# CHEMICAL ENGINEERING

## VOLUME TWO

### UNIT OPERATIONS

**J. M. COULSON**

*Emeritus Professor of Chemical Engineering*
*University of Newcastle-upon-Tyne*

and

**J. F. RICHARDSON**

*University College of Swansea*

with

**J. R. BACKHURST and J. H. HARKER**

*University of Newcastle-upon-Tyne*

**THIRD EDITION**
**(SI units)**

**PERGAMON PRESS**

OXFORD · NEW YORK · TORONTO · SYDNEY · PARIS · FRANKFURT

| U.K. | Pergamon Press Ltd., Headington Hill Hall, Oxford OX3 0BW, England |
| U.S.A | Pergamon Press Inc., Maxwell House, Fairview Park, Elmsford, New York 10523, U.S.A. |
| CANADA | Pergamon of Canada Ltd., 75 The East Mall, Toronto, Ontario, Canada |
| AUSTRALIA | Pergamon Press (Aust.) Pty. Ltd., 19a Boundary Street, Rushcutters Bay, N.S.W. 2011, Australia |
| FRANCE | Pergamon Press SARL, 24 rue des Ecoles, 75240 Paris, Cedex 05, France |
| FEDERAL REPUBLIC OF GERMANY | Pergamon Press GmbH, 6242 Kronberg/Taunus, Pferdstrasse 1, Federal Republic of Germany |

First edition 1955; Second (revised) impression 1956; Third (revised) impression 1959; Fourth (revised) impression 1960; Fifth impression 1962; Second edition 1968; Reprinted 1976; Third edition (SI units) 1978

**British Library Cataloguing in Publication Data**

Chemical engineering.
Vol. 2: Unit operations.—3rd ed. (SI units).
1. Chemical engineering
I. Title II. Coulson, John Metcalfe
III. Richardson, John Francis IV. Backhurst,
John Rayner V. Harker, John Hadlett
660.2      TP155      78-40157

ISBN 0-08-018090-6 (Hard cover)
ISBN 0-08-022919-0 (Flexicover)

*Printed in Great Britain by William Clowes (Beccles) Ltd., Beccles, Suffolk.*

# Contents

CONTENTS

# *Preface to Third Edition*

IN producing a third edition, we have taken the opportunity, not only of updating the material but also of expressing the values of all the physical properties and characteristics of the systems in the SI System of units, as has already been done in Volumes 1 and 3. The SI system, which is described in detail in Volume 1, is widely adopted in Europe and is now gaining support elsewhere in the world. However, because some readers will still be more familiar with the British system, based on the foot, pound and second, the old units have been retained as alternatives wherever this can be done without causing confusion.

The material has, to some extent, been re-arranged and the first chapter now relates to the characteristics of particles and their behaviour in bulk, the blending of solids, and classification according to size or composition of material. The following chapters describe the behaviour of particles moving in a fluid and the effects of both gravitational and centrifugal forces and of the interactions between neighbouring particles. The old chapter on centrifuges has now been eliminated and the material dispersed into the appropriate parts of other chapters. Important applications which are considered include flow in granular beds and packed columns, fluidisation, transport of suspended particles, filtration and gas cleaning. An example of the updating which has been carried out is the addition of a short section on fluidised bed combustion, potentially the most important commercial application of the technique of fluidisation. In addition, we have included an entirely new section on flocculation, which has been prepared for us by Dr. D. J. A. Williams of University College, Swansea, to whom we are much indebted.

Mass transfer operations play a dominant role in chemical processing and this is reflected in the continued attention given to the operations of solid–liquid extraction, distillation, gas absorption and liquid–liquid extraction. The last of these subjects, together with material on liquid–liquid mixing, is now dealt within a single chapter on liquid–liquid systems, the remainder of the material which appeared in the former chapter on mixing having been included earlier under the heading of solids blending. The volume concludes with chapters on evaporation, crystallisation and drying.

Volumes 1, 2 and 3 form an integrated series with the fundamentals of fluid flow, heat transfer and mass transfer in the first volume, the physical operations of chemical engineering in this, the second volume, and in the third volume, the basis of chemical and biochemical reactor design, some of the physical operations which are now gaining in importance and the underlying theory of both process control and computation. The solutions to the problems listed in Volumes 1 and 2 are now available as Volumes 4 and 5 respectively. Furthermore, an additional volume in the series is in course of preparation and will provide an introduction to chemical engineering design and indicate how the principles enunciated in the earlier volumes can be translated into chemical plant.

We welcome the collaboration of J. R. Backhurst and J. H. Harker as co-authors in the preparation of this edition, following their assistance in the editing of the latest edition of Volume 1 and their authorship of Volumes 4 and 5. We also look forward to the appearance of R. K. Sinnott's volume on chemical engineering design.

# *Preface to Second Edition*

THIS text deals with the physical operations used in the chemical and allied industries. These operations are conveniently designated "unit operations" to indicate that each single operation, such as filtration, is used in a wide range of industries, and frequently under varying conditions of temperature and pressure.

Since the publication of the first edition in 1955 there has been a substantial increase in the relevant technical literature but the majority of developments have originated in research work in government and university laboratories rather than in industrial companies. As a result, correlations based on laboratory data have not always been adequately confirmed on the industrial scale. However, the section on absorption towers contains data obtained on industrial equipment and most of the expressions used in the chapters on distillation and evaporation are based on results from industrial practice.

In carrying out this revision we have made substantial alteration to Chapters 1, 5, 6, 7, 12, 13 and 15* and have taken the opportunity of presenting the volume paged separately from Volume 1. The revision has been possible only as the result of the kind co-operation and help of Professor J. D. Thornton (Chapter 12), Mr. J. Porter (Chapter 13), Mr. K. E. Peet (Chapter 10) and Dr. B. Waldie (Chapter 1), all of the University at Newcastle, and Dr. N. Dombrowski of the University of Leeds (Chapter 15). We want in particular to express our appreciation of the considerable amount of work carried out by Mr. D. G. Peacock of the School of Pharmacy, University of London. He has not only checked through the entire revision but has made numerous additions to many chapters and has overhauled the index.

We should like to thank the companies who have kindly provided illustrations of their equipment and also the many readers of the previous edition who have made useful comments and helpful suggestions.

Chemical engineering is no longer confined to purely physical processes and the unit operations, and a number of important new topics, including reactor design, automatic control of plants, biochemical engineering, and the use of computers for both process design and control of chemical plant will be covered in a forthcoming Volume 3 which is in course of preparation.

Chemical engineering has grown in complexity and stature since the first edition of the text, and we hope that the new edition will prove of value to the new generation of university students as well as forming a helpful reference book for those working in industry.

In presenting this new edition we wish to express our gratitude to Pergamon Press who have taken considerable trouble in coping with the technical details.

J. M. COULSON
J. F. RICHARDSON

* N.B. Chapter numbers are altered in the current (third) edition.

# *Preface to First Edition*

In presenting Volume 2 of *Chemical Engineering*, it has been our intention to cover what we believe to be the more important unit operations used in the chemical and process industries. These unit operations, which are mainly physical in nature, have been classified, as far as possible, according to the underlying mechanism of the transfer operation. In only a few cases is it possible to give design procedures when a chemical reaction takes place in addition to a physical process. This difficulty arises from the fact that, when we try to design such units as absorption towers in which there is a chemical reaction, we are not yet in a position to offer a thoroughly rigorous method of solution. We have not given an account of the transportation of materials in such equipment as belt conveyors or bucket elevators, which we feel lie more distinctly in the field of mechanical engineering.

In presenting a good deal of information in this book, we have been much indebted to facilities made available to us by Professor Newitt, in whose department we have been working for many years. The reader will find a number of gaps, and a number of principles which are as yet not thoroughly developed. Chemical engineering is a field in which there is still much research to be done, and, if this work will in any way stimulate activities in this direction, we shall feel very much rewarded. It is hoped that the form of presentation will be found useful in indicating the kind of information which has been made available by research workers up to the present day. Chemical engineering is in its infancy, and we must not suppose that the approach presented here must necessarily be looked upon as correct in the years to come. One of the advantages of this subject is that its boundaries are not sharply defined.

Finally, we should like to thank the following friends for valuable comments and suggestions: Mr. G. H. Anderson, Mr. R. W. Corben, Mr. W. J. De Coursey, Dr. M. Guter, Dr. L. L. Katan, Dr. R. Lessing, Dr. J. Rasbash, Dr. H. Sawistowski, Dr. W. Smith, Mr. D. Train, Mr. M. E. O'K. Trowbridge, Mr. F. E. Warner and Dr. W. N. Zaki.

# Acknowledgements

THE authors and publishers acknowledge with thanks the assistance given by the following companies and individuals in providing illustrations and data for this volume and giving their permission for reproduction. Everyone was most helpful and some firms went to considerable trouble to provide exactly what was required. We are extremely grateful to them all.

N.E.I. International Combustion Ltd. for Figs. 1.2, 1.29, 1.30, 1.32, 1.33, 1.39, 2.12, 2.24–2.27, 2.30, 2.31, 2.35, 7.11, 7.12, 9.20–9.22, 9.25, 9.26, 14.21, 16.28.

Simon Solitec Ltd. for Figs. 1.11, 1.12.

Baker Perkins Ltd. for Figs. 1.18, 2.8, 2.32, 2.33, 13.48, 13.49.

Buss–Hamilton Ltd. for Figs. 1.19, 16.31, 16.32.

Dorr–Oliver Co. Ltd. for Figs. 1.21, 1.31, 5.18, 10.5, 10.11.

R. O. Stokes & Co. Ltd. for Fig. 1.24.

The Denver Equipment Co. Ltd. for Figs. 1.25–1.27.

Wilfley Mining Machinery Co. Ltd. for Fig. 1.28.

Davies Magnet Works Ltd. for Figs. 1.35, 1.37.

Simon Carves Ltd. for Fig. 1.40.

Hadfields Ltd. for Figs. 2.3, 2.16, 2.18.

Messrs. Edgar Allen & Co. Ltd. for Figs. 2.4, 2.6, 2.10, 2.11, 2.15.

Sturtevant Engineering Co. Ltd. for Figs. 2.9, 2.20, 8.5, 8.11, 8.17, 8.18.

Kek Ltd. for Fig. 2.13.

J. Harrison Carter Ltd. for Fig. 2.14.

Babcock & Wilcox Ltd. for Fig. 2.23.

The Bradley Pulverizer Co. Ltd. for Figs. 2.29, 9.27.

Premier Colloid Mills for Figs. 2.36, 2.37.

McGraw–Hill Book Co. for Fig. 3.3.

Norton Chemical Process Products Division for Figs. 4.10–4.17, Table 4.3, Figs. 4.26, 4.29, 4.31, 12.21, 12.22.

Mass Transfer Ltd. for Fig. 4.18, Table 4.3, Table 12.4.

Butterworths Scientific Publications for Fig. 4.20.

I.C.I. Pollution Control Systems Ltd. for Fig. 4.21.

Thomas Broadbent & Sons Ltd. for Figs. 5.24, 5.25, 5.27, 5.28.

Alfa–Laval Ltd. for Figs. 5.29, 5.31, 13.50.

Pennwalt Ltd. for Figs. 5.32, 5.33, 5.35, 5.36.

Institution of Mechanical Engineers for Fig. 8.11.

The Power Gas Corporation for Figs. 8.21, 15.16, 15.17.

F. H. Schule GmbH, Hamburg for Fig. 9.6.

Johnson–Progress Ltd. for Figs. 9.7–9.12.

The Mirrlees Watson Co. Ltd. for Figs. 9.13, 9.15.

Amafilter (UK) Ltd. for Figs. 9.16–9.19.

The Metafiltration Co. Ltd. for Figs. 9.28–9.30.

Stream-Line Filters Ltd. for Fig. 9.31.

Messrs. Rose, Downs & Thompson Ltd. for Fig. 10.4.

APV Mitchell Ltd. for Figs. 11.49–11.52, 12.17.

Davy Powergas Ltd. for Figs. 13.36, 13.37.

Swenson Evaporator Co. for Figs. 14.19, 14.20, 14.22, 15.13.

Messrs. George Scott & Son (London) Ltd. for Fig. 14.24.

The Editor and Publishers of *Chemical and Process Engineering* for Figs. 14.27, 14.32, 14.33.

The A.P.V. Co. Ltd. for Figs. 14.28–14.30.

Woodhall–Duckham Ltd. for Fig. 15.12.

The Girdler Company for Fig. 15.15.

Messrs. Dunford and Elliott (Sheffield) Ltd. for Fig. 16.12.

Buflovak Equipment Div. of Blaw–Knox Co. Ltd. for Figs. 16.18, 16.19, and Table 16.3.

Dr. N. Dombrowski for Figs. 16.20, 16.24–16.26.

# *Introduction*

THE understanding of the design and construction of chemical plant is frequently regarded as the essence of chemical engineering. Starting from the original conception of the process by the chemist, it is necessary to understand the chemical, physical and many of the engineering features in order to develop the laboratory process on an industrial scale. In this volume we shall be mainly concerned with the physical nature of the processes that take place in industrial units, and, in particular, with determining the factors that influence the rate of transfer of material. The basic principles of these operations, namely fluid dynamics, and heat and mass transfer, have been discussed in Volume 1, and it is the application of these principles that forms the main part of this volume.

Throughout what are conveniently regarded as the process industries, there are many physical operations that are common to a number of the individual industries, and, as explained in Volume 1, these are regarded as unit operations. Thus, the separation of solids from a suspension by filtration, the separation of liquids by distillation, and the removal of water by evaporation and drying are typical operations of this kind. The problem of designing a distillation unit for the fermentation industry, the petroleum industry or the organic chemical industry is, in principle, the same, and it is mainly in the details of construction that the differences will occur. The concentration of solutions by evaporation is again a typical operation that is basically similar in the handling of sugar, or salt, or fruit juices, though there will be differences in the most suitable arrangement. This form of classification has been used here, but we have grouped the operations according to the mechanism of the transfer operation, so that the diffusion processes of distillation, absorption and liquid–liquid extraction are taken in successive chapters, and the operations involving solids in fluids are considered together. In examining many of these unit operations, we shall find that the rate of heat transfer or the nature of the fluid flow is the governing feature. The transportation of a solid or a fluid stream is another instance of the importance of understanding fluid dynamics.

One of the difficult problems of design is that of maintaining conditions of similarity between laboratory units and the larger industrial plants. Thus, if a mixture is to be maintained at a certain temperature during the course of an exothermic reaction, then on the laboratory scale there is rarely any real difficulty in maintaining isothermal conditions. On the other hand, in a large reactor the ratio of the external surface to the volume— which is inversely proportional to the linear dimension of the unit—is in most cases of a different order, and the problem of removing the heat of reaction becomes a major item in design. Some of the general problems associated with scaling up are considered in Chapter 13 on liquid–liquid systems, and particular features occur in many chapters. Again, the introduction and removal of the reactants may present difficult problems on the large scale, especially if they contain corrosive liquids or abrasive solids. The general tendency with industrial units is to provide a continuous process, frequently involving a series of stages. Thus, exothermic reactions may be carried out in a series of reactors with interstage cooling between them.

The planning of a process plant will involve the determining of the most economic method, and later the most economic arrangement of the unit operations used in the process. This amounts to designing a chemical process to provide the best combination of capital and operating costs. We have not in this volume considered the question of costs in any great detail, but we have aimed at indicating the conditions under which various types of units will operate in the most economical manner. Without a thorough knowledge of the physical principles involved in the various operations, it will not be possible to select the most suitable one for a given process. This aspect of the design can be considered by taking one or two simple illustrations of separation processes. The particles in a solid–solid system may be separated, first according to size, and secondly according to the material. Generally, sieving is the most satisfactory method of classifying relatively coarse materials according to size, but the method is impracticable for very fine particles and a form of settling process is generally used. In the first of these processes, we are directly utilising the size of the particle as the basis for the separation, and, in the second, we are dependent on the variation with size of the behaviour of particles in a fluid. A mixed material can also be separated into its components by means of settling methods, because the shape and density of particles also affect their behaviour in a fluid. Other methods of separation depend on differences in surface properties (froth flotation), magnetic properties (magnetic separation), and on differences in solubility in a solvent (leaching). For the separation of miscible liquids, the three commonly used methods are:

1. Distillation—depends on difference in volatility.
2. Liquid–liquid extraction—depends on difference in solubility in a liquid solvent.
3. Refrigeration—depends on differences in melting point.

This problem of selecting the most appropriate operation will be further complicated by such factors as the concentration of liquid which gives rise to crystals. Thus, in the separation of a mixture of ortho-, meta-, and para-mononitrotoluenes, the decision must be made as to whether it is better to carry out the separation by distillation followed by crystallisation, or in the reverse order. The same kind of consideration will arise when we are concentrating a solution of a solid; we must decide whether to stop the evaporation process when a certain concentration of solid has been reached and then to proceed with filtration followed by drying, or whether to continue the concentration by evaporation to such an extent that we can leave out the filtration stage and go straight on to drying.

In many operations, for instance in a distillation column, it is necessary to understand the fluid dynamics of the unit, as well as the heat and mass transfer relationships. These factors are frequently interdependent in a complex manner, and it is essential to consider the individual contributions of each of the mechanisms. Again, in a chemical reaction the final rate of the process may be governed by a heat transfer process or by the chemical kinetics, and it is essential to decide which is the controlling factor; this problem is discussed in Volume 3.

Some indication of the method of using the information on unit operations can be obtained by considering the example of a sulphuric acid plant. First it is necessary to select the process and the raw materials which are to be used. It will be supposed that the process involves burning sulphur to sulphur dioxide, followed by oxidation to sulphur trioxide in what is known as the contact process, and that it is proposed to employ rock sulphur, oxygen from the atmosphere, and water. Alternative sources of sulphur include iron pyrites and gypsum (calcium sulphate), but, although these are considerably cheaper than rock sulphur, it is necessary to install additional equipment for cleaning the gas. Secondly, we must study the kinetics of the chemical reactions. Here we are concerned basically with the following three reactions:

(i) $S + O_2 = SO_2$ ($\Delta H = -16{,}940 \text{ kJ/kmol}$).
(ii) $2SO_2 + O_2 = 2SO_3$ ($\Delta H = -11{,}350 \text{ kJ/kmol}$).
(iii) $SO_3 + H_2O = H_2SO_4$ ($\Delta H = -7{,}600 \text{ kJ/kmol}$)

The overall reaction is therefore

$$2S + 3O_2 + 2H_2O \rightarrow 2H_2SO_4.$$

Each of the reactions is exothermic, and the successful and economic operation of the plant depends on the removal and subsequent utilisation of the heats of reaction. The first reaction is a simple gas-phase combustion which goes virtually to completion under adiabatic conditions, and it is not necessary therefore to remove heat in the combustion chamber itself. The second reaction does not proceed sufficiently rapidly except in the presence of a catalyst, generally vanadium pentoxide dispersed on a filler, and it becomes self-supporting at temperatures above about 675 K. Because the reaction is strongly exothermic, the equilibrium point is adversely affected by rise in temperature, and, because this effect is very marked, the reaction should be carried out as nearly as possible under isothermal conditions. The reaction rate is high in the presence of the catalyst, and the reaction can be regarded as proceeding to completion at each stage. The composition of the final gas can be calculated as a function of temperature, and therefore the composition of the equilibrium mixture at each adiabatic stage can be determined because all the heat of reaction appears in the products. The temperature of the outlet gases from each successive reactor will be lower, and therefore the $SO_3$ content will rise. The third reaction involves absorption of $SO_3$ in water and an immediate chemical reaction to produce $H_2SO_4$. As this reaction is exothermic, there would be a considerable rise in temperature and mist formation if it were carried out in this way. In practice, the heat capacity of the liquid phase is increased by absorbing the $SO_3$ in 98 per cent $H_2SO_4$ rather than in water.

It can be seen from these considerations that a thorough understanding of the physical processes taking place in the individual units of chemical plant is an essential requirement in the training of a chemical engineer. It is only by understanding the mechanisms, and then the method of constructing the equipment, that it will be possible to obtain a complete picture, including the cost of the operation.

# CHAPTER 1

# Particulate Solids

## 1.1. INTRODUCTION

In Volume 1, consideration has been given to the behaviour of fluids, both liquids and gases, with particular reference to their flow properties and their heat and mass transfer characteristics. Once the composition, temperature and pressure of the fluid have been specified, its relevant physical properties, such as density, viscosity, thermal conductivity and molecular diffusivity, are defined. In the early chapters of this volume consideration will be given to the properties and behaviour of systems containing solid particles. Such systems are generally more complicated, not only because of the complex geometrical arrangements which are possible, but also because of the basic problem of defining completely the physical state of the material.

The three most important characteristics of an individual particle are its composition, its size and its shape. The composition will determine such properties as its density and conductivity provided that it is completely uniform. In many cases, however, the particle will be porous or may consist of a continuous matrix in which are distributed small particles of a second material. The particle size is important in that it affects properties such as the surface per unit volume and the rate at which the particle will settle in a fluid. The particle shape may be regular, such as spherical or cubical, or it may be irregular like, for instance, a piece of broken glass. Regular shapes are capable of precise definition by mathematical equations; irregular shapes are not and the properties of irregular particles are usually expressed in terms of some particular characteristics of a regular shaped particle.

On the industrial scale, one is concerned with handling large quantities of particles and frequently needs to define the system as a whole. Thus, in place of particle size, one needs to know the distribution of particle sizes in the mixture and to be able to define a mean size which in some way represents the behaviour of the particulate mass as a whole. Important operations relating to systems of particles include storage in hoppers, flow through orifices and pipes, metering of flow. Frequently, it will be necessary to reduce the size of the particles, or alternatively to form them into aggregates or sinters. Sometimes it may be desired to mix two or more solids and at other times there may be a requirement to separate a mixture into its components or according to the sizes of the particles.

In some cases the interaction between the particles and the surrounding fluid will be of little significance, but at other times it can have a dominating effect on the behaviour of the system. Thus, in filtration or the flow of fluids through beds of granular particles, the characterisation of the porous mass as a whole is the principal feature and the resistance to flow is dominated by the size and shape of the free space between the particles. In these cases, the particles are in physical contact with their neighbours and there is little relative movement between the particles. However, in processes such as the sedimentation of particles in a liquid, each particle is completely surrounded by fluid and is free to move relative to the other particles. Only very simple cases are capable of a precise theoretical analysis and Stokes' law, which gives the drag on an isolated spherical particle due to its motion relative to the surrounding fluid at very low velocities, is outstandingly the most important theoretical relation in this area of study. In fact, very many empirical laws are based on the concept of defining correction factors to be applied to Stokes' law.

## 1.2. PARTICLE CHARACTERISATION

### 1.2.1. Single Particles

The particle of simplest shape is the sphere, in that because of its symmetry, the question of orientation does not have to be considered, i.e. the particle looks exactly the same from whatever direction it is viewed and behaves in the same manner in a fluid, irrespective of its orientation. No other particle has this characteristic. Frequently, the size of a particle of irregular shape is defined in terms of the size of an equivalent sphere. However, the particle is represented by a sphere of different size according to the property selected. Some of the important sizes of equivalent spheres are as follows:

(a) The sphere of the same volume as the particle.
(b) The sphere of the same surface as the particle.
(c) The sphere of the same surface per unit volume as the particle.
(d) The sphere of the same area as the particle when projected on to a plane perpendicular to its direction of motion.
(e) The sphere of the same projected area as the particle, as viewed from above, when lying in its position of maximum stability (e.g. on a microscope slide).
(f) The sphere which will just pass through the same size of square aperture as the particle (as on a screen).
(g) The sphere with the same settling velocity as the particle in a specified fluid.

Several definitions depend on the measurement of a particle in a particular orientation. Thus the Feret diameter is the distance apart of two parallel planes which just touch the particle. Clearly this depends on the orientation of the particle relative to the planes.

Several definitions depend on the measurement of a particle in a particular orientation. Thus Feret's statistical diameter is the mean length of the distance apart of two parallel lines which are tangential to the particle in an arbitrarily fixed direction and irrespective of the orientation of the particles relative to the planes (see Fig. 1.1).

A measure of particle shape which is frequently used is the sphericity, $\psi$, defined by the relation:

$$\psi = \frac{\text{surface area of sphere of same volume as particle}}{\text{surface area of particle}} \tag{1.1}$$

Another method of indicating shape is to use the factor by which the cube of the size of the particle must be multiplied to give the volume; in this case the particle size is usually defined by method (e) given above.

Other properties of the particle which may at times be important are: whether it is crystalline or amorphous; whether it is porous; and the properties of its surface, including roughness and presence of adsorbed films.

Hardness may also be important if the particle is to be subjected to heavy loading.

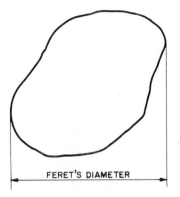

FIG. 1.1. Feret's diameter

## 1.2.2. Measuring Techniques

A wide range of measuring techniques is available both for single particles and for systems of particles. Each method is applicable, in practice, to a finite range of sizes and gives a particular equivalent size, dependent on the nature of the method. Reference should be made to standard works[1,2] on particle size measurement for details of the methods, but the principles of some of the chief methods are given below together with an indication of the size range to which they are applicable.

1. *Sieving* ($> 50 \, \mu m$). Sieve analysis may be carried out using a nest of sieves, each lower sieve being of smaller aperture size. Generally, sieve series are arranged so that the ratio of aperture sizes on consecutive sieves is 2, $2^{1/2}$ or $2^{1/4}$ according to the closeness of sizing which is required. The sieves can either be mounted on a vibrator (Fig. 1.2), which should be designed to give a degree of vertical movement in addition to the horizontal vibration, or may be hand shaken. Whether or not a particle passes through an aperture depends not only upon its size, but also on the probability that it will be presented at the required orientation at the surface of the screen. The sizing is based purely on the linear dimensions of the particle and the lower limit of size which can be used is determined by two principal factors. The first is that the proportion of free space on the screen surface becomes very small as the size of the aperture is reduced. The second is that attractive forces between particles become larger at small particle sizes and consequently the particles tend to stick together and block the screen. Sieves are available in a number of standard series. There are several standard series of screens; the sizes of the openings are determined by the thickness of wire used. In the U.K. British Standard (B.S.) screens are made in sizes from 300-mesh upwards, but they are too fragile for some work. The Institute of Mining and Metallurgy (I.M.M.) screens are more robust, with the thickness of the wire approximately equal to the size of the apertures. The Tyler series, which is standard in the United States, is intermediate between the two British series. Details of the three series of screens are given in Table 1.1, together with the American Society for Testing Materials series and a French standard.

The efficiency of screening is defined as the ratio of the weight of material which passes the screen to that which is capable of passing; this will differ according to the size of the material. It may be assumed that the rate of passage of particles of a given size through the screen is proportional to the number or weight of particles of that size on the screen at any

FIG. 1.2. Rotap sieve shaker

instant. Thus, if $w$ is the mass of particles of a particular size on the screen at a time $t$:

$$\frac{dw}{dt} = -kw \qquad (1.2)$$

where $k$ is a constant for a given size and shape of particle and for a given screen.

Thus, the mass of particles $(w_1 - w_2)$ passing the screen in time $t$ is given by

$$\ln\frac{w_2}{w_1} = -kt$$
$$\therefore \quad w_2 = w_1 e^{-kt} \qquad (1.3)$$

If the screen contains a large proportion of material just a little larger than the maximum size of particle which will pass, its capacity is considerably reduced. Screening is generally continued either for a predetermined time or until the rate of screening falls off to a certain fixed value.

Screening can be carried out either wet or dry. In wet screening, the material is washed evenly over the screen and clogging is prevented; further, the small particles are washed off the surface of the large ones. It has the obvious disadvantage, however, that it may be necessary to dry the material afterwards. With dry screening, the material is sometimes brushed lightly over the screen so as to form a thin even sheet. Care should always be taken to ensure that agitation is not so vigorous that size reduction occurs, because screens are usually quite fragile and easily damaged by rough treatment. In general, the larger and the more abrasive the solids the more robust the screen should be.

TABLE 1.1. *Standard Sieve Sizes*

**British fine mesh (B.S.S. 410)[3,4]**

| Sieve no. | Nominal aperture in. | μm |
|---|---|---|
| 300 | 0·0021 | 53 |
| 240 | 0·0026 | 66 |
| 200 | 0·0030 | 76 |
| 170 | 0·0035 | 89 |
| 150 | 0·0041 | 104 |
| 120 | 0·0049 | 124 |
| 100 | 0·0060 | 152 |
| 85 | 0·0070 | 178 |
| 72 | 0·0083 | 211 |
| 60 | 0·0099 | 251 |
| 52 | 0·0116 | 295 |
| 44 | 0·0139 | 353 |
| 36 | 0·0166 | 422 |
| 30 | 0·0197 | 500 |
| 25 | 0·0236 | 600 |
| 22 | 0·0275 | 699 |
| 18 | 0·0336 | 853 |
| 16 | 0·0395 | 1003 |
| 14 | 0·0474 | 1204 |
| 12 | 0·0553 | 1405 |
| 10 | 0·0660 | 1676 |
| 8 | 0·0810 | 2057 |
| 7 | 0·0949 | 2411 |
| 6 | 0·1107 | 2812 |
| 5 | 0·1320 | 3353 |

**I.M.M.[3]**

| Sieve no. | Nominal aperture in. | μm |
|---|---|---|
| 200 | 0·0025 | 63 |
| 150 | 0·0033 | 84 |
| 120 | 0·0042 | 107 |
| 100 | 0·0050 | 127 |
| 90 | 0·0055 | 139 |
| 80 | 0·0062 | 157 |
| 70 | 0·0071 | 180 |
| 60 | 0·0083 | 211 |
| 50 | 0·0100 | 254 |
| 40 | 0·0125 | 347 |
| 30 | 0·0166 | 422 |
| 20 | 0·0250 | 635 |
| 16 | 0·0312 | 792 |
| 12 | 0·0416 | 1056 |
| 10 | 0·0500 | 1270 |
| 8 | 0·0620 | 1574 |
| 5 | 0·1000 | 2540 |

**U.S. Tyler[3]**

| Sieve no. | Nominal aperture in. | μm |
|---|---|---|
| 325 | 0·0017 | 43 |
| 270 | 0·0021 | 53 |
| 250 | 0·0024 | 61 |
| 200 | 0·0029 | 74 |
| 170 | 0·0035 | 89 |
| 150 | 0·0041 | 104 |
| 115 | 0·0049 | 125 |
| 100 | 0·0058 | 147 |
| 80 | 0·0069 | 175 |
| 65 | 0·0082 | 208 |
| 60 | 0·0097 | 246 |
| 48 | 0·0116 | 295 |
| 42 | 0·0133 | 351 |
| 35 | 0·0164 | 417 |
| 32 | 0·0195 | 495 |
| 28 | 0·0232 | 589 |
| 24 | 0·0276 | 701 |
| 20 | 0·0328 | 833 |
| 16 | 0·0390 | 991 |
| 14 | 0·0460 | 1168 |
| 12 | 0·0550 | 1397 |
| 10 | 0·0650 | 1651 |
| 9 | 0·0780 | 1981 |
| 8 | 0·0930 | 2362 |
| 7 | 0·1100 | 2794 |
| 6 | 0·1310 | 3327 |
| 5 | 0·1560 | 3962 |
| 4 | 0·1850 | 4699 |

**U.S. A.S.T.M.[5]**

| Sieve no. | Nominal aperture in. | μm |
|---|---|---|
| 325 | 0·0017 | 44 |
| 270 | 0·0021 | 53 |
| 230 | 0·0024 | 61 |
| 200 | 0·0029 | 74 |
| 170 | 0·0034 | 88 |
| 140 | 0·0041 | 104 |
| 120 | 0·0049 | 125 |
| 100 | 0·0059 | 150 |
| 80 | 0·0070 | 177 |
| 70 | 0·0083 | 210 |
| 60 | 0·0098 | 250 |
| 50 | 0·0117 | 297 |
| 45 | 0·0138 | 350 |
| 40 | 0·0165 | 420 |
| 35 | 0·0197 | 500 |
| 30 | 0·0232 | 590 |
| 25 | 0·0280 | 710 |
| 20 | 0·0331 | 840 |
| 18 | 0·0394 | 1000 |
| 16 | 0·0469 | 1190 |
| 14 | 0·0555 | 1410 |
| 12 | 0·0661 | 1680 |
| 10 | 0·0787 | 2000 |
| 8 | 0·0937 | 2380 |
| 7 | 0·1110 | 2839 |
| 6 | 0·1320 | 3360 |
| 5 | 0·1570 | 4000 |
| 4 | 0·1870 | 4760 |

**French (A.F.N.O.R.)[5]**

| Sieve no. | Nominal aperture in. | μm |
|---|---|---|
| 17 | 0·0015 | 40 |
| 18 | 0·0019 | 50 |
| 19 | 0·0024 | 61 |
| 20 | 0·0031 | 80 |
| 21 | 0·0039 | 100 |
| 22 | 0·0049 | 125 |
| 23 | 0·0063 | 160 |
| 24 | 0·0079 | 200 |
| 25 | 0·0098 | 250 |
| 26 | 0·0124 | 315 |
| 27 | 0·0157 | 400 |
| 28 | 0·0197 | 500 |
| 29 | 0·0248 | 630 |
| 30 | 0·0315 | 800 |
| 31 | 0·0394 | 1000 |
| 32 | 0·0492 | 1250 |
| 33 | 0·0630 | 1600 |
| 34 | 0·0787 | 2000 |
| 35 | 0·0984 | 2500 |
| 36 | 0·1240 | 3150 |
| 37 | 0·1570 | 4000 |
| 38 | 0·1970 | 5000 |

2. *Microscopic analysis* (1–100 μm). Microscopic examination permits measurement of the projected area of the particle and also enables an assessment to be made of its two-dimensional shape. In general, the third dimension will not be capable of estimation except using special stereomicroscopes. The apparent size of particle is compared with that of circles engraved on a graticule in the eyepiece (Fig. 1.3). Automatic methods of scanning have now been developed. By using the electron microscope, the lower limit of size can be reduced to about 0·001 μm.

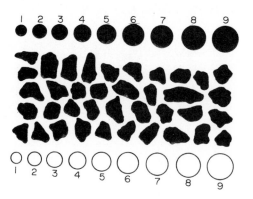

FIG. 1.3. Particle profiles and comparison circles

3. *Sedimentation and elutriation methods* (>1 μm). These methods depend on the fact that the terminal falling velocity of a particle in a fluid increases with size. Sedimentation methods are of two main types. In the first, the pipette method, samples are abstracted from the settling suspension at a fixed horizontal level at intervals of time; each sample will contain a representative sample of the suspension, with the exception of particles larger than a critical size, all of which will have settled below the level of the sampling point. In the second, the sedimentation balance, particles settle on an immersed balance pan which is continuously weighed. The largest particles will be deposited preferentially and consequently the rate of increase of weight will fall off progressively as particles settle out.

Sedimentation analyses must be carried out at concentrations which are sufficiently low for interactive effects between particles to be negligible so that their terminal falling velocities can be taken as equal to those of isolated particles. Careful temperature control (preferably to ±0·1 K) is necessary to suppress convection currents. The lower limit of particle size is set by the increasing importance of Brownian motion for progressively smaller particles. However, it is possible to replace gravitational forces by centrifugal forces and this reduces the lower size limit to about 0·05 μm.

The elutriation method is really a reverse sedimentation process in which the particles are dispersed in an upward flowing stream of fluid. All particles with terminal falling velocities less than the upward velocity of the fluid will be carried away. A complete size analysis can be obtained by using successively higher fluid velocities. Figure 1.4 shows the standard elutriator (B.S. 893)[6] for particles with settling velocities between 7 and 70 mm/s.

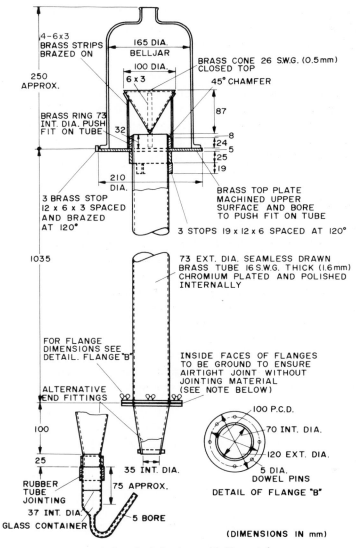

FIG. 1.4. Standard elutriator with 70-mm tube

4. *The Coulter counter* (1–100 μm). The particles are suspended in an electrolyte and are forced through a small orifice. The voltage difference between two electrodes on either side of the orifice is continuously measured. Because the electrical resistance is affected by the presence of the particle, and the magnitude of the change is a function of the size of the particle, the number and size of the particles passing through the orifice can be obtained from the record of voltage differences.

5. *Permeability methods* (> 1 μm). These methods depend on the fact that at low flowrates the flow through a packed bed is directly proportional to the pressure difference, the proportionality constant being proportional to the square of the specific surface

(surface:volume ratio) of the powder. From this method it is possible therefore to obtain the diameter of the sphere with the same specific surface as the powder. The reliability of the method is dependent upon the care with which the sample of powder is packed. Further details are given in Chapter 4.

6. *Light-scattering methods* (0·5–50 μm). These methods depend on the light scattering by particles in a direction at right angles to the incident-beam. The fraction of the incident light scattered is directly proportional to the concentration of the particles and to the cube of their diameter. It is critically dependent on the wavelength of the incident radiation, being inversely proportional to its fourth power.

### 1.2.3. Particle Size Distribution

Most particulate systems which are of practical interest consist of particles of a wide range of sizes and it is necessary to be able to give a quantitative indication of the mean size and of the spread of sizes. The results of a size analysis can most conveniently be represented by means of a *cumulative weight fraction curve*, in which the proportion of particles $(x)$ smaller than a certain size $(d)$ is plotted against that size $(d)$. In most practical determinations of particle size, the size analysis will be obtained as a series of steps, each step representing the proportion of particles lying within a certain small range of size. From these results a cumulative size distribution can be built up and this can then be approximated by a smooth curve provided that the size intervals are sufficiently small. A typical curve for size distribution on a cumulative basis is shown in Fig. 1.5. This curve rises from zero to unity over the range from the smallest to the largest particle size present.

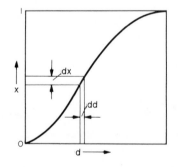

FIG. 1.5. Size distribution curve—cumulative basis

The distribution of particle sizes can be seen more readily by plotting a *size frequency curve* such as that shown in Fig. 1.6 in which the slope $(dx/dd)$ of the cumulative curve (Fig. 1.5) is plotted against particle size $(d)$. The most frequently occurring size is then shown by the maximum of the curve. For naturally occurring materials the curve will generally have a single peak. For mixtures of particles, there may be as many peaks as components in the mixture. Again, if the particles are formed by crushing larger particles, the curve may have two peaks, one characteristic of the material and the other characteristic of the equipment.

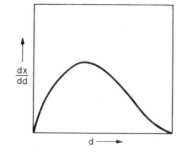

FIG. 1.6. Size distribution curve—frequency basis

## 1.2.4. Mean Particle Size

It is often desirable to express the particle size of a powder in terms of a single linear dimension. For coarse particles Bond[7,8] has somewhat arbitrarily chosen the size of the opening through which 80 per cent of the material will pass. This size $d_{80}$ is a useful rough comparative measure for the size of material which has been through a crusher.

It should be remembered that a mean size will describe only one particular characteristic of the powder and it is important to decide what that characteristic is before the mean is calculated. Thus, it may be desirable to define the size of particle such that its weight or its surface or its length is the mean value for all the particles in the system. In the following treatment, it will be assumed that each of the particles has the same shape.

Consider a unit mass of particles consisting of $n_1$ particles of characteristic dimension $d_1$, constituting a mass fraction $x_1$, $n_2$ particles of size $d_2$, etc.

Then

$$x_1 = n_1 k_1 d_1^3 \rho_s \tag{1.4}$$

and

$$\Sigma x_1 = 1 = \rho_s k_1 \Sigma(n_1 d_1^3) \tag{1.5}$$

Thus

$$n_1 = \frac{1}{\rho_s k_1} \frac{x_1}{d_1^3} \tag{1.6}$$

If the size distribution can be represented by a continuous function, then:

$$\mathrm{d}x = \rho_s k_1 d^3 \, \mathrm{d}n$$

i.e.

$$\frac{\mathrm{d}x}{\mathrm{d}n} = \rho_s k_1 d^3 \tag{1.7}$$

and

$$\int_0^1 \mathrm{d}x = 1 = \rho_s k_1 \int d^3 \, \mathrm{d}n \tag{1.8}$$

where $\rho_s$ is the density of the particles, and
$k_1$ is a constant whose value depends on the shape of the particle.

*Mean Sizes Based on Volume*

The mean abscissa in Fig. 1.5 is defined as the *volume mean diameter $d_v$*, or as the *weight mean diameter*. Then

$$d_v = \frac{\int_0^1 d \, dx}{\int_0^1 dx} = \int_0^1 d \, dx. \tag{1.9}$$

Expressing this relation in finite difference form:

$$d_v = \frac{\Sigma(d_1 x_1)}{\Sigma x_1} = \Sigma(x_1 d_1) \tag{1.10}$$

Writing in terms of particle numbers, rather than mass fractions:

$$d_v = \frac{\rho_s k_1 \Sigma(n_1 d_1^4)}{\rho_s k_1 \Sigma(n_1 d_1^3)} = \frac{\Sigma(n_1 d_1^4)}{\Sigma(n_1 d_1^3)} \tag{1.11}$$

Another mean size based on volume is the *mean volume diameter $d_v'$*. If all the particles are of diameter $d_v'$, then the total volume of particles is the same as in the mixture.
   Thus:

$$k_1 d_v'^3 \Sigma n_1 = \Sigma(k_1 n_1 d_1^3)$$

i.e.

$$d_v' = \sqrt[3]{\frac{\Sigma(n_1 d_1^3)}{\Sigma n_1}} \tag{1.12}$$

Substituting from equation 1.6:

$$d_v' = \sqrt[3]{\frac{\Sigma x_1}{\Sigma(x_1/d_1^3)}} = \sqrt[3]{\frac{1}{\Sigma(x_1/d_1^3)}} \tag{1.13}$$

*Mean Sizes Based on Surface*

If in Fig. 1.5, instead of fraction of total mass, the surface in each fraction is plotted against size, a similar curve is obtained but the mean abscissa $d_s$ is then the *surface mean diameter*.
   Then

$$d_s = \frac{\Sigma[(n_1 d_1)S_1]}{\Sigma(n_1 S_1)} = \frac{\Sigma(n_1 k_2 d_1^3)}{\Sigma(n_1 k_2 d_1^2)} = \frac{\Sigma(n_1 d_1^3)}{\Sigma(n_1 d_1^2)} \tag{1.14}$$

where $S_1 = k_2 d_1^2$, $k_2$ being a constant whose value depends on particle shape. $d_s$ is also known as the *Sauter mean diameter* and is the diameter of the particle with the same specific surface as the powder.
   Substituting for $n_1$ from equation 1.6:

$$d_s = \frac{\Sigma x_1}{\Sigma\left(\dfrac{x_1}{d_1}\right)} = \frac{1}{\Sigma\left(\dfrac{x_1}{d_1}\right)} \tag{1.15}$$

The *mean surface diameter* is defined as the size of particle $d'_s$ which is such that if all the particles are of this size, the total surface will be the same as in the mixture.

Thus

$$k_2 d'^2_s \Sigma n_1 = \Sigma(k_2 n_1 d_1^2)$$

i.e.

$$d'_s = \sqrt{\frac{\Sigma(n_1 d_1^2)}{\Sigma n_1}} \qquad (1.16)$$

Substituting for $n_1$:

$$d'_s = \sqrt{\frac{\Sigma(x_1/d_1)}{\Sigma(x_1/d_1^3)}} \qquad (1.17)$$

*Mean Dimensions Based on Length*

Similarly, a *length mean diameter* may be defined as:

$$d_l = \frac{\Sigma[(n_1 d_1)d_1]}{\Sigma(n_1 d_1)} = \frac{\Sigma(n_1 d_1^2)}{\Sigma(n_1 d_1)}$$

$$= \frac{\Sigma\left(\dfrac{x_1}{d_1}\right)}{\Sigma\left(\dfrac{x_1}{d_1^2}\right)} \qquad (1.18)$$

A *mean length diameter* or arithmetic mean diameter may also be defined by the relation:

$$d'_l \Sigma n_1 = \Sigma(n_1 d_1)$$

$$d'_l = \frac{\Sigma(n_1 d_1)}{\Sigma n_1}$$

$$= \frac{\Sigma\left(\dfrac{x_1}{d_1^2}\right)}{\Sigma\left(\dfrac{x_1}{d_1^3}\right)} \qquad (1.19)$$

**Example 1.1**

The size analysis of a powdered material on a weight basis is represented by a straight line from 0 per cent weight at 1 μm particle size to 100 per cent weight at 101 μm particle size (Fig. 1.7). Calculate the surface mean diameter of the particles constituting the system.

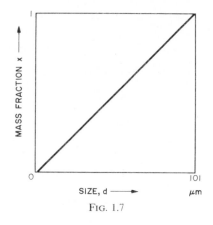

FIG. 1.7

**Solution**

From equation 1.15, the surface mean diameter is given by:

$$d_s = \frac{1}{\Sigma\left(\dfrac{x_1}{d_1}\right)}$$

Since the size analysis is represented by the continuous curve:

$$d = 100x + 1$$

$$d_s = \frac{1}{\displaystyle\int_0^1 \frac{dx}{d}}$$

$$= \frac{1}{\displaystyle\int_0^1 \frac{dx}{100x + 1}}$$

$$= \frac{100}{\ln 101}$$

$$= 21.7 \, \mu m$$

## 1.3. PARTICULATE SOLIDS IN BULK

### 1.3.1. General Characteristics

The properties of solids in bulk will be a function of the properties of the individual particles including their shapes and sizes and size distribution, and of the way in which the particles interact with one another. Furthermore, by the very nature of a particulate material, it is always interspersed with a fluid, generally air, and the interaction between the fluid and the particles may have a considerable effect on the behaviour of the bulk material. Particulate solids present considerably greater problems than fluids in storage, in removal at a controlled rate from storage, and when introduced into vessels or reactors where they become involved in a process. Although there has recently been a considerable amount of research on the properties and behaviour of solids in bulk, there is still a considerable lack of understanding of all the factors determining their behaviour.

One of the most important characteristics of any particulate mass is its voidage, the fraction of the total volume which is made up of the free space between the particles which is filled with fluid. Clearly a low voidage corresponds to a high density of packing of the particles. The way in which the particles pack depends not only on their physical properties, including shape and size distribution, but also on the way in which the particulate mass has been introduced to its particular location. In general, isometric particles, which have approximately the same linear dimension in each of the three principal directions, will pack more densely than long thin particles or plates. The more rapidly material is poured on to a surface or into a vessel the more densely will it pack; if it is then subjected to vibration further consolidation may occur. The packing density or voidage is important in that it determines the bulk density of the material, and hence the volume taken up by a given mass; it affects the tendency for agglomeration of the particles and it critically influences the resistance which the material offers to the percolation of fluid through it—as, for example, in filtration (see Chapter 9).

## 1.3.2. Agglomeration

Because it is necessary in chemical plant to transfer material from storage to process, it is important to know how the particulate material will flow. If a significant amount of the material is in the form of particles smaller than $10\,\mu m$ or if the particles deviate substantially from isometric form, it can be inferred that the flow characteristics will be poor; if the particles tend to agglomerate again poor flow properties may be expected. Agglomeration arises from interaction between particles, as a result of which they adhere to one another to form clusters. The main mechanisms giving rise to agglomeration are as follows:

(1) *Mechanical interlocking.* This can occur particularly if the particles are long and thin in shape in which case large masses may become completely interlocked.

(2) *Surface attraction.* Surface forces, including van der Waal's forces, may give rise to substantial bonds between particles, particularly where particles are very fine ($<10\,\mu m$) with the result that their surface per unit volume is high. In general, freshly formed surface, such as that resulting from particle fracture, gives rise to high surface forces.

(3) *Plastic welding.* When irregular particles are in contact the forces between the particles will be borne on extremely small surfaces and the very high pressures developed may give rise to plastic welding.

(4) *Electrostatic attraction.* Particles may become charged as they are fed into equipment and significant electrostatic charges may be built up, particularly on fine solids.

(5) *Effect of moisture.* Moisture may have two effects. First, it will tend to collect near the points of contact between particles and give rise to surface tension effects. Secondly, it may dissolve a little of the solid, which then acts as a bonding agent on subsequent evaporation.

(6) *Temperature fluctuations* give rise to changes in particle structure and to greater cohesiveness.

### 1.3.3. Resistance to Shear and Tensile Forces

A particulate mass may offer a significant resistance to both shear and tensile forces and this is specially marked when there is a significant amount of agglomeration. However, even in non-agglomerating powders there is some resistance to relative movement between the particles and it is always necessary for the bed to dilate (i.e. for the particles to move apart) to some extent before internal movement can take place. The greater the density of packing, the higher will be this resistance to shear and tension.

The resistance of a particulate mass to shear may be measured in a shear cell such as that described by Jenike[9,10]. The powder is contained in a shallow cylindrical cell, axis vertical, which is split horizontally. The lower half of the cell is fixed and the upper half is subjected to a shear force which is applied slowly and continuously measured. The shearing is carried out for a range of normal loads and the relationship between shear force and normal force is measured to give the shear strength at different degrees of compaction.

A method of measuring tensile strength has been developed at the Warren Spring Laboratory by Ashton *et al.*[11] who also used a cylindrical cell but this time split diametrically. One half of the cell is fixed and the other, which is movable, is connected to a spring, the other end of which is driven at a slow constant speed. A slowly increasing force is thus exerted on the powder compact and the point at which failure occurs determines the tensile strength; this has been shown to depend on the degree of compaction of the powder.

The magnitude of the shear and tensile strength of the powder has a considerable effect on the way in which the powder will flow, and particularly on the way in which it will discharge from a storage hopper through an outlet nozzle.

### 1.3.4. Angles of Repose and of Friction

A rapid method of assessing the behaviour of a particulate mass is to measure its *angle of repose*. If solid is poured from a nozzle on to a plane surface it will form an approximately conical heap and the angle between the sloping side of the cone and the horizontal is the angle of repose; when determined in this manner it is sometimes referred to as the *dynamic angle of repose* or the *poured angle*. In practice, the heap will not be exactly conical and there will be irregularities in the sloping surface. Furthermore, there will be a tendency for large particles to roll down from the top and collect at the base, thus giving a greater angle at the top and a smaller angle at the bottom.

The angle of repose may also be measured using a plane sheet to which has been stuck a layer of particles from the powder. Loose powder is then poured on to the sheet which is then tilted until the powder slides. The angle of slide is known as the *static angle of repose* or the *drained angle*.

Angles of repose vary from about 20° with free-flowing solids to about 60° with solids with poor flow characteristics. In extreme cases of highly agglomerated solids, angles of repose up to nearly 90° can be obtained. Generally, material which is free of particles smaller than 100 μm has a low angle of repose.

Powders with low angles of repose tend to pack rapidly to give a high packing density almost immediately. If the angle of repose is large a loose structure is formed initially and the material subsequently consolidates if subjected to vibration.

An angle which is similar to the static angle of repose is the *angle of slide* which is measured in the same manner as the drained angle except that the surface is smooth and is not coated with a layer of particles.

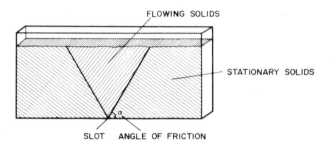

FIG. 1.8. Angle of friction—flow through slot

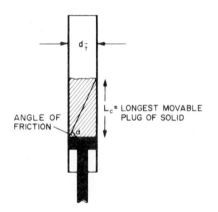

FIG. 1.9. Angle of friction—tube test

A measure of the frictional forces within the particulate mass is the *angle of friction*. This can be measured in a number of ways, two of which will be described. In the first, the powder is contained in a two-dimensional bed (Fig. 1.8) with transparent walls and is allowed to flow out through a slot in the centre of the base. It will be found that a triangular wedge of material in the centre flows out leaving stationary material at the outside. The angle between the cleavage between stationary and flowing material and the horizontal is the angle of friction. A simple alternative method[12] of measuring the angle of friction employs a vertical tube, open at the top, with a loosely fitting piston in the base (Fig. 1.9). With small quantities of solid in the tube, the piston will freely move upwards, but when a certain critical amount is exceeded no force, however large, will force the solids upwards in the tube. With the largest movable core of solids in the tube, the ratio of its length to diameter is the tangent of the angle of friction.

The angle of friction is important in its effect on the design of bins and hoppers. If the pressure at the base of a column of solids is measured as a function of depth, it is found to increase approximately linearly with height up to a certain critical point beyond which it remains constant. A typical curve is shown in Fig. 1.10. The point of discontinuity on the curve is given by:

$$L_c/d_t = \tan \alpha \qquad (1.20)$$

For heights greater than $L_c$ the weight of additional solids is supported by frictional forces at the walls of the hopper. It can thus be seen that hoppers must be designed to resist considerable pressures due to the solids acting on the walls.

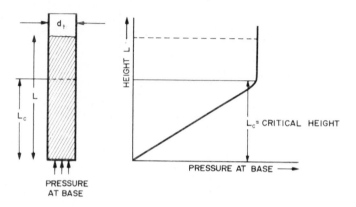

FIG. 1.10. Angle of friction—pressure at base of column

### 1.3.5. Flow of Solids in Hoppers

Solids may be stored in heaps or in sacks but the subsequent handling problems may be serious for large-scale operations. Frequently, therefore, the solids are stored in hoppers which are usually circular or rectangular in cross-section with conical or tapering sections at the bottom. The hopper will be filled at the top and it should be noted that, if there is an appreciable size distribution of the particles, some segregation may occur during filling with the larger particles tending to roll to the outside of the piles in the hopper.

Discharge from the hopper takes place through an aperture at the bottom of the cone and it is common to experience difficulties in obtaining a regular, or sometimes any, flow. Bridging of particles may take place and sometimes stable arches may form inside the hopper and, although these can usually be broken down by vibrators attached to the walls, problems of persistent blockage are not unknown. A further problem which is commonly encountered is that of "piping" or "rat-holing" in which the central core of material is discharged leaving a stagnant surrounding mass of solids. As a result some solids may be retained for long periods in the hopper and may deteriorate. Ideally, "mass flow" is required in which the solids are in plug flow and move downwards *en masse* in the hopper. The residence time of all particles in the hopper will then be the same.

In general, tall thin hoppers give better flow characteristics than short fat ones and the use of long small-angle conical sections at the base is advantageous. The nature of the surface of the hopper is important and smooth surfaces give improved discharge characteristics; monel metal cladding of steel is frequently used.

### 1.3.6. Flow of Solids through Orifices

The discharge rate of solid particles is usually controlled by the size of the orifice or aperture at the base of the hopper, though sometimes screw feeders or rotating table feeders may be incorporated to encourage an even flowrate.

The flow of solids through an orifice depends on the ability of the particles to dilate in the region of the aperture. Flow will occur if the shear force exerted by the super-incumbent material exceeds the shear strength of the powder near the outlet.

The rate of discharge of solids through the outlet orifice is substantially independent of the depth of solids in the hopper, provided this exceeds about four times the diameter of the hopper, and is proportional to the effective diameter of the orifice, raised to the power 2·5. The effective diameter is the actual orifice diameter less a correction which is equal to between 1 and $1\frac{1}{2}$ times the particle diameter.

Brown[13] has developed an equation for flow through an orifice, by assuming that the plug of solids issuing from the orifice has a minimum total potential plus kinetic energy. This gives:

$$G = \frac{\pi}{4} \rho_s d_{\text{eff}}^{2\cdot5} g^{0\cdot5} \left\{ \frac{1 - \cos \beta}{2 \sin^3 \beta} \right\}^{0\cdot5} \tag{1.21}$$

where  $G$  is the mass flowrate,

$\rho_s$  is the density of the solid particles,

$d_{\text{eff}}$  is the effective diameter of the orifice (orifice–particle diameter),

$g$  is the acceleration due to gravity, and

$\beta$  is the acute angle between the cone wall and the horizontal.

The attachment of a discharge pipe of the same diameter as the orifice immediately beneath it has been found to increase the flowrate, particularly of fine solids. Thus, in one case, with a pipe with a length to diameter ratio of 50, the discharge rate of a fine sand could be increased by 50 per cent and that of a coarse sand by 15 per cent. Another method of increasing the discharge rate of fine particles is to fluidise the particles in the neighbourhood of the orifice by the injection of air (see Chapter 6 for detailed discussion of fluidisation).

### 1.3.7. Conveying of Solids

The variety of needs in connection with the conveying of solids has led to the development of a wide range of equipment. This includes:

(a) *Gravity chutes*—down which the solids fall under the action of gravity.

(b) *Air slides*—where the particles, which are maintained partially suspended in a channel by the upward flow of air through a porous distributor, flow at a small angle to the horizontal.

(c) *Belt conveyors*—where the solids are conveyed horizontally, or at small angles to the horizontal, on a continuous moving belt.

(d) *Screw conveyors*—in which the solids are moved along a pipe or channel by a rotating helical impeller—shown in Fig. 1.11 as a screw lift elevator.

(e) *Bucket elevators*—in which the particles are carried upwards in buckets attached to a continuously moving vertical belt as illustrated in Fig. 1.12.

(f) *Vibrating conveyors*—in which the particles are subjected to an asymmetric

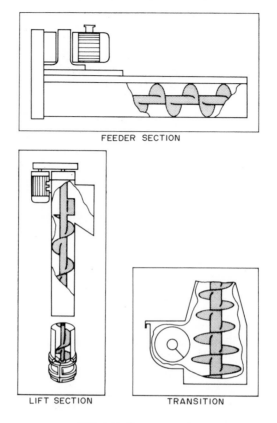

FEEDER SECTION

LIFT SECTION                    TRANSITION

FIG. 1.11. Screw conveyor

vibration and travel in a series of steps over a table. During the forward stroke of the table the particles are carried forward in contact with it, but the acceleration in the reverse stroke is so high that the table slips under the particles. With fine powders, vibration of sufficient intensity results in a fluid-like behaviour.

(g) *Pneumatic/hydraulic conveying installations*—in which the particles are transported in a stream of air/water; their operation is described in Chapter 7.

### 1.3.8. Measurement and Control of Solids Flowrate

The flowrate of solids can be measured either as they leave the hopper or as they are conveyed. In the former case, the hopper may be supported on load cells so that a continuous record of the weight of contents may be obtained as a function of time. Alternatively, the level of the solids in the hopper may be continuously monitored using transducers covered by flexible diaphragms flush with the walls of the hopper; the diaphragm responds to the presence of the solids and thus indicates whether there are solids present at a particular level.

Continuous weighing of a section of the belt on a belt conveyor provides a means of continuously recording the flowrate of solids. Over a portion of its travel the belt passes over support pulleys resting on load cells.

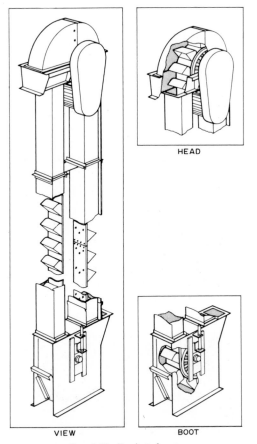

HEAD

VIEW    BOOT

FIG. 1.12. Bucket elevator

The rate of feed of solids may be controlled using screw feeders, rotating tables or vibrating feeders such as magnetically vibrated troughs. Volumetric rates may be controlled by regulating the speeds of rotation of star feeders or rotary vaned valves.

## 1.4. BLENDING OF SOLID PARTICLES

In the mixing of solid particles, three mechanisms may be involved:

(a) convective mixing, in which groups of particles are moved from one position to another,

(b) diffusion mixing, where the particles are distributed over a freshly developed interface, and

(c) shear mixing, where slipping planes are formed.

These three mechanisms will occur to varying extents in different kinds of mixers and with different kinds of particles. A trough mixer with a ribbon spiral will give almost pure convective mixing, but a simple barrel-mixer will give mainly a form of diffusion mixing. These mixers are illustrated in Figs. 1.18 and 1.19 later in this chapter.

The mixing of pastes is discussed in the chapter on Non-Newtonian Technology in Volume 3.

### 1.4.1. The Degree of Mixing

It is difficult to express the degree of mixing, but any index should be related to the properties of the required mix, should be easy to measure, and should be suitable for a variety of different mixers. When dealing with solid particles, the statistical variation in composition among samples withdrawn at any time from a mix is commonly used as a measure of the degree of mixing. The standard deviation $s$ (i.e. the square root of the mean

FIG. 1.13. Uniform mosaic

FIG. 1.14. Overall but not point uniformity in mix

of the squares of the individual deviations) or the variance $s^2$ is generally used. A particulate material cannot attain the perfect mixing that is possible with two fluids, for the best that can be obtained will be a degree of randomness in which two similar particles may well be side by side. No amount of mixing will lead to the formation of a uniform mosaic (Fig. 1.13), but only to a condition, such as Fig. 1.14, where there is an overall uniformity but not point uniformity. For a completely random mix of uniform particles distinguishable, say, only by colour, Lacey[14,15] has shown that:

$$s_r^2 = \frac{p(1-p)}{n} \tag{1.22}$$

where $s_r^2$ is the variance for the mixture, $p$ is the overall proportion of particles of one colour, and $n$ is the number of particles in each sample.

This at once brings out the importance of the size of the sample in relation to the size of the particles. In an incompletely randomised material, $s^2$ will be greater, and in a completely unmixed system (indicated by suffix 0) it can be shown that:

$$s_0^2 = p(1-p) \tag{1.23}$$

which is independent of the number of particles. Only a definite number of samples can in practice be taken from a mixture, and hence $s$ will itself be subject to random errors. The above analysis has been extended to systems containing particles of different sizes by Buslik[16].

When a material is partly mixed, then the degree of mixing can be represented by some term $M$, and several methods have been suggested for expressing $M$ in terms of measurable quantities. If $s$ is obtained from examination of a large number of samples then, as suggested by Lacey, $M$ may be defined as being equal to $s_r/s$ or, as more recently suggested by Kramers[17], $M = (s_0 - s)/(s_0 - s_r)$, where, as before, $s_0$ is the value of $s$ for the unmixed material. This form of expression is useful in that $M = 0$ for an unmixed material and 1 for a completely randomised material ($s = s_r$). If $s^2$ is used instead of $s$, then the above expression may be modified to give:

$$M = \frac{s_0^2 - s^2}{s_0^2 - s_r^2} \tag{1.24}$$

or

$$1 - M = \frac{s^2 - s_r^2}{s_0^2 - s_r^2} \tag{1.25}$$

For diffusive mixing, $M$ will be independent of sample size provided the sample is small. For convective mixing, Kramers has suggested that for groups of particles randomly distributed, each group behaves as a unit containing $n_g$ particles. As mixing proceeds $n_g$ becomes smaller. The number of groups will then be $n_p/n_g$, where $n_p$ is the number of particles in each sample. Applying equation 1.22:

$$s^2 = \frac{p(1-p)}{n_p/n_g} = n_g s_r^2$$

and this gives

$$1 - M = \frac{n_g s_r^2 - s_r^2}{n_p s_r^2 - s_r^2} = \frac{n_g - 1}{n_p - 1} \tag{1.26}$$

Thus, with convective mixing, $1 - M$ will depend on the sample size.

### 1.4.2. The Rate of Mixing

Expressions for the rate of mixing can be developed for any one of the possible mechanisms. Since mixing involves obtaining an equilibrium condition of uniform randomness, the relation between $M$ and time might be expected to be of the general form:

$$M = 1 - e^{-ct} \tag{1.27}$$

where $c$ is some constant depending on the nature of the particles and the physical action of the mixer.

Imagine a cylindrical vessel (Fig. 1.15) in which one substance **A** is poured into the bottom and the second **B** on top. Then the boundary surface between the two materials is a minimum. The process of mixing consists in making some of **A** enter the space occupied by **B**, and some of **B** enter the lower section originally filled by **A**. We may consider this as the diffusion of **A** across the initial boundary into **B**, and of **B** into A. This process continues until there is a maximum degree of dispersion, and a maximum interfacial area between the two materials. This type of process is somewhat akin to that of diffusion, and tentative use may be made of the relationship given by Fick's Law (see chapter 11 and chapter 8 in Volume 1). This law is applied in the following way.

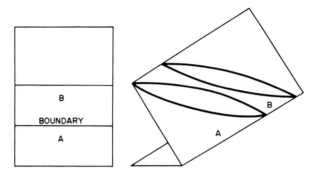

FIG. 1.15. Initial arrangement in mixing test

Let $a$ be the area of the interface per unit volume of the mix, and $a_m$ be the maximum interfacial surface per unit volume that can be obtained. Then

$$\frac{\mathrm{d}A}{\mathrm{d}t} = c(a_m - a) \tag{1.28}$$

and

$$a = a_m(1 - \mathrm{e}^{-ct}) \tag{1.29}$$

Suppose that, after any time $t$, a number of samples are removed from the mix, and are examined to see how many contain both components. If a sample contains both components, it will contain an element of the interfacial surface. If $Y$ is the fraction of the samples containing the two materials in approximately the proportion in the whole mix, then:

$$Y = 1 - \mathrm{e}^{-ct} \tag{1.30}$$

Coulson and Maitra[18] have examined the mixing of a number of pairs of materials in a simple drum mixer, and have expressed their results in the form of a plot of $\ln(100/X)$ vs. $t$, where $X$ is the percentage of the samples that is unmixed, i.e. $Y = 1 - (X/100)$. They were able to show that the constant $c$ depends on:
  (a) the total volume of the material,
  (b) the inclination of the drum,
  (c) the speed of rotation of the drum,
  (d) the particle size of each component,
  (e) the density of each component, and
  (f) the relative volume of each component.

Whilst the precise values of $c$ are only of value for the particular mixer under examination, they do bring out the effect of these variables. Thus, Fig. 1.16 shows the effect of speed of rotation, and the best results are obtained when the mixture is just not taken round by centrifugal action. If fine particles are put in at the bottom and coarse on the top,

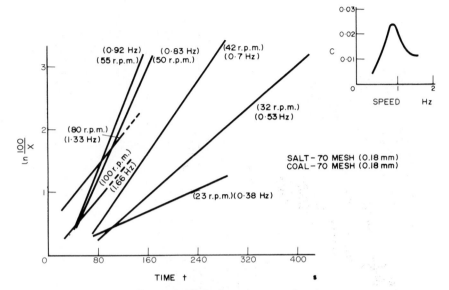

FIG. 1.16. Effect of speed on rate of mixing

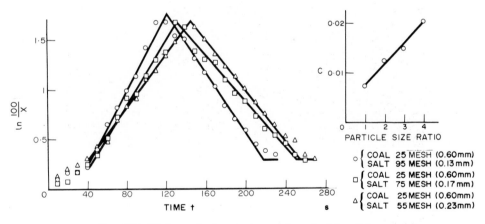

FIG. 1.17. Mixing and subsequent separation of solid particles

then on rotation no mixing occurs, the coarse remaining on top. If the coarse particles are put at the bottom and the fine on top, then on rotation mixing occurs to an appreciable extent, but on further rotation $Y$ falls, and the coarse particles settle out on the top. This is shown in Fig. 1.17 which shows a maximum degree of mixing. Again, with particles of the same size but differing density, the denser migrate to the bottom and the lighter to the top. Thus, if the lighter particles are put in at the bottom and the heavier at the top, rotation of the drum will give improved mixing to a definite value, after which the heavier particles will settle to the bottom. This kind of analysis of simple mixers has been made by Blumberg and Maritz[19] and by Kramers[17], but the results of their experiments are little more than qualitative.

FIG. 1.18. The Oblicone blender

Brothman *et al.*[20] have given an alternative theory for the process of mixing of solid particles, which has been described as a shuffling process. In their view, as mixing takes place there will be an increasing chance that a sample of a given size intercepts more than one part of the interface, and hence the number of mixed samples will increase less rapidly than the surface. This is further described by Herdan[1]. The application of some of these ideas to continuous mixing has been attempted by Danckwerts[21,22].

Two different types of industrial mixers are illustrated in Figs. 1.18 and 1.19. The Oblicone blender, which is available in sizes from 0·3 to 3 m, is used where ease of cleaning and 100 per cent discharge is essential, e.g. in the pharmaceutical, food or metal powder industries.

The Buss kneader is a continuous mixing or kneading machine in which the characteristic feature is that a reciprocating motion is superimposed on the rotation of the kneading screw, resulting in the interaction of specially profiled screw flights with rows of kneading teeth in the casing as shown in Fig. 1.19.

The motion of the screw causes the kneading teeth to pass between the flights of the screw at each stroke backward and forward. In this manner, a positive exchange of material occurs both in axial and radial directions, resulting in a homogeneous distribution of all components in a short casing length and with a short residence time.

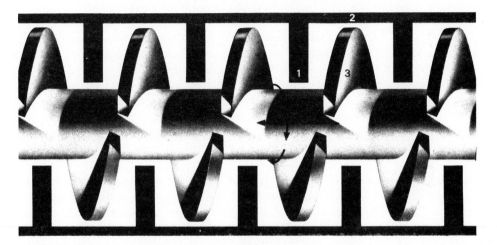

FIG. 1.19. The Buss kneader
1, Kneading tooth. 2, Kneader casing. 3, Screw flight

## 1.5. CLASSIFICATION OF SOLID PARTICLES

### 1.5.1. Introduction

The problem of separating solid particles according to their physical properties arises on a large scale in the mining industry, where it is necessary to separate the valuable constituents in a mineral from the adhering gangue, as it is called, which is usually of a lower density. In this case, it is first necessary to crush the material so that each individual particle contains only one constituent. The coal mining industry also is faced with a similar problem in coal washing plants in which dirt is separated from the clean coal. The chemical industry is more usually concerned with separating a single material, for instance, the product from a size reduction plant, into a number of size fractions, or to obtain a uniform material for incorporation in a system in which a chemical reaction takes place. As similar problems are involved in separating a mixture into its constituents and into size fractions, the two processes will here be considered together.

Separation depends on the selection of a process in which the behaviour of the material is influenced to a very marked degree by some physical property. Thus, if a material is to be separated into various size fractions, a sieving method may be used because this process depends primarily on the size of the particles, though other physical properties such as the shape of the particles and their tendency to agglomerate may also be involved. Other methods of separation depend on the differences in the behaviour of the particles in a moving fluid, and in this case the size and the density of the particles are the most important factors and shape is of secondary importance. Other processes make use of differences in electrical or magnetic properties of the materials or in their surface properties.

Generally, large particles are separated into size fractions by means of screens, and small particles, which would clog the fine apertures of the screen or for which it would be impracticable to make the openings sufficiently fine, are separated in a fluid. Fluid separation is commonly used for separating a mixture of two materials but magnetic, electrostatic and froth flotation methods are also used where appropriate.

Considerable development has taken place recently in techniques for size separation in the sub-sieve range. The emphasis has been on techniques lending themselves to automatic working[23]. Many of the methods of separation and much of the equipment have been developed for use in the mining and metallurgical industries[24].

Most processes which depend on differences in the behaviour of particles in a stream of fluid separate materials according to their terminal falling velocities (see Chapter 3), which in turn depend primarily on density and size and to a lesser extent on shape. Thus, in many cases it is possible to use the method to separate a mixture of two materials into its constituents, or to separate a mixture of particles of the same material into a number of size fractions.

Suppose that it is desired to separate particles of a relatively dense material **A** (density $\rho_A$) from particles of a less dense material **B**. If the size range is large, the terminal falling velocities of the largest particles of **B** (density $\rho_B$) may be greater than those of the smallest particles of **A**, and therefore a complete separation will not be possible. The maximum range of sizes that can be separated is calculated from the ratio of the sizes of the particles of the two materials which have the same terminal falling velocities. It will be shown in Chapter 3 that this condition is given by equation 3.25 as:

$$\frac{d_B}{d_A} = \left\{\frac{\rho_A - \rho}{\rho_B - \rho}\right\}^s \tag{1.31}$$

where $s = \frac{1}{2}$ when Stokes' Law applies, and
$\quad\quad s = 1$ when Newton's Law applies.

It is seen that this size range becomes wider with increase in the density of the separating fluid and, when the fluid has the same density as the less dense material, complete separation is possible whatever the relative sizes. Although water is the most commonly used fluid, a specific gravity greater than unity is obtained if hindered settling takes place. Frequently, the fluid density is increased artificially by forming a suspension of small particles of a very dense solid, such as galena, ferro-silicon, magnetite, sand or clay, in the water. These suspensions have effective specific gravities up to about 3·5. Alternatively zinc chloride or calcium chloride solutions may be used. On the small scale liquids with a range of densities are produced by mixing benzene and carbon tetrachloride.

**Example 1.2**

It is desired to separate into two pure fractions a mixture of quartz and galena of a size range from 0·015 mm to 0·065 mm by the use of a hindered settling process. What is the minimum apparent density of the fluid that will give this separation? Density of galena $= 7500\,\text{kg/m}^3$. Density of quartz $= 2650\,\text{kg/m}^3$.

## Solution

Assume shapes of galena and quartz particles are similar. From equation 1.31, the required density of fluid when Stokes' Law applies is given by:

$$\frac{0.065}{0.015} = \left\{\frac{7500 - \rho}{2650 - \rho}\right\}^{0.5}$$

∴
$$\rho = 2377\,\text{kg/m}^3$$

Required density of fluid when Newton's Law applies is given by:

$$\frac{0.065}{0.015} = \frac{7500 - \rho}{2650 - \rho}$$

∴
$$\rho = 1196\,\text{kg/m}^3$$

Thus the required density of the fluid is between 1196 and $2377\,\text{kg/m}^3$.

If the particles are allowed to settle in the fluid for only a very short time, they will not attain their terminal falling velocities and a better degree of separation can be obtained. A particle of material **A** will have an initial acceleration $g[1 - (\rho/\rho_A)]$, because there is no fluid friction when the relative velocity is zero. Thus the initial velocity is a function only of density and is unaffected by size and shape. A very small particle of the denser material will therefore always commence settling at a greater rate than a large particle of the less dense material. Theoretically, therefore, it should be possible to separate materials completely, irrespective of the size range, provided that the periods of settling are sufficiently short. In practice, the required periods will often be so short that it is impossible to make use of this principle alone, but a better degree of separation can be obtained if the particles are not allowed to become fully accelerated. As the time of settling increases the larger particles of the less dense material catch up and overtake the small heavy particles.

Size separation equipment in which particles move in a fluid stream will now be described. Most of the plant utilises the difference in the terminal falling velocities of the particles; in the hydraulic jig, however, the particles are allowed to settle for only very brief periods at a time, and this equipment can therefore be used when the size range of the material is large.

### 1.5.2. Gravity Settling

*The Settling Tank*

The material is introduced in suspension into a tank containing a relatively large volume of water moving at a low velocity as shown in Fig. 1.20. The particles soon enter the slowly moving water and, because the small particles settle at a lower rate, they are carried further forward before they reach the bottom of the tank; the very fine particles are carried away in the liquid overflow. Receptacles at various distances from the inlet collect different grades of particles according to their terminal falling velocities, the particles of high terminal falling velocity collecting near the inlet. The positions at which the particles

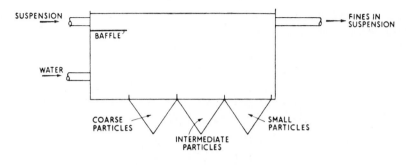

FIG. 1.20. Gravity settling tank

are collected can be calculated on the assumption that they rapidly reach their terminal falling velocities, and attain the same horizontal velocity as the fluid.

If the material is introduced in solid form down a chute, the position at which the particles are deposited will be determined by the rate at which they lose the horizontal component of their velocities. The larger particles will therefore be carried further forward and the smaller particles will be deposited near the inlet. The position at which the particles are deposited can be calculated using the relations given in Chapter 3.

### The Elutriator

Material can be separated by means of an elutriator, which consists of a vertical tube up which fluid is passed at a controlled velocity. The particles are introduced, often through a side tube, and the smaller particles are carried over in the fluid stream while the large particles settle against the upward current. Further size fractions can be collected if the overflow from the first tube is passed vertically upwards through a second tube of greater cross-section; any number of such tubes can be arranged in series.

### The Spitzkasten

This plant consists of a series of vessels of conical shape arranged in series. A suspension of the material is fed into the top of the first vessel and the larger particles settle, while the smaller ones are carried over in the liquid overflow and enter the top of a second conical vessel of greater cross-sectional area. The bottoms of the vessels are fitted with wide diameter outlets, and a stream of water can be introduced near the outlet so that the particles have to settle against a slowly rising stream of liquid. The size of the smallest particle which is collected in each of the vessels is influenced by the upward velocity at the bottom outlet of each of the vessels. The size of each successive vessel is increased, partly because the amount of liquid to be handled includes all the water used for classifying in the previous vessels, and partly because it is desired to reduce, in stages, the surface velocity of the fluid flowing from one vessel to the next. The Spitzkasten thus combines the principles used in the settling tank and in the elutriator.

The size of the material collected in each of the units is determined by the rate of feeding of suspension, the upward velocity of the liquid in the vessel and the diameter of the vessel.

FIG. 1.21. Dorrco sizer

The equipment can also be used for separating a mixture of materials into its constituents, provided that the size range is not large. The individual units can be made of wood or sheet metal.

The Dorrco sizer, illustrated in Fig. 1.21, works on the same principle but has a number of compartments trapezoidal in section. It is suitable for use with materials finer than about 4-mesh and it works at high concentrations in order to obtain the advantages of hindered settling.

### The Double Cone Classifier

This classifier (Fig. 1.22) consists of a conical vessel, with a second hollow cone of greater angle arranged apex downwards inside it so that there is an annular space of

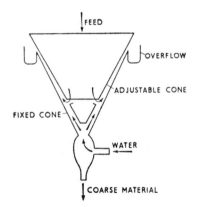

FIG. 1.22. Double cone classifier

approximately constant cross-section between the two cones. The bottom portion of the inner cone is cut away and its position relative to the outer cone can be regulated by a screw adjustment. Water is passed in an upward direction, as in the Spitzkasten, and overflows into a launder arranged round the whole of the periphery of the outer cone. The material to be separated is fed in suspension to the centre of the inner cone, and the liquid level is maintained slightly higher than the overflow level, so that there is a continuous flow of liquid downwards in the centre cone. The particles are therefore brought into the annular space where they are subjected to a classifying action; the smaller particles are carried away in the overflow, and the larger particles settle against the liquid stream and are taken off at the bottom.

### The Rake Classifier

Several forms of classifier exist in which the material of lower settling velocity is carried away in a liquid overflow, and the material of higher settling velocity is deposited on the bottom of the equipment and is dragged upwards against the flow of the liquid, by some mechanical means. During the course of the raking action, the solids are turned over so that any small particles trapped under larger ones are brought to the top again.

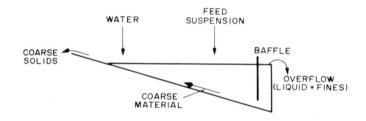

FIG. 1.23. Principle of operation of Dorr classifier

In the Dorr classifier, which consists of a shallow rectangular tank inclined to the horizontal, the feed is introduced in the form of a suspension near the middle of the tank and water for classifying is added at the upper end (see Fig. 1.23). The liquid, together with the material of low terminal falling velocity, flows down the tank under a baffle and is then discharged over an overflow weir. The heavy material settles to the bottom and is then dragged upwards by means of a rake; it is thus separated from the liquid and is discharged at the upper end of the tank. The Dorr classifier is extensively used in conjunction with ball mills where it is often necessary to separate a large amount of fines from the oversize material. Arrangement of a number of Dorr classifiers in series makes it possible for a material to be separated into several size fractions.

Numerous other classifiers of a similar type are made; they include the Stokes double-acting classifier (Fig. 1.24), in which a more uniform discharge of the heavy solids is obtained by the use of two sets of rakes suitably staggered.

The Akins classifier and the Denver cross-flow classifier (Fig. 1.25) are similar in action but employ a trough which is semicircular in cross-section, and the material which settles to the bottom is continuously moved to the upper end by means of a rotating helical scraper.

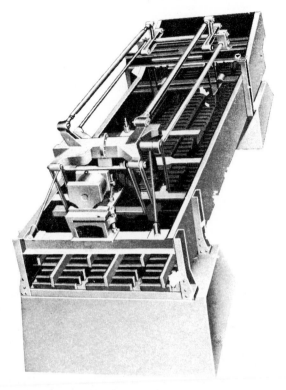

FIG. 1.24. Stokes double-acting rake classifier

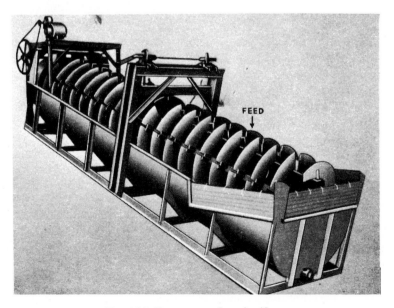

FIG. 1.25. Denver cross-flow classifier

*The Bowl Classifier*

The bowl classifier, which is used for fine materials, consists of a shallow bowl with a concave bottom (Fig. 1.26). The suspension is fed into the centre of the bowl near the liquid surface, and the liquid and the fine particles are carried in a radial direction and pass into the overflow, an open launder running round the whole of the periphery of the bowl at the top. The heavier or larger material settles to the bottom and is raked towards the outlet at the centre. The classifier has a large overflow area, and consequently high volumetric rates of flow of liquid can be used without producing a high linear velocity at the overflow. Its action will be seen to be similar to that of a thickener which is effecting incomplete clarification.

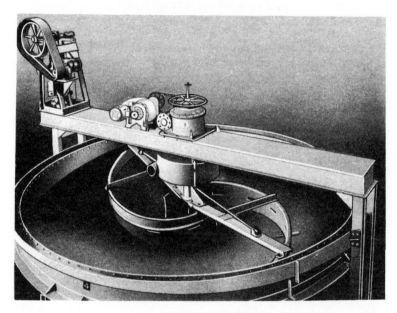

FIG. 1.26.  Bowl classifier

*The Hydraulic Jig*

The hydraulic jig operates by allowing material to settle for brief periods so that the particles do not attain their terminal falling velocities, and is therefore suitable for separating materials of wide size range into their constituents. The material to be separated is fed dry, or more usually in suspension, over a screen and is subjected to a pulsating action by liquid which is set in oscillation by means of a reciprocating plunger. The particles on the screen constitute a suspension of high concentration and therefore the advantages of hindered settling are obtained. The jig usually consists of a rectangular-section tank with a tapered bottom, divided into two portions by a vertical baffle. In one section, the plunger operates in a vertical direction; the other incorporates the screen over which the separation is carried out. In addition, a stream of liquid is fed to the jig during the upward stroke.

The particles on the screen are brought into suspension during the downward stroke of the plunger (Fig. 1.27a). As the water passes upwards the bed opens up, starting at the top, and thus tends to rise *en masse*. During the upward stroke the input of water is adjusted so that there is virtually no flow through the bed (Fig. 1.27b). During this period differential settling takes place and the denser material tends to collect near the screen and the lighter material above it. After a short time the material becomes divided into three strata, the bottom layer consisting of the large particles of the heavy material, the next of large particles of the lighter material together with small particles of the heavy material, and the top stratum of small particles of the light material. Large particles wedge at an earlier stage than small ones and therefore the small particles of the denser material are able to fall through the spaces between the larger particles of the light material. Many of these small particles then fall through the supporting gauze.

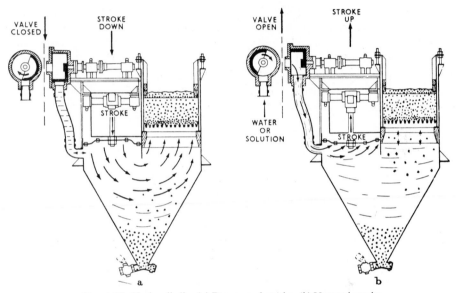

FIG. 1.27. Hydraulic jig: (*a*) Downward stroke, (*b*) Upward stroke

Four separate fractions are obtained from the jig and the successful operation of the plant depends on their rapid removal. The small particles of the heavy material which have fallen through the gauze are taken off at the bottom of the tank. The larger particles of each of the materials are retained on the gauze in two layers: the denser material at the bottom and the less dense material on top. These two fractions are removed through gates at the side of the jig. The remaining material, consisting of small particles of the light material, is carried away in the liquid overflow.

*Riffled Tables*

The riffled table, of which the Wilfley table (Fig. 1.28) is a typical example, consists of a flat table which is inclined at an angle of about 3° to the horizontal. Running approximately parallel to the top edge is a series of slats, or riffles as they are termed, about 6 mm in height. The material to be separated is fed on to one of the top corners and a

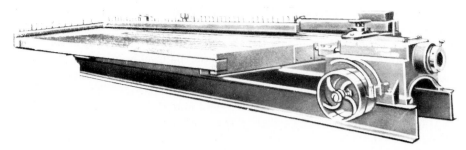

FIG. 1.28. Wilfley riffled table

reciprocating motion, consisting of a slow movement in the forward direction and a very rapid return stroke, causes it to move across the table. The particles also tend to move downwards under the combined action of gravity and of a stream of water which is introduced along the top edge, but are opposed by the riffles behind which the smaller particles or the denser material tend to be held. Thus, the large particles and the less dense material are carried downwards, and the remainder is carried parallel to the riffles. In many cases, each riffle is tapered with the result that a number of fractions can be collected.

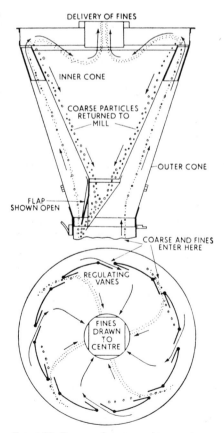

FIG. 1.29. Raymond vacuum air separator

Riffled tables can be used for separating materials down to about 50 μm in size, provided that the difference in the densities is large.

### 1.5.3. Centrifugal Separators

The use of cyclone separators for the removal of suspended dust particles from gases will be referred to in Chapter 8. By suitable choice of operating conditions, it is possible to use centrifugal methods for the classification of solid particles according to their terminal falling velocities. Figure 1.29 shows a typical air separation unit which is similar in construction to the double cone-classifier. The solids are fed to the bottom of the annular space between the cones and are carried upwards in the air stream and enter the inner cone through a series of ports fitted with adjustable vanes. As will be seen in the diagram, the suspension enters approximately tangentially and is therefore subjected to the action of centrifugal force. The coarse solids are thrown outwards against the walls and fall to the bottom under the action of gravity, while the small particles are removed by means of an exhaust fan. This type of separator is widely used for separating the oversize material from the product from a ball mill, and is suitable for materials as fine as 50 μm.

A mechanical air separator is shown in Fig. 1.30. The material is introduced at the top through the hollow shaft and falls on to the rotating disc which throws it outwards. Very

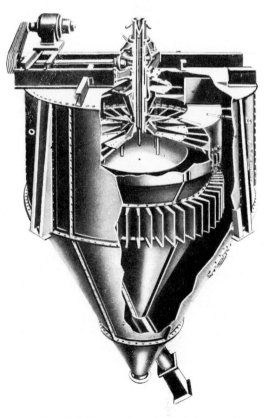

FIG. 1.30. Raymond mechanical air separator

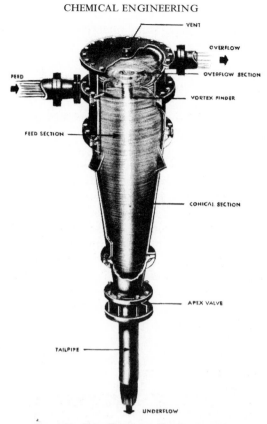

VENT

OVERFLOW

OVERFLOW SECTION

FEED

VORTEX FINDER

FEED SECTION

CONICAL SECTION

APEX VALVE

TAILPIPE

UNDERFLOW

FIG. 1.31. DorrClone centrifugal separator

large particles fall into the inner cone and the remainder are lifted by the air current produced by the rotating vanes above the disc. Because a rotary motion has been imparted to the air stream, the coarser particles are thrown to the walls of the inner cone and, together with the very large particles, are withdrawn from the bottom. The fine particles remain in suspension and are carried down the space between the two cones and are collected from the bottom of the outer cone. The air returns to the inner cone through a series of deflector vanes, and blows any fines from the surfaces of the larger particles as they fall. The size at which the "cut" is made is controlled by the number of fixed and rotating blades and the speed of rotation.

Considerable interest has recently been shown in the use of hydraulic cyclone separators and the results of preliminary laboratory investigations have been reported by Fern[25] and by Kelsall[26,27]. The movement of aluminium particles has been studied, and it appears that the flow pattern is similar to that obtained in the cyclone air separator. The main interest in this plant is for the washing of coal, but it is now being used in the metallurgical industry. A commercial model, the DorrClone, is illustrated in Fig. 1.31.

### 1.5.4. Sieves or Screens

Sieves or screens are used industrially on a large scale for the separation of particles according to their sizes, and on the small scale for the production of closely graded materials and for the carrying out of size analyses. The method is applicable for particles of size down to about 50 μm, but not for very fine materials because of the difficulty of producing accurately woven fine gauzes of sufficient strength and because the screens become clogged. Woven wire cloth is generally used for fine sizes and perforated plates for the larger meshes. Some large industrial screens are formed either from a series of parallel rods or from H-shaped links bolted together, though square or circular openings are more usual.

The only large screen that is hand operated is the grizzly. This has a plane screening surface composed of longitudinal bars up to 3 m long and fixed in a rectangular framework. In some cases, a reciprocating motion is imparted to alternate bars so as to

Fig. 1.32. Hummer electromagnetic screen

reduce the risk of clogging. The grizzly is usually inclined at an angle to the horizontal; the greater the angle the greater is the throughput but the lower is the screening efficiency. If the grizzly is used for wet screening, a very much smaller angle is employed. In some screens the longitudinal bars are replaced by a perforated plate.

The mechanically operated screens are vibrated by means of an electromagnetic device (Fig. 1.32), or mechanically (Fig. 1.33). In the former case the screen itself is vibrated, and in the latter the whole assembly. Because very rapid accelerations and retardations are produced, the power consumption and the wear on the bearings are high. These screens are sometimes mounted in a multi-deck fashion with the coarsest screen on the top, either horizontally or inclined at angles up to 45°. With the horizontal machine, the vibratory motion fulfils the additional function of moving the particles across the screen.

The screen area which is required for a given operation cannot be predicted without testing the material under similar conditions on the small plant. In particular, the tendency of the material to clog the screening surface can only be determined by experiment.

FIG. 1.33. Tyrock mechanical screen

A very large mechanically operated screen is the trommel (Fig. 1.34), which consists of a slowly rotating perforated cylinder with its axis at a slight angle to the horizontal. The material to be screened is fed in at the top and gradually moves down the screen and passes over apertures of gradually increasing size, with the result that all the material has to pass over the finest screen. There is therefore a tendency for blockage of the apertures by the large material and for oversize particles to be forced through; further, the relatively fragile fine screen is subjected to the abrasive action of the large particles. These difficulties are obviated to some extent by arranging the screens in the form of concentric cylinders with the coarsest in the centre. The disadvantage of all screens of this type is that only a small fraction of the screening area is in use at any one time. The speed of rotation of the trommel should not be high enough for the material to be carried completely round in contact with the screening surface. The lowest speed at which this occurs is known as the critical speed

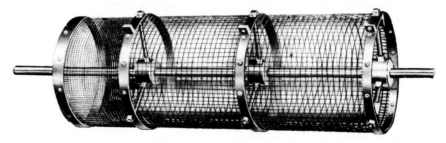

FIG. 1.34. Trommel

and is analogous to the critical speed of the ball mill; speeds of between one-third and a half of the critical speed are usually recommended. In a modified form of the trommel, the screen surfaces are in the form of truncated cones. Such screens are mounted with their axes horizontal and the material flows away from the apex of the cone.

### 1.5.5. Magnetic Separators

In the magnetic separator, material is passed through the field of an electromagnet which causes the retention or retardation of the magnetic constituent. It is important that the material should be supplied as a thin sheet in order that all the particles are subjected to a field of the same intensity and so that the free movement of individual particles is not impeded. The two main types of equipment are as follows:

(a) Eliminators are used for the removal of small quantities of magnetic material from the charge to a plant. They are frequently employed, for example, for the removal of stray pieces of scrap iron from the feed to crushing equipment. A common type of eliminator is a magnetic pulley (Fig. 1.35) incorporated in a belt conveyor so that the non-magnetic material is discharged in the normal manner but the magnetic material adheres to the belt and falls off from the underside.

FIG. 1.35. Magnetic pulley

(b) Concentrators are used for the separation of magnetic ores from the accompanying mineral matter. The Ball–Norton machine (Fig. 1.36), which is a typical concentrator, employs two horizontally staggered belt conveyors running parallel, one above the other. The material is fed as a thin sheet to the lower belt and is conveyed under the second belt where it is subjected to the action of a magnetic field. The non-magnetic material is discharged in the normal manner, but the magnetic material adheres to the lower side of the upper belt. The time for which the material is subjected to the magnetic field can be varied by modifying the speed of the belts or by altering the overlap.

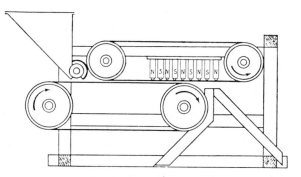

FIG. 1.36.  Ball–Norton magnetic separator

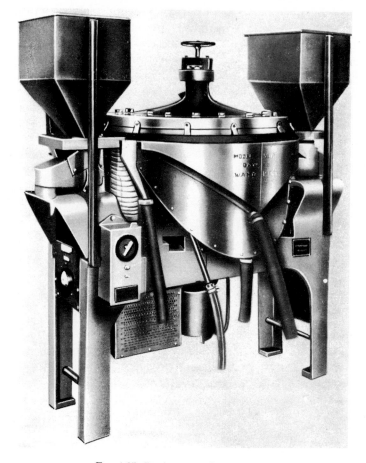

FIG. 1.37.  Davies magnetic separator

Another concentrator is the Davies magnetic separator (Fig. 1.37) in which material is fed in a thin sheet over a magnetic feed plate and the magnetic constituent is picked up by rotating electromagnets. The current is automatically switched off as the electromagnets pass over a collecting bin into which the magnetic material then falls. The non-magnetic constituent passes directly over the feed plate into a collecting hopper.

### 1.5.6. Electrostatic Separators

Electrostatic separators, in which differences in the electrical properties of the materials are exploited in order to effect a separation, are now sometimes used with small quantities of fine material. The solids are fed from a hopper on to a rotating drum, which is either charged or earthed, and an electrode bearing the opposite charge is situated at a small distance from the drum (Fig. 1.38). The point at which the material leaves the drum is determined by the charge it acquires, and by suitable arrangement of the collecting bins (*A, B, C*) a sharp classification can be obtained.

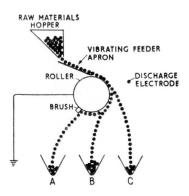

FIG. 1.38. Electrostatic separator

### 1.5.7. Flotation

*Froth Flotation*

Separation of a mixture using flotation methods depends on differences in the surface properties of the materials involved. If the mixture is suspended in an aerated liquid, the gas bubbles will tend to adhere preferentially to one of the constituents—the one which is more difficult to wet by the liquid—and its effective density may be reduced to such an extent that it will rise to the surface. If a suitable frothing agent is added to the liquid, the particles will be held in the surface by means of the froth until they can be discharged over a weir. Froth flotation is widely used in the metallurgical industries where, generally, the ore is difficult to wet and the residual earth is readily wetted.

The process depends on the existence, or development, of a selective affinity of one of the constituents for the envelopes of the gas bubbles. In general, this affinity must be induced, and the reagents which increase the angle of contact between the liquid and one of the materials are known as promoters and collectors. Promoters are selectively adsorbed on the surface of one material and form a monomolecular layer; use of excess material destroys the effect. Concentrations of the order of 0·05 kg/Mg of solids are usually required. A commonly used promoter is sodium ethyl xanthate which has the formula

$$\begin{matrix} NaS-\!\!\!\diagdown \\ \phantom{NaS}\diagup\!\!C-O-C_2H_5 \\ S=\!\!\!\diagup \end{matrix}$$

In solution this material ionises giving positively charged sodium ions and negatively charged xanthate ions. The polar end of the xanthate ion is adsorbed and the new surface of the particles is therefore made up of the non-polar part of the radicle, so that the contact angle with water is increased. A large number of other materials are used in particular cases; for instance, other xanthates and diazoaminobenzene. Collectors are materials which form surface films on the particles; these films are rather thicker than those produced by promoters, and collectors therefore have to be added in higher concentrations—about 0·5 kg/Mg of solids. Pine oil is a commonly used collector, and petroleum compounds are frequently used though they often form very greasy froth which is then difficult to disperse.

In many cases, it is necessary to modify the surface of the other constituent so that it does not adsorb the collector or promoter. This is effected by means of materials known as modifiers which are either adsorbed on the surface of the particles or react chemically at the surface, and thereby prevent the adsorption of the collector or promoter. Mineral acids, alkalis, and salts are frequently used for this purpose.

An essential requirement of the process is the production of a froth of sufficient stability to retain the particles of the one constituent in the surface so that they can be discharged over the overflow weir. On the other hand, the froth should not be so stable that it is difficult to break down at a later stage. The frothing agent should reduce the interfacial tension between the liquid and gas, and the quantities required are less when the frothing agent is adsorbed at the interface. Liquid soaps, soluble oils, cresol and alcohols, in the range between amyl and octyl alcohols, are frequently used as frothing agents. In many cases, the stability of the foam is increased by adding a stabiliser—usually a mineral oil— which increases the viscosity of the liquid film forming the bubble, and therefore reduces the rate at which liquid can drain from the envelope. Pine oil is sometimes used as a frothing agent; it produces a stable froth and acts as a collector, in addition.

The reagents which are used in froth flotation are usually specific in their action and the choice of suitable reagents is usually made as a result of tests on small-scale equipment. Mixtures of more than two components can be separated in stages using different reagents at each stage. Sometimes the behaviour of a system is considerably influenced by change in the pH of the solution, and it is then possible to remove materials at successive stages by progressive alterations in the pH. In general, froth flotation processes can be used for particles between about 5 and 250 μm in size. The tendency of various materials to respond to froth flotation methods is given by Doughty[28].

### The Agitair Flotation Machine

The Agitair is an example of a typical froth flotation unit used in the mining industry and is illustrated in Fig. 1.39. The pulp is introduced through a feed box and is distributed over the entire width of the first cell. Circulation of the pulp through each cell is such that

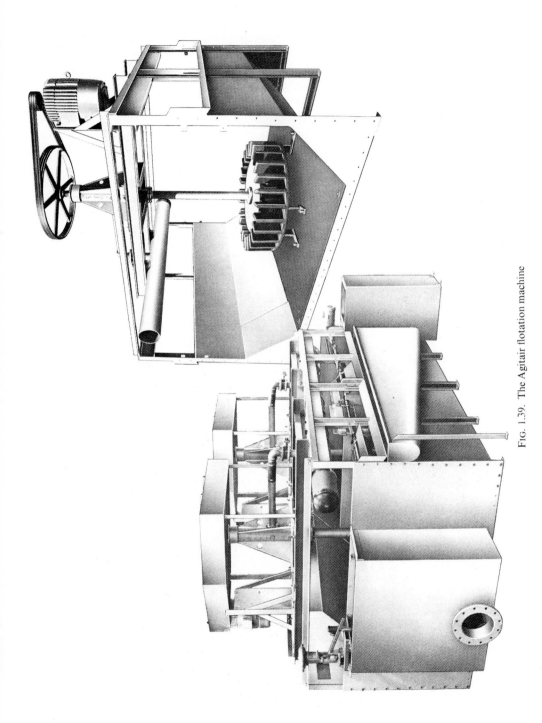

FIG. 1.39. The Agitair flotation machine

as the pulp comes into contact with the impeller it is subjected to intense agitation and aeration. Low pressure air for this purpose is introduced down the hollow impeller shaft. It is thoroughly disseminated throughout the pulp in the form of minute bubbles when it enters the impeller/stabiliser zone, thus assuring maximum contact with the solids.

The flow pattern of the aerated pulp from the impeller is downwards and outwards. As the aerated pulp leaves the impeller, contact is made with the stabiliser which peels the masses of bubbles from the impeller fingers. The stabiliser then directs them to the surface where the valuable solids are collected and discharged at the froth overflow. In addition, the stabiliser eliminates dead spots in the agitation zone and prevents accumulation of sands or granular materials in the corners or on the bottom of the cells. Total cell volume is thus utilised at all times.

Each cell is provided with an individually controlled air valve. Air pressure can be set between 106 and 115 kN/m$^2$ depending on the depth and size of the machine and the pulp density. Adjustment is achieved by opening the valve until there is a "boiling action" or breaking through of pulp into the froth, when the valves should be gradually closed until the froth surface becomes quiet. Typical energy requirements for this machine range from 70 W to 3·7 kW for airflow rates of $1·4 \times 10^{-3}$ to 0·2 m$^3$/s respectively.

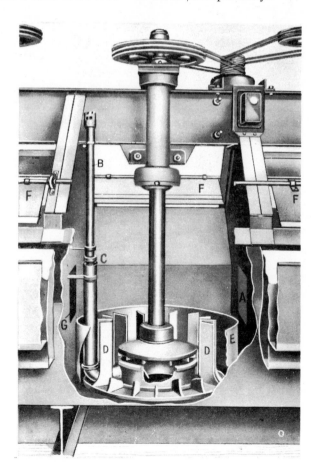

A froth flotation cell used for coal washing is illustrated in Fig. 1.40. The suspension, containing about 20 per cent of solids together with the necessary reagents, enters at *A*, and the rotating impeller draws in air through the pipe *B*, together with some suspension. The air volume is controlled by a valve at the inlet, and the volume of suspension by an adjustable cone at *C*. The aerated suspension passes through the baffles *D* which destroy the vortex, and then is thrown against a rubber-covered wear ring *E*. The froth rises to the surface and is driven over the froth weirs by means of rotating paddles *F*. The rejects are discharged through the port *G*, either to waste or to a further unit. In the latter case, the flow is maintained by allowing a small fall in liquor level from cell to cell.

## Electroflotation

Electroflotation represents an interesting recent development of the flotation process for the treatment of dilute suspensions[29]. Gas bubbles are generated electrolytically within the suspension, and attach themselves to the suspended particles which then rise to the surface. Because the bubbles are very small, they have a high surface:volume ratio and are therefore very effective in suspensions of fine particles.

The method has been developed particularly for the treatment of dilute industrial effluents, including suspensions and colloids, especially those containing small quantities of organic materials. It allows the dilute suspension to be separated into a concentrated slurry and a clear liquid and, at the same time, permits oxidation of unwanted organic materials at the positive electrode.

Typically, the electrodes will be of lead dioxide on a titanium substrate in the form of horizontal perforated plates, the distance apart, usually 5–40 mm, depending on the conductivity of the liquid. A potential difference of 5 to 10 V may be applied to give

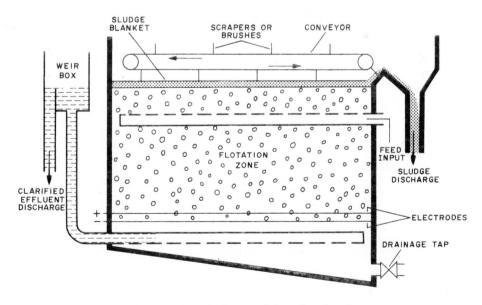

FIG. 1.41. Schematic diagram of electroflotation plant

current densities of the order of $100 \, A/m^2$. Frequently, the conductivity of the suspension itself is adequate, but it may be necessary to add ionic materials, such as sodium chloride or sulphuric acid. Electrode fouling can usually be prevented by periodically reversing the polarity of the electrodes. Occasionally, consumable iron or aluminium anodes may be used because the ions released into the suspension may then assist flocculation of the suspended solids.

The electroflotation plant usually consists of a steel or concrete tank with a sloping bottom as shown in Fig. 1.41; liquid depth may typically be about 1 m. Since the flotation process is much faster than sedimentation with fine particles, flotation can be achieved with much shorter retention times—usually about 1 hour (3·6 ks)—and the land area required may be only about one-eighth of that for a sedimentation tank.

The floated sludge, which frequently contains 95 per cent of the solids, forms a blanket on the liquid surface which can be continuously removed by means of slowly moving brushes or scrapers mounted across the top of the tank.

## 1.6. FURTHER READING

DALLAVALLE, J. M.: *Micromeritics*, 2nd edn. (Pitman, 1948).
HERDAN, G.: *Small Particle Statistics*, 2nd edn. (Butterworths, London, 1960).
STERBACEK, Z. and TAUSK, P.: *Mixing in the Chemical Industry* (translated from Czech by K. MAYER) (Pergamon Press, 1965).
TAGGART, A. F.: *Handbook of Mineral Dressing* (Wiley, 1945).

## 1.7. REFERENCES

1. HERDAN, G.: *Small Particle Statistics*, 2nd edn. (Butterworths, 1960).
2. DALLAVALLE, J. M.: *Micromeritics*, 2nd edn. (Pitman, 1948).
3. ROSE, J. W. and COOPER, J. R. (Eds.) *Technical Data on Fuel*, 7th edn. (British National Committee, World Energy Conference, 1977).
4. B.S. 410: British Standard 410 (1962) Specification for test sieves.
5. International Combustion Ltd., Derby. Private communication (1977).
6. B.S. 893: British Standard 893 (1940). The method of testing dust extraction plant and the emission of solids from chimneys of electric power stations.
7. BOND, F. C.: *Brit. Chem. Eng.* **8** (1963) 631. Some recent advances in grinding theory and practice.
8. BOND, F. C.: *Chem. Eng. Albany* **71**, No. 6 (16 Mar. 1964) 134. Costs of process equipment.
9. JENIKE, A. W., ELSEY, P. J., and WOOLLEY, R. H.: *Proc. Amer. Soc. Test. Mat.* **60** (1960) 1168. Flow properties of bulk solids.
10. JENIKE, A. W.: *Trans. I. Chem. E.* **40** (1962) 264. Gravity flow of solids.
11. ASHTON, M. D., FARLEY, R., and VALENTIN, F. H. H.; *J. Sci. Instrum.* **41** (1964) 763. An improved apparatus for measuring the tensile strength of powders.
12. ZENZ, F. A. and OTHMER, D. F.: *Fluidization and Fluid Particle Systems* (Reinhold, 1960).
13. BROWN, R. L.: *Nature* **191**, No. 4.787 (1961) 458. Minimum energy theorem for flow of dry granules through apertures.
14. LACEY, P. M. C.: *Trans. Inst. Chem. Eng.* **21** (1943) 53. The mixing of solid particles.
15. LACEY, P. M. C.: *J. Appl. Chem.* **4** (1954) 257. Developments in the theory of particle mixing.
16. BUSLIK, D.: *Bull. Am. Soc. Testing Mat.* **165** (1950) 66. Mixing and sampling with special reference to multi-sized granular particles.
17. KRAMERS, H.: Private communication.
18. COULSON, J. M. and MAITRA, N. K.: *Ind. Chemist* **26** (1950) 55. The mixing of solid particles.
19. BLUMBERG, R. and MARITZ, J. S.: *Chem. Eng. Sci.* **2** (1953) 240. Mixing of solid particles.
20. BROTHMAN, A., WOLLAN, G. N., and FELDMAN, S. M.: *Chem. Met. Eng.* **52** (iv) (1945) 102. New analysis provides formula to solve mixing.
21. DANCKWERTS, P. V.: *Appl. Scient. Res.* **3A** (1952) 279. The definition and measurement of some characteristics of mixtures.
22. DANCKWERTS, P. V.: *Research (London)* **6** (1953) 355. Theory of mixtures and mixing.
23. WORK, L. T.: *Ind. Eng. Chem.* **57**, No. 11 (Nov. 1965) 97. Annual review—size reduction (21 refs.).

24. TAGGART, A. F.: *Handbook of Mineral Dressing* (Wiley, 1945).
25. FERN, K. A.: *Trans. Inst. Chem. Eng.* **30** (1952) 82. The cyclone as a separating tool in mineral dressing.
26. KELSALL, D. F.: *Trans. Inst. Chem. Eng.* **30** (1952) 87. A preliminary study of the motion of solid particles in a hydraulic cyclone.
27. KELSALL, D. F.: *Chem. Eng. Sci.* **2** (1953) 254. A further study of the hydraulic cyclone.
28. DOUGHTY, F. T. C.: *Mining Magazine* **81** (1949) 268. Floatability of minerals.
29. KUHN, A. T.: The electrochemical treatment of aqueous effluent streams, in *Electrochemistry of Cleaner Environments*, ed. J. O'M. BOCKRIS (Plenum Press, 1972).

## 1.8. NOMENCLATURE

| | | Units in SI System | Dimensions in MLT |
|---|---|---|---|
| $a$ | Interfacial area per unit volume between constituents | $m^2/m^3$ | $L^{-1}$ |
| $a_m$ | Maximum interfacial area per unit volume | $m^2/m^3$ | $L^{-1}$ |
| $c$ | Coefficient of $t$ | $1/s$ | $T^{-1}$ |
| $d$ | Particle size | m | $L$ |
| $d_A, d_B$ | Equivalent spherical diameters of particles of **A**, **B** | m | $L$ |
| $d_{eff}$ | Effective diameter of orifice | m | $L$ |
| $d_l$ | Length mean diameter | m | $L$ |
| $d_s$ | Surface mean diameter (Sauter mean) | m | $L$ |
| $d_t$ | Tube diameter | m | $L$ |
| $d_v$ | Volume mean diameter | m | $L$ |
| $d_{80}$ | Size of opening through which 80 per cent of material passes | m | $L$ |
| $d'_l$ | Mean length diameter | m | $L$ |
| $d'_s$ | Mean surface diameter | m | $L$ |
| $d'_v$ | Mean volume diameter | m | $L$ |
| $G$ | Mass rate of flow of solids | kg/s | $MT^{-1}$ |
| $g$ | Acceleration due to gravity | $m/s^2$ | $LT^{-2}$ |
| $k$ | Constant in sieving equation | $1/s$ | $T^{-1}$ |
| $k_1$ | Constant depending on particle shape (volume/$d^3$) | — | — |
| $L$ | Length of solids plug | m | $L$ |
| $L_c$ | Critical length of solids plug | m | $L$ |
| $M$ | Degree of mixing | — | — |
| $n$ | Number of particles per unit mass | $1/kg$ | $M^{-1}$ |
| $n_g$ | Number of particles in a group | — | — |
| $n_p$ | Number of particles in a sample | — | — |
| $p$ | Overall proportion of particles of one type | — | — |
| $s$ | An index | — | — |
| $s$ | Standard deviation | — | — |
| $s_r$ | Standard deviation of random mixture | — | — |
| $s_0$ | Standard deviation of unmixed material | — | — |
| $t$ | Time | s | $T$ |
| $w$ | Mass of particles on screen at time $t$ | kg | $M$ |
| $X$ | Percentage of samples in which particles are not mixed | — | — |
| $x$ | Proportion of particles smaller than size $d$ | — | — |
| $Y$ | Fraction of samples in which particles are mixed | — | — |
| $\alpha$ | Angle of friction | — | — |
| $\beta$ | Angle between cone wall and horizontal | — | — |
| $\rho$ | Density of fluid | $kg/m^3$ | $ML^{-3}$ |
| $\rho_A, \rho_B$ | Density of material **A**, **B** | $kg/m^3$ | $ML^{-3}$ |
| $\rho_s$ | Density of particle | $kg/m^3$ | $ML^{-3}$ |
| $\psi$ | Sphericity of particle | — | — |

# CHAPTER 2

# Size Reduction of Solids

## 2.1. INTRODUCTION

In the chemical industry, size reduction is usually carried out in order to increase the surface because, in most reactions involving solid particles, the rate is directly proportional to the area of contact with a second phase. Thus the rate of combustion of solid particles is proportional to the area presented to the gas, though a number of secondary factors may also be involved such as, for example, the free flow of gas may be impeded because of the higher resistance to flow of a bed of small particles. Again in leaching, not only is the rate of extraction increased by virtue of the increased area of contact between the solvent and the solid, but the distance the solvent has to penetrate into the particles in order to gain access to the more remote pockets of solute is also reduced. This factor is also important in the drying of porous solids, where reduction in size causes both an increase in area and a reduction in the distance the moisture must travel within the particles in order to reach the surface; in this case the capillary forces acting on the moisture are also affected.

There are a number of other reasons for effecting size reduction. Thus it may be necessary to break a material into very small particles in order to separate two constituents, especially where one is dispersed in small isolated pockets. Further, the properties of a material may be considerably influenced by the particle size; for example, the chemical reactivity of fine particles is greater than that of coarse particles, and the colour and covering power of a pigment is considerably affected by the size of the particles. Again, far more intimate mixing of solids can be achieved if the particle size is small.

## 2.2. MECHANISM OF SIZE REDUCTION

The mechanism of the process of size reduction is extremely complex, though in recent years a number of attempts have been made at a more detailed analysis of the problem. If a single lump of material is subjected to a sudden impact, it will generally break so as to yield a few relatively large particles and a number of fine particles, with relatively few particles of intermediate size. If the energy in the blow is increased, the larger particles will be of a rather smaller size and more numerous and, whereas the number of fine particles will be appreciably increased, their size will not be much altered. It therefore appears that the size of the fine particles is closely connected with the internal structure of the material, and the size of the larger particles is more closely connected with the process by which the size reduction is effected.

This effect is well illustrated by a series of experiments on the grinding of coal in a small mill, carried out by Heywood[1]. The results are shown in Fig. 2.1 in which the distribution of particle size in the product is shown as a function of the number of revolutions of the mill. The initial size distribution shows a single mode corresponding to a relatively coarse size, but as the degree of crushing is gradually increased this mode progressively decreases in magnitude and a second mode develops at a particular size. This process continues until the first mode has completely disappeared. Here the second mode is characteristic of the material and is known as the persistent mode, and the first is known as the transitory mode.

The energy required to effect size reduction is related to the internal structure of the material and the process really consists of two parts, first opening up any small fissures which are already present and secondly forming new surface. A material such as coal contains a number of small cracks and tends first to break along these, and therefore the

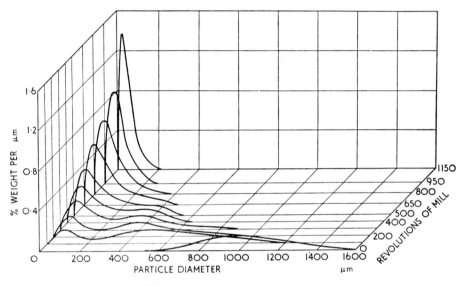

FIG. 2.1. Effect of progressive grinding on size distribution

large pieces are broken up more readily than the small ones. Further, since a very much greater increase in surface results from crushing a given quantity of fine as opposed to coarse material, fine grinding requires very much more power.

From the point of view of energy utilisation, size reduction is a very inefficient process and only between 0.1 and 2.0 per cent of the energy supplied to the machine appears as increased surface energy in the solids. The efficiency of the process is very much influenced by the manner in which the load is applied and its magnitude, but in addition the nature of the force exerted is also very important, for example, whether it is predominantly a compressive, an impact or a shearing force. If the applied force is insufficient for the elastic limit to be exceeded, the material is compressed (if compressive force is used) and energy is stored in the particle. When the load is removed, the particle expands again to its original condition without doing useful work; the energy appears as heat and no size reduction is effected. A somewhat greater force, however, will cause the particle to fracture, and in order to obtain the most effective utilisation of energy the force should be only slightly in excess of the crushing strength of the material. The surface of the particles will generally be of a very irregular nature so that the force is initially taken on the high spots with the result that very high stresses and temperatures may be set up locally in the material. As soon as a small amount of breakdown of material takes place, the point of application of the force will alter. The exact method by which fracture occurs is not known but it is suggested by Piret[2] that the compressive force produces small flaws in the material. If the energy concentration exceeds a certain critical value, these flaws will grow rapidly and will generally branch and the particles will break up. The rate of application of the force is important because there is generally a time lag between attainment of maximum load and fracture. Thus, a rather smaller force will cause fracture provided it is maintained for a sufficient time. This is a phenomenon similar to the ignition lag which is obtained with a combustible gaseous mixture. Here the interval between introducing the ignition source and the occurrence of ignition is a function of the temperature of the source, and when it is near the minimum ignition temperature delays of several seconds may be encountered. The greater the rate at which the load is applied, the less effectively is the energy utilised and the higher is the proportion of fine material which is produced. The efficiency of utilisation of energy as supplied by a falling weight has been compared with that of energy applied slowly by means of hydraulic pressure; up to three or four times more surface can be produced per unit of energy if it is applied by the latter method. Piret suggests that there is a close similarity between the crushing operation and a chemical reaction. In both cases a critical energy level must be exceeded before the process will start, and in both cases time is an important variable.

## 2.3. ENERGY FOR SIZE REDUCTION

It is impossible to estimate accurately the amount of energy required in order to effect a size reduction of a given material but a number of empirical laws have been put forward. The two earliest laws were due to Kick[3] and Rittinger[4] and a third law due to Bond[5, 6] has since been put forward. These three laws can all be derived from the basic differential equation:

$$\frac{dE}{dL} = -CL^p \tag{2.1}$$

This equation states that the energy $dE$ required to effect a small change $dL$ in the size of unit mass of material is a simple power function of the size. If $p$ is put equal to $-2$, integration gives:

$$E = C\left(\frac{1}{L_2} - \frac{1}{L_1}\right)$$

Writing $C = K_R f_c$, where $f_c$ is the crushing strength of the material, Rittinger's Law, which was first postulated in 1867, is obtained:

$$E = K_R f_c\left(\frac{1}{L_2} - \frac{1}{L_1}\right) \tag{2.2}$$

Since the surface of unit mass of material is proportional to $1/L$, the interpretation of this law is that the energy required for size reduction is directly proportional to the increase in surface.

On the other hand, if $p$ is equal to $-1$, then:

$$E = C \ln \frac{L_1}{L_2}$$

and writing $C = K_K f_c$:

$$E = K_K f_c \ln \frac{L_1}{L_2} \tag{2.3}$$

which is known as Kick's Law. This supposes that the energy required is directly related to the reduction ratio $L_1/L_2$ and that the energy required to crush a given amount of material from a 50 mm to a 25 mm size is the same as that required to reduce the size from 12 mm to 6 mm. In equations 2.2 and 2.3, $K_R$ and $K_K$ are known respectively as Rittinger's constant and Kick's constant. It should be noted that neither is dimensionless.

Neither of the above two laws permits an accurate calculation of the energy requirements. Rittinger's Law is applicable mainly to that part of the process where new surface is being created and holds most accurately for fine grinding where the increase in surface per unit mass of material is large. Kick's Law, however, more closely relates to the energy required to effect elastic deformation before fracture occurs, and is more accurate than Rittinger's Law for coarse crushing where the amount of surface produced is considerably less.

Bond has suggested a law intermediate between Rittinger's and Kick's Laws, where by putting $p = -3/2$ in equation 2.1:

$$E = 2C \left( \frac{1}{L_2^{1/2}} - \frac{1}{L_1^{1/2}} \right)$$

$$= 2C \sqrt{\frac{1}{L_2}} \left\{ 1 - \frac{1}{q^{1/2}} \right\} \tag{2.4}$$

where $\qquad q = \dfrac{L_1}{L_2}$

the reduction ratio. Writing $C = 5E_i$

$$E = E_i \sqrt{\frac{100}{L_2}} \left\{ 1 - \frac{1}{q^{1/2}} \right\} \tag{2.5}$$

Bond terms $E_i$ the work index and expresses it as the amount of energy required to reduce unit mass of material from an infinite particle size to a size $L_2$ of 100 μm (i.e. $q = \infty$). The size of material is taken as the size of the square hole through which 80 per cent of it will pass. Expressions for the work index are given in the original papers[5, 6] for various types of materials and various forms of size reduction equipment.

Austin and Klimpel[7] have reviewed these three laws and their applicability, whilst Cutting[8] described laboratory work to assess grindability using rod mill tests.

### 2.3.1. Energy Utilisation

One of the first important investigations into the distribution of the energy fed into a crusher was carried out by Owens[9] who concluded that energy was utilised in the following ways:

(a) In producing elastic deformation of the particles before fracture occurs.
(b) In producing inelastic deformation which results in size reduction.
(c) In causing elastic distortion of the equipment.
(d) In friction between particles, and between particles and the machine.
(e) In noise, heat and vibration in the plant, and
(f) In friction losses in the plant itself.

Owens estimated that only about 10 per cent of the total power is usefully employed.

During the course of an investigation by the U.S. Bureau of Mines[10], in which a drop weight type crusher was employed, it was found that the increase in surface was directly proportional to the input of energy but the rate of application of the load was an important factor.

This conclusion was substantiated in a more recent investigation of the power consumption in a size reduction process which is reported in three papers by Kwong and others[11,12,13]. A sample of material was crushed by putting it in a cavity in a steel mortar, placing a steel plunger over the sample and dropping a steel ball of known weight on the plunger over the sample from a measured height. Bouncing of the ball was prevented by three soft aluminium cushion wires under the mortar, and these wires were calibrated so that the energy absorbed by the system could be determined from their deformation. Losses in the plunger and ball were assumed to bear a fixed relationship to the energy absorbed by the wires, and the energy actually used for size reduction was then obtained as the difference between the energy of the ball on striking the plunger and the energy absorbed. Surfaces were measured by a water or air permeability method or by gas adsorption; the latter method gave a value approximately double that obtained from the former, indicating that, in these experiments, the internal surface was approximately the same as the external surface. The experimental results showed that, provided the new surface did not exceed about 400 cm$^2$/g, the new surface produced was directly proportional to the energy input. For a given energy input it was independent of

(a) the velocity of impact,
(b) the weight and arrangement of the sample,
(c) the initial particle size, and
(d) the moisture content of the sample.

Between 30 and 50 per cent of the energy of the ball on impact was absorbed by the material but no indication was obtained of how this was utilised. An extension of the range of the experiments, so that up to 120 m$^2$ of new surface was produced per kilogram of material, showed that the linear relationship between energy and new surface no longer held rigidly. However, in further experiments in which the crushing was effected slowly, using a hydraulic press, it was found that the linear relationship still held for the larger surface increases.

In order to determine the efficiency of the surface production process, experiments were carried out with sodium chloride and it was found that 90 J was required to produce a square metre of new surface. As the theoretical value of the surface energy of sodium chloride is only 0·08 J/m$^2$, the efficiency of the process is about 0·1 per cent. More recently, Zeleny and Piret[14] have reported calorimetric studies on the crushing of glass and quartz. They found a fairly constant energy requirement of 77 J/m$^2$ of new surface created, compared with a surface-energy value of less than $5 \text{ J/m}^2$. In some cases over 50 per cent of the energy supplied was used to produce plastic deformation of the steel crusher surfaces.

As indicated here, the apparent efficiency of the size reduction operation depends on the type of equipment used. Thus, for instance, a ball mill is rather less efficient than a drop weight type of crusher because of the ineffective collisions that take place.

Further work[2] on the crushing of quartz showed that more surface was created per unit of energy with single particles than with a collection of particles. This appears to be attributable to the fact that the crushing strength of apparently identical particles may vary by a factor as large as 20, and it is necessary to provide a sufficient energy concentration to crush the strongest particle.

## 2.4. METHODS OF OPERATING CRUSHERS

There are two distinct methods of feeding material to a crusher. The first is known as "free crushing" and involves feeding the material at a comparatively low rate so that the product can readily escape; its time of residence in the machine is therefore short and the production of appreciable quantities of undersize material is avoided. The second method is known as "choke feeding". In this case, the machine is kept full of material and discharge of the product is impeded so that the material remains in the crusher for a longer period. This results in a higher degree of crushing, but the capacity of the machine is reduced and energy consumption is high because of the cushioning action produced by the accumulated product. This method is therefore used only when a comparatively small amount of materials is to be crushed and when it is desired to complete the whole of the size reduction in one operation.

If the plant is operated, as in "choke feeding", so that the material is passed only once through the equipment, the process is known as "open circuit grinding". If, on the other hand, the product contains material which is insufficiently crushed, it may be necessary to separate the product and return the oversize material for a second crushing. This system is generally to be preferred and is known as "closed circuit grinding". A flow-sheet for a typical closed circuit grinding process, in which a coarse crusher, an intermediate crusher and a fine grinder are used, is shown in Fig. 2.2. In many plants, the product is continuously removed, either by allowing the material to fall on to a screen or by

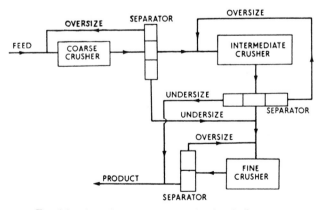

FIG. 2.2. Flow diagram for closed circuit grinding system

subjecting it to the action of a stream of fluid: the small particles are carried away and the oversize material falls back to be crushed again.

The desirability of using a number of size reduction units when the particle size is to be considerably reduced arises from the fact that it is not generally economical to effect a large reduction ratio in a single machine. The equipment is usually roughly divided into classes as given in Table 2.1 according to the size of the feed and the product:

TABLE 2.1. *Classification of Size Reduction Equipment*

|  | Feed size | Product size |
|---|---|---|
| Coarse crushers | 1500–40 mm (60–1$\frac{1}{2}$ in.) | 50–5 mm (2–$\frac{1}{4}$ in.) |
| Intermediate crushers | 50–5 mm (2–$\frac{1}{4}$ in.) | 5–0·1 mm ($\frac{1}{4}$ in.–200-mesh) |
| Fine crushers | 5–2 mm ($\frac{1}{4}$–$\frac{1}{8}$ in.) | 0·1 mm (about 200-mesh) |
| Colloid mills | 0·2 mm (80-mesh) | down to 0·01 μm ($10^{-6}$ in.) |

A greater size reduction ratio can be obtained in fine crushers than in coarse crushers.

The equipment can also be classified, to some extent, according to the nature of the force which is applied but, as generally a number of forces are involved, it is a less convenient basis.

Grinding can be carried out either wet or dry but wet grinding is generally applicable only with low speed mills. The advantages of wet grinding are

(a) The power consumption is less by about 20–30 per cent.
(b) The capacity of the plant is increased.
(c) Removal of the product is facilitated and the amount of fines is reduced.
(d) Dust formation is eliminated.
(e) The solids are more easily handled.

Against this, the wear on the grinding medium will generally be about 20 per cent greater, and it may be necessary to dry the product.

The separator indicated in Fig. 2.2 is usually either a cyclone type or a mechanical air separator. Cyclone separators, the theory of operation and application of which are fully discussed in the context of gas cleaning in Chapter 8, may be used. Alternatively, a *whizzer* type of air separator such as the Raymond air separator which appears in Fig. 1.30 is often included as an integral part of the mill as shown in the examples of the Lopulco and Raymond mills in Figs. 2.24 and 2.26 later in this chapter. Oversize particles drop down the inner case and are returned directly to the mill whilst the fine material is removed as a separate product stream.

## 2.5. NATURE OF THE MATERIAL TO BE CRUSHED

The choice of machine for a given crushing operation will be influenced by the nature of the product required and the quantity and size of material to be handled. The more important properties of the feed apart from its size are as follows:

*Hardness.* The hardness of the material affects the power consumption and the wear on the machine. With hard and abrasive materials it is necessary to use a low-speed machine and to protect the bearings from the abrasive dusts that are produced. Pressure lubrication is recommended. Materials are arranged in order of increasing hardness in the Mohr scale. The first four items rank as soft and the remainder as hard.

Mohr Scale of Hardness

| | | |
|---|---|---|
| 1. Talc | 5. Apatite | 8. Topaz |
| 2. Rock salt or gypsum | 6. Felspar | 9. Carborundum |
| 3. Calcite | 7. Quartz | 10. Diamond. |
| 4. Fluorspar | | |

*Structure.* Normal granular materials such as coal, ores and rocks can be effectively crushed employing the normal forces of compression, impact, etc. With fibrous materials it is necessary to effect a tearing action.

*Moisture content.* It is found that materials do not flow well if they contain between about 5 and 50 per cent of moisture; under these conditions the material tends to cake together in the form of balls. Grinding can be carried out satisfactorily, in general, outside these limits.

*Crushing strength.* The power required for crushing is almost directly proportional to the crushing strength of the material.

*Friability.* The friability of the material is its tendency to fracture during normal handling. In general, a crystalline material will break along well-defined planes and the power required for crushing will increase as the particle size is reduced.

*Stickiness.* A sticky material will tend to clog the grinding equipment and therefore it should be ground in a plant that can be readily cleaned.

*Soapiness.* This is, in general, a measure of the coefficient of friction of the surface of the material. If the coefficient of friction is low the crushing may be more difficult.

*Explosive materials* must be ground wet or in the presence of an inert atmosphere.

*Materials yielding dusts that are harmful to the health* must be ground under conditions where the dust is not allowed to escape.

Work[15] has presented a guide to equipment selection based on size and abrasiveness of material.

## 2.6. TYPES OF CRUSHING EQUIPMENT

The most important coarse, intermediate and fine crushers may be classified as in Table 2.2.

TABLE 2.2. *Crushing Equipment*

| Coarse crushers | Intermediate crushers | Fine crushers |
|---|---|---|
| Blake jaw crusher | Crushing rolls | Buhrstone mill |
| Dodge jaw crusher | Disc crusher | Roller mill |
| Gyratory crusher | Edge runner mill | Raymond mill |
| | Conical crusher | Griffin mill |
| | Stamp battery | Centrifugal ball mill |
| | Hammer mill | Ring roller mill |
| | Single roll crusher | Ball mill |
| | Pin mill | Tube mill |
| | Edge runner mill | Hardinge mill |
| | Squirrel cage disintegrator | Lopulco mill |

The features of these crushers will now be considered in some detail.

### 2.6.1. Coarse Crushers

*The Blake Jaw Crusher*

The Blake jaw crusher (Fig. 2.3) has a fixed jaw and a moving jaw pivoted at the top. The crushing faces themselves are formed either of manganese steel or of chilled cast iron and must be carefully fitted because they are brittle; the risk of breakage is reduced by grinding the back surface flat or packing with lead. Since the maximum movement of the jaw is at the bottom, there will be little tendency for the machine to clog though some uncrushed material may fall through and have to be returned to the crusher. Further, the maximum

FIG. 2.3. Blake jaw crusher

pressure will be exerted on the large material which is introduced at the top. The machine is usually protected so that it is not damaged if lumps of metal inadvertently enter, by making one of the toggle plates in the driving mechanism relatively weak so that, if any large stresses are set up, this is the first part to fail. Easy renewal of the damaged part is then possible.

Blake crushers are made with jaw widths varying from about 50 mm to 1·2 m and the running speed varies from about 1·5 to 6 Hz, the smaller machines running at the higher speeds. The speed of operation should not be so high that a large quantity of fines is produced as a result of material being repeatedly crushed because it cannot escape sufficiently quickly. The angle of nip, the angle between the jaws, is usually about 30°.

Because the crushing action is intermittent, the loading on the machine is uneven and the crusher therefore incorporates a heavy flywheel.

*The Dodge Jaw Crusher*

In the Dodge crusher (Fig. 2.4) the moving jaw is pivoted at the bottom. The minimum movement is thus at the bottom and a more uniform product is obtained, but the crusher is less widely used because of its tendency to choke. The large opening at the top enables it to take very large feed and to effect a large size reduction. This crusher is usually made in smaller sizes than the Blake crusher, because of the high fluctuating stresses that are produced in the members of the machine.

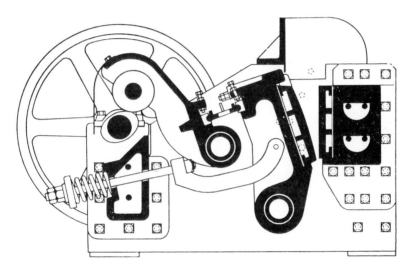

FIG. 2.4. Dodge crusher

*The Gyratory Crusher*

The gyratory crusher (Fig. 2.5) employs a crushing head, in the form of a truncated cone, mounted on a shaft, the upper end of which is held in a flexible bearing, whilst the lower end is driven eccentrically so as to describe a circle. The crushing action takes place round the whole of the cone and, since the maximum movement is at the bottom, the characteristics of the machine are similar to those of the Blake crusher. As the crusher is continuous in action, the fluctuations in the stresses are smaller than in the jaw crushers and the power consumption is lower. It has a large capacity per unit area of grinding surface, particularly if it is used to produce a small size reduction. However, it will not take such a large size of feed as a jaw crusher, but it will give a rather finer and more uniform product. Because the capital cost is high, the crusher is suitable only where large quantities of material are to be handled.

The jaw crushers and the gyratory crusher both employ a predominantly compressive force.

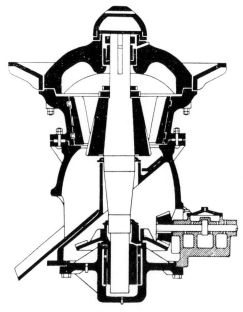

FIG. 2.5. Gyratory crusher

## Other Coarse Crushers

Friable materials, such as coal, can be broken up without the application of large forces and therefore less robust plant can be used. A common form of coal breaker consists of a large hollow cylinder with perforated walls. The axis is at a small angle to the horizontal and the feed is introduced at the top. The cylinder is rotated and the coal is lifted by means of arms attached to the inner surface and then falls against the cylindrical surface. The coal

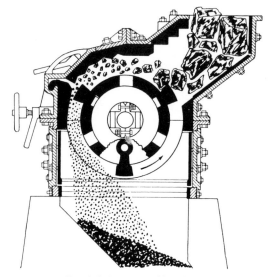

FIG. 2.6. Rotary coal breaker

breaks by impact and passes through the perforations as soon as the size has been sufficiently reduced. This type of equipment is less expensive and has a higher throughput than the jaw or gyratory crusher. Another coarse rotary breaker, similar in action to the hammer mill described later, is shown in Fig. 2.6.

### 2.6.2. Intermediate Crushers

*The Stamp Battery*

The stamp battery (Fig. 2.7) has been used extensively for moderately fine crushing in the past but it has now been very largely superseded by more efficient equipment. It consists of a number of heavy stamps—up to 500–1000 kg in weight—which are raised mechanically and allowed to fall under gravity on to the material to be crushed. The larger the number of stamps in the battery, the more even is the load on the driving mechanism. The solids are usually fed in as a suspension in water so that the product can be continuously removed from the system.

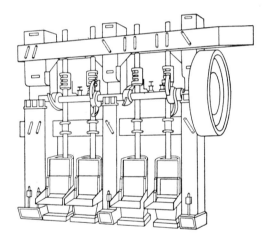

Fig. 2.7. Stamp battery

*The Edge Runner Mill*

In the edge runner mill (Fig. 2.8) a heavy cast iron or granite wheel, or muller as it is called, is mounted on a horizontal shaft which is rotated in a horizontal plane in a heavy pan; alternatively, the muller remains stationary and the pan is rotated. In some cases the mill incorporates two mullers. Material is fed to the centre of the pan and is worked outwards by the action of the muller whilst a scraper continuously removes material that has adhered to the sides of the pan, and returns it to the crushing zone. In many models the outer rim of the bottom of the pan is perforated, so that the product can be removed continuously as soon as its size has been sufficiently reduced. The mill can be operated wet or dry and is used extensively for the grinding of paints, clays and sticky materials.

FIG. 2.8.  Edge runner mill

## The End Runner Mill

The end runner mill (Fig. 2.9) is usually made in small laboratory sizes and consists of a cast iron or porcelain mortar which is rotated so that grinding takes place against a cylindrical pestle mounted with its axis vertical. Material is continuously scraped from the sides of the mortar with a doctor knife. A fine product can usually be obtained.

FIG. 2.9.  End runner mill

*The Hammer Mill*

The hammer mill is an impact mill employing a high speed rotating disc, to which are fixed a number of hammer bars which are swung outwards by centrifugal force. Two industrial models are illustrated in Figs. 2.10 and 2.11 and a laboratory model in Fig. 2.12.

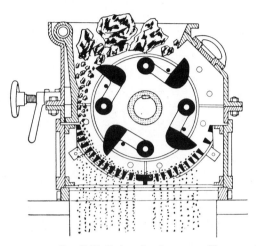

FIG. 2.10. Swing claw hammer mill

FIG. 2.11. Swing hammer mill

The material is fed in, either at the top or at the centre, and is thrown out centrifugally and crushed by being beaten between the hammer bars, or against breaker plates fixed around the periphery of the cylindrical casing. The material is beaten until it is small enough to fall through the screen which forms the lower portion of the casing. Since the hammer bars are hinged the presence of any hard material does not cause damage to the equipment. The bars are readily replaced when they are worn out. The machine is suitable for the crushing of both brittle and fibrous materials, and, in the latter case, it is usual to employ a screen with cutting edges. The hammer mill is suitable for hard materials but since a large amount of fines is produced, it is advisable to employ positive pressure lubrication to the bearings in order to prevent the entry of dust. The size of the product is regulated by the size of the screen and the speed of rotation.

A number of similar machines are made, and in some the hammer bars are rigidly fixed in position. Since a large current of air is produced, the dust must be separated in a cyclone separator or a bag filter.

### The Pin-type Mill

The Kek mill (Fig. 2.13) is a form of pin mill and consists of two horizontal steel plates with vertical projections on their near faces. The upper disc is stationary whilst the lower disc is rotated at high speed. The material is fed in through a hopper to the centre of the upper disc, and is thrown outwards by centrifugal action and broken against the projections. The mill gives a fairly uniform fine product with little dust and is extensively used with chemicals, fertilisers and other materials that are non-abrasive and easily broken. Control of the size of the product is effected by means of the speed and the spacing of the projections.

FIG. 2.13. Kek mill
*E*, Fixed pins. *F*, Rotating pins. *G*, Drive

FIG. 2.14. Bar mill

The squirrel cage disintegrator or bar mill (Fig. 2.14) is similar in action but employs vertical discs with horizontal bars. It is used with friable materials, such as coal and limestone, and with fibrous materials.

### The Single Roll Crusher

The single roll crusher (Fig. 2.15) consists of a toothed crushing roll which rotates close to a breaker plate. The material is crushed by compression and shearing between the two surfaces. It is used extensively in crushing coal. In the model shown in Fig. 2.16, the coal is crushed in three stages.

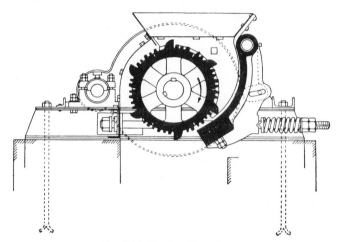

FIG. 2.15.  Single roll crusher

FIG. 2.16.  Three-stage breaker

*Crushing Rolls*

Two rolls, one in adjustable bearings, rotate in opposite directions and the clearance between them can be adjusted according to the size of feed and the required size of product (Figs. 2.17 and 2.18). The machine is protected, by spring loading, against damage from

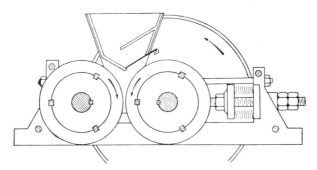

FIG. 2.17. Crushing rolls

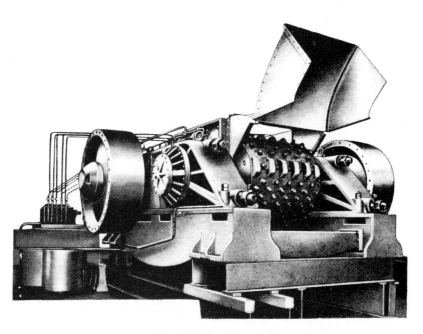

FIG. 2.18. Crushing rolls

very hard material. Generally, one roll is driven directly and the other by friction with the solids. The crushing rolls, which may vary from a few centimetres up to about 1·2 m in diameter, are suitable for effecting only a small size reduction in a single operation and it is therefore common to employ a number of pairs of rolls in series, one above the other. Tyres with either smooth or ridged surfaces are keyed to the rolls. The capacity is usually between one-tenth and one-third of that calculated on the assumption that a continuous ribbon of the material forms between the rolls.

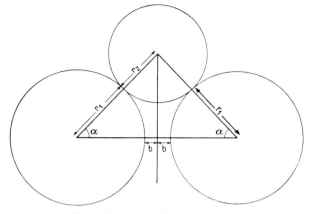

FIG. 2.19. Particle fed to crushing rolls

Figure 2.19 shows an idealised system where a spherical or cylindrical particle of radius $r_2$ is being fed to crushing rolls of radius $r_1$. $2\alpha$ is the angle of nip, the angle between the two common tangents to the particle and each of the rolls, and $2b$ is the distance between the rolls. It will be seen from the geometry of the system that the angle of nip is given by

$$\frac{r_1 + b}{r_2 + r_1} = \cos\alpha \qquad (2.6)$$

For steel rolls, the angle of nip will be not greater than about $32°$.

Crushing rolls are extensively used for crushing oil seeds and in the gunpowder industry and are also suitable for abrasive materials. They are simple in construction and do not give a large percentage of fines.

*Conical Crushers*

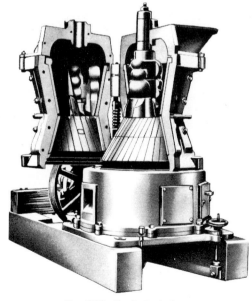

FIG. 2.20. Conical grinder

Conical crushers are now replacing many of the other types of intermediate crusher. They are similar in construction to the gyratory crusher, though they will not take such a coarse feed and give a very much finer product; they operate at rather higher speeds. In some conical grinders (Fig. 2.20), the crushing head is given a rotating motion instead of an eccentric motion. For efficient operation, the conical grinders should be supplied with dry and uniformly sized material and are therefore most suited to closed circuit grinding; they give free discharge of product.

### The Symons Disc Crusher

The disc crusher (Fig. 2.21) employs two saucer-shaped discs mounted on horizontal shafts, one of which is rotated and the other is mounted in an eccentric bearing so that the two crushing faces continuously approach and recede. The material is fed into the centre between the two discs and the product is discharged by centrifugal action as soon as it is fine enough to escape through the opening between the faces.

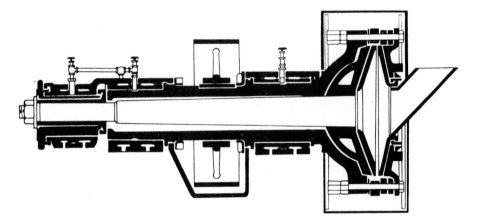

FIG. 2.21. Symons disc crusher

### 2.6.3. Fine Crushers

#### The Buhrstone Mill

The Buhrstone mill (Fig. 2.22) is one of the oldest forms of fine crushing equipment though it has, very largely, been superseded now by roller mills. Grinding takes place between two heavy horizontal wheels, one of which is stationary and the other driven. The surface of the stones is carefully dressed so that the material is continuously worked from the centre to the circumference of the wheels. Size reduction takes place by a shearing action between the edges of the grooves on the two grinding stones. This equipment has been used in the past for the grinding of grain, pigments for paints, pharmaceuticals, cosmetics and printer's ink but is now used only where the quantity of material is very small.

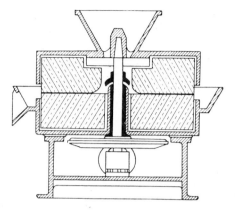

FIG. 2.22. Buhrstone mill

## Roller Mills

The roller mill consists of a pair of rollers that rotate at different speeds (e.g. 3:1 ratio) in
. opposite directions. As in crushing rolls, one of the rollers is held in a fixed bearing whereas
the other has an adjustable spring-loaded bearing. Since the rollers rotate at different
speeds, size reduction is effected by a combination of compressive and shear forces. The
roller mill is now extensively used in the flour milling industry and for the manufacture of
pigments for paints.

## Centrifugal Attrition Mills

*The Babcock mill.* This mill (Fig. 2.23) consists of a series of pushers which cause heavy
cast iron balls to rotate against a bull ring like a ball race, and the pressure of the balls on
the bull ring is produced by pressure applied from above. Material fed into the mill falls on
the bull ring, and the product is continuously removed in an upward stream of air which
carries the ground material between the blades of the classifier, which is shown towards the
top of the photograph; the oversize material falls back and is reground. The air stream is
produced by means of an external blower and may involve a considerable additional power
consumption. The fineness of the product is controlled by the rate of feeding and the air
velocity. This machine is used mainly for preparation of pulverised coal and sometimes for
cement.

*The Lopulco mill or ring roll pulveriser.* These machines (Fig. 2.24) are being
manufactured in large numbers at the present time for the production of industrial minerals
such as limestone and gypsum. A slightly concave circular bull ring rotates at high speed and
the feed is thrown outwards by centrifugal action under the crushing rollers, which are
shaped like truncated cones (Fig. 2.25). The rollers are spring-loaded and the strength of the
springs determines the grinding force available. There is always a clearance between the
rollers and the bull ring so that there is no wear on the grinding heads if the plant is operated
light, and quiet operation is obtained. The product is continuously removed by means of a
stream of air produced by an external fan and is carried into a separator fitted above the
grinding mechanism. In this separator, the cross-sectional area for flow is gradually

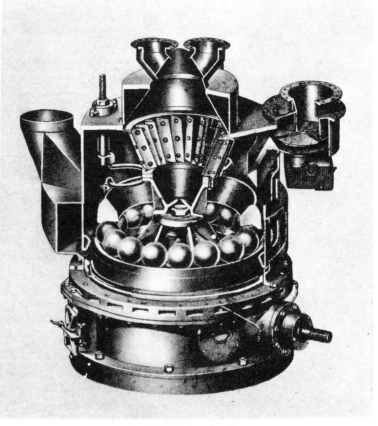

FIG. 2.23. Babcock mill

increased and, as the air velocity fails, the oversize material falls back and is ground again. The product is separated in a cyclone separator from the air which is then recycled. Chemicals, dyes, cements, and phosphate rocks are also ground in the Lopulco mill. As the risk of sparking is reduced by the maintenance of a clearance between the grinding media, the mill can be used for grinding explosive materials.

*The Raymond mill.* The Raymond mill (Fig. 2.26) is slightly less economical in operation than the Lopulco mill, but will give a rather finer and more uniform product. A central shaft driven by a bevel gear carries a yoke at the top and terminates in a foot-step bearing at the bottom. On the yoke are pivoted a number of heavy arms (Fig. 2.27) carrying the rollers which are thrown outwards by centrifugal action and bear on a circular bull ring. Both the rollers and the bull ring are readily replaceable. The material, which is introduced by means of an automatic feed device, is forced on to the bull ring by means of a plough which rotates on the central shaft. The ground material is removed by means of an air current, as in the case of the Lopulco mill, and the oversize material falls back and is again brought on to the bull ring by the plough.

As the mill operates at high speeds, it is not suitable for use with abrasive materials; nor will it handle materials that soften during milling. Though the power consumption and

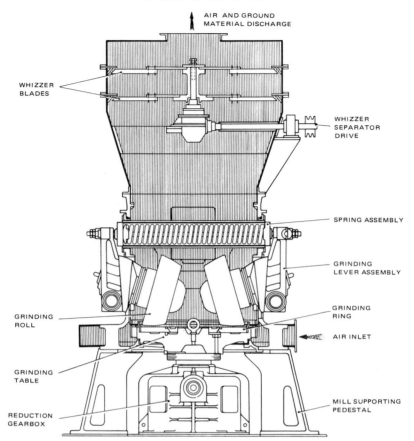

FIG. 2.24. Section through Lopulco mill with whizzer classifier

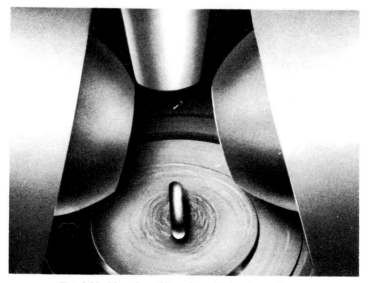

FIG. 2.25. Plan view of Lopulco mill showing rollers

maintenance costs are low, this machine does not compare favourably with the Lopulco mill under most conditions. It has the added disadvantage that wear will take place if the machine is run without any feed, because no clearance is maintained between the grinding heads and the bull ring. Besides being used for the preparation of pulverised coal, the Raymond mill is used extensively in cement and pottery manufacture. In the pottery industry, sizing of the raw materials has to be carried out within very fine limits, possibly

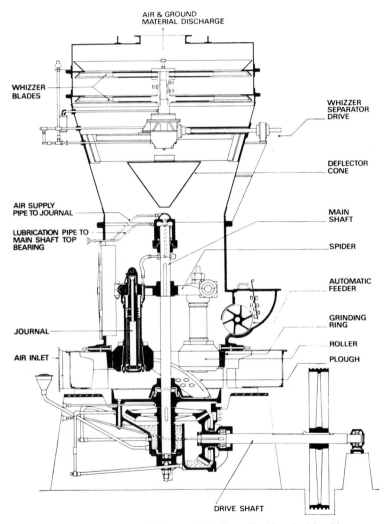

FIG. 2.26. Sectional arrangement of a Raymond 3 roller mill with a double whizzer separator

between 55 and 65 µm, and the Raymond mill is capable of achieving this. A comparison between the Lopulco and Raymond mills has shown that, whereas the Lopulco mill would give a product 98 per cent of which was below 50 µm in size, the Raymond mill would give 100 per cent below this size; in the latter case, however, the power consumption is considerably higher. Some typical figures for power consumption with the Lopulco mill are given in Table 2.3.

Fig. 2.27. Crushing heads of Raymond mill

TABLE 2.3. *Typical Power Requirements for Grinding*

| Material | Mill size | Product fineness | Output | | Power consumption | |
|---|---|---|---|---|---|---|
| | | | tonne/hr | kg/s | HP hr/ton (mill and fan installed) | MJ/Mg |
| Gypsum | LM12 | 99%—200-mesh B.S.S. | 4·5 | 1·3 | 50 | 134 |
| | LM14 | 99%—100-mesh B.S.S. | 11·2 | 3·2 | 29 | 78 |
| | LM16 | 80%—100-mesh B.S.S. | 27·5 | 7·8 | 18 | 48 |
| Limestone | LM12 | 70%—200-mesh B.S.S. | 11·0 | 3·1 | 20·5 | 55 |
| | LM14 | 80%—200-mesh B.S.S. | 11·5 | 3·2 | 28 | 75 |
| | LM16 | 99%—200-mesh B.S.S. | 8·0 | 2·3 | 62 | 166 |
| Phosphate | LM12 | 75%—100-mesh B.S.S. | 12 | 3·4 | 18·5 | 50 |
| (Morocco) | LM14 | 90%—100-mesh B.S.S. | 13 | 3·7 | 25 | 67 |
| | LM16 | 90%—100-mesh B.S.S. | 19 | 5·4 | 23 | 62 |
| | LM16/3 | 90%—100-mesh B.S.S. | 35 | 9·9 | 20 | 54 |
| Phosphate (Nauru) | LM16/3 | 97%—100-mesh B.S.S. | 27 | 7·6 | 26 | 70 |
| Coal | LM12 | 96%—100-mesh B.S.S. | 9·4 | 2·7 | 21 | 56 |
| | LM14 | 96%—100-mesh B.S.S. | 13·4 | 3·8 | 21 | 56 |
| | LM16 | 96%—100-mesh B.S.S. | 20 | 5·6 | 21 | 56 |

100-mesh B.S.S. $\equiv 150\,\mu m$      200-mesh B.S.S. $\equiv 75\,\mu m$

*The crushing force.* For a Raymond mill let

$M$ = the mass of the grinding head,
$m$ = the mass per unit length of the arm,
$\omega$ = the angular speed of rotation,
$\theta$ = the angle between the arm and the vertical,
$l$ = the length of the arm,
$c$ = the radius of the yoke, and
$R$ = the normal reaction of the bull ring on the grinding head.

Considering one arm rotating under uniform conditions, the sum of the moments of all the forces, acting on the roller and arm, about the point of support $O$ will be zero (Fig. 2.28).

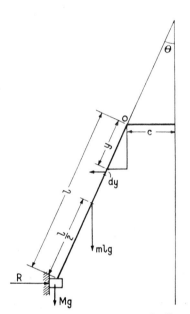

FIG. 2.28. Forces in Raymond mill

For a length $dy$ of the rod at a distance $y$ from $O$:

mass of the element $= m\,dy$
centrifugal force acting on element $= m\,dy(c + y\sin\theta)\omega^2$
moment of the force about $O$ $= -m\,dy(c + y\sin\theta)\omega^2 y\cos\theta$
Total moment of the whole arm $= -m\omega^2\cos\theta(\tfrac{1}{2}cl^2 + \tfrac{1}{3}l^3\sin\theta)$
Moment of the centrifugal force acting on the grinding head
$\qquad\qquad = -M\omega^2 l\cos\theta(c + l\sin\theta)$
Moment of the weight of the arm $= \tfrac{1}{2}ml^2 g\sin\theta$
Moment of the weight of the grinding head $= Mgl\sin\theta$
Moment of the normal reaction of the bull ring on the grinding head $= Rl\cos\theta$
Total moment about $O = ml^2 g\sin\theta + M\,gl\sin\theta + Rl\cos\theta$
$\qquad\qquad - ml^2\omega^2\cos\theta\,(\tfrac{1}{2}c + \tfrac{1}{3}l\sin\theta) - M\omega^2 l\cos\theta\,(c + l\sin\theta) = 0$

Thus     $R = M'\omega^2(\tfrac{1}{2}c + \tfrac{1}{3}l\sin\theta) + M\omega^2(c + l\sin\theta) - \tfrac{1}{2}M'g\tan\theta - Mg\tan\theta$

(where $M' = ml =$ mass of arm)

$$= M'(\tfrac{1}{2}\omega^2 c + \tfrac{1}{3}l\omega^2\sin\theta - \tfrac{1}{2}g\tan\theta) + M(\omega^2 c + \omega^2 l\sin\theta - g\tan\theta) \qquad (2.7)$$

*The Griffin mill.* The Griffin mill (Fig. 2.29) is similar to the Raymond mill except that it employs only one grinding head and the separation of the product is effected using a screen.

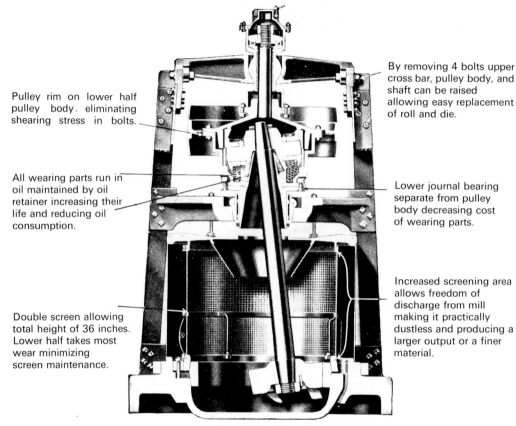

Pulley rim on lower half pulley body. eliminating shearing stress in bolts.

By removing 4 bolts upper cross bar, pulley body, and shaft can be raised allowing easy replacement of roll and die.

All wearing parts run in oil maintained by oil retainer increasing their life and reducing oil consumption.

Lower journal bearing separate from pulley body decreasing cost of wearing parts.

Double screen allowing total height of 36 inches. Lower half takes most wear minimizing screen maintenance.

Increased screening area allows freedom of discharge from mill making it practically dustless and producing a larger output or a finer material.

Fig. 2.29. Griffin mill

### The Ball Mill

In its simplest form, the ball mill consists of a rotating hollow cylinder, partially filled with balls, with its axis either horizontal or at a small angle to the horizontal. The material to be ground may be fed in through a hollow trunnion at one end and the product leaves through a similar trunnion at the other end. The outlet is normally covered with a coarse screen to prevent the escape of the balls. Figure 2.30 shows a section of a ball mill which is an example of the Hardinge mill which is discussed later in this chapter.

The inner surface of the cylinder is usually lined with an abrasion-resistant material (Fig. 2.31), such as manganese steel, stoneware or rubber. Less wear takes place in rubber-lined mills, and the coefficient of friction between the balls and the cylinder is greater than with steel or stoneware linings. The balls are therefore carried further in contact with the cylinder and thus drop on to the feed from a greater height. In some cases, lifter bars are fitted to the inside of the cylinder. A new type of ball mill is now being used to an increasing extent. The mill is vibrated instead of being rotated and the rate of passage of material is controlled by the slope of the mill.

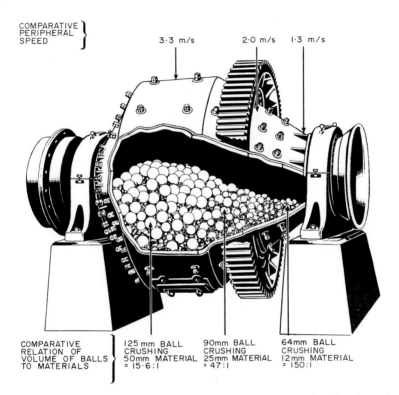

FIG. 2.30. Cut-away view of the conical ball mill showing how energy is proportioned to the work required

The ball mill is used for the grinding of a wide range of materials, including coal, pigments, and felspar for pottery, and will take feed up to about 50 mm in size. The efficiency of grinding increases with the hold-up in the mill until the voids between the balls are filled. Further increase in the quantity then lowers the efficiency again.

The balls are usually made of flint or steel and occupy between 30 and 50 per cent of the volume of the mill. The diameter of ball used will vary between 12 mm and 125 mm and the optimum diameter is approximately proportional to the square root of the size of the feed, the proportionality constant being a function of the nature of the material.

During grinding, the balls themselves wear and are constantly replaced by new ones so that the mill contains balls of various ages, and hence of various sizes. This is advantageous since the large balls deal effectively with the feed and the small ones are responsible for giving a fine product. The maximum rate of wear of steel balls, using very abrasive materials, is about 0·3 kg per Mg of material for dry grinding, and 1–1·5 kg/Mg for wet grinding. The normal charge of balls amounts to about 5 Mg/m³. In small mills where very fine grinding is required, pebbles are often used in place of balls (Fig. 2.32).

In the compound mill, the cylinder is divided into a number of compartments by vertical perforated plates. The material flows axially along the mill and can pass from one

FIG. 2.31. View of inside of ball mill showing liners

FIG. 2.32. Double unit pebble mill

compartment to the next only when its size has been reduced to less than that of the perforations in the plate. Each compartment is supplied with balls of a different size; the large balls are at the entry end and thus operate on the feed material, whilst the small balls come into contact with the material immediately before it is discharged. This results in economical operation and the formation of a uniform product.

### Factors Influencing the Size of the Product

(a) *The rate of feed.* With high rates of feed, less size reduction is effected since the material is in the mill for a shorter time.

(b) *The properties of the feed material.* The larger the feed the larger will be the product under given operating conditions. A smaller size reduction is obtained with a hard material.

(c) *Weight of balls.* A heavy charge of balls produces a fine product. The weight of the charge can be increased either by increasing the number of balls or by using a material of higher density. Since optimum grinding conditions are usually obtained when the bulk volume of the balls is equal to 50 per cent of the volume of the mill, variation of the weight of balls is normally effected by the use of materials of different densities.

(d) *The diameter of the balls.* Small balls facilitate the production of fine material but they do not deal so effectively with the larger particles in the feed. The limiting size reduction obtained with a given size of balls is known as the free grinding limit. For most economical operation, the smallest possible balls should be used.

(e) *The slope of the mill.* Increase in the slope of the mill increases the capacity of the plant because the retention time is reduced but a coarser product is obtained.

(f) *Discharge freedom.* Increasing the freedom of discharge of the product has the same effect as increasing the slope. In some mills, the product is discharged through openings in the lining.

(g) *The speed of rotation of the mill.* At low speeds of rotation, the balls simply roll over one another and little crushing action is obtained. At slightly higher speeds, they are projected short distances across the mill, and at still higher speeds they are thrown greater distances and considerable wear of the lining of the mill takes place. At very high speeds, the balls are carried right round in contact with the sides of the mill and little relative movement or grinding takes place again. The minimum speed at which the balls are carried round in this manner is called the critical speed of the mill and, under these conditions, there will be no resultant force acting on the ball when it is situated in contact with the lining of the mill in the uppermost position, i.e. the centrifugal force will be exactly equal to the weight of the ball. If the mill is rotating at the critical angular velocity $\omega_c$,

$$r\omega_c^2 = g$$

*i.e.*

$$\omega_c = \sqrt{\frac{g}{r}} \tag{2.8}$$

Here $r$ is the radius of the mill less that of the particle. It is found that the optimum speed is between one-half and three-quarters of the critical speed. Figure 2.33 illustrates conditions in a ball mill operating at the correct rate.

(h) *The level of material in the mill.* Power consumption is reduced by maintaining a low level of material in the mill, and this can be controlled most satisfactorily by fitting a suitable discharge opening for the product. If the level of material is raised, the cushioning action is increased and power is wasted by the production of an excessive quantity of undersize material.

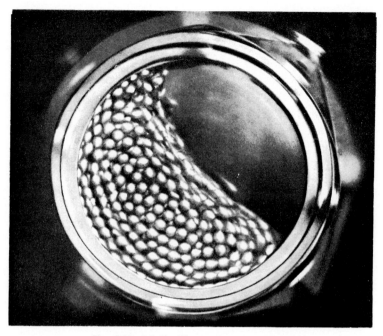

FIG. 2.33. A ball mill operating at correct speed

### Advantages of the Ball Mill

(a) The mill can be used wet or dry but wet grinding facilitates the removal of the product.

(b) The costs of installation and of power are low.

(c) The ball mill can be used with an inert atmosphere and therefore can be used for the grinding of certain explosive materials.

(d) The grinding medium is cheap.

(e) The mill is suitable for materials of all degrees of hardness.

(f) It can be used for batch or continuous operation.

(g) It can be used for open or closed circuit grinding. With open circuit grinding a wide range of particle size is obtained in the product. With closed circuit grinding, the use of an external separator can be obviated by continuous removal of the product by means of a current of air or through a screen (Fig. 2.34).

### The Tube Mill

The tube mill is similar to the ball mill in construction and operation, but the ratio of length to the diameter is usually 3 or 4:1, as compared with 1 or $1\frac{1}{2}$ for the ball mill. The mill is filled with pebbles, rather smaller in size than the balls used in the ball mill, and the inside of the mill is so shaped that a layer of pebbles becomes trapped in it to form a self-renewing lining. The characteristics of the two mills are similar but the material remains longer in the tube mill because of its greater length, and a finer product is therefore obtained.

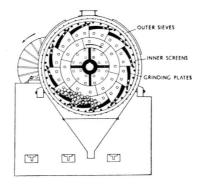

FIG. 2.34. End view of ball mill showing screens

*The Rod Mill*

In the rod mill, high carbon steel rods about 50 mm diameter and extending the whole length of the mill are used in place of balls. This mill gives a very uniform fine product and power consumption is low, but it is not suitable for very tough materials and the feed should not exceed about 25 mm in size. Worn rods must be removed from time to time and replaced by new ones, which are rather cheaper than balls. It is particularly useful with sticky materials which would hold the balls together in aggregates because the greater weight of the rods causes them to pull apart again.

*The Hardinge Mill*

The Hardinge mill (Fig. 2.35) is a ball mill in which the balls segregate themselves according to size. The main portion of the mill is cylindrical, like the ordinary ball mill, but the outlet end is conical and tapers towards the discharge point. The large balls collect in the cylindrical portion while the smaller balls, in order of decreasing size, locate themselves in the conical portion as shown earlier in Fig. 2.30. The material is therefore crushed by the action of successively smaller balls, and the mill is thus similar in characteristics to the compound ball mill. It is not known exactly how balls of different sizes segregate but it is suggested that, if the balls are initially mixed, the large ones will attain a slightly higher falling velocity and therefore strike the sloping surface of the mill before the smaller ones, and then run down towards the cylindrical section. The mill has an advantage over the compound ball mill in that the large balls are raised to the greatest height and therefore are able to exert the maximum force on the feed. As the size of the material is reduced, smaller forces are needed to cause fracture and it is therefore unnecessary to raise the smaller balls so high. The capacity of the Hardinge mill is higher than that of a ball mill of similar size and it gives a finer and more uniform product with a

FIG. 2.35. The Hardinge mill

lower consumption of power, but it is difficult to select an optimum speed because of the variation in feed diameter. It is extensively used for the grinding of such materials as cement, fuels, carborundum, silica, talc, slate and barytes.

### 2.6.4. Vibration Mills

One of the main limitations of the standard ball or tube mill is that it must be operated below the critical speed given by equation 2.8. If the effective value of the acceleration due to gravity could be increased a higher operating speed would be possible. The rotation of the mill simultaneously about a vertical and a horizontal axis has been used to simulate the effect of an increased gravitational acceleration, but clearly such techniques are applicable only to small machines.

By imparting a vibrating motion to a mill, either by the rotation of out-of-balance weights or by the use of electro-mechanical devices, accelerations many times the gravitational acceleration can be imparted to the machine. The body of the machine is generally supported on powerful springs and caused to vibrate in a vertical direction; vibration frequencies of between 6 and 60 Hz are common. In some machines the grinding takes place in two stages, the material falling from an upper to a lower chamber when its size has been reduced below a certain value.

The vibration mill has a very much higher capacity than a conventional mill of the same size and consequently either smaller equipment can be used or a much greater throughput obtained. Vibration mills lend themselves particularly to incorporation in continuous grinding systems.

### 2.6.5. Colloid Mills

Colloidal suspensions, emulsions and solid dispersions are produced by means of colloid mills or dispersion mills. Droplets of particles of sizes less than one micron may be formed, and solids suspensions consisting of discrete solid particles are obtainable with feed material of approximately 100-mesh size (50 µm). A typical colloid mill is shown in Fig. 2.36.

Fig. 2.36. A colloid mill

The mill consists of a flat rotor and stator (see Fig. 2.37) manufactured in a chemically inert synthetic abrasive material, and the mills can be set to operate at clearances from virtually zero to 1·25 mm, though in practice the maximum clearance used is about 0·3 mm. When duty demands, steel working surfaces can be fitted, and in such cases the minimum setting between rotor and stator must be 0·50 to 0·75 mm, otherwise "pick up" between the steel surfaces will occur.

The gap setting between rotor and stator is not necessarily in direct proportion to the droplet size or particle size of the end product. The thin film of material continually passing between the working surfaces is subjected to a high degree of shear, and consequently the energy absorbed within this film is frequently sufficient to reduce the dispersed phase to a particle size far smaller than the gap setting used. The rotor speed will vary with the physical size of the mill and the clearance necessary to achieve the desired

Fig. 2.37. Rotor and stator of a colloid mill

result, but peripheral speeds of the rotor of between 18 m/s and 36 m/s are usual. The required operating conditions and size of mill can only be found by experiment.

Some of the energy imparted to the film of material does, of course, appear in the form of heat, and a jacketed shroud is frequently fitted round the periphery of the working surfaces so that some of the heat may be removed by coolant. This jacket can also be used for circulation of a heating medium to maintain a desired temperature of the material being processed.

In all colloid mills, the power consumption is very high and the material should therefore be ground as fine as possible before it is fed to the mill.

### 2.6.6. Fluid Energy Mills

Another form of mill which does not give quite such a fine product is the jet pulveriser in which the solid is pulverised in jets of high pressure superheated steam or compressed air, supplied at pressures up to $3.5 \, \text{MN/m}^2$. The pulverising takes place in a shallow cylindrical chamber with a number of jets arranged tangentially at equal intervals around the circumference. The solid is thrown to the outside walls of the chamber and the fine particles are formed by the shearing action resulting from the differential velocities within the fluid streams. The jet pulveriser will give a product with a particle size between 1 and 10 μm.

The microniser is probably the best known of this type of pulveriser, which effects comminution by bombarding the particles of material against each other. Pre-ground material, of about 500 μm size, is fed into a shallow circular grinding chamber (which may be horizontal or vertical) the periphery of which is fitted with a number of jets, equally spaced, and arranged tangentially to a common circle.

Gaseous fluid (compressed air at approximately $800 \, \text{kN/m}^2$ or superheated steam at pressures from 0·8 to 1·6 MN/m² and temperatures ranging from 480 K to 810 K) issue through these jets, thereby promoting high-speed reduction of the contents of the grinding chamber, with turbulence and bombardment of the particles against each other. An intense centrifugal classifying action within the grinding chamber causes the coarser particles to concentrate towards the periphery of the chamber whilst the finer particles leave the chamber, with the fluid, through the central opening.

The majority of applications for fluid energy mills are for producing powders in the sub-sieve range in the order of 20 µm and less, and it should be mentioned that the power consumption per kg is proportionately higher than for conventional milling systems which grind to a top size of about 44 µm.

A section through a typical microniser is shown in Fig. 2.38.

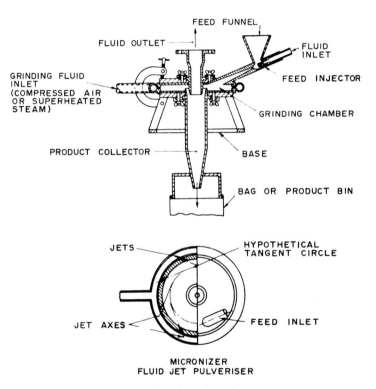

FIG. 2.38. Section through a microniser

Another pulveriser in this category is the Wheeler fluid energy mill which is in the form of a vertical loop. The pre-ground feed material is injected towards the bottom of the loop in which are situated the nozzles for admitting the compressed air or superheated steam. Reduction is by bombardment of the particles against each other and classification is effected by arranging for the fluid to leave the circulating gas stream through vanes which are situated just downstream of the top of the loop and on the inner face of the loop, i.e. against the centrifugal force. Oversize particles continue their downward path with the circulating fluid and re-enter the reduction chamber for further grinding.

## 2.7. FURTHER READING

BOND, F. C.: *Brit. Chem. Eng.* **6** (1961) 378–85, 543–8. Crushing and grinding calculations.
*Comminution*, ed. V. C. MARSHALL (I. Chem. E., London, 1974).
LOWRISON, G. C.: *Crushing and Grinding* (Butterworths, London, 1975).
WORK, L. T.: *Ind. Eng. Chem.* **57** (1965) 11, 97–99. Annual review—size reduction (21 refs.).

## 2.8. REFERENCES

1. HEYWOOD, H.: *J. Imp. Coll. Eng. Soc.* **6** (1950–2) 26. Some notes on grinding research.
2. PIRET, E. L.: *Chem. Eng. Prog.* **49** (1953) 56. Fundamental aspects of crushing.
3. KICK, F.: *Das Gesetz der proportionalen Widerstande und seine Anwendungen* (Leipzig, 1885).
4. VON RITTINGER, P. R.: *Lehrbuch der Aufbereitungskunde in ihrer neuesten Entwicklung und Ausbildung systematisch dargestellt* (Ernst und Korn, 1867).
5. BOND, F. C.: *Min. Engng, N.Y.* **4** (1952) 484. Third theory of comminution.
6. BOND, F. C.: *Chem. Eng., Albany* **59** (Oct. 1952) 169. New grinding theory aids equipment selection.
7. AUSTIN, L. G. and KLIMPEL, R. R.: *Ind. Eng. Chem.* **56**, No. 11 (Nov. 1964) 18–29. The theory of grinding operations (53 refs.).
8. CUTTING, G. W.: *Chem. Engnr.* (Oct. 1977) 702–4. Grindability assessments using laboratory rod mill tests.
9. OWENS, J. S.: *Trans. Inst. Min. Met.* **42** (1933) 407. Notes on power used in crushing ore, with special reference to rolls and their behaviour.
10. GROSS, J.: *U.S. Bur. Mines Bull.* **402** (1938). Crushing and grinding.
11. KWONG, J. N. S., ADAMS, J. T., JOHNSON, J. F., and PIRET, E. L.: *Chem. Eng. Prog.* **45** (1949) 508. Energy–new surface relationship in crushing. I. Application of permeability methods to an investigation of the crushing of some brittle solids.
12. ADAMS, J. T., JOHNSON, J. F., and PIRET, E. L.: *Chem. Eng. Prog.* **45** (1949) 655. Energy–new surface relationship in the crushing of solids. II. Application of permeability measurements to an investigation of the crushing of halite.
13. JOHNSON, J. F., AXELSON, J., and PIRET, E. L.: *Chem. Eng. Prog.* **45** (1949) 708. Energy–new surface relationship in the crushing of solids. III. Application of gas adsorption measurements to an investigation of crushing of quartz.
14. ZELENY, R. A. and PIRET, E. L.: *Ind. Eng. Chem. Process Design and Development* **1,** No. 1 (Jan. 1962) 37–41. Dissipation of energy in single particle crushing.
15. WORK, L. T.: *Ind. Eng. Chem.* **55**, No. 2 (Feb. 1963) 56–58. Trends in particle size technology.

## 2.9. NOMENCLATURE

| | | Units in SI System | Dimensions in **MLT** |
|---|---|---|---|
| $b$ | Half distance between crushing rolls | m | **L** |
| $C$ | A coefficient | — | $\mathbf{L}^{1-p}\mathbf{T}^{-2}$ |
| $c$ | Radius of yoke of Raymond mill | m | **L** |
| $E$ | Energy per unit mass | J/kg | $\mathbf{L}^2\mathbf{T}^{-2}$ |
| $E_i$ | Work index | J/kg | $\mathbf{L}^2\mathbf{T}^{-2}$ |
| $f_c$ | Crushing strength of material | N/m$^2$ | $\mathbf{ML}^{-1}\mathbf{T}^{-2}$ |
| $g$ | Acceleration due to gravity | m/s$^2$ | $\mathbf{LT}^{-2}$ |
| $K_K$ | Kick's constant | m$^3$/kg | $\mathbf{M}^{-1}\mathbf{L}^3$ |
| $K_R$ | Rittinger's constant | m$^4$/kg | $\mathbf{M}^{-1}\mathbf{L}^4$ |
| $L$ | Characteristic linear dimension | m | **L** |
| $l$ | Length of arm of Raymond mill | m | **L** |
| $M$ | Mass of crushing head in Raymond mill | kg | **M** |
| $M'$ | Mass of arm of Raymond mill | kg | **M** |
| $m$ | Mass per unit length of arm of Raymond mill | kg/m | $\mathbf{ML}^{-1}$ |
| $p$ | A constant used as an index | — | — |
| $q$ | Size reduction $L_1/L_2$ | — | — |
| $R$ | Normal reaction | N | $\mathbf{MLT}^{-2}$ |
| $r$ | Radius of ball mill less radius of particle | m | **L** |

|        |                                          | Units in SI System | Dimensions in **MLT** |
|--------|------------------------------------------|--------------------|-----------------------|
| $r_1$  | Radius of crushing rolls                 | m                  | **L**                 |
| $r_2$  | Radius of particle in feed               | m                  | **L**                 |
| $y$    | Distance along arm from point of support | m                  | **L**                 |
| $\alpha$ | Half angle of nip                      | —                  | —                     |
| $\theta$ | Angle between axis and vertical        | —                  | —                     |
| $\rho_s$ | Density of solid material              | kg/m$^3$           | **ML**$^{-3}$         |
| $\omega$ | Angular velocity                       | 1/s                | **T**$^{-1}$          |
| $\omega_c$ | Critical speed of rotation of ball mill | 1/s              | **T**$^{-1}$          |

CHAPTER 3

# Motion of Particles in a Fluid

## 3.1. INTRODUCTION

Many processes for the separation of particles of various sizes and shapes depend on the variation in the behaviour of the particles when they are subjected to the action of a moving fluid. Further, many of the methods for the determination of the sizes of particles in the sub-sieve ranges involve relative motion between the particles and a fluid.

The flow problems considered in Volume 1 were unidirectional, with the fluid flowing along a pipe or channel, and the effect of an obstruction has been discussed only in so far as it causes an alteration in the forward velocity of the fluid. In the present chapter, the force exerted on a body as a result of the flow of fluid past it will be considered and, as in general the fluid will be diverted all round it, the more complex problem of three-dimensional flow exists. The flow of fluid relative to an infinitely long cylinder, a spherical particle and a non-spherical particle will now be considered. The problem is essentially the same whether the fluid or the particle is moving. In the last part of the chapter the motion of particles in a centrifugal field will be examined.

## 3.2. FLOW PAST A CYLINDER

The flow of fluid past an infinitely long cylinder, in a direction perpendicular to its axis, will be considered in the first instance because this involves only two-directional flow, there being no flow parallel to the axis. For a non-viscous fluid flowing past a cylinder, as shown in Fig. 3.1, the velocity and direction of flow will vary round the circumference. Thus at $A$ and $D$ the fluid is brought to rest and at $B$ and $C$ the velocity is at a maximum. Since the fluid is non-viscous, there will be no drag, and an infinite velocity gradient will exist at the surface of the cylinder. If the fluid is incompressible and the cylinder is small, the sum of the kinetic energy and the pressure energy is constant at all points on the surface. The kinetic energy is a maximum at $B$ and $C$ and zero at $A$ and $D$, so that the pressure falls from $A$ to $B$ and from $A$ to $C$ and rises again from $B$ to $D$ and from $C$ to $D$, the pressure at $A$ and $D$ being the same. There will therefore be no net force exerted by the fluid on the cylinder. It is found that, whereas the predicted pressure variation for a non-viscous fluid agrees well with the results obtained with a viscous fluid over the front face, very considerable differences occur at the rear face.

It has already been shown in Volume 1, Chapter 9 that, when a viscous fluid flows over a surface, the fluid is retarded in the boundary layer which is formed near the surface and that the boundary layer increases in thickness with increase in distance from the leading edge. If the pressure is falling in the direction of flow, the retardation of the fluid will be less and the boundary layer will be thinner in consequence. On the other hand, if the pressure is rising, there will be a greater retardation and the thickness of the boundary layer will

86

increase more rapidly. The force acting on the fluid at some point in the boundary layer may then be sufficient to bring it to rest or to cause flow in the reverse direction with the result that an eddy current is set up. A region of reverse flow then exists near the surface where the boundary layer has separated (Fig. 3.2). The velocity rises from zero at the surface to a maximum negative value and falls again to zero; it then increases in the positive direction until it reaches the main stream velocity at the edge of the boundary layer, as seen in Fig. 3.2. At $PQ$ the velocity in the $X$-direction is zero and the direction of flow in the eddies must be in the $Y$-direction.

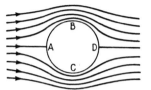

FIG. 3.1. Flow round a cylinder

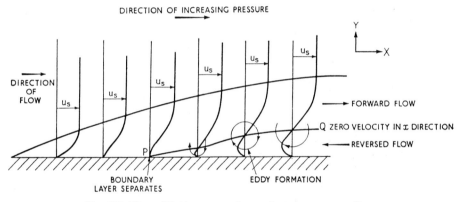

FIG. 3.2. Flow of fluid over a surface against a pressure gradient

For the flow of a viscous fluid past the cylinder, the pressure decreases from $A$ to $B$ and from $A$ to $C$ so that the boundary layer is thin and the flow is similar to that obtained with a non-viscous fluid: from $B$ to $D$ and from $C$ to $D$ the pressure is rising and therefore the boundary layer rapidly thickens with the result that it tends to separate from the surface. If separation occurs, eddies are formed in the wake of the cylinder and energy is thereby dissipated and an additional force, known as form drag, is set up. Thus, on the forward surface of the cylinder, the pressure distribution is similar to that obtained with the ideal fluid of zero viscosity; but on the rear surface, the boundary layer is thickening rapidly and pressure variations are very different in the two cases.

The total force on a body immersed in a moving fluid is made up of two components, the viscous drag or skin friction and the form drag. At low rates of flow no separation of the boundary layer takes place and the whole of the drag results from skin friction. As the velocity is increased, separation of the boundary layer occurs and the skin friction forms a gradually decreasing proportion of the total drag. However, if the velocity of the fluid is very high, or if turbulence is artificially induced, the flow within the boundary layer will

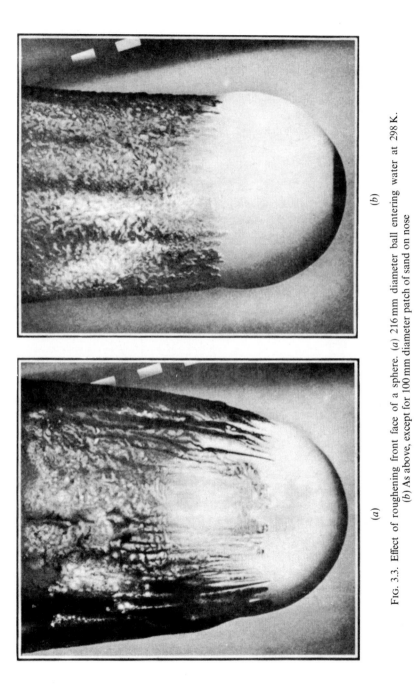

(b)

(a)

Fig. 3.3. Effect of roughening front face of a sphere. (a) 216 mm diameter ball entering water at 298 K. (b) As above, except for 100 mm diameter patch of sand on nose

change from streamline to turbulent before separation takes place. Since the rate of transfer of momentum through a fluid in turbulent motion is much greater than that in a fluid flowing under streamline conditions, separation is less likely to occur, because the fast-moving fluid outside the boundary layer is able to keep the fluid within the boundary layer moving in the forward direction. If separation does occur, it will take place nearer to $D$ in Fig. 3.1, and the resulting eddies will be smaller and the total drag will be reduced.

Turbulence may arise either from an increased fluid velocity or from artificial roughening of the forward face of the immersed body. Prandtl roughened the forward face of a sphere by fixing a hoop to it with the result that the drag was considerably reduced. Further experiments have been carried out in which sand particles have been stuck to the front face (see Fig. 3.3). The tendency for separation and hence the magnitude of the form drag are also dependent on the shape of the body.

Conditions of flow relative to a spherical particle are similar to those relative to the cylinder except that the flow pattern is three-directional. The method of calculating the force on a sphere is discussed in the following section.

## 3.3. THE DRAG FORCE ON A SPHERICAL PARTICLE

The only practically important case for which the drag on an immersed body has been calculated from purely theoretical considerations is that of a sphere moving at a low velocity in an infinite extent of continuous fluid. In 1851 Stokes[1] obtained the formula:

$$F = 3\pi\mu du \tag{3.1}$$

where $F$ is the drag force,
$\quad \mu$ is the viscosity of the fluid,
$\quad d$ is the diameter of the sphere, and
$\quad u$ is the velocity of the fluid relative to the particle.

The experimental data for the drag force exerted on a spherical particle by a moving fluid are conveniently expressed over a wide range of velocities in a manner similar to that used for flow in a pipe, by plotting a dimensionless group $R'/\rho u^2$ against a modified Reynolds number $Re'$, in which the linear dimension is the diameter of the sphere, and the velocity is the relative velocity between the fluid and the particle. $Re'$ is then equal to $ud\rho/\mu$, where $\rho$ is the density of the fluid. $R'$ is the force exerted on unit projected area of particle in a plane at right angles to its direction of motion: this area is thus equal to that of a circle of the same diameter as the sphere.

Values of $R'/\rho u^2$ are given as a function of $Re'$ in Table 3.1 and $\log R'/\rho u^2$ is plotted against $\log Re'$ in Figs. 3.4 and 3.5. The curve can be divided into four regions as follows:

*Region (a)* $(10^{-4} < Re' < 0.2)$. The equation of the curve approximates to:

$$\frac{R'}{\rho u^2} = 12 \, Re'^{-1} \tag{3.2}$$

i.e. on the logarithmic scale it is a straight line of slope $-1$. It will be shown later that equations 3.1 and 3.2 are consistent.

*Region (b)* $(0.2 < Re' < 500{-}1000)$. Various attempts have been made to derive theoretical expressions which can be used at higher values of $Re'$. For instance, Goldstein[2] has given an infinite series for the drag force on a spherical particle. This can be used for

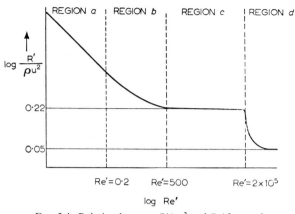

Fig. 3.4. Relation between $R'/\rho u^2$ and $Re'$ for a sphere

values of $Re'$ up to about 2, but at higher values, the series either converges very slowly or diverges. His relation is:

$$\frac{R'}{\rho u^2} = \frac{12}{Re'}\left\{1 + \frac{3}{16}Re' - \frac{19}{1280}Re'^2 + \frac{71}{20480}Re'^3 - \frac{30179}{34406400}Re'^4\right.$$
$$\left. + \frac{122519}{560742400}Re'^5 - \ldots\right\} \quad (3.3)$$

From this expression it is seen that equation 3.2 is in error by nearly 4 per cent at a value of $Re'$ of 0.2.

Oseen[3] takes the first two terms only of the series in $Re'$ and his formula can be used for values of $Re'$ up to 0.1. It therefore becomes:

$$\frac{R'}{\rho u^2} = \frac{12}{Re'}\left\{1 + \frac{3}{16}Re'\right\} \quad (3.4)$$

The slope of the curve gradually changes from $-1$ to 0 as $Re'$ changes from 0.2 to 500 or 1000. Schiller and Naumann[4] have given the following empirical equation to cover this range:

$$\frac{R'}{\rho u^2} = 12\,Re'^{-1}(1 + 0.15\,Re'^{0.687}) \quad (3.5)$$

Equation 3.5 is reliable for values of $Re'$ up to about 800. Some typical values of $R'/\rho u^2$ calculated from this formula are as follows:

$$Re' = 0.2 \qquad R'/\rho u^2 = 12\,Re'^{-1} \times 1.05$$
$$Re' = 2.0 \qquad R'/\rho u^2 = 12\,Re'^{-1} \times 1.24$$
$$Re' = 500 \qquad R'/\rho u^2 = 0.28$$
$$Re' = 800 \qquad R'/\rho u^2 = 0.24$$

Various other approximate formulae have been given for the drag force for this region. Allen[5] has considered the curve as having a mean slope of $-1/2$ and suggested that the data can be represented by taking $R'/\rho u^2$ as proportional to $Re'^{-1/2}$.

Table 3.1. $R'/\rho u^2$, $(R'/\rho u^2)\,Re'^2$ and $(R'/\rho u^2)\,Re'^{-1}$ as a Function of $Re'$

| $Re'$ | $R'/\rho u^2$ | $(R'/\rho u^2)\,Re'^2$ | $(R'/\rho u^2)\,Re'^{-1}$ |
|---|---|---|---|
| $10^{-3}$ | 12,000 | | |
| $2 \times 10^{-3}$ | 6,000 | | |
| $5 \times 10^{-3}$ | 2,400 | | |
| $10^{-2}$ | 1,200 | $1\cdot20 \times 10^{-1}$ | $1\cdot20 \times 10^5$ |
| $2 \times 10^{-2}$ | 600 | $2\cdot40 \times 10^{-1}$ | $3\cdot00 \times 10^4$ |
| $5 \times 10^{-2}$ | 240 | $6\cdot00 \times 10^{-1}$ | $4\cdot80 \times 10^3$ |
| $10^{-1}$ | 124 | $1\cdot24$ | $1\cdot24 \times 10^3$ |
| $2 \times 10^{-1}$ | 63 | $2\cdot52$ | $3\cdot15 \times 10^2$ |
| $5 \times 10^{-1}$ | 26·3 | $6\cdot4$ | $5\cdot26 \times 10$ |
| $10^0$ | 13·8 | $1\cdot38 \times 10$ | $1\cdot38 \times 10$ |
| $2 \times 10^0$ | 7·45 | $2\cdot98 \times 10$ | $3\cdot73$ |
| $5 \times 10^0$ | 3·49 | $8\cdot73 \times 10$ | $7\cdot00 \times 10^{-1}$ |
| $10$ | 2·08 | $2\cdot08 \times 10^2$ | $2\cdot08 \times 10^{-1}$ |
| $2 \times 10$ | 1·30 | $5\cdot20 \times 10^2$ | $6\cdot50 \times 10^{-2}$ |
| $5 \times 10$ | 0·768 | $1\cdot92 \times 10^3$ | $1\cdot54 \times 10^{-2}$ |
| $10^2$ | 0·547 | $5\cdot47 \times 10^3$ | $5\cdot47 \times 10^{-3}$ |
| $2 \times 10^2$ | 0·404 | $1\cdot62 \times 10^4$ | $2\cdot02 \times 10^{-3}$ |
| $5 \times 10^2$ | 0·283 | $7\cdot08 \times 10^4$ | $5\cdot70 \times 10^{-4}$ |
| $10^3$ | 0·221 | $2\cdot21 \times 10^5$ | $2\cdot21 \times 10^{-4}$ |
| $2 \times 10^3$ | 0·22 | $8\cdot8 \ \times 10^5$ | $1\cdot1 \ \times 10^{-4}$ |
| $5 \times 10^3$ | 0·22 | $5\cdot5 \ \times 10^6$ | $4\cdot4 \ \times 10^{-5}$ |
| $10^4$ | 0·22 | $2\cdot2 \ \times 10^7$ | $2\cdot2 \ \times 10^{-5}$ |
| $2 \times 10^4$ | 0·22 | | |
| $5 \times 10^4$ | 0·22 | | |
| $10^5$ | 0·22 | | |
| $2 \times 10^5$ | 0·05 | | |
| $5 \times 10^5$ | 0·05 | | |
| $10^6$ | 0·05 | | |
| $2 \times 10^6$ | 0·05 | | |
| $5 \times 10^6$ | 0·05 | | |
| $10^7$ | 0·05 | | |

When $0\cdot2 < Re' < 1000$, $\qquad R'/\rho u^2 = 12\,Re'^{-1}(1 + 0\cdot15\,Re'^{0\cdot687})$

When $Re' < 0\cdot2$, $\qquad\qquad R'/\rho u^2 = 12\,Re'^{-1}$

$$(R'/\rho u^2)\,Re'^2 = 12\,Re'$$

$$(R'/\rho u^2)\,Re'^{-1} = 12\,Re'^{-2}$$

When $Re' > 1000$ (but $< 10^5$), $\quad R'/\rho u^2 = 0\cdot22$

$$(R'/\rho u^2)\,Re'^2 = 0\cdot22\,Re'^2$$

$$(R'/\rho u^2)\,Re'^{-1} = 0\cdot22\,Re'^{-1}$$

When $Re' > $ ca. $10^5$, $\qquad\qquad R'/\rho u^2 = 0.05$

*Region (c)*. The value of $R'/\rho u^2$ remains approximately constant at 0·22 for values of $Re'$ between about 500–1000 and $2 \times 10^5$. The data can therefore be represented by the equation:

$$\frac{R'}{\rho u^2} = 0\cdot22 \qquad (3.6)$$

*Region (d)*. When $Re'$ exceeds about $2 \times 10^5$, the flow in the boundary layer changes from streamline to turbulent and the separation takes place nearer to the rear of the sphere; the drag force is decreased considerably and:

$$\frac{R'}{\rho u^2} = 0\cdot05 \qquad (3.7)$$

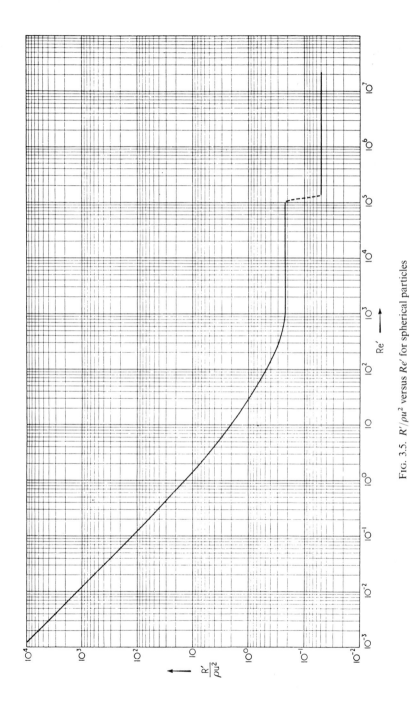

Fig. 3.5. $R'/\rho u^2$ versus $Re'$ for spherical particles

The curve shown in Figs. 3.4 and 3.5 is really continuous and its division into four regions is only a convenient method by which a series of equations can be assigned to it. In region (a) it is assumed that skin friction is solely responsible for the drag, and in regions (c) and (d) that only form drag is acting. In fact, over the whole range of Reynolds numbers both skin friction and form drag are present to some extent but the two are of comparable magnitude only in region (b). The exact divisions between regions (a) and (b), and between regions (b) and (c), depend on the limits at which form drag and skin friction respectively are assumed to become significant. Thus the upper limit for equation 3.2 is set by the accuracy required. As this is a simple equation to use, it is often applied for Reynolds numbers up to about 2, but the calculated values of $R'/\rho u^2$ are then far less accurate over this extended range.

Although $R'/\rho u^2$ is shown exactly constant at 0·22 in region (c), it is often reported that the value gradually rises as the Reynolds number increases, but the accuracy with which the relations can be applied seldom warrants a refinement of this sort.

### 3.3.1. Total Force on the Particle

In region (a)

$$R' = 12\rho u^2 \frac{\mu}{ud\rho} = \frac{12u\mu}{d} \tag{3.8}$$

The projected area of the particle $= \frac{1}{4}\pi d^2$. Thus the total force on the particle is given by:

$$F = \frac{12u\mu}{d}\pi d^2/4 = 3\pi\mu du \tag{3.9}$$

This is the expression originally obtained by Stokes[1] (equation 3.1).

In region (b), from equation 3.4:

$$R' = \frac{12u\mu}{d}(1+0·15\,Re'^{0·687}) \tag{3.10}$$

and therefore

$$F = 3\pi\mu du(1+0·15\,Re'^{0·687}) \tag{3.11}$$

In region (c):

$$R' = 0·22\rho u^2 \tag{3.12}$$

and

$$F = 0·22\rho u^2\,\tfrac{1}{4}\pi d^2 = 0·055\pi d^2\rho u^2 \tag{3.13}$$

This relation is often known as Newton's Law.

In region (d):

$$R' = 0·05\rho u^2 \tag{3.14}$$

$$F = 0·0125\pi d^2\rho u^2 \tag{3.15}$$

If a spherical particle is allowed to settle in a fluid under the action of the earth's gravitational field, it will increase in velocity until the accelerating force is exactly balanced by the resistance force. Although this state is approached exponentially, the effective acceleration period is generally of short duration for very small particles. If this terminal falling velocity is such that the corresponding value of $Re'$ is less than 0·2, the drag force on the particle will be given by equation 3.9. If the corresponding value of $Re'$ lies between 0·2 and 500, the drag force will be given approximately by Schiller and Naumann's equation (3.11). It should be noted, however, that if the particle has started from rest, the drag force will be given by equation 3.9 until $Re'$ exceeds 0·2. Again if the terminal falling velocity corresponds to a value of $Re'$ greater than about 500, the drag on the particle will be given by equation 3.13. Under terminal falling conditions, velocities are rarely high enough for $Re'$ to approach $10^5$, with the small particles generally used in the chemical industry. The accelerating force due to gravity:

$$= \tfrac{1}{6}\pi d^3 (\rho_s - \rho) g \tag{3.16}$$

where $\rho_s$ is the density of the solid.

The terminal falling velocity $u_0$ corresponding to region $(a)$ is given by:

$$\tfrac{1}{6}\pi d^3 (\rho_s - \rho) g = 3\pi \mu d u_0$$

$$u_0 = \frac{d^2 g}{18\mu}(\rho_s - \rho) \tag{3.17}$$

The terminal falling velocity corresponding to region $(c)$ is given by:

$$\tfrac{1}{6}\pi d^3 (\rho_s - \rho) g = 0.055\pi d^2 \rho u_0^2$$

i.e.

$$u_0^2 = 3dg\frac{(\rho_s - \rho)}{\rho} \tag{3.18}$$

In the expressions which have been given for the drag force and the terminal falling velocity it has been assumed:

(a) that the settling is not affected by the presence of other particles in the fluid. This condition is known as "free settling"; when the interference of other particles is appreciable, the process is known as "hindered settling";

(b) that the walls of the containing vessel do not exert an appreciable retarding effect; and

(c) that the fluid can be considered as a continuous medium, i.e. that the particle is large compared with the mean free path of the molecules of the fluid, otherwise the particles may occasionally "slip" between the molecules and thus attain a velocity higher than the calculated one.

From equations 3.17 and 3.18, it is seen that terminal falling velocity of a particle in a given fluid becomes greater as both particle size and density are increased. For a particle of material **A** of diameter $d_A$ and density $\rho_A$ if Stokes' Law is applicable the terminal falling velocity $u_{0A}$ is given by equation 3.17 as:

$$u_{0A} = \frac{d_A^2 g}{18\mu}(\rho_A - \rho) \tag{3.19}$$

Similarly for a particle of material **B**:

$$u_{0B} = \frac{d_B^2 g}{18\mu}(\rho_B - \rho) \tag{3.20}$$

The condition for the two terminal velocities to be equal is then:

$$\frac{d_B}{d_A} = \left\{\frac{\rho_A - \rho}{\rho_B - \rho}\right\}^{1/2} \tag{3.21}$$

If Newton's Law is applicable, equation 3.18 is applicable and:

$$u_{0A}^2 = \frac{3d_A g(\rho_A - \rho)}{\rho} \tag{3.22}$$

and

$$u_{0B}^2 = \frac{3d_B g(\rho_B - \rho)}{\rho} \tag{3.23}$$

For equal settling velocities:

$$\frac{d_B}{d_A} = \frac{\rho_A - \rho}{\rho_B - \rho} \tag{3.24}$$

In general, the relationship for equal settling velocities is:

$$\frac{d_B}{d_A} = \left(\frac{\rho_A - \rho}{\rho_B - \rho}\right)^s \tag{3.25}$$

$s = \frac{1}{2}$ for the Stokes' Law region, $s = 1$ for Newton's Law and $1/2 < s < 1$ for the intermediate region.

A number of small correction factors may be applied to account for the effect of the walls and for the effect of "slip". Ladenburg[6] states that the velocity of fall in a vessel of diameter $d_t$ must be multiplied by $[1 + 2\cdot4(d/d_t)]$ to give the velocity in an infinite medium. This is only approximate for values of $d/d_t$ approaching unity. He has introduced another similar factor, $[1 + 1\cdot7(d/dL')]$ which must be introduced when the particle is at a comparatively small distance $L'$ from the bottom of the container. According to Cunningham[7], the calculated falling velocity of particles below $0\cdot1$ μm in diameter must be multiplied by $[1 + (J\lambda/d)]$ to account for "slip", where $\lambda$ is the mean free path. Davies[8] gives $J$ equal to $1\cdot764 + 0\cdot562e^{-0\cdot785d/\lambda}$.

The behaviour of very small particles is also affected by Brownian motion. The molecules of the fluid bombard each particle in a random manner; if the particle is small, the net resultant force acting at any instant may be large enough to cause a change in its direction of motion. This effect has been studied by Davies[9], who has developed an expression for the combined effects of gravitation and Brownian motion on particles suspended in a fluid.

### 3.3.2. Effect of Motion of the Fluid

If the fluid is moving relative to some surface other than that of the particle, there will be a superimposed velocity distribution and the drag on the particle may be altered. Thus, if the particle is situated at the axis of a vertical tube up which fluid is flowing in streamline motion, the velocity near the particle will be twice the mean velocity because of the parabolic velocity profile in the fluid. The drag force is then determined by the difference in the velocities of the fluid and the particle at the axis.

The effect of turbulence in the fluid stream has been studied[9] by suspending a particle on a thread at the centre of a vertical pipe up which water was passed under conditions of turbulent flow. The upper end of the thread was attached to a lever fixed on a coil free to rotate in the field of an electromagnet. By passing a current through the coil it was possible to bring the level back to a null position. After calibration the current required could be related to the force acting on the sphere.

The results were expressed as the friction factor $(R'/\rho u^2)$, which was found to have a constant value of 0·40 for particle Reynolds numbers $(Re')$ over the range from 3000 to 9000, and for tube Reynolds numbers $(Re)$ from 12,000 to 26,000. Thus the value of $R'/\rho u^2$ has been approximately doubled as a result of turbulence in the fluid.

By surrounding the particle with a fixed array of similar particles on a hexagonal spacing the effect of neighbouring particles was measured. The results are discussed in Chapter 5.

Rowe and Henwood[10] made similar studies by supporting a spherical particle 12·7 mm diameter in water at the end of a 100 mm length of fine nichrome wire. The force exerted by the water when flowing in a 150 mm square duct was calculated from the measured deflection of the wire. The experiments were carried out at low Reynolds numbers with respect to the duct ($<1200$) corresponding to between 32 and 96 relative to the particle. The experimental values of the drag force were about 10 per cent higher than those calculated from the Schiller and Naumann equation. The work was then extended to cover the measurement of the force on a particle surrounded by an assemblage of particles and this is described in Chapter 5.

If $Re'$ is of the order of $10^5$, the drag on the sphere may be reduced if the fluid stream is turbulent. The flow in the boundary layer changes from streamline to turbulent and the formation of eddies in the wake of the particle is reduced. The higher the turbulence of the fluid, the lower is the value of $Re'$ at which the transition from region $(c)$ to region $(d)$ occurs. The value of $Re'$ at which $R'/\rho u^2$ is 0·15 is known as the *turbulence number* and is taken as an indication of the turbulence of the fluid.

### 3.3.3. Terminal Falling Velocities

Expressions have already been given for the terminal falling velocities of spherical particles but each is applicable over only a limited range of values of the Reynolds number $Re'$. If the value of $Re'$ is known approximately, it is possible to select the correct expression. In general, however, $Re_0 = u_0 d\rho/\mu$ will be unknown because it is a function of the unknown terminal velocity and a different approach must be used. The resistance force per unit projected area of the particle under terminal falling conditions $R'_0$ is given by:

$$R'_0 \tfrac{1}{4}\pi d^2 = \tfrac{1}{6}\pi d^3(\rho_s - \rho)g$$

i.e.

$$R'_0 = \tfrac{2}{3}d(\rho_s - \rho)g \tag{3.26}$$

and

$$\frac{R_0'}{\rho u_0^2} = \frac{2dg}{3\rho u_0^2}(\rho_s - \rho) \tag{3.27}$$

The curve of $R'/\rho u^2$ against $Re'$ cannot be used to calculate $u_0$ because both $Re_0'$ and $R_0'/\rho u_0^2$ are themselves functions of $u_0$, whereas the dimensionless group $(R_0'/\rho u_0^2)Re_0'^2$ does not involve $u_0$ since:

$$\frac{R_0'}{\rho u_0^2}\frac{u_0^2 d^2 \rho^2}{\mu^2} = \frac{2dg(\rho_s - \rho)}{3\rho u_0^2}\frac{u_0^2 d^2 \rho^2}{\mu^2}$$

$$= \frac{2d^3 \rho g(\rho_s - \rho)}{3\mu^2} \tag{3.28}$$

The group

$$\frac{d^3 \rho g(\rho_s - \rho)}{\mu^2}$$

is known as the Galileo number $Ga$.

Thus

$$\frac{R_0'}{\rho u_0^2}Re_0'^2 = \frac{2}{3}Ga \tag{3.29}$$

Using equations 3.2, 3.5 and 3.6 to express $R'/\rho u^2$ in terms of $Re'$ over the appropriate range of $Re'$:

$$Ga = 18Re_0' \qquad (Ga < 3.6) \tag{3.30}$$

$$Ga = 18Re_0' + 2.7Re_0'^{1.687} \quad (3.6 < Ga < \text{ca. } 10^5) \tag{3.31}$$

$$Ga = \tfrac{1}{3}Re_0'^2 \qquad (Ga > \text{ca. } 10^5). \tag{3.32}$$

$(R_0'/\rho u_0^2)Re_0'^2$ can be evaluated if the properties of the fluid and the particle are known. In Table 3.2, values of $\log Re'$ are given as a function of $\log\{(R'/\rho u^2)Re'^2\}$ and the data are represented in graphical form in Fig. 3.6 (data are taken from tables given by Heywood[11]. In order to determine the terminal falling velocity of a particle, $(R_0'/\rho u_0^2)Re_0'^2$ is evaluated and the corresponding value of $Re_0'$, and hence of the terminal velocity, is found either from Table 3.2 or from Fig. 3.6.

TABLE 3.2. *Values of* $\log Re'$ *as a Function of* $\log\{(R'/\rho u^2)Re'^2\}$ *for Spherical Particles*

| $\log\{(R'/\rho u^2)Re'^2\}$ | 0·0 | 0·1 | 0·2 | 0·3 | 0·4 | 0·5 | 0·6 | 0·7 | 0·8 | 0·9 |
|---|---|---|---|---|---|---|---|---|---|---|
| $\bar{2}$ | | | | | | | | 3·620 | 3·720 | 3·819 |
| $\bar{1}$ | $\bar{3}$·919 | $\bar{2}$·018 | $\bar{2}$·117 | $\bar{2}$·216 | $\bar{2}$·315 | $\bar{2}$·414 | $\bar{2}$·513 | $\bar{2}$·612 | $\bar{2}$·711 | $\bar{2}$·810 |
| 0 | $\bar{2}$·908 | $\bar{1}$·007 | $\bar{1}$·105 | $\bar{1}$·203 | $\bar{1}$·301 | $\bar{1}$·398 | $\bar{1}$·495 | $\bar{1}$·591 | $\bar{1}$·686 | $\bar{1}$·781 |
| 1 | $\bar{1}$·874 | $\bar{1}$·967 | 0·008 | 0·148 | 0·236 | 0·324 | 0·410 | 0·495 | 0·577 | 0·659 |
| 2 | 0·738 | 0·817 | 0·895 | 0·972 | 1·048 | 1·124 | 1·199 | 1·273 | 1·346 | 1·419 |
| 3 | 1·491 | 1·562 | 1·632 | 1·702 | 1·771 | 1·839 | 1·907 | 1·974 | 2·040 | 2·106 |
| 4 | 2·171 | 2·236 | 2·300 | 2·363 | 2·425 | 2·487 | 2·548 | 2·608 | 2·667 | 2·725 |
| 5 | 2·783 | 2·841 | 2·899 | 2·956 | 3·013 | 3·070 | 3·127 | 3·183 | 3·239 | 3·295 |

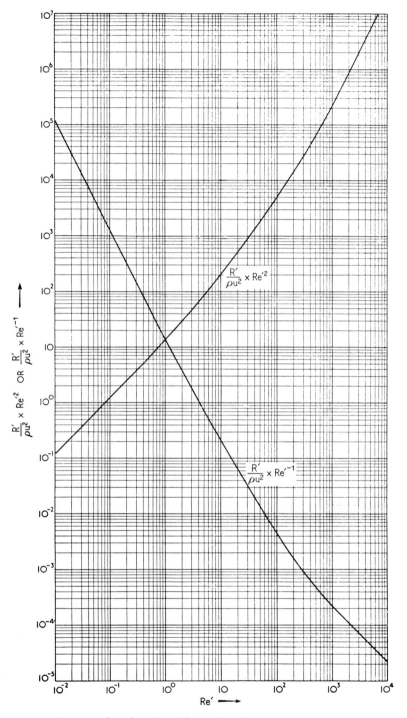

FIG. 3.6. $(R'/\rho u^2)\,Re'^2$ and $(R'/\rho u^2)\,Re'^{-1}$ versus $Re'$ for spherical particles

A similar difficulty is encountered in calculating the size of a sphere having a given terminal falling velocity, since $Re_0'$ and $R_0'/\rho u^2$ are both functions of the diameter $d$ of the particle. This calculation is similarly facilitated by the use of another combination, $(R_0'/\rho u_0^2)Re_0'^{-1}$, which is independent of diameter:

$$\frac{R_0'}{\rho u_0^2} Re_0'^{-1} = \frac{2\mu g}{3\rho^2 u_0^3}(\rho_s - \rho) \tag{3.33}$$

Log $Re'$ is given as a function of $\log\{(R'/\rho u^2)Re'^{-1}\}$ in Table 3.3 and the functions are plotted in Fig. 3.6. The diameter of a sphere of known terminal falling velocity can then be calculated by evaluating $(R_0'/\rho u_0^2)Re_0'^{-1}$, and then finding the corresponding value of $Re_0'$, from which the diameter can be calculated.

TABLE 3.3. *Values of log Re' as a Function of log $\{(R'/\rho u^2)Re'^{-1}\}$ for Spherical Particles*

| $\log\{(R'/\rho u^2)Re'^{-1}\}$ | 0·0 | 0·1 | 0·2 | 0·3 | 0·4 | 0·5 | 0·6 | 0·7 | 0·8 | 0·9 |
|---|---|---|---|---|---|---|---|---|---|---|
| $\bar{5}$ | | | | | | | | | | 3·401 |
| $\bar{4}$ | 3·316 | 3·231 | 3·148 | 3·065 | 2·984 | 2·903 | 2·824 | 2·745 | 2·668 | 2·591 |
| $\bar{3}$ | 2·517 | 2·443 | 2·372 | 2·300 | 2·231 | 2·162 | 2·095 | 2·027 | 1·961 | 1·894 |
| $\bar{2}$ | 1·829 | 1·763 | 1·699 | 1·634 | 1·571 | 1·508 | 1·496 | 1·383 | 1·322 | 1·260 |
| $\bar{1}$ | 1·200 | 1·140 | 1·081 | 1·022 | 0·963 | 0·904 | 0·846 | 0·788 | 0·730 | 0·672 |
| 0 | 0·616 | 0·560 | 0·505 | 0·449 | 0·394 | 0·339 | 0·286 | 0·232 | 0·178 | 0·125 |
| 1 | 0·072 | 0·019 | $\bar{1}$·969 | $\bar{1}$·919 | $\bar{1}$·865 | $\bar{1}$·811 | $\bar{1}$·760 | $\bar{1}$·708 | $\bar{1}$·656 | $\bar{1}$·605 |
| 2 | $\bar{1}$·554 | $\bar{1}$·503 | $\bar{1}$·452 | $\bar{1}$·401 | $\bar{1}$·350 | $\bar{1}$·299 | $\bar{1}$·249 | $\bar{1}$·198 | $\bar{1}$·148 | $\bar{1}$·097 |
| 3 | $\bar{1}$·047 | $\bar{2}$·996 | $\bar{2}$·946 | $\bar{2}$·895 | $\bar{2}$·845 | $\bar{2}$·794 | $\bar{2}$·744 | $\bar{2}$·694 | $\bar{2}$·644 | $\bar{2}$·594 |
| 4 | $\bar{2}$·544 | $\bar{2}$·493 | $\bar{2}$·443 | $\bar{2}$·393 | $\bar{2}$·343 | $\bar{2}$·292 | | | | |

## Example 3.1

What will be the settling velocity of a spherical steel particle, 0·40 mm diameter, in an oil of specific gravity 0·82 and viscosity $10\,\text{mN s/m}^2$? The specific gravity of steel is 7·87.

## Solution

For a sphere

$$\frac{R_0'}{\rho u_0^2} Re_0'^2 = \frac{2d^3}{3\mu^2}\rho(\rho_s - \rho)g \qquad \text{(equation 3.28)}$$

$$= \frac{2 \times 0.0004^3 \times 820(7870-820)9.81}{3(10 \times 10^{-3})^2}$$

$$= 24.2$$

$$\log_{10} 24.2 = 1.384$$
$$\log_{10} Re_0' = 0.222 \text{ (from Table 3·2)}$$
$$Re_0' = 1.667$$

$$u_0 = \frac{1.667 \times 10 \times 10^{-3}}{820 \times 0.0004}$$

$$= 0.051\,\text{m/s or } \underline{\underline{51\,\text{mm/s}}}$$

## 3.4. NON-SPHERICAL PARTICLES

### 3.4.1. Effect of Shape and Orientation on Drag

A spherical particle is unique in that it presents the same surface to the oncoming fluid whatever its orientation. For non-spherical particles, on the other hand, the orientation must be specified before the drag force can be calculated. The experimental data for the drag can be correlated in the same way as for the sphere by plotting the dimensionless group $R'/\rho u^2$ against the Reynolds number, $Re' = ud'\rho/\mu$, using logarithmic coordinates, and a separate curve is obtained for each shape of particle and for each orientation. In the above groups, $R'$ is taken, as before, as the resistance force per unit area of particle, projected on to a plane perpendicular to the direction of flow. $d'$ is defined as the diameter of the circle having the same area as the projected area of the particle and is therefore a function of the orientation, as well as the shape, of the particle.

The curve for $R'/\rho u^2$ against $Re'$ can be divided into four regions, $(a)$, $(b)$, $(c)$ and $(d)$, as before. In region $(a)$ the flow is entirely streamline and, although no theoretical expressions have been developed for the drag on the particle, the practical data suggest that a law of the form:

$$\frac{R'}{\rho u^2} = K\,Re'^{-1} \tag{3.34}$$

is applicable. The constant $K$ varies somewhat according to the shape and orientation of the particle but always has a value of about 12. In this region, a particle falling freely in the fluid under the action of gravity will normally move with its longest surface nearly parallel to the direction of motion.

At higher values of $Re'$, the linear relation between $R'/\rho u^2$ and $Re'^{-1}$ no longer holds and the slope of the curve gradually changes until $R'/\rho u^2$ becomes independent of $Re'$ in region $(c)$. Region $(b)$ represents transition conditions and commences at a lower value of $Re'$, and a correspondingly higher value of $R'/\rho u^2$, than in the case of the sphere: a freely falling particle will tend to change its orientation as the value of $Re'$ changes and some instability may be apparent. In region $(c)$ the particle will tend to fall so that it is presenting the maximum possible surface to the oncoming fluid. Typical values of $R'/\rho u^2$ for non-spherical particles in region $(c)$ are as follows:

*Thin rectangular plates, arranged with their planes perpendicular to the direction of motion.*

| | |
|---|---|
| Length/breadth = 1–5 | $R'/\rho u^2 = 0\cdot6$ |
| 20 | $0\cdot75$ |
| $\infty$ | $0\cdot95$ |

*Cylinders with axes parallel to direction of motion.*

| | |
|---|---|
| Length/diameter = 1 | $R'/\rho u^2 = 0\cdot45$ |

*Cylinders with axes perpendicular to the direction of motion.*

| | |
|---|---|
| Length/diameter = 1 | $R'/\rho u^2 = 0\cdot3$ |
| 5 | $0\cdot35$ |
| 20 | $0\cdot45$ |
| $\infty$ | $0\cdot6$ |

It should be noted that all the above values of $R'/\rho u^2$ are higher than the value of $0\cdot22$ for the sphere.

### 3.4.2. Terminal Falling Velocities

Heywood[11] has developed an approximate method for calculating the terminal falling velocity of a non-spherical particle, or for calculating its size from its terminal falling velocity. The method is an adaptation of his method for spheres.

A mean projected diameter of the particle $d_p$ is defined as the diameter of a circle having the same area as the particle when viewed from above and lying in its most stable position. Heywood selected this particular dimension because it is easily measured by microscopic examination.

If $d_p$ is the mean projected diameter, the mean projected area is $\pi d_p^2/4$ and the volume is $k'd_p^3$, where $k'$ is a constant whose value depends on the shape of the particle. For a spherical particle, $k'$ is equal to $\pi/6$. For rounded isometric particles (i.e. particles in which the dimension in three mutually perpendicular directions is approximately the same) $k'$ is about 0·5, but for angular particles $k'$ is about 0·4. For most minerals $k'$ lies between 0·2 and 0·5.

The method of calculating the terminal falling velocity consists in evaluating $(R_0'/\rho u^2)\,Re_0'^2$, using $d_p$ as the characteristic linear dimension of the particle and $\frac{1}{4}\pi d_p^2$ as the projected area in a plane perpendicular to the direction of motion. The corresponding value of $Re_0'$ is then found from Table 3.2 or Fig. 3.6, which refer to spherical particles, and a correction is then applied to the value of $\log Re_0'$ to account for the deviation from spherical shape. Values of this correction factor, which is a function both of $k'$ and of $(R'/\rho u^2)Re'^2$, are given in Table 3.4. A similar procedure is adopted for calculating the size of a particle of given terminal velocity (see Table 3.5).

The method is only approximate because it is assumed that $k'$ completely defines the shape of the particle, whereas there are many different shapes of particle for which the value of $k'$ is the same. Further it assumes that the diameter $d_p$ is the same as the mean projected diameter $d'$. This is very nearly so in regions (b) and (c), but in region (a) the particle tends to settle so that the longest face is parallel to the direction of motion and some error may therefore be introduced in the calculation, as indicated by Heiss and Coull[12].

For the non-spherical particle:

$$\text{total drag force} = R_0'\tfrac{1}{4}\pi d_p^2 = (\rho_s - \rho)gk'd_p^3$$

TABLE 3.4. *Logarithmic Corrections to log Re' as a Function of* $\log\{(R'/\rho u^2)\,Re'^2\}$ *for Non-spherical Particles*

| $\log\{(R'/\rho u^2)Re'^2\}$ | $k' = 0\cdot4$ | $k' = 0\cdot3$ | $k' = 0\cdot2$ | $k' = 0\cdot1$ |
|---|---|---|---|---|
| $\bar{2}$ | −0·022 | −0·002 | +0·032 | +0·131 |
| $\bar{1}$ | −0·023 | −0·003 | +0·030 | +0·131 |
| 0 | −0·025 | −0·005 | +0·026 | +0·129 |
| 1 | −0·027 | −0·010 | +0·021 | +0·122 |
| 2 | −0·031 | −0·016 | +0·012 | +0·111 |
| 2·5 | −0·033 | −0·020 | 0·000 | +0·080 |
| 3 | −0·038 | −0·032 | −0·022 | +0·025 |
| 3·5 | −0·051 | −0·052 | −0·056 | −0·040 |
| 4 | −0·068 | −0·074 | −0·089 | −0·098 |
| 4·5 | −0·083 | −0·093 | −0·114 | −0·146 |
| 5 | −0·097 | −0·110 | −0·135 | −0·186 |
| 5·5 | −0·109 | −0·125 | −0·154 | −0·224 |
| 6 | −0·120 | −0·134 | −0·172 | −0·255 |

TABLE 3.5. *Logarithmic Corrections to log Re′ as a Function of* $\{log\,(R'/\rho u^2)\,Re'^{-1}\}$ *for Non-spherical Particles*

| $\log\{(R'/\rho u^2)\,Re'^{-1}\}$ | $k' = 0{\cdot}4$ | $k' = 0{\cdot}3$ | $k' = 0{\cdot}2$ | $k' = 0{\cdot}1$ |
|---|---|---|---|---|
| $\bar{4}$ | $+0{\cdot}185$ | $+0{\cdot}217$ | $+0{\cdot}289$ | |
| $\bar{4}{\cdot}5$ | $+0{\cdot}149$ | $+0{\cdot}175$ | $+0{\cdot}231$ | |
| $\bar{3}$ | $+0{\cdot}114$ | $+0{\cdot}133$ | $+0{\cdot}173$ | $+0{\cdot}282$ |
| $\bar{3}{\cdot}5$ | $+0{\cdot}082$ | $+0{\cdot}095$ | $+0{\cdot}119$ | $+0{\cdot}170$ |
| $\bar{2}$ | $+0{\cdot}056$ | $+0{\cdot}061$ | $+0{\cdot}072$ | $+0{\cdot}062$ |
| $\bar{2}{\cdot}5$ | $+0{\cdot}038$ | $+0{\cdot}034$ | $+0{\cdot}033$ | $-0{\cdot}018$ |
| $\bar{1}$ | $+0{\cdot}028$ | $+0{\cdot}018$ | $+0{\cdot}007$ | $-0{\cdot}053$ |
| $\bar{1}{\cdot}5$ | $+0{\cdot}024$ | $+0{\cdot}013$ | $-0{\cdot}003$ | $-0{\cdot}061$ |
| 0 | $+0{\cdot}022$ | $+0{\cdot}011$ | $-0{\cdot}007$ | $-0{\cdot}062$ |
| 1 | $+0{\cdot}019$ | $+0{\cdot}009$ | $-0{\cdot}008$ | $-0{\cdot}063$ |
| 2 | $+0{\cdot}017$ | $+0{\cdot}007$ | $-0{\cdot}010$ | $-0{\cdot}064$ |
| 3 | $+0{\cdot}015$ | $+0{\cdot}005$ | $-0{\cdot}012$ | $-0{\cdot}065$ |
| 4 | $+0{\cdot}013$ | $+0{\cdot}003$ | $-0{\cdot}013$ | $-0{\cdot}066$ |
| 5 | $+0{\cdot}012$ | $+0{\cdot}002$ | $-0{\cdot}014$ | $-0{\cdot}066$ |

Thus
$$\frac{R'_0}{\rho u_0^2} = \frac{4k'd_p g}{\pi \rho u_0^2}(\rho_s - \rho) \tag{3.35}$$

$$\frac{R'_0}{\rho u_0^2}Re_0'^2 = \frac{4k'\rho d_p^3 g}{\mu^2 \pi}(\rho_s - \rho) \tag{3.36}$$

and
$$\frac{R'_0}{\rho u_0^2}Re_0'^{-1} = \frac{4k'\mu g}{\pi \rho^2 u_0^3}(\rho_s - \rho) \tag{3.37}$$

Provided $k'$ is known, the appropriate dimensionless group can be evaluated and the terminal falling velocity, or diameter, calculated.

### Example 3.2

What will be the settling velocities of mica plates, 1 mm thick and ranging in area from 6 to 600 mm², in an oil of specific gravity 0·82 and viscosity 10 mN s/m²? The specific gravity of mica is 3·0.

### Solution

| | smallest particles | largest particles |
|---|---|---|
| $A'$ | $6 \times 10^{-6}\,\text{m}^2$ | $6 \times 10^{-4}\,\text{m}^2$ |
| $d_p$ | $\sqrt{(4 \times 6 \times 10^{-6}/\pi)} = 2{\cdot}76 \times 10^{-3}\,\text{m}$ | $\sqrt{(4 \times 6 \times 10^{-4}/\pi)} = 2{\cdot}76 \times 10^{-2}\,\text{m}$ |
| $d_p^3$ | $2{\cdot}103 \times 10^{-8}\,\text{m}^3$ | $2{\cdot}103 \times 10^{-5}\,\text{m}^3$ |
| volume | $6 \times 10^{-9}\,\text{m}^3$ | $6 \times 10^{-7}\,\text{m}^3$ |
| $k'$ | $0{\cdot}285$ | $0{\cdot}0285$ |

$$\left(\frac{R'_0}{\rho u^2}\right)Re_0'^2 = \frac{4k'}{\mu^2 \pi}(\rho_s - \rho)\rho d_p^3 g \qquad \text{(equation 3.36)}$$

$$= (4 \times 0{\cdot}285/\pi \times 0{\cdot}01^2)(3000 - 820)820 \times 2{\cdot}103 \times 10^{-8} \times 9{\cdot}81$$
$$= 1340 \text{ for smallest particles and}$$
$$\quad 134{,}000 \text{ for largest particles.}$$

| | smallest particles | largest particles | |
|---|---|---|---|
| $\log\left\{\dfrac{R_0'}{\rho u_0^2}Re_0'^2\right\}$ | 3·127 | 5·127 | |
| $\log Re_0'$ | 1·581 | 2·857 | (from Table 3.2) |
| Correction from Table 3.4 | −0·038 | −0·300 | (estimated) |
| Corrected $\log Re_0'$ | 1·543 | 2·557 | |
| $Re_0'$ | 34·9 | 361 | |
| $u_0$ | 0·154 m/s | 0·159 m/s | |

Thus it is seen that all the mica particles settle at approximately the same velocity.

### 3.5. MOTION OF BUBBLES AND DROPS

The drag force acting on a gas bubble or a liquid droplet will not, in general, be the same as that acting on a rigid particle of the same shape and size because circulating currents are set up inside the bubble. The velocity gradient at the surface is thereby reduced and the drag force is therefore less than for the rigid particle. Hadamard[13] showed that, if the effects of surface energy are neglected, the terminal falling velocity of a drop, as calculated from Stokes' Law, must be multiplied by a factor $Q$, to account for the internal circulation, such that:

$$Q = \frac{3\mu + 3\mu_l}{2\mu + 3\mu_l} \tag{3.38}$$

where $\mu$ is the viscosity of the continuous fluid and $\mu_l$ is the viscosity of the fluid forming the drop or bubble. This expression applies only in the range for which Stokes' Law is valid.

If $\mu_l/\mu$ is large, $Q$ approaches unity, but if $\mu_l/\mu$ is small, $Q$ approaches a value of 1·5. Thus the effect of circulation is small when a liquid drop falls in a gas but is large when a gas bubble rises in a liquid. If the fluid within the drop is very viscous, the amount of energy which has to be transferred in order to induce circulation is large and circulation effects are therefore small.

Hadamard's work was later substantiated by Bond[14] and by Bond and Newton[15] who showed that equation 3.38 is valid provided that surface tension forces do not play a large role. With very small droplets, the surface tension forces tend to nullify the tendency for circulation, and the droplet falls at a velocity close to that of a solid sphere.

Drops and bubbles are, in addition, subject to deformation because of the differences in the pressures acting on various parts of the surface. Thus, when a drop is settling in a still fluid, both the hydrostatic and the impact pressures will be greater on the forward face than on the rear face and will tend to flatten the drop, whereas the viscous drag will tend to elongate it. Deformation of the drop is opposed by the surface tension forces so that very small drops retain their spherical shape, whereas large drops may be considerably deformed and the resistance to their motion thereby increased. For drops above a certain size, the deformation is so great that the drag force increases at the same rate as the volume, and the terminal falling velocity therefore becomes independent of size.

Garner has shown the importance of circulation within a drop in determining the coefficient of mass transfer between the drop and the surrounding medium. The critical Reynolds number at which circulation commences has been shown[16] to increase at a rate proportional to the logarithm of viscosity of the liquid constituting the drop and to increase with interfacial tension. The circulation rate may be influenced by mass transfer because of the effect of concentration of diffusing material on both the interfacial tension and on the viscosity of the surface layers. As a result of circulation the falling velocity may be up to 50 per cent greater than for a rigid sphere whereas oscillation of the drop between oblate and prolate forms will reduce the velocity of fall[17]. Terminal falling velocities of droplets have also been calculated by Hamielec and Johnson[18] from approximate velocity profiles at the interface and the values so obtained compare well with experimental values for droplet Reynolds numbers up to 80.

## 3.6. ACCELERATING MOTION OF A PARTICLE IN THE GRAVITATIONAL FIELD

The motion of a particle through a fluid can be traced provided that the value of the drag factor $R'/\rho u^2$ for a given value of the Reynolds number is fixed. The behaviour of a particle undergoing acceleration or retardation has been the subject of a very large number of investigations, which have been critically reviewed by Torobin and Gauvin[19]. The results of different workers are not consistent but it is shown that the drag factor is often related not only to the Reynolds number but also to the number of particle diameters traversed by the particle since the initiation of the motion. Originally it had been suggested that the particle behaved as if its mass had been increased by an amount proportional to the mass of displaced fluid, but this concept appears to be useful only over a very limited range of conditions. It is generally more satisfactory to regard the drag factor as modified by the effects of acceleration. In most cases it is found that the rate of acceleration is also important and that the effect of acceleration is less at high Reynolds numbers. A number of experimental investigations do contradict these conclusions, however.

In the following treatment it will be assumed that the steady-state drag factors can be used during the accelerating motion of a sphere and the limitations of this assumption must be borne in mind.

Consider the motion of a particle of mass $m$ in the earth's gravitational field. At some time $t$ the particle will be moving at an angle $\alpha$ to the horizontal with a velocity $u$ (Fig. 3.7). This velocity $u$ can then be resolved into two components, $\dot{x}$ and $\dot{y}$, in the horizontal and vertical directions. ($\dot{x}$ and $\ddot{x}$ will be taken to denote the first and second derivatives of the

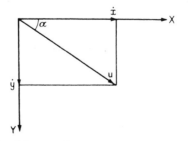

Fig. 3.7. Two-dimensional motion of particle

displacement $x$ in the $X$-direction with respect to time, and $\dot{y}$ and $\ddot{y}$ the corresponding derivatives of $y$.)

Now

$$\cos\alpha = \frac{\dot{x}}{u} \tag{3.39}$$

$$\sin\alpha = \frac{\dot{y}}{u} \tag{3.40}$$

and

$$u = \sqrt{\dot{x}^2 + \dot{y}^2} \tag{3.41}$$

There are two forces acting on the body:

(1) In the vertical direction, the apparent weight of the particle,

$$mg\left(1 - \frac{\rho}{\rho_s}\right)$$

(2) The drag force which is equal to $R'A'$ and acts in such a direction as to oppose the motion of the particle. Its direction therefore changes as $\alpha$ changes. Here $A'$ is the projected area of the particle on a plane at right angles to the direction of motion and its value varies with the orientation of the particle in the fluid. The drag force can be expressed by the relation

$$F = \frac{R'}{\rho u^2}\rho u^2 A' \tag{3.42}$$

This has a component in the $X$-direction of

$$\frac{R'}{\rho u^2}\rho u^2 A' \cos\alpha = \frac{R'}{\rho u^2}A'\rho\dot{x}\sqrt{\dot{x}^2 + \dot{y}^2}$$

and in the $Y$-direction of

$$\frac{R'}{\rho u^2}\rho u^2 A' \sin\alpha = \frac{R'}{\rho u^2}A'\rho\dot{y}\sqrt{\dot{x}^2 + \dot{y}^2}$$

The equations of motion in the $X$- and $Y$-directions are therefore:

$$m\ddot{x} = -\frac{R'}{\rho u^2}\rho A'\dot{x}\sqrt{\dot{x}^2 + \dot{y}^2} \tag{3.43}$$

and

$$m\ddot{y} = -\frac{R'}{\rho u^2}\rho A'\dot{y}\sqrt{\dot{x}^2 + \dot{y}^2} + mg\left(1 - \frac{\rho}{\rho_s}\right) \tag{3.44}$$

assuming the velocities $\dot{x}$ and $\dot{y}$ to be positive. If $\dot{x}$ or $\dot{y}$ is negative the drag force acts in the opposite direction and the corresponding term bears a positive sign. In general direct integration is not possible because the equation of motion in the $X$-direction involves the velocity in the $Y$-direction and vice versa, but for the following cases we can obtain a solution:

(1) flow under conditions where form drag is negligible;
(2) unidimensional flow, when the effect of gravity can be neglected; and
(3) unidimensional flow in the vertical direction, under the action of gravity.

### 3.6.1. Motion of a Particle (Form Drag Negligible)

Under these conditions, $R'/\rho u^2$ is given by the relation:

$$\frac{R'}{\rho u^2} = K\, Re'^{-1} \qquad\qquad \text{(equation 3.34)}$$

If this value is substituted in equations 3.43 and 3.44, then:

$$\ddot{x} = -\frac{A'\dot{x}\rho}{m}\frac{K\mu}{ud'\rho}\sqrt{\dot{x}^2+\dot{y}^2} = -\frac{K\mu}{md'}A'\dot{x} \qquad\qquad (3.45)$$

and

$$\ddot{y} = -\frac{K\mu}{md'}A'\dot{y}+g\left(\frac{\rho_s-\rho}{\rho_s}\right) \qquad\qquad (3.46)$$

since

$$u = \sqrt{\dot{x}^2+\dot{y}^2} \qquad\qquad \text{(equation 3.41)}$$

Thus, in this particular case, the motions in two mutually perpendicular directions can be considered independently. Further, it is seen that if $\dot{x}$ or $\dot{y}$ is negative, the resistance term becomes positive and the equation of motion is still valid.

For the motion of a spherical particle, $K = 12$, so that:

$$\ddot{x} = -\frac{\dfrac{\pi}{4}d^2 12\mu}{\dfrac{\pi}{6}d^3\rho_s d}\,\dot{x}$$

$$= -18\frac{\mu}{d^2\rho_s}\dot{x} \qquad\qquad (3.47)$$

$$= -a\dot{x} \qquad\qquad (3.48)$$

and

$$\ddot{y} = -\frac{\dfrac{\pi}{4}d^2 12\mu}{\dfrac{\pi}{6}d^3\rho_s d}\,\dot{y}+\left(1-\frac{\rho}{\rho_s}\right)g$$

$$= -18\frac{\mu}{d^2\rho_s}\dot{y}+\left(1-\frac{\rho}{\rho_s}\right)g \qquad\qquad (3.49)$$

$$= -a\dot{y}+b \qquad\qquad (3.50)$$

(where

$$a = 18\frac{\mu}{d^2\rho_s}$$                                                          (3.51)

and

$$b = \left(1 - \frac{\rho}{\rho_s}\right)g$$                                              (3.52)

are constant for a given fluid and particle).

These equations can now be integrated independently.

$$\ddot{y} = -a\dot{y} + b$$                                             (equation 3.50)

Integrating with respect to $t$,

$$\dot{y} = -ay + bt + \text{constant}.$$

The axes will now be so chosen that the particle is at the origin at time $t = 0$. Let the initial component of the velocity of the particle in the $Y$-direction be $v$. Thus, when $t = 0$, $y = 0$ and $\dot{y} = v$, and the constant $= v$, i.e.

$$\dot{y} + ay = bt + v$$

$$\therefore \quad e^{at}\dot{y} + e^{at}ay = (bt + v)e^{at}$$

Thus,

$$e^{at}y = (bt + v)\frac{e^{at}}{a} - \int b\frac{e^{at}}{a}\,dt$$

$$= (bt + v)\frac{e^{at}}{a} - \frac{b}{a^2}e^{at} + \text{constant}$$

When $t = 0$, $y = 0$, and the constant $= \dfrac{b}{a^2} - \dfrac{v}{a}$.

Thus,

$$y = \frac{b}{a}t + \frac{v}{a} - \frac{b}{a^2} + \left(\frac{b}{a^2} - \frac{v}{a}\right)e^{-at}$$                    (3.53)

where

$$a = 18\frac{\mu}{d^2\rho_s} \quad \text{and} \quad b = \left(1 - \frac{\rho}{\rho_s}\right)g.$$

It should be noted that $b/a = u_0$, the terminal falling velocity of the particle. This equation enables the displacement of the particle in the $Y$-direction to be calculated at any time $t$. For the displacement in the $X$-direction:

$$\ddot{x} = -a\dot{x}$$                                                  (equation 3.48)

$$\dot{x} = -ax + \text{constant}$$

Let $w$ be the component of the velocity of the particle in the $X$-direction at time $t = 0$.

Then, when $t = 0$, $x = 0$ and $\dot{x} = w$. The constant therefore is equal to $w$. Hence

$$\dot{x} + ax = w$$

$$\therefore \quad x\,e^{at} = \frac{w}{a}e^{at} + \text{constant}$$

When $t = 0$, $x = 0$, and the constant $= -w/a$,

$$\therefore \quad x\,e^{at} = \frac{w}{a}(e^{at} - 1)$$

i.e.

$$x = \frac{w}{a}(1 - e^{-at}) \tag{3.54}$$

where

$$a = \frac{18\mu}{d^2 \rho_s}.$$

Thus the displacement in the $X$-direction can also be calculated for any time $t$.

By eliminating $t$ between equations 3.53 and 3.54, a relation between the displacements in the $X$- and $Y$-directions is obtained. Equations of this form are useful for calculating the trajectories of particles in size-separation equipment.

From equation 3.54,

$$e^{-at} = 1 - \frac{ax}{w}$$

and

$$t = -\frac{1}{a}\ln\left(1 - \frac{ax}{w}\right) \tag{3.55}$$

Substituting in equation 3.53 gives:

$$y = \frac{b}{a}\left\{-\frac{1}{a}\ln\left(1 - \frac{ax}{w}\right)\right\} + \frac{v}{a} - \frac{b}{a^2} + \left(\frac{b}{a^2} - \frac{v}{a}\right)\left(1 - \frac{ax}{w}\right)$$

$$= -\frac{b}{a^2}\ln\left(1 - \frac{ax}{w}\right) - \frac{bx}{aw} + \frac{vx}{w} \tag{3.56}$$

The values of $a$ and $b$ can now be substituted and the final relation is:

$$y = -\frac{g\rho_s(\rho_s - \rho)d^4}{324\mu^2}\left\{\ln\left(1 - \frac{18\mu x}{w\rho_s d^2}\right) + \frac{18\mu x}{w\rho_s d^2}\left(1 - \frac{18v\mu}{d^2(\rho_s - \rho)g}\right)\right\} \tag{3.57}$$

### 3.6.2. Unidimensional Motion in the Absence of Gravitation

When the resistance force is very large compared with the gravitational force, the latter may be neglected. The only force then acting on the body is that due to the resistance of the fluid and the direction of motion will not alter. This condition is met when the densities of the particle and of the fluid are almost the same, when the particle is very small and when the viscosity of the fluid is high.

Consider the motion of a particle in the $X$-direction under conditions where the effect of gravity can be neglected. Since $\dot{y} = 0$, then from equation 3.43 for $\dot{x}$ positive:

$$m\ddot{x} = -\frac{R'}{\rho u^2} A'\rho\dot{x}^2 \tag{3.58}$$

For a sphere

$$\ddot{x} = -\frac{R'}{\rho u^2}\frac{\frac{1}{4}\pi d^2\rho}{\frac{1}{6}\pi d^3\rho_s}\dot{x}^2$$

$$= -\frac{R'}{\rho u^2}\frac{1\cdot5\rho}{d\rho_s}\dot{x}^2 \tag{3.59}$$

### Resistance due to Skin Friction only

Under these conditions of motion:

$$\frac{R'}{\rho u^2} = 12\,Re'^{-1} \tag{equation 3.2}$$

and

$$\ddot{x} = -\frac{18\mu}{d^2\rho_s}\dot{x} \tag{equation 3.47}$$

This equation has already been integrated to give:

$$x = \frac{w}{a}(1-\mathrm{e}^{-at}) \tag{equation 3.54}$$

where $a = (18\mu/d^2\rho_s)$ and $w$ is the initial velocity.

Equation 3.54 is applicable if $Re'$ is less than about 0·2, and for positive or negative values of $w$.

### Resistance due to Form Drag only

If $Re'$ exceeds about 500, $R'/\rho u^2$ attains a constant value of 0·22. Then:

$$\ddot{x} = -\frac{0\cdot33}{d}\frac{\rho}{\rho_s}\dot{x}^2 \tag{3.60}$$

$$= -c\dot{x}^2 \tag{3.61}$$

where

$$c = \frac{0\cdot33}{d}\frac{\rho}{\rho_s} \tag{3.62}$$

Thus

$$\frac{d\dot{x}}{\dot{x}^2} = -c\,dt$$

Integrating,

$$-\dot{x}^{-1} = -ct + \text{constant}$$

When $t = 0$, $\dot{x} = w$, and therefore the constant $= -w^{-1}$. Thus

$$\dot{x}^{-1} = ct + w^{-1}$$

$$\therefore \quad \frac{dt}{ct + w^{-1}} = dx$$

and

$$x = \frac{1}{c}\ln\left(t + \frac{1}{wc}\right) + \text{constant}$$

When $t = 0$, $x = 0$ and the constant $= -\frac{1}{c}\ln\frac{1}{wc}$. Thus:

$$x = \frac{1}{c}\ln\frac{t + (1/wc)}{1/wc}$$

$$= \frac{1}{c}\ln(wct + 1) \qquad (3.63)$$

where

$$c = \frac{0 \cdot 33\rho}{d\rho_s}$$

If $w$ is negative,

$$x = -\frac{1}{c}\ln(1 - wct) \qquad (3.64)$$

*Transition Region*

When $Re'$ lies between about $0 \cdot 2$ and 500, $R'/\rho u^2$ cannot readily be expressed as a simple function of $Re'$. A graphical solution of the problem is therefore necessary.

For a spherical particle when $w$ is positive:

$$\ddot{x} = -\frac{R'}{\rho u^2}\frac{1 \cdot 5}{d}\frac{\rho}{\rho_s}\dot{x}^2 \qquad \text{(equation 3.59)}$$

Now

$$\dot{x} = Re'\frac{\mu}{d\rho}$$

and thus

$$\frac{\mu}{d\rho}\frac{dRe'}{dt} = -\frac{R'}{\rho u^2}\frac{1\cdot 5}{d}\frac{\rho}{\rho_s}\frac{\mu^2}{d^2\rho^2}Re'^2$$

i.e.

$$\frac{dRe'}{dt} = -\frac{1\cdot 5\mu}{d^2\rho_s}\frac{R'}{\rho u^2}Re'^2$$

so that

$$t = -\frac{2d^2\rho_s}{3\mu}\int_{Re'_1}^{Re'_2}\frac{1}{(R'/\rho u^2)\,Re'^2}\,dRe' \qquad (3.65)$$

If $w$ is negative,

$$t = +\frac{2d^2\rho_s}{3\mu}\int_{Re'_1}^{Re'_2}\frac{1}{(R'/\rho u^2)\,Re'^2}\,dRe' \qquad (3.66)$$

This expression can be integrated graphically between the limits $Re'_1$ and $Re'_2$ in order to obtain $t$ as a function of $Re'$. The corresponding values of $\dot{x}$ are then calculated; $x$ is then obtained in terms of $t$ by means of a further graphical integration. This method can be used over any range of values of $Re'$ but, since graphical integrations are rather tedious, it is better to use the analytical expressions already derived for conditions corresponding to $Re'$ below 0·2, or greater than 500, and to confine the application of the graphical method to the intervening region. Schiller and Naumann's equation (3.5) can be used for calculating the values of $R'/\rho u^2$ required for the graphical integration.

### 3.6.3. Vertical Motion of a Particle

For vertical motion, the equations will be different for upward and downward motion. If the particle is initially moving downwards (taken as $\dot{y}$ positive), the gravitational force and the fluid resistance will act in opposite directions; the particle will then always move in the same sense. On the other hand, if the particle is initially moving upwards, the gravitational and resistance forces will both act downwards so that, in due course, it will be brought to rest and will then move in the opposite sense. As the equations of motion are different for the two senses, care must be exercised in the application of the expressions which are obtained.

Consider the equation of motion for a particle moving vertically downwards. From equation 3.44:

$$m\ddot{y} = -\frac{R'}{\rho u^2}A'\rho\dot{y}^2 + mg\left(1 - \frac{\rho}{\rho_s}\right) \qquad (3.67)$$

i.e.

$$\ddot{y} = -\frac{R'}{\rho u^2}\frac{A'}{m}\rho\dot{y}^2 + g\left(1 - \frac{\rho}{\rho_s}\right) \qquad (3.68)$$

For a spherical particle, $A' = \pi d^2/4$ and $m = \pi d^3 \rho_s/6$. Thus:

$$\ddot{y} = -\frac{R'}{\rho u^2} \frac{1 \cdot 5}{d} \frac{\rho}{\rho_s} \dot{y}^2 + g\left(1 - \frac{\rho}{\rho_s}\right) \tag{3.69}$$

*Resistance due to Skin Friction only*

Under these conditions of motion:

$$\frac{R'}{\rho u^2} = 12\, Re'^{-1} \tag{equation 3.2}$$

and

$$\ddot{y} = -\frac{18}{d^2 \rho_s} \mu \dot{y} + g\left(1 - \frac{\rho}{\rho_s}\right) \tag{3.70}$$

This equation, which is applicable whether the particle is moving upwards or downwards, has already been integrated to give:

$$y = \frac{b}{a}t + \frac{v}{a} - \frac{b}{a^2} + \left(\frac{b}{a^2} - \frac{v}{a}\right)e^{-at} \tag{equation 3.53}$$

where

$$a = 18\frac{\mu}{d^2 \rho_s} \quad \text{and} \quad b = \left(1 - \frac{\rho}{\rho_s}\right)g.$$

*Resistance due to Form Drag only*

When the value of $Re'$ exceeds about 500:

$$\frac{R'}{\rho u^2} = 0 \cdot 22 \tag{equation 3.6}$$

Thus

$$\ddot{y} = -\frac{0 \cdot 33}{d} \frac{\rho}{\rho_s} \dot{y}^2 + g\left(1 - \frac{\rho}{\rho_s}\right) \tag{3.71}$$

$$= -c\dot{y}^2 + b \tag{3.72}$$

where

$$c = \frac{0 \cdot 33}{d} \frac{\rho}{\rho_s} \tag{equation 3.62}$$

and

$$b = g\left(1 - \frac{\rho}{\rho_s}\right) \tag{equation 3.52}$$

If the particle is moving upwards, $\dot{y}$ is negative and the drag force is positive. Thus, for this condition:

$$\ddot{y} = c\dot{y}^2 + b \qquad (3.73)$$

Equations 3.72 and 3.73 will now be integrated separately.

*Motion downwards:*

$$\ddot{y} = -c\dot{y}^2 + b \qquad \text{(equation 3.72)}$$

$$\therefore \quad \frac{d\dot{y}}{b - c\dot{y}^2} = dt$$

$$\therefore \quad \frac{d\dot{y}}{f^2 - \dot{y}^2} = c\,dt$$

where

$$f = \sqrt{\frac{b}{c}} = \sqrt{\frac{d(\rho_s - \rho)g}{0\cdot 33\rho}} \qquad (3.74)$$

Integrating,

$$\frac{1}{2f}\ln\frac{f + \dot{y}}{f - \dot{y}} = ct + \text{constant}$$

When $t = 0$, $\dot{y} = v$, say, and therefore the constant $= \dfrac{1}{2f}\ln\dfrac{f + v}{f - v}$.

$$\therefore \quad \frac{1}{2f}\ln\left(\frac{f + \dot{y}}{f - \dot{y}}\right)\left(\frac{f - v}{f + v}\right) = ct$$

$$\therefore \quad \left(\frac{f + \dot{y}}{f - \dot{y}}\right)\left(\frac{f - v}{f + v}\right) = e^{2fct}$$

$$\therefore \quad -1 + \frac{2f}{f - \dot{y}} = \frac{f + v}{f - v}e^{2fct}$$

$$\therefore \quad f - \dot{y} = \frac{2f}{1 + \dfrac{f + v}{f - v}e^{2fct}}$$

$$\therefore \quad y = ft - 2f\int\frac{dt}{1 + \dfrac{f + v}{f - v}e^{2fct}}$$

$$= ft - 2fI \text{ (say)}$$

where

$$I = \int\frac{dt}{1 + je^{pt}} \quad \left(\text{where } j = \frac{f + v}{f - v} \quad \text{and} \quad p = 2fc\right)$$

Put

$$s = 1 + j e^{pt}$$

Then

$$ds = pj e^{pt} dt = p(s-1) dt$$

Thus

$$I = \int \frac{ds}{ps(s-1)}$$

$$= \frac{1}{p} \int \left\{ \frac{1}{s-1} - \frac{1}{s} \right\} ds$$

$$= \frac{1}{p} \ln \frac{s-1}{s} + \text{constant}$$

$$= \frac{1}{p} \ln \frac{j e^{pt}}{j e^{pt} + 1} + \text{constant}$$

$$= \frac{1}{2fc} \ln \frac{1}{1 + \dfrac{f-v}{f+v} e^{-2fct}} + \text{constant}$$

Thus,

$$y = ft - \frac{1}{c} \ln \frac{1}{1 + \dfrac{f-v}{f+v} e^{-2fct}} + \text{constant}$$

When $t = 0$, $y = 0$ and therefore:

$$\text{constant} = \frac{1}{c} \ln \frac{1}{1 + \dfrac{f-v}{f+v}} = \frac{1}{c} \ln \frac{f+v}{2f}$$

$$\therefore \quad y = ft + \frac{1}{c} \ln \frac{f+v}{2f} \left\{ 1 + \frac{f-v}{f+v} e^{-2fct} \right\}$$

i.e.

$$y = ft + \frac{1}{c} \ln \frac{1}{2f} \left\{ f + v + (f-v) e^{-2fct} \right\} \tag{3.75}$$

where

$$c = \frac{0.33}{d} \frac{\rho}{\rho_s} \tag{equation 3.62}$$

and

$$f = \sqrt{\frac{d(\rho_s - \rho)g}{0.33\rho}} \tag{equation 3.74}$$

*Upward motion:*

$$\ddot{y} = c\dot{y}^2 + b \qquad \text{(equation 3.73)}$$

$$\therefore \quad \frac{d\dot{y}}{b + c\dot{y}^2} = dt$$

$$\therefore \quad \frac{d\dot{y}}{f^2 + \dot{y}^2} = c\,dt$$

where

$$f = \sqrt{\frac{b}{c}} = \sqrt{\frac{d(\rho_s - \rho)g}{0\cdot33\rho}} \qquad \text{(equation 3.74)}$$

Integrating,

$$\frac{1}{f}\tan^{-1}\frac{\dot{y}}{f} = ct + \text{constant}$$

When $t = 0$, $\dot{y} = v$, say, and the constant $= 1/f \tan^{-1} v/f$; in this case, $v$ is a negative quantity. Thus:

$$\frac{1}{f}\tan^{-1}\frac{\dot{y}}{f} = ct + \frac{1}{f}\tan^{-1}\frac{v}{f}$$

$$\therefore \quad \frac{\dot{y}}{f} = \tan\left(fct + \tan^{-1}\frac{v}{f}\right)$$

$$\therefore \quad y = \frac{f}{-fc}\ln\cos\left(fct + \tan^{-1}\frac{v}{f}\right) + \text{constant}$$

When $t = 0$, $y = 0$, so that:

$$\text{constant} = \frac{1}{c}\ln\cos\tan^{-1}\frac{v}{f}$$

$$\therefore \quad y = -\frac{1}{c}\ln\frac{\cos\left(fct + \tan^{-1}\frac{v}{f}\right)}{\cos\tan^{-1}\frac{v}{f}}$$

$$= -\frac{1}{c}\ln\frac{\cos fct \cos\tan^{-1}\frac{v}{f} - \sin fct \sin\tan^{-1}\frac{v}{f}}{\cos\tan^{-1}\frac{v}{f}}$$

i.e.

$$y = -\frac{1}{c}\ln\left(\cos fct - \frac{v}{f}\sin fct\right) \qquad \text{(3.76)}$$

where

$$c = \frac{0\cdot33}{d}\frac{\rho}{\rho_s} \qquad \text{(equation 3.62)}$$

and

$$f = \sqrt{\frac{d(\rho_s - \rho)g}{0 \cdot 33\rho}}$$                    (equation 3.74)

*Transition Region*

The relation between $y$ and $t$ can also be obtained graphically, though the process is more tedious than that of using the analytical solution, appropriate to the particular case in question. When $Re'$ lies between 0·2 and 500, there is no analytical solution to the problem and the graphical method must be used.

When the spherical particle is moving downwards, i.e. when its velocity is positive:

$$\ddot{y} = -\frac{3}{2d}\frac{\rho}{\rho_s}\frac{R'}{\rho u^2}\dot{y}^2 + \left(1 - \frac{\rho}{\rho_s}\right)g$$                    (equation 3.69)

$$\therefore \quad \frac{\mu}{\rho d}\frac{dRe'}{dt} = -\frac{3}{2d}\frac{\rho}{\rho_s}\frac{R'}{\rho u^2}Re'^2\frac{\mu^2}{d^2\rho^2} + \left(1 - \frac{\rho}{\rho_s}\right)g$$

$$\therefore \quad t = \int_{Re'_1}^{Re'_2} \frac{dRe'}{\dfrac{d\rho(\rho_s - \rho)g}{\mu\rho_s} - \dfrac{3\mu}{2d^2\rho_s}\dfrac{R'}{\rho u^2}Re'^2}$$                    (3.77)

If the particle is moving upwards, the corresponding expression for $t$ is:

$$t = \int_{Re'_1}^{Re'_2} \frac{dRe'}{\dfrac{d\rho(\rho_s - \rho)g}{\mu\rho_s} + \dfrac{3u}{2d^2\rho_s}\dfrac{R'}{\rho u^2}Re'^2}$$                    (3.78)

From the above equations, $Re'$ can be obtained as a function of $t$. The velocity $\dot{y}$ can then be calculated. By means of a second graphical integration, the displacement $y$ can be found at any time $t$.

In using the various relations which have been obtained, it must be remembered that the law of motion of the particle will change as the relative velocity between the particle and the fluid changes.

If, for instance, a particle is initially moving upwards with a velocity $v$, so that the corresponding value of $Re'$ is greater than about 500, the relation between $y$ and $t$ will be given by equation 3.76. The velocity of the particle will progressively decrease and, when $Re'$ is less than 500, the motion is obtained by application of equation 3.78. The upward velocity will then fall still further until $Re'$ falls below 0·2. While the particle is moving under these conditions, its velocity will fall to zero and will then gradually increase in the downward direction. The same equation (3.53) can be applied for the whole of the time the Reynolds group is less than 0·2, irrespective of sense. Then for higher downward velocities, the particle motion is given by equations 3.77 and 3.75.

Unidimensional motion in the vertical direction, under the action of gravity, occurs frequently in elutriation and other size separation equipment. This equipment was described in Chapter 1.

**Example 3.3**

A material of specific gravity 2·5 is fed to a size separation plant where the separating fluid is water which rises with a velocity of 1·2 m/s. The upward vertical component of the velocity of the particles is 6 m/s. How far will an approximately spherical particle, 6 mm diameter, rise relative to the walls of the plant before it comes to rest relative to the fluid?

**Solution**

Initial velocity of particle relative to fluid, $v = (6{·}0 - 1{·}2) = 4{·}8 \text{ m/s}$

$$Re' = \frac{6 \times 10^{-3} \times 4{·}8 \times 1000}{1 \times 10^{-3}}$$

$$= 28{,}800$$

When the particle has been retarded to such a velocity that $Re' = 500$, the minimum value for which equation 3.76 is applicable:

$$\dot{y} = (4{·}8 \times 500/28{,}800) = 0{·}083 \text{ m/s}$$

When $Re'$ is greater than 500, the relation between the displacement of the particle $y$ and the time $t$ is

$$y = -\frac{1}{c} \ln \left( \cos fct - \frac{v}{f} \sin fct \right) \qquad \text{(equation 3.76)}$$

where

$$c = \frac{0{·}33}{d} \frac{\rho}{\rho_s} = (0{·}33/6 \times 10^{-3})(1000/2500) = 22{·}0 \qquad \text{(equation 3.62)}$$

$$f = \sqrt{\frac{d(\rho_s - \rho)g}{0{·}33\rho}} = \sqrt{[(6 \times 10^{-3} \times 1500 \times 9{·}81)/(0{·}33 \times 1000)]} = 0{·}517 \qquad \text{(equation 3.74)}$$

and

$$v = -4{·}8 \text{ m/s}$$

Thus

$$y = -\frac{1}{22{·}0} \ln \left( \cos 0{·}517 \times 22t + \frac{4{·}8}{0{·}517} \sin 0{·}517 \times 22t \right)$$

$$= -0{·}0455 \ln (\cos 11{·}37t + 9{·}28 \sin 11{·}37t)$$

$$\dot{y} = -\frac{0{·}0455(-11{·}37 \sin 11{·}37t + 9{·}28 \times 11{·}37 \cos 11{·}37t)}{\cos 11{·}37t + 9{·}28 \sin 11{·}37t}$$

$$= -\frac{0{·}517(9{·}28 \cos 11{·}37t - \sin 11{·}37t)}{\cos 11{·}37t + 9{·}28 \sin 11{·}37t}$$

The time taken for the velocity of the particle relative to the fluid to fall from 4·8 m/s to 0·083 m/s is given by

$$-0.083 = -\frac{0.517(9.28 \cos 11.37t - \sin 11.37t)}{\cos 11.37t + 9.28 \sin 11.37t}$$

i.e.

$$\cos 11.37t + 9.28 \sin 11.37t = -6.23 \sin 11.37t + 57.8 \cos 11.37t$$

$$\therefore \quad \sin 11.37t = 3.66 \cos 11.37t$$

$$\therefore \quad \cos 11.37t = 0.264$$

$$\therefore \quad \sin 11.37t = \sqrt{1 - 0.264^2} = 0.965 \qquad (3.79)$$

The distance moved by the particle relative to the fluid during this period is therefore given by

$$y = -0.0455 \ln (0.264 + 9.28 \times 0.965)$$

$$= -0.101 \text{ m}$$

If equation 3.76 were applied for a relative velocity down to zero, the time taken for the particle to come to rest would be given by

$$9.28 \cos 11.37t = \sin 11.37t$$

$$\therefore \quad \cos 11.37t = 0.107$$

and

$$\sin 11.37t = \sqrt{1 - 0.107^2} = 0.994$$

The corresponding distance the particle moves relative to the fluid is then given by

$$y = -0.0455 \ln (0.107 + 9.28 \times 0.994)$$

$$= -0.102 \text{ m}$$

i.e. the particle moves only a very small distance with a velocity of less than 0·083 m/s.

If form drag were neglected for all velocities less than 0·083 m/s, the distance moved by the particle would be given by

$$y = \frac{b}{a}t + \frac{v}{a} - \frac{b}{a^2} + \left(\frac{b}{a^2} - \frac{v}{a}\right)e^{-at} \qquad \text{(equation 3.53)}$$

and

$$\dot{y} = \frac{b}{a} - \left(\frac{b}{a} - v\right)e^{-at}$$

where

$$a = 18\frac{\mu}{d^2 \rho_s} = \frac{18 \times 0.001}{0.006^2 \times 2500} = 0.20 \qquad \text{(equation 3.51)}$$

$$b = \left(1 - \frac{\rho}{\rho_s}\right)g = (1 - 1000/2500)9.81 = 5.89 \qquad \text{(equation 3.52)}$$

$$b/a = 29.43$$

$$v = -0.083 \text{ m/s}$$

Thus

$$y = 29 \cdot 43t - \left(\frac{0 \cdot 083}{0 \cdot 20} + \frac{29 \cdot 43}{0 \cdot 20}\right)(1 - e^{-0 \cdot 20t})$$

$$= 29 \cdot 43t - \frac{29 \cdot 51}{0 \cdot 20}(1 - e^{-0 \cdot 20t})$$

$$\therefore \quad \dot{y} = 29 \cdot 43 - 29 \cdot 51e^{-0 \cdot 20t}$$

When the particle comes to rest in the fluid, $\dot{y} = 0$ and

$$e^{-0 \cdot 20t} = 29 \cdot 43/29 \cdot 51$$

$$\therefore \quad t = 0 \cdot 0141 \text{ s}$$

The corresponding distance moved by the particle is given by

$$y = 29 \cdot 43 \times 0 \cdot 0141 - \frac{29 \cdot 51}{0 \cdot 20}(1 - e^{-0 \cdot 20 \times 0 \cdot 0141})$$

$$= 0 \cdot 41442 - 0 \cdot 41550 = -0 \cdot 00108 \text{ m}$$

Thus whether the resistance force is calculated by equation 3.9 or equation 3.13, the particle moves a negligible distance with a velocity relative to the fluid of less than $0 \cdot 083$ m/s. Further, the time is also negligible and thus the fluid also has moved through only a very small distance.

It can therefore be taken that the particle moves through $0 \cdot 102$ m before it comes to rest in the fluid. The time taken for the particle to move this distance is given by equation 3.79, on the assumption that the drag force corresponds to that given by equation 3.13. The time is therefore given by

$$\cos 11 \cdot 37t = 0 \cdot 264 \qquad \text{(from equation 3.79)}$$

$$\therefore \quad 11 \cdot 37t = 1 \cdot 304$$

$$t = 0 \cdot 115 \text{ s}$$

The distance travelled by the fluid in this time $= 1 \cdot 2 \times 0 \cdot 115$

$$= 0 \cdot 138 \text{ m}$$

Thus the total distance moved by the particle relative to the walls of the plant

$$= 0 \cdot 102 + 0 \cdot 138 = 0 \cdot 240 \text{ m}$$

$$\text{or } \underline{\underline{240 \text{ mm}}}$$

## 3.7. MOTION OF PARTICLES IN A CENTRIFUGAL FIELD

In most practical cases where a particle is moving in a fluid under the action of a centrifugal field, gravitational effects will be comparatively small and can be neglected. The equation of motion for the particles will be similar to that for motion in the gravitational field, except that the gravitational acceleration $g$ must be replaced by the centrifugal acceleration $r\omega^2$, where $r$ is the radius of rotation and $\omega$ is the angular velocity. It will be noted, however, that in this case the acceleration is a function of the position $r$ of the particle.

For a spherical particle in a fluid, the equation of motion for the Stokes' Law region is:

$$\frac{\pi}{6}d^3(\rho_s-\rho)r\omega^2 - 3\pi\mu d\frac{dr}{dt} = \frac{\pi}{6}d^3\rho_s\frac{d^2r}{dt^2} \tag{3.80}$$

Thus

$$\frac{d^2r}{dt^2} + \frac{18\mu}{d^2\rho_s}\frac{dr}{dt} - \frac{\rho_s-\rho}{\rho_s}\omega^2 r = 0 \tag{3.81}$$

i.e.

$$\frac{d^2r}{dt^2} + a\frac{dr}{dt} - nr = 0 \tag{3.82}$$

The solution of equation 3.82 takes the form:

$$r = B_1 e^{-[a/2+\sqrt{a^2/4+n}]t} + B_2 e^{-[a/2-\sqrt{a^2/4+n}]t} \tag{3.83}$$

$$= e^{-at/2}\{B_1 e^{-kt} + B_2 e^{kt}\} \tag{3.84}$$

where

$$a = \frac{18\mu}{d^2\rho_s}$$

$$n = \left(1-\frac{\rho}{\rho_s}\right)\omega^2$$

$$k = \sqrt{a^2/4+n}$$

Equation 3.84 requires the specification of two boundary conditions in order that the constants $B_1$ and $B_2$ may be evaluated.

Suppose that the particle starts ($t = 0$) at a radius $r_1$ with zero velocity $(dr/dt) = 0$. From equation 3.84:

$$\frac{dr}{dt} = e^{-at/2}\{-kB_1 e^{-kt} + kB_2 e^{kt}\} - \frac{a}{2}e^{-at/2}\{B_1 e^{-kt} + B_2 e^{kt}\}$$

$$= e^{-at/2}\left\{\left(k-\frac{a}{2}\right)B_2 e^{kt} - \left(k+\frac{a}{2}\right)B_1 e^{-kt}\right\} \tag{3.85}$$

Substituting the boundary conditions into equations 3.84 and 3.85:

$$r_1 = B_1 + B_2$$

$$0 = B_2\left(k-\frac{a}{2}\right) - B_1\left(k+\frac{a}{2}\right)$$

Thus

$$\frac{B_1}{B_2} = \frac{k - a/2}{k + a/2}$$

$$r_1 = \left(\frac{k - a/2}{k + a/2} + 1\right) B_2 = \left(\frac{2k}{k + a/2}\right) B_2$$

$$B_2 = \frac{k + a/2}{2k} r_1$$

and

$$B_1 = \frac{k - a/2}{2k} r_1$$

Thus

$$r = e^{-at/2} \left\{ \frac{k - a/2}{2k} r_1 e^{-kt} + \frac{k + a/2}{2k} r_1 e^{kt} \right\}$$

i.e.

$$\frac{r}{r_1} = e^{-at/2} \left\{ \cosh kt + \frac{a}{2k} \sinh kt \right\} \tag{3.86}$$

Hence $r/r_1$ can be directly calculated at any value of $t$, but a numerical solution is required to determine $t$ for any particular $r/r_1$ value.

If the effects of particle acceleration can be neglected, equation 3.82 simplifies to:

$$a \frac{dr}{dt} - nr = 0$$

Direct integration gives:

$$\ln \frac{r}{r_1} = \frac{n}{a} t = \frac{d^2 (\rho_s - \rho)\omega^2}{18\mu} t \tag{3.87}$$

Thus the time taken for a particle to move to a radius $r$ from an initial radius $r_1$ is given by:

$$t = \frac{18\mu}{d^2 \omega^2 (\rho_s - \rho)} \ln \frac{r}{r_1} \tag{3.88}$$

If the particle is initially situated in the liquid surface ($r_1 = r_i$) the time taken to reach the wall of the basket ($r = R$) is given by:

$$t = \frac{18\mu}{d^2 \omega^2 (\rho_s - \rho)} \ln \frac{R}{r_i}$$

If $h$ is the thickness of the liquid layer at the walls:

$$h = R - r_i$$

Then

$$\ln R/r_i = \ln R/(R - h) = -\ln (1 - h/R)$$

$$= \frac{h}{R} + \left(\frac{h}{R}\right)^2 + \ldots$$

If $h$ is small compared with $R$:

$$\ln\frac{R}{r_i} = \frac{h}{R}$$

Equation 3.88 then becomes:

$$t = \frac{18\mu h}{d^2\omega^2(\rho_s - \rho)R} \tag{3.89}$$

For the Newton's Law region, the equation of motion is:

$$\frac{\pi}{6}d^3(\rho_s - \rho)r\omega^2 - 0.22\frac{\pi}{4}d^2\rho\left(\frac{dr}{dt}\right)^2 = \frac{\pi}{6}d^3\rho_s\frac{d^2r}{dt^2}$$

This equation can be solved numerically only. However, if the acceleration term can be neglected:

$$\left(\frac{dr}{dt}\right)^2 = 3d\omega^2\left(\frac{\rho_s - \rho}{\rho}\right)r$$

Thus

$$r^{-1/2}\frac{dr}{dt} = \left\{3d\omega^2\left(\frac{\rho_s - \rho}{\rho}\right)\right\}^{1/2} \tag{3.90}$$

On integration:

$$2(r^{1/2} - r_1^{1/2}) = \left\{3d\omega^2\frac{\rho_s - \rho}{\rho}\right\}^{1/2}t$$

or

$$t = \left[\frac{\rho}{3d\omega^2(\rho_s - \rho)}\right]^{1/2}2(r^{1/2} - r_1^{1/2}) \tag{3.91}$$

## 3.8. FURTHER READING

DALLAVALLE, J. M.: *Micromeritics*, 2nd edn. (Pitman, 1948).
HERDAN, G.: *Small Particle Statistics*, 2nd edn. (Butterworths, London, 1960).
ORR, C.: *Particulate Technology* (Macmillan, New York, 1966).

## 3.9. REFERENCES

1. STOKES, G. G.: *Trans. Cam. Phil. Soc.* **9** (1851) 8. On the effect of the internal friction of fluids on the motion pendulum.
2. GOLDSTEIN, S.: *Proc. Roy. Soc.* **A123** (1929) 225. The steady flow of viscous fluid past a fixed spherical obstacle at small Reynolds numbers.
3. OSEEN, C. W.: *Ark. Mat. Astr. Fys.* **9,** No. 16 (1913) 1–15. Über den Gültigkeitsbereich der Stokesschen Widerstandsformel.
4. SCHILLER, L. and NAUMANN, A.: *Z. Ver. deut. Ing.* **77** (1933) 318. Über die grundlegenden Berechnungen bei der Schwerkraftaufbereitung.
5. ALLEN, H. S.: *Phil. Mag.* **1** (1900) 323. On the motion of a sphere in a viscous fluid.
6. LADENBURG, R.: *Ann. Phys.* **23** (1907) 447. Über den Einfluss von Wänden auf die Bewegung einer Kügel in einer reibenden Flüssigkeit.

7. CUNNINGHAM, E.: *Proc. Roy. Soc.* **A83** (1910) 357. Velocity of steady fall of spherical particles.

8. DAVIES, C. N.: *Proc. Phys. Soc.* **57** (1945) 259. Definitive equations for the fluid resistance of spheres.

9. DAVIES, C. N.: *Proc. Roy. Soc.* **A200** (1949) 100. The sedimentation and diffusion of small particles.

10. ROWE, P. N. and HENWOOD, G. N.: *Trans. Inst. Chem. Eng.* **39** (1961) 43. Drag forces in a hydraulic model of a fluidised bed. Part 1.

11. HEYWOOD, H.: *J. Imp. Coll. Chem. Eng. Soc.* **4** (1948) 17. Calculation of particle terminal velocities.

12. HEISS, J. F. and COULL, J.: *Chem. Eng. Prog.* **48** (1952) 133. The effect of orientation and shape on the settling velocity of non-isometric particles in a viscous medium.

13. HADAMARD, J.: *Comptes rendus* **152** (1911) 1735. Mouvement permanent lent d'une sphère liquide et visqueuse dans un liquide visqueux.

14. BOND, W. N.: *Phil. Mag.*, 7th. ser. **4** (1927) 889. Bubbles and drops and Stokes' law.

15. BOND, W. N. and NEWTON, D. A.: *Phil. Mag.* 7th ser. **5** (1928) 794. Bubbles, drops and Stokes' law (Paper 2).

16. GARNER, F. H. and SKELLAND, A. H. P.: *Trans. Inst. Chem. Eng.* **29** (1951) 315. Liquid–liquid mixing as affected by the internal circulation within drops.

17. GARNER, F. H. and SKELLAND, A. H. P.: *Chem. Eng. Sci.* **4** (1955) 149. Some factors affecting drop behaviour in liquid–liquid systems.

18. HAMIELEC, A. E. and JOHNSON, A. I.: *Can. J. Chem. Eng.* **40** (1962) 41. Viscous flow around fluid spheres at intermediate Reynolds numbers.

19. TOROBIN, L. B. and GAUVIN, W. H.: *Can. J. Chem. Eng.* **38** (1959) 129, 167, 224. Fundamental aspects of solids–gas flow. Part I: Introductory concepts and idealized sphere-motion in viscous regime. Part II: The sphere wake in steady laminar fluids. Part III: Accelerated motion of a particle in a fluid.

## 3.10. NOMENCLATURE

| | | Units in SI System | Dimensions in **MLT** |
|---|---|---|---|
| $A'$ | Projected area of particle in plane perpendicular to direction of motion | | |
| $a$ | $18\mu/d^2\rho_s$ | 1/s | $T^{-1}$ |
| $B_1, B_2$ | Coefficients in equation 3.83 | m | **L** |
| $b$ | $[1-(\rho/\rho_s)]g$ | m/s² | $LT^{-2}$ |
| $c$ | $0.33\rho/d\rho_s$ | 1/m | $L^{-1}$ |
| $d$ | Diameter of sphere or characteristic dimension of particle | m | **L** |
| $d_p$ | Mean projected diameter of particle | m | **L** |
| $d_t$ | Diameter of tube or vessel | m | **L** |
| $d'$ | Linear dimension of particle | m | **L** |
| $F$ | Total force on particle | N | $MLT^{-2}$ |
| $f$ | $\sqrt{b/c}$ | m/s | $LT^{-1}$ |
| $g$ | Acceleration due to gravity | m/s² | $LT^{-2}$ |
| $h$ | Thickness of liquid layer | m | **L** |
| $I$ | $\int dt/(1+je^{pt})$ | s | **T** |
| $J$ | Constant for given fluid and particle | — | — |
| $j$ | $(f+v)/(f-v)$ | — | — |
| $K$ | Constant for given shape and orientation of particle | — | — |
| $k$ | $\sqrt{a^2/4+n}$ | 1/s | $T^{-1}$ |
| $k'$ | Constant for calculating volume of particle | — | — |
| $L'$ | Distance of particle from bottom of container | m | **L** |
| $m$ | Mass of particle | kg | **M** |
| $n$ | $[1-(\rho/\rho_s)]\omega^2$ | 1/s² | $T^{-2}$ |
| $p$ | $2fc$ | 1/s | $T^{-1}$ |
| $Q$ | Correction factor for velocity of bubble | — | — |
| $R$ | Radius of basket | m | **L** |
| $R'$ | Resistance per unit projected area of particle | N/m² | $ML^{-1}T^{-2}$ |
| $R'_0$ | Resistance per unit projected area of particle at free falling condition | N/m² | $ML^{-1}T^{-2}$ |
| $r$ | Radius of rotation | m | **L** |
| $r_i$ | Radius of inner surface of liquid | m | **L** |
| $s$ | $1+je^{pt}$ | — | — |
| $t$ | Time | s | **T** |
| $u$ | Velocity of fluid relative to particle | m/s | $LT^{-1}$ |
| $u_0$ | Terminal falling velocity of particle | m/s | $LT^{-1}$ |

| | | Units in SI System | Dimensions in **MLT** |
|---|---|---|---|
| $v$ | Initial component of velocity of particle in $Y$-direction | m/s | $\mathbf{LT^{-1}}$ |
| $w$ | Initial component of velocity of particle in $X$-direction | m/s | $\mathbf{LT^{-1}}$ |
| $x$ | Displacement of particle in $X$-direction at time $t$ | m | $\mathbf{L}$ |
| $\dot{x}$ | Velocity of particle in $X$-direction at time $t$ | m/s | $\mathbf{LT^{-1}}$ |
| $\ddot{x}$ | Acceleration of particle in $X$-direction at time $t$ | m/s$^2$ | $\mathbf{LT^{-2}}$ |
| $y$ | Displacement of particle in $Y$-direction at time $t$ | m | $\mathbf{L}$ |
| $\dot{y}$ | Velocity of particle in $Y$ direction at time $t$ | m/s | $\mathbf{LT^{-1}}$ |
| $\ddot{y}$ | Acceleration of particle in $Y$-direction at time $t$ | m/s$^2$ | $\mathbf{LT^{-2}}$ |
| $\alpha$ | Angle between direction of motion of particle and horizontal | — | — |
| $\lambda$ | Mean free path | m | $\mathbf{L}$ |
| $\mu$ | Viscosity of fluid | N s/m$^2$ | $\mathbf{ML^{-1}T^{-1}}$ |
| $\mu_l$ | Viscosity of fluid in drop or bubble | N s/m$^2$ | $\mathbf{ML^{-1}T^{-1}}$ |
| $\rho$ | Density of fluid | kg/m$^3$ | $\mathbf{ML^{-3}}$ |
| $\rho_s$ | Density of solid | kg/m$^3$ | $\mathbf{ML^{-3}}$ |
| $\omega$ | Angular velocity | rad/s | $\mathbf{T^{-1}}$ |
| $Re'$ | Reynolds number $ud\rho/\mu$ or $ud'\rho/\mu$ | — | — |
| $Re'_0$ | Reynolds number $u_0 d\rho/\mu$ | — | — |

# CHAPTER 4

# Flow of Fluids through Granular Beds and Packed Columns

## 4.1. INTRODUCTION

The flow of fluids through beds composed of stationary granular particles is a frequent occurrence in the chemical industry and therefore expressions are needed to predict pressure drop across beds due to the resistance caused by the presence of the particles. For example, in fixed bed catalytic reactors, such as $SO_2$–$SO_3$ converters, and drying columns containing silica gel or molecular sieves, gases are passed through a bed of particles. In the case of gas absorption into a liquid, the gas flows upwards against a falling liquid stream, the fluids being contained in a vertical column packed with suitable particles. In the filtration of a suspension, liquid flows at a relatively low velocity through the spaces between the particles which have been retained by the filter medium and, as a result of the continuous deposition of solids, the resistance to flow increases progressively throughout the operation. Filtration through a fixed bed takes place on a very large scale, for example in water treatment through sand filters. In all these instances it is necessary to estimate the size of the equipment required, and design expressions are required for the drop in pressure for a fluid flowing through a packing, either alone or as a two-phase system. The corresponding expressions for fluidised beds are discussed in Chapter 6. The drop in pressure for flow through a bed of small particles provides a convenient method for obtaining a measure of the external surface area of a powder, for example cement or pigment.

The flow of either a single phase through a bed of particles or the more complex flow of two fluid phases is approached by using the ideas developed in Volume 1 for the flow of an incompressible fluid through regular pipes or ducts. It will be found, however, that the problem is not in practice capable of complete analytical solution and the use of experimental data obtained for many different systems is essential. Later in the chapter some aspects of the design of industrial packed columns involving countercurrent flow of liquids and gases are described.

## 4.2. FLOW OF A SINGLE FLUID THROUGH A GRANULAR BED

### 4.2.1. Darcy's Law and Permeability

The first experimental work on the subject was carried out by Darcy[1] in 1830 in Dijon when he examined the rate of flow of water from the local fountains through beds of sand of various thicknesses. He showed that the average velocity, as measured over the whole area of the bed, was directly proportional to the driving pressure and inversely proportional to the thickness of the bed. This relation, often termed Darcy's Law, has subsequently been confirmed by a number of workers and can be written as follows:

$$u = K\frac{(-\Delta P)}{l} \tag{4.1}$$

where   $-\Delta P$ is the pressure drop across the bed,
$\phantom{where}$   $l$ is the thickness of the bed,
$\phantom{where}$   $u$ is the average velocity of flow of the fluid, defined as $(1/A)(dV/dt)$,
$\phantom{where}$   $A$ is the total cross-sectional area of the bed,
$\phantom{where}$   $V$ is the volume of fluid flowing in time $t$, and
$\phantom{where}$   $K$ is a constant depending on the physical properties of the bed and fluid.

The linear relation between the rate of flow and the pressure difference leads one to suppose that the flow was streamline (Volume 1, Chapter 3). This would be expected because the Reynolds number for the flow through the pore spaces in a granular material is low, since both the velocity of the fluid and the width of the channels are normally small. The resistance to flow then arises mainly from viscous drag. Equation 4.1 can then be expressed as:

$$u = \frac{K(-\Delta P)}{l} = B\frac{(-\Delta P)}{\mu l} \tag{4.2}$$

where $\mu$ is the viscosity of the fluid and $B$ is termed the permeability coefficient for the bed, and depends only on the properties of the bed.

The value of the permeability coefficient is frequently used to give an indication of the ease of passing a fluid through, say, a bed of particles or a filter medium. Some values of $B$ for various packings[2] are shown in Table 4.1 and it can be seen that $B$ can vary over a wide range of values. It should be noted that these values of $B$ apply only to the laminar flow regime.

### 4.2.2. Specific Surface and Voidage

The general structure of a bed of particles can often be characterised by the specific surface area of the bed $S_B$ and the fractional voidage of the bed $e$.

$S_B$ is the surface area presented to the fluid per unit volume of bed when the particles are packed in a bed. Its units are (length)$^{-1}$.

$e$ is the fraction of the volume of the bed not occupied by solid material and is termed the fractional voidage, voidage, or porosity. It is dimensionless. Thus the fractional volume of the bed occupied by solid material is $(1-e)$.

$S$ is the specific surface area of the particles and is the surface area of a particle divided by its volume. Its units are again (length)$^{-1}$. For a sphere, for example:

$$S = \frac{\pi d^2}{\pi(d^3/6)} = \frac{6}{d} \tag{4.3}$$

It can be seen that $S$ and $S_B$ are not equal due to the voidage occurring when the particles are packed into a bed. If point contact occurs between particles so that only a very small fraction of surface area is lost by overlapping, then:

$$S_B = S(1-e) \tag{4.4}$$

Some values of $S$ and $e$ for different beds of particles are listed in Table 4.1. Values of $e$ much higher than those shown in Table 4.1, sometimes up to about 0·95, are possible in beds of fibres[3] and some ring packings. For a given shape of particle $S$ increases as the particle size is reduced as is shown in Table 4.1.

TABLE 4.1. *Properties of Beds of Some Regular-shaped Materials*[2]

| | Solid constituents | | Porous mass | |
|---|---|---|---|---|
| No. | Description | Specific surface area $S$ $(m^2/m^3)$ | Fractional voidage, $e$ | $B$ $(m^2)$ |
| | **Spheres** | | | |
| 1 | $\frac{1}{32}$ in. diam. (0·794 mm) | 7600 | 0·393 | $6\cdot2 \times 10^{-10}$ |
| 2 | $\frac{1}{16}$ in. diam. (1·588 mm) | 3759 | 0·405 | $2\cdot8 \times 10^{-9}$ |
| 3 | $\frac{1}{8}$ in. diam. (3·175 mm) | 1895 | 0·393 | $9\cdot4 \times 10^{-9}$ |
| 4 | $\frac{1}{4}$ in. diam. (6·35 mm) | 948 | 0·405 | $4\cdot9 \times 10^{-8}$ |
| 5 | $\frac{5}{16}$ in. diam. (7·94 mm) | 756 | 0·416 | $9\cdot4 \times 10^{-8}$ |
| | **Cubes** | | | |
| 6 | $\frac{1}{8}$ in. (3·175 mm) | 1860 | 0·190 | $4\cdot6 \times 10^{-10}$ |
| 7 | $\frac{1}{8}$ in. (3·175 mm) | 1860 | 0·425 | $1\cdot5 \times 10^{-8}$ |
| 8 | $\frac{1}{4}$ in. (6·35 mm) | 1078 | 0·318 | $1\cdot4 \times 10^{-8}$ |
| 9 | $\frac{1}{4}$ in. (6·35 mm) | 1078 | 0·455 | $6\cdot9 \times 10^{-8}$ |
| | **Hexagonal prisms** | | | |
| 10 | $\frac{3}{16}$ in. $\times \frac{3}{16}$ in. thick (4·76 mm $\times$ 4·76 mm) | 1262 | 0·355 | $1\cdot3 \times 10^{-8}$ |
| 11 | $\frac{3}{16}$ in. $\times \frac{3}{16}$ in. thick (4·76 mm $\times$ 4·76 mm) | 1262 | 0·472 | $5\cdot9 \times 10^{-8}$ |
| | **Triangular pyramids** | | | |
| 12 | $\frac{1}{4}$ in. length $\times$ 0·113 in. ht. (6·35 mm $\times$ 2·87 mm) | 2410 | 0·361 | $6\cdot0 \times 10^{-9}$ |
| 13 | $\frac{1}{4}$ in. length $\times$ 0·113 in. ht. (6·35 mm $\times$ 2·87 mm) | 2410 | 0·518 | $1\cdot9 \times 10^{-8}$ |
| | **Cylinders** | | | |
| 14 | $\frac{1}{8}$ in. diam. $\times \frac{1}{8}$ in. (3·175 mm $\times$ 3·175 mm) | 1840 | 0·401 | $1\cdot1 \times 10^{-8}$ |
| 15 | $\frac{1}{8}$ in. diam. $\times \frac{1}{4}$ in. (3·175 mm $\times$ 6·35 mm) | 1585 | 0·397 | $1\cdot2 \times 10^{-8}$ |
| 16 | $\frac{1}{4}$ in. diam. $\times \frac{1}{4}$ in. (6·35 mm $\times$ 6·35 mm) | 945 | 0·410 | $4\cdot6 \times 10^{-8}$ |
| | **Plates** | | | |
| 17 | $\frac{1}{4}$ in. $\times \frac{1}{4}$ in. $\times \frac{1}{32}$ in. (6·35 mm $\times$ 6·35 mm $\times$ 0·794 mm) | 3033 | 0·410 | $5\cdot0 \times 10^{-9}$ |
| 18 | $\frac{1}{4}$ in. $\times \frac{1}{4}$ in. $\times \frac{1}{16}$ in. (6·35 mm $\times$ 6·35 mm $\times$ 1·59 mm) | 1984 | 0·409 | $1\cdot1 \times 10^{-8}$ |
| | **Discs** | | | |
| 19 | $\frac{1}{8}$ in. diam. $\times \frac{1}{16}$ in. (3·175 mm $\times$ 1·59 mm) | 2540 | 0·398 | $6\cdot3 \times 10^{-9}$ |
| | **Porcelain Berl saddles** | | | |
| 20 | 0·236 in. (6 mm) | 2450 | 0·685 | $9\cdot8 \times 10^{-8}$ |
| 21 | 0·236 in. (6 mm) | 2450 | 0·750 | $1\cdot73 \times 10^{-7}$ |
| 22 | 0·236 in. (6 mm) | 2450 | 0·790 | $2\cdot94 \times 10^{-7}$ |
| 23 | 0·236 in. (6 mm) | 2450 | 0·832 | $3\cdot94 \times 10^{-7}$ |
| 24 | Lessing rings (6 mm) | 5950 | 0·870 | $1\cdot71 \times 10^{-7}$ |
| 25 | Lessing rings (6 mm) | 5950 | 0·889 | $2\cdot79 \times 10^{-7}$ |

As $e$ is increased, flow through the bed becomes easier and so the permeability coefficient $B$ increases; a relation between $B$, $e$, and $S$ is developed in a later section of this chapter. If the particles are randomly packed, then $e$ should be approximately constant throughout the bed and the resistance to flow the same in all directions. Often near containing walls, $e$ is higher, and corrections for this should be made if the particle size is a significant fraction of the size of the containing vessel. This correction is discussed in more detail later.

### 4.2.3. General Expressions for Flow through Beds in Terms of Carman–Kozeny Equations

*Streamline Flow—Kozeny Equation*

Many attempts have been made to obtain general expressions for pressure drop and mean velocity for flow through packings in terms of voidage and specific surface, as these quantities are often known or can be measured. Alternatively, measurements of the pressure drop, velocity, and voidage provide a convenient way of measuring the surface areas of some particulate materials, as described later.

The analogy between streamline flow through a tube and streamline flow through the pores in a bed of particles is a useful starting point for deriving a general expression.

From Volume 1, the equation for streamline flow through a circular tube is:

$$u_t = \frac{d_t^2}{32\mu} \frac{(-\Delta P)}{l_t} \tag{4.5}$$

where $\mu$ is the viscosity of the fluid,

$u_t$ is the mean velocity of the fluid,

$d_t$ is the diameter of the tube, and

$l_t$ is the length of the tube.

If the free space in the bed is assumed to consist of a series of tortuous channels, equation 4.5 for flow through a bed may be rewritten as:

$$u_1 = \frac{d_m'^2}{K'\mu} \frac{(-\Delta P)}{l'} \tag{4.6}$$

where $d_m'$ is some equivalent diameter of the pore channels,

$K'$ is a dimensionless constant whose value depends on the structure of the bed,

$l'$ is the length of channel, and

$u_1$ is the average velocity through the pore channels.

It should be noted that $u_1$ and $l'$ in equation 4.6 now represent conditions in the pores and are not the same as $u$ and $l$ in equations 4.1 and 4.2. However, it is a reasonable assumption that $l'$ is directly proportional to $l$. Also, Dupuit[4] related $u$ and $u_1$ by the following argument:

In a cube of side $X$, the volume of free space is $eX^3$ so that the mean cross-sectional area for flow is the free volume divided by the height, or $eX^2$. The volume flowrate through this cube is $uX^2$, so that the average linear velocity through the pores, $u_1$, is given by:

$$u_1 = \frac{uX^2}{eX^2} = \frac{u}{e} \tag{4.7}$$

Although equation 4.7 is reasonably true for random packings, it does not apply to all regular packings. Thus with a bed of spheres arranged in cubic packing, $e = 0.476$, but the fractional free area varies continuously, from 0.215 in a plane across the diameters to unity between successive layers.

For equation 4.6 to be generally useful, an expression is needed for $d_m'$, the equivalent diameter of the pore space. Kozeny[5,6] proposed that $d_m'$ could be taken as:

$$d_m' = \frac{e}{S_B} = \frac{e}{S(1-e)} \tag{4.8}$$

where $e/S_B = \dfrac{\text{volume of voids filled with fluid}}{\text{wetted surface area of the bed}}$

$\qquad\qquad = \dfrac{\text{cross-sectional area normal to flow}}{\text{wetted perimeter}}$

The hydraulic mean diameter for such a flow passage has been shown in Volume 1, Chapter 3 to equal:

$$4 \times \frac{\text{cross-sectional area}}{\text{wetted perimeter}}$$

then it is seen that:

$$\frac{e}{S_B} = \tfrac{1}{4} \times \text{hydraulic mean diameter}$$

Then taking $u_1 = u/e$ and $l' \propto l$, equation 4.6 becomes:

$$u = \frac{1}{K''} \frac{e^3}{S_B^2 \mu} \frac{1}{l} (-\Delta P)$$

$$= \frac{1}{K''} \frac{e^3}{S^2 (1-e)^2} \frac{1}{\mu} \frac{(-\Delta P)}{l} \qquad (4.9)$$

$K''$ is generally known as Kozeny's constant and a commonly accepted value for $K''$ is 5. However, as will be shown later, $K''$ is dependent on porosity, particle shape, and other factors. Comparison with equation 4.2 shows that $B$ the permeability coefficient is given by:

$$B = \frac{1}{K''} \frac{e^3}{S^2 (1-e)^2} \qquad (4.10)$$

*Streamline and Turbulent Flow*

Equation 4.9 applies to streamline flow conditions, but Carman[7] and others have extended the analogy with pipe flow to cover both streamline and turbulent flow conditions through packed beds. In this treatment a modified Reynolds number $Re_1$ is plotted against a modified friction factor $R_1/\rho u_1^2$. This is analogous to plotting $Re$ against $R/\rho u^2$ for flow through a pipe as in Volume 1, Chapter 3.

The modified Reynolds number $Re_1$ is obtained by taking the same velocity and characteristic linear dimension $d'_m$ as were used in deriving equation 4.9. Thus:

$$Re_1 = \frac{u}{e} \frac{e}{S(1-e)} \frac{\rho}{\mu}$$

$$= \frac{u\rho}{S(1-e)\mu} \qquad (4.11)$$

The friction factor, which is plotted against the modified Reynolds number, is $R_1/\rho u_1^2$, where $R_1$ is the component of the drag force per unit area of particle surface in the direction of motion. $R_1$ can be related to the properties of the bed and pressure gradient as follows. Consider the forces acting on the fluid in a bed of unit cross-sectional area and thickness $l$. The volume of particles in the bed is $l(1-e)$ and therefore the total surface is $Sl(1-e)$. Thus the resistance force $= R_1 Sl(1-e)$. This force on the fluid must be equal to that produced by a pressure difference of $\Delta P$ across the bed. Then, since the free cross-section of fluid is equal to $e$:

$$(-\Delta P)e = R_1 Sl(1-e)$$

and

$$R_1 = \frac{e}{S(1-e)}\frac{(-\Delta P)}{l} \tag{4.12}$$

Thus

$$\frac{R_1}{\rho u_1^2} = \frac{e^3}{S(1-e)}\frac{(-\Delta P)}{l}\frac{1}{\rho u^2} \tag{4.13}$$

Carman found that when $R_1/\rho u_1^2$ was plotted against $Re_1$ using logarithmic coordinates, his data for the flow through randomly packed beds of solid particles could be correlated approximately by a single curve (curve $A$, Fig. 4.1), whose general equation was:

$$\frac{R_1}{\rho u_1^2} = 5Re_1^{-1} + 0.4Re_1^{-0.1} \tag{4.14}$$

The form of equation 4.14 is similar to 4.15 proposed by Forchheimer[8] who suggested that the resistance to flow should be considered in two parts: that due to the viscous drag at the surface of the particles, and that due to loss in turbulent eddies and at the sudden changes in the cross-section of the channels. Thus:

$$(-\Delta P) = \alpha u + \alpha' u^n \tag{4.15}$$

The first term in this equation will predominate at low rates of flow where the losses are mainly attributable to skin friction, and the second term will become significant at high flowrates and in very thin beds where the enlargement and contraction losses become very important. At very high flowrates the effects of viscous forces are negligible.

From equation 4.14 it can be seen that for values of $Re_1$ less than about 2, the second term is small and, approximately:

$$\frac{R_1}{\rho u_1^2} = 5Re_1^{-1} \tag{4.16}$$

Taking $R_1/\rho u_1^2$ from equation 4.13 it can be shown that the equation for the straight line part of the plot for $Re_1 < 2$ is in fact equation 4.9 taking the Kozeny constant $K''$ as 5.0.

As the value of $Re_1$ increases from about 2 to 100, the second term in equation 4.14 becomes more significant and the slope of the plot gradually changes from $-1.0$ to about $-\frac{1}{4}$. Above $Re_1$ of 100 the plot is approximately straight. The change from complete streamline flow to complete turbulent flow is very gradual because flow conditions are not the same in all the pores. Thus, the flow starts to become turbulent in the larger pores, and subsequently in successively smaller pores as the value of $Re_1$ increases. It is probable that the flow never becomes completely turbulent since some of the passages may be so small that streamline conditions prevail even at high flowrates.

Rings, which as described later are often used in industrial packed columns, tend to deviate from the generalised curve $A$ on Fig. 4.1 particularly at high values of $Re_1$.

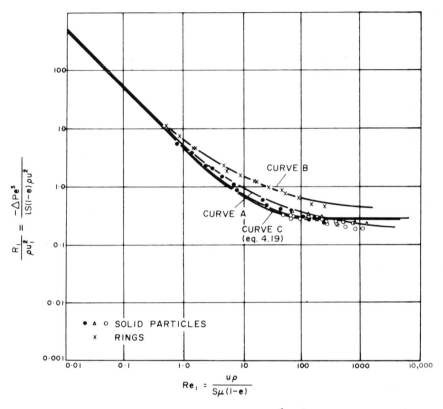

FIG. 4.1. Carman's graph of $R_1/\rho u_1^2$ against $Re_1$

Sawistowski[9] has compared the results obtained for flow of fluids through beds of hollow packings (see later) and has noted that equation 4.14 gives a consistently low result for these materials. He proposed:

$$\frac{R_1}{\rho u_1^2} = 5Re_1^{-1} + Re_1^{-0.1} \tag{4.17}$$

This equation is plotted as curve $B$ in Fig. 4.1.

For flow through ring packings which as described later are often used in industrial packed columns, Ergun[10] has obtained a good semi-empirical correlation for pressure drop as follows:

$$\frac{-\Delta P}{l} = 150\frac{(1-e)^2}{e^3}\frac{\mu u}{d^2} + 1\cdot75\frac{(1-e)}{e^3}\frac{\rho u^2}{d} \tag{4.18}$$

Writing $d = 6/S$ (from equation 4.3):

$$\frac{-\Delta P}{Sl\rho u^2}\frac{e^3}{1-e} = 4{\cdot}17\frac{\mu S(1-e)}{\rho u} + 0{\cdot}29$$

i.e.

$$\frac{R_1}{\rho u_1^2} = 4{\cdot}17 Re_1^{-1} + 0{\cdot}29 \tag{4.19}$$

This equation is plotted as curve $C$ in Fig. 4.1. The form of equation 4.19 is somewhat similar to that of equations 4.14 and 4.15, in that the first term represents viscous drag losses which are most significant at low velocities and the second term represents kinetic energy losses which become more significant at high velocities. The equation is thus applicable over a wide range of velocities and was found by Ergun to correlate experimental data well for values of $Re_1/(1-e)$ from 1 to over 2000.

### Dependence of $K''$ on Structure of Bed

*Tortuosity.* Although it was implied in the derivation of equation 4.9 that a single value of the Kozeny constant $K''$ applied to all packed beds, in practice this assumption does not hold.

Carman[7] has shown that:

$$K'' = \left(\frac{l'}{l}\right)^2 \times K_0 \tag{4.20}$$

where $(l'/l)$ is the tortuosity and is a measure of the fluid path length through the bed compared with the actual depth of the bed,

$K_0$ is a factor which depends on the shape of the cross-section of a channel through which fluid is passing.

For streamline fluid flow through a circular pipe where Poiseuille's equation applies (Volume 1, Chapter 3), $K_0$ is equal to 2·0, and for streamline flow through a rectangle where the ratio of the lengths of the sides is 10:1, $K_0 = 2{\cdot}65$. Carman[11] has listed values of $K_0$ for other cross-sections. From equation 4.20 it can be seen that if, say, $K_0$ were constant, then $K''$ would increase with increase in tortuosity. The reason for $K''$ being near to 5·0 for many different beds is probably that changes in tortuosity from one bed to another have been compensated by changes in $K_0$ in the opposite direction.

*Wall effect.* In a packed bed the particles will not pack as closely in the region near the wall as in the centre of the bed, so that the actual resistance to flow in a bed of small diameter is less than it would be in an infinite container for the same flowrate per unit area of bed cross-section. A correction factor $f_w$ for this effect has been determined experimentally by Coulson[12] as:

$$f_w = \left(1 + \tfrac{1}{2}\frac{S_c}{S}\right)^2 \tag{4.21}$$

where $S_c$ is the surface of the container per unit volume of bed.

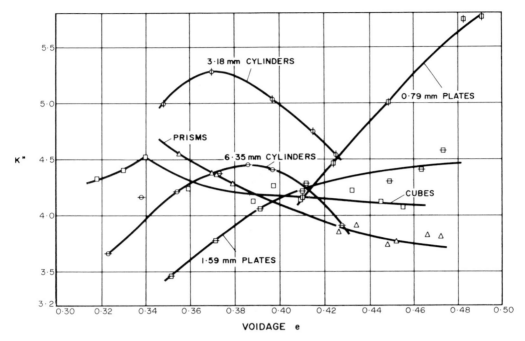

FIG. 4.2. Variation of Kozeny's constant $K''$ with voidage for various shapes

Equation 4.9 then becomes:

$$u = \frac{1}{K''} \frac{e^3}{S^2(1-e)^2} \frac{1}{\mu} \frac{(-\Delta P)}{l} f_w \qquad (4.22)$$

The values of $K''$ shown on Fig. 4.2 apply to equation 4.22.

*Non-spherical particles.* Coulson[12] and Wyllie and Gregory[13] have each determined values of $K''$ for particles of many different sizes and shapes, including prisms, cubes, and plates.

Some of these values for $K''$ are shown in Fig. 4.2 where it is seen that they lie between 3 and 6 with the extreme values only occurring with thin plates. This variation of $K''$ with plates probably arises, not only from the fact that area contact is obtained between the particles, but also because the plates tend to give greater tortuosities. For normal granular materials Kihn[14] and Pirie[15] have found that $K''$ is reasonably constant and does not vary so widely as the $K''$ values on Fig. 4.2 for extreme shapes.

*Spherical particles.* Equation 4.22 has been tested with spherical particles over a wide range of sizes and $K''$ has been found[16,12] to be about $4\cdot8 \pm 0\cdot3$.

For beds composed of spheres of mixed sizes the porosity of the packing can change very rapidly if the smaller spheres are able to fill the pores between the larger ones. Thus Coulson[12] found that, with a mixture of spheres of size ratio 2:1, a bed behaves much in accordance with equation 4.17 but, if the size ratio is 5:1 and the smaller particles form less than 30 per cent by volume of the larger ones, then $K''$ falls very rapidly, emphasising that only for uniform sized particles can bed behaviour be predicted with confidence.

*Beds with high voidage.* Spheres and particles which are approximately isometric do not pack to give beds with voidages in excess of about 0·6. With fibres and some ring packings, however, values of $e$ near unity can be obtained and for these high values $K''$ rises rapidly. Some values are given in Table 4.2.

TABLE 4.2. *Experimental Values of K'' for Beds of High Porosity*

| Voidage $e$ | Experimental value of $K''$ | | |
|---|---|---|---|
| | Brinkman[3] | Davies[32] | Silk fibres Lord[17] |
| 0·5 | 5·5 | | |
| 0·6 | 4·3 | | |
| 0·8 | 5·4 | 6·7 | 5·35 |
| 0·9 | 8·8 | 9·7 | 6·8 |
| 0·95 | 15·2 | 15·3 | 9·2 |
| 0·98 | 32·8 | 27·6 | 15·3 |

Deviations from the Kozeny equation (4.9) become more pronounced in these beds of fibres as the voidage increases, because the nature of the flow changes from one of channel flow to one in which the fibres behave as a series of obstacles in an otherwise unobstructed passage. The flow pattern is also different in expanded fluidised beds and the Kozeny equation does not apply there either. As fine spherical particles move far apart in a fluidised bed Stokes' Law can be applied whereas the Kozeny equation leads to no such limiting resistance. This problem is further discussed by Carman[11].

*Effect of bed support.* The structure of the bed, and hence $K''$, is markedly influenced by the nature of the support. For example, the initial condition in a filtration may affect the whole of a filter cake. Figure 4.3 shows the difference in orientation of two beds of cubical particles. The importance of the packing support should not be overlooked in considering the drop in pressure through the column since the support may itself form an important resistance, and by orientating the particles as indicated may also affect the total pressure drop.

## The Application of Carman–Kozeny Equations

Equations 4.9 and 4.14, which involve $e/S_B$ as a measure of the effective pore diameter, are developed from a relatively sound theoretical basis and are recommended for beds of small particles when they are nearly spherical in shape. The correction factor for wall effects, given by equation 4.21, should be included where appropriate. With larger particles which will frequently be far from spherical in shape, the correlations are not so reliable. As shown in Fig. 4.1, deviations can occur for rings at higher values of $Re_1$. Efforts to correct for non-sphericity, though frequently useful, are not universally effective, and in such cases it will often be more rewarding to use correlations (such as equation 4.19) which are based on experimental data for large packings.

## Use of Kozeny Equation for Measurement of Particle Surface

The Kozeny equation relates the drop in pressure through a bed to the specific surface of the material and can therefore be used as a means of calculating $S$ from measurements of the drop in pressure. This method is strictly only suitable for beds of uniformly packed particles and it is not a suitable method for measuring the size distribution of particles in the subsieve range. A convenient form of apparatus has been developed by Lea and

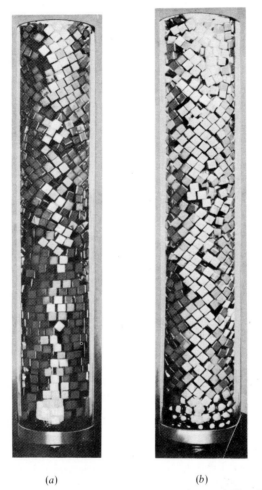

(a)                                    (b)

FIG. 4.3. Packing of cubes, stacked on (a) plane surface, and (b) on bed of spheres

Nurse[18] and is shown diagrammatically in Fig. 4.4. In this apparatus air or other suitable gas flows through the bed contained in a cell (25 mm dia., 87 mm deep), and the pressure drop is obtained from $h_1$ and the gas flow rate from $h_2$.

If the diameters of the particles are below about 5 μm, then slip will occur and this must be allowed for. This factor is discussed by Carman and Malherbe[19].

The method has been successfully developed for measurement of the surface area of cement and for such materials as pigments, fine metal powders, pulverised coal, and fine fibres.

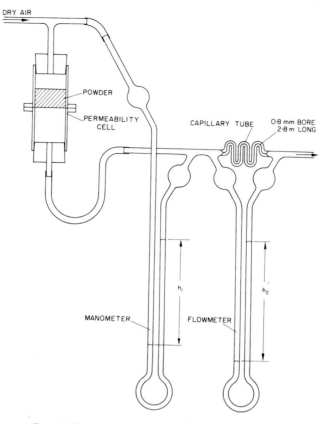

FIG. 4.4. Permeability apparatus of Lea and Nurse[18]

### 4.2.4. Molecular Flow

In the relations which have been given earlier, it is assumed that the fluid can be regarded as a continuum and that there is no slip between the wall of the capillary and the fluid layers in contact with it. However, when conditions are such that the mean free path of the molecules of a gas is a significant fraction of the capillary diameter, the flowrate at a given value of the pressure gradient becomes greater than the predicted value. If the mean free path exceeds the capillary diameter, the flowrate becomes independent of the viscosity and the process is one of diffusion. Whereas these considerations apply only at very low pressures in normal tubes, in fine-pored materials the pore diameter and the mean free path may be of the same order of magnitude even at atmospheric pressure.

### 4.3. DISPERSION

Dispersion is the general term which is used to describe the various types of self-induced mixing process which can occur during the flow of a fluid through a pipe or vessel. The effects of dispersion are particularly important in packed beds, though they are also present under the simple flow conditions which exist in a straight tube or pipe. Dispersion can arise from the effects of molecular diffusion or as the result of the flow pattern existing within the fluid. An important consequence of dispersion is that the flow in a packed bed reactor deviates from plug flow with an important effect on the characteristics of the reactor.

It is of interest first to consider what is happening in pipe flow. Random molecular movement gives rise to a mixing process which can be described by Fick's Law. If concentration differences exist, the rate of transfer of a component is proportional to the product of the molecular diffusivity and the concentration gradient. If the fluid is in laminar flow, a parabolic velocity profile is set up over the cross-section and the fluid at the centre moves with twice the mean velocity in the pipe. This can give rise to dispersion since elements of fluid will take different times to traverse the length of the pipe according to their radial positions. When the fluid leaves the pipe, elements that have been within the pipe for very different periods of time will be mixed together. Thus, if the concentration of a tracer material in the fluid is suddenly changed, the effect will first be seen in the outlet stream after an interval required for the fluid at the axis to traverse the length of the pipe. Then as time increases the effect will be evident in the fluid issuing at progressively greater distances from the centre. Because the fluid velocity approaches zero at the pipe wall, the fluid near the wall will reflect the change over only a very long period.

If the fluid in the pipe is in turbulent flow, the effects of molecular diffusion will be supplemented by the action of the turbulent eddies and a much higher rate of transfer of material will occur within the fluid. Because the turbulent eddies also give rise to momentum transfer, the velocity profile is much flatter and the dispersion due to the effects of the different velocities of the fluid elements will be correspondingly less.

In a packed bed, the effects of dispersion will generally be greater than in a straight tube. The fluid is flowing successively through constrictions in the flow channels and then through broader passages or cells. Radial mixing readily takes place in the cells because the fluid enters them with an excess of kinetic energy, much of which is converted into rotational motion within the cells. Furthermore, the velocity profile is continuously changing within the fluid as it proceeds through the bed. Wall effects can be important in a packed bed because the bed voidage will be higher near the wall and flow will occur preferentially in that region.

At low rates of flow the effects of molecular diffusion predominate and cell mixing contributes relatively little to the dispersion. At high rates, on the other hand, a realistic model is presented by considering the bed to consist of a series of mixing cells, the dimension of each of which is of the same order as the size of the particles forming the bed. Whatever the mechanism, however, the rate of dispersion can be conveniently described by means of a dispersion coefficient. The process is generally anisotropic, except at very low flowrates, that is the dispersion rate is different in the longitudinal and radial directions, and therefore separate dispersion coefficients $D_L$ and $D_R$ are generally used to represent the behaviour in the two directions. The process is normally linear, with the rate of dispersion proportional to the product of the corresponding dispersion coefficient and concentration gradient. The principal factors governing dispersion in packed beds are discussed in a critical review by Gunn[20].

The differential equation for dispersion in a cylindrical bed of voidage $e$ may be obtained by taking a material balance over an annular element of height $\delta l$, inner radius $r$, and outer radius $r + \delta r$ (Fig. 4.5). On the basis of a dispersion model it is seen that if $C$ is

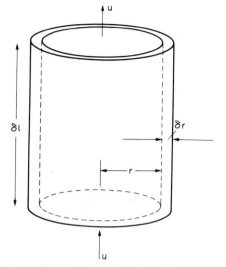

FIG. 4.5. Dispersion in packed beds

concentration of a reference material as a function of axial position $l$, radial position $r$, time $t$, and $D_L$ and $D_R$ are the axial and radial dispersion coefficients, then:

Rate of entry of reference material due to flow in axial direction:

$$= u(2\pi r \, \delta r)C$$

Corresponding efflux rate:

$$= u(2\pi r \, \delta r)\left(C + \frac{\partial C}{\partial l}\delta l\right)$$

Net accumulation rate in element due to flow in axial direction:

$$= u(2\pi r \, \delta r)\frac{\partial C}{\partial l}\delta l$$

Rate of diffusion in axial direction across inlet boundary:

$$= -(2\pi r \, \delta r \, e)D_L\frac{\partial C}{\partial l}$$

Corresponding rate at outlet boundary:

$$= -(2\pi r \, \delta r \, e)D_L\left(\frac{\partial C}{\partial l} + \frac{\partial^2 C}{\partial l^2}\delta l\right)$$

Net accumulation rate due to diffusion from boundaries in axial direction:

$$= (2\pi r \, \delta r \, e)D_L\frac{\partial^2 C}{\partial l^2}\delta l$$

Diffusion in radial direction at radius $r$:

$$= (2\pi r \, \delta l \, e)D_R\frac{\partial C}{\partial r}$$

Corresponding rate at radius $r + \delta r$:

$$= [2\pi(r + \delta r)\delta l\, e]D_R\left[\frac{\partial C}{\partial r} + \frac{\partial^2 C}{\partial r^2}\delta r\right]$$

Net accumulation rate due to diffusion from boundaries in radial direction:

$$= -[2\pi r\,\delta l\, e]D_R\frac{\partial C}{\partial r} + [2\pi(r + \delta r)\delta l\, e]D_R\left(\frac{\partial C}{\partial r} + \frac{\partial^2 C}{\partial r^2}\delta r\right)$$

$$= 2\pi\,\delta l\, eD_R\left[\frac{\partial C}{\partial r}\delta r + r\,\delta r\frac{\partial^2 C}{\partial r^2} + (\delta r)^2\frac{\partial^2 C}{\partial r^2}\right]$$

$$= 2\pi\,\delta l\, eD_R\left[\delta r\frac{\partial}{\partial r}\left(r\frac{\partial C}{\partial r}\right)\right] \text{ (ignoring the last term)}$$

Thus, the total accumulation rate:

$$= (2\pi r\,\delta r\,\delta l)e\frac{\partial C}{\partial t}$$

$$= -u2\pi r\,\delta r\frac{\partial C}{\partial l}\delta l + 2\pi r\,\delta r\, eD_L\frac{\partial^2 C}{\partial l^2}\delta l + 2\pi\,\delta l\, eD_R\,\delta r\frac{\partial}{\partial r}\left(r\frac{\partial C}{\partial r}\right)$$

i.e.

$$\frac{\partial C}{\partial t} + \frac{1}{e}u\frac{\partial C}{\partial l} = D_L\frac{\partial^2 C}{\partial l^2} + \frac{1}{r}D_R\frac{\partial}{\partial r}\left(r\frac{\partial C}{\partial r}\right) \tag{4.23}$$

Longitudinal dispersion coefficients can be readily obtained by injecting a pulse of tracer into the bed in such a way that radial concentration gradients are eliminated, and measuring the change in shape of the pulse as it passes through the bed. Since $\partial C/\partial r$ is then zero, equation 4.23 becomes:

$$\frac{\partial C}{\partial t} + \frac{u}{e}\frac{\partial C}{\partial l} = D_L\frac{\partial^2 C}{\partial l^2} \tag{4.24}$$

Values of $D_L$ can be calculated from the change in shape of a pulse of tracer as it passes between two locations in the bed, and a typical procedure is described by Edwards and Richardson[21]. Gunn and Pryce[22], on the other hand, imparted a sinusoidal variation to the concentration of tracer in the gas introduced into the bed. The results obtained by a number of workers are shown in Fig. 4.6 as a Peclet number $Pe(ud/eD_L)$ plotted against the particle Reynolds number ($Re' = ud\rho/\mu$).

For gases, at low Reynolds numbers ($<1$), the Peclet number increases linearly with Reynolds number, giving:

$$\frac{ud}{eD_L} = K\frac{ud\rho}{\mu} = K.Sc^{-1}\frac{ud}{D} \tag{4.25}$$

or

$$\frac{D_L}{D} = \text{constant}, \gamma \approx 0.7 \tag{4.26}$$

since $Sc$, the Schmidt number, is approximately constant for gases and the voidage of a randomly packed bed is usually about 0.4. This is consistent with the hypothesis that, at

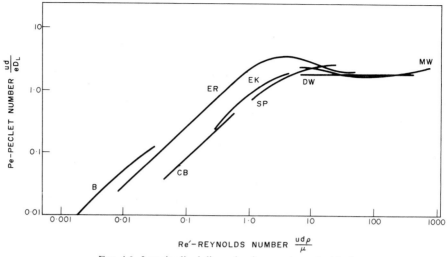

FIG. 4.6. Longitudinal dispersion in gases in packed beds

ER —Edwards and Richardson[21]
B  —Blackwell et al.[25]
CB —Carberry and Bretton[26]
DW —De Maria and White[27]
MW—McHenry and Wilhelm[28]
SP —Sinclair and Potter[29]
EK —Evans and Kenney[30] $N_2/He : N_2/H_2$

low Reynolds numbers, molecular diffusion predominates. The factor 0·7 is a tortuosity factor which allows for the fact that the molecules must negotiate a tortuous path because of the presence of the particles.

At Reynolds numbers greater than about 10 the Peclet number becomes approximately constant, giving:

$$D_L \approx \tfrac{1}{2}\frac{u}{e}d \qquad (4.27)$$

This equation is predicted by the mixing cell model, and turbulence theories put forward by Aris and Amundson[23] and by Prausnitz[24].

In the intermediate range of Reynolds numbers, the effects of molecular diffusivity and of macroscopic mixing are approximately additive, and the dispersion coefficient is given by an equation of the form:

$$D_L = \gamma D + \tfrac{1}{2}\frac{ud}{e} \qquad (4.28)$$

However, the two mechanisms interact and molecular diffusion can reduce the effects of convective dispersion. This can be understood by the fact that with streamline flow in a tube molecular diffusion will tend to smooth out the concentration profile arising from the velocity distribution over the cross-section. Similarly radial dispersion can give rise to lower values of longitudinal dispersion than predicted by equation 4.28. As a result the curves of Peclet versus Reynolds number tend to pass through a maximum as shown in Fig. 4.6.

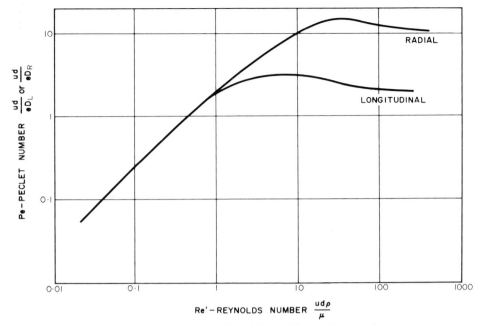

FIG. 4.7. Longitudinal and radial mixing coefficients for argon in air

A comparison of the effects of axial and radial mixing is seen from Fig. 4.7, in which are shown results obtained by Gunn and Pryce[22] for dispersion of argon into air. The values of $D_L$ were obtained as indicated earlier and $D_R$ was determined by injecting a steady stream of tracer at the axis and measuring the radial concentration gradient across the bed. It is seen that molecular diffusion dominates at low Reynolds numbers, with both the axial and radial dispersion coefficients $D_L$ and $D_R$ equal to approximately 0·7 times the molecular diffusivity. At high Reynolds numbers, however, the ratio of the longitudinal dispersion coefficient to the radial dispersion coefficient approaches a value of about 5, i.e. at high Reynolds numbers:

$$D_L/D_R \approx 5 \qquad (4.29)$$

The experimental results for dispersion coefficients in gases show that they can be satisfactorily represented as Peclet number expressed as a function of particle Reynolds number, and that similar correlations are obtained, irrespective of the gases used. However, it might be expected that the Schmidt number would be an important variable but it is not possible to test this hypothesis with gases as the values of Schmidt number are all approximately the same and equal to about unity.

With liquids, however, the Schmidt number is variable and it is generally about three orders of magnitude greater than for a gas. Results for longitudinal dispersion available in the literature, and plotted in Fig. 4.8, show that over the range of Reynolds numbers studied ($10^{-2} < Re' < 10^3$) the Peclet number shows little variation and is of the order of unity. Comparison of these results with the corresponding ones for gases (Fig. 4.6) shows that the effect of molecular diffusion in liquids is insignificant at Reynolds numbers up to unity. This difference can be attributed to the very different magnitudes of the Schmidt numbers.

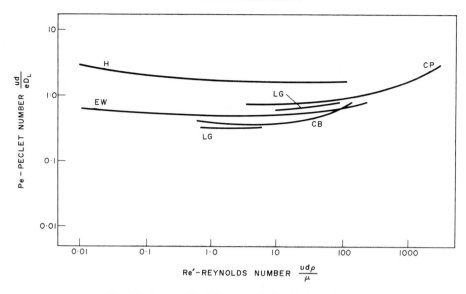

FIG. 4.8. Longitudinal dispersion in liquids in packed beds

CP —Cairns and Prausnitz[31]
CB —Carberry and Bretton[26]
EW—Ebach and White[33]
H  —Hiby[34]
LG —Liles and Geankoplis[35]

## 4.4. HEAT TRANSFER IN PACKED BEDS

For heat and mass transfer through a stationary or streamline fluid to a single spherical particle it has been šhown (Volume 1, Chapter 7) that the heat and mass transfer coefficients reach limiting low values given by:

$$Nu' = Sh' = 2 \tag{4.30}$$

where $Nu'(= hd/k)$ and $Sh'(= h_D d/D)$ are the Nusselt and Sherwood numbers with respect to the fluid.

Kramers[36] has shown that, for conditions of forced convection, the heat transfer coefficient can be represented by the following equation:

$$Nu' = 2 \cdot 0 + 1 \cdot 3 Pr^{0 \cdot 15} + 0 \cdot 66 Pr^{0 \cdot 31} Re'^{0 \cdot 5} \tag{4.31}$$

where $Re'$ is the particle Reynolds number $ud\rho/\mu$, and
        $Pr$ is the Prandtl number $C_p \mu/k$.

This expression has been obtained on the basis of experimental results obtained with fluids of Prandtl numbers ranging from 0·7 to 380.

For natural convection, Ranz and Marshall[37] have given:

$$Nu' = 2 \cdot 0 + 0 \cdot 6 Pr^{1/3} Gr'^{1/4} \tag{4.32}$$

where $Gr'$ is the Grashof number (see Volume 1, Chapter 7).

Results for packed beds are much more difficult to obtain because the driving force cannot be measured very readily. Gupta and Thodos[38] suggest that the $j$-factor for heat transfer, $j_h$ (see Volume 1, Chapter 8), forms the most satisfactory basis of correlation for experimental results:

$$ej_h = 2 \cdot 06 Re'^{-0 \cdot 575} \tag{4.33}$$

where  $e$ is the voidage of the bed,

$j_h = St' Pr^{2/3}$, and

$St' =$ Stanton number $h/C_p \rho u$.

The $j$-factors for heat and mass transfer, $j_h$ and $j_d$, are found to be equal, and therefore equation 4.33 can also be used for the calculation of mass transfer rates.

Reproducible correlations for the heat transfer coefficient between a fluid flowing through a packed bed and the cylindrical wall of the container are very difficult to obtain. The main difficulty here is that a wide range of packing conditions can occur in the vicinity of the walls. However, the results quoted by Zenz and Othmer[39] suggest that

$$Nu \propto Re'^{0 \cdot 7 - 0 \cdot 9} \tag{4.34}$$

It will be noted that in this expression the Nusselt number with respect to the tube wall $Nu$ is related to the Reynolds number with respect to the particle $Re'$.

## 4.5. PACKED COLUMNS

Packed columns are very widely used to bring about intimate contact between two immiscible or partially miscible fluids, either a gas and a liquid, or two liquids. Countercurrent flow is normally used with the gas or lighter liquid entering at the bottom and the second fluid at the top. An example of the liquid–gas system is an absorption process where a soluble gas is scrubbed from a mixture of gases by means of a liquid (Fig. 4.9). In a packed column used for distillation, the more volatile component of, say, a binary mixture is progressively transferred to the vapour phase and the less volatile condenses out in the liquid. Packed columns have also been used extensively for liquid–liquid extraction processes where a solute is transferred from one solvent to another (Chapter 12). Some principles involved in the design and operation of packed columns will be

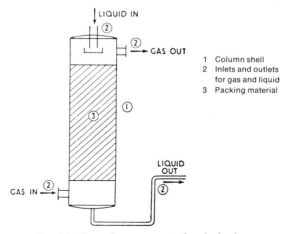

FIG. 4.9. General arrangement of packed column

illustrated by consideration of columns for gas absorption. In this chapter the construction of the column and the flow characteristics will be dealt with, whereas the magnitude of the mass transfer coefficients will be discussed later in Chapters 10, 11, and 12.

In order to obtain a good rate of transfer per unit volume of the tower, a packing is selected which will promote a high interfacial area between the two phases and a high degree of turbulence in the fluids. Usually increased area and turbulence are achieved at the expense of increased capital cost and/or pressure drop, and a balance must be made between these factors when arriving at an economic design.

### 4.5.1. General Description

The construction of packed towers is relatively straightforward. The shell of the column may be constructed from metal, ceramics, glass, or plastics material, or from metal with a corrosion-resistant lining. The column should be mounted truly vertically to help uniform liquid distribution. Detailed information on the mechanical design and mounting of industrial scale column shells is given by Brownell and Young[40], Molyneux[41] and in B.S. 1500[42].

The bed of packing rests on a support plate which should be designed to have at least 75 per cent free area for the passage of the gas so as to offer as low a resistance as possible. The simplest support is a grid with relatively widely spaced bars on which a few layers of large Raschig or partition rings are stacked. One such arrangement is shown in Fig. 4.10. The gas injection plate[43] shown in Fig. 4.11 is designed to provide separate passageways for gas and liquid so that they need not vie for passage through the same opening. This is achieved by providing the gas inlets to the bed at a point above the level at which liquid leaves the bed.

At the top of the packed bed a liquid distributor of suitable design provides for the uniform irrigation of the packing which is necessary for satisfactory operation. Four examples of different distributors are shown in Fig. 4.12.

(a) A simple orifice type. This type gives very fine distribution but must be correctly sized for a particular duty and should not be used where there is any risk of the holes plugging.

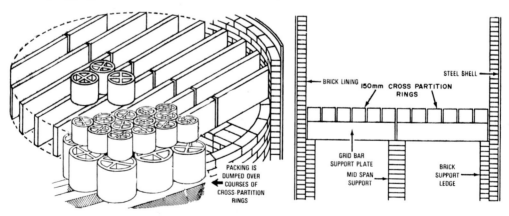

FIG. 4.10. Grid bar supports for packed towers

(b) Notched chimney type of distributor, which has a good range of flexibility for the medium and upper flowrates, and is not prone to blockage.

(c) Notched trough distributor which is specially suitable for the larger sizes of tower. Because of its large free area it is also suitable for the higher gas rates.

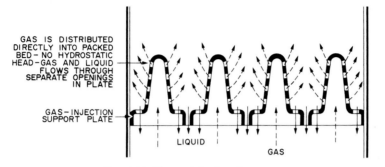

GAS IS DISTRIBUTED DIRECTLY INTO PACKED BED – NO HYDROSTATIC HEAD – GAS AND LIQUID FLOWS THROUGH SEPARATE OPENINGS IN PLATE

GAS-INJECTION SUPPORT PLATE

LIQUID    GAS

FIG. 4.11. The gas injection plate

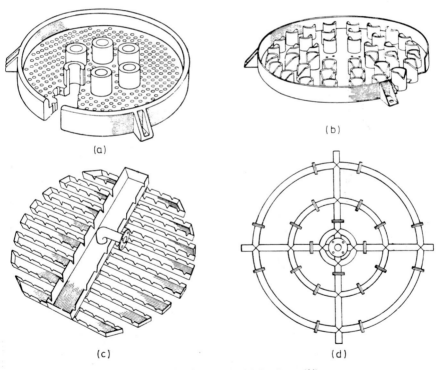

(a)

(b)

(c)

(d)

FIG. 4.12. Types of liquid distributor[44]

(d) Perforated ring type of distributor for use with absorption columns where high gas rates and relatively small liquid rates are encountered. This type is especially suitable where pressure loss must be minimised. For the larger size of tower, where installation through manholes is necessary, it may be made up in flanged sections.

Uniform liquid flow is essential if the best use is to be made of the packing and, if the tower is high, re-distributing plates are necessary. These plates are needed at intervals of about $2\frac{1}{2}$–3 column diameters for Raschig rings and about 5–10 column diameters for Pall rings, but are usually not more than 6 m apart[45]. A "hold-down" plate is often placed at the top of a packed column to minimise movement and breakage of the packing caused by surges in flowrates. The gas inlet should also be designed for uniform flow over the cross-section and the gas exit should be separate from the liquid inlet. Further details on internal fittings are given by Leva[43].

Columns for both absorption and distillation vary in diameter from about 25 mm for small laboratory purposes to over 4·5 m for large industrial operations; these industrial columns may be up to about 30 m in height. Columns may operate at pressures ranging from high vacuum to high pressure, the optimum pressure depending on both the chemical and the physical properties of the system.

### 4.5.2. Packings

Packings can be divided into three main classes—broken solids, shaped packings, and grids. Broken solids are the cheapest form and are used in sizes from about 10 mm to 100 mm according to the size of the column. Although they frequently form a good corrosion-resistant material they are not as satisfactory as shaped packings either in regard to liquid flow or to effective surface offered for transfer. The packing should be of as uniform size as possible so as to produce a bed of uniform characteristics with a desired voidage.

The most commonly used packings in chemical plant were Raschig rings, Pall rings, Lessing rings, and Berl saddles. Newer packings include Intalox and Super Intalox saddles, Hy-Pak, and Mini rings and, because of their high performance characteristics and low pressure drop, these packings now account for a large share of the market. They are illustrated in Figs. 4.13 to 4.18. Most of the packings mentioned are available in a wide range of materials such as ceramics, metals, glass, plastics, carbon, and sometimes rubber. Ceramic packings are resistant to corrosion and comparatively cheap, but are heavy and may require a stronger packing support and foundations. The smaller metal rings are also available made from wire mesh, and these give much-improved transfer characteristics. A

FIG. 4.13. Ceramic and metallic Raschig rings and Lessing rings (made in sizes from 150 mm to 15 mm)

FIG. 4.14. Ceramic, plastic and steel Pall rings

non-porous solid should be used if there is any risk of crystal formation in the pores when the packing dries, as this can give rise to serious damage to the packing elements. However, some plastics are not very good because they are not wetted by many liquids. Channelling, i.e. non-uniform distribution of liquid across the column cross-section, is much less marked with shaped packings, and their resistance to flow is much less. Shaped packings also give a more effective surface per unit volume because surface contacts are

FIG. 4.15. Hy-Pak metal packing

FIG. 4.16. Super Intalox packing

reduced to a minimum and the film flow is much improved compared with broken solids. On the other hand, the shaped packings are more expensive, particularly when small sizes are used. The voidage obtainable with these packings varies from about 0·45 to 0·95. Ring packings are either dumped into a tower, dropped in small quantities, or may be individually stacked if 75 mm or larger in size. To obtain high and uniform voidage and to prevent breakage, it is often found better to dump the packings into a tower full of liquid. Stacked packings as indicated in Fig. 4.10 have the advantage that the flow channels are vertical and there is much less tendency for the liquid to flow to the walls than with random packings. The properties of some commonly used industrial packings are shown in Table 4.3.

The size of packing used influences the height and diameter of a column and the pressure drop and cost of packing. Generally, as the packing size is increased, the cost per unit volume of packing and the pressure drop per unit height of packing are reduced, and the mass transfer efficiency is reduced. Reduced mass transfer efficiency results in a taller column being needed so that the overall column cost is not always reduced by increasing the packing size. Normally, in a column in which the packing is randomly arranged the packing size should not exceed one-eighth of the column diameter. Above this size, liquid distribution—and hence mass transfer efficiency—deteriorate rapidly. Since cost per unit volume of packing does not fall much for sizes above 50 mm whereas efficiency continues to fall, there is seldom any advantage in using packings much larger than 50 mm in a randomly packed column.

For laboratory purposes a number of special packings have been developed which are, in general, too expensive for large diameter towers. Dixon packings, which are Lessing rings made from wire mesh, and Knitmesh, a fine wire mesh packing, are shown in Fig. 4.19. These packings give very great interfacial areas and, if they are flooded with liquid before operation, all of the surface is active so that the transfer characteristics are very good even at low liquid rates. The volume of liquid held up in such a packing is low and the pressure drop is also low. Some of these high efficiency woven wire packings have been used in columns up to 500 mm diameter.

Grid packings, which are relatively easy to fabricate, are usually used in columns of square section (Fig. 4.20). They may be made from wood, plastics, carbon, or ceramic

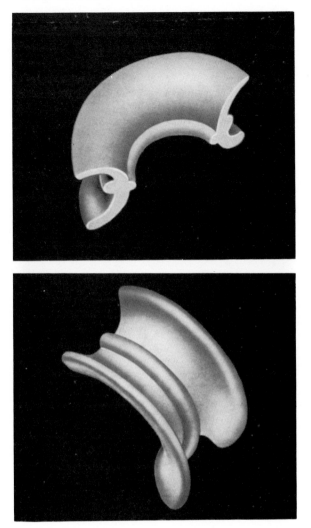

FIG. 4.17.  The Intalox saddle

FIG. 4.18.  Mini ring

TABLE 4.3. *Design Data for Various Packings*

| | Size in. | Size mm | Wall thickness in. | Wall thickness mm | Number /ft³ | Number /m³ | Density lb/ft³ | Density kg/m³ | Contact surface $S_B$ ft²/ft³ | Contact surface $S_B$ m²/m³ | Free space % | Packing factor F |
|---|---|---|---|---|---|---|---|---|---|---|---|---|
| **Ceramic Raschig Rings** | 0·25 | 6·35 | 0·03 | 0·76 | 85,600 | 3,022,600 | 60 | 961 | 242 | 794 | 62 | 1600 |
| | 0·38 | 9·65 | 0·05 | 1·27 | 24,700 | 872,175 | 61 | 977 | 157 | 575 | 67 | 1000 |
| | 0·50 | 12·7 | 0·07 | 1·78 | 10,700 | 377,825 | 55 | 881 | 112 | 368 | 64 | 640 |
| | 0·75 | 19·05 | 0·09 | 2·29 | 3090 | 109,110 | 50 | 801 | 73 | 240 | 72 | 255 |
| | 1·0 | 25·4 | 0·14 | 3·56 | 1350 | 47,670 | 42 | 673 | 58 | 190 | 71 | 160 |
| | 1·25 | 31·75 | | | 670 | 23,660 | 46 | 737 | | | 71 | 125 |
| | 1·5 | 38·1 | | | 387 | 13,665 | 43 | 689 | | | 73 | 95 |
| | 2·0 | 50·8 | 0·25 | 6·35 | 164 | 5790 | 41 | 657 | 29 | 95 | 74 | 65 |
| | 3·0 | 76·2 | | | 50 | 1765 | 35 | 561 | | | 78 | 36 |
| **Metal Raschig Rings** | 0·25 | 6·35 | 0·03125 | 0·794 | 88,000 | 3,107,345 | 133 | 2131 | | | 72 | 700 |
| | 0·38 | 9·65 | 0·03125 | 0·794 | 27,000 | 953,390 | 94 | 1506 | | | 81 | 390 |
| | 0·50 | 12·7 | 0·03125 | 0·794 | 11,400 | 402,540 | 75 | 1201 | 127 | 417 | 85 | 300 |
| | 0·75 | 19·05 | 0·03125 | 0·794 | 3340 | 117,940 | 52 | 833 | 84 | 276 | 89 | 185 |
| | 0·75 | 19·05 | 0·0625 | 1·59 | 3140 | 110,875 | 94 | 1506 | | | 80 | 230 |
| | 1·0 | 25·0 | 0·03125 | 0·794 | 1430 | 50,494 | 39 | 625 | 63 | 207 | 92 | 115 |
| | 1·0 | 25·0 | 0·0625 | 1·59 | 1310 | 46,260 | 71 | 1137 | | | 86 | 137 |
| | 1·25 | 31·75 | 0·0625 | 1·59 | 725 | 25,600 | 62 | 993 | | | 87 | 110 |
| | 1·5 | 38·1 | 0·0625 | 1·59 | 400 | 14,124 | 49 | 785 | | | 90 | 83 |
| | 2·0 | 50·8 | 0·0625 | 1·59 | 168 | 5932 | 37 | 593 | 31 | 102 | 92 | 57 |
| | 3·0 | 76·2 | 0·0625 | 1·59 | 51 | 1800 | 25 | 400 | 22 | 72 | 95 | 32 |
| **Carbon Raschig Rings** | 0·25 | 6·35 | 0·0625 | 1·59 | 85,000 | 3,001,410 | 46 | 737 | 212 | 696 | 55 | 1600 |
| | 0·50 | 12·7 | 0·0625 | 1·59 | 10,600 | 374,290 | 27 | 433 | 114 | 374 | 74 | 410 |
| | 0·75 | 19·05 | 0·125 | 3·175 | 3140 | 110,875 | 34 | 545 | 75 | 246 | 67 | 280 |
| | 1·0 | 25·0 | 0·125 | 3·175 | 1325 | 46,787 | 27 | 433 | 57 | 187 | 74 | 160 |
| | 1·25 | 31·75 | | | 678 | 23,940 | 31 | 496 | | | 69 | 125 |
| | 1·5 | 38·1 | | | 392 | 13,842 | 34 | 545 | | | 67 | 130 |
| | 2·0 | 50·8 | 0·250 | 6·35 | 166 | 5862 | 27 | 433 | 29 | 95 | 74 | 65 |
| | 3·0 | 76·2 | 0·312 | 7·92 | 49 | 1730 | 23 | 368 | 19 | 62 | 78 | 36 |
| **Metal Pall Rings** | 0·625 | 15·9 | 0·018 | 0·46 | 5950 | 210,098 | 37 | 593 | 104 | 341 | 93 | 70 |
| | 1·0 | 25·4 | 0·024 | 0·61 | 1400 | 49,435 | 30 | 481 | 64 | 210 | 94 | 48 |
| | 1·25 | 31·75 | 0·030 | 0·76 | 375 | 13,240 | 24 | 385 | 39 | 128 | 95 | 28 |
| | 2·0 | 50·8 | 0·036 | 0·915 | 170 | 6003 | 22 | 353 | 31 | 102 | 96 | 20 |
| | 3·5 | 76·2 | 0·048 | 1·219 | 33 | 1165 | 17 | 273 | 20 | 65·6 | 97 | 16 |

N.B. weights are for mild steel; multiply by 1·105, 1·12, 1·37, 1·115 for stainless steel, copper, aluminium, and monel respectively

Metal Pall Rings
N.B. weights are for mild steel

| Packing | Size | mm | | | | | | | | | | |
|---|---|---|---|---|---|---|---|---|---|---|---|---|
| Plastic Pall Rings N.B. weights are for polypropylene | 0·625 | 15·9 | 0·03 | 0·762 | 6050 | 213,630 | 7·0 | 112 | 104 | 341 | 87 | 97 |
| | 1·0 | 25·4 | 0·04 | 1·016 | 1440 | 50,848 | 5·5 | 88 | 63 | 207 | 90 | 52 |
| | 1·5 | 38·1 | 0·04 | 1·016 | 390 | 13,770 | 4·75 | 76 | 39 | 128 | 91 | 40 |
| | 2·0 | 50·8 | 0·06 | 1·524 | 180 | 6356 | 4·25 | 68 | 31 | 102 | 92 | 25 |
| | 3·5 | 88·9 | 0·06 | 1·524 | 33 | 1165 | 4·0 | 64 | 26 | 85 | 92 | 16 |
| Ceramic Intalox Saddles | 0·25 | 6·35 | | | 117,500 | 4,149,010 | 54 | 865 | | | 65 | 725 |
| | 0·38 | 9·65 | | | 49,800 | 1,758,475 | 50 | 801 | | | 67 | 330 |
| | 0·50 | 12·7 | | | 18,300 | 646,186 | 46 | 737 | | | 71 | 200 |
| | 0·75 | 19·05 | | | 5640 | 199,150 | 44 | 705 | | | 73 | 145 |
| | 1·0 | 25·4 | | | 2150 | 75,918 | 42 | 673 | 59 | 194 | 73 | 92 |
| | 1·5 | 38·1 | | | 675 | 23,835 | 39 | 625 | | | 76 | 52 |
| | 2·0 | 50·8 | | | 250 | 8828 | 38 | 609 | | | 76 | 40 |
| | 3·0 | 76·2 | | | 52 | 1836 | 36 | 577 | | | 79 | 22 |
| Plastic Super Intalox | No. 1 | | | | 1620 | 57,200 | 6·0 | 96 | 63 | 207 | 90 | 33 |
| | No. 2 | | | | 190 | 6710 | 3·75 | 60 | 33 | 108 | 93 | 21 |
| | No. 3 | | | | 42 | 1483 | 3·25 | 52 | 27 | 88·6 | 94 | 16 |
| Ceramic Super Intalox | No. 1 | | | | 1490 | 52,613 | 39 | 625 | 76 | 249 | 79 | 60 |
| | No. 2 | | | | 180 | 6356 | 37 | 593 | 32 | 105 | 81 | 30 |
| Hy-Pak N.B. weights are for mild steel | No. 1 | | | | 850 | 30,014 | 19 | 304 | | | 96 | 43 |
| | No. 2 | | | | 107 | 3778 | 14 | 224 | | | 97 | 18 |
| | No. 3 | | | | 31 | 1095 | 13 | 208 | | | 97 | 15 |
| Plastic Cascade Mini Rings | No. 1 | | | | | | | | | | | 25 |
| | No. 2 | | | | | | | | | | | 15 |
| | No. 3 | | | | | | | | | | | 12 |
| Metal Cascade Mini Rings | No. 0 | | | | | | | | | | | 55 |
| | No. 1 | | | | | | | | | | | 34 |
| | No. 2 | | | | | | | | | | | 22 |
| | No. 3 | | | | | | | | | | | 14 |
| | No. 4 | | | | | | | | | | | 10 |
| Ceramic Cascade Mini Rings | No. 2 | | | | | | | | | | | 38 |
| | No. 3 | | | | | | | | | | | 24 |
| | No. 5 | | | | | | | | | | | 18 |

N.B. The packing factor F replaces the term $S_B/e^3$ which appears later in this chapter. Use of the given value of F in Fig. 4.29 permits more predictable performance of designs incorporating packed beds since the values quoted are derived from operating characteristics of the packings rather than from their physical dimensions.

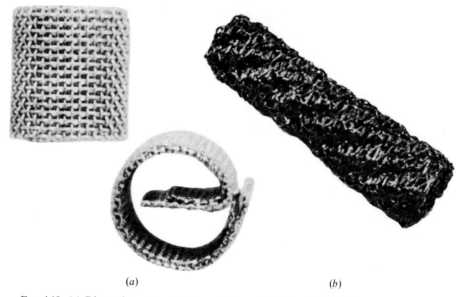

(a)                                              (b)

FIG. 4.19. (<i>a</i>) Dixon rings (3 mm) enlarged view, (<i>b</i>) Knitmesh Multifil packing for 25 mm diameter column

materials, and, because of the relatively large spaces between the individual grids, they give low pressure drops. Further advantages lie in their ease of assembly, their ability to accept fluids with suspended solids, and their ease of wetting even at very low liquid rates. The main problem is that of obtaining good liquid distribution since, at high liquid rates, the liquid tends to cascade from one grid to the next without being broken up into fine droplets which are desirable for a high interfacial surface.

Plastic grid-type packings have been the subject of development in the U.K. by I.C.I. Limited, initially with the aim of producing an ideal biological filter medium for use in effluent treatment[46]. This resulted in the design of a PVC packing known as Flocor. The properties of this packing have been fully described by Askew[47]. More recently, a further development by the same Company has produced a similar product, Flocool, shown in Fig. 4.21 for use in the more familiar chemical engineering packed tower applications. Based on a standard module size of $1.14 \times 0.62 \times 0.61$ m, this packing is available in two types with specific areas of 60 and 90 $m^2/m^3$ with densities of 19·3 and 28·9 $kg/m^3$ respectively.

Further discussion on the merits of these various packings on the basis of their contribution to mass transfer rates is given in Chapters 11 and 12.

### 4.5.3. Fluid Flow in Packed Columns

*Pressure Drop*

It is important to be able to predict the drop in pressure for the flow of the two fluid streams through a packed column. Earlier in this chapter the drop in pressure arising from the flow of a single phase through granular beds was considered and the same general form of approach is usefully adopted for the flow of two fluids through packed columns. It was

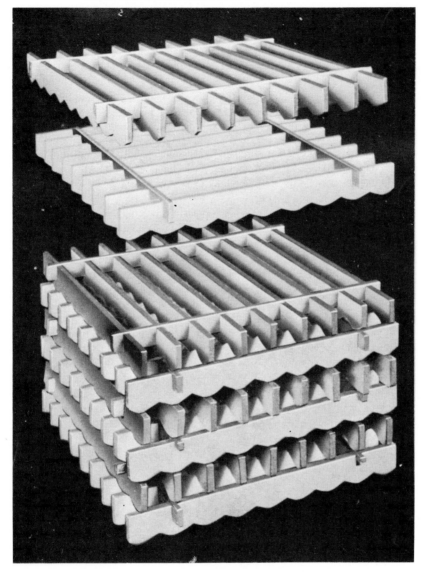

FIG. 4.20. Grid packings

noted that the expressions for flow through ring-type packings are less reliable than those for flow through beds of solid particles; for the typical absorption column there is no very accurate expression, but there are several correlations that are useful for design purposes. In the majority of cases the gas flow is turbulent and the general form of the relation between the drop in pressure $-\Delta P$ and the volumetric gas flowrate per unit area of column $u_G$ is shown on curve $A$ of Fig. 4.22. $-\Delta P$ is then proportional to about $u_G^{1.8}$, in agreement with the slope of curve $A$ of Fig. 4.1 at higher Reynolds numbers. If, in addition to the gas flow, liquid flows down the tower, the passage of the gas is not significantly affected at low liquid rates and the pressure drop line is similar to line $A$, but for a given

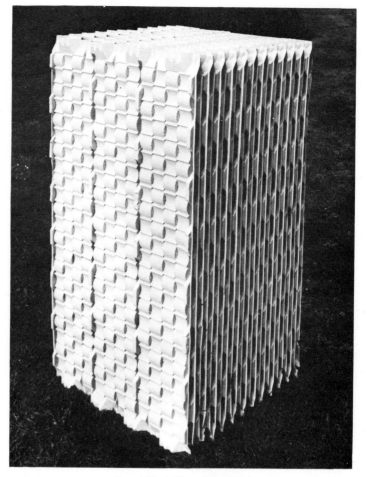

Fig. 4.21. A Flocool module

value of $u_G$ the value of $-\Delta P$ is somewhat increased. When the gas rate reaches a certain value, then the pressure drop rises very much more quickly and is proportional to $u_G^{2.5}$ as shown by the section $XY$ on curve $C$. Over this section the liquid flow is interfering with the gas flow and the hold-up of liquid is progressively increasing. The free space in the packings is therefore being continuously taken up by the liquid, and thus the resistance to flow rises quickly. At gas flows beyond $Y$, $-\Delta P$ rises very steeply and the liquid is held up in the column. The point $X$ is known as the loading point, and point $Y$ as the flooding point for the given liquid flow. If the flowrate of liquid is increased, a similar plot $D$ is obtained in which the loading point is achieved at a lower gas rate but at a similar value of $-\Delta P$. Whilst it is advantageous to have a reasonable hold-up in the column as this promotes interphase contact, it is not practicable to operate under flooding conditions, and columns are best operated over the section $XY$. Since this is a section with a relatively short range in gas flow, the safe practice is to design for operation at the loading point $X$. It is of interest to note that, if a column is flooded and then allowed to drain, the value of $-\Delta P$ for a given gas flow is increased over that for an entirely dry packing as

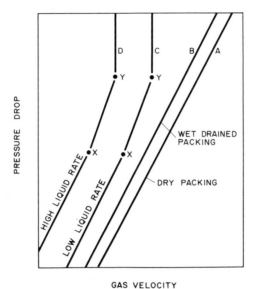

FIG. 4.22. Pressure drops in wet packings (logarithmic axes)

shown by curve $B$. Rose and Young[48] correlated their experimental pressure drop data for Raschig rings by the following equation:

$$-\Delta P_w = -\Delta P_d\left(1 + \frac{3 \cdot 30}{d_n}\right) \tag{4.35}$$

where $-\Delta P_w$ is the pressure drop across the wet drained column,
$\quad -\Delta P_d$ is the pressure drop across the dry column, and
$\quad\quad d_n$ is the nominal size of the Raschig rings in mm.
This effect will thus be most significant for small packings.

There are several ways of calculating the pressure drop across a packed column when gas and liquid are flowing simultaneously and the column is operating below the loading point.

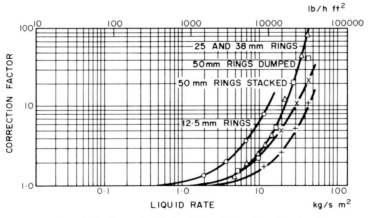

FIG. 4.23. Correction factor for liquid flow (Raschig rings)

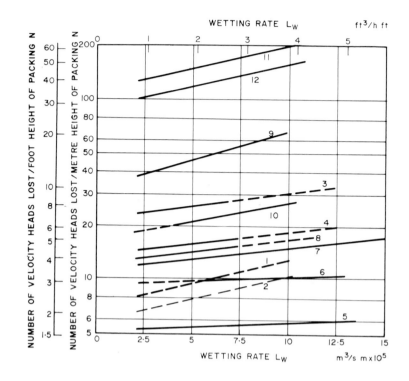

FIG. 4.24. Pressure drop through grids and stacked rings.
Curves shown in broken lines are based on estimated figures

Plain grids:
1   25 mm × 25 mm × 1·6 mm
2   25 mm × 50 mm × 1·6 mm
3   25 mm × 25 mm × 6·4 mm
4   25 mm × 50 mm × 6·4 mm

Serrated grids:
5   100 mm × 100 mm × 12·5 mm
6   50 mm × 50 mm × 9·5 mm
7   38 mm × 38 mm × 4·8 mm

Stacked stoneware rings:
8   100 mm × 100 mm × 9·5 mm
9   75 mm × 75 mm × 9·5 mm
10  75 mm × 75 mm × 6·4 mm
11  50 mm × 50 mm × 6·4 mm
12  50 mm × 50 mm × 4·8 mm

Random metal rings:
13  50 mm × 50 mm × 1·6 mm
14  25 mm × 25 mm × 1·6 mm
15  12·5 mm × 12·5 mm × 0·8 mm

Random stoneware rings:
16  75 mm × 75 mm × 9·5 mm
17  50 mm × 50 mm × 6·4 mm
18  50 mm × 50 mm × 4·8 mm
19  38 mm × 38 mm × 4·8 mm
20  25 mm × 25 mm × 2·4 mm
22  12·5 mm × 12·5 mm × 1·6 mm

Random carbon rings:
23  50 mm × 50 mm × 6·4 mm*
24  25 mm × 25 mm × 4·8 mm
25  12·5 mm × 12·5 mm × 1·6 mm†

Quartz:
29  50 mm
30  12·5–32 mm

* Probably same as 17
† Probably same as 22

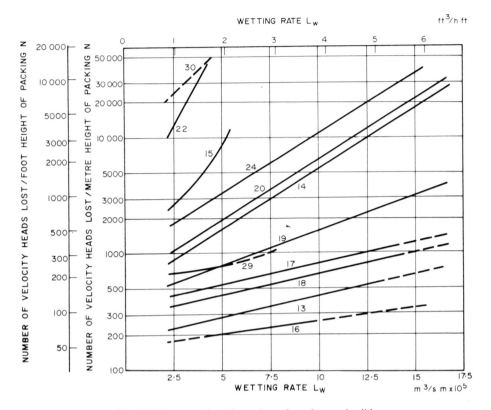

FIG. 4.25. Pressure drop through random rings and solids.
Curves shown in broken lines are based on estimated figures
(For key to numbered curves see Fig. 4.24)

One approach is to calculate the pressure drop for gas flow only and then multiply this pressure drop by a factor which accounts for the effect of the liquid flow. Equation 4.17 may be used for predicting the pressure drop for the gas only, and then the pressure drop with gas and liquid flowing is obtained by using the correction factors for the liquid flow rate shown on Figs. 4.23. This graph of correction factors is taken from Sherwood and Pigford[49] where other examples are given.

Another approach is that of Morris and Jackson[50] who have arranged experimental data for a wide range of solid ring and grid packings into a graphical form convenient for calculation and are included here as they cover a number of unusual packings for which data are scarce. The graphs are shown on Figs. 4.24 and 4.25. The number of velocity heads $N$ lost per unit height of packing is found from the graph for the appropriate value of the wetting rate (see equation 4.37) and then $N$ is substituted in the equation:

$$-\Delta P = \tfrac{1}{2} N \rho_G u_G^2 l \qquad (4.36)$$

where  $-\Delta P$ = pressure drop,

$\rho_G$ = gas density,

$u_G$ = gas velocity, based on the empty column cross-sectional area, and

$l$ = height of packing.

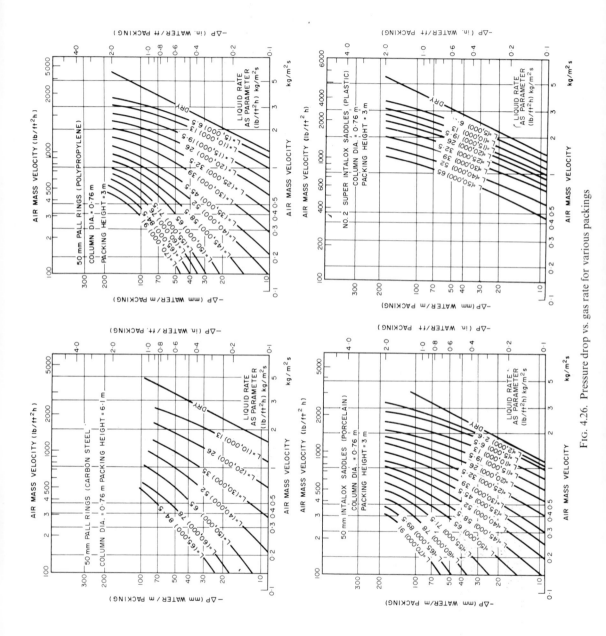

FIG. 4.26. Pressure drop vs. gas rate for various packings

Equation 4.36 is in consistent units. For example, writing $\rho_G$ in kg/m³, $u_G$ in m/s, $l$ in m, and $N$ in 1/m, $-\Delta P$ is calculated in N/m².

An empirical correlation of experimental data for pressure drop has also been presented by Leva[51], and by Eckert *et al.*[52] for Pall rings.

Where the data are available, the most accurate method of obtaining the pressure drop for flow through a bed of packing is from the manufacturer's own literature. This is usually presented as a logarithmic plot of gas rate against pressure drop, with a parameter of liquid flowrate on the graphs. Typical curves for four packings are shown as Fig. 4.26.

It should be stressed that all of these methods apply only to conditions at or below the loading point ($X$ on Fig. 4.22). If applied to conditions above the loading point the calculated pressure drop would be too low. It is therefore necessary first to check whether

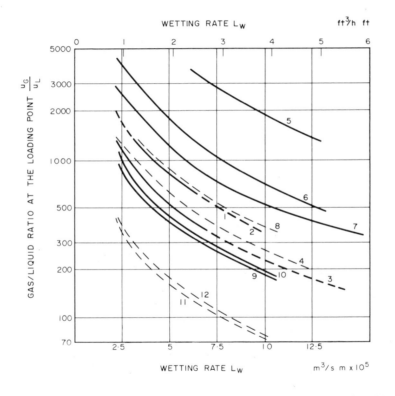

FIG. 4.27. Relation between gas/liquid ratio and wetting rate at the loading point for grids and stacked rings.
Curves shown in broken lines are based on estimated figures

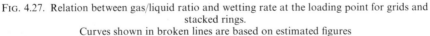

| Plain grids: | Serrated grids: | Stacked stoneware rings: |
|---|---|---|
| 1  25 mm × 25 mm × 1·6 mm | 5  100 mm × 100 mm × 12·5 mm | 8  100 mm × 100 mm × 9·5 mm |
| 2  25 mm × 50 mm × 1·6 mm | | 9  75 mm × 75 mm × 9·5 mm |
| 3  25 mm × 25 mm × 6·4 mm | 6  50 mm × 50 mm × 9·5 mm | 10  75 mm × 75 mm × 6·4 mm |
| 4  25 mm × 50 mm × 6·4 mm | 7  38 mm × 38 mm × 4·8 mm | 11  50 mm × 50 mm × 6·4 mm |
| | | 12  50 mm × 50 mm × 4·8 mm |

the column is operating at or below the loading point, and methods of predicting loading points are discussed in the next section.

### Loading and Flooding Points

The loading and flooding points have been shown on Fig. 4.22. There is no completely generalised expression for calculating the onset of loading, but one of the following semi-empirical correlations will often be adequate. Morris and Jackson[50] have given their results in the form of plots of $\psi(u_G/u_L)$ at the loading rate for various wetting rates $L_W$ ($m^3/s\, m$) (see equation 4.37) and their data for grids and rings are given in Fig. 4.27. $u_G$ and $u_L$ are average gas and liquid velocities based on the empty column and $\psi(= \sqrt{\rho_G/\rho_A})$ is a gas density correction factor, in which $\rho_A$ is the density of air at 293 K.

A useful graphical correlation for flooding rates was first presented by Sherwood *et al.*[53] and later developed by Lobo *et al.*[54] for random-dumped packings as shown in Fig. 4.28 in which $\dfrac{u_G^2 S_B}{ge^3}\dfrac{\rho_G}{\rho_L}\left(\dfrac{\mu_L}{\mu_w}\right)^{0.2}$ is plotted against $\dfrac{L}{G}\sqrt{\dfrac{\rho_G}{\rho_L}}$

where $u_G$ is the velocity of the gas, calculated over the whole cross-section of the bed,

$S_B$ is the surface of the packing per unit volume of bed, $S(1-e)$,

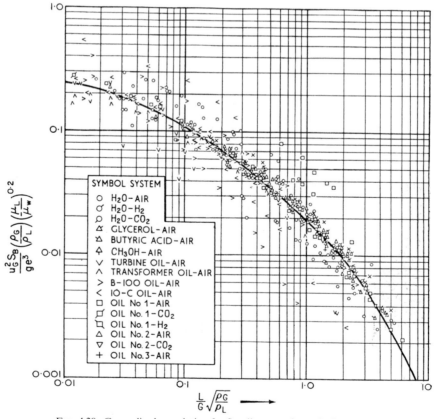

FIG. 4.28. Generalised correlation for flooding rates in packed towers

$g$ is the acceleration due to gravity,
$L$ is the mass rate of flow of the liquid,
$G$ is the mass rate of flow of the gas, and
$\mu_w$ is the viscosity of water at 293 K (approximately 1 mN s/m$^2$).

Suffix $G$ refers to the gas and suffix $L$ to the liquid.

The area inside the curve represents possible conditions of operation. In the above expressions, the ratios $\rho_G/\rho_L$ and $\mu_L/\mu_w$ have been introduced so that the relationship can be applied for a wide range of liquids and gases.

### The Generalised Pressure Drop Correlation

The generalised pressure drop correlation[55] has been developed as a practical aid to packed tower design and incorporates flowrates, physical properties of the fluid, a wide range of packings and pressure drop on one chart presented here as Fig. 4.29. A line representing flooding would lie above the top curve in Fig. 4.29 and hence the correlation may be used with safety in design procedures. The packing factor $F$ which is employed in the correlation is a modification of the specific surface of the packing which is used in Fig. 4.28. Values of $F$ are included in Table 4.3 for all packings. In practice, a pressure drop is selected for a given duty and use is made of the correlation to determine the gas flowrate per unit area $G'$ from which the tower diameter may be calculated. The method is discussed in further detail in Chapter 11 where the mass transfer properties of packings are also considered.

### Liquid Distribution

Provision of a packing with a high surface area per unit volume may not result in good contacting of gas and liquid unless the liquid is distributed uniformly over the surface of the packing. The need for liquid distribution and redistribution and correct packing size was stressed earlier. The effective wetted area decreases as the liquid rate is decreased and, for a given packing, there is a minimum liquid rate for effective use of the surface area of the packing. A useful measure of the effectiveness of wetting of the available area is the wetting rate $L_W$ defined as:

Volumetric liquid rate per unit cross-sectional area of column
——————————————————————————————————
Packing surface area per unit volume of column

i.e.
$$L_W = \frac{L}{A\rho_L S_B}$$

$$= \frac{u_L}{S_B} \tag{4.37}$$

Thus the wetting rate is analogous to the volumetric liquid rate per unit length of circumference in a wetted wall column (Chapter 12) where the liquid flows down the surface of a cylinder. If the liquid rate were too low, a continuous liquid film would not be formed around the circumference of the cylinder and some of the area would be ineffective.

Similar effects occur in a packed column although the flow patterns and arrangement of the surfaces are then obviously much more complex. Morris and Jackson[50] have recommended minimum wetting rates of $2 \times 10^{-5}$ m$^3$/s m for rings between 25 mm and

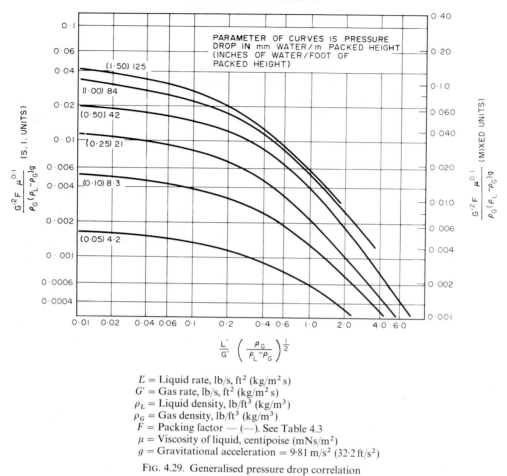

$L'$ = Liquid rate, lb/s, ft$^2$ (kg/m$^2$ s)
$G'$ = Gas rate, lb/s, ft$^2$ (kg/m$^2$ s)
$\rho_L$ = Liquid density, lb/ft$^3$ (kg/m$^3$)
$\rho_G$ = Gas density, lb/ft$^3$ (kg/m$^3$)
$F$ = Packing factor — (—). See Table 4.3
$\mu$ = Viscosity of liquid, centipoise (mNs/m$^2$)
$g$ = Gravitational acceleration = 9·81 m/s$^2$ (32·2 ft/s$^2$)

FIG. 4.29. Generalised pressure drop correlation

75 mm diameter and grids of pitch less than 50 mm, and $3\cdot3 \times 10^{-5}$ m$^3$/s m for larger packings. Other workers have quoted lower values. If the minimum wetting rate cannot be achieved in normal operation, then flooding the column at intervals will help to wet the packing. Alternatively, part of the liquid may be recirculated but the advantage of countercurrent flow may then be lost.

The distribution of liquid over packings has been studied experimentally by many workers and, for instance, Tour and Lerman[56,57] showed that for a single point feed the distribution is given by:

$$Q_x = c \exp(-a^2 x^2) \tag{4.38}$$

where $Q_x$ is the fraction of the liquid collected at a distance $x$ from the centre and $c$ and $a$ are constants depending on the packing arrangement. Norman[58] and Manning and Cannon[59] and others have shown that this mal-distribution is one cause of falling transfer coefficients with tall towers.

Mayo et al.[60] found the wetted area of a packing by running coloured water through a packing of random rings formed from waxed paper and measuring the area dyed. The

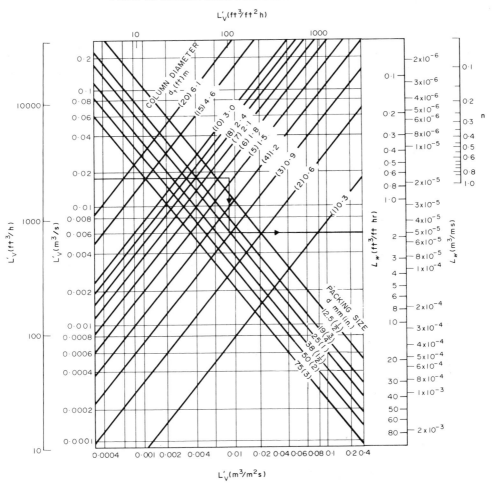

FIG. 4.30. Nomograph for the estimation of the degree of wetting in a packed column

percentage of the geometrical area wetted increased with liquid rates but fell off lower down the tower. Thus, with 75 mm rings at a depth of 0·6 m, the fraction wetted ranged from 34 to 50 per cent with increasing liquid flow. The effective wetted area for mass transfer had been found by Shulman et al.[61] by measuring the vaporisation rate of Raschig rings made from naphthalene in a current of air.

A nomograph which relates liquid rate, tower diameter and packing size is reproduced in Fig. 4.30[62]. The wetting rate $L_W$ may be obtained as an absolute value from the inner right-hand axis or as *total wetting* from the outer scale. A value of total wetting exceeding unity on that scale indicates that the packing is satisfactorily wet. This figure is discussed further in Chapter 12 and in Chapter 11 for its application to distillation in packed columns. It should be noted that many organic liquids have favourable wetting properties and wetting may be effective at much lower rates, though materials such as plastics and polished stainless steel are difficult to wet. Figure 4.30 does, however, represent the best available data on the subject of wetting.

*Hold-up*

In many instances in the industrial application of packed columns it is desirable to know the volumetric hold-up of the liquid phase in the column. This information might be needed, for example, if the liquid were involved in a chemical reaction or if a control system

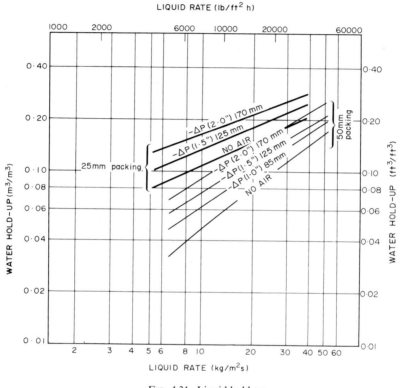

Fig. 4.31. Liquid hold-up
ceramic Super Intalox saddles
(Pressure drop in in. w.g./ft or mm w.g./m)

for the column were being designed. For gas–liquid systems the hold-up of liquid $H_w$ for conditions below the loading point has been found[43] to vary approximately as the 0·6 power of the liquid rate and for rings and saddles is given approximately by:

$$H_w = 0.143 \left(\frac{L'}{d}\right)^{0.6} \tag{4.39}$$

where   $L'$ is the liquid flowrate in kg/m² s,
        $d$ is the equivalent diameter of the packing in mm, and
        $H_w$ is the hold-up in m³ liquid per m³ of column.
Thus with 25 mm Raschig rings, $L'$ of 1·0 kg/m² s and $d = 20$ mm, $H_w$ has a value of 0·021 m³/m³.

Manufacturers' data are often available on hold-up for specific packings and typical figures are shown in Fig. 4.31 for 25 mm and 50 mm saddles.

*Economic Design of Packed Columns*

In designing industrial scale packed columns a balance must be made between the capital cost of the column and ancillary equipment on the one side and the running costs on the other. Generally, reducing the diameter of the column will reduce the capital cost but will increase the cost of pumping the gas through the column due to the increased pressure drop.

For columns operating at atmospheric and sub-atmospheric pressure and where the mass transfer rate is controlled by transfer through the gas film, Morris and Jackson[50] have calculated ranges of economic gas velocities for various packings under specified conditions. In selecting a gas velocity, and hence column cross-sectional area, it is necessary to check that the liquid rate is above the minimum wetting rate as discussed in the previous section. Selection of the appropriate packing will help in achieving the minimum wetting rate. The loading condition should also be calculated to ensure that the column would not be operating above this condition.

For columns operating at high pressures, the capital cost of the column shell becomes more significant and it may be more economic to operate at gas velocities above the loading conditions. Morris and Jackson[50] suggest a gas velocity about 75–80 per cent of the flooding velocity for normal systems, and less than 40 per cent of the flooding rate if foaming is likely to occur. The height of the column would have to be taken into consideration in making an economically optimum design. The height is usually determined by the mass transfer duty of the columns and mass transfer rates per unit height of packing. Mass transfer in packed columns is discussed in Chapter 11.

*Vacuum Columns*

Sawistowski[9] has shown that the curve in Fig. 4.28 can be converted to a straight line by plotting:

$$\ln\left\{\ln\left[\frac{u_G^2 S_B}{ge^3}\left(\frac{\rho_G}{\rho_L}\right)\left(\frac{\mu_L}{\mu_w}\right)^{0\cdot2}\right]\right\} \text{ against } \ln\left\{\frac{L}{G}\sqrt{\frac{\rho_G}{\rho_L}}\right\}$$

The equation of the curve is thus found to be:

$$\ln\left\{\frac{u_G^2 S_B}{ge^3}\left(\frac{\rho_G}{\rho_L}\right)\left(\frac{\mu_L}{\mu_w}\right)^{0\cdot2}\right\} = -4\left(\frac{L}{G}\right)^{1/4}\left(\frac{\rho_G}{\rho_L}\right)^{1/8} \tag{4.40}$$

When a column is operating under reduced pressure and the pressure drop is of the same order of magnitude as the absolute pressure, it is not immediately obvious whether the onset of flooding will be determined by conditions at the top or the bottom of the column. If $G_F$ is the gas flowrate under flooding conditions in the column:

$$G_F = u_G \rho_G A \tag{4.41}$$

Substituting in equation 4.40:

$$\ln\left\{\frac{G_F^2 S_B}{A^2 ge^3}\frac{1}{\rho_G \rho_L}\left(\frac{\mu_L}{\mu_w}\right)^{0\cdot2}\right\} = -4\left(\frac{L}{G}\right)^{1/4}\left(\frac{\rho_G}{\rho_L}\right)^{1/8} \tag{4.42}$$

For a column operating at a given reflux ratio, $L/G$ is constant and the only variables over the length of the column are, now, the minimum flooding rate $G_F$ and the gas density $\rho_G$. In order to find the condition for a minimum or maximum value of $G_F \, \mathrm{d}(G_F^2)/\mathrm{d}\rho_G$ is obtained from equation 4.42 and equated to zero. Thus:

$$\left(\frac{G_F^2}{\rho_G'}\right)^{-1} \left\{ G_F^2 (-\rho_G)^{-2} + \rho_G^{-1} \frac{\mathrm{d}G_F^2}{\mathrm{d}\rho_G} \right\} = -4 \left(\frac{L}{G}\right)^{1/4} \rho_L^{-1/8} (\tfrac{1}{8}\rho_G^{-7/8}) \tag{4.43}$$

This gives $L/G \sqrt{\rho_G/\rho_L} = 16$, when $\mathrm{d}G_F^2/\mathrm{d}\rho_G = 0$. As the second differential coefficient is negative at this point, $G_F^2$ is a maximum.

A value of $L/G \sqrt{\rho_G/\rho_L} = 16$ is well in excess of the normal operating range of the column (especially of a distillation column operating at reduced pressure) as seen in Fig. 4.28. Thus, over the whole operating range of a column, the value of $G'$ which just gives rise to flooding increases with gas density and hence with the absolute pressure. The tendency for a column to flood will always be greater therefore at the low pressure end, i.e. at the top.

Calculation of the pressure drop and flooding rates is particularly important for vacuum columns, in which the pressure may increase severalfold from the top to the bottom of the column. When a heat-sensitive liquid is distilled, the maximum temperature, and hence pressure, at the bottom of the column is limited and hence the vapour rate must not exceed a certain value. In a vacuum column, the throughput is very low because of the high specific volume of the vapour, and the liquid reflux rate is generally so low that the liquid flow has little effect on the pressure drop. The pressure drop can be calculated by applying equation 4.13 over a differential height and integrating. Thus:

$$-\frac{\mathrm{d}P}{\mathrm{d}l} = \left(\frac{R_1}{\rho u_1^2}\right) S \rho_G u^2 \frac{1-e}{e^3} \tag{4.44}$$

Writing $G' = \rho_G u$ and $P/\rho_G = P_0/\rho_0$ for isothermal operation, where $\rho_0$ is the vapour density at some arbitrary pressure $P_0$:

$$-\frac{\mathrm{d}P}{\mathrm{d}l} = \left(\frac{R_1}{\rho u_1^2}\right) S \frac{(1-e)}{e^3} \frac{G'^2}{P} \frac{P_0}{\rho_0} \tag{4.45}$$

The Reynolds number, and hence $R_1/\rho u_1^2$, will remain approximately constant over the column.

Integrating,

$$P_1^2 - P_2^2 = 2 \left(\frac{R_1}{\rho u_1^2}\right) \frac{(1-e)}{e^3} \frac{SG'^2 P_0}{\rho_0} l \tag{4.46}$$

It will be noted that when the pressure at the top of the column is small compared with that at the bottom, the pressure drop is directly proportional to the vapour rate. Use may also be made of the generalised pressure-drop correlation to calculate the required pressure drop as shown in Example 4.1.

## Example 4.1

A column 0·6 m diameter and 4 m tall, packed with 25 mm ceramic Raschig rings, is used in a gas absorption process carried out at atmospheric pressure and at 293 K. If the liquid and gas can be assumed to have the properties of water and air, and their flowrates are 6·5 and 0·5 kg/m² s respectively, what will be the pressure drop across the column? How much can the liquid rate be increased before the column floods?

**Solution**

*Properties of Gas*

Density $= (29/22\cdot4) \times (273/293) = 1\cdot206\,\text{kg/m}^3$.
Viscosity $= 0\cdot018\,\text{mN s/m}^2 = 1\cdot8 \times 10^{-5}\,\text{N s/m}^2$.
Kinematic viscosity $= 1\cdot8 \times 10^{-5}/1\cdot206 = 1\cdot49 \times 10^{-5}\,\text{m}^2/\text{s}$.
Gas velocity (superficial) $= 0\cdot5/1\cdot206 = 0\cdot415\,\text{m/s}$.

*Pressure Drop over Packing*

Carman's method:

Using equation 4.17, $R_1/\rho u_1^2 = 5Re_1^{-1} + Re_1^{-0\cdot1}$

$$Re_1 = \frac{u\rho}{S\mu(1-e)} = \frac{0\cdot415}{1\cdot49 \times 10^{-5} \times 190} = 147 \quad \text{(Table 4.3 gives } S_B = S(1-e) = 190\,\text{m}^{-1})$$

Thus
$$\frac{R_1}{\rho u_1^2} = \frac{5}{147} + \frac{1\cdot0}{147^{0\cdot1}} = 0\cdot0339 + 0\cdot607 = 0\cdot641$$

Thus
$$\frac{-\Delta P e^3}{l S(1-e)\rho u^2} = 0\cdot641$$

i.e.
$$-\Delta P = \frac{0\cdot641 \times 4 \times 190 \times 1\cdot206 \times 0\cdot415^2}{0\cdot71^3} \quad (e = 0\cdot71.\ \text{Table 4.3})$$
$$= 282\cdot7\,\text{N/m}^2$$

The above value of pressure drop must be multiplied by $1\cdot8$ to allow for the effect of the liquid flow (Fig. 4.23).
Thus, value of pressure drop over irrigated packing is
$$282\cdot7 \times 1\cdot8 = 509\,\text{N/m}^2$$

Morris and Jackson's method:

$$\text{Wetting rate } L_W = \frac{u_L}{S_B} = \frac{6\cdot5}{1000} \cdot \frac{1}{190} = 0\cdot000034\,\text{m}^3/\text{s m}$$

From Fig. 4.25, number of velocity heads lost $= N = 1445$.
Using equation 4.36:
$$-\Delta P = \tfrac{1}{2}N\rho_G u_G^2 l$$
$$= \tfrac{1}{2} \times 1445 \times 1\cdot206 \times 0\cdot415^2 \times 4 = 600\cdot3\,\text{N/m}^2.$$

Using the generalised pressure-drop correlation (Fig. 4.29):
$$\rho_L = 1000\,\text{kg/m}^3$$
$$G' = 0\cdot5\,\text{kg/m}^2\,\text{s}$$
$$F = 160 \quad \text{(Table 4.3)}$$

$$\therefore \quad \frac{G'^2 F\mu^{0\cdot1}}{\rho_G(\rho_L - \rho_G)g} = \frac{0\cdot5^2 \times 160 \times 1\cdot0^{0\cdot1}}{1\cdot206(1000 - 1\cdot206)9\cdot81} = 0\cdot0034$$

$$\frac{L'}{G'}\sqrt{\frac{\rho_G}{\rho_L - \rho_G}} = \frac{6\cdot5}{0\cdot5}\sqrt{\frac{1\cdot206}{1000 - 1\cdot206}} = 0\cdot434$$

From Fig. 4.29, pressure drop $= 15\,\text{mm water/m of packing}$

$$\therefore \quad -\Delta P = 15 \times 4 = 60\,\text{mm H}_2\text{O} \equiv 588\,\text{N/m}^2$$

*Flooding Condition*

$$\frac{u_G^2 S_B}{g e^3} \frac{\rho_G}{\rho_L}\left(\frac{\mu_L}{\mu_w}\right)^{0\cdot2} = \frac{0\cdot415^2 \times 190 \times 1\cdot206}{9\cdot81 \times 0\cdot71^3 \times 1000} = 0\cdot112$$

$$\ln\left[\frac{u_G^2 S_B}{g e^3}\frac{\rho_G}{\rho_L}\left(\frac{\mu_L}{\mu_w}\right)^{0\cdot2}\right] = -4\cdot49$$

From equation 4.40:

$$-4\cdot49 = -4\left(\frac{L}{G}\right)^{1/4}\left(\frac{\rho_G}{\rho_L}\right)^{1/8}$$

$$\therefore \quad \frac{L}{G} = (1\cdot123)^4 \sqrt{\frac{1000}{1\cdot206}} = \frac{L'}{G'}$$

$$\therefore \quad L' = 0\cdot5 \times 1\cdot59 \times 28\cdot8 = 22\cdot9\,\text{kg/m}^2\,\text{s}$$

## 4.6. FURTHER READING

BACKHURST, J. R. and HARKER, J. H.: *Process Plant Design* (Heinemann Educational Books, London, 1973).
LEVA, M.: *Tower Packings and Packed Tower Design* (U.S. Stoneware Co., 1958).
MORRIS, G. A. and JACKSON, J.: *Absorption Towers* (Butterworths, 1953).
NORMAN, W. S.: *Absorption, Distillation and Cooling Towers* (Longmans, London, 1961).

## 4.7. REFERENCES

1. DARCY, H. P. G.: *Les Fontaines publiques de la ville de Dijon. Exposition et application à suivre et des formules à employer dans les questions de distribution d'eau* (Victor Dalamont, 1856).
2. CREMER, H. W. and DAVIES, T.: *Chemical Engineering Practice*, Vol. 2 (Butterworths, 1956).
3. BRINKMAN, H. C.: *Appl. Scient. Res.* **1A** (1948) 81–86. On the permeability of media consisting of closely packed porous particles.
4. DUPUIT, A. J. E. J.: *Études théoriques et pratiques sur le mouvement des eaux* (1863).
5. KOZENY, J.: *Sber. Akad. Wiss. Wien* (Abt. IIa) **136** (1927) 271–306. Über kapillare Leitung des Wassers im Boden (Aufstieg, Versicherung, und Anwendung auf die Bewässerung).
6. KOZENY, J.: *Z. Pfl.-Ernähr. Düng. Bodenk.* **28A** (1933) 54–56. Über Bodendurchlässigkeit.
7. CARMAN, P. C.: *Trans. Inst. Chem. Eng.* **15** (1937) 150–66. Fluid flow through granular beds.
8. FORCHHEIMER, P.: *Hydraulik* (Teubner, 1930).
9. SAWISTOWSKI, H.: *Chem. Eng. Sci.* **6** (1957) 138. Flooding velocities in packed columns operating at reduced pressures.
10. ERGUN, S.: *Chem. Eng. Prog.* **48** (1952) 89–94. Fluid flow through packed columns.
11. CARMAN, P. C.: *Flow of Gases Through Porous Media* (Butterworths, 1956).
12. COULSON, J. M.: *Trans. Inst. Chem. Eng.* **27** (1949) 237–57. The flow of fluids through granular beds; effect of particle shape and voids in streamline flow.
13. WYLLIE, M. R. J. and GREGORY, K. R.: *Ind. Eng. Chem.* **47** (1955) 1379–88. Fluid flow through unconsolidated porous aggregates—effect of porosity and particle shape on Kozeny–Carman constants.
14. KIHN, E.: University of London, Ph.D. Thesis (1939). Streamline flow of fluids through beds of granular materials.
15. PIRIE, J. M.: in discussion of COULSON[12].
16. MUSKAT, M. and BOTSET, H. G.: *Physics* **1** (1931) 27–47. Flow of gas through porous material.
17. LORD, E.: *J. Text. Inst.* **46** (1951) T191. Air flow through plugs of textile fibres. Part I. General flow relations.

18. LEA, F. M. and NURSE, R. W.: *Trans. Inst. Chem. Eng.* **25** (1947). Supplement: Symposium on Particle Size Analysis, pp. 47–63. Permeability methods of fineness measurement.
19. CARMAN, P. C. and MALHERBE, P. LE R.: *J. Soc. Chem. Ind. Trans.* **69** (1950) 134T–143T. Routine measurement of surface of paint pigments and other fine powders. I.
20. GUNN, D. J.: *Chem. Engnr.* **219** (1968) CE153. Mixing in packed and fluidised beds.
21. EDWARDS, M. F. and RICHARDSON, J. F.: *Chem. Eng. Sci.* **22** (1968) 109. Gas dispersion in packed beds.
22. GUNN, D. J. and PRYCE, C.: *Trans. Inst. Chem. Eng.* **47** (1969) T341. Dispersion in packed beds.
23. ARIS, R. and AMUNDSON, N. R.: *A.I.Ch.E.Jl.* **3** (1957) 280. Some remarks on longitudinal mixing or diffusion in fixed beds.
24. PRAUSNITZ, J. M.: *A.I.Ch.E.Jl.* **4** (1958) 14M. Longitudinal dispersion in a packed bed.
25. BLACKWELL, R. J., RAYNE, J. R., and TERRY, M. W.: *J. Petrol. Technol.* **11** (1959) 1–8. Factors influencing the efficiency of miscible displacement.
26. CARBERRY, J. J. and BRETTON, R. H.: *A.I.Ch.E.Jl.* **4** (1958) 367. Axial dispersion of mass in flow through fixed beds.
27. DE MARIA, F. and WHITE, R. R.: *A.I.Ch.E.Jl.* **6** (1960) 473. Transient response study of gas flowing through irrigated packing.
28. McHENRY, K. W. and WILHELM, R. H.: *A.I.Ch.E.Jl.* **3** (1957) 83. Axial mixing of binary gas mixtures flowing in a random bed of spheres.
29. SINCLAIR, R. J. and POTTER, O. E.: *Trans. I. Chem. E.* **43** (1965) 3. Dispersion of gas in flow through a bed of packed solids.
30. EVANS, E. V. and KENNEY, C. N.: *Trans. I. Chem. E.* **44** (1966) T189. Gaseous dispersion in packed beds at low Reynolds numbers.
31. CAIRNS, E. J. and PRAUSNITZ, J. M.: *Chem. Eng. Sci.* **12** (1960) 20. Longitudinal mixing in packed beds.
32. DAVIES, C. N.: in discussion of HUTCHISON, H. P., NIXON, I. S., and DENBIGH, K. G.: *Discns. Faraday Soc.* **3** (1948) 86–129. The thermosis of liquids through porous materials.
33. EBACH, E. E. and WHITE, R. R.: *A.I.Ch.E. Jl.* **4** (1958) 161. Mixing of fluids flowing through beds of packed solids.
34. HIBY, J. W.: *Proceedings of the Symposium on Interaction between Fluids and Particles. I. Chem. E. London* (1962) 312. Longitudinal and transverse mixing during single phase flow through granular beds.
35. LILES, A. W. and GEANKOPLIS, C. J.: *A.I.Ch.E.Jl.* **6** (1960) 591. Axial diffusion of liquids in packed beds and end effects.
36. KRAMERS, H.: *Physica, 's Grav.* **12** (1946) 61. Heat transfer from spheres to flowing media.
37. RANZ, W. E. and MARSHALL, W. R.: *Chem. Eng. Prog.* **48** (1952) 141, 173. Evaporation from drops.
38. GUPTA, A. S. and THODOS, G.: *A.I.Ch.E.Jl.* **9** (1963) 751. Direct analogy between mass and heat transfer to beds of spheres.
39. ZENZ, F. A. and OTHMER, D. F.: *Fluidization and Fluid-particle Systems* (Reinhold, 1960).
40. BROWNELL, L. E. and YOUNG, E. H.: *Process Equipment Design, Vessel Design* (Chapman & Hall, 1959).
41. MOLYNEUX, F.: *Chemical Plant Design*, Vol. 1 (Butterworths, 1963).
42. B.S. 1500 Pt. 1. *Fusion Welded Pressure Vessels* (British Standards Institution, London, 1958).
43. LEVA, M.: *Tower Packings and Packed Tower Design* (U.S. Stoneware Co., 1953).
44. Norton Chemical Process Products Div., Box 350, Akron, Ohio; Hydronyl Ltd., King St., Fenton, Stoke-on-Trent, U.K.
45. ECKERT, J. S.: *Chem. Eng. Prog.* **57**, No. 9 (1961) 54. Design techniques for designing packed towers.
46. CHIPPERFIELD, P. N. J.: Preprint of paper presented at 5th Effluent and Water Treatment Convention, March 14–17, 1967 (1967). The development, use and future of plastics in biological treatment.
47. ASKEW, M. W.: *Proc. Biochem.* **1** (1966) 483 and **2** (1967) 31. Aspects of the use of modern materials in the treatment of sewage and industrial wastes.
48. ROSE, H. E. and YOUNG, P. H.: *Proc. Inst. Mech. Eng.* **1B** (1952) 114. Hydraulic characteristics of packed towers operating under countercurrent flow conditions.
49. SHERWOOD, T. K. and PIGFORD, R. L.: *Absorption and Extraction* (McGraw-Hill, 1952).
50. MORRIS, G. A. and JACKSON, J.: *Absorption Towers* (Butterworths, 1953).
51. LEVA, M.: *Chem. Eng. Prog.* Symp. Ser. No. 10, **50** (1954) 51–59. Flow through irrigated dumped packings. Pressure drop, loading, flooding.
52. ECKERT, J. S., FOOTE, E. H., and HUNTINGTON, R. L.: *Chem. Eng. Prog.* **54**, No. 1 (Jan. 1958) 70–5. Pall rings—new type of tower packing.
53. SHERWOOD, T. K., SHIPLEY, G. H., and HOLLOWAY, F. A. L.: *Ind. Eng. Chem.* **30** (1938) 765–9. Flooding velocities in packed columns.
54. LOBO, W. E., FRIEND, L., HASHMALL, F., and ZENZ, F.: *Trans. Am. Inst. Chem. Eng.* **41** (1945) 693–710. Limiting capacity of dumped tower packings.
55. ECKERT, J. S.: *Chem. Eng. Prog.* **59**, No. 5 (1963) 76. Tower packings—comparative performance.
56. TOUR, R. S. and LERMAN, F.: *Trans. Am. Inst. Chem. Eng.* **35** (1939) 709–18. An improved device to demonstrate the laws of frequency distribution, with special reference to liquid flow in packed towers.

57. Tour, R. S. and Lerman, F.: *Trans. Am. Inst. Chem. Eng.* **35** (1939) 719–42. The unconfined distribution of liquid in tower packing.
58. Norman, W. S.: *Trans. Inst. Chem. Eng.* **29** (1951) 226–39. The performance of grid-packed towers.
59. Manning, R. E. and Cannon, M. R.: *Ind Eng. Chem.* **49** (1957) 347–9. Distillation improvement by control of phase channelling in packed columns.
60. Mayo, F., Hunter, T. G., and Nash, A. W.: *J. Soc. Chem. Ind. Trans.* **54** (1935) 375T–385T. Wetted surface in ring-packed towers.
61. Shulman, H. L., Ullrich, C. F., Proulx, A. Z., and Zimmerman, J. O.: *A.I.Ch.E.Jl.* **1** (1955) 253–8. Performance of packed columns. II. Wetted and effective-interfacial areas, gas- and liquid-phase mass transfer rates.
62. Tower Packings, Hydronyl Limited, Stoke-on-Trent, England.

# 4.8. NOMENCLATURE

| | | Units in SI System | Dimensions in **MLTθ** |
|---|---|---|---|
| $A$ | Total cross-sectional area of bed or column | $m^2$ | $L^2$ |
| $B$ | Permeability coefficient (equation 4.2) | $m^2$ | $L^2$ |
| $C$ | Concentration | $kg/m^3$ | $ML^{-3}$ |
| $C_p$ | Specific heat at constant pressure | $J/kg\,K$ | $L^2T^{-2}\theta^{-1}$ |
| $D$ | Molecular diffusivity | $m^2/s$ | $L^2T^{-1}$ |
| $D_L$ | Axial dispersion coefficient | $m^2/s$ | $L^2T^{-1}$ |
| $D_R$ | Radial dispersion coefficient | $m^2/s$ | $L^2T^{-1}$ |
| $d$ | Diameter of particle | $m$ | $L$ |
| $d_t$ | Diameter of tube or column | $m$ | $L$ |
| $d'_m$ | Equivalent diameter of pore space $= e/S_B$ as used by Kozeny | $m$ | $L$ |
| $d_n$ | Nominal packing size (e.g. diameter for a Raschig ring) | $m$ | $L$ |
| $e$ | Fractional voidage of bed of particles or packing | — | — |
| $F$ | Packing factor | — | — |
| $f_w$ | Wall correction factor (equation 4.21) | — | — |
| $G$ | Gas mass flowrate | $kg/s$ | $MT^{-1}$ |
| $G_F$ | Gas mass flowrate under flooding conditions | $kg/s$ | $MT^{-1}$ |
| $G'$ | Gas mass velocity | $kg/s\,m^2$ | $ML^{-2}T^{-1}$ |
| $g$ | Acceleration due to gravity | $m^2/s$ | $LT^{-2}$ |
| $h$ | Heat transfer coefficient | $W/m^2\,K$ | $MT^{-3}\theta^{-1}$ |
| $h_D$ | Mass transfer coefficient | $m/s$ | $LT^{-1}$ |
| $H_w$ | Liquid hold-up in bed, volume of liquid per unit volume of bed | — | — |
| $j_d$ | $j$-factor for mass transfer | — | — |
| $j_h$ | $j$-factor for heat transfer | — | — |
| $K$ | Constant in flow equations 4.1 and 4.2 | $m^3\,s/kg$ | $M^{-1}L^3T$ |
| $K'$ | Dimensionless constant in equation 4.6 | — | — |
| $K''$ | Kozeny constant in equation 4.9 | — | — |
| $K_0$ | Shape factor in equation 4.20 | — | — |
| $k$ | Thermal conductivity | $W/m\,K$ | $MLT^{-3}\theta^{-1}$ |
| $l$ | Length of bed or height of column packing | $m$ | $L$ |
| $l'$ | Length of flow passage through bed | $m$ | $L$ |
| $l_t$ | Length of circular tube | $m$ | $L$ |
| $L$ | Liquid mass flowrate | $m$ | $MT^{-1}$ |
| $L'$ | Liquid mass velocity | $m$ | $ML^{-2}T^{-1}$ |
| $L_v$ | Volumetric liquid rate | $m^3/s$ | $L^3T^{-1}$ |
| $L'_v$ | Volumetric liquid rate per unit area | $m/s$ | $LT^{-1}$ |
| $L_W$ | Wetting rate ($u_L/S_B$) | $m^2/s$ | $L^2T^{-1}$ |
| $N$ | Number of velocity heads lost through unit height of bed (equation 4.36) | $1/m$ | $L^{-1}$ |
| $P$ | Pressure | $N/m^2$ | $ML^{-1}T^{-2}$ |
| $-\Delta P$ | Pressure drop across bed or column | $N/m^2$ | $ML^{-1}T^{-2}$ |
| $-\Delta P_d$ | Pressure drop across bed of dry packing | $N/m^2$ | $ML^{-1}T^{-2}$ |
| $-\Delta P_w$ | Pressure drop across bed of wet packing | $N/m^2$ | $ML^{-1}T^{-2}$ |
| $Q_x$ | Fraction of liquid collected at distance $x$ from centre line of packing in equation 4.38 | — | — |
| $R$ | Drag force per unit area of tube wall | $N/m^2$ | $ML^{-1}T^{-2}$ |

|  |  | Units in SI System | Dimensions in $\mathbf{MLT\theta}$ |
|---|---|---|---|
| $R_1$ | Drag force per unit area of particles | N/m$^2$ | $\mathbf{ML^{-1}T^{-2}}$ |
| $r$ | Radius | m | $\mathbf{L}$ |
| $S$ | Surface area per unit volume of particle or packing | m$^2$/m$^3$ | $\mathbf{L^{-1}}$ |
| $S_B$ | Surface area per unit volume of bed (specific surface) | m$^2$/m$^3$ | $\mathbf{L^{-1}}$ |
| $S_c$ | Surface area of container per unit volume of bed | m$^2$/m$^3$ | $\mathbf{L^{-1}}$ |
| $t$ | Time | s | $\mathbf{T}$ |
| $u$ | Average fluid velocity based on cross-sectional area $A$ of empty column | m/s | $\mathbf{LT^{-1}}$ |
| $u_t$ | Mean velocity of fluid in tube | m/s | $\mathbf{LT^{-1}}$ |
| $u_1$ | Mean velocity in pore channel | m/s | $\mathbf{LT^{-1}}$ |
| $V$ | Volume of fluid flowing in time $t$ | m$^3$ | $\mathbf{L^3}$ |
| $X$ | Side of cube | m | $\mathbf{L}$ |
| $x$ | Distance | m | $\mathbf{L}$ |
| $\gamma$ | Coefficient of $D$ in equation 4.26 | — | — |
| $\mu$ | Fluid viscosity | N s/m$^2$ | $\mathbf{ML^{-1}T^{-1}}$ |
| $\rho$ | Density of fluid | kg/m$^3$ | $\mathbf{ML^{-3}}$ |
| $\psi$ | Density correction factor $\sqrt{(\rho_G/\rho_A)}$ | — | — |
| $Gr'$ | Grashof number (particle) (see Volume 1, Chapter 7) | — | — |
| $Nu$ | Nusselt number for tube wall $(hd_t/k)$ | — | — |
| $Nu'$ | Nusselt number (particle) $(hd/k)$ | — | — |
| $Pe$ | Peclet number $(ud/eD_L)$ or $(ud/eD_R)$ | — | — |
| $Pr$ | Prandtl number $(C_p\mu/k)$ | — | — |
| $Re$ | Reynolds number for flow through tube $(ud_t\rho/\mu)$ | — | — |
| $Re_1$ | Modified Reynolds number based on pore size as used by Carman (equation 4.11) | — | — |
| $Re'$ | Modified Reynolds number based on particle size $(ud\rho/\mu)$ | — | — |
| $Sc$ | Schmidt number | — | — |
| $Sh'$ | Sherwood number (particle) $(h_D d/D)$ | — | — |
| $St'$ | Stanton number (particle) $(h/C_p\rho u)$ | — | — |
| Suffix | $A$ refers to air at 293 K | — | — |
|  | $G$ refers to gas |  |  |
|  | $L$ refers to liquid |  |  |
|  | $w$ refers to water at 293 K |  |  |
|  | 0 refers to standard conditions |  |  |

# CHAPTER 5

# *Sedimentation*

## 5.1. INTRODUCTION

In Chapter 3 consideration has been given to the forces acting on an isolated particle moving relative to a fluid and it has been seen that the frictional drag can be expressed in terms of a friction factor which is in turn a function of the particle Reynolds number. If the particle is settling in the gravitational field, it rapidly reaches its terminal falling velocity when the frictional force has become equal to the net gravitational force. In a centrifugal field the particle may reach a very much higher velocity because the centrifugal force may be many thousands of times greater than the gravitational force.

In practice, the concentrations of suspensions used in industry will usually be high enough for there to be significant interaction between the particles, and the frictional force exerted at a given velocity of the particles relative to the fluid may be greatly increased as a result of modifications of the flow pattern, and *hindered settling* takes place. As a corollary, the sedimentation rate of a particle in a concentrated suspension may be considerably less than its terminal falling velocity under *free settling* conditions when the effects of mutual interference are negligible. In this chapter the behaviour of concentrated suspensions in a gravitational and a centrifugal field will be discussed and the equipment used industrially for concentrating or *thickening* such suspensions will be described.

It is important to note that suspensions of fine particles tend to behave rather differently from coarse suspensions in that a high degree of flocculation may occur as a result of the very high specific surface of the particles. For this reason fine and coarse suspensions are considered separately and the factors giving rise to flocculation are discussed in the final section of the chapter.

Although the sedimentation rate of particles tends to increase steadily as the concentration of the suspension is increased, it has been shown by Kaye and Boardman[1] that particles in very dilute suspensions may settle at velocities up to 1·5 times the normal terminal falling velocities, due to the formation of clusters of particles which settle in well-defined streams. This effect is important when particle size is determined by a method involving the measurement of the settling velocity of particles in dilute concentration, but is not significant with concentrated suspensions.

## 5.2. GRAVITATIONAL SEDIMENTATION

### 5.2.1. Fine Suspensions

The sedimentation of metallurgical slimes has been studied by Coe and Clevenger[2], who concluded that a concentrated suspension may settle in one of two different ways. In the first, after an initial brief acceleration period, the interface between the clear liquid and the suspension moves downwards at a constant rate and a layer of sediment builds up at the bottom of the container. When this interface approaches the layer of sediment, its rate of fall decreases until the "critical settling point" is reached when a direct interface is formed between the sediment and the clear liquid. Further sedimentation then results solely from a consolidation of the sediment, with liquid being forced upwards around the solids which are then forming a loose bed with the particles in contact with one another. Since the flow area is gradually being reduced, the rate progressively diminishes. In Fig. 5.1*a*, a stage in the sedimentation process is illustrated. *A* is clear liquid, *B* is suspension of the original concentration, *C* is a layer through which the concentration gradually

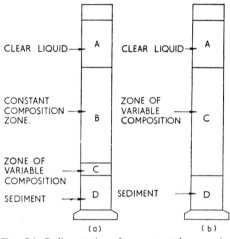

FIG. 5.1. Sedimentation of concentrated suspensions.
(*a*) Type 1 settling, (*b*) Type 2 settling

increases, and *D* is sediment. The sedimentation rate remains constant until the upper interface corresponds with the top of zone *C* and it then falls until the critical settling point is reached when both zones *B* and *C* will have disappeared. A second and rather less common mode of sedimentation (Fig. 5.1*b*) is obtained when the range of particle size is very great. The sedimentation rate progressively decreases throughout the whole operation because there is no zone of constant composition, and zone *C* extends from the top interface to the layer of sediment.

The main reasons for the modification of the settling rate of particles in a concentrated suspension are as follows:

(a) The large particles are settling relative to a suspension of smaller ones so that the effective density and viscosity of the fluid are increased.

(b) The upward velocity of the fluid displaced during settling is appreciable in a concentrated suspension and the apparent settling velocity is less than the actual velocity relative to the fluid.

(c) The velocity gradients in the fluid close to the particles are increased as a result of the change in the area and shape of the flow spaces.

(d) The smaller particles tend to be dragged downwards by the motion of the large particles and therefore accelerated.

(e) Because the particles are closer together in a concentrated suspension, flocculation is more marked in an ionised solvent and the effective size of the small particles is increased.

If the range of particle size is not more than about 6:1, a concentrated suspension settles with a sharp interface and all the particles fall at the same velocity. This is in contrast with the behaviour of a dilute suspension, for which the rates of settling of the particles can be calculated by the methods given in Chapter 3, and where the settling velocity is greater for the large particles. The two types of settling are often referred to as *sludge line settling* and *selective settling* respectively. The overall result is that in a concentrated suspension the large particles are retarded and the small ones accelerated.

A number of attempts have been made to predict the apparent settling velocity of a concentrated suspension. In 1926 Robinson[3] suggested a modification of Stokes' Law and used the density ($\rho_c$) and viscosity ($\mu_c$) of the suspension in place of the properties of the fluid.

Thus

$$u_c = \frac{K'' d^2 (\rho_s - \rho_c) g}{\mu_c} \tag{5.1}$$

where $K''$ is a constant.

The effective buoyancy force is readily calculated since:

$$\rho_s - \rho_c = \rho_s - \{\rho_s(1-e) + \rho e\} = e(\rho_s - \rho) \tag{5.2}$$

where $e$ is the voidage of the suspension.

Robinson determined the viscosity of the suspension $\mu_c$ experimentally, but it may be obtained approximately from the following formula of Einstein[4]:

$$\mu_c = \mu(1 + k''C) \tag{5.3}$$

where $k''$ is a constant for a given shape of particle (2·5 for spheres),
$\quad$ $C$ is the volumetric concentration of particles, and
$\quad$ $\mu$ is the viscosity of the fluid.

This equation holds for values of $C$ up to 0·02. For more concentrated suspensions, Vand[5] gives the equation:

$$\mu_c = \mu e^{k''C/(1 - a'C)} \tag{5.4}$$

in which $a'$ is a second constant, equal to $\frac{39}{64}$ (0·609) for spheres.

Steinour[6], who studied the sedimentation of small uniform particles, adopted a similar approach, using the viscosity of the fluid, the density of the suspension and a function of the voidage of the suspension to take account of the character of the flow spaces, and obtained an expression for the velocity of the particle relative to the fluid $u_p$ as follows:

$$u_p = \frac{d^2(\rho_s - \rho_c)g}{18\mu} f(e) \tag{5.5}$$

Since the fraction of the area available for flow of the displaced fluid is $e$, its upward velocity is $u_c(1-e)/e$ so that:

$$u_p = u_c + u_c \frac{1-e}{e} = \frac{u_c}{e} \tag{5.6}$$

From his experiments on the sedimentation of tapioca in oil Steinour has found:

$$f(e) = 10^{-1 \cdot 82(1-e)} \tag{5.7}$$

Substituting in equation 5.5, from equations 5.2, 5.6 and 5.7:

$$u_c = \frac{e^2 d^2 (\rho_s - \rho) g}{18\mu} 10^{-1 \cdot 82(1-e)} \tag{5.8}$$

Hawksley[7] also used a similar method and gave:

$$u_p = \frac{u_c}{e} = \frac{d^2(\rho_s - \rho_c)g}{18\mu_c} \tag{5.9}$$

In each of the above cases, it is correctly assumed that the upthrust acting on the particles is determined by the density of the suspension rather than that of the fluid. The use of an effective viscosity, however, is valid only for a large particle settling in a fine suspension. For the sedimentation of uniform particles the increased drag is attributable to a steepening of the velocity gradients rather than to a change in viscosity.

The rate of sedimentation of a suspension of fine particles is difficult to predict because of the large number of factors involved. Thus, for instance, the presence of an ionised solute in the liquid and the nature of the surface of the particles will affect the degree of flocculation and hence the mean size and density of the flocs. The flocculation of a suspension is usually completed quite rapidly so that it is not possible to detect an increase in the sedimentation rate in the early stages after the formation of the suspension. Most fine suspensions flocculate readily in tap water and it is generally necessary to add a deflocculating agent to maintain the particles individually dispersed; the factors involved in flocculation are discussed later. A further factor influencing the sedimentation rate is the degree of agitation of the suspension. Gentle stirring may produce accelerated settling if the suspension behaves as a non-Newtonian fluid in which the apparent viscosity is a function of the rate of shear. The change in apparent viscosity can probably be attributed to the re-orientation of the particles. The effect of stirring is, however, most marked on the consolidation of the final sediment, in which "bridge formation" by the particles can be prevented by gentle stirring. During these final stages of consolidation of the sediment, liquid is being squeezed out through a bed of particles which are gradually becoming more tightly packed.

A number of empirical equations have been obtained for the rate of sedimentation of suspensions, as a result of tests carried out in vertical tubes. For a given solid and liquid, the chief factors which affect the process are the height of the suspension, the diameter of the containing vessel, and the volumetric concentration. An attempt at bringing together the results obtained under a variety of conditions has been made by Wallis[8].

*Height of Suspension*

The height of suspension does not generally affect either the rate of sedimentation or the consistency of the sediment ultimately obtained. If, however, the position of the sludge line is plotted as a function of time for two different initial heights of slurry, curves of the form shown in Fig. 5.2 are obtained in which the ratio $OA' : OA''$ is everywhere constant. Thus, if the curve is obtained for any one initial height, the curves can be drawn for any other height.

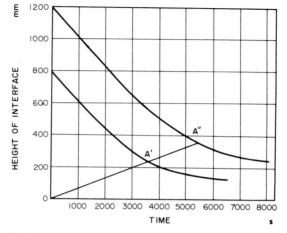

FIG. 5.2. Effect of height on sedimentation of 3 per cent (volume) suspension of calcium carbonate

## Diameter of Vessel

If the ratio of the diameter of the vessel to the diameter of the particle is greater than about 100, the walls of the container appear to have no effect on the rate of sedimentation. For smaller values, the sedimentation rate may be reduced because of the retarding influence of the walls.

## Concentration of Suspension

As already indicated, the higher the concentration, the lower is the rate of fall of the sludge line because the greater is the upward velocity of the displaced fluid and the steeper are the velocity gradients in the fluid. Typical curves for the sedimentation of a suspension of precipitated calcium carbonate in water are shown in Fig. 5.3, and in Fig. 5.4 the mass

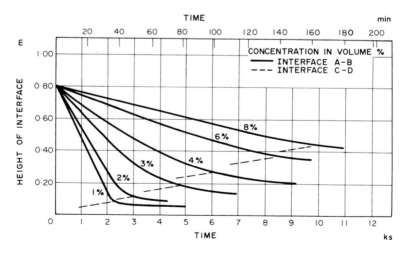

FIG. 5.3. Effect of concentration on sedimentation of calcium carbonate suspensions

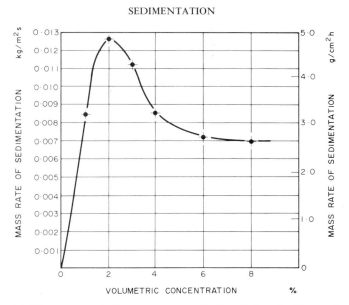

FIG. 5.4. Effect of concentration on mass rate of sedimentation of calcium carbonate

rate of sedimentation ($kg/m^2 s$) is plotted against the concentration. This curve shows a maximum corresponding to a volumetric concentration of about 2 per cent. Egolf and McCabe[9], Work and Kohler[10], and others have given empirical expressions for the rate of sedimentation at the various stages, but these are generally applicable over a narrow range of conditions and involve constants which need to be determined experimentally for each suspension.

The final consolidation of the sediment is the slowest part of the process because the displaced fluid has to flow through the small spaces between the particles. As consolidation occurs, the rate falls off because the resistance to the flow of liquid progressively increases. The porosity of the sediment is smallest at the bottom because the compressive force due to the weight of particles is greatest and because the lower portion was formed at an earlier stage in the sedimentation process. The rate of sedimentation during this period is given approximately by the expression:

$$-\frac{dH}{dt} = b(H - H_\infty)$$
(5.10)

where $H$ is the height of the sludge line at time $t$,
$H_\infty$ is the final height of the sediment, and
$b$ is constant for a given suspension.

The time taken for the sludge line to fall from a height $H_c$, corresponding to the critical settling point, to a height $H$ is given by:

$$-bt = \ln(H - H_\infty) - \ln(H_c - H_\infty)$$
(5.11)

Thus, if $\ln(H - H_\infty)$ is plotted against $t$, a straight line of slope $-b$ is obtained.

The value of $H_\infty$ is determined largely by the surface film of liquid adhering to the particles.

*Shape of Vessel*

Provided that the walls of the vessel are vertical and the cross-sectional area does not vary with depth, the shape of the vessel has little effect on the sedimentation rate. However, if parts of the walls of the vessel face downwards, as in an inclined tube, or if part of the cross-section is obstructed for a portion of the height, the effect on the sedimentation process may be considerable.

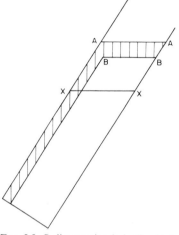

FIG. 5.5. Sedimentation in inclined tube

Pearce[11] studied the effect of a downward-facing surface by considering an inclined tube as shown in Fig. 5.5. Starting with a suspension reaching a level $AA$, suppose that the sludge line falls to a new level $BB$. Then material will tend to settle out from the whole of the shaded area. This configuration is not stable and the system tends to adjust itself so that the sludge line takes up a new level $XX$, the volume corresponding to the area $AAXX$ being equal to that corresponding to the shaded area. By applying this principle it is seen that it is possible to obtain an accelerated rate of settling in an inclined tank by inserting a series of inclined plates.

The effect of a non-uniform cross-section was considered by Robins[12], by studying the effect of reducing the area in part of the vessel by immersing a solid body, as shown in Fig. 5.6. Suppose that the cross-sectional area, sedimentation velocity, and fractional volumetric concentration are $C$, $u_c$, and $A$ below the obstruction, and $C'$, $u'_c$, and $A'$ at the horizontal level of the obstruction. If $\psi$ and $\psi'$ are the corresponding rates of deposition of solids per unit area, then:

$$A\psi = ACu_c \tag{5.12}$$

and

$$A'\psi' = A'C'u'_c \tag{5.13}$$

For continuity at the bottom of the obstruction:

$$\psi' = \frac{A}{A'}\psi \tag{5.14}$$

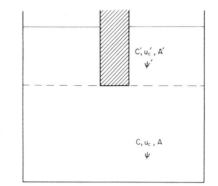

FIG. 5.6. Sedimentation in partially obstructed vessel

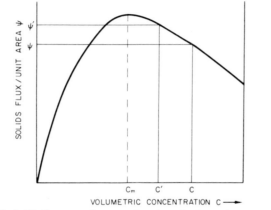

FIG. 5.7. Solids flux per unit area as function of volumetric concentration

A curve of $\psi$ versus $C$ will have the same general form as Fig. 5.4 and a typical curve is given in Fig. 5.7. If the concentration $C$ is appreciably greater than the value $C_m$ at which $\psi$ is a maximum, $C'$ will be less than $C$ and the system will be stable. On the other hand, if $C$ is less than $C_m$, $C'$ will be greater than $C$ and mixing will take place because of the greater density of the upper portion of the suspension. The range of values of $C$ for which equation 5.14 is in practice satisfied, and for which mixing currents are absent, may be very small.

### 5.2.2. Coarse Suspensions

Richardson and Zaki[13] have studied the sedimentation of uniform particles ($>100$ μm), sufficiently large for anomalous viscosity effects and flocculation to be negligible; experimental work on sedimentation and fluidisation is used to establish the effects of concentration and of the walls of the container on sedimentation rates.

The drag $R_t$ per unit projected area of spherical particle settling at its terminal falling velocity $u_0$ is a function of the density and viscosity of the fluid, the diameter of the particle, the terminal falling velocity, and the ratio of the diameter of the particle ($d$) to that of the containing vessel ($d_t$). Thus:

$$R_t = f\left(\rho, \mu, u_0, d, \frac{d}{d_t}\right) \tag{5.15}$$

For the isolated spherical particle:

$$R_t \frac{\pi}{4} d^2 = \frac{\pi}{6} d^3 (\rho_s - \rho) g \qquad (5.16)$$

For the sedimentation of a particle in a suspension, $R_t$ will alter to $R_t'$ because the upthrust is equal to the weight of displaced suspension and:

$$R_t' \frac{\pi}{4} d^2 = \frac{\pi}{6} d^3 (\rho_s - \rho_c) g = \frac{\pi}{6} d^3 e(\rho_s - \rho) g \qquad \text{(from equation 5.2)}$$

Thus:

$$R_t' = e R_t \qquad (5.17)$$

In this case, $R_t'$ will be a function of the velocity of the particle relative to the fluid $_p$, and also a function of the voidage $e$ which determines the flow pattern and the area available for the flow of the displaced fluid. Thus

$$R_t' = f(\rho, \mu, u_p, d, e, d/d_t) \qquad (5.18)$$

Rearranging equations 5.15 and 5.18 to give $u_0$ and $u_p$ explicitly:

$$u_0 = f\left(R_t, \rho, \mu, d, \frac{d}{d_t}\right) \qquad (5.19)$$

and

$$u_p = f\left(R_t', \rho, \mu, d, e, \frac{d}{d_t}\right) \qquad (5.20)$$

$$= f\left(R_t, \rho, \mu, d, e, \frac{d}{d_t}\right) \qquad (5.21)$$

since $R_t'$ is a function only of $R_t$ and $e$. Dividing:

$$\frac{u_p}{u_0} = f\left(R_t, \rho, \mu, d, e, \frac{d}{d_t}\right) \qquad (5.22)$$

Since the left-hand side of equation 5.22 is dimensionless the right-hand side must also be. The only dimensionless combinations of the variables $R_t$, $\rho$, $\mu$ and $d$ involve the group $R_t d^2 \rho / \mu^2$, and therefore:

$$\frac{u_p}{u_0} = f\left(\frac{R_t d^2 \rho}{\mu^2}, e, \frac{d}{d_t}\right) \qquad (5.23)$$

Now $(R_t d^2 \rho / \mu^2) = (R_t / \rho u_0^2)(u_0^2 d^2 \rho^2 / \mu^2)$, and since $R_t / \rho u_0^2$ is a unique function of $(u_0 d \rho / \mu)(= Re'_0)$ for a spherical particle (see Chapter 3),

$$\frac{u_p}{u_0} = f\left(Re'_0, e, \frac{d}{d_t}\right) \qquad (5.24)$$

The measured sedimentation velocity $u_c$ is given in terms of the velocity of the particles relative to the fluid by:

$$u_c = e u_p \qquad \text{(equation 5.6)}$$

Thus

$$\frac{u_c}{u_0} = f\left(Re'_0, e, \frac{d}{d_t}\right) \tag{5.25}$$

At low velocities $(Re'_0 < 0.2)$ where the drag force is attributable entirely to skin friction, from equation 3.1:

$$R_t \frac{\pi}{4} d^2 = 3\pi\mu d u_0 \tag{5.26}$$

and $R_t$ is independent of the density of the fluid.

Again when the Reynolds number $Re'_0$ exceeds 500, skin friction is negligible and:

$$R_t \propto \rho u^2 \tag{5.27}$$

$R_t$ is then independent of $\mu$ and $d$.

Neither $R_t$, $\mu$ and $d$, nor $R_t$ and $\rho$ can be arranged in the form of a dimensionless group, and therefore the ratio $u_c/u_0$ is independent of $Re'_0$, when $Re'_0$ is less than 0.2 or greater than 500, i.e.

$$\frac{u_c}{u_0} = f\left(e, \frac{d}{d_t}\right) \tag{5.28}$$

The results for the sedimentation of uniform spheres, and for the fluidisation experiments referred to in Chapter 6, were plotted as $\log u_c$ versus $\log e$ (e.g. as in Fig. 5.8) for each suspension and a linear relationship was obtained so that:

$$\log u_c = n \log e + \log u_i \tag{5.29}$$

where $\log u_i$ is the intercept corresponding to infinite dilution $(e = 1)$. Thus

$$\frac{u_c}{u_i} = e^n \tag{5.30}$$

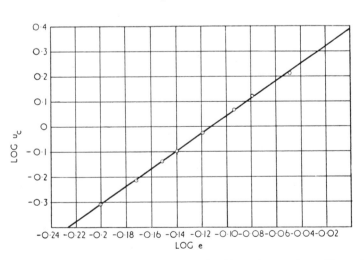

FIG. 5.8. Relation between rate of sedimentation and voidage of suspensions of 1 mm glass balls in bromoform

$u_i$ corresponds to the falling velocity of the suspension in a tube of diameter $d_t$ at infinite dilution and therefore

$$\frac{u_i}{u_0} = f\left(\frac{d}{d_t}\right) \quad \text{only} \tag{5.31}$$

Thus from equations 5.25, 5.30, and 5.31:

$$e^n = f\left(Re'_0, e, \frac{d}{d_t}\right)$$

and since $n$ is independent of $e$:

$$n = f\left(Re'_0, \frac{d}{d_t}\right) \tag{5.32}$$

For $Re'_0 < 0.2$, and for $Re'_0 > 500$, from equation 5.28:

$$n = f\left(\frac{d}{d_t}\right) \quad \text{only} \tag{5.33}$$

The experimental values of $n$ have been plotted against $d/d_t$ in Fig. 5.9 and are correlated by a family of straight lines. As expected, a single curve represents the data for all values of $Re'_0$ less than 0.2, and another single curve for all values of $Re'_0$ greater than 500. At intermediate values, $n$ is a function of $Re'_0$. From an analysis of the curves, the following equations give values of $n$:

For $\qquad\qquad 0 < Re'_0 < 0.2 \qquad n = 4.6 + 20\dfrac{d}{d_t}$ $\qquad\qquad\qquad$ (5.34)

$\qquad\qquad 0.2 < Re'_0 < 1 \qquad n = \left(4.4 + 18\dfrac{d}{d_t}\right) Re'_0{}^{-0.03}$ $\qquad$ (5.35)

$\qquad\qquad 1 < Re'_0 < 200 \qquad n = \left(4.4 + 18\dfrac{d}{d_t}\right) Re'_0{}^{-0.1}$ $\qquad$ (5.36)

$\qquad\qquad 200 < Re'_0 < 500 \qquad n = 4.4 Re'_0{}^{-0.1}$ $\qquad\qquad\qquad$ (5.37)

$\qquad\qquad Re'_0 > 500 \qquad n = 2.4$ $\qquad\qquad\qquad\qquad$ (5.38)

It was found that $u_0$ and $u_i$ were equal for the sedimentation experiments, and for the fluidisation tests (see Chapter 6), i.e.

$$\log u_0 = \log u_i + \frac{d}{d_t} \tag{5.39}$$

The calculation of the sedimentation rate in terms of the free-falling velocity therefore involves obtaining the index $n$ by inserting the values of $d/d_t$ and $Re'_0$ in the appropriate equation (5.34–5.38 inclusive) and substituting for $u_i$ in terms of $u_0$. The sedimentation velocity $u_c$ is then obtained from equation 5.30. For non-spherical particles, the index $n$, as calculated above, must be multiplied by a factor of $1.1 k'^{0.165}$, where $k'$ is the shape factor as defined in Chapter 3: this has been established only at values of $Re'_0$ greater than 500.

The use of equations 5.34 to 5.38 to calculate the value of $n$ in equation 5.30 presupposes a knowledge of $Re'_0$ which is proportional to the free-falling velocity of a single particle in the fluid. It is therefore frequently more convenient to work in terms of the Galileo number which is directly calculable from the properties of the fluid and of the particles.

For an isolated particle settling in the gravitational field, the dimensionless group $R_t/\rho u_0^2 . Re_0'^2$ is obtained as follows:

$$\frac{R_t}{\rho u_0^2} = \frac{(\pi/6)d^3(\rho_s-\rho)g}{(\pi/4)d^2 \rho u_0^2} = \frac{2}{3}dg\frac{\rho_s-\rho}{\rho u_0^2} \qquad (5.40)$$

and

$$\frac{R_t}{\rho u_0^2} . Re_0'^2 = \left[\frac{2}{3}dg\frac{\rho_s-\rho}{\rho u_0^2}\right]\left[\frac{u_0^2 d^2 \rho^2}{\mu^2}\right] = \frac{2}{3}\frac{d^3 g}{\mu^2}(\rho_s-\rho)\rho = \frac{2}{3}Ga \qquad (5.41)$$

where $Ga = (d^3 g/\mu^2)(\rho_s-\rho)\rho$ is the Galileo number.

Equations 5.34 to 5.38 can then be written:

$$0 < Ga < 3.6 \qquad n = 4.6 + 20d/d_t \qquad (5.42)$$

$$3.6 < Ga < 21 \qquad n = (4.8 + 20d/d_t)Ga^{-0.03} \qquad (5.43)$$

$$21 < Ga < 2.4 \times 10^4 \qquad n = (5.5 + 23d/d_t)Ga^{-0.075} \qquad (5.44)$$

$$2.4 \times 10^4 < Ga < 8.3 \times 10^4 \qquad n = 5.5 Ga^{-0.075} \qquad (5.45)$$

$$Ga > 8.3 \times 10^4 \qquad n = 2.4 \qquad (5.46)$$

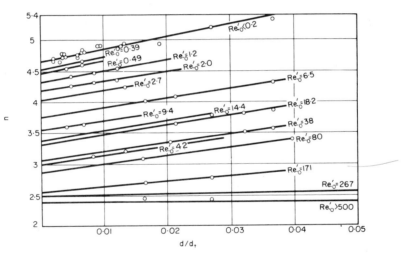

FIG. 5.9. $n$ for various values of $Re_0'$ and $d/d_t$

Garside and Al-Dibouni[14] suggest that the exponent "$n$" in equation 5.30 is most conveniently calculated using the following equation:

$$\frac{5.1 - n}{n - 2.7} = 0.1 Re_0'^{0.9} \qquad (5.47)$$

They suggest that particle interaction effects are underestimated by equation 5.30, particularly at high voidages ($e > 0.9$) and for the turbulent region, and that for a given value of voidage ($e$) it predicts too high a value for the sedimentation (or fluidisation) velocity ($u_c$). They prefer the following equation which, they claim, overall gives a higher degree of accuracy:

$$\frac{u_c/u_0 - e^{5\cdot14}}{e^{2\cdot68} - u_c/u_0} = 0\cdot06\, Re'_0 \tag{5.48}$$

However, equation 5.48 is less easy to use than equation 5.30 and it is doubtful whether it gives any significantly better accuracy, there being evidence[15] that with viscous oils equation 5.30 is more satisfactory.

### 5.2.3. Solids Flux in Batch Sedimentation

In a sedimenting suspension, the sedimentation velocity $u_c$ is a function of fractional volumetric concentration $C$, and the mass rate of sedimentation per unit area or flux $\psi$ is equal to the product $u_c C$.

Thus

$$\psi = u_c C = u_c(1-e) \tag{5.49}$$

Then, if the relation between settling velocity and concentration can be expressed in terms of a terminal falling velocity $(u_0)$ for the particles, substituting for $u_0$ using equation 5.30 gives:

$$\psi = u_0\, e^n\, (1-e) \tag{5.50}$$

From the form of the function, it is seen that $\psi$ should have a maximum at some value of $e$ lying between 0 and 1.

Now

$$\frac{d\psi}{de} = u_0\{ne^{n-1}(1-e)+e^n(-1)\} = u_0\{ne^{n-1}-(n+1)e^n\} \tag{5.51}$$

When $d\psi/de$ is zero:

$$ne^{n-1}(1-e)-e^n = 0$$

or

$$n(1-e) = e$$

and

$$e = \frac{n}{1+n} \tag{5.52}$$

If $n$ ranges from 2·4 to 4·6 as for suspensions of uniform spheres, the maximum flux should occur at a voidage between 0·71 and 0·82 (volumetric concentration 0·29 to 0·18). Furthermore, there will be a point of inflexion if $d^2\psi/de^2$ is zero for real values of $e$. Differentiating equation 5.51:

$$\frac{d^2\psi}{de^2} = u_0\{n(n-1)e^{n-2}-(n+1)ne^{n-1}\} \tag{5.53}$$

When $d^2\psi/de^2 = 0$:

$$n-1-(n+1)e = 0$$

or

$$e = \frac{n-1}{n+1} \tag{5.54}$$

If $n$ ranges from 2·4 to 4·6, there should be a point of inflexion in the curve and it will occur at values of $e$ between 0·41 and 0·65 corresponding to very high concentrations ($C = 0·59$ to 0·35). It will be noted that the point of inflexion occurs at a value of voidage $e$ below that at which the mass rate of sedimentation $\psi$ is a maximum. For coarse particles ($n \approx 2·4$), the point of inflexion is not of practical interest, since the concentration at which it occurs ($C \approx 0·59$) corresponds to a packed bed rather than a suspension.

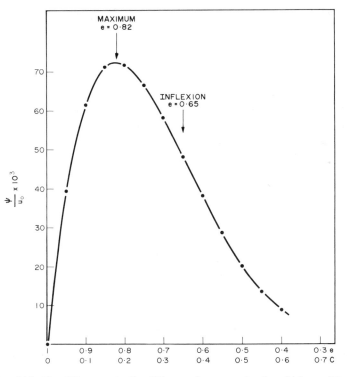

FIG. 5.10. Flux ($\psi$)–concentration ($C$) curve for suspension for which $n = 4·6$

The form of variation of flux ($\psi$) with voidage ($e$) and volumetric concentration ($C$) is shown in Fig. 5.10 for an "$n$" value of 4·6. This corresponds to the sedimentation of uniform spheres for which the free-falling velocity is given by Stokes' Law. It should be compared with Fig. 5.4 obtained for a flocculated suspension of calcium carbonate.

### 5.2.4. The Kynch Theory of Sedimentation

The behaviour of concentrated suspensions during sedimentation has been analysed by Kynch[16], largely using considerations of continuity. The basic assumptions which are made are as follows:

(a) particle concentration is uniform across any horizontal layer,
(b) wall effects can be ignored,
(c) there is no differential settling of particles as a result of differences in shape, size, or composition,

(d)  the velocity of fall of particles depends only on the local concentration of particles,

(e)  the initial concentration is either uniform or increases towards the bottom of the suspension, and

(f)  the sedimentation velocity tends to zero as the concentration approaches a limiting value corresponding to that of the sediment layer deposited at the bottom of the vessel.

If at some horizontal level where the volumetric concentration of particles is $C$ the sedimentation velocity is $u_c$, the volumetric rate of sedimentation per unit area or flux is given by:

$$\psi = Cu_c \tag{5.55}$$

Then a material balance taken between a height $H$ above the bottom at which the concentration is $C$ and the mass flux is $\psi$ and a height $H+\mathrm{d}H$ where the concentration is $C+(\partial C/\partial H)\mathrm{d}H$ and the mass flux is $\psi + (\partial\psi/\partial H)\mathrm{d}H$ gives:

$$\left\{ \left( \psi + \frac{\partial\psi}{\partial H}\mathrm{d}H \right) - \psi \right\}\mathrm{d}t = \frac{\partial}{\partial t}(C\mathrm{d}H)\,\mathrm{d}t$$

i.e.

$$\frac{\partial\psi}{\partial H} = \frac{\partial C}{\partial t} \tag{5.56}$$

Now

$$\frac{\partial\psi}{\partial H} = \frac{\partial\psi}{\partial C}\cdot\frac{\partial C}{\partial H} = \frac{\mathrm{d}\psi}{\mathrm{d}C}\cdot\frac{\partial C}{\partial H} \quad\text{(since } \psi \text{ depends only on } C) \tag{5.57}$$

Thus

$$\frac{\partial C}{\partial t} - \frac{\mathrm{d}\psi}{\mathrm{d}C}\cdot\frac{\partial C}{\partial H} = 0 \tag{5.58}$$

In general, the concentration of particles will be a function of position and time and thus:

$$C = \mathrm{f}(H, t)$$

and

$$\mathrm{d}C = \frac{\partial C}{\partial H}\mathrm{d}H + \frac{\partial C}{\partial t}\mathrm{d}t$$

Conditions of constant concentration are therefore defined by the relation:

$$\frac{\partial C}{\partial H}\mathrm{d}H + \frac{\partial C}{\partial t}\mathrm{d}t = 0$$

Thus

$$\frac{\partial C}{\partial H} = -\frac{\partial C}{\partial t}\bigg/\frac{\mathrm{d}H}{\mathrm{d}t} \tag{5.59}$$

Substituting in equation 5.58 gives the following relation for constant concentration:

$$\frac{\partial C}{\partial t} - \frac{\mathrm{d}\psi}{\mathrm{d}C}\left\{ -\frac{\partial C}{\partial t}\bigg/\frac{\mathrm{d}H}{\mathrm{d}t} \right\} = 0$$

i.e.

$$-\frac{\mathrm{d}\psi}{\mathrm{d}C} = \frac{\mathrm{d}H}{\mathrm{d}t} = u_w \tag{5.60}$$

Since equation 5.60 refers to a constant concentration, $d\psi/dC$ is constant and $u_w$ $= (dH/dt)$ is therefore also constant for any given concentration and is the velocity of propagation of a zone of constant concentration $C$. Thus lines of constant slope, on a plot of $H$ versus $t$, will refer to zones of constant composition each of which will be propagated at a constant rate, dependent only on the concentration. Then since $u_w = -(d\psi/dC)$ (equation 5.60) when $d\psi/dC$ is negative (as it is at volumetric concentrations greater than $0.18$ in Fig. 5.10), $u_w$ is positive and the wave will propagate upwards; at lower concentrations $d\psi/dC$ is positive and the wave will propagate downwards. Thus, waves originating at the base of the sedimentation column will propagate upwards to the suspension interface if $d\psi/dC$ is negative but will be prevented from propagating if $d\psi/dC$ is positive because of the presence of the base. Although Kynch's arguments can be applied to any suspension in which the initial concentration increases continuously from top to bottom, consideration will be confined to suspensions initially of uniform concentration.

In an initially uniform suspension of concentration $C_0$, the interface between the suspension and the supernatant liquid will fall at a constant rate until a zone of composition, greater than $C_0$, has propagated from the bottom to the free surface. The sedimentation rate will then fall off progressively as zones of successively greater concentrations reach the surface, until eventually sedimentation will cease when the $C_{max}$ zone reaches the surface. Now this assumes that the propagation velocity decreases progressively with increase of concentration.

However, if zones of higher concentration propagate at velocities greater than those of lower concentrations, they will automatically overtake them, giving rise to a sudden discontinuity in concentration. In particular, if the propagation velocity of the suspension of maximum possible concentration $C_{max}$ exceeds that of all the intermediate concentrations between $C_0$ and $C_{max}$, sedimentation will take place at a constant rate, corresponding to the initial uniform concentration $C_0$, and will then cease abruptly as the concentration at the interface changes from $C_0$ to $C_{max}$.

Now since the propagation velocity $u_w$ is equal to $-(d\psi/dC)$, the sedimentation behaviour will be affected by the shape of the curve of $\psi$ versus $C$. If this is consistently concave to the time-axis, $d\psi/dC$ will become increasingly negative as $C$ increases, $u_w$ will increase monotonically and consequently there will be a discontinuity because the rate of propagation of a zone of concentration $C_{max}$ exceeds that for all lower concentrations; this is the condition referred to in the previous paragraph. On the other hand, if there is a point of inflexion in the curve (as at $C = 0.35$ in Fig. 5.10), the propagation rate will increase progressively up to the condition given by this point of inflexion (concentration $C_i$) and will then decrease as the concentration is further increased. There will again be a discontinuity, but this time when the wave corresponding to concentration $C_i$ reaches the interface; then the sedimentation rate will fall off gradually as zones of successively higher concentration reach the interface, and sedimentation will finally cease when the concentration at the interface reaches $C_{max}$.

It is possible to apply this analysis to obtain the relationship between flux of solids and concentration over the range where $-(d\psi/dC)$ is decreasing with increase of concentration using the results of a single sedimentation test. For the suspension whose flux–concentration curve is given by Fig. 5.10, this condition is met at concentrations exceeding the value $C_i$ at which the point of inflexion occurs on the curve. By taking a suspension of initial concentration $C_0(\ C_i)$, it is possible to obtain the $\psi - C$ curve over the

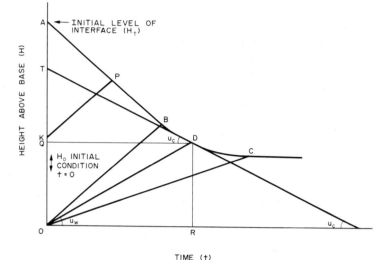

FIG. 5.11. Construction for Kynch theory

concentration range $C_0$ to $C_{max}$ from a single experiment. Figure 5.11 shows a typical sedimentation curve for such a suspension. The $H$-axis represents the initial condition $(t = 0)$ and such lines such as $KP$, $OB$ represent constant concentrations and have slopes of $u_w (= dH/dt)$. Lines from all points between $A$ and $O$ corresponding to the top and bottom of the suspension, respectively, will be parallel because the concentration is constant and their location and slope are determined by the initial concentration of the suspension. As solids become deposited at the bottom the concentration there will rapidly rise to the maximum possible value $C_{max}$ (ignoring the effects of possible sediment consolidation) and the line $OC$ represents the line of constant concentration $C_{max}$. Other lines, such as $OD$ of greater slope, all originate at the base of the suspension and correspond to intermediate concentrations.

Consider a line such as $KP$ which refers to the propagation of a wave corresponding to the initial uniform composition from an initial position $K$ in the suspension. This line terminates on the curve $ABDC$ at $P$ which is the position of the top interface of the suspension at time $t$. The location of $P$ is determined by the fact that $KP$ represents the upward propagation of a zone of constant composition at a velocity $u_w$ through which particles are falling at a sedimentation velocity $u_c$. Thus the total volume of particles passing per unit area through the plane in time $t$ is given by:

$$V = C_0(u_c + u_w)t$$

Since $P$ corresponds to the surface of the suspension, $V$ must be equal to the total volume of particles which was originally above the level indicated by $K$.

Thus

$$C_0(u_c + u_w)t = C_0(H_t - H_0)$$

or

$$(u_c + u_w)t = H_t - H_0 \tag{5.61}$$

Because the concentration of particles is initially uniform and the sedimentation rate is a function solely of the particle concentration, the line $APB$ will be straight, having a slope $(-dH/dt)$ equal to $u_c$.

After point $B$, the sedimentation curve has a decreasing negative slope reflecting the increasing concentration of solids at the interface. Line $OD$ represents the locus of points of some concentration $C$, where $C_0 < C < C_{max}$. It corresponds to the propagation of a wave at a velocity $u_w$ from the bottom of the suspension. Thus when the wave reaches the interface, point $D$, all the particles in the suspension must have passed through the plane of the wave. Thus considering unit area:

$$C(u_c + u_w)t = C_0 H_t \qquad (5.62)$$

In Fig. 5.11:

$$H_t = OA$$

By drawing a tangent to the curve $ABDC$ at $D$, the point $T$ is located.
  Then

$$u_c t = QT \quad \text{(since } -u_c \text{ is the slope of the curve at } D)$$

$$u_w t = RD = OQ \quad \text{(since } u_w \text{ is the slope of line } OD)$$

and

$$(u_c + u_w)t = OT$$

Thus the concentration $C$ corresponding to the line $OD$ is given by:

$$C = C_0 \frac{OA}{OT} \qquad (5.63)$$

and the corresponding solids flux is given by:

$$\psi = Cu_c = \frac{OA}{OT} . u_c \qquad (5.64)$$

Thus by drawing the tangent at a series of points on the curve $BDC$ and measuring the corresponding slope $-u_c$ and intercept $OT$, it is possible to establish the solids flux $\psi$ for any concentration $C$ $(C_i < C < C_{max})$.

It has already been shown that for coarse particles (low values of $n$) the point of inflexion does not occur at a concentration which would be obtained in practice in a suspension, and therefore the particles will settle throughout at a constant rate until an interface forms between the clear liquid and the sediment when sedimentation will abruptly cease. As the value of $n$ increases, that is as the settling velocity of the particles decreases, the point of inflexion occurs at progressively lower concentrations and with a highly flocculated suspension with a high value of $n$ may occur at a very low volumetric concentration. In these circumstances, there will be a wide range of concentrations for which the constant rate sedimentation is followed by a period of falling rate.

### 5.2.5. Comparison of Sedimentation with Flow through Fixed Beds

Meikle[17] has shown that at high concentrations the results of sedimentation and fluidisation experiments can be represented in a manner similar to that used by Carman[18] for fixed beds (see°Chapter 4). Using the interstitial velocity and a linear dimension given by the reciprocal of the surface of particles per unit volume of fluid, the Reynolds number is defined as:

$$Re_1 = \frac{(u_c/e)e/[(1-e)S]\rho}{\mu} = \frac{u_c\rho}{S\mu(1-e)} \qquad (5.65)$$

The friction group $\phi''$ is defined in terms of the resistance force per unit particle surface $(R_1)$ and the interstitial velocity.

Thus

$$\phi'' = \frac{R_1}{\rho(u_c/e)^2} \qquad (5.66)$$

Now a force balance on an element of system containing unit volume of particles gives:

$$R_1 S = (\rho_s - \rho_c)g = e(\rho_s - \rho)g \qquad (5.67)$$

$$\therefore \quad \phi'' = \frac{e^3(\rho_s - \rho)g}{S\rho u_c^2} \qquad (5.68)$$

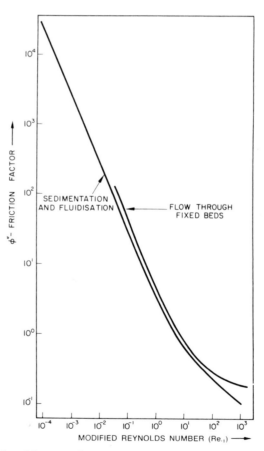

FIG. 5.12. Correlation of data on sedimentation and fluidisation and comparison with results for flow through fixed granular beds

In Fig. 5.12, $\phi''$ is plotted against $Re_1$ using results obtained in experiments on sedimentation and fluidisation. On a logarithmic scale a linear relation is obtained, for values of $Re_1$ less than 1, with the following equation:

$$\phi'' = 3\cdot36\,Re_1^{-1} \tag{5.69}$$

Results for flow through a fixed bed are also shown. They can be represented by:

$$\phi'' = 5\,Re_1^{-1} \tag{equation 4.16}$$

Because equation 4.16 is applicable to low Reynolds numbers at which the flow is streamline, it appears that the flow of fluid at high concentrations of particles in a sedimenting or fluidised system is also streamline. The resistance to flow in the latter case appears to be about 30 per cent lower, presumably because the particles are free to move relative to one another.

Equation 5.69 can be arranged to give:

$$u_c = \frac{e^3}{3\cdot36\,(1-e)}\frac{(\rho_s-\rho)g}{S^2\mu} \tag{5.70}$$

In the Stokes' Law region, from equation 3.17:

$$u_0 = \frac{d^2 g}{18\mu}(\rho_s-\rho) = \frac{2g}{S^2\mu}(\rho_s-\rho) \tag{5.71}$$

Dividing:

$$\frac{u_c}{u_0} = \frac{e^3}{6\cdot7\,(1-e)} \tag{5.72}$$

From equations 5.30 and 5.34:

$$\frac{u_c}{u_0} = e^{4\cdot6} \quad \text{(for } Re_1 < 0\cdot2) \tag{5.73}$$

The functions $e^3/[6\cdot7(1-e)]$ and $e^{4\cdot6}$ are both plotted as a function of $e$ in Fig. 5.13, from which it is seen that they correspond closely for voidages less than 0·75.

The application of the relations obtained for monodisperse systems to fine suspensions containing particles of a wide range of sizes was examined by Shabi[19] who studied the behaviour of aqueous suspensions of zirconia of particle size ranging from less than 1 to 20 μm. A portion of the solid forming the suspension was irradiated and the change in concentration with time at various depths below the surface was followed by means of a Geiger–Müller counter. It was found that selective settling took place initially and that the effect of concentration on the falling rate of a particle was the same as in a monodisperse system, provided that the total concentration of particles of all sizes present was used in the calculation of the correction factor. Agglomeration was found to occur rapidly at high concentrations, and the conclusions were therefore applicable only to the initial stages of settling.

### 5.2.6. Model Experiments

The effect of surrounding a particle with an array of similar particles has been studied by Meikle[20] and by Rowe and Henwood[21]. In both investigations, the effect of the proximity and arrangement of fixed neighbouring particles on the drag was examined.

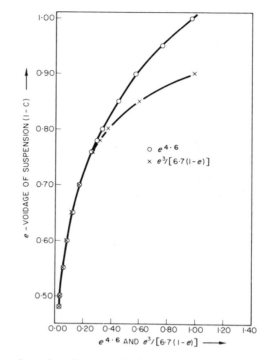

FIG. 5.13. Comparison of two functions of voidage used as correction factors for sedimentation velocities

Meikle surrounded his test particle by a uniplanar hexagonal arrangement of identical spheres which were held in position by means of rods passing through glands in the tube walls, as shown in Fig. 5.14. This arrangement permitted the positioning of the spheres to be altered without interfering with the flow pattern in the hexagonal space. For each arrangement of the spheres, the drag force was found to be proportional to the square of the liquid velocity. The effective voidage $e$ of the system was calculated by considering a hexagonal cell located symmetrically about the test sphere, of depth equal to the diameter of the particle, as shown in Fig. 5.15. The drag force $F_c$ on a particle of volume $v$ settling in a suspension of voidage $e$ is given by:

$$F_c = v(\rho_s - \rho_c)g = ev(\rho_s - \rho)g \qquad (5.74)$$

Thus, for any voidage the drag force on a sedimenting particle can be calculated, and the corresponding velocity required to produce this force on a particle at the same voidage in the model is obtained from the experimental results. All the experiments were carried out at particle Reynolds numbers greater than 500, and under these conditions the observed sedimentation velocity is given by equations 5.30 and 5.34 as:

$$\frac{u_c}{u_0} = e^{2\cdot4} \qquad (5.75)$$

Writing $u_c = eu_p$, where $u_p$ is the velocity of the particle relative to the fluid:

$$\frac{u_p}{u_0} = e^{1\cdot4} \qquad (5.76)$$

In Fig. 5.16 the velocity of the fluid relative to the particle, as calculated for the model experiments and from equation 5.76, is plotted against voidage. It will be seen that reasonable agreement is obtained at voidages between 0·45 and 0·90, indicating that the model does fairly closely represent the conditions in a suspension.

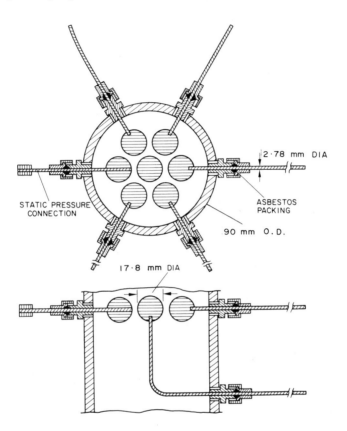

FIG. 5.14. Arrangement of hexagonal spacing of particles in tube

Rowe and Henwood[21] made assemblies of particles from cast blocks of polythene. They found that the force on a single particle was increased by an order of magnitude as a result of surrounding it with a close assembly. They also found that adjacent spheres tend to repel one another, that surfaces facing downstream tend to expel particles whereas those facing upstream attract particles. They offer this as an explanation for the stable upper surface and rather diffuse lower surface of a bubble in a fluidised bed.

### 5.2.7. Sedimentation of Two-component Mixtures

*Particles of Different Size but Same Density*

Several workers[22,23,24] have studied the sedimentation of suspensions formed of particles of two different sizes, but of the same densities. The large particles have higher

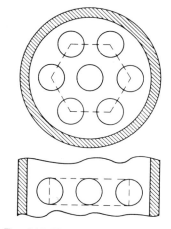

FIG. 5.15. Hexagonal cell of particles

settling velocities than the small ones and therefore four zones form, in order, from the top downwards:

(a)  Clear liquid.
(b)  Suspension of the fine particles.
(c)  Suspension of mixed sizes.
(d)  Sediment layer.

For relatively coarse particles, the rates of all of the interface between (a) and (b) and between (b) and (c) can be calculated approximately if the relation between sedimentation velocity and voidage, or concentration, is given by equation 5.30. Richardson and Shabi[19] have shown that, in a suspension of particles of mixed sizes, it is the total concentration which controls the sedimentation rate of each species.

Consider a suspension of a mixture of large particles of terminal falling velocity $u_{0L}$ and of small particles of terminal falling velocity $u_{0S}$; the fractional volumetric concentrations are $C_L$ and $C_S$, respectively. Suppose that the value of $n$ in equation 5.30 is the same for each

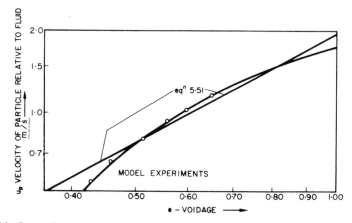

FIG. 5.16. Comparison of velocity in model required to produce a drag force equal to buoyant weight of particle, with velocity determined experimentally

particle. For each of the spheres therefore settling on its own:

$$\frac{u_{cL}}{u_{0L}} = e^n \tag{5.77}$$

and

$$\frac{u_{cS}}{u_{0S}} = e^n \tag{5.78}$$

Then considering the velocities of the particles relative to the fluid, $u_{cL}/e$ and $u_{cS}/e$, respectively:

$$u_{cL}/e = u_{0L}e^{n-1} \tag{5.79}$$

and

$$\frac{u_{cS}}{e} = u_{0S}e^{n-1} \tag{5.80}$$

When the particles of two sizes are settling together, the upflow of displaced fluid is caused by the combined effects of the sedimentation of the large and small particles. If this upward velocity is $u_F$, the sedimentation rates $u_{ML}$ and $u_{MS}$ will be obtained by deducting $u_F$ from the velocities relative to the fluid. Thus:

$$u_{ML} = u_{0L}e^{n-1} - u_F \tag{5.81}$$

and

$$u_{MS} = u_{0S}e^{n-1} - u_F \tag{5.82}$$

Then, since the volumetric flow of displaced fluid upwards must be equal to the total volumetric flowrate of particles downwards:

$$u_F e = (u_{0L}e^{n-1} - u_F)C_L + (u_{0S}e^{n-1} - u_F)C_S$$

$$\therefore \quad u_F = e^{n-1}(u_{0L}C_L + u_{0S}C_S) \tag{5.83}$$

$$(\text{since } e + C_L + C_S = 1)$$

Then substituting from equation 5.83 into equations 5.81 and 5.82:

$$u_{ML} = e^{n-1}[u_{0L}(1 - C_L) - u_{0S}C_S] \tag{5.84}$$

and

$$u_{MS} = e^{n-1}[u_{0S}(1 - C_S) - u_{0L}C_L] \tag{5.85}$$

Equation 5.84 gives the rate of fall of the interface between zones (b) and (c); i.e. it is the apparent rate of settling of the zone of mixed particles.

The velocity of fall $u_F$ of the interface between zones (a) and (b) is the sedimentation rate of the suspension composed only of fine particles and will therefore depend on the free-falling velocity $u_{0S}$ and the concentration $C_f$ of this zone. Now:

$$C_f = \frac{\text{Volumetric rate at which solids are entering zone}}{\text{Total volumetric growth rate of zone}}$$

$$= \frac{(u_{ML} - u_{MS})C_S}{u_{ML} - u_F} \tag{5.86}$$

Thus

$$u_F = u_{0S}(1 - C_f)^n \tag{5.87}$$

$u_F$ and $C_f$ are determined by solving equations 5.86 and 5.87 simultaneously.

*Particles of Equal Terminal Falling Velocities*

By studying suspensions containing two different solid components, it is possible to obtain a fuller understanding of the process of sedimentation of a complex mixture. Meikle[17] investigated the sedimentation characteristics of suspensions of glass ballotini and polystyrene particles in a 22 per cent by weight ethanol–water mixture. The free-falling velocity, and the effect of concentration on sedimentation rate, were identical for each of the two solids alone in the liquid.

The properties of the components were as given in Table 5.1.

TABLE 5.1. *Properties of Solids and Liquids in Sedimentation of Two-component Mixtures*

|  | Glass ballotini | Polystyrene | 22% ethanol in water |
|---|---|---|---|
| Density (kg/m³) | $\rho_B = 1921$ | $\rho_P = 1045$ | $\rho = 969$ |
| Particle size (μm) | 71·1 | 387 | — |
| Viscosity (mN s/m²) | — | — | 1·741 |
| Free-falling velocity $(u_0)$ (mm/s) | 3·24 | 3·24 | — |

The sedimentation of mixtures containing equal volumes of the two solids was then studied, and it was found that segregation of the two components tended to take place, the degree of segregation increasing with concentration. This arises because the sedimentation velocity of an individual particle in the suspension is different from its free-falling velocity, first because the buoyancy force is greater, and, secondly, because the flow pattern is different. For a monodisperse suspension of either constituent of the mixture, the effect of flow pattern as determined by concentration is the same, but the buoyant weights of the two species of particles are altered in different proportions. The settling velocity of a particle of polystyrene or ballotini in the mixture ($u_{PM}$ or $u_{BM}$) can be written in terms of its free-falling velocity ($u_{P0}$ or $u_{B0}$) in the following way:

$$u_{PM} = u_{P0}\frac{\rho_P - \rho_c}{\rho_P - \rho}f(e) \tag{5.88}$$

and

$$u_{BM} = u_{B0}\frac{\rho_B - \rho_c}{\rho_B - \rho}f(e) \tag{5.89}$$

Here $f(e)$ represents the effects of concentration, other than those associated with an alteration of buoyancy arising from the fact that the suspension has a higher density than the liquid. In a uniform suspension, the density of the suspension is given by:

$$\rho_c = \rho e + \frac{1-e}{2}(\rho_B + \rho_P) \tag{5.90}$$

Substitution of the numerical values of the densities in equations 5.88, 5.89 and 5.90 gives:

$$u_{PM} = u_{P0}(13·35e - 12·33)f(e) \tag{5.91}$$

and

$$u_{BM} = u_{B0}(0·481 + 0·520e)f(e) \tag{5.92}$$

Noting that $u_{P0}$ and $u_{B0}$ are equal, it is seen that the rate of fall of the polystyrene becomes progressively less than that of the ballotini as the concentration is increased. When $e = 0.924$, the polystyrene particles should remain suspended in the mixed suspension, and settling should occur only when the ballotini have separated out from that region. At lower values of $e$, the polystyrene particles should rise in the mixture. At a voidage of $0.854$, the polystyrene should rise at the rate at which the ballotini are settling, so that there will then be no net displacement of liquid. When the ballotini are moving at the higher rate, the net displacement of liquid will be upwards and the zone of polystyrene suspension will become more dilute. When the polystyrene is moving at the higher rate, the converse is true.

Because there is no net displacement of liquid at a voidage of $0.854$, corresponding to a total volumetric concentration of particles of 14·6 per cent, the concentration of each zone of suspension after separation should be equal. This has been confirmed within the limits of experimental accuracy. It has also been confirmed experimentally that upward movement of the polystyrene particles does not take place at concentrations less than 8 per cent ($e > 0.92$).

On the above basis, the differences between the velocities of the two types of particles should increase with concentration and consequently the concentration of the two separated zones should become increasingly different. This is found to be so at volumetric concentrations up to 20 per cent. At higher concentrations, the effect is not observed, probably because the normal corrections for settling velocities cannot be applied when particles are moving in opposite directions in suspensions of high concentration.

The behaviour of a suspension of high concentration ($e < 0.85$) is shown in Fig. 5.17 where it is seen that the glass ballotini settle out completely before any deposition of polystyrene particles occurs.

These experiments clearly show that the tendency for segregation to occur in a two-component mixture becomes progressively greater as the concentration is increased. This behaviour is in distinct contrast to that observed in the sedimentation of a suspension of multi-sized particles of a given material, when segregation becomes less as the concentration is increased.

## 5.2.8. The Thickener

The thickener is the industrial plant in which the concentration of a suspension is increased by sedimentation, with the formation of a clear liquid. In most cases, the concentration of the suspension is high and hindered settling takes place. Thickeners may operate as batch or continuous units, and consist of relatively shallow tanks from which the clear liquid is taken off at the top and the thickened liquor at the bottom.

In order to obtain the largest possible throughput from a thickener of given size, the rate of sedimentation should be as high as possible. In many cases, the rate may be artificially increased by the addition of small quantities of an electrolyte, which causes precipitation of colloidal particles and the formation of flocs. The suspension is also frequently heated because this lowers the viscosity of the liquid, and encourages the larger particles in the suspension to grow in size at the expense of the more soluble small particles. Further, the thickener frequently incorporates a slow stirrer, which causes a reduction in the apparent viscosity of the suspension and also aids in the consolidation of the sediment.

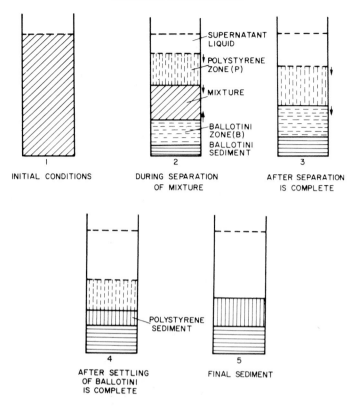

FIG. 5.17. Settling behaviour of mixture consisting of equal volumes of polystyrene and ballotini
at volumetric concentrations exceeding 15 per cent

FIG. 5.18. Dorr thickener—single tray

The batch thickener usually consists of a cylindrical tank with a conical bottom. After sedimentation has proceeded for an adequate time, the thickened liquor is withdrawn from the bottom and the clear liquid is taken off through an adjustable offtake pipe from the upper part of the tank. The conditions prevailing in the batch thickener are similar to those in the ordinary laboratory sedimentation tube, and during the initial stages there will generally be a zone in which the concentration of the suspension is the same as that in the feed.

The continuous thickener, such as the Dorr thickener (Fig. 5.18), consists of a large diameter shallow tank with a flat bottom. The liquor is fed in at the centre, at a depth of from 0·3 to 1 m below the surface of the liquid, with as little disturbance as possible. The thickened liquor is continuously removed through an outlet at the bottom, and any solids which are deposited on the floor of the tank are directed towards the outlet by means of a slowly rotating rake mechanism incorporating scrapers. The rake is hinged so that the arms fold up automatically if the torque exceeds a certain value; this prevents it from being damaged if it is overloaded. Because of the action of the rake, a greater degree of thickening is obtained than with the batch plant. The clarified liquid is continuously removed from an overflow which runs round the whole of the upper edge of the tank. The solids are therefore moving continuously downwards, and then inwards towards the thickened liquor outlet; the liquid is moving upwards and radially outwards (Fig. 5.19). In general, there will be no region of constant composition in the continuous thickener.

The thickener has a twofold function. First it must produce a clarified liquid, and therefore the upward velocity of the liquid must, at all times, be less than the settling velocity of the particles. Thus, for a given throughput, the clarifying capacity is determined by the diameter of the tank. Secondly, the thickener is required to produce a given degree of thickening of the suspension. This is controlled by the time of residence of the particles in the tank, and hence by the depth below the feed inlet.

The satisfactory operation of the thickener as a clarifier depends upon the existence of a zone of negligible solid content towards the top. In this zone conditions approach those under which free settling takes place, and the rate of sedimentation of any particles which have been carried to this height is therefore sufficient for them to settle against the upward current of liquid. If this upper zone is too shallow, some of the smaller particles may escape in the liquid overflow. The volumetric rate of flow of liquid upwards through the

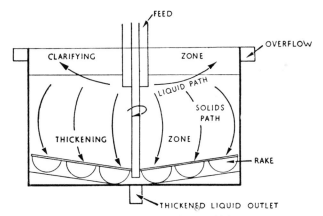

FIG. 5.19. Flow in continuous thickener

clarification zone is equal to the difference between the rate of feed of liquid in the slurry and the rate of removal in the underflow. Thus the required concentration of solids in the underflow, as well as the throughput, determines the conditions in the clarification zone.

In a continuous sedimentation tank, therefore, allowance must be made for the downward velocity $u_r$ arising from the removal of the underflow.

In the settling zone of the tank of area $A$, the volumetric rate of sedimentation of solids $q$ is given by:

$$q = (u_c + u_r)AC \qquad (5.93)$$

where $u_c$ is the sedimentation velocity at a concentration $C$ in a batch system.

In the underflow, the corresponding relation is:

$$q = (u_u + u_r)AC_u \qquad (5.94)$$

where $C_u$ is the concentration in the underflow and $u_u$ is the sedimentation velocity at a concentration $C_u$ in a batch system.

From equation 5.94:

$$u_r = \frac{q}{AC_u} - u_u \qquad (5.95)$$

Substituting in equation 5.93:

$$q = (u_c + \frac{q}{AC_u} - u_u)AC$$

$$\therefore \quad \frac{q}{A} = \frac{u_c - u_u}{(1/C) - (1/C_u)} \qquad (5.96)$$

If the sedimentation rate in the underflow is small compared with that in the settling zone:

$$\frac{q}{A} = \frac{u_c C}{1 - (C/C_u)} \qquad (5.97)$$

Thus if suspension of concentration $C$ is fed to the tank at a volumetric rate $Q$:

$$\text{Solids input} = QC = \frac{Au_c C}{1 - (C/C_u)} \qquad \text{(from equation 5.97)}$$

Thus:

$$\frac{C}{C_u} = 1 - \frac{Au_c}{Q} \qquad (5.98)$$

The liquid flowrate in the overflow $Q_0$ is the difference between the feed rate to the thickener and the rate at which it leaves with underflow.

Thus:

$$Q_0 = Q(1 - C) - QC \frac{1 - C_u}{C_u}$$

or

$$\frac{Q_0}{Q} = 1 - \frac{C}{C_u} \qquad (5.99)$$

Now the area required in a thickener for clarification is given by equation 5.98 from which:

$$A = \frac{Q}{u_c}\left(1 - \frac{C}{C_u}\right)$$   (5.100)

This can usefully be rearranged in terms of the mass ratio of liquid to solid in the feed ($Y$) and the corresponding value ($U$) in the underflow. Now

$$Y = \frac{1-C}{C}\frac{\rho}{\rho_s} \quad \text{and} \quad U = \frac{1-C_u}{C_u}\frac{\rho}{\rho_s}$$

Then

$$C = \frac{1}{1+Y(\rho_s/\rho)} \qquad C_u = \frac{1}{1+U(\rho_s/\rho)}$$

and

$$A = \frac{Q}{u_c}\left\{1 - \frac{1+U(\rho_s/\rho)}{1+Y(\rho_s/\rho)}\right\}$$

$$= \frac{Q(Y-U)C\rho_s}{u_c\rho}$$   (5.101)

The values of $A$ should be calculated for the whole range of concentrations present in the thickener, and the design should then be based on the maximum value so obtained.

## Example 5.1

A slurry containing 5 kg of water per kg of solids is to be thickened to a sludge containing 1·5 kg of water per kg of solids in a continuous operation. Laboratory tests using five different concentrations of the slurry yielded the following results:

| Concentration (kg water/kg solid) | 5·0 | 4·2 | 3·7 | 3·1 | 2·5 |
|---|---|---|---|---|---|
| Rate of sedimentation (mm/s) | 0·20 | 0·12 | 0·094 | 0·070 | 0·050 |

Calculate the minimum area of a thickener to effect the separation of 1·33 kg of solids per second.

## Solution

*Basis of 1 kg solids*

$$\text{Mass rate of feed of solids} = 1\cdot33 \text{ kg/s}$$

1·5 kg water is carried away in underflow, balance in overflow. $V = 1\cdot5$

| $Y$ concentration | $Y-U$ water to overflow | $u_c$ sedimentation rate (m/s) | $\dfrac{Y-U}{u_c}$ (s/m) |
|---|---|---|---|
| 5·0 | 3·5 | $2\!\cdot\!00 \times 10^{-4}$ | $1\!\cdot\!75 \times 10^{4}$ |
| 4·2 | 2·7 | $1\!\cdot\!20 \times 10^{-4}$ | $2\!\cdot\!25 \times 10^{4}$ |
| 3·7 | 2·2 | $0\!\cdot\!94 \times 10^{-4}$ | $2\!\cdot\!34 \times 10^{4}$ |
| 3·1 | 1·6 | $0\!\cdot\!70 \times 10^{-4}$ | $2\!\cdot\!29 \times 10^{4}$ |
| 2·5 | 1·0 | $0\!\cdot\!50 \times 10^{-4}$ | $2\!\cdot\!00 \times 10^{4}$ |

Maximum value of $\dfrac{Y-U}{u_c} = 2\!\cdot\!34 \times 10^{4}$ s/m.

From equation 5.101:

$$A = \frac{Y-U}{u_c}\frac{QC\rho_s}{\rho}$$

Now $QC\rho_s = 1\!\cdot\!33$ kg/s and taking $\rho$ as 1000 kg/m³,

$$A = 2\!\cdot\!34 \times 10^{4}\,\frac{1\!\cdot\!33}{1000}$$

$$= 31\!\cdot\!2\ \text{m}^2$$

The concentration of solids in the underflow will be a function of the depth of the tank below the feed point and the time of residence of the solids in the thickener. In most cases, the time of compression of the sediment will be large compared with the time taken for the critical settling conditions to be reached for any portion of the slurry. The time required to concentrate the sediment after it has reached the critical condition can be determined approximately by allowing a sample of the slurry at its critical composition to settle in a vertical glass tube, and measuring the time taken for the interface between the sediment and the clear liquid to fall to such a level that the concentration is that required in the underflow from the thickener. The use of data so obtained assumes that the average concentration in the sediment in the laboratory test is the same as that which would be obtained in the thickener after the same time. This is not quite so because, in the thickener, the various parts of the sediment have been under compression for different times. Further, it assumes that the time taken for the sediment to increase in concentration by a given amount is independent of its depth.

An approximate value for the depth of the thickening zone is then found by adding the volume of the liquid in the sediment to the corresponding volume of solid and dividing by the area, which has already been calculated in order to determine the clarifying capacity of the thickener. The required depth of the thickening region is thus:

$$\left\{\frac{Wt_R}{A\rho_s} + W\frac{t_R}{A\rho}X\right\} = \frac{Wt_R}{A\rho_s}\left(1 + \frac{\rho_s}{\rho}X\right) \tag{5.102}$$

where   $t_R$ is the required time of retention of the solids, as determined experimentally,
  $W$ is the mass rate of feed of solids to the thickener,
  $X$ is the average value of the mass ratio of liquid to solids in the thickening portion, and
  $\rho$ and $\rho_s$ are the densities of the liquid and solid respectively.

This method of design is only approximate and therefore, in a large tank, about 1 metre should be added to the calculated depth as a safety margin and to allow for the depth required for the suspension to reach the critical concentration. In addition, the bottom of the tanks may be slightly pitched to assist the flow of material towards the thickened liquor outlet. The diameter of the tank is usually large compared with its depth and therefore a large ground area is required. Space is often saved by arranging a number of thickeners vertically above one another. They operate as separate units but a common central shaft is utilised to drive the rakes and stirrers. Figure 5.18 shows a single tray thickener which incorporates two units in this way.

Thickeners may be anything from a few metres to several hundred metres in diameter. The small ones are made of wood or metal and the rakes rotate at about 0·02 Hz (1 r.p.m.). The very large thickeners generally consist of large concrete tanks, and the stirrers and rakes are driven by means of traction motors which drive on a rail running round the whole circumference; the speed of rotation may be as low as 0·002 Hz (0·1 r.p.m.).

## 5.3. CENTRIFUGAL SEPARATION

### 5.3.1. Introduction

The rate of sedimentation of particles in a fluid will be very much greater in a centrifugal field than in the gravitational field. The ratio of the centrifugal to the gravitational acceleration, $r\omega^2/g$, is a measure of the separating power of a centrifuge; this ratio may have a value as high as $10^4$. Not only does the use of a centrifuge permit very much more rapid sedimentation but it does also provide the possibility of effecting separations which would be quite impossible under gravitational conditions. For example, a colloid or emulsion may be quite stable under ordinary gravitational conditions where the dispersive forces such as those due to Brownian Motion are very much greater than the gravitational force. However, in a centrifuge the colloid or emulsion may break down completely because the vastly greater centrifugal forces will be able to overcome the effects of the dispersive forces. Similarly, it is possible to obtain a very much drier solid by centrifugal action than by draining under gravity, because the surface tension forces in the fine pores which cause retention of moisture can be overcome by the centrifugal but not by the gravitational force.

### 5.3.2. Behaviour of Liquid in Centrifuge Basket

Most centrifuges are constructed with the axis of rotation either vertical or horizontal, though in some cases they may be arranged with the axis at an inclined angle in order to facilitate discharge of solids. Generally, the speed of operation will be such that the centrifugal force considerably exceeds the gravitational force and the liquid surface will be concentric with the wall of the basket and in a simple bowl the inner surface of the liquid will be cylindrical.

However, it is of interest to consider the effect of speed of rotation on the shape of the liquid surface when the axis is vertical. Any element of liquid is subjected to a force $g$ per unit mass in the vertical direction and $r\omega^2$ in the radial direction, when it is rotating at a radius $r$ and angular velocity $\omega$. For an element situated at the free surface, the resultant

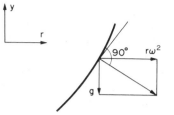

FIG. 5.20. Element of surface of liquid

force must be perpendicular to the surface (Fig. 5.20). Then the slope of the surface at a height $y$ and radius $r$ will be given by:

$$\frac{dy}{dr} = \frac{\text{radial component of force}}{\text{vertical component of force}} = \frac{r\omega^2}{g} \qquad (5.103)$$

On integration:

$$y = \frac{\omega^2}{2g}r^2 + \text{constant}$$

At the bottom of the basket $(y = 0)$, $r$ has some value $r_0$, say. Then:

$$y = \frac{\omega^2}{2g}(r^2 - r_0^2) \qquad (5.104)$$

This is the equation of a parabola and therefore the surface of the liquid is part of a paraboloid of revolution. However, for all practical purposes, the curvature of the surface may be neglected.

### 5.3.3. Fluid Pressure

When a liquid is rotated in a centrifuge basket, there will be a radial pressure gradient arising from the centrifugal force contributed by each liquid element. Suppose that the basket of radius $R$ and its contents are rotating at a constant angular velocity $\omega$, with no slip between the basket and the liquid.

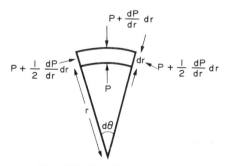

FIG. 5.21. Centrifugal pressure

Consider an element of liquid, of length $dy$ in the axial direction, of inner radius $r$ and outer radius $r + dr$, subtending an angle $d\theta$ at the centre. If $\rho$ is the density of the liquid:

$$\text{Mass of element} = (r\,d\theta)(dy)(dr)\rho$$
$$\text{Centrifugal force exerted} = [(r\,d\theta)(dy)(dr)\rho]r\omega^2$$
$$= \rho\omega^2 r^2\,d\theta\,dy\,dr$$

If the pressure in the fluid is $P$ at radius $r$ and $P + (dP/dr)\,dr$ at radius $r + dr$, the mean pressure on the side boundaries of the element will be $P + \frac{1}{2}(dP/dr)\,dr$ as shown in Fig. 5.21. The net force in the radial direction attributable to the pressure distribution is therefore:

$$P(r\,d\theta\,dy) - \left(P + \frac{dP}{dr}\,dr\right)[(r+dr)\,d\theta\,dy] + 2\left(P + \tfrac{1}{2}\frac{dP}{dr}\,dr\right)(dr\,dy)\sin\frac{d\theta}{2}$$

$$= -\frac{dP}{dr}\,dr\,r\,d\theta\,dy$$

(ignoring terms of smaller order of magnitude). A force balance therefore gives:

$$\rho\omega^2 r^2\,d\theta\,dy\,dr - (dP/dr)\,dr\,r\,d\theta\,dy = 0$$

Thus:
$$\frac{dP}{dr} = \rho\omega^2 r \qquad\qquad (5.105)$$

Then, since at the inner radius $r_i$ of the liquid surface the centrifugal pressure is zero, the centrifugal pressure at the walls of the basket of radius $R$ will be obtained by integration of equation 5.105 to give:

$$P = \tfrac{1}{2}\rho\omega^2(R^2 - r_i^2) \qquad\qquad (5.106)$$

### 5.3.4. Sedimentation in a Centrifugal Field

Centrifuges are extensively used for separating fine solids from suspension in a liquid. As a result of the far greater separating power compared with that available using gravity, it becomes feasible readily to separate fine solids and even colloids. Furthermore, it is possible to break down emulsions and to separate dispersions of fine liquid droplets. The same considerations apply as to suspensions of fine solids, though in this case the suspended phase is in the form of liquid droplets which will coalesce following separation. Centrifuges may be used for batch operation when dealing with small quantities of suspension, but on the large scale arrangements must be made for the continuous removal of the separated constituents. There is a clear difference here in the methods required according to whether the disperse phase is a solid or a liquid. If it is a solid, the material will be deposited on the walls of the basket and some mechanical means of discharge must be used. When it is a liquid each phase can be discharged separately using a suitably designed overflow weir system.

Because centrifuges are normally used for separating fine particles and droplets, it is necessary to consider only the Stokes' Law region in calculating the drag between the particle and the liquid. If the thickness of the liquid layer at the wall of the basket is small compared woth the radius of the basket, the particles can be assumed to be moving at their equilibrium velocities when rotating at a radius equal to that of the basket ($=R$).

If the suspension is sufficiently dilute for the effects of particle interaction to be small, a force balance then gives the settling velocity $dr/dt$ as follows for a spherical particle of diameter $d$:

$$\frac{\pi}{6}d^3(\rho_s-\rho)r\omega^2 = 3\pi\,\mu d\frac{dr}{dt}$$

Thus:

$$\frac{dr}{dt} = \frac{d^2\omega^2(\rho_s-\rho)r}{18\mu} \tag{5.107}$$

Then if the depth of liquid at the wall is $h(h\ll R)$, the time $t_R$ taken for a particle to travel from the inner surface of the liquid to the basket wall, i.e. the time taken for the collection of the particle in the most unfavourable position when introduced into the basket, is given by:

$$t_R = \frac{h}{dr/dt} \tag{5.108}$$

Then, if it is assumed that all the particles are spheres of diameter $d$ rotating at a radius equal to the radius of the basket ($R$):

$$t_R = \frac{18\mu h}{d^2\omega^2(\rho_s-\rho)R} \tag{5.109}$$

(from equation 5.107, cf. equation 3.89).

If the volumetric rate of feed of suspension to the centrifuge is $Q$ and $V'$ is the volumetric capacity of the basket:

$$t_R = \frac{V'}{Q} \tag{5.110}$$

The rate of feed of suspension must be adjusted to give at least the required residence time. If $L$ is the depth of the basket of radius $R$ and $r_i$ is the inner radius of the liquid surface (which is determined by the setting of the overflow weir):

$$V' = \frac{\pi}{4}(R^2 - r_i^2)L \tag{5.111}$$

and therefore the depth, $L$, of the bowl determines the retention time.

Thus from equations 5.109 and 5.110:

$$Q = \frac{d^2\omega^2(\rho_s-\rho)RV'}{18\mu h} \tag{5.112}$$

$$= \frac{d^2(\rho_s-\rho)g}{18\mu}\frac{R\omega^2 V'}{hg}$$

$$= u_0\frac{R\omega^2 V'}{hg} = u_0\Sigma \text{ (say)} \tag{5.113}$$

where $u_0$ is the terminal falling velocity of the particle. Now $\Sigma$ in equation 5.113 is independent of the properties of the fluid–particle system and is dependent only on the operating conditions of the centrifuge. It is equal to the cross-sectional area of a gravity settling tank with the same clarifying capacity as the centrifuge and therefore comparison of the values of $\Sigma$ for different machines provides a method of comparing their performances.

If the thickness ($h$) of the layer of liquid at the walls of the basket is not small compared with the basket radius ($R$) and $r_i$ is the inner radius of the liquid in the basket, it is necessary to use equation 5.107 to give the residence time ($V'/Q$). Assuming that the particle is situated in the most unfavourable location, i.e. at the inner surface of the liquid, $r_i$ the required residence time, $V'/Q$ is obtained by integrating equation 5.107 between the limits $r = r_i$ and $r = R$ to give:

$$t_R = \frac{V'}{Q} = \frac{18\mu}{d^2\omega^2(\rho_s-\rho)}\ln\frac{R}{r_i} \tag{5.114}$$

$$\therefore \quad Q = \frac{d^2(\rho_s-\rho)g}{18\mu}\frac{\omega^2 V'}{g\ln(R/r_i)} \tag{5.115}$$

In this case:

$$\Sigma = \frac{\omega^2 V'}{g\ln(R/r_i)} \tag{5.116}$$

A similar analysis can be carried out with various geometrical arrangements of the bowl of the centrifuge. Thus, for instance, for a disc machine the value of $\Sigma$ is very much greater than for a cylindrical bowl of the same size. Values of $\Sigma$ for different arrangements are quoted by Hayter[25] and by Trowbridge[26] who have a slightly different definition in that they chose a cut-point such that only half the particles of the specified size are removed from the suspension.

### 5.3.5. Separation of Two Immiscible Liquids

The problem of separating two immiscible liquids continuously in a centrifuge can most simply be understood by first considering the problem of a gravity separator. Figure 5.22

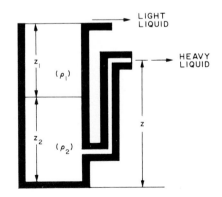

FIG. 5.22. Gravity separation of two immiscible liquids

shows two immiscible liquids which have separated under the action of gravity. Then the height of the overflow pipe for discharge of the denser liquid in a continuous flow system must be such that the combined hydrostatic pressure due to the two liquids in the separator exactly balances the hydrostatic pressure of the denser liquid in the overflow pipe. Thus

$$z\rho_2 g = z_2\rho_2 g + z_1\rho_1 g$$

i.e.

$$z = z_2 + z_1 \frac{\rho_1}{\rho_2} \tag{5.117}$$

For the centrifuge it is necessary to position the overflow on the same principle (Fig. 5.23). In this case the radius of the weir for the less dense liquid will correspond to the radius of the inner surface of the liquid $r_i$ in the basket. That of the outer weir, $r_w$, will be such that the pressure developed at the wall of the basket (radius $R$) by the heavy liquid alone as it flows over the weir is equal to that due to the two liquids within the basket.

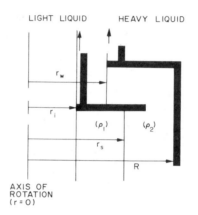

FIG. 5.23. Separation of two immiscible liquids in centrifuge

Thus, obtaining the pressures from equation 5.106, denoting the densities of the light and heavy liquids by $\rho_1$ and $\rho_2$ respectively and the radius of the interface between the two liquids in the basket as $r_s$:

$$\tfrac{1}{2}\rho_2\omega^2 (R^2 - r_w^2) = \tfrac{1}{2}\rho_2\omega^2 (R^2 - r_s^2) + \tfrac{1}{2}\rho_1\omega^2 (r_s^2 - r_i^2)$$

$$\therefore \quad \frac{r_s^2 - r_i^2}{r_s^2 - r_w^2} = \frac{\rho_2}{\rho_1} \tag{5.118}$$

If $Q_1$ and $Q_2$ are the volumetric rates of feed of the light and heavy liquids respectively, on the assumption that there is no slip between the liquids and the basket:

$$\frac{Q_1}{Q_2} = \frac{r_s^2 - r_i^2}{R^2 - r_s^2} \tag{5.119}$$

Equation 5.119 enables the value of $r_s$ to be calculated for a given operating condition. The radius $r_i$ of the inner surface of the liquid fixes the volumetric holdup of the basket and for a given feed rate of liquid this determines the average retention time $t_R$ in the centrifuge which must be adequate for separation of the two liquids to take place. The required value of $t_R$, and hence of $Q_1$ and $Q_2$, may be calculated from a form of equation 5.114 applied to each of the two liquids, and the radius of the overflow weir $r_w$ for the heavy liquid is obtained from equation 5.118.

### 5.3.6. Centrifugal Equipment

Centrifuges range from the simple batch machines used in the laboratory or for drying crystals to highly automated equipment in which elaborate arrangements are made for the continuous removal of the separated materials.

*Simple Bowl Centrifuges*

Most small batch centrifuges are mounted with their axes vertical and, because of the possibility of uneven loading of the machine, the basket is normally supported in bearings

FIG. 5.24. Underdriven centrifuge

either above or below, but not in both positions, so that a certain degree of flexibility is provided. In the underdriven machine, where the drive and bearings are underneath (Fig. 5.24), access to the basket is easier and the material is normally discharged from the top. In the overdriven centrifuge (Fig. 5.25), in which the basket is suspended, a flap valve can be incorporated in the bottom of the basket for easy discharge of the solids. This latter type of machine is the more versatile and can cope more readily with uneven loading.

Centrifuges with imperforate baskets are used either for producing an accelerated separation of solid particles in a liquid, or for separating mixtures of two liquids. In the former case, the solids are deposited on the walls of the basket and the liquid is removed through an overflow or skimming tube. The suspension is continuously fed in until a suitable depth of solids has been built up on the walls; this deposit is then removed either by hand or by a mechanical scraper. With the basket mounted about a horizontal axis,

FIG. 5.25. Overdriven centrifuge

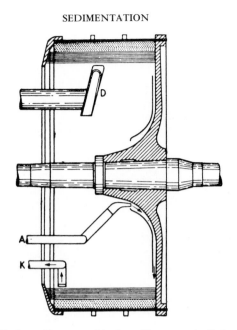

Fig. 5.26. Horizontally mounted basket with automatic discharge of solids.
*A*—Feed,   *D*—Cutter,   *K*—Skimming tube

solids are more readily discharged because they can be allowed to fall directly into a chute. In the centrifuges shown in Figs. 5.26 and 5.27, the liquid is taken off through a skimming tube and the solids are removed by a cutter which operates with the machine running at full speed; a considerable saving of time and energy is thereby achieved. The solids may be

Fig. 5.27. Horizontal centrifuge with automatic discharge of solids

Fig. 5.28. Inclined centrifuge

washed before discharge if desired. A similar machine with an inclined axis is shown in Fig. 5.28.

Perforated baskets are used when relatively large particles are to be separated from a liquid, as for example in the separation of mother liquor and drying crystals. The mother liquor passes through the bed of particles and then through the perforations in the basket. When the centrifuge is used for filtration, a coarse gauze is laid over the inner surface of the basket and the filter cloth rests on the gauze; space is thus provided behind the cloth for the filtrate to flow to one of the perforations.

When a mixture of liquids is to be separated the denser liquid collects near the walls and the lighter liquid forms an inner layer. Overflow weirs are arranged so that the two constituents are continuously removed; the design of the weirs has already been considered.

### Disc Centrifuges

For a given rate of feed to the centrifuge, the degree of separation obtained will depend on the thickness of the liquid layer formed at the walls of the basket and on the total depth of the basket, as these factors will control the time the mixture remains in the machine. A high degree of separation is therefore obtained with a long basket of small diameter, but the speed required is then very high. The introduction of conical discs in the bowl, as illustrated in Fig. 5.29, enables the liquid stream to be split into a large number of very thin layers in a bowl of much greater diameter. The separation of a mixture of water and dirt from a relatively low density oil takes place as shown in Fig. 5.30, with the dirt and water collecting close to the undersides of the discs and moving radially outwards, and with the

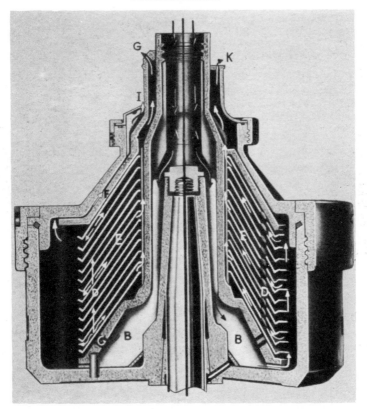

FIG. 5.29. Bowl with conical discs (left-hand side for separating liquids, right-hand side for separating solid from liquid)

oil moving inwards along the top sides. This bowl can be run at a very much lower speed and its size is very much smaller, as seen from Fig. 5.31. The separation of two liquids in a disc-type bowl is illustrated in the left-hand side of Fig. 5.29. Liquid enters through the distributor *AB*, passes through *C*, and is distributed between the discs *E* through the holes *D*. The denser liquid is taken off through *F* and *I* and the lighter liquid through *G*.

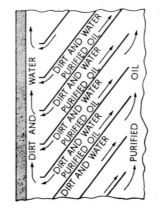

FIG. 5.30. Separation of water and dirt from oil in disc bowl

Fig. 5.31. Two bowls of equal capacity; with discs (left) and without discs (right)

A disc-type bowl is often used for the separation of fine solids from a liquid and its construction is shown in the right-hand side of Fig. 5.29. Here there is only one liquid outlet $K$, and the solids are retained in the space between the ends of the discs and the walls of the basket.

### Valve Nozzle Centrifuges

The continuous removal of solids from the centrifuge basket can be effected by fitting a number of discharge nozzles around the periphery of the basket. The centrifuge is operated in such a way that solids are ejected with sufficient liquid to enable them to flow. The frequency of opening of the valves and the duration for which they remain open can be controlled independently or, alternatively, the valves may be actuated automatically according to the level of the solids which have built up. Valve nozzles can be used on simple bowl or disc centrifuges (Fig. 5.32) and are suitable for applications where a solid is to be separated from a liquid or there are two liquids containing suspended solids.

Centrifuges of this type are used in the processing of yeast, starch, meat, fish products and fruit juices. They form essential components of the process of impulse rendering for the extraction of oils and fats from cellular materials. The raw material, consisting of bones, animal fat, fish offal, or vegetable seeds, is first disintegrated and then, after a preliminary gravitational separation, the final separation of water, oil, and suspended solids is carried out in a number of valve nozzle centrifuges.

### Scroll-type Centrifuges

In the scroll type of centrifuge the mixture is fed to the machine through a hollow shaft which discharges near one end of the basket; with thick suspensions the flow is assisted by

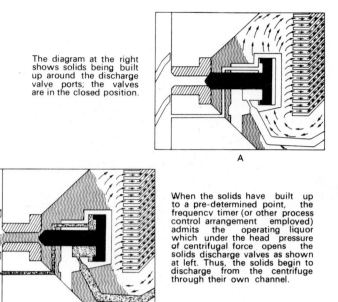

The diagram at the right shows solids being built up around the discharge valve ports; the valves are in the closed position.

A

When the solids have built up to a pre-determined point, the frequency timer (or other process control arrangement employed) admits the operating liquor which under the head pressure of centrifugal force opens the solids discharge valves as shown at left. Thus, the solids begin to discharge from the centrifuge through their own channel.

B

The duration timer is so adjusted that it stops the flow of operating liquor, thus closing the valve, before all the solids have been discharged. This prevents loss of the mother liquor with the solids. The operating liquor passes out of the valve and the centrifuge bowl through its own controlled bleed line. Thus, the operating liquor does not come in contact with either the solids discharge or the liquid being centrifuged.

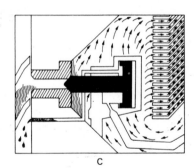

C

FIG. 5.32. Operating of valve nozzle centrifuge

means of a screw mechanism. A spiral scroll rotates at a speed slightly different from that of the basket and causes the solids deposited on the wall to move steadily along in an axial direction away from the inlet. It operates at high speed giving accelerations up to 3000 $g$, but the speed differential is not sufficient to cause interference with the separation. The time for which material remains in the machine is directly proportional to its length, and therefore this type of centrifuge is usually relatively long and of small diameter. It can readily be adapted for operation at high pressures. The axis of the centrifuge is usually horizontal though vertical mounting is sometimes used. The basket is either cylindrical or in the form of a truncated cone (Fig. 5.33), in which case the feed is introduced at the large diameter end. The conical form is preferred where the prime requirement is the dryness of the solids, and the cylindrical form where liquor clarity is of overriding importance. The basket and scroll may be driven by a single motor using a differential gear, but in some models separate drives are employed so that the differential speed can be varied. The use of separate drives has the additional advantage that the basket and scroll will rotate at the same rate if the mechanism tends to become overloaded.

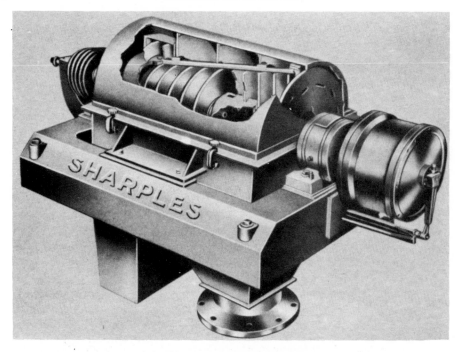

FIG. 5.33. Continuous conical centrifuge

Machines of this type are made in a wide range of sizes and can be fitted with perforated or imperforate baskets. The liquid is taken off through an overflow or through the perforations. When this centrifuge is used as a filter, the filter cake can be washed and dried as it passes along the basket.

*Pusher-type Centrifuges*

This type of centrifuge is used for the separation of suspensions and is fitted either with a perforated or imperforate basket. The feed is introduced through a conical funnel and the cake is formed in the space between the flange and the end of the basket. The solids are intermittently moved along the surface of the basket by means of a reciprocating pusher. The pusher comes forward and returns immediately but waits until a further layer of solids has been built up before advancing again. In this machine the thickness of filter cake cannot exceed the distance between the surface of the basket and the flange of the funnel. The liquid either passes through the holes in the basket or, in the case of an imperforate basket, is taken away through an overflow. The solids are washed by means of a spray, as shown in Fig. 5.34.

A form of pusher-type centrifuge which is particularly suitable for filtering slurries of low concentrations is shown in Fig. 5.35. A perforated pusher cone gently accelerates the feed and secures a large amount of preliminary drainage near the apex of the cone. The solids from the partially concentrated suspension are then evenly laid on the cylindrical surface and the risk of the solids being washed out of the basket is minimised.

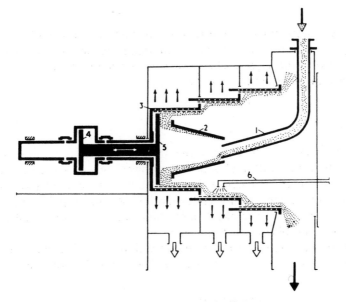

FIG. 5.34. Pusher-type centrifuge.
1, Inlet. 2, Inlet funnel. 3, Basket. 4, Piston. 5, Pusher disc. 6, Washing spray

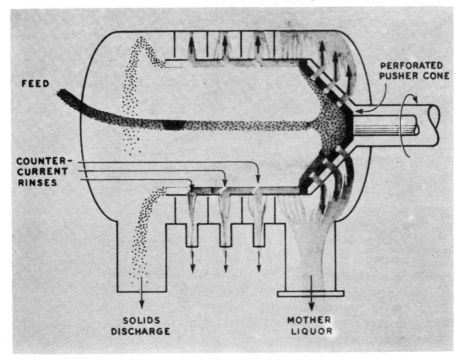

FIG. 5.35. Pusher centrifuge for low concentration slurries

## Statifuge

This machine is well suited to the removal of small quantities of solids from large volumes of liquid. It has a stationary bowl containing a revolving disc stack which is also effective in clarifying the liquid. This system is considerably cheaper than the more conventional arrangements in many cases.

## Supercentrifuge

Because, for a given separating power, the stress in the wall is a minimum for machines of small radius, machines with high separating powers generally use very tall baskets of small diameters. A typical centrifuge (Fig. 5.36) would consist of a basket about 100 mm diameter and 1 m long incorporating baffles to bring the liquid rapidly up to speed. Speeds up to 1000 Hz are used to give accelerations 50,000 times the gravitational acceleration. A wide range of materials of construction can be used.

The position of the liquid interface is determined by balancing centrifugal forces as in Fig. 5.23. The lip (of radius $r_w$) over which the denser liquid leaves the bowl is part of a

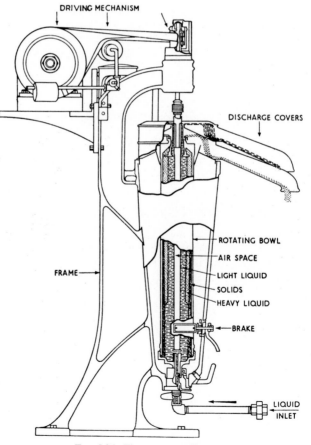

FIG. 5.36. The supercentrifuge

removable ring. Various sizes may be fitted to provide for the separation of liquids of various relative densities.

Often the material fed to these machines contains traces of denser solids in addition to the two liquid phases. These solids are deposited on the inner wall of the bowl, and the machine is dismantled periodically to remove them. A common application is the removal of water and suspended solids from lubricating oil.

The supercentrifuge is used for clarification of oils and fruit juices and for the removal of oversize and undersize particles from pigmented liquids. The liquid is continuously discharged but the solids are retained in the bowl and must be removed periodically.

*Ultracentrifuge*

For separation of colloidal particles and for breaking down emulsions, the ultracentrifuge is used. It operates at speeds up to 1600 Hz and produces a force of as much as 500,000 times the force of gravity. The basket is usually driven by means of a small air turbine. The ultracentrifuge is often run either at low pressures or in an atmosphere of hydrogen in order to reduce frictional losses, and a fivefold increase in the maximum speed can be attained by this means.

## 5.4. FLOCCULATION

### 5.4.1. Introduction

The tendency of the particulate phase of colloidal dispersions to aggregate is an important physical property which finds practical application in solid–liquid separation processes, such as sedimentation and filtration. The aggregation of colloids is known as coagulation, or flocculation. Particles dispersed in liquid media collide due to their relative motion; and stability (i.e. stability against aggregation) of the dispersion is determined by the interaction between particles during these collisions. Attractive and repulsive forces can be operative between the particles; these forces may react in different ways depending on environmental conditions (salt concentration, pH). The commonly occurring forces between colloidal particles are van der Waals forces, electrostatic forces and forces due to adsorbed macromolecules. In the absence of macromolecules, aggregation is largely due to van der Waals attractive forces, whereas stability is due to repulsive interaction between similarly charged electrical double-layers.

### 5.4.2. The Electrical Double-layer

Most particles acquire a surface electric charge when in contact with a polar medium. Ions of opposite charge (counter-ions) in the medium are attracted towards the surface and ions of like charge (co-ions) are repelled, and this process, together with the mixing tendency due to thermal motion, results in the creation of an electrical double-layer which comprises the charged surface and a neutralising excess of counter-ions over co-ions distributed in a diffuse manner in the polar medium. The quantitative theory of the electrical double-layer, which deals with the distribution of ions and the magnitude of electric potentials, is beyond the scope of this text but an understanding of it is essential in an analysis of colloid stability[27,28].

For present purposes the electrical double-layer is represented in terms of Stern's model (Fig. 5.37) wherein the double-layer is divided into two parts separated by a plane (Stern plane) located at a distance of about one hydrated-ion radius from the surface. The potential changes from $\psi_0$ (surface) to $\psi_\delta$ (Stern potential) in the Stern layer and decays to zero in the diffuse double-layer; quantitative treatment of the diffuse double-layer follows the Gouy–Chapman theory[29,30].

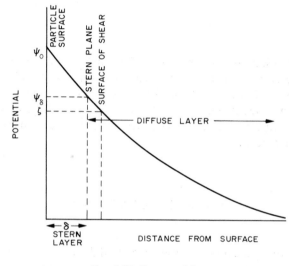

FIG. 5.37. Stern model

$\psi_\delta$ can be estimated from electrokinetic measurements (e.g. electrophoresis, streaming potential). In such measurements surface and liquid move tangentially with respect to each other. For example, in electrophoresis the liquid is stationary and the particles move under the influence of an applied electric field. A thin layer (few molecules thick) of liquid moves together with the particle so that the actual hydrodynamic boundary between the moving unit and the stationary liquid is a *slipping plane* inside the solution. The potential at the slipping plane is termed the *zeta potential*, $\zeta$ (Fig. 5.37).

Lyklema[31] considers that the slipping plane can be identified with the Stern plane so that $\psi_\delta \simeq \zeta$; thus, since the surface potential $\psi_0$ is inaccessible, zeta potentials find practical application in the calculation of $V_R$ (see equation 5.121). In practice, electrokinetic measurements must be carried out with considerable care if reliable estimates of $\zeta$ are to be made[32].

### 5.4.3. Interactions between Particles

The interplay of forces between particles in lyophobic sols may be interpreted in terms of the theory of Derjaguin and Landau[33] and Verwey and Overbeek[27]. Their theory (DLVO theory) considers that the potential energy of interaction between a pair of particles consists of two components:

(a) a repulsive component $V_R$ arising from the overlap of the electrical double-layers;

(b) a component $V_A$ due to van der Waals attraction arising from electromagnetic effects.

These were considered to be additive so that the total potential energy of interaction $V_T$ is given by:

$$V_T = V_R + V_A \qquad (5.120)$$

In general, the calculation of $V_R$ is complex[34] but a useful approximation for identical spheres of radius $a$ is given by[27]:

$$V_R = \frac{64\pi \, an_i \, KT\gamma^2 \, e^{-\kappa H_s}}{\kappa^2} \qquad (5.121)$$

where

$$\gamma = \frac{\exp \left(Ze_c \psi_\delta / 2KT\right) - 1}{\exp \left(Ze_c \psi_\delta / 2KT\right) + 1} \qquad (5.122)$$

and

$$\kappa = \left(\frac{2e_c^2 n_i Z^2}{\varepsilon KT}\right)^{1/2} \qquad (5.123)$$

For identical spheres with $H_s \leqslant 10\text{--}20\,\text{mm}$ (100–200 Å) and when $H_s \ll a$, the energy of attraction $V_A$ is given by the approximate expression[27]:

$$V_A = -\frac{\mathscr{A}a}{12H_s} \qquad (5.124)$$

where $\mathscr{A}$ is the Hamaker[35] constant whose value depends on the nature of the material of the particles. The presence of liquid between particles reduces $V_A$ and an effective Hamaker constant is calculated from:

$$\mathscr{A} = (\mathscr{A}_2^{1/2} - \mathscr{A}_1^{1/2})^2 \qquad (5.125)$$

where subscripts 1 and 2 refer to dispersion medium and particles respectively. Equation 5.124 is based on the assumption of complete additivity of intermolecular interactions; this assumption is avoided in the theoretical treatment of Lifshitz[36] which is based on macroscopic properties of materials[37]. Tables of $\mathscr{A}$ are available in the literature[38]; values are generally found to lie in the range 0·1 to $10 \times 10^{-20}$ J.

The general form of $V_T$ versus distance of separation between particle surfaces $H_s$ is shown schematically in Fig. 5.38. At very small distances of separation repulsion due to overlapping electron clouds (Born repulsion)[27] predominates and consequently a deep minimum (primary minimum) occurs in the potential energy curve. For smooth surfaces this limits the distance of closest approach ($H_{smin}$) to $\sim 0.4\,\text{nm}$ (4 Å). Aggregation of particles occurring in this primary minimum, e.g. aggregation of lyophobic sols in the presence of NaCl, is termed *coagulation*[37].

At high surface potentials, low ionic strengths and intermediate distances the electrical repulsion term is dominant and so a maximum (primary maximum) occurs in the potential energy curve. At larger distances of separation $V_R$ decays more rapidly than $V_A$

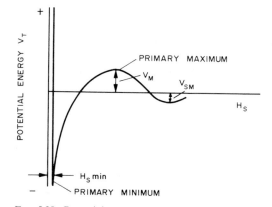

FIG. 5.38. Potential energy as a function of separation

and a secondary minimum appears. If the potential energy maximum is large compared with the thermal energy $KT$ ($\sim 4\cdot2 \times 10^{-2}$ J) of the particles the system should be stable, otherwise the particles would coagulate. The height of this energy barrier to coagulation depends upon the magnitude of $\psi_\delta$ (and $\zeta$) and upon the range of the repulsive forces (i.e. upon $1/\kappa$). If the depth of the secondary minimum is large compared with $KT$, it should produce a loose, easily reversible form of aggregation which is termed flocculation; this term also describes aggregation of particles in the presence of polymers[37] (see Section 5.4.6). It is of interest to note that both $V_A$ and $V_R$ increase as particle radius $a$ becomes larger and thus $V_M$ (Fig. 5.38) would be expected to increase with the sol becoming more stable; also if $a$ increases then $V_{SM}$ increases and may become large enough to produce "secondary minimum" flocculation.

### 5.4.4. Coagulation Concentrations

Coagulation concentrations are the electrolyte concentrations required just to coagulate a sol. Clearly $V_M$ (Fig. 5.38) must be reduced, preferably to zero, to allow coagulation. This can be achieved by increasing the ionic strength of the solution, thus increasing $\kappa$ and thereby reducing $V_R$ in equation 5.121. The addition of salts with multivalent ions (e.g. $Al^{3+}$, $Ca^{2+}$, $Fe^{3+}$) is most effective because of the effect of charge number $Z$ and $\kappa$ (equation 5.123). Taking as a criterion that $V_T = 0$ and $dV_T/dH_s = 0$ for the same value of $H_s$, it is readily shown[27] that the coagulation concentration $c_c$ is given by:

$$c_c = \frac{9\cdot75\,B^2\,\varepsilon^3\,K^5\,T^5\,\gamma^4}{e_c^2 N_A \mathscr{A}^2 Z^6} \tag{5.126}$$

where $B = 3\cdot917 \times 10^{39}$ $1/C^2$. At high values of surface potentials $\gamma \approx 1$ and equation 5.126 predicts that the coagulation concentration should be inversely proportional to the sixth power of the valency $Z$. Thus coagulation concentrations of indifferent electrolytes whose counter-ions have charge numbers 1, 2, 3 should be in the ratio $100:1\cdot6:0\cdot13$. It should be noted that if an ion is specifically adsorbed on the particles $\psi_\delta$ can be drastically reduced and coagulation effected without any great increase in ionic strength; for instance, minute traces of certain hydrolysed metal ions can cause coagulation of negatively charged particles[39]. In such cases charge reversal often occurs and the particles can be restabilised if excess coagulant is added.

### 5.4.5. Kinetics of Coagulation

The rate of coagulation of particles in a liquid depends on the frequency of collisions between particles due to their relative motion. When this motion is due to Brownian movement coagulation is termed *perikinetic*; when the relative motion is caused by velocity gradients coagulation is termed *orthokinetic*.

Modern analyses of perikinesis and orthokinesis take account of hydrodynamic forces as well as interparticle forces. In particular, the frequency of binary collisions between spherical particles has received considerable attention[40-43].

The frequency of binary encounters during perikinesis is determined by considering the process as that of diffusion of spheres (radius $a_2$) and number concentration $n_2$ towards a central reference sphere of radius $a_1$, whence the frequency of collision $I$ is given by[40]:

$$I = \frac{4\pi D_{12}^{(\infty)} n_2 (a_1 + a_2)}{1 + \dfrac{a_2}{a_1} \displaystyle\int_{1+(a_2/a_1)}^{\infty} (D_{12}^{(\infty)}/D_{12}) \exp(V_T/KT)\dfrac{ds}{s^2}} \tag{5.127}$$

where $s = a_r/a_1$ and the coordinate $a_r$ has its origin at the centre of sphere 1.

Details of $D_{12}$, the relative diffusivity between unequal particles, are given by Spielman[40] who illustrates the dependence of $D_{12}$ on the relative separation $a_r$ between particle centres. At infinite separation (where hydrodynamic effects vanish):

$$D_{12} = D_{12}^{(\infty)} = D_1 + D_2 \tag{5.128}$$

where $D_1$ and $D_2$ are absolute diffusion coefficients given by the Stokes–Einstein equation.

$$D_1 = KT/(6\pi\mu a_1)$$
$$D_2 = KT/(6\pi\mu a_2) \tag{5.129}$$

When long-range particle interactions and hydrodynamic effects are ignored equation 5.127 becomes equivalent to the solution of Smoluchowski who obtained the collision frequency $I_s$ as:

$$I_s = 4\pi D_{12}^{(\infty)} n_2 (a_1 + a_2) \tag{5.130}$$

and who assumed an attractive potential only given by:

$$V_A = -\infty \quad H_s \leqslant (a_1 + a_2)$$
$$V_A = 0 \quad H_s > (a_1 + a_2) \tag{5.131}$$

Thus:
$$I = \alpha_p I_s \tag{5.132}$$

where the ratio $\alpha_p$ is the reciprocal of the denominator in equation 5.127; values of $\alpha_p$ are tabulated by Spielman[40].

Assuming an attractive potential only given by equation 5.131 Smoluchowski showed that the frequency of collisions per unit volume between particles of radii $a_1$ and $a_2$ in the presence of a laminar shear gradient $\dot{\gamma}$ is given by:

$$J_s = \frac{4}{3} n_1 n_2 (a_1 + a_2)^3 \dot{\gamma} \qquad (5.133)$$

Analyses[43] of the orthokinetic encounters between equi-sized spheres have shown that as with perikinetic encounters equation 5.133 can be modified to include a ratio $\alpha_0$ to give the collision frequency $J$ as:

$$J = \alpha_0 J_s \qquad (5.134)$$

where $\alpha_0$, which is a function of $\dot{\gamma}$, corrects the Smoluchowski relation for hydrodynamic interactions and interparticle forces. Zeichner and Schowalter[43] graphically present $\alpha_0^{-1}$ as a function of a dimensionless parameter $N_F (= 6\pi\mu a^3 \dot{\gamma}/\mathscr{A})$ for the condition $V_R = 0$, whence it is possible to show that, for values of $N_F > 10$, $J$ is proportional to $\dot{\gamma}$ raised to the 0·77 power instead of the first power as given by equation 5·133.

Perikinetic coagulation is normally too slow for economic practical use in such processes as wastewater treatment, and orthokinetic coagulation is often used to produce rapid growth of aggregate or floc size. In such situations floc–floc collisions occur under non-uniform turbulent flow conditions. A rigorous analysis of the kinetics of coagulation under these conditions is not available at present. A widely used method of evaluating a mean shear gradient in such practical situations is given by Camp and Stein[44], viz.:

$$\dot{\gamma} = [\mathbf{P}/\mu]^{1/2} \qquad (5.135)$$

where $\mathbf{P}$ = power input/unit volume of fluid.

### 5.4.6. Effect of Polymers on Stability

The stability of colloidal dispersions is strongly influenced by the presence of adsorbed polymers. Sols can be stabilised or destabilised depending on a number of factors including the relative amounts of polymer and sol, the mechanism of adsorption of polymer and the method of mixing polymer and dispersion[45]. Adsorption of polymer on to colloidal particles may increase their stability by decreasing $V_A$[46,47], increasing $V_R$[48] or by introducing a *steric* component of repulsion $V_S$[49,50].

Flocculation is readily produced by linear homopolymers of high molecular weight. Although they may be non-ionic they are commonly polyelectrolytes; polyacrylamides and their derivatives are widely used in practical situations[51]. Flocculation by certain high molecular weight polymers can be interpreted in terms of a *bridging* mechanism; polymer molecules may be long and flexible enough to adsorb on to several particles. The precise nature of the attachment between polymer and particle surface depends on the nature of the surfaces of particle and polymer and on the chemical properties of the solution. Various types of interaction between polymer segments and particle surfaces may be envisaged. In the case of polyelectrolytes, the strongest of these interactions would be ionic association between a charged site on the surface and an oppositely charged polymer segment, e.g. polyacrylic acid and positively charged silver iodide particles[52–55].

Polymers may show an optimum flocculation concentration which depends on molecular weight and concentration of solids in suspension. Overdosing with flocculant may lead to restabilisation[56], as a consequence of particle surfaces becoming saturated with polymer. Optimum flocculant concentrations may be determined by a range of techniques including sedimentation rate, sedimentation volume, filtration rate and clarity of supernatant liquid[57].

### 5.4.7. Effect of Flocculation on Sedimentation

In a flocculated (or coagulated) suspension the aggregates of fine particles or flocs are the basic structural units and in a low shear rate process, such as gravity sedimentation, their settling rates and sediment volumes depend largely on volumetric concentration of floc and on interparticle forces. The type of settling behaviour exhibited by flocculated suspensions depends largely on the initial solids concentration and chemical environment. Two kinds of batch settling curve are frequently seen. At low initial solids concentration the flocs may be regarded as discrete units consisting of particles and immobilised fluid. The flocs settle initially at a constant settling rate but as they accumulate on the bottom of the vessel they deform under the weight of the overlying flocs. The curves shown earlier in Fig. 5.3 for calcium carbonate suspensions relate to this type of sedimentation; when the solids concentration is very high the maximum settling rate is not immediately reached and may increase with increasing initial height of suspension[58]. Such behaviour appears to be characteristic of structural flocculation associated with a continuous network of flocs extending to the walls of the vessel. Quantitative treatment of both forms of sedimentation has been given by Michaels and Bolger[58]; in particular the constant velocity characteristic of the first type of behaviour has been interpreted in terms of a modified form of equation 5.30 involving the free-falling velocity of a characteristic floc and floc volume concentration.

## 5.5. FURTHER READING

AMBLER, C. M.: *Chem. Eng. Prog.* **48** (1952) 150. The evaluation of centrifuge performance.
AMBLER, C. M.: *Ind. Eng. Chem.* **53** (1961) 430. Centrifugation equipment. Theory.
DALLAVALLE, J. M.: *Micromeritics*, 2nd edn. (Pitman, 1948).
FLOOD, J. E.: *Ind. Eng. Chem.* **53** (1961) 489. Centrifugation.
IVES, K. J. (ed.): *The Scientific Basis of Filtration* (Noordhoff, Leyden, 1975).
ORR, C.: *Particulate Technology* (Macmillan, New York, 1966).
OTTEWILL, R. H.: Particulate dispersions. In *Colloid Science* 2, 173–219 (The Chemical Society, London, 1973).
SMITH, A. L. (ed.): *Particle Growth in Suspensions* (Academic Press, 1973).
SMITH, J. C.: *Ind. Eng. Chem.* **53** (1961) 439. Centrifugation equipment. Applications.

## 5.6. REFERENCES

1. KAYE, B. H. and BOARDMAN, R. P.: Third Congress of the European Federation of Chemical Engineering (1962). *Symposium on the Interaction between Fluids and Particles* 17. Cluster formation in dilute suspensions.
2. COE, H. S. and CLEVENGER, G. H.: *Trans. Am. Inst. Min. Met. Eng.* **55** (1916) 356. Methods for determining the capacities of slime-settling tanks.
3. ROBINSON, C. S.: *Ind. Eng. Chem.* **18** (1926) 869. Some factors influencing sedimentation.
4. EINSTEIN, A.: *Ann. Phys.* **19** (1906) 289–306. Eine neue Bestimmung der Molekuldimensionen.
5. VAND, V.: *J. Phys. Coll. Chem.* **52** (1948) 277. Viscosity of solutions and suspensions.
6. STEINOUR, H. H.: *Ind. Eng. Chem.* **36** (1944) 618, 840, and 901. Rate of sedimentation.
7. HAWKSLEY, P. G. W.: *Inst. of Phys. Symposium* (1950) 114. The effect of concentration on the settling of suspensions and flow through porous media.

8. WALLIS, G. B.: Third Congress of the European Federation of Chemical Engineering (1962). *Symposium on the Interaction between Fluids and Particles*, 9. A simplified one-dimensional representation of two-component vertical flow and its application to batch sedimentation.

9. EGOLF, C. B. and MCCABE, W. L.: *Trans. Am. Inst. Chem. Eng.* **33** (1937) 620. Rate of sedimentation of flocculated particles.

10. WORK, L. T. and KOHLER, A. S.: *Trans. Am. Inst. Chem. Eng.* **36** (1940) 701. The sedimentation of suspensions.

11. PEARCE, K. W.: Third Congress of the European Federation of Chemical Engineering (1962). *Symposium on the Interaction between Fluids and Particles*, 30. Settling in the presence of downward-facing surfaces.

12. ROBINS, W. H. M.: Third Congress of the European Federation of Chemical Engineering (1962). *Symposium on the Interaction between Fluids and Particles*, 26. The effect of immersed bodies on the sedimentation of suspensions.

13. RICHARDSON, J. F. and ZAKI, W. N.: *Trans. Inst. Chem. Eng.* **32** (1954) 35. Sedimentation and fluidisation: Part I.

14. GARSIDE, J. and AL-DIBOUNI, M. R.: *Ind. Eng. Chem. Proc. Des. Dev.* **16** (1977) 206. Velocity–voidage relationships for fluidization and sedimentation in solid–liquid systems.

15. KHAN, A. R.: Private communications. Work in progress.

16. KYNCH, G. J.: *Trans. Faraday Soc.* **48** (1952) 166. A theory of sedimentation.

17. RICHARDSON, J. F. and MEIKLE, R. A.: *Trans. Inst. Chem. Eng.* **39** (1961) 348. Sedimentation and fluidisation. Part III. The sedimentation of uniform fine particles and of two-component mixtures of solids.

18. CARMAN, P. C.: *Trans. Inst. Chem. Eng.* **15** (1937) 150. Fluid flow through granular beds.

19. RICHARDSON, J. F. and SHABI, F. A.: *Trans. Inst. Chem. Eng.* **38** (1960) 33. The determination of concentration distribution in a sedimenting suspension using radioactive solids.

20. RICHARDSON, J. F. and MEIKLE, R. A.: *Trans. Inst. Chem. Eng.* **39** (1961) 357. Sedimentation and fluidisation. Part IV. Drag force on individual particles in an assemblage.

21. ROWE, P. N. and HENWOOD, G. N.: *Trans. Inst. Chem. Eng.* **39** (1961) 43. Drag forces in a hydraulic model of a fluidised bed. Part 1.

22. SMITH, T. N.: *Trans. Inst. Chem. Eng.* **45** (1967) T311. The differential sedimentation of particles of various species.

23. LOCKETT, M. J. and AL-HABBOOBY, H. M.: *Trans. Inst. Chem. Eng.* **51** (1973) 281. Differential settling by size of two particle species in a liquid.

24. MIRZA, S.: University of Wales M.Sc. thesis (1978). Sedimentation of polydiapase suspensions.

25. HAYTER, A. J.: *J. Soc. cosmet. Chem.* (1962) 152. Progress in centrifugal separations.

26. TROWBRIDGE, M. E. O'K.: *Chem. Engr. London* No. **162** (Aug. 1962) A73. Problems in the scaling-up of centrifugal separation equipment.

27. VERWEY, E. J. W. and OVERBEEK, J. TH. G.: *Theory of the Stability of Lyophobic Colloids* (Elsevier, Amsterdam, 1948).

28. SPARNAAY, M. J.: *The Electrical Double Layer* (Pergamon Press, Oxford, 1972).

29. SHAW, D. J.: *Introduction to Colloid and Surface Chemistry* (Butterworths, London, 1970).

30. SMITH, A. L.: Electrical phenomena associated with the solid–liquid interface, in *Dispersion of Powders in Liquids*, ed. G. D. PARFITT (Applied Science Publishers, London, 1973).

31. LYKLEMA, J.: *J. Coll. and Interface Sci.* **58** (1977) 242. Water at interfaces: A colloid chemical approach.

32. WILLIAMS, D. J. A. and WILLIAMS, K. P.: *J. Coll. and Interface Sci.* **65** (1978) 79. Electrophoresis and zeta potential of kaolinite.

33. DERJAGUIN, B. V. and LANDAU, L.: *Acta Physicochim.* **14** (1941) 663. Theory of the stability of strongly charged lyophobic solutions and of the adhesion of strongly charged particles in solution of electrolytes.

34. GREGORY, J.: Interfacial phenomena, in *The Scientific Basis of Filtration*, ed. K. J. IVES (Noordhoff, Leyden, 1975).

35. HAMAKER, H. C.: *Physica* **4** (1937) 1058. The London–van der Waals attraction between spherical particles.

36. LIFSHITZ, E. M.: *Soviet Physics. J.E.T.P.* **2** (1956) 73. Theory of molecular attractive forces between solids.

37. OTTEWILL, R. H.: Particulate dispersions. In *Colloid Science* **2** (The Chemical Society, London, 1973) 173.

38. VISSER, J.: *Adv. Colloid Interface Sci.* **3** (1972) 331. On Hamaker constants: a comparison between Hamaker constants and Lifshitz–van der Waals constants.

39. MATIJEVIC, E., JANAUER, G. E. and KERKER, M.: *J. Coll. Sci.* **19** (1964) 333. Reversal of charge of hydrophobic colloids by hydrolyzed metal ions, I. Aluminium nitrate.

40. SPIELMAN, L. A.: *J. Coll. and Interface Sci.* **33** (1970) 562. Viscous interactions in Brownian coagulation.

41. HONIG, E. P., ROEBERSEN, G. J., and WIERSEMA, P. H.: *J. Coll. and Interface Sci.* **36** (1971) 97. Effect of hydrodynamic interaction on the coagulation rate of hydrophobic colloids.

42. VAN DE VEN, T. G. M. and MASON, S. G.: *J. Coll. and Interface Sci.* **57** (1976) 505. The microrheology of colloidal dispersions, IV. Pairs of interacting spheres in shear flow.

43. ZEICHNER, G. R. and SCHOWALTER, W. R.: *A.I.Ch.E.Jl* **23** (1977) 243. Use of trajectory analysis to study stability of colloidal dispersions in flow fields.

44. CAMP, T. R. and STEIN, P. C.: *J. Boston Soc. Civ. Eng.* **30** (1943) 219. Velocity gradients and internal work in fluid motion.
45. LA MER, V. K. and HEALY, T. W.: *Rev. Pure Appl. Chem.* (Australia) **13** (1963) 112. Adsorption–flocculation reactions of macromolecules at the solid–liquid interface.
46. VOLD, M. J.: *J. Coll. Sci.* **16** (1961) 1. The effect of adsorption on the van der Waals interaction of spherical particles.
47. OSMOND, D. W. J., VINCENT, B., and WAITE, F. A.: *J. Coll. and Interface Sci.* **42** (1973) 262. The van der Waals attraction between colloid particles having adsorbed layers, I. A re-appraisal of the "Vold" effect.
48. VINCENT, B.: *Adv. Coll. Interface Sci.* **4** (1974) 193. The effect of adsorbed polymers on dispersion stability.
49. NAPPER, D. H.: *J. Coll. and Interface Sci.* **58** (1977) 390. Steric stabilisation.
50. DOBBIE, J. W., EVANS, R. E., GIBSON, D. V., SMITHAM, J. B. and NAPPER, D. H.: *J. Coll. and Interface Sci.* **45** (1973) 557. Enhanced steric stabilisation.
51. O'GORMAN, J. V. and KITCHENER, J. A.: *Int. J. Min. Proc.* **1** (1974) 33. The flocculation and dewatering of Kimberlite clay slimes.
52. WILLIAMS, D. J. A. and OTTEWILL, R. H.: *Kolloid-Z. u. Z. Polymere* **243** (1971) 141. The stability of silver iodide solutions in the presence of polyacrylic acids of various molecular weights.
53. GREGORY, J.: *Trans. Faraday Soc.* **65** (1969) 2260. Flocculation of polystyrene particles with cationic polyelectrolytes.
54. GREGORY, J.: *J. Coll. and Interface Sci.* **55** (1976) 35. The effect of cationic polymers on the colloid stability of latex particles.
55. GRIOT, O. and KITCHENER, J. A.: *Trans. Faraday Soc.* **61** (1965) 1026. Role of surface silanol groups in the flocculation of silica by polycrylamide.
56. LA MER, V. K.: *Disc. Faraday Soc.* **42** (1966) 248. Filtration of colloidal dispersions flocculated by anionic and cationic polyelectrolytes.
57. SLATER, R. W. and KITCHENER, J. A.: *Disc. Faraday Soc.* **42** (1966) 267. Characteristics of flocculation of mineral suspensions by polymers.
58. MICHAELS, A. S. and BOLGER, J. C.: *Ind. Eng. Chem. Fundamentals* **1** (1962) 24. Settling rates and sediment volumes of flocculated kaolinite suspensions.

## 5.7. NOMENCLATURE

| | | Units in SI System | Dimensions in $M, L, T, \theta, A$ |
|---|---|---|---|
| $A$ | Cross-sectional area of vessel or tube | m² | $L^2$ |
| $A'$ | Cross-sectional area at level of obstruction | m² | $L^2$ |
| $\mathscr{A}$ | Hamaker constant | J | $ML^2T^{-2}$ |
| $a$ | Particle radius | m | $L$ |
| $a_r$ | Separation distance between particle centres | m | $L$ |
| $a'$ | Constant in Vand's equation (5.4) | — | — |
| $B$ | Constant in equation 5.126 ( $= 3.917 \times 10^{39} \, C^{-2}$ ) | $C^{-2}$ | $T^{-2}A^{-2}$ |
| $b$ | Constant in equation 5.10 | $s^{-1}$ | $T^{-1}$ |
| $C$ | Fractional volumetric concentration | — | — |
| $C_0$ | Initial uniform concentration | — | — |
| $C_f$ | Concentration of fines in upper zone | — | — |
| $C_i$ | Concentration corresponding to point of inflexion on $\psi$–$C$ curve | — | — |
| $C_m$ | Value of $C$ at which $\psi$ is a maximum | — | — |
| $C_{max}$ | Concentration corresponding to sediment layer | — | — |
| $C_u$ | Concentration of underflow in continuous thickener | — | — |
| $C'$ | Value of $C$ at constriction | — | — |
| $c_c$ | Critical coagulation concentration | kmol/m³ | $ML^{-3}$ |
| $D_1, D_2$ | Absolute diffusivities | m²/s | $L^2T^{-1}$ |
| $D_{12}$ | Relative diffusivity | m²/s | $L^2T^{-1}$ |
| $d$ | Diameter of sphere or equivalent spherical diameter | m | $L$ |
| $d_t$ | Diameter of tube or vessel | m | $L$ |
| $e$ | Voidage of suspension | — | — |
| $e_c$ | Elementary charge ( $= 1.60 \times 10^{-19} \, C$ ) | C | $TA$ |
| $F_c$ | Drag force on particle of volume $v$ | N | $MLT^{-2}$ |
| $g$ | Acceleration due to gravity | m/s² | $LT^{-2}$ |
| $H$ | Height in suspension above base | m | $L$ |
| $H_0$ | Value of $H$ at $t = 0$ | m | $L$ |

|  |  | Units in SI System | Dimensions in M, L, T, $\theta$, A |
|---|---|---|---|
| $H_c$ | Height corresponding to critical settling point | m | L |
| $H_s$ | Separation between particle surfaces | m | L |
| $H_t$ | Initial total depth of suspension | m | L |
| $H_\infty$ | Final height of sediment | m | L |
| $h$ | Depth of liquid at basket wall | m | L |
| $I$ | Collision frequency calculated from equation 5.127 | 1/s | $T^{-1}$ |
| $I_s$ | Collision frequency calculated from equation 5.131 | 1/s | $T^{-1}$ |
| $J$ | Collision frequency calculated from equation 5.134 | $m^{-3}s^{-1}$ | $L^{-3}T^{-1}$ |
| $J_s$ | Collision frequency calculated from equation 5.133 | $m^{-3}s^{-1}$ | $L^{-3}T^{-1}$ |
| $K$ | Boltzmann's constant ($=1\cdot38 \times 10^{-23}$ J/K) | J/K | $ML^2T^{-2}\theta^{-1}$ |
| $K''$ | Constant in Robinson's equation (5.1) | — | — |
| $k'$ | Shape factor | — | — |
| $k''$ | Constant in Einstein's equation (5.3) | — | — |
| $L$ | Depth of basket of centrifuge | m | L |
| $N_A$ | Avogadro's constant ($=6\cdot0225 \times 10^{-20}$ kmol$^{-1}$) | 1/kmol | $M^{-1}$ |
| $N_F$ | Dimensionless parameter ($6\pi\mu a^3\dot{\gamma}/\mathscr{A}$) | — | — |
| $n$ | Index in equations 5.29 and 5.30 | — | — |
| $n_i$ | Number of ions per unit volume | $1/m^3$ | $L^{-3}$ |
| $n_1, n_2$ | Number of particles per unit volume | $1/m^3$ | $L^{-3}$ |
| $P$ | Pressure | $N/m^2$ | $ML^{-1}T^{-2}$ |
| $\mathbf{P}$ | Power input per unit volume of fluid | $W/m^3$ | $ML^{-1}T^{-3}$ |
| $Q$ | Volumetric feed rate of suspension to continuous thickener of centrifuge | $m^3/s$ | $L^3T^{-1}$ |
| $Q_0$ | Volumetric flowrate of overflow from continuous thickener | $m^3/s$ | $L^3T^{-1}$ |
| $q$ | Volumetric rate of sedimentation of solids | $m^3/s$ | $L^3T^{-1}$ |
| $R$ | Radius of centrifuge basket | m | L |
| $R_t$ | Drag force per unit projected area of isolated spherical particle | $N/m^2$ | $ML^{-1}T^{-2}$ |
| $R_t'$ | Drag force per unit projected area of spherical particle in suspension | $N/m^2$ | $ML^{-1}T^{-2}$ |
| $R'$ | Resistance per unit area of spherical particle | $N/m^2$ | $ML^{-1}T^{-2}$ |
| $R_1$ | Drag force per unit surface of particle | $N/m^2$ | $ML^{-1}T^{-2}$ |
| $r$ | Radius of rotation in centrifuge | m | L |
| $r_0$ | Radius of liquid surface at bottom of centrifuge basket | m | L |
| $r_i$ | Radius of inner surface of liquid | m | L |
| $r_s$ | Radius of interface between liquids in centrifuge | m | L |
| $r_w$ | Radius of outer weir in centrifuge | m | L |
| $S$ | Specific surface | $m^2/m^3$ | $L^{-1}$ |
| $s$ | Dimensionless separation distance ($=a_r/a_1$) | — | — |
| $T$ | Absolute temperature | K | $\theta$ |
| $t$ | Time | s | T |
| $t_R$ | Residence time | s | T |
| $U$ | Mass ratio of liquid to solid in underflow from continuous thickener | — | — |
| $u$ | Velocity | m/s | $LT^{-1}$ |
| $u_0$ | Free-falling velocity of particle | m/s | $LT^{-1}$ |
| $u_c$ | Sedimentation velocity of particle in suspension | m/s | $LT^{-1}$ |
| $u_c'$ | Sedimentation velocity at level of obstruction | m/s | $LT^{-1}$ |
| $u_F$ | Velocity of displaced fluid | m/s | $LT^{-1}$ |
| $u_f$ | Velocity of sedimentation of small particles in upper layer | m/s | $LT^{-1}$ |
| $u_i$ | Sedimentation velocity at infinite dilution | m/s | $LT^{-1}$ |
| $u_p$ | Velocity of particle relative to fluid | m/s | $LT^{-1}$ |
| $u_r$ | Downward velocity due to removal of underflow | m/s | $LT^{-1}$ |
| $u_u$ | Sedimentation velocity at concentration $C_u$ of underflow | m/s | $LT^{-1}$ |
| $u_w$ | Velocity of propagation of concentration wave | m/s | $LT^{-1}$ |
| $V$ | Volume of particles passing plane per unit area | m/s | $LT^{-1}$ |
| $V_A$ | van der Waals attraction energy | J | $ML^2T^{-2}$ |
| $V_R$ | Electrical repulsion energy | J | $ML^2T^{-2}$ |
| $V_T$ | Total potential energy of interaction | J | $ML^2T^{-2}$ |
| $V'$ | Volumetric capacity of centrifuge basket | $m^3$ | $L^3$ |

|  |  | Units in SI System | Dimensions in M, L, T, $\theta$, A |
|---|---|---|---|
| $v$ | Volume of particle | m³ | $L^3$ |
| $W$ | Mass rate of feed of solids | kg/s | $MT^{-1}$ |
| $X$ | Average value of mass ratio of liquid to solids | — | — |
| $Y$ | Mass ratio of liquid to solids in feed to continuous thickener | — | — |
| $y$ | Height above centrifuge basket | m | $L$ |
| $Z$ | Valence of ion | — | — |
| $z$ | Hydrostatic height | m | $L$ |
| $\alpha_0$ | Ratio defined by equation 5.134 | — | — |
| $\alpha_p$ | Ratio defined by equation 5.132 | — | — |
| $\gamma$ | Constant defined by equation 5.122 | — | — |
| $\dot{\gamma}$ | Shear rate | $s^{-1}$ | $T^{-1}$ |
| $\delta$ | Stern layer thickness | m | $L$ |
| $\varepsilon$ | Permittivity | $s^4 A^2/kg\,m^3$ | $M^{-1}L^{-3}T^4A^2$ |
| $\kappa$ | Debye–Hückel parameter (equation 5.123) | $m^{-1}$ | $L^{-1}$ |
| $\zeta$ | Electrokinetic or zeta potential | V | $ML^2T^{-3}A^{-1}$ |
| $\mu$ | Viscosity of fluid | $Ns/m^2$ | $ML^{-1}T^{-1}$ |
| $\mu_c$ | Viscosity of suspension | $Ns/m^2$ | $ML^{-1}T^{-1}$ |
| $\rho$ | Density of fluid | kg/m³ | $ML^{-3}$ |
| $\rho_0$ | Density of suspension | kg/m³ | $ML^{-3}$ |
| $\rho_s$ | Density of solid | kg/m³ | $ML^{-3}$ |
| $\Sigma$ | Capacity factor for centrifuge defined by equation 5.113 | m² | $L^2$ |
| $\theta$ | Angle subtended at centre of rotation | — | — |
| $\phi''$ | Friction factor defined by equation 5.66 | — | — |
| $\psi$ | Mass rate of sedimentation per unit area | $kg/m^2\,s$ | $ML^{-2}T^{-1}$ |
| $\psi'$ | Mass rate of sedimentation per unit area at level of obstruction | $kg/m^2\,s$ | $ML^{-2}T^{-1}$ |
| $\psi_\delta$ | Stern plane potential | V | $ML^2T^{-3}A^{-1}$ |
| $\psi_0$ | Surface potential | V | $ML^2T^{-3}A^{-1}$ |
| $\omega$ | Angular velocity | rad/s | $T^{-1}$ |
| $Ga$ | Galileo number ($d^3 g(\rho_s - \rho)\rho/\mu^2$) | — | — |
| $Re'$ | Reynolds number ($u d\rho/\mu$) | — | — |
| $Re'_0$ | Reynolds number for particle under terminal falling conditions ($u_0 d\rho/\mu$) | — | — |
| $Re_1$ | Bed Reynolds number ($u_c \rho/S\mu(1-e)$) | — | — |

Suffix
| | |
|---|---|
| $B$ | Glass ballotini particles |
| $L$ | Large particles |
| $M$ | Mixture |
| $P$ | Polystyrene particles |
| $S$ | Small particles |
| 0 | Value under free-falling conditions |
| 1 | Light liquid |
| 2 | Heavy liquid |

CHAPTER 6

# *Fluidisation*

## 6.1. CHARACTERISTICS OF FLUIDISED SYSTEMS

### 6.1.1. Properties of Gas–Solid and Liquid–Solid Systems

If a fluid is passed downwards through a bed of solids, no relative movement between the particles takes place, unless the initial orientation is unstable. If the flow is streamline, the pressure drop across the bed will be directly proportional to the rate of flow but at higher rates it will rise more rapidly. The expressions given in Chapter 4 can be used for the calculation of the pressure drop under these conditions[1].

If the fluid passes upwards through the bed, the pressure drop will be the same as for downward flow at low rates, but when the frictional drag on the particles becomes equal to their apparent weight (actual weight less buoyancy), the particles become rearranged so that they offer less resistance to the flow of fluid and the bed starts to expand. This process continues as the velocity is increased, with the total frictional force remaining equal to the weight of the particles, until the bed has assumed the loosest stable form of packing. If the velocity is then increased still further, the individual particles separate from one another and become freely supported in the fluid and the bed is said to be *fluidised*. Further increase in the velocity causes the particles to separate still further from one another, and the pressure difference remains approximately equal to the weight per unit area of the bed.

Up to this stage the system behaves in a similar way whether the fluid is a liquid or gas, but at high fluid velocities when the expansion of the bed is large there is a fairly sharp distinction between the behaviour in the two cases. With a liquid, the bed continues to expand as the velocity is increased and it maintains its uniform character, with the amount of agitation of the particles increasing progressively. This type of fluidisation is known as *particulate fluidisation*. With a gas, however, uniform fluidisation is obtained only at relatively low velocities. At high velocities two separate *phases* are formed; the continuous phase which is often referred to as the *dense* or *emulsion* phase, and the discontinuous phase known as the *lean* or *bubble* phase. The fluidisation is then said to be *aggregative*. Gas bubbles pass through a high-density fluidised bed with the result that the system closely resembles a boiling liquid, with the lean phase corresponding to the vapour and the dense or continuous phase to the liquid. The bed is then often referred to as a *boiling bed*, as opposed to a *quiescent bed* at low flow rates. Thus as the flow of the gas is increased, its velocity relative to the particles in the dense phase may not change appreciably, and it has been shown that the flow relative to the particles can as a result remain streamline even at very high overall rates of flow. If the rate of passage of gas is high, and if the bed is deep, coalescence of the bubbles takes place, and in a narrow vessel slugs of gas occupying the whole cross-section may be produced. These slugs of gas alternate with slugs of fluidised solids which are carried upwards and subsequently collapse, causing the solids to fall back again.

It has been suggested[2] that the Froude group $u_{mf}^2/gd$ gives a criterion from which the type of fluidisation can be predicted.

Here

$u_{mf}$   is the minimum velocity of flow, calculated over the whole cross-section of the bed, at which fluidisation takes place,

$d$   is the diameter of the particles, and

$g$   is the acceleration due to gravity.

At values less than unity, particulate fluidisation occurs and, at higher values, aggregative fluidisation takes place. Much lower values of the Froude number are generally obtained with liquids because the velocity required to produce fluidisation is less. A theoretical justification for using the Froude group to distinguish between particulate and aggregative fluidisation is provided in the work of Jackson[3], which will be referred to later.

Although the possibility of forming fluidised beds had been known for many years, the subject remained of academic interest until the adoption of fluidised catalysts by the petroleum industry for the cracking of heavy hydrocarbons and for the synthesis of fuels from natural gas or from carbon monoxide and hydrogen. The fluidised bed in many ways behaves as a fluid of the same density as that of the solids and fluid combined. It transmits hydrostatic forces, and solid objects float if their densities are lower than that of the bed. Intimate mixing takes place and heat transfer within the bed is very rapid, and therefore uniform temperatures are quickly attained throughout the system. The easy control of temperature is the feature which has led to the use of fluidised solids for strongly exothermic processes, and where close control of temperature is important.

In order to understand the properties of a fluidised system it is necessary to study the flow patterns of the solids and the fluid. In this connection the mode of formation and behaviour of fluid bubbles is of particular importance because they can account for the flow of a large proportion of the fluid in a gas–solid system.

In any study of the properties of a fluidised system, it is necessary to select conditions which are reproducible, and the lack of agreement between the results of many workers, particularly those relating to heat transfer, is largely attributable to the existence of widely different conditions within the bed. One of the prime requirements is that the fluidisation shall be of a good quality; that is to say, that the bed shall be free from irregularities and channelling. It must be accepted that many solids will never give good fluidisation, particularly those whose shape is appreciably non-isometric, and those which form soft particles which readily form agglomerates. Furthermore, when the solid is capable of giving good quality fluidisation, the fluid must be evenly distributed at the bottom of the bed and this necessitates the provision of a distributor giving rise to a pressure drop at least equal to that across the bed. This condition is much more readily satisfied in a small laboratory apparatus than in large-scale industrial equipment.

As already mentioned, systems fluidised with a liquid do not tend to give rise to bubbles, whereas those fluidised with a gas tend to produce bubbles when the flow rate appreciably exceeds the minimum for fluidisation. In an attempt to improve the reproducibility of conditions within the bed, much of the earlier research work with gas fluidised systems was carried out at gas velocities sufficiently low for bubble formation to be absent. In recent years, however, it has been recognised that bubbles normally tend to form in such systems, that they exert an important influence on the flow pattern of both gas and solids, and that the behaviour of individual bubbles can be predicted quite accurately.

### 6.1.2. Effect of Fluid Velocity on Pressure Gradient

The relation between the superficial velocity $u_c$ of the fluid (calculated over the whole cross-section of the containing vessel) and the pressure gradient is shown in Fig. 6.1. If the flow is streamline a straight line of slope unity is obtained at low velocities. At the fluidising point, the pressure gradient begins to fall because the porosity of the bed increases; this fall continues until the velocity is high enough for transport of the material to take place; the pressure gradient then starts to increase again because the frictional drag of the fluid at the walls of the tube starts to become significant.

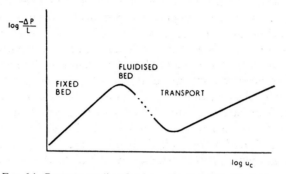

FIG. 6.1. Pressure gradient in bed as function of fluid velocity

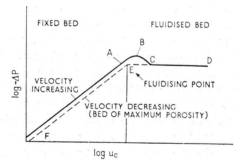

FIG. 6.2. Pressure drop over fixed and fluidised beds

If the pressure across the whole bed instead of the pressure gradient is plotted against velocity, again using logarithmic coordinates (Fig. 6.2), a linear relation is again obtained up to the point where expansion of the bed takes place ($A$), but the slope of the curve then gradually diminishes as the bed expands. As the velocity is increased, the pressure drop passes through a maximum value ($B$) and then falls slightly and attains an approximately constant value, independent of fluid velocity ($CD$). If the velocity of flow is reduced again, the bed contracts until it reaches the condition where the particles are just resting on one another ($E$); it then has the maximum stable porosity for a fixed bed of the particles in question. If the velocity is further decreased, the bed then remains in this condition provided that it is not shaken. The pressure drop ($EF$) in this reformed fixed bed is then less than that originally obtained at the same velocity. If the velocity were now increased again, it would be expected that the new curve ($FE$) would be retraced and that the slope would suddenly change from 1 to 0 at the fluidising point; this condition is difficult to produce, however, because the bed tends to become consolidated again unless it is completely free from vibration. In the absence of channelling, the shape and size of the particles alone determine both the maximum porosity and the pressure drop at fluidisation for a bed of given dimensions. In an ideal fluidised bed the pressure drop corresponding to *ECD* is equal to the buoyant weight per unit area. In practice, it may deviate appreciably from this value, and not remain constant, as a result of channelling. Point *B* lies above *CD* because the frictional forces between the particles have to be overcome before rearrangement can take place.

The velocity corresponding to the fluidising point is readily calculated from the equations given in Chapter 4. The pressure drop through the bed is then equal to its apparent weight per unit area, and the porosity at the onset of fluidisation is the maximum which the fixed bed can attain. van Heerden *et al.*[4] have suggested that the porosity term be eliminated from the above equations for the drop in pressure through a fixed bed by combining it with the numerical constant, because both the constant and the maximum porosity are functions solely of particle shape.

In a fluidised bed, the total frictional force on the particles must be equal to the effective weight of the bed. Thus in a bed of unit cross-sectional area, depth *l*, and porosity *e*:

$$-\Delta P = (1-e)(\rho_s - \rho)lg \qquad (6.1)$$

where $-\Delta P$ is the pressure drop across the bed and
$\quad\quad g$ is the acceleration due to gravity.

This relation applies from the initial expansion of the bed until transport of solids takes place. There may be some discrepancy between the calculated and measured minimum velocities for fluidisation. This may be attributable to channelling, as a result of which the drag force acting on the bed is reduced, to the action of electrostatic forces in the case of gaseous fluidisation (these are particularly important in the case of sands), to agglomeration which is often considerable with small particles, or to friction between the fluid and the walls of the containing vessel: this last factor is of greatest importance with beds of small diameters. Leva *et al.*[5] have introduced a term $(G_F - G_E)/G_F$, a fluidisation efficiency, in which $G_F$ is the flowrate of fluid required to produce fluidisation and $G_E$ is the rate which will cause an initial expansion of the bed.

If flow conditions within the bed are streamline, the relation between fluid velocity $u_c$ and voidage *e* is obtained by substituting from equation 4.9 into equation 6.1 to give, for spherical particles:

$$u_c = 0.0055 \frac{e^3}{1-e} d^2 \frac{(\rho_s - \rho)g}{\mu} \qquad (6.2)$$

### 6.1.3. Minimum Fluidising Velocity

From equation 6.2, it is clear that the minimum fluidising velocity is very dependent on the voidage of the bed. At the point of incipient fluidisation, the minimum fluidising velocity $u_{mf}$ is obtained by substituting the appropriate value of voidage $e_{mf}$ in the equation. Thus:

$$u_{mf} = 0.0055 \frac{e_{mf}^3}{1-e_{mf}} \frac{d^2(\rho_s - \rho)g}{\mu} \qquad (6.3)$$

Values of $e_{mf}$ will vary considerably from powder to powder, but a typical value for a bed of isometric particles is 0·4. Substituting this value:

$$(u_{mf})_{e_{mf}=0·4} = 0·0059\left(\frac{d^2(\rho_s-\rho)g}{\mu}\right) \tag{6.4}$$

For larger particles, where the flow in the bed at the point of incipient fluidisation is no longer streamline, it is necessary to use a more general equation, such as the Ergun equation, for the pressure drop through the bed. Applying the Ergun equation (4.18) at the incipient fluidisation point:

$$\frac{-\Delta P}{l} = 150\frac{(1-e_{mf})^2}{e_{mf}^3}\frac{\mu u_{mf}}{d^2} + 1·75\frac{(1-e_{mf})}{e_{mf}^3}\frac{\rho u_{mf}^2}{d} \tag{6.5}$$

Substituting for $-\Delta P/l$ from equation 6.1 and multiplying both sides by $\rho d^3/\mu^2(1-e_{mf})$:

$$\frac{\rho(\rho_s-\rho)gd^3}{\mu^2} = 150\frac{1-e_{mf}}{e_{mf}^3}\frac{u_{mf}\,d\rho}{\mu} + \frac{1·75}{e_{mf}^3}\left(\frac{u_{mf}\,d\rho}{\mu}\right)^2 \tag{6.6}$$

Equation 6.6 can be solved to give $u_{mf}$ explicitly. However, it is convenient to work in terms of dimensionless groups, noting that:

$\dfrac{\rho(\rho_s-\rho)gd^3}{\mu^2}$ is the Galileo number, $Ga$ (see page 97),

$\dfrac{u_{mf}\,d\rho}{\mu}$ is a form of Reynolds number and will be designated $Re'_{mf}$.

Equation 6.6 then becomes:

$$Ga = 150\frac{1-e_{mf}}{e_{mf}^3}Re'_{mf} + \frac{1·75}{e_{mf}^3}Re'^2_{mf} \tag{6.7}$$

Again taking a typical value of 0·4 for $e_{mf}$, and rearranging:

$$Re'^2_{mf} + 51·4Re'_{mf} - 0·0366\,Ga = 0 \tag{6.8}$$

Thus

$$(Re'_{mf})_{e_{mf}=0·4} = 25·7\{\sqrt{1+5·53\times10^{-5}Ga} - 1\} \tag{6.9}$$

and

$$u_{mf} = \frac{\mu}{d\rho}Re'_{mf} \tag{6.10}$$

Work already reported on sedimentation[6] has suggested that where the particles are free to adjust their orientations with respect to one another, as in sedimentation and fluidisation, the equations obtained for pressure drop in fixed beds overestimate the pressure drop. For streamline flow the Carman–Kozeny constant has a value of 3·36 instead of 5. Substituting this lower value, equation 6.2 becomes:

$$(u_{mf})_{e_{mf}=0·4} = 0·0089\frac{d^2(\rho_s-\rho)g}{\mu} \tag{6.11}$$

It is probable that the Ergun equation also overpredicts pressure drop for fluidised systems but no experimental evidence is available on the basis of which the values of the coefficients may be amended.

The minimum fluidising velocity $u_{mf}$ can be expressed in terms of the free-falling velocity $u_0$ of the particles in the fluid. The Ergun equation (equation 6.5) relates the Galileo number $Ga$ to the Reynolds number $Re'_{mf}$ in terms of the voidage $e_{mf}$ at the incipient fluidisation point.

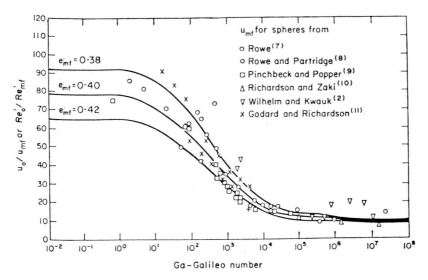

FIG. 6.3. Ratio of terminal falling to minimum fluidising velocity

$$\left( \frac{u_0}{u_{mf}} = \frac{Re'_0}{Re'_{mf}} \right)$$

The Reynolds number $u_0 d\rho/\mu \; (= Re'_0)$ involving the terminal falling velocity $u_0$ of the particle in the fluid can also be expressed in terms of the Galileo number (see page 97) but no single expression covers the whole range of Reynolds numbers and it is most satisfactory to cover each of three zones of $Ga$ separately to give:

$$Ga = 18 \, Re'_0 \qquad\qquad (Ga < 3 \cdot 6) \qquad\qquad (6.12)$$

$$Ga = 18 \, Re'_0 + 2 \cdot 7 \, Re'^{1 \cdot 687}_0 \qquad (3 \cdot 6 < Ga < 10^5) \qquad\qquad (6.13)$$

$$Ga = \tfrac{1}{3} Re'^2_0 \qquad\qquad (Ga > 10^5) \qquad\qquad (6.14)$$

Thus, it is possible for any value of $e_{mf}$ to calculate $Re'_0/Re'_{mf} \; (= u_0/u_{mf})$ as a function of $Ga$. In Fig. 6.3, $Re'_0/Re'_{mf}$ is plotted against $Ga$ and curves are given for values of $e_{mf}$ of 0.42, 0.40, and 0.38. It is of interest to note that the range of velocities over which particulate fluidisation can be obtained is very much greater in the streamline than in the turbulent region ($\sim 80$ compared with $\sim 10$-fold range). In addition to the calculated curves, some experimental results of various workers for spherical particles are also given. The values of $e_{mf}$ for this experimental work are not known, but it will be noted that, in general, the points straddle the curves for $0.38 < e_{mf} < 0.42$.

## 6.2. LIQUID–SOLID SYSTEMS

### 6.2.1. Bed Expansion

Liquid-fluidised systems are generally characterised by the regular expansion of the bed which takes place as the velocity increases from the minimum fluidisation velocity to the terminal falling velocity of the particles. The general relation between velocity and volumetric concentration or voidage is found to be similar to that existing between sedimentation velocity and concentration for particles in a suspension. The two systems are hydrodynamically similar in that in the fluidised bed the particles undergo no net movement and are maintained in suspension by the upward flow of liquid, whereas in the sedimenting suspension the particles move downwards and the only flow of liquid is the upward flow of that liquid which is displaced by the settling particles. Zaki[10] showed that, for sedimentation or fluidisation of uniform particles, cf. equation 5.30,

$$\frac{u_c}{u_i} = e^n = (1 - C)^n \tag{6.15}$$

where    $u_c$ is the observed sedimentation velocity or the empty tube fluidisation velocity,
$u_i$ is the corresponding value at infinite dilution,
$e$ is the voidage of the system,
$C$ is the volumetric fractional concentration of solids, and
$n$ is an index.

$u_i$ was found to correspond closely to the free-falling velocity $u_0$ for sedimentation and to be related to $u_0$ by the following expression for fluidisation:

$$\log_{10} u_0 = \log_{10} u_i + \frac{d}{d_t} \tag{6.16}$$

where $d/d_t$ is the ratio of the particle to tube diameters.

For spherical particles, the index $n$ was found by dimensional analysis to be defined by the Reynolds number $Re'_0$ $(= u_0\, d\rho/\mu)$ and the ratio $d/d_t$. Furthermore, it was shown that, provided the Reynolds number was less than 0·2 or greater than 500, it would not affect the value of $n$. Experimentally determined values of the index $n$ are correlated by equations 5.34 to 5.38 given earlier.

Thus, if the voidage $e$ of the bed is plotted against the velocity of the fluid $u_c$ (using logarithmic coordinates) (Fig. 6.4), the curve can be divided approximately into two straight lines joined by a short curve. At low velocities the voidage remains constant corresponding to the fixed bed, and for the fluidised state there is a linear relation between $\log u_c$ and $\log e$. The curve shown refers to the fluidisation of steel spheres in water. It should be noted that whereas, in the absence of channelling, the pressure drop across a bed of a given expansion is directly proportional to its depth, the fluidising velocity is independent of depth. A relation similar to equation 6.15 has also been obtained by Lewis and Bowerman[12].

It is now possible to examine further the variation of the index $n$ in equation 6.15 for the expansion of particulately fluidised systems. Neglecting effects due to the container wall:

$$(u_c/u_0) = e^n \qquad \text{(from equation 6.15)}$$

i.e.

$$n = \log(u_c/u_0)/\log e \tag{6.17}$$

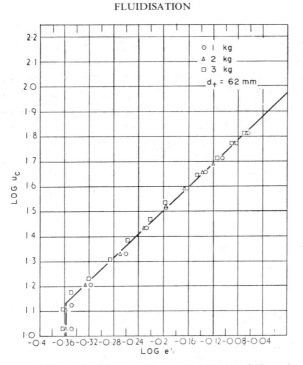

FIG. 6.4. Relation between fluid velocity ($u_c$) and voidage for fluidisation of 6·4 mm steel spheres in water[10]

On the assumption that equations 6.15 and 6.17 apply at the point of incipient fluidisation:

$$n = \frac{\log\,(u_{mf}/u_0)}{\log\,e_{mf}} = \frac{-\log\,(Re_0'/Re_{mf}')}{\log\,e_{mf}} \tag{6.18}$$

Values of $n$, calculated from the curves in Fig. 6.3, are plotted against Galileo number $Ga$ in Fig. 6.5. It will be noted that there is good agreement between the calculated values and the experimental points at the extremities ($n = 4·6$ and $n = 2·3$) but that in the intermediate region, the theory consistently overpredicts the value of $n$. This is probably attributable to the fact that the constants in the Ergun equation apply to fixed beds and that, in the intermediate range of Galileo numbers, in particular, the effect of particle reorientation is not taken into account. In this region the Ergun equation is giving an overestimate of the pressure drop at a given velocity.

When the bed consists of particles with a significant size range stratification occurs, with the largest particles forming a bed of low voidage near the bottom and the smallest particles a bed of high voidage near the top. If the particles are in the form of sharp-cut size fractions, segregation will be virtually complete with what is in effect a number of fluidised beds of different voidage, one above the other. If the size range is small, there will be a continuous variation in both composition and concentration of particles throughout the depth of the bed.

Leva *et al.*[5], Cathala[14], and others prefer to plot $e^3/(1-e)^2$ versus $\log u_c$ because $e^3/(1-e)^2$ is the function of voidage appearing in the Kozeny equation (4.9) for the streamline flow of a fluid through a granular bed. As already indicated, Meikle[6] has shown that the results for high concentrations can be expressed in a similar manner to those for flow through a fixed bed.

The limitations of equation 6.15 were pointed out by Lawther and Berglin[15] who fluidised lead shot with water. Smith[16] and Bailey[17] have also investigated the behaviour of dense particles and have stated that deviations from equation 6.15 follow a systematic pattern and increase with the density of the particles and the height of the bed, and as the particle size is reduced. The deviations from equation 6.15 are accompanied by an unevenness in the fluidisation, and it is observed that circulating eddies appear within the fluid from which the particles are separated out by centrifugal action. Thus, density fluctuations occur throughout the bed, but these are frequently of a completely different

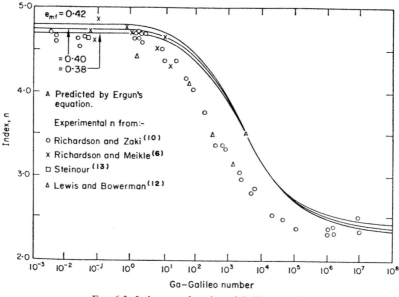

FIG. 6.5. Index, $n$, as function of Galileo number

character from bubbles though true bubbles are sometimes formed with lead or tungsten particles fluidised in water.

Anderson[18] has shown that, in very narrow diameter beds, bands of low particle concentration develop within a liquid fluidised bed. These bands tend to propagate upwards through the bed, increasing in amplitude as they do so.

In some circumstances, particularly where the particle to tube diameter ratio is significant, bridging may take place at high concentrations. This is clearly seen in the relationship between pressure difference and velocity and in that between concentration and velocity.

From experiments[17] on the fluidisation by downflow of particles of density lower than that of the liquid, it is found that the relation between flowrate and concentration follows equation 6.15 very closely. It is noticeable that the particles undergo very little movement in the bed, moving only a few millimetres in a second.

### 6.2.2. Liquid and Solid Mixing

Kramers et al.[19] have studied longitudinal dispersion in the liquid in a fluidised bed composed of glass spheres of 0·5 mm and 1 mm diameter. A step change was introduced by feeding a normal solution of potassium chloride into the system; the concentration at the top of the bed was measured as a function of time by means of a small conductivity cell. On the assumption that the flow pattern could be regarded as longitudinal diffusion superimposed on piston flow, an eddy longitudinal diffusivity was calculated. It was found to range from 1 to 10 cm$^2$/s, increasing with both voidage and particle size.

The movement of individual particles in a liquid–solid fluidised bed has been measured by Handley et al.[20], Carlos[21,22], and Latif[23,24]. In all cases, the method involved fluidising transparent particles in a liquid of the same refractive index so that the whole system became transparent. The movement of coloured tracer particles, whose other physical properties were identical to those of the bed particles, could then be followed photographically.

Handley fluidised soda glass particles using methyl benzoate, and obtained information on the flow pattern of the solids and the distribution of vertical velocity components of the particles. He found that a bulk circulation of solids was superimposed on their random movement. Particles normally tended to move upwards in the centre of the bed and downwards at the walls; this circulation pattern was less marked in regions remote from the distributor.

Carlos and Latif fluidised glass particles in dimethyl phthalate, and analysed their ciné film with a Benson–Lehner analyser which produced data of the coordinates of the tracer particles directly on to cards which could then be used as the input for a computer. Vertical, radial, and tangential velocity components and total velocities were obtained for the particles. When plotted as a histogram, the total velocity distribution was found to be of the same form as that predicted by the kinetic theory for the molecules in a gas. A typical result is shown in Fig. 6.6[21]. Effective diffusion or mixing coefficients for the particles were

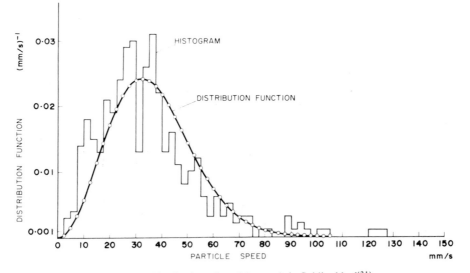

FIG. 6.6. Distribution of particle speeds in fluidised bed[21]

then calculated from the product of the mean velocity and mean free path of the particles using the simple kinetic theory.

Solids mixing was also studied by Carlos[22] in the same apparatus, starting with a bed composed of transparent particles and a layer of tracer particles at the base of the bed. The concentration of particles in a control zone was then determined at various intervals of time after the commencement of fluidisation. The mixing process was found to be described by a diffusion-type equation; this was then used to calculate the mixing coefficient. A comparison of the values of mixing coefficient obtained by the two methods then enabled the persistence of velocity factor to be calculated. A typical value of the mixing coefficient was $15\,\mathrm{cm^2/s}$ for 9 mm glass ballotini fluidised at a velocity of twice the minimum fluidising velocity.

Latif[23,24] was able to represent the circulation currents of the particles in a fluidised bed, by plotting stream functions for the particles on the assumption that the particles could be regarded as behaving as a continuum. A typical result for the fluidisation of 6-mm glass particles by dimethyl phthalate is shown in Fig. 6.7; in this case the velocity has been adjusted to give a bed voidage of 0.65. Because the bed is symmetrical about its axis, the pattern over only a radial slice is shown. It will be noted that the circulation patterns are concentrated mainly in the lower portion of the bed, with particles moving upwards in the

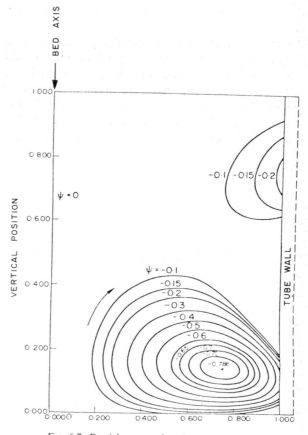

FIG. 6.7.   Particle stream functions $\psi$ $(e = 0.65)$

centre and downwards at the walls. As the bed voidage is decreased, the circulation patterns tend to occupy progressively smaller portions of the bed, but there is a tendency for a small reverse circulation pattern to develop in the upper regions of the bed.

## 6.3. GAS–SOLID SYSTEMS

### 6.3.1. Bed Characteristics

In systems fluidised by means of a gas, only a limited expansion of the continuous phase in the bed usually takes place. With fine materials, such as the silica–alumina microspheres used as petroleum cracking catalysts, a 30 per cent expansion may be obtained and this will occur at velocities of about three times the minimum fluidisation velocity. With coarser solids a much smaller expansion is obtained and bubble formation commences at lower gas rates. During this period of even expansion of the bed, equation 6.15 is closely followed[25]. In fact, deviations from this equation are always associated with the tendency for a second phase to appear in the bed.

The ease with which a fluidised bed can be formed varies with the physical properties of the solid. Thus, silica–alumina catalysts are very readily fluidised and can be maintained in that condition by the gas entrained when the bed of material is gently stirred. Most materials become fluidised only during the constant upward passage of gas, however. Large particles require relatively high rates of flow of gas because their terminal falling velocities are high, and because the pressure drop is relatively low when a fluid flows through a bed of large particles.

The uniformity of the fluidised bed is a function of the size and the surface properties of the solids. Thus some materials tend to agglomerate, especially when the particles are less than about 50 μm in diameter, and channelling takes place. Occasionally agglomeration is so marked that fluidisation does not occur at all and, when the pressure drop exceeds the weight per unit area of the particles, the whole bed is raised like a solid piston; this structure subsequently collapses and large aggregates of particles are produced.

The effect of stirring on the properties of the fluidised system has been studied by a number of workers. If the bed is stirred using an agitator with blades so arranged as to lift the particles, an increase in volume takes place and fluidisation can be effected by the use of a smaller rate of flow of gas. Various attempts have been made to measure a property analogous to the viscosity of the bed. Thus Matheson, Herbst, and Holt[26] measured the torque required to rotate a paddle at 3 rev/s (3 Hz) in a fluidised bed. They found that it was almost directly proportional to the linear size of the particles, and that it increased with increase in the density of the material. Increased aeration considerably reduced the torque required. Diekman and Forsythe[27] found that the most uniform fluidisation was obtained with those materials which showed only a small increase in apparent viscosity as the flowrate was reduced.

The tendency for the gas bubbles in a fluidised bed to coalesce and form alternate slugs of gas and fluidised solid is known to increase as the ratio of the depth of the bed to its diameter is increased. Further, the tendency for slugging is known to be dependent on the properties of the particles. Matheson et al.[26] determined the maximum value of the ratio $l/d_t$ for which slugging did not take place for a number of systems. They then plotted this ratio against the apparent viscosity of the bed and obtained a curve which approximated to a straight line. It was found that the tendency for slugging was least with those materials which formed beds whose resistance to shear was small, and thus slugging increased with the particle size of the material and its density, i.e. with terminal falling velocity.

In general, as the gas velocity is increased the voidage of the fluidised bed changes very little. A small expansion of the fluidised bed occurs at velocities just above the minimum fluidising velocity, but any further increase in gas rate results in the formation of bubbles which tend to increase both in number and in size as the flowrate is increased. The importance of the bubbles, not only because they transmit a large proportion of the gas through the bed, but also because they affect the overall flow pattern of gas and solids, is now well recognised, and research into the behaviour of bubbles is proceeding in a number of establishments.

### 6.3.2. Properties of Bubbles in the Bed

Research at Cambridge has involved a study of the formation of bubbles at orifices in a fluidised bed, including measurement of their size, the conditions under which they will coalesce with one another, and their rate of rise in the bed. Thus Davidson et al.[28] injected air from an orifice into a fluidised bed composed of particles of sand (0·3–0·5 mm) and glass ballotini (0·15 mm) fluidised by air at a velocity just above the minimum for fluidisation. By varying the depth of the injection point from the free surface they were able to show that the injected bubble rises through the bed with a constant velocity, which is dependent only on the volume of the bubble. Furthermore, this velocity of rise corresponds with that of a spherical cap bubble in an inviscid liquid of zero surface tension, as determined from the equation of Davies and Taylor[29].

$$u_b = 0.792 \, V_B^{1/6} g^{1/2} \qquad (6.19)$$

Thus the velocity of rise is apparently independent of the velocity of the fluidising air and of the properties of the particles making up the bed. Equation 6.19 is applicable provided that the density of the gas in the bubbles can be neglected in comparison with the density of the bed. Otherwise, the expression must be multiplied by a factor of $(1 - \rho/\rho_c)^{1/2}$.

In a second paper[30], these results are applied to the problem of explaining why gas and liquid fluidised systems behave differently. Photographs of bubbles in beds of lead shot fluidised with air and with water have shown that an injected bubble is stable in the former case but tends to collapse in the latter. It has already been mentioned, in any case, that the water–lead system tends to give rise to inhomogeneities, and it is therefore interesting to note that bubbles as such are apparently not stable. As the bubble rises in a bed, internal circulation currents are set up because of the shear stresses existing at the boundary of the bubble. These circulation velocities will be of the same order of magnitude as the rising velocity of the bubble. If the circulation velocity is appreciably greater than the falling velocity of the particles, the bubble will tend to draw in particles from the wake and will therefore tend to be destroyed. On the other hand, if the rising velocity is lower, particles will not be drawn in at the wake and the bubble will be stable.

As a first approximation, therefore, a bubble is assumed to be stable if its rising velocity is less than the free-falling velocity of the particles, and therefore for any system the limiting size of stable bubble can be calculated using equation 6.19. If this is of the same order of size as the particle diameter, the bubble will not readily be detected. On the other hand, if it is more than about ten times the particle diameter it will be visible and the system will be seen to contain stable bubbles. On the basis of this argument, large bubbles are generally stable in gases, whereas in liquids the largest size of stable bubble is comparable with the diameter of the particles. It should be possible to achieve fluidisation free of bubbles with very light particles by using gas of high density. Leung[31] succeeded in reaching this condition by fluidising hollow phenolic microballoons at pressures of about 4·5 MN/m². Furthermore, it was found possible to form stable bubbles with glycerine–water mixtures and lead shot of particle size equal to 0·77 mm. This transitional region has also been studied by Simpson and Rodger[32].

Harrison and Leung[33] have shown that the frequency of formation of bubbles at an orifice (size range 1·2–25 mm) is independent of the bed depth, the flowrate of gas and the properties of the particles constituting the continuous phase, but depends on the injection rate of gas, tending to a frequency of 18–21 per second at high flowrates.

In a further paper[34], they have shown that a wake extends for about 1·1 bubble diameter behind each rising bubble. If a second bubble follows in this wake, its velocity is increased by an amount equal to the velocity of the leading bubble, and in this way coalescence takes place.

The differences between liquid and gas fluidised systems have also been studied theoretically by Jackson[3] who showed that small discontinuities would tend to grow in a fluidised bed, but that the rate of growth would be greater in a gas–solid system.

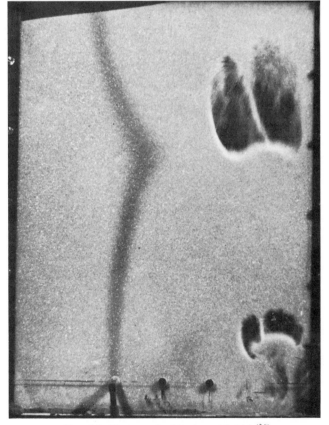

FIG. 6.8. Photograph of tracer and bubble[36]

Rowe, Wace, and Burnett[35,36] at Harwell have examined the influence of gas bubbles on the flow of gas in their vicinity. By constructing a thin bed 300 mm wide, 375 mm deep, and only 25 mm across, it was possible to take photographs through the perspex wall showing the behaviour of a thin filament of nitrogen dioxide gas injected into the bed. In a bed composed of 0·20 mm ballotini, it was found that the filament tended to be drawn towards a rising bubble (see Fig. 6.8), and through it if sufficiently close. This establishes that there is a flow of gas from the continuous phase into the bubble and out through the roof of the bubble again, and that the gas tends to flow in definite streamlines.

As a result, the gas is accelerated towards the bubble and is given a horizontal velocity component; as a consequence, the gas velocity in the continuous phase close to the bubble is reduced.

The pressure distribution round a stationary bubble was measured by inserting a gauze sphere 50 mm in diameter in the bed and measuring the pressure throughout the bed using the pressure inside the sphere as a datum (Fig. 6.9). It was found that near the bottom of the bubble the pressure was less than that remote from it at the same horizontal level, and that the situation was reversed, though to a smaller degree, towards the top of the bubble. Although the pressure distribution would be somewhat modified in a moving bubble, the model serves qualitatively to explain the observed flow patterns of the tracer gas.

When the rate of rise of the bubble exceeds the velocity of the gas in the continuous phase, the gas leaving the top of the bubble is recycled and enters the base again. As a result the gas in the bubble comes into contact with only those solid particles which immediately surround the bubble. Davidson[37] has analysed this problem and has shown that if the

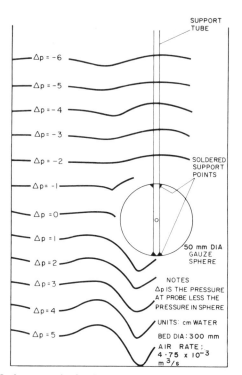

FIG. 6.9. Isobars round a fixed gauze sphere in a bed of mixed sand[36]

inertia of the gas is neglected, the diameter $d_c$ of the cloud of recycling gas surrounding a bubble of diameter $d_b$ is given by:

$$\frac{d_c}{d_b} = \left\{ \frac{\alpha+2}{\alpha-1} \right\}^{1/3}$$ (6.20)

where $\alpha$ is the ratio of the linear velocity of the gas in the bubble to that in the emulsion phase, i.e.

$$\alpha = \frac{eu_b}{u_c}$$ (6.21)

The corresponding expression for a thin, essentially two-dimensional bed is

$$\frac{d_c}{d_b} = \left\{ \frac{\alpha+1}{\alpha-1} \right\}^{1/2}$$ (6.22)

Rowe, Partridge, and Lyall[38] have studied the behaviour of the cloud surrounding a bubble using $NO_2$ gas as a tracer. They discovered that Davidson's theory consistently overestimated the size of the cloud. A more sophisticated theory developed by Murray[39], however, predicted the observed size of the cloud much more closely over the forward face, but neither theory satisfactorily represented conditions in the wake of the bubble. The gas cloud was found to break off at intervals, and the gas so detached became dispersed in the emulsion phase. Furthermore, by X-ray photographs it was shown that the bubbles would from time to time become split by fingers of particles falling through the roof of the bubble.

In most gas–solid systems used in practice, the particles are fine and the rising velocities of the individual bubbles are considerably greater than the velocity of the gas in the continuous phase. Thus, $\alpha$ in equation 6.21 is considerably greater than unity. Furthermore, the bubbles will tend to flow in well-defined paths at velocities considerably in excess of the rising velocity of individual bubbles. The whole pattern is, in practice, very complex because of the large number of bubbles present simultaneously, and of the size range of the bubbles.

The work of Rowe and Henwood[40] on the drag force exerted by a fluid on a particle, already referred to on page 193, showed why a bubble tended to be stable during its rise. Surfaces containing particles which face downstream, corresponding to the bottom of a bubble, tend to expel particles and therefore are diffuse. Surfaces facing upstream tend to attract particles and thus the top surface of a bubble will be sharp. Particles in a close packed array were found to be subjected to a force 68·5 times greater than that on an isolated particle for the same relative velocity. It should therefore be possible to evaluate the minimum fluidising velocity as the fluid velocity at which the drag force acting on a single isolated particle would be equal to 1/68·5 of its buoyant weight. Values calculated on this basis agree well with experimental determinations. A similar method of calculating minimum fluidising velocities has also been proposed by Davies[25] who gives a mean value of 71·3 instead of 68·5 for the drag force factor.

From Fig. 6.3, it is seen that for $e_{mf}$ equal to 0·40, $u_0/u_{mf}$ is about 78 at low values of Galileo number and about 9 for high values. In the first case, the drag on the particle is directly proportional to velocity and in the latter case to the square of the velocity; thus the force on a particle in a fluidised bed of voidage 0·4 is about 80 times that on an isolated particle for the same velocity.

### 6.3.3. Gas and Solid Mixing

*Flow Pattern of Solids*

The movement of the solid particles in a gas-fluidised bed has been studied by a number of workers using a variety of techniques, but very few satisfactory quantitative studies have been made. In general, it has been found that a high degree of mixing is usually present, and that for most practical purposes the solids in a bed can be regarded as completely mixed. One of the earliest quantitative investigations was that of Bart[41] who added some particles impregnated with salt to a bed 32 mm diameter composed of spheres of cracking catalyst. The tracer particles were fed continuously to the middle of the bed and a sample was removed at the top. Massimilla and Bracale[42] fluidised glass beads in a bed of diameter 100 mm and showed that, in the absence of bubbles, solids mixing could be represented by a diffusional type of equation. Most other workers, including Tailby and Cocquerel[43], report almost complete mixing of solids, and fail to account for any deviations by assuming a diffusional model.

May[44] studied the flow patterns of solids in beds up to 1·5 m in diameter by introducing radioactive particles near the top of the bed, and monitoring the appearance of radioactivity at different levels by means of a series of scintillation counters. With small beds, he found that the solids mixing could be represented as a diffusional type of process, but in large-diameter beds there appeared to be an overall circulation superimposed. Values of longitudinal diffusivity ranged from 45 cm$^2$/s in a bed of 0·3 m diameter to about 150 cm$^2$/s in a bed 1·5 m in diameter. May also investigated the residence time distribution of gas in a fluidised bed by causing a step-change in the concentration of helium in the fluidising air and following the change in concentration in the outlet air using a mass spectrometer.

Sutherland[45] used nickel particles as tracers in a bed of copper shot. The nickel particles were virtually identical to the copper ones in all the relevant physical properties but, being magnetic, could be readily separated again from the bed. To the top of a fluidised bed (140 mm diameter and up to 2 m deep) of copper spheres was added about 1 per cent of nickel particles. Mixing was found to be rapid but results were difficult to interpret because the flow patterns were complex. It was established, however, that vertical mixing was less in a bed which was tapered towards the base. In this case the formation of bubbles was reduced, because the gas velocity was maintained nearly constant by increasing the flow area in the direction in which the pressure was falling. Gas bubbles appear to play a very large part in causing circulation of the solids, but the type of flow obtained will depend very much on the geometry of the system.

Rowe and Partridge[46] have studied photographically the movement of solids produced by the rise of a single bubble. As the bubble forms and rises it gathers a wake of particles, and then draws up a spout of solids behind it as shown in Fig. 6.10. The wake grows by the addition of particles and then becomes so large that it sheds a fragment of itself, usually in the form of a complete ring. This process is repeated at intervals as the bubble rises (see Fig. 6.11). A single bubble causes roughly half its own volume of particles to be raised in the spout through a distance of half its diameter, and one-third of its volume in the wake through a bubble diameter. The overall effect is to move a quantity equal to a bubble volume through one and a half diameters. In a fluidised bed there will normally be many bubbles present simultaneously and their interaction results in a very complex process giving rise to a high degree of mixing.

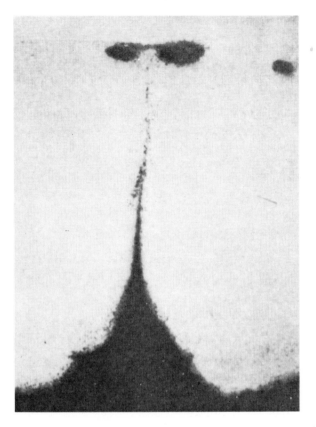

FIG. 6.10. Photograph of the solids displacement caused by a single bubble[46]

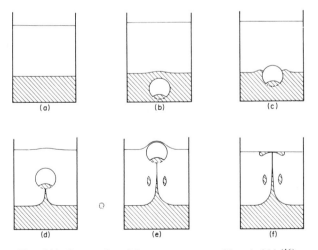

FIG. 6.11. Stages of particle movement caused by a bubble[46]

*Flow Pattern of Gas*

The existence and the movement of gas bubbles in a fluidised bed exerts an appreciable influence on the flow pattern within the bed. A number of studies have been made of gas-flow patterns, but these have, in most cases, suffered from the disadvantage of having been carried out in small apparatus where the results tend to be specific to the equipment used.

Some measurements have been made in industrial equipment. Thus Askins *et al.*[47] measured the gas composition in a catalytic regenerator, and concluded that much of the gas passed through the bed in the form of bubbles. Danckwerts *et al.*[48] used a tracer injection technique, also in a catalytic regenerator, to study the residence time distribution of gas, and concluded that the flow was approximately piston-type and that the amount of back-mixing was small. Gilliland and co-workers[49,50,51] carried out a series of laboratory-scale investigations, but their results were not conclusive. Generally our knowledge of flow patterns within the bed is poor; they are very much dependent on the scale and geometry of the equipment and on the quality of fluidisation obtained. However, the gas-flow pattern can be regarded as a combination of piston-type flow, of back-mixing, and of by-passing. Back-mixing appears to be associated primarily with the movement of the solids and to be relatively unimportant at high gas rates. This is confirmed by the studies of Lanneau[52].

### 6.3.4. Transfer between Continuous and Bubble Phases

There is no well-defined boundary between the bubble and continuous phase though, at the same time, the bubble exhibits many properties in common with a gas bubble in a liquid. If the gas in the bubble did not interchange with the gas in the continuous phase, there would be the possibility of a large proportion of the gas in the bed effectively by-passing the continuous phase and not coming into contact with the solids. This would obviously have a serious effect where a chemical reaction involving the solids was being carried out. The rate of interchange between the bubble and continuous phases has been studied by Szekely[53] and by Davies[25].

In a preliminary investigation, Szekely[53] injected bubbles of air containing a known concentration of carbon tetrachloride vapour into a bed (diameter 86 mm) of silica–alumina microspheres fluidised by air. By assuming that the vapour at the outside of the bubble was instantaneously adsorbed on the particles and by operating with very low concentrations of adsorbed material, it was possible to determine the transfer rate between the two phases from the vapour concentrations in the issuing gas.

By studying beds of differing depths, it was established that most of the mass transfer occurs during the formation of the bubble rather than during its rise. Mass transfer coefficients were calculated on the assumption that the bubbles are spherical during formation and rise as spherical cap bubbles. Values of the order of 20 or 30 mm/s were obtained. The variation of the average value of the coefficient with depth is shown in Fig. 6.12; at low bed depths the coefficient is low because the gas is able to bypass the bed during the formation of the bubble.

Davies[25] has found that, with a range of fine materials, a bubble injected into a fluidised bed will tend to grow as a result of a net transfer of gas to it from the continuous phase. The effect becomes progressively less marked as the minimum fluidising velocity of the system increases. It is found that the growth in bubble volume from $V_1$ to $V_2$ in a height

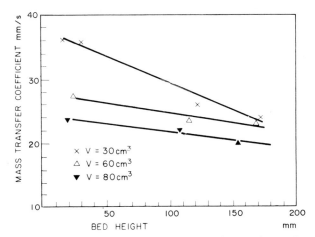

FIG. 6.12. Mass transfer coefficient between bubble and continuous phase as function of bed
height for various bubble volumes[92]

$\Delta z$ of bed for a system with a minimum fluidising velocity $u_{mf}$ is given by:

$$\ln \frac{V_2}{V_1} = \frac{\Delta z}{K} \qquad (6.23)$$

$K$ represents the distance the bubble must travel for its volume to increase by a factor of e. Typically, $K$ has a value of 900 mm for cracker catalyst of mean particle size 55 $\mu$m, fluidised at a velocity of $2 \cdot 3\, u_{mf}$.

As a result of gas flow into the bubble, the mean residence time of the gas in such systems is reduced because the bubble rises more rapidly than the gas in the continuous phase. Thus the injection of a single bubble into the bed will initially cause the bed to expand by an amount equal to the volume of the bubble. However, when this bubble has broken the surface of the bed, the bed volume decreases to a value less than its initial value. If the value of $u_c$ is only slightly in excess of $u_{mf}$, the gas in a small injected bubble may, however, become dispersed throughout the continuous phase so that no bubble appears at the surface of the bed.

Equation 6.23 enables the net increase in volume of bubble to be calculated. The bubble grows because more gas enters through the base than leaves through the cap. By injecting bubbles of carbon dioxide which is not adsorbed by the particles and analysing the concentration in the continuous phase at various heights in the bed, it has been possible to determine the actual transfer from the bubble when rising in a bed of cracker catalyst fluidised at an air velocity of $u_{mf}$. For bubbles in the size range of 80–250 cm$^3$, the transfer velocity through the roof of the bubble is constant at about 20 mm/s, which compares very closely with the results of Szekely[53] using a quite different technique and assumptions.

### 6.3.5. Beds of Particles of Mixed Sizes

The most uniform fluidisation would be expected when all the particles are approximately the same size so that there is no great difference in their yerminal falling velocities. However, the presence of a very small quantity of fines is often found to improve the quality of fluidisation of gas–solid systems, though if fines are present in the bed bubble formation may occur at a lower velocity. If the sizes differ appreciably, elutriation occurs and the smaller particles are continuously removed from the system. If the particles forming the bed are initially of the same size, fines will often be produced as a result of mechanical attrition or as a result of breakage due to high thermal stresses. Further, if the particles themselves take part in a chemical reaction, their sizes may alter as a result of the elimination of part of the material (e.g. during carbonisation or combustion). Any fines which are produced should be recovered using a cyclone separator. Final traces of fine material can then be eliminated with an electrostatic precipitator (see Chapter 8).

Leva[54] measured the rate of elutriation from a bed composed of particles of two different sizes fluidised in air. He found that, if the height of the containing vessel above the top of the bed was small, the rate of elutriation was high, but if it was greater than a certain value the rate was not affected. This was attributable to the fact that the small particles were expelled from the bed with a velocity higher than the equilibrium value in the unobstructed tube above the bed, because the linear velocity of the fluid in the bed is much higher than that in the empty tube. The experiments showed that the concentration of the fine particles in the bed varied with the time of elutriation according to a law of the form

$$C = C_0 e^{-Mt} \tag{6.24}$$

where    $C$ is the concentration of particles at time $t$,
        $C_0$ is the initial concentration of particles, and
        $M$ is a constant.

The bed containing particles of different sizes behaves in a similar manner to a mixture of liquids of different volatilities. Thus the finer particles, when associated with the fluidising medium, correspond to the lower boiling liquid and are more readily elutriated; their rate of removal from the system and the degree of separation are affected by the height of the reflux column. Further, a law analogous to Henry's Law for the solubility of gases in liquids is obeyed, with the concentration of solids of given size in the bed bearing, at equilibrium, a constant relation to the concentration of solids in the gas which is passing upwards through the bed. Thus, if clean gas is passed upwards through a bed containing fine particles, these particles are continuously removed. On the other hand, if a dust-laden gas is passed through the bed, particles will be deposited until the equilibrium condition is reached. Fluidised beds are therefore sometimes used for removing suspended dusts and mist droplets from gases. They have the advantage that the resistance to flow is in many cases less than in equipment employing a fixed filter medium. Clay particles and small glass beads are used for the removal of sulphuric acid mists.

For fluidisation with a liquid, a bed of particles of mixed sizes will become sharply stratified with the small particles on top and the large ones at the bottom. The pressure drop is that which would be expected for each of the layers in series. If the size range is small, however, no appreciable segregation will occur.

## 6.4. HEAT TRANSFER TO A BOUNDARY SURFACE

### 6.4.1. Mechanisms Involved

The good heat transfer properties of fluidised systems have led to their adoption in circumstances where close control of temperature is required. The presence of the particles in a fluidised system results in an increase of up to one-hundredfold in the heat transfer coefficient, as compared with the value obtained with a gas alone at the same velocity. In a liquid-fluidised system the increase is not so marked.

Many experimental investigations of heat transfer between a gas-fluidised system and a heat transfer surface have been made, and the agreement between the correlations given by different workers is very poor, differences of 1 or even 2 orders of magnitude occurring at times. The reasons for these large discrepancies appear to be associated with the critical dependence of heat transfer coefficients on the geometry of the system, on the quality of fluidisation, and consequently on the flow patterns obtained. Much of the work was carried out before any real understanding existed of the nature of the flow patterns within the bed. However, there is almost universal agreement that the one property which has virtually no influence on the process is the thermal conductivity of the solids.

Three main mechanisms have been suggested for the improvement in the heat transfer coefficients brought about by the presence of the solids. First, the particles, whose heat capacity per unit volume is many times greater than that of the gas, act as heat-transferring agents. As a result of their rapid movement within the bed they pass from the bulk of the bed to the layers of gas in close contact with the heat transfer surface, exchanging heat at this point and returning to the body of the bed again. Because of their short time of residence and their high heat capacity, they change little in temperature, and this fact, coupled with the extremely short physical contact time of the particle with the surface, accounts for the unimportance of their thermal conductivity. The second mechanism which has been suggested is the erosion of the laminar sub-layer at the heat transfer surface by the particles, and the consequent reduction in its effective thickness. The third mechanism, suggested by Mickley and Fairbanks[55], is that "packets" of particles move to the heat transfer surface, and an unsteady state heat transfer process takes place.

### 6.4.2. Liquid–Solid Systems

It is of interest to study the heat transfer characteristics of liquid–solid fluidised systems, in which the heat capacity per unit volume of the solids is of the same order as that of the fluid. The first investigation with such a system was made by Lemlich and Caldas[56], but most of their results were obtained in the transitional region between streamline and turbulent flow and are therefore difficult to assess. Mitson[57] and Smith[16] both measured heat transfer coefficients for systems in which a number of different solids were fluidised by water in a 50 mm diameter brass tube, fitted with an annular heating jacket. The apparatus was equipped with thermocouples which measured the wall and liquid temperatures at various points along the heating section. Heat transfer coefficients at the tube wall were calculated and were plotted against volumetric concentration as shown in Fig. 6.13 for gravel. They were found to increase with concentration and to pass through a maximum at a volumetric concentration of 25–30 per cent. Since in a liquid–solid fluidised system there is a unique relation between concentration and velocity, as the concentration is increased the velocity necessarily falls, and the heat transfer coefficient for liquid alone at the corresponding velocity shows a continuous decrease as the concentration is increased. The difference in the two values, namely the increase in coefficient attributable to the presence of the particles $h-h_l$, is plotted against concentration in Fig. 6.14 for ballotini; these curves also pass through a maximum.

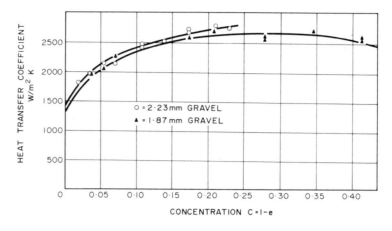

FIG. 6.13. Heat transfer coefficients for gravel particles fluidised in water[16]

Experimental results in the region of turbulent flow can conveniently be correlated in terms of the specific heat $c_s$ of the solid (kJ/kg K) by the equation:

$$h - h_l = 24 \cdot 4 \,(1 + 1 \cdot 71 \, c_s^{2 \cdot 12})(1 - e)^m \left(\frac{u_c}{e}\right)^{1 \cdot 15} \tag{6.25}$$

Here the film coefficients are expressed in kW/m² K and the fluidising velocity $u_c$ is in m/s. The value of the index $m$ is given by:

$$m = 0 \cdot 079 \left(\frac{u_i \, d\rho}{\mu}\right)^{0 \cdot 36} \tag{6.26}$$

The maximum value of the ratio of the coefficient for the fluidised system to that for liquid alone at the same velocity is about 3.

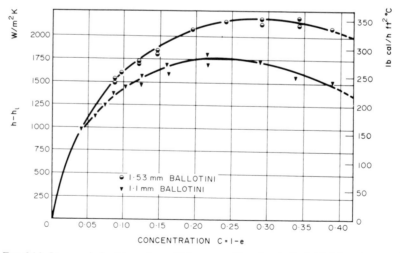

FIG. 6.14. Increase in heat transfer coefficient caused by glass ballotini fluidised in water [16]

In a modified system in which a suspension of solids was conveyed through the heat transfer section, the heat transfer coefficient was greater than that obtained with liquid alone, but was lower than that obtained at the same concentration in a fluidised system. Similar conclusions were reached by Jepson, Poll, and Smith[58] who measured the heat transfer to a suspension of solids in gas.

More recently[59-62] heat-transfer coefficients to liquid–solid systems have been measured using small electrically heated surfaces immersed in the bed. The temperature of the element may be obtained from its electrical resistance, provided that the temperature coefficient of resistance is known. The heat supplied is obtained from the measured applied voltage and resistance and is equal to $V^2/R$

where    $V$ is the voltage applied across the element, and
            $R$ is its resistance.

The energy $Q$ given up by the element to the bed in which it is immersed can be expressed as the product of the heat transfer coefficient $h$, the area $A$ and the temperature difference between the element and the bed $(T_E - T_B)$. Thus:

$$Q = hA(T_E - T_B) \qquad (6.27)$$

At equilibrium, the energy supplied must be equal to that given up to the bed and:

$$\frac{V^2}{R} = hA(T_E - T_B) \qquad (6.28)$$

Thus the heat-transfer coefficient can be obtained from the slope of the straight line connecting $V^2$ to $T_E$. This method has been successfully applied for measuring the heat transfer coefficient from a small heated surface, 25 mm square, to a wide range of fluidised systems. Khan[61] measured the coefficient for heat transfer between the 25 mm square surface and fluidised beds formed in a tube of 100 mm diameter. Uniform particles of sizes between 3 mm and 9 mm were fluidised by liquids consisting of mixtures of kerosene and lubricating oil whose viscosities ranged from 1·55 to 940 mN s/m². The heat transfer coefficient increases as the voidage, and hence the velocity, is increased and passes through a maximum; this effect has been noted by most of the previous workers in the area. The voidage $e_{max}$ at which the maximum heat transfer coefficient occurs becomes progressively greater as the viscosity of the liquid is increased.

The results of this work, together with those obtained earlier by Davies[62] and by Richardson, Romani, and Shakiri[60], who fluidised uniform particles with dimethyl phthalate, have been correlated by expressing the particle Nusselt number ($Nu' = hd/k$) in terms of the Prandtl number of the fluid ($Pr$), the particle Reynolds number ($Re'$) and the bed voidage ($e$) to give (see Fig. 6.15):

$$Nu' = (0{\cdot}033\,Re' + 1{\cdot}88\,Re'^{0{\cdot}43})\,Pr^{0{\cdot}37}(1-e)^{0{\cdot}725} \qquad (6.29)$$

An alternative form of correlation, which was equally satisfactory, involved the Reynolds number for the bed ($Re_1 = u_c\rho/S\mu(1-e)$) and the ratio of the particle to column diameter ($d/d_t$) (see Fig. 6.16):

$$\frac{e\,Nu'}{Re_1\,Pr^{0{\cdot}422}} = (0{\cdot}0080 + 0{\cdot}27\,Re_1^{-0{\cdot}53})(d/d_t)^{0{\cdot}05} \qquad (6.30)$$

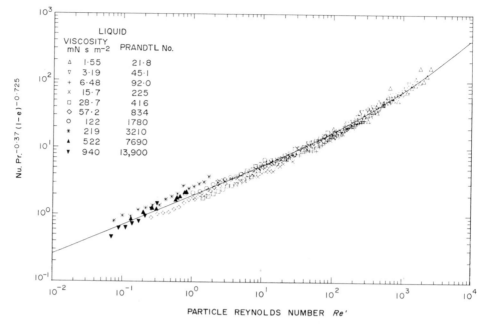

FIG. 6.15. Correlation of heat transfer coefficients to immersed surface in terms of particle Reynolds number $Re'$ $(= u_c d\rho/\mu)$

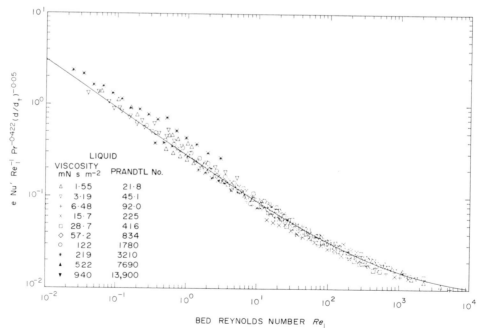

FIG. 6.16. Correlation of heat transfer coefficients to immersed surface in terms of bed Reynolds number $Re_1 [= u_c\rho/S\mu(1-e)]$

Although parameters in equations 6.29 and 6.30 were varied over a wide range ($20 < Pr < 14{,}000$) ($0.5 < e < 0.9$) ($0.01 < Re' < 3000$) and ($0.01 < Re_1 < 3000$), the specific heat capacities and the thermal conductivities of the liquids were almost constant, as they are for most organic liquids, and the dimensions of the surface and of the tube were not varied. Nevertheless, for the purposes of comparison with other results, it is useful to work in terms of dimensionless groups.

The maximum values of the heat-transfer coefficient, and of the corresponding Nusselt numbers, can be predicted satisfactorily from equation 6.29 by differentiating with respect to voidage and putting the derivative equal to zero.

### 6.4.3. Gas–Solid Systems

With gas–solid systems, the heat-transfer coefficient to a surface is very much dependent on the geometrical arrangement and the quality of fluidisation; furthermore, in many cases the temperature measurements are suspect. Leva[63] has plotted the heat transfer coefficient for a bed, composed of silica sand particles of diameter 0·15 mm fluidised in air, as a function of gas rate, using the correlations put forward as a result of ten different investigations as shown in Fig. 6.17. For a gas flow of 0·3 kg/m²s, the values of the coefficient ranged from about 75 W/m² K when calculated by the formula of Levenspiel and Walton[64] to about 340 when Vreedenberg's[65] expression was used. One of the most reliable equations appears to be that of Dow and Jakob[66], which gives a value of about 200 for the above case. Their equation is:

$$\frac{h d_t}{k} = 0.55 \left(\frac{d_t}{l}\right)^{0.65} \left(\frac{d_t}{d}\right)^{0.17} \left\{\frac{(1-e)\rho_s c_s}{e \rho c_p}\right\}^{0.25} \left(\frac{u_c d_t \rho}{\mu}\right)^{0.80} \tag{6.31}$$

where     $h$ is the heat transfer coefficient,
           $k$ is the thermal conductivity of the gas,
           $d$ is the diameter of the particle,
           $d_t$ is the diameter of the tube,
           $l$ is the depth of the bed,
           $e$ is the voidage of the bed,
           $\rho_s$ is the density of the solid,
           $\rho$ is the density of the gas,
           $c_s$ is the specific heat of the solid,
           $c_p$ is the specific heat of the gas at constant pressure,
           $\mu$ is the viscosity of the gas, and
           $u_c$ is the superficial velocity based on the empty tube.

In other words, the Nusselt number with respect to the tube $Nu$ ($= h d_t/k$) is expressed as a function of four dimensionless groups: the ratio of tube diameter to length, the ratio of tube to particle diameter, the ratio of the heat capacity per unit volume of the solid to that of the fluid, and the tube Reynolds number, $Re$ ($= u_c d_t \rho/\mu$). It is doubtful whether it is possible to give a more reliable relation than this with our present state of knowledge. At this stage it would seem to be more important to study the mechanism of transfer than to add yet another equation to the literature.

The mechanism of heat transfer to a surface has been studied in a fixed and a fluidised bed by Botterill and co-workers[67,68,69,70]. They constructed an apparatus in which

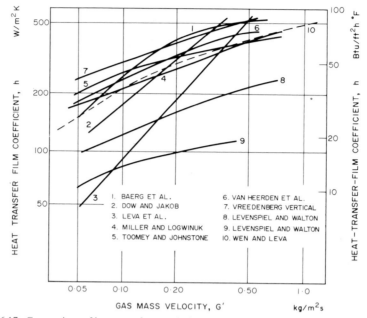

FIG. 6.17. Comparison of heat transfer correlations. Silica sand (0·15 mm) fluidised in air[63]

particle replacement at a heat transfer surface was obtained by means of a rotating stirrer with blades close to the surface. A steady-state system was employed using an annular bed, with heat supplied at the inner surface and removed at the outer wall by means of a jacket. The average residence time of the particle at the surface in a fixed bed was calculated, assuming the stirrer to be perfectly efficient, and in a fluidised bed it was shown that the mixing effects of the gas and the stirrer were additive. The heat transfer process was considered to be one of unsteady-state thermal conduction during the period of residence of the particle and its surrounding layer of fluid at the surface. Virtually all the heat passing between the particle and the surface does so through the intervening fluid as there is only point contact between the particle and the surface. It is shown by calculation that the heat transfer rate falls off in an exponential manner with time and that, with gases, the heat transfer is confined to the region surrounding the point of contact. With liquids, however, the thermal conductivities and diffusivities of the two phases are comparable, and appreciable heat flow occurs over a more extended region. Thus, the heat transfer process is shown to be capable of being broken down into a series of unsteady state stages, at the completion of each of which the particle with its attendant fluid layer becomes mixed again with the bulk of the bed. This process is very similiar to that assumed in the penetration theory of mass transfer proposed by Higbie[71] and by Danckwerts[72], and described in Volume 1. It shows how the heat-carrying effect of the particles and their disruption of the laminar sub-layer at the surface must be considered as component parts of a single mechanism. Reasonable agreement is obtained by Botterill *et al.* between the practical measurements of heat transfer coefficient and the calculated values.

Mickley, Fairbanks, and Hawthorn[73] made measurements of heat transfer between an air-fluidised bed 104 mm diameter and a concentric heater 6·4 mm diameter and 600 mm long, divided into six 100 mm lengths each independently controllable. The mean value of

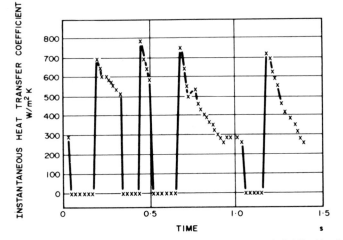

FIG. 6.18. Instantaneous heat transfer coefficients for glass beads fluidised in air[73]

the heat transfer coefficient was found to fall with increase in height, and this was probably attributable to an increased tendency for slugging. The instantaneous value of the heat transfer coefficient was found by replacing a small segment of one of the heaters (arc 120° and depth 7·9 mm) by a piece of platinum foil of low heat capacity with a thermocouple behind it. The coefficient was found to fluctuate from < 50 to about 600 W/m² K. A typical curve is shown (Fig. 6.18) from which it is seen that the coefficient rises rapidly to a peak and then falls off slowly before there is a sudden drop in value as a bubble passes the foil. It is suggested that heat transfer takes place by unsteady-state conduction to a packet of particles which are then displaced by a gas slug. Coefficients were calculated on the assumption that heat flowed for a mean exposure time into a mixture whose physical and thermal properties corresponded with those of an element of the continuous phase, and that the heat flow into the gas slug could be neglected. The fraction of the time during which the surface was in contact with a gas slug ranged from zero to 40 per cent with an easily fluidised solid, but appeared to continue increasing with gas rate to much higher values for a solid which gave uneven fluidisation. Thus, it is easy on this basis to see why very variable values of heat transfer coefficients have been obtained by different workers. For good heat transfer, rapid replacement of the solids is required, without an appreciable coverage of the surface by gas alone.

The method previously described for the determination of heat transfer coefficients in liquid–solid fluidised beds in which a small immersed element is electrically heated has been successfully used for gas–solid systems. Figure 6.19 shows the way in which the heat transfer coefficient varies as gas velocity is increased. The heat transfer coefficient is plotted against multiples of the minimum fluidising velocity $(u_c/u_{mf})$ for glass particles (9 μm) fluidised with air at atmospheric pressure. Region $AB$, corresponding to the fixed bed, is characterised by a low, almost constant, coefficient. At point $B$, corresponding to the minimum fluidising velocity, the coefficient starts to rise rapidly due to the circulation of the solids; it then reaches a maximum $C$, after which the coefficient falls again. This behaviour can be attributed to two effects occurring simultaneously as the gas flowrate is increased; first there is a progressively greater rate of circulation of solids, but secondly

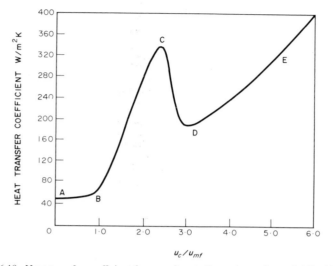

FIG. 6.19. Heat transfer coefficient from surface to 9 $\mu$m glass spheres fluidised by air

there is an increasing tendency for gas bubbles to cover the surface. Once the gas flowrate corresponding to the maximum has been reached, the effects of gas blanketing of the surface start to dominate as velocity is raised further. However, at a still higher flowrate the curve passes through a minimum $D$ and then for higher gas rates there is a steady increase in coefficient. At these very high gas rates $E$ there is very rapid replacement of material at the surface.

Because the process depends on the transfer of heat between fluid and particle, both near the heat transfer surface and again in the body of the bed where thermal equilibrium is rapidly attained, it is important to know what factors control the transfer between fluid and particle. Because mass transfer is a similar process, information on mass transfer rates will also be helpful in interpreting results for heat transfer.

## 6.5. MASS AND HEAT TRANSFER BETWEEN FLUID AND PARTICLES

### 6.5.1. Introduction

The calculation of coefficients for the transfer of heat or mass between the particles and the fluid stream necessitates a knowledge of the heat or mass flow, the interfacial area, and the driving force expressed either as a temperature or a concentration difference. Many of the earlier investigations were unsatisfactory in that one or more of these variables was inaccurately determined. This applied particularly to the driving force which was frequently based on completely erroneous assumptions concerning the nature of the flow in the bed.

One of the main problems of making measurements of transfer coefficients is that equilibrium is rapidly attained between particles and fluidising medium. This difficulty has in some cases been obviated by the use of very shallow beds. Furthermore, in measurements of mass transfer, the methods of analysis have been inaccurate and the particles used have frequently been of such a nature that it has not been possible to obtain fluidisation of good quality.

### 6.5.2. Mass Transfer between Fluid and Particles

Bakhtiar[74] adsorbed toluene and iso-octane vapours from a vapour-laden air stream on to the surface of synthetic alumina microspheres and followed the change of concentration of the outlet gas with time, using a sonic gas analyser. He found that equilibrium was attained between outlet gas and solids in all cases, and therefore transfer coefficients could not be calculated. However, he was able to follow the progress of the adsorption process.

A material balance over the bed at any time $t$ after the commencement of the experiment gives:

$$G_m(y_0 - y) = \frac{d}{dt}(WF) \tag{6.32}$$

where   $G_m$ is the molar rate of flow of gas,
  $W$ is the mass of solids in the bed,
  $F$ is the number of moles of vapour adsorbed on unit mass of solid, and
  $y(y_0)$ is the mol fraction of vapour in the outlet (inlet) stream.

Equation 6.32 is given on the basis of complete mixing of the solids in the bed. If the adsorption isotherm is linear, the concentration $F$ of adsorbed material can be related to the composition $y^*$ of the vapour with which it is in equilibrium by the relation:

$$F = f + by^* \tag{6.33}$$

Now, if equilibrium is reached between the outlet gas and the solids and if none of the gas by-passes the bed in the form of bubbles:

$$y = y^* \tag{6.34}$$

Thus, substituting from equations 6.33 and 6.34 into equation 6.32:

$$G_m(y_0 - y) = \frac{d}{dt}\{W(f + by)\}$$

$$\frac{G_m}{Wb}(y_0 - y) = \frac{dy}{dt} \tag{6.35}$$

Rearranging equation 6.35 and integrating:

$$\ln\left(1 - \frac{y}{y_0}\right) = -\frac{G_m}{Wb}t \tag{6.36}$$

Here time $t$ is reckoned from the instant at which vapour starts to appear in the outlet gas from the bed.

If the hypotheses put forward, namely complete mixing of the solids, no by-passing by the gas, and equilibrium between outlet gas and solids, are correct, the plot of $\ln(1 - y/y_0)$ against $(G_m/W)t$ should give straight lines of slope equal to $-1/b$. In all the experiments on adsorption, and in the converse experiments in which desorption was effected, straight

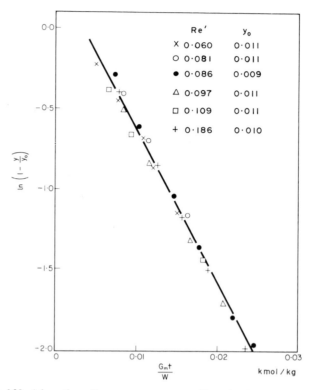

FIG. 6.20. Adsorption of iso-octane vapour on silica–alumina microspheres[74]

lines were obtained and the hypotheses were borne out. A typical curve for the adsorption of iso-octane vapour is given in Fig. 6.20.

Szekely[75] modified the system so that equilibrium was not achieved at the outlet. He used thin beds and low concentrations of vapour, so that the slope of the adsorption isotherm was greater. Particles of charcoal of different pore structures, and of silica gel, were fluidised by means of air or hydrogen containing a known concentration of carbon tetrachloride or water vapour. A small glass apparatus was used so that it could be readily dismantled, and the adsorption process was followed by weighing the bed at intervals. The inlet concentration was known and the outlet concentration was determined as a function of time from a material balance, using the information obtained from the periodic weighing of the apparatus. The driving force was then obtained at the inlet and the outlet of the bed, on the assumption that the solids were completely mixed and that the partial pressure of vapour at their surface was given by the adsorption isotherm.

At any height $z$ above the bottom of the bed, the mass transfer rate per unit time, on the assumption of piston flow of gas, is given by:

$$\mathrm{d}N_A = h_D \Delta C \, a' \, \mathrm{d}z \tag{6.37}$$

where $a'$ is the transfer area per unit height of bed.

Integrating over the whole depth of the bed:

$$N_A = h_D a' \int_0^Z \Delta C \, \mathrm{d}z \tag{6.38}$$

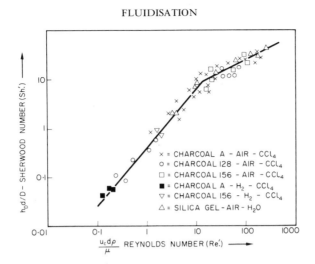

FIG. 6.21. Sherwood number as a function of Reynolds number for adsorption experiments[75]

The integration can be carried out only if the variation of driving force throughout the depth of the bed can be estimated. It was not possible to make measurements of the concentration profiles within the bed, but as the value of $\Delta C$ did not vary greatly from the inlet to the outlet, no serious error was introduced by using the logarithmic mean value $\Delta C_{lm}$.

Thus:

$$N_A \approx h_D a' Z \, \Delta C_{lm} \tag{6.39}$$

Values of mass transfer coefficients were calculated using equation 6.39, and it was found that the coefficient progressively became less as each experiment proceeded and as the solids became saturated. This effect was attributed to the gradual build up of the resistance to transfer in the solids. In all cases the transfer coefficient was plotted against the relative saturation of the bed, and the values were extrapolated back to zero relative saturation, corresponding to the commencement of the experiment. These maximum extrapolated values were then correlated by plotting the corresponding value of the Sherwood number $(Sh' = (h_D d/D))$ against the particle Reynolds number $(Re' = (u_c d\rho/\mu))$ to give two lines as shown in Fig. 6.21, which could be represented by the following equations:

$$(0.1 < Re' < 15) \qquad \frac{h_D d}{D} = Sh' = 0.37 \, Re'^{1.2} \tag{6.40}$$

$$(15 < Re' < 250) \qquad \frac{h_D d}{D} = Sh' = 2.01 \, Re'^{0.5} \tag{6.41}$$

These correlations were applicable to all the systems employed, provided that the initial maximum values of the transfer coefficients were used. This suggests that the extrapolation gives the true gas-film coefficient. This is borne out by the fact that the coefficient remained unchanged for a considerable period when the pores were large, but fell off extremely rapidly with solids with a fine pore structure. It was not possible, however, to relate the behaviour of the system quantitatively to the pore size distribution.

A study of mass transfer between a liquid and a particle forming part of an assemblage of particles was made by Mullin and Treleaven[76], who subjected a sphere of benzoic acid to the action of a stream of water. For a fixed sphere, or a sphere free to circulate in the liquid, the mass transfer coefficient was given by:

$$(50 < Re' < 700) \qquad Sh' = 0.94 \, Re'^{\,1/2} \, Sc^{1/3} \qquad (6.42)$$

The presence of adjacent spheres caused an increase in the coefficient because the turbulence was thereby increased. The effect became progressively greater as the concentration increased, but the results were not influenced by whether or not the surrounding particles were free to move. This suggested that the transfer coefficient should be the same in a fixed or a fluidised bed.

The results of earlier work by Chu, Kalil, and Wetteroth[77] had suggested that transfer coefficients were similar in fixed and fluidised beds. Apparent differences at low Reynolds numbers were probably attributable to the fact that there could be appreciable back-mixing of fluid in the fluidised bed.

### 6.5.3. Heat Transfer between Fluid and Particles

The difficulty of measuring heat transfer coefficients is shown by the fact that many workers failed to measure any temperature difference between gas and solid in a fluidised bed. Frequently, an incorrect area for transfer was assumed, since it was not appreciated that thermal equilibrium existed everywhere in a fluidised bed, except within a thin layer immediately above the gas distributor. Kettenring, Manderfield, and Smith[78] and Heertjes and McKibbins[79] measured heat transfer coefficients for the evaporation of water from particles of alumina or silica gel fluidised by heated air. In the former work there were probably considerable errors arising from the conduction of heat along the leads of the thermocouples used for measuring the gas temperature. Heertjes found that any temperature gradient was confined to the bottom part of the bed. He used a suction thermocouple for measuring gas temperatures, but this probably caused some disturbance to the flow pattern in the bed. Frantz[80] has reviewed many of the recent investigations.

Ayers[81] used a steady-state system in which spherical particles were fluidised in a rectangular bed by means of hot air. A continuous flow of solids was maintained across the bed, and the particles on leaving the system were cooled and then returned to the bed.

Temperature gradients within the bed were measured using a fine thermocouple assembly, with a junction formed by welding together 40-gauge wires of copper and constantan. The thermojunction leads were held in an approximately isothermal plane to minimise the effect of heat conduction. After steady-state conditions had been reached, it was found that the temperature gradient was confined to a shallow zone, not more than 2·5 mm deep at the bottom of the bed. Elsewhere, the temperature was uniform and equilibrium existed between the gas and the solids. A typical temperature profile is shown in Fig. 6.22.

At any height $z$ above the bottom of the bed, the heat transfer rate between the particles and the fluid, on the assumption of complete mixing of the solids and piston flow of the gas, is given by:

$$dQ = h\Delta T a' \, dz \qquad (6.43)$$

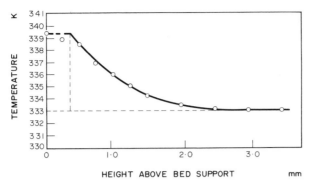

FIG. 6.22. Vertical temperature gradient in fluidised bed[81]

Integrating,

$$Q = ha' \int_0^z \Delta T \, dz \tag{6.44}$$

In equation 6.44, $Q$ could be obtained from the change in temperature of the gas stream, and $a'$ (the area for transfer per unit height of bed) from measurements of the surface of solids in the bed. In this case, the integration could be carried out graphically, because the relation between $\Delta T$ and $z$ had been obtained from the readings of the thermocouple. The temperature recorded by the thermocouple was assumed to be the gas temperature, and if the solids were completely mixed their temperature would be the same as that of the gas in the upper portion of the bed.

The results for the heat transfer coefficient were satisfactorily correlated by equation 6.45 (see Fig. 6.23). Since the resistance to heat transfer in the solid could be neglected compared with that in the gas, the coefficients which were calculated were gas-film coefficients:

$$Nu' = \frac{hd}{k} = 0.054 \left(\frac{u_c d \rho}{e \mu}\right)^{1.28} \tag{6.45}$$

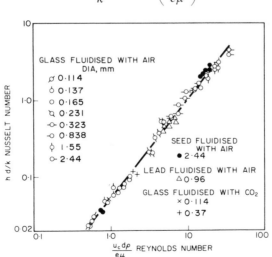

FIG. 6.23. Correlation of experimental results for heat transfer to particles fluidised bed[81]

Taking an average value of 0·57 for the voidage of the bed, this equation can be rewritten:

$$Nu' = 0.11\, Re'^{1.28} \tag{6.46}$$

This equation was found to be applicable for values of $Re'$ from 0·25 to 18.

### 6.5.4. Analysis of Results for Heat and Mass Transfer to Particles

A comparison of equations 6.40 and 6.46 shows that similar forms of equation describe the processes of heat and mass transfer. However, the values of the coefficients are different in the two cases. This appears to be attributable largely to the fact that the average value for the Prandtl number ($Pr$) in the heat transfer work was lower than the value of the Schmidt number ($Sc$) in the mass transfer experiments.

It is convenient to express results for experiments on heat transfer and mass transfer to particles in the form of $j$-factors. If the concentration of the diffusing component is small, the $j$-factor for mass transfer may be defined by the relationship:

$$j_d = \frac{h_D}{u_c} Sc^{0.67} \tag{6.47}$$

where  $h_D$ is the mass transfer coefficient,

  $u_c$ is the fluidising velocity,

  $Sc$ is the Schmidt number ($\mu/\rho D$),

  $\mu$ is the fluid viscosity,

  $\rho$ is the fluid density, and

  $D$ is the diffusivity of the transferred component in the fluid.

The corresponding relation for heat transfer is:

$$j_h = \frac{h}{c_p \rho u_c} Pr^{0.67} \tag{6.48}$$

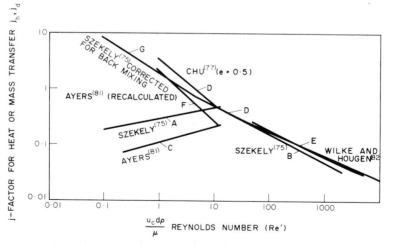

Fig. 6.24. Heat and mass transfer results expressed as $j$-factors

where    $h$ is the heat transfer coefficient,
$\quad\quad c_p$ is the specific heat of the fluid at constant pressure,
$\quad\quad Pr$ is the Prandtl number $(c_p\mu/k)$, and
$\quad\quad k$ is the thermal conductivity of the fluid.

The significance of $j$-factors has been discussed in detail in Volume 1.

Rearranging equations 6.40, 6.41, and 6.46 in the form of 6.47 and 6.48, and substituting mean values of 2·0 and 0·7 respectively for $Sc$ and $Pr$, gives equations 6.49, 6.50, and 6.51 respectively:

$(0·1 < Re' < 15)$

$$j_d = \frac{Sh'}{Sc\,Re'}\,Sc^{0·67} = 0·37\,Re'^{\,0·2}\,Sc^{-0·33} = 0·29\,Re'^{\,0·2} \tag{6.49}$$

$(15 < Re' < 250)$

$$j_d = \frac{Sh'}{Sc\,Re'}\,Sc^{0·67} = 2·01\,Re'^{\,-0·5}\,Sc^{-0·33} = 1·59\,Re'^{\,-0·5} \tag{6.50}$$

$(0·25 < Re' < 18)$

$$j_h = \frac{Nu'}{Pr\,Re'}\,Pr^{0·67} = 0·11\,Re'^{\,0·28}\,Pr^{-0·33} = 0·13\,Re'^{\,0·28} \tag{6.51}$$

These relations are plotted in Fig. 6.24 as lines $A$, $B$, and $C$ respectively.

Resnick and White[83] fluidised naphthalene crystals of five different size ranges (between 1000 and 250 $\mu$m) in air, hydrogen, and carbon dioxide at a temperature of 298 K. The gas was passed through a sintered disc, which served as the bed support, at rates between 0·01 and 1·5 kg/m² s. Because of the nature of the surface and of the shape of the particles, uneven fluidisation would have been obtained. The rate of vaporisation was

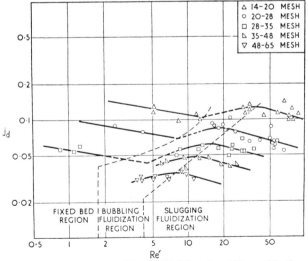

[14–20 mesh $\approx$ 1000 $\mu$m, 20–28 mesh $\approx$ 700 $\mu$m, 28–35 mesh $\approx$ 500 $\mu$m, 35–48 mesh $\approx$ 350 $\mu$m, 48–65 mesh $\approx$ 250 $\mu$m]

FIG. 6.25. $j$-factor, $j_d$, for transfer of naphthalene vapour to air in fixed and fluidised beds[83]

determined by a gravimetric analysis of the outlet gas, and mass transfer coefficients were calculated. These were expressed as $j$-factors and plotted against Reynolds number $Re'$ $(=u_c d\rho/\mu)$ in Fig. 6.25. It will be seen that separate curves were obtained for each size fraction of particles, but that each curve was of the same general shape, showing a maximum in the fluidisation region, roughly at the transition between bubbling and slugging conditions.

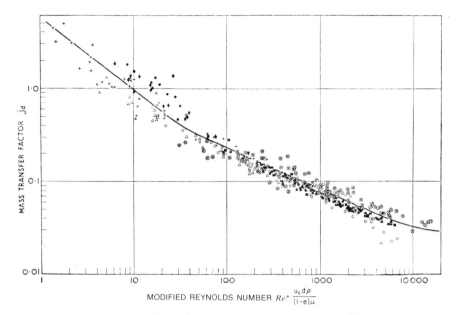

FIG. 6.26. $j$-factor, $j_d$, for fixed and fluidised beds[77]

| Symbol | System | Schmidt No. | Type of particle | State of bed | Ref. |
|---|---|---|---|---|---|
| $\otimes$ | Naphthalene–air | 2·57 | Spheres, cylinders | Fixed, fluidised | 77 |
| ■ | Water–air | 0·60 | Spheres, cylinders | Fixed | 82 |
| ○ | 2-naphthol–water | 1400 | Modified spheres | Fixed, fluidised | 84 |
| ● | Isobutyl alcohol–water | 866 | Spheres | Fixed | 85 |
| ⌀ | Methyl ethyl ketone–water | 776 | Spheres | Fixed | 85 |
| + | Salicylic acid–benzene | 368 | Modified spheres | Fixed | 86 |
| × | Succinic acid–$n$-butyl alcohol | 690 | Modified spheres | Fixed | 86 |
| △ | Succinic acid–acetone | 164 | Modified spheres | Fixed | 86 |

Chu, Kalil, and Wetteroth[77] obtained an improved quality of fluidisation by coating spherical particles with naphthalene, but it is probable that some attrition occurred. Experiments were carried out with particles ranging in size from 0·75 to 12·5 mm and voidages from 0·25 to 0·97. Fixed beds were also used. Again, it was found that particle size was a parameter in the relation between $j$-factor and Reynolds number. However, when plotted (Fig. 6.26) against a modified Reynolds number $Re^*$ $(=(u_c d\rho/(1-e)\mu))$ a single correlation was obtained. Furthermore, it was possible to represent with a single curve the results of a number of workers, obtained in fixed and fluidised beds with both liquids and gases as the fluidising media. A range of 0·6–1400 in Schmidt number was covered. It will be noted that the results for fluidised systems are confined to values obtained at relatively high values of the Reynolds number. The curve could be represented approximately by the equations:

$$(1 < Re^* < 30) \qquad j_d = 5\cdot7\,Re^{*-0\cdot78} \tag{6.52}$$

$$(30 < Re^* < 5000) \qquad j_d = 1\cdot77\,Re^{*-0\cdot44} \tag{6.53}$$

These two relations are also shown as curve $D$ in Fig. 6.24 for a voidage equal to 0·5.

A number of other workers have measured mass transfer rates. Thus McCune and Wilhelm[84] studied transfer between naphthol particles and water in fixed and fluidised beds. Hsu and Molstad[87] absorbed carbon tetrachloride vapour on activated carbon particles in very shallow beds which were sometimes less than one particle diameter deep. Wilke and Hougen[82] dried Celite particles (size range approximately 3–19 mm) in a fixed bed by means of a stream of air, and found that their results could be represented by:

$$(50 < Re' < 250) \qquad j_d = 1\cdot82\,Re'^{-0\cdot51} \tag{6.54}$$

$$(Re' > 350) \qquad j_d = 0\cdot99\,Re'^{-0\cdot41} \tag{6.55}$$

These relations are plotted as curve $E$ in Fig. 6.24.

It will be seen that the general trend of the results of different workers is similar but that the agreement is not good. In most cases a direct comparison of results is not possible because the experimental data are not available in the required form.

The importance of flow pattern on the experimental results is clearly apparent, and the reasons for discrepancies between the results of different workers are largely attributable to the rather different characters of the fluidised systems. It is of particular interest to note that, at high values of the Reynolds number when the effects of back-mixing are unimportant, similar results are obtained in fixed and fluidised beds. This conclusion was also reached by Mullin and Treleaven[76] in their experiments with models.

There is apparently an inherent anomaly in the heat and mass transfer results in that, at low Reynolds numbers, the Nusselt and Sherwood numbers (Figs. 6.23 and 6.21) are very low, and substantially below the theoretical minimum value of 2 for transfer by thermal conduction or molecular diffusion to a spherical particle when the temperature or concentration difference is spread over an infinite distance. (See Volume 1.) The most probable explanation is that at low Reynolds numbers there is appreciable back-mixing of gas associated with the circulation of the solids. If this could be represented as a diffusional type of process with a longitudinal diffusivity of $D_L$, the basic equation for the heat transfer process would become:

$$D_L c_p \rho \frac{\mathrm{d}^2 T}{\mathrm{d}z^2} - u_c c_p \rho \frac{\mathrm{d}T}{\mathrm{d}z} - ha(T - T_s) = 0 \tag{6.56}$$

Equation 6.56 can be obtained in exactly the same way as equation 4.24, but with the addition of the last term which represents the transfer of sensible heat from the gas to the solids. The time derivative is zero because a steady-state process is considered.

On integration, equation 6.56 gives a relation between $h$ and $D_L$, and $h$ can only be evaluated if $D_L$ is known. If it is assumed that at low Reynolds numbers the value of the Nusselt or Sherwood numbers approaches the theoretical minimum value of 2, it is possible to estimate the values of $D_L$ at low Reynolds numbers, and then to extrapolate these values over the whole range of Reynolds numbers used. This provides a means of recalculating all the results using equation 6.56. When this is done, it is found that the

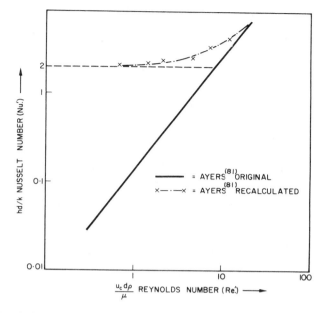

FIG. 6.27. Recalculated values of Nusselt number, taking into account effects of back-mixing[75]

results for low Reynolds numbers are substantially modified and the anomaly is eliminated, whereas the effect at high Reynolds numbers is small. Recalculated values of Nusselt number for heat transfer experiments are shown in Fig. 6.27. This confirms the view already expressed, and borne out by the work of Lanneau[52], that back-mixing is of importance only at low flow rates.

Recalculated values of $j_h$ and $j_d$ obtained from the results of Ayers[81] and Szekely[75] are shown in Fig. 6.24 as curves $F$ and $G$. It will be seen that the curves $B$, $D$, $E$, $F$, and $G$ follow the same trend.

Cornish[88] has given further consideration to the minimum possible value of the Nusselt number in a multiple particle system. He regards an individual particle as a source and the remote fluid as the sink, and shows that values of Nusselt number less than 2 can then be obtained. In a fluidised system, however, the inter-particle fluid is usually regarded as the sink and under these circumstances the theoretical lower limit of 2 for the Nusselt number applies. Zabrodsky[89] has also discussed the fallacy of Cornish's argument.

## 6.6. SUMMARY OF THE PROPERTIES OF FLUIDISED BEDS

Understanding and knowledge relating to fluidised systems is increasing at a very high rate, and as many as a hundred papers may appear in any given year. It must be understood, therefore, that any summarised picture of our present state of knowledge will rapidly become outdated.

Fluidised beds can be roughly divided into two classes. In the first, there is a uniform dispersion of the particles within the fluid and the bed expands in a regular manner as the fluid velocity is increased. This behaviour, termed *particulate fluidisation*, is exhibited by most liquid–solid systems, the only important exceptions being those composed of fine particles of high density. It is also exhibited by certain gas–solid systems over a very small range of velocities just in excess of the minimum fluidising velocity—particularly where the particles are approximately spherical and have very low free-falling velocities. In particulate fluidisation the rate of movement of the particles is comparatively low, and the fluid is predominantly in piston-type flow with some back-mixing, particularly at low flowrates. Overall turbulence normally exists in the system.

In the other form of fluidisation, *aggregative fluidisation*, two phases are present in the bed—a continuous or emulsion phase, and a discontinuous or bubble phase. This is the pattern normally encountered with gas–solid systems. Bubbles tend to form at gas rates above the minimum fluidising rate and grow as they rise through the bed. The bubbles grow because the hydrostatic pressure is falling, as a result of coalescence with other bubbles, and by flow of gas from the continuous to the bubble phase. The rate of rise of the bubble is approximately proportional to the one-sixth power of its volume. If the rising velocity of the bubble exceeds the free-falling velocity of the particles, it will tend to draw in particles at its wake and to destroy itself. There is therefore a maximum stable bubble size in a given system. If this exceeds about 10 particle diameters, the bubble will be obvious and aggregative fluidisation will exist; this is the usual condition with a gas–solid system. Otherwise, the bubble will not be observable and particulate fluidisation will occur. In aggregative fluidisation, the flow of the fluid in the continuous phase is predominantly streamline.

In a gas–solid system, the gas distributes itself between the bubble phase and the continuous phase which generally has a voidage a little greater than at the point of incipient fluidisation. If the rising velocity of the bubbles is less than that of the gas in the continuous phase, it behaves as a rising void through which the gas will tend to flow preferentially. If the rising velocity exceeds the velocity in the continuous phase—and this is the usual case—the gas in the bubble is continuously recycled through a cloud surrounding the bubble. Partial by-passing therefore occurs and the gas comes into contact with only a limited quantity of solids. However, the gas cloud surrounding the bubble detaches itself from time to time.

The bubbles appear to be responsible for a large amount of mixing of the solids. A rising bubble draws up a spout of particles behind it and carries a wake of particles equal to about one-third of the volume of the bubble and wake together. This wake detaches itself at intervals. Again, of course, the pattern in a bed containing a large number of bubbles must be very much more complex.

One of the most important properties of the fluidised bed is its good heat transfer characteristics. For a liquid–solid system, the presence of the particles may increase the coefficient by a factor of 2 or 3. In a gas–solid system, the factor may be about two orders of magnitude, the coefficient being raised by the presence of the particles from a value for the gas to one normally associated with a liquid. The improved heat transfer is associated with the movement of the particles between the main body of the bed and the heat transfer surface. The particles act as heat-transferring elements and bring material at the bulk temperature in close proximity to the heat transfer surface. A rapid circulation therefore gives a high heat transfer coefficient. In a gas–solid system, therefore, the amount of bubbling within the bed should be sufficient to give adequate mixing, but at the same time should not be sufficient to cause an appreciable blanketing of the heat transfer surface by gas.

## 6.7. APPLICATIONS OF THE FLUIDISED SOLIDS TECHNIQUE

### 6.7.1. General

The use of the fluidised solids technique was developed very largely by the petroleum and chemical industries, for processes where the very high heat transfer coefficients and the high degree of uniformity of temperature within the bed enabled the development of processes which would otherwise be impracticable. Fluidised solids are now used quite extensively in many industries where it is desirable to bring about intimate contact between small solid particles and a gas stream. In many cases it is possible to produce the same degree of contact between the two phases with a very much lower pressure drop over the system. Drying of finely divided solids is now carried out in a fluidised system, and some carbonisation and gasification processes are now in operation. Fluidised beds are employed in gas purification work, in the removal of suspended dusts and mists from gases, in lime burning and in the manufacture of phthalic anhydride.

### 6.7.2. Fluidised Bed Catalytic Cracking[90]

The existence of a large surplus of high boiling material after the distillation of crude oil led to the introduction of a cracking process to convert these materials into compounds of lower molecular weight and lower boiling point—in particular into petroleum spirit. The cracking was initially carried out using a fixed catalyst, but local variations in temperature in the bed led to a relatively inefficient process, and the deposition of carbon on the surface of the catalyst particles necessitated taking the catalyst bed out of service periodically so that the carbon could be burned off. Many of these difficulties are obviated by the use of a fluidised catalyst, since it is possible continuously to remove catalyst from the reaction vessel and to supply regenerated catalyst to the plant. The high heat transfer coefficients for fluidised systems account for the very uniform temperatures within the reactors and make it possible to control conditions very closely. The fluidised system has one serious drawback: some longitudinal mixing occurs and this gives rise to a number of side reactions.

A diagram of the plant used for the catalytic cracking process is given in Fig. 6.28. Hot oil vapour, containing the required amount of regenerated catalyst, is introduced into the reactor which is at a uniform temperature of about 775 K. As the velocity of the vapour falls in the reactor, because of the greater cross-sectional area for flow, a fluidised bed is formed and the solid particles are maintained in suspension. The vapours escape from the top, and the flowrate is such that the vapour remains in the reactor for about 20 s. It is necessary to provide a cyclone separator at the gas outlet to remove entrained catalyst particles and droplets of heavy oil. The vapour from the cracked material then passes to the fractionating unit, whilst the catalyst particles and the heavy residue are returned to the bed. Some of the catalyst is continuously removed from the bottom of the reactor and, together with any fresh catalyst which is required, is conveyed in a stream of hot air into the regenerator where the carbon deposit and any adhering film of heavy oil is burned off at about 875 K: in the regenerator the particles are again suspended as a fluidised bed. The hot gases leave the regenerator through a cyclone separator, from which the solids return to the bed, and then flow to waste through an electrostatic precipitator which removes any

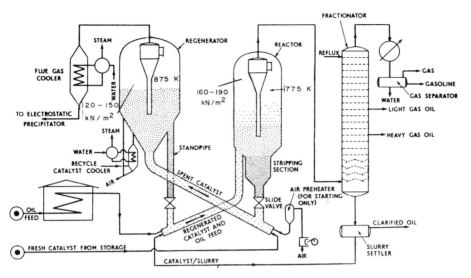

FIG. 6.28. Fluid catalytic cracking plant

very fine particles which are still in suspension. The temperature in the regenerator remains constant to within about 3 deg K, even where the fluidised bed is as much as 6 m deep and 15 m in diameter. Catalyst is continuously returned from the regenerator to the reactor by introducing it into the supply of hot vapour. The complete time cycle for the catalyst material is about 600 s. By this process a product consisting of between about 50 per cent and 75 per cent of petroleum spirit of high octane number is obtained. The quality of the product can be controlled by the proportion of catalyst which is used and the exact temperature in the reactor.

More recent experimental work has shown that much of the cracking takes place in the transfer line in which the regenerated catalyst is conveyed into the reactor in the stream of oil vapour. The chemical reaction involved is very fast, and the performance of the reactor is not sensitive to the hydrodynamic conditions.

From the diagram of the catalytic cracking plant in Fig. 6.28 it will be noted that there is a complete absence of moving parts in the reactor and the regenerator. The relative positions of components are such that the catalyst is returned to the reactor under the action of gravity.

### 6.7.3. Applications in the Chemical Industry

Fluidised catalysts are also used in the synthesis of high-grade fuels from mixtures of carbon monoxide and hydrogen obtained either by coal carbonisation or from partial oxidation of methane. An important application in the chemical industry is the oxidation of naphthalene to phthalic anhydride; this process is discussed by Riley[91]. The kinetics of this reaction are much slower than those of catalytic cracking, and considerable difficulties have been experienced in correctly designing the system.

Purely physical operations are also frequently carried out in a fluidised bed. Thus, fluidised bed dryers (see Chapter 16) are successfully used, frequently for heat-sensitive materials which must not be subjected to elevated temperatures for prolonged periods.

The design of a fluidised bed for the carrying out of an exothermic reaction involving a long reaction time has been considered by Rowe[92], and, as an example, he has considered the reaction between gaseous hydrogen fluoride and solid uranium dioxide to give solid uranium tetrafluoride and water vapour. This is a complex reaction which, as an approximation, however, can be considered as first order with respect to hydrogen fluoride. Here the problem is to obtain the required time of contact between the two phases in the most economical method. The amount of gas in the bubbles must be sufficient to give an adequate heat transfer coefficient, but the gas in the bubbles does, however, have a shorter contact time with the solids because of its greater velocity of rise. In order to increase the contact time of the gas, the bed can be made deep, but this results in a large pressure drop. The bubble size may be reduced by the incorporation of baffles and this is frequently an effective manner both of increasing the contact time and of permitting a more reliable scale-up from small scale experiments[93]. The most effective control is obtained by careful selection of the particle size of the solids. If the particle size is increased, there will be a higher gas flow in the emulsion phase and less in the bubble phase. Thus the ratio of bubble to emulsion phase gas velocity will be reduced, and the size of the gas cloud will increase. However, if the particle size is increased too much, there will be insufficient bubble phase to give good mixing. In practice, an overall gas flowrate of about twice that required for incipient fluidisation will frequently be suitable.

### 6.7.4. Fluidised Bed Combustion

An important application of fluidisation which has attracted considerable interest in recent years is fluidised bed combustion. The combustible material is held in a fluidised bed of inert material and the air for combustion is the fluidising gas. The system is being developed for steam raising on a very large scale for electricity generation and for incineration of domestic refuse.

The particular features of fluidised combustion of coal which have given rise to the current interest are first its suitability for use with very low-grade coals, including those with very high ash contents and sulphur concentrations, and secondly the very low concentrations of sulphur dioxide which can be attained in the stack gases. This situation arises from the very much lower bed temperatures ($\sim 1200\,K$) than those obtained in conventional grate-type furnaces, and the possibility of reacting the sulphur in the coal with limestone or dolomite to enable its discharge as part of the ash.

Much of the basic research, the development studies and the design features of large-scale fluidised bed combustors is discussed in the Proceedings of the Symposium on Fluidised Bed Combustion organised by the Institute of Fuel in 1975[94]. Pilot scale furnaces with ratings up to 0·5 MW have been operated and large-scale furnaces up to 30 MW are in course of construction.

The bed material normally consists initially of an inert material, such as sand or ash, of particle size between 500 and 1500 μm. This gradually becomes replaced by ash from the coal and additives used for sulphur removal. Ash is continuously removed from the bottom of the bed and, in addition, there is a considerable carryover by elutriation and this flyash must be collected in cyclone separators. Bed depths are usually kept below about 0·6 m in order to limit power requirements.

Coal has a lower density than the bed material and therefore tends to float, but in a vigorously bubbling bed it can become well mixed with the remainder of the material and the degree of mixing determines the number of feed points which are required. Generally the combustible material does not exceed about 5 per cent of the total inventory of the bed. The maximum size of coal which has been successfully used is 25 mm; this is about two orders of magnitude greater than the particle size in pulverised fuel. The ability to use coal directly from the colliery eliminates the need for pulverising equipment, with its high capital and operating costs.

The coal, when it first enters the bed, gives up its volatiles and the combustion process thus involves both vapour and char. Mixing of the volatiles with air in the bed is not usually very good with the result that there is considerable flame burning above the surface of the bed. Because the air rates are chosen to give vigorously bubbling beds, much of the oxygen for combustion must pass from the bubble phase to the dense phase before it can react with the char. Then in the dense phase there will be a significant diffusional resistance to transfer of oxygen to the surface of the particles. The combustion process is, as a result, virtually entirely diffusion-controlled at temperatures above about 1120 K. The reaction is one of oxidation to carbon monoxide near the surface of the particles and subsequent reaction to carbon dioxide. Despite the diffusional limitations, up to 90 per cent utilisation of the inlet oxygen may be achieved. The residence time of the gas in the bed is of the order of a second, whereas the coal particles may remain in the bed for many minutes. There is generally a significant amount of carryover char in the flyash which is then usually recycled to a burner operating at a rather higher temperature than the main bed. An additional advantage of using a large size of feed coal is that the proportion carried over is correspondingly small.

Fluidised bed furnaces can operate in the range 1075–1225 K but most operate close to 1175 K. Some of the tubes are immersed in the bed and others are above the free surface; heat transfer to the immersed tubes is good. Tube areas are usually 6–10 $m^2/m^3$ of furnace and transfer coefficients usually range from 300 to 500 $W/m^2 K$; the radiation component of heat transfer is highly important. Heat releases in large furnaces are about $10^6$ $W/m^3$ of furnace.

One of the major advantages of fluidised bed combustion of coal is that it is possible to absorb the sulphur dioxide which is formed. Generally limestone or dolomite is added and it breaks up in the bed to yield calcium oxide or magnesium and calcium oxide, which then react with the sulphur dioxide as follows:

$$CaO + SO_2 + \tfrac{1}{2}O_2 \rightarrow CaSO_4$$

It is possible to regenerate the solid in a separate reactor using a reducing gas consisting of hydrogen and carbon monoxide. There is some evidence that the reactivity of the limestone or dolomite is improved by the addition of chloride, but its use is not generally favoured because of corrosion problems. Fluidised combustion gives less pollution also because less oxides of nitrogen are formed.

Corrosion and erosion of the tubes immersed in the bed are at a low level. However, there is evidence that the addition of limestone or dolomite causes some sulphide penetration. The chief operating danger is corrosion by chlorine.

Experimental and pilot scale work has been carried out on pressurised operation and plants have been operated up to 600 $kN/m^2$, and in one case up to 1 $MN/m^2$ pressure. Pressure operation permits the use of smaller beds. The fluidising velocity required to produce a given condition in the bed is largely independent of pressure and thus the mass rate of feed of oxygen to the bed is approximately linearly related to the pressure. It is practicable to use deeper beds for pressure operation. Because of the low temperature of operation of fluidised beds the ash is friable and relatively non-erosive, so that the combustion products can be passed directly through a gas turbine. The combination in this way of a gas turbine is an essential feature of the economic operation of pressurised combustors. Generally, it is better to use dolomite in place of limestone as an absorbent for sulphur dioxide, because the higher pressures of carbon dioxide lead to inhibition of the breakdown of calcium carbonate to oxide.

It appears likely that fluidised bed combustion of coal may, in the near future, be one of the most important applications of fluidised systems and it may well be that a large proportion of the new coal-fired generating stations will incorporate fluidised bed combustors.

## 6.8. FURTHER READING

BOTTERILL, J. S. M.: *Fluid Bed Heat Transfer* (Academic Press, 1975).

DALLAVALLE, J. M.: *Micromeritics*, 2nd edn. (Pitman, 1943).

DAVIDSON, J. F. and HARRISON, D.: *Fluidized Particles* (Cambridge, 1963).

DAVIDSON, J. F. and HARRISON, D.: *Fluidization* (Academic Press, 1971).

HALOW, J. S. (ed.): *Fluidization Theories and Applications*. A.I.Ch.E. Symposium Series 161 (1977).

KEAIRNS, D. L. (ed.): *Fluidized Bed Fundamentals and Applications*. A.I.Ch.E. Symposium Series 128 (1973).

KEAIRNS, D. L. (ed.): *Fluidization and Fluid-Particle Systems*. A.I.Ch.E. Symposium Series 141 (1974).

KEAIRNS, D. L. (ed.): *Fluidization Technology*. Proc. Int. Fluidisation Conf., Pacific Grove, California, 1975.

LEVA, M.: *Fluidization* (McGraw-Hill, 1959).

OSTERGAARD, K.: *Studies of Gas–Liquid Fluidisation*. Thesis (Technical University of Denmark, 1968).

ZABRODSKY, S. S.: *Hydrodynamics and Heat Transfer in Fluidized Beds* (The M.I.T. Press, 1966).

ZANDI, I. (ed.): *Advances in Solid–Liquid Flow in Pipes and its Application* (Pergamon Press, Oxford, 1971).

ZENZ, F. A. and OTHMER, D. F.: *Fluidization and Fluid-Particle Systems* (Reinhold, 1960).

## 6.9. REFERENCES

1. DAVIDSON, J. F. and HARRISON, D.: *Fluidized Particles* (Cambridge, 1963).
2. WILHELM, R. H. and KWAUK, M.: *Chem. Eng. Prog.* **44** (1948) 201. Fluidization of solid particles.
3. JACKSON, R.: *Trans. Inst. Chem. Eng.* **41** (1963) 13, 22. The mechanisms of fluidised beds. Part 1. The stability of the state of uniform fluidisation. Part 2. The motion of fully developed bubbles.
4. VAN HEERDEN, C., NOBEL, P., and VAN KREVELEN, D. W.: *Chem. Eng. Sci.* **1** (1951) 37, 51. Studies in fluidisation. 1. The critical mass velocity. 2. Heat transfer.
5. LEVA, M., GRUMMER, M., WEINTRAUB, M., and POLLCHIK, M.: *Chem. Eng. Prog.* **44** (1948) 619. Fluidization of non-vesicular particles.
6. RICHARDSON, J. F. and MEIKLE, R. A.: *Trans. Inst. Chem. Eng.* **39** (1961) 348. Sedimentation and fluidisation. Part III. The sedimentation of uniform fine particles and of two-component mixtures of solids.
7. ROWE, P. N.: *Trans. Inst. Chem. Eng.* **39** (1961) 175. Drag forces in a hydraulic model of a fluidised bed—Part II.
8. ROWE, P. N. and PARTRIDGE, B. A.: *Trans. Inst. Chem. Eng.* **43** (1965) T157. An X-ray study of bubbles in fluidised beds.
9. PINCHBECK, P. H. and POPPER, F.: *Chem. Eng. Sci.* **6** (1956) 57. Critical and terminal velocities in fluidization.
10. RICHARDSON, J. F. and ZAKI, W. N.: *Trans. Inst. Chem. Eng.* **32** (1954) 35. Sedimentation and fluidisation. Part 1.
11. GODARD, K. E. and RICHARDSON, J. F.: *Proceedings of the Tripartite Chem. Engineering Conference, Montreal* (1968). Symposium of Fluidisation 126. The behaviour of bubble-free fluidised beds.
12. LEWIS, E. W. and BOWERMAN, E. W.: *Chem. Eng. Prog.* **48** (1952) 603. Fluidization of solid particles in liquids.
13. STEINOUR, H. H.: *Ind. Eng. Chem.* **36** (1944) 618, 840, 901. Rate of sedimentation.
14. CATHALA, J.: *Chem. Eng. Sci.* **2** (1953) 273. Fluidisation and research methods of chemical engineering science.

15. LAWTHER, K. P. and BERGLIN, C. L. W.: United Kingdom Atomic Energy Authority Report, A.E.R.E.CE/R 2360 (1957). Fluidisation of lead shot with water.

16. RICHARDSON, J. F. and SMITH, J. W.: *Trans. Inst. Chem. Eng.* **40** (1962) 13. Heat transfer to liquid fluidised systems and to suspensions of coarse particles in vertical transport.

17. BAILEY, C.: reported in Davidson, J. F. and Harrison, D.: *Fluidization* (Academic Press, 1971) 38.

18. ANDERSON, T. B.: University of Edinburgh, Ph.D. thesis (1967). The dynamics of fluidised beds—with particular reference to the stability of the fluidised state.

19. KRAMERS, H., WESTERMANN, M. D., DE GROOT, J. H., and DUPONT, F. A. A.: Third Congress of the European Federation of Chemical Engineering (1962). *The Interaction between Fluids and Particles* 114. The longitudinal dispersion of liquid in a fluidised bed.

20. HANDLEY, D., DORAISAMY, A., BUTCHER, K. L., and FRANKLIN, N. L.: *Trans. Inst. Chem. Eng.* **44** (1966) T260. A study of the fluid and particle mechanics in liquid-fluidised beds.

21. CARLOS, C. R. and RICHARDSON, J. F.: *Chem. Eng. Sci.* **22** (1967) 705. Particle speed distribution in a fluidised system.

22. CARLOS, C. R.: University of Wales, Ph.D. thesis (1967). Solids mixing in fluidised beds.

23. LATIF, B. A. J.: University of Wales, Ph.D. thesis (1971). Variation of particle velocities and concentration in liquid–solid fluidised beds.

24. LATIF, B. A. J. and RICHARDSON, J. F.: *Chem. Eng. Sci.* **27** (1972) 1933. Circulation patterns and velocity distributions for particles in a liquid fluidised bed.

25. DAVIES, L. and RICHARDSON, J. F.: *Trans. Inst. Chem. Eng.* **44** (1966) T293. Gas interchange between bubbles and the continuous phase in a fluidised bed.

26. MATHESON, G. L., HERBST, W. A., and HOLT, P. H.: *Ind. Eng. Chem.* **41** (1949) 1099–1104. Characteristics of fluid–solid systems.

27. DIEKMAN, R. and FORSYTHE, W. L.: *Ind. Eng. Chem.* **45** (1953) 1173. Laboratory prediction of flow properties of fluidised solids.

28. DAVIDSON, J. F., PAUL, R. C., SMITH, M. J. S., and DUXBURY, H. A.: *Trans. Inst. Chem. Eng.* **37** (1959) 323. The rise of bubbles in a fluidised bed.

29. DAVIES, R. M. and TAYLOR, G. I.: *Proc. Roy. Soc.* **A200** (1950) 375. The mechanics of large bubbles rising through extended liquids and liquids in tubes.

30. HARRISON, D., DAVIDSON, J. F., and DE KOCK, J. W.: *Trans. Inst. Chem. Eng.* **39** (1961) 202. On the nature of aggregative and particulate fluidisation.

31. LEUNG, L. S.: University of Cambridge, Ph.D. thesis (1961). Bubbles in fluidised beds.

32. SIMPSON, H. C. and RODGER, B. W.: *Chem. Eng. Sci.* **16** (1961) 153. The fluidization of light solids by gases under pressure and heavy solids by water.

33. HARRISON, D. and LEUNG, L. S.: *Trans. Inst. Chem. Eng.* **39** (1961) 409. Bubble formation at an orifice in a fluidised bed.

34. HARRISON, D. and LEUNG, L. S.: Third Congress of the European Federation of Chemical Engineering (1962). *The Interaction between Fluids and Particles* 127. The coalescence of bubbles in fluidised beds.

35. ROWE, P. N. and WACE, P. F.: *Nature* **188** (1960) 737. Gas-flow patterns in fluidised beds.

36. WACE, P. F. and BURNETT, S. T.: *Trans. Inst. Chem. Eng.* **39** (1961) 168. Flow patterns in gas-fluidised beds.

37. DAVIDSON, J. F.: *Trans. Inst. Chem. Eng.* **39** (1961) 230. In discussion of Symposium on Fluidisation.

38. ROWE, P. N., PARTRIDGE, B. A., and LYALL, E.: *Chem. Eng. Sci.* **19** (1964) 973; **20** (1965) 1151. Cloud formation around bubbles in gas fluidised beds.

39. MURRAY, J. D.: *J. Fluid Mech.* **21** (1965) 465; **22** (1965) 57. On the mathematics of fluidization. Part 1. Fundamental equations and wave propagation. Part 2. Steady motion of fully developed bubbles.

40. ROWE, P. N. and HENWOOD, G. N.: *Trans. Inst. Chem. Eng.* **39** (1961) 43. Drag forces in a hydraulic model of a fluidised bed. Part 1.

41. BART, R.: Massachusetts Institute of Technology, Sc.D. thesis (1950). Mixing of fluidized solids in small diameter columns.

42. MASSIMILLA, L. and BRACALE, S.: *La Ricerca Scientifica* **27** (1957) 1509. Il mesolamento della fase solida nei sistemi: Solido-gas fluidizzati, liberi e frenati.

43. TAILBY, S. R. and COCQUEREL, M. A. T.: *Trans. Inst. Chem. Eng.* **39** (1961) 195. Some studies of solids mixing in fluidised beds.

44. MAY, W. G.: *Chem. Eng. Prog.* **55** (1959) 49. Fluidized bed reactor studies.

45. SUTHERLAND, K. S.: *Trans. Inst. Chem. Eng.* **39** (1961) 188. Solids mixing studies in gas fluidised beds. Part 1. A preliminary comparison of tapered and non-tapered beds.

46. ROWE, P. N. and PARTRIDGE, B. A.: Third Congress of the European Federation of Chemical Engineering (1962). *The Interaction between Fluids and Particles* 135. Particle movement caused by bubbles in a fluidised bed.

47. ASKINS, J. W., HINDS, G. P., and KUNREUTHER, F.: *Chem. Eng. Prog.* **47** (1951) 401. Fluid catalyst–gas mixing in commercial equipment.

48. DANCKWERTS, P. V., JENKINS, J. W., and PLACE, G.: *Chem. Eng. Sci.* **3** (1954) 26. The distribution of residence times in an industrial fluidised reactor.
49. GILLILAND, E. R. and MASON, E. A.: *Ind. Eng. Chem.* **41** (1949) 1191. Gas and solid mixing in fluidized beds.
50. GILLILAND, E. R. and MASON, E. A.: *Ind. Eng. Chem.* **44** (1952) 218. Gas mixing in beds of fluidized solids.
51. GILLILAND, E. R., MASON, E. A., and OLIVER, R. C.: *Ind. Eng. Chem.* **45** (1953) 1177. Gas-flow patterns in beds of fluidized solids.
52. LANNEAU, K. P.: *Trans. Inst. Chem. Eng.* **38** (1960) 125. Gas–solids contacting in fluidised beds.
53. SZEKELY, J. Third Congress of the European Federation of Chemical Engineering (1962). *The Interaction between Fluids and Particles* 197. Mass transfer between the dense phase and lean phase in a gas–solid fluidised system.
54. LEVA, M.: *Chem. Eng. Prog.* **47** (1951) 39. Elutriation of fines from fluidized systems.
55. MICKLEY, H. S. and FAIRBANKS, D. F.: *A.I.Ch.E.Jl.* **1** (1955) 374. Mechanism of heat transfer to fluidized beds.
56. LEMLICH, R. and CALDAS, I.: *A.I.Ch.E.Jl.* **4** (1958) 376. Heat transfer to a liquid fluidized bed.
57. RICHARDSON, J. F. and MITSON, A. E.: *Trans. Inst. Chem. Eng.* **36** (1958) 270. Sedimentation and fluidisation. Part II. Heat transfer from a tube wall to a liquid-fluidised system.
58. JEPSON, G., POLL, A., and SMITH, W.: *Trans. Inst. Chem. Eng.* **41** (1963) 207. Heat transfer from gas to wall in a gas–solids transport line.
59. ROMANI, M. N. and RICHARDSON, J. F.: *Letters in Heat and Mass Transfer* **1** (1974) 55. Heat transfer from immersed surfaces to liquid-fluidized beds.
60. RICHARDSON, J. F., ROMANI, M. N., SHAKIR., K. J.: *Chem. Eng. Sci.* **31** (1976) 619. Heat transfer from immersed surfaces in liquid fluidised beds.
61. KHAN, A. R.: University of Wales, Ph.D. thesis (1978). Heat transfer from immersed surfaces to liquid-fluidised beds.
62. DAVIES, R.: University of Wales, Ph.D. thesis (1975). Local heat transfer in liquid–solid fluidised beds.
63. LEVA, M.: *Fluidization* (McGraw-Hill, 1959).
64. LEVENSPIEL, O. and WALTON, J. S.: *Proc. Heat Transf. Fluid Mech. Inst. Berkeley, California* (1949) 139–46. Heat transfer coefficients in beds of moving solids.
65. VREEDENBERG, H. A.: *J. Appl. Chem.* **2** (1952) S26. Heat transfer between fluidised beds and vertically inserted tubes.
66. DOW, W. M. and JAKOB, M.: *Chem. Eng. Prog.* **47** (1951) 637. Heat transfer between a vertical tube and a fluidized air–solid mixture.
67. BOTTERILL, J. S. M., REDISH, K. A., ROSS, D. K., and WILLIAMS, J. R.: Third Congress of the European Federation of Chemical Engineering (1962). *The Interaction between Fluids and Particles* 183. The mechanism of heat transfer to fluidised beds.
68. BOTTERILL, J. S. M. and WILLIAMS, J. R.: *Trans. Inst. Chem. Eng.* **41** (1963) 217. The mechanism of heat transfer to gas-fluidised beds.
69. BOTTERILL, J. S. M., BRUNDRETT, G. W., CAIN, G. L., and ELLIOTT, D. E.: *Chem. Eng. Prog.* Symp. Ser. No. 62, **62** (1966) 1. Heat transfer to gas-fluidized beds.
70. WILLIAMS, J. R.: University of Birmingham, Ph.D. thesis (1962). The mechanism of heat transfer to fluidised beds.
71. HIGBIE, R.: *Trans. Am. Inst. Chem. Eng.* **31** (1935) 365. The rate of absorption of a pure gas into a still liquid during periods of exposure.
72. DANCKWERTS, P. V.: *Ind. Eng. Chem.* **43** (1951) 1460. Significance of liquid-film coefficients in gas absorption.
73. MICKLEY, H. S., FAIRBANKS, D. F., and HAWTHORN, R. D.: *Chem. Eng. Prog.* Symp. Ser. No. 32, **57** (1961) 51. The relation between the transfer coefficient and thermal fluctuations in fluidized-bed heat transfer.
74. RICHARDSON, J. F. and BAKHTIAR, A. G.: *Trans. Inst. Chem. Eng.* **36** (1958) 283. Mass transfer between fluidised particles and gas.
75. RICHARDSON, J. F. and SZEKELY, J. *Trans. Inst. Chem. Eng.* **39** (1961) 212. Mass transfer in a fluidised bed.
76. MULLIN, J. W. and TRELEAVEN, C. R.: Third Congress of the European Federation of Chemical Engineering (1962). *The Interaction between Fluids and Particles* 203. Solids–liquid mass transfer in multi-particulate systems.
77. CHU, J. C., KALIL, J., and WETTEROTH, W. A.: *Chem. Eng. Prog.* **49** (1953) 141. Mass transfer in a fluidized bed.
78. KETTENRING, K. N., MANDERFIELD, E. L., and SMITH, J. M.: *Chem. Eng. Prog.* **46** (1950) 139. Heat and mass transfer in fluidized systems.
79. HEERTJES, P. M. and MCKIBBINS, S. W.: *Chem. Eng. Sci.* **5** (1956) 161. The partial coefficient of heat transfer in a drying fluidized bed.
80. FRANTZ, J. F.: *Chem. Eng. Prog.* **57** (1961) 35. Fluid-to-particle heat transfer in fluidized beds.
81. RICHARDSON, J. F. and AYERS, P: *Trans. Inst. Chem. Eng.* **37** (1959) 314. Heat transfer between particles and a gas in a fluidised bed.

82. WILKE, C. R. and HOUGEN, O. A.: *Trans. Am. Inst. Chem. Eng.* **41** (1945) 445. Mass transfer in the flow of gases through granular solids extended to low modified Reynolds Numbers.
83. RESNICK, W. and WHITE, R. R.: *Chem. Eng. Prog.* **45** (1949) 377. Mass transfer in systems of gas and fluidized solids.
84. MCCUNE, L. K. and WILHELM, R. H.: *Ind. Eng. Chem.* **41** (1949) 1124. Mass and momentum transfer in solid–liquid system. Fixed and fluidized beds.
85. HOBSON, M. and THODOS, G.: *Chem. Eng. Prog.* **45** (1949) 517. Mass transfer in flow of liquids through granular solids.
86. GAFFNEY, B. J. and DREW, T. B.: *Ind. Eng. Chem.* **42** (1950) 1120. Mass transfer from packing to organic solvents in single phase flow through a column.
87. HSU, C. T. and MOLSTAD, M. C.: *Ind. Eng. Chem.* **47** (1955) 1550. Rate of mass transfer from gas stream to porous solid in fluidized beds.
88. CORNISH, A. R. H.: *Trans. Inst. Chem. Eng.* **43** (1965) T332. Note on minimum possible rate of heat transfer from a sphere when other spheres are adjacent to it.
89. ZABRODSKY, S. S.: *J. Heat and Mass Transfer* **10** (1967) 1793. On solid-to-fluid heat transfer in fluidized systems.
90. WINDERBANK, C. S.: *J. Imp. Coll. Chem. Eng. Soc.* **4** (1948) 31. The fluid catalyst technique in modern petroleum refining.
91. RILEY, H. L.: *Trans. Inst. Chem. Eng.* **37** (1959) 305. Design of fluidised reactors for naphthalene oxidation: a review of patent literature.
92. ROWE, P. N.: *Soc. Chem. Ind. Fluidisation* (1964) 15. A theoretical study of a batch reaction in a gas-fluidised bed.
93. VOLK, W., JOHNSON, C. A., and STOTLER, H. H.: *Chem. Eng. Prog. Symp. Ser.* No. 38, **58** (1962) 38. Effect of reactor internals on quality of fluidization.
94. INSTITUTE OF FUEL: Symposium Series No. 1: *Fluidised Combustion* **1** (1975).

## 6.10 NOMENCLATURE

| | | Units in SI System | Dimensions in M, L, T, $\theta$, A |
|---|---|---|---|
| $A$ | Cross-sectional area of bed | m$^2$ | $\mathbf{L^2}$ |
| $a$ | Area for transfer per unit volume of bed | m$^2$/m$^3$ | $\mathbf{L^{-1}}$ |
| $a'$ | Area for transfer per unit height of bed | m$^2$/m | $\mathbf{L}$ |
| $b$ | Slope of adsorption isotherm | — | — |
| $C$ | Fractional volumetric concentration of solids | — | — |
| $C_0$ | Value of $C$ at $t = 0$ | — | — |
| $\Delta C$ | Driving force expressed as a molar concentration difference | kmol/m$^3$ | $\mathbf{ML^{-3}}$ |
| $\Delta C_{lm}$ | Logarithmic mean value of $C$ | kmol/m$^3$ | $\mathbf{ML^{-3}}$ |
| $c_p$ | Specific heat of gas at constant pressure | J/kg K | $\mathbf{L^2 T^{-2} \theta^{-1}}$ |
| $c_s$ | Specific heat of solid particle | J/kg K | $\mathbf{L^2 T^{-2} \theta^{-1}}$ |
| $D$ | Gas phase diffusivity | m$^2$/s | $\mathbf{L^2 T^{-1}}$ |
| $D_L$ | Longitudinal diffusivity | m$^2$/s | $\mathbf{L^2 T^{-1}}$ |
| $d$ | Particle diameter or diameter of sphere with same surface as particle | m | $\mathbf{L}$ |
| $d_b$ | Bubble diameter | m | $\mathbf{L}$ |
| $d_c$ | Cloud diameter | m | $\mathbf{L}$ |
| $d_t$ | Tube diameter | m | $\mathbf{L}$ |
| $e$ | Voidage | — | — |
| $e_{mf}$ | Voidage corresponding to minimum fluidising velocity | — | — |
| $e_{max}$ | Voidage corresponding to maximum heat transfer coefficient | — | — |
| $F$ | Moles of vapour adsorbed on unit mass of solid | kmol/kg | — |
| $f$ | Intercept of adsorption isotherm | — | — |
| $G$ | Mass flowrate of fluid | kg/s | $\mathbf{MT^{-1}}$ |
| $G_E$ | Mass flowrate of fluid to cause initial expansion of bed | kg/s | $\mathbf{MT^{-1}}$ |
| $G_F$ | Mass flowrate of fluid to initiate fluidisation | kg/s | $\mathbf{MT^{-1}}$ |
| $G_m$ | Molar rate of flow of gas | kmol/s | $\mathbf{MT^{-1}}$ |
| $G'$ | Mass flowrate of fluid per unit area | kg/m$^2$s | $\mathbf{ML^{-2}T^{-1}}$ |
| $g$ | Acceleration due to gravity | 9·81 m/s$^2$ | $\mathbf{LT^{-2}}$ |
| $h$ | Heat transfer coefficient | W/m$^2$ K | $\mathbf{MT^{-3}\theta^{-1}}$ |
| $h_D$ | Mass transfer coefficient | m/s | $\mathbf{LT^{-1}}$ |
| $h_l$ | Heat transfer coefficient for liquid alone at same rate as in bed | W/m$^2$ K | $\mathbf{MT^{-3}\theta^{-1}}$ |

| | | Units in SI System | Dimensions in $\mathbf{M}$, $\mathbf{L}$, $\mathbf{T}$, $\theta$, $\mathbf{A}$ |
|---|---|---|---|
| $j_d$ | $j$-factor for mass transfer | — | — |
| $j_h$ | $j$-factor for heat transfer | — | — |
| $K$ | Distance travelled by bubble to increase its volume by a factor e | m | $\mathbf{L}$ |
| $k$ | Thermal conductivity of fluid | W/m K | $\mathbf{MLT^{-3}\theta^{-1}}$ |
| $l$ | Depth of fluidised bed | m | $\mathbf{L}$ |
| $M$ | Constant in equation 6.24 | 1/s | $\mathbf{T^{-1}}$ |
| $m$ | Index of $(1-e)$ in equation 6.26 | — | — |
| $N_A$ | Molar rate of transfer of diffusing component | kmol/s | $\mathbf{MT^{-1}}$ |
| $n$ | Index of $e$ in equation 6.15 | — | — |
| $P$ | Pressure | N/m$^2$ | $\mathbf{ML^{-1}T^{-2}}$ |
| $-\Delta P$ | Pressure drop across bed | N/m$^2$ | $\mathbf{ML^{-1}T^{-2}}$ |
| $Q$ | Rate of transfer of heat | W | $\mathbf{ML^2T^{-3}}$ |
| $R$ | Electrical resistance | W/A$^2$ | $\mathbf{ML^2T^{-2}A^{-2}}$ |
| $S$ | Specific surface of particles | m$^2$/m$^3$ | $\mathbf{L^{-1}}$ |
| $T$ | Temperature of gas | K | $\theta$ |
| $T_B$ | Bed temperature | K | $\theta$ |
| $T_E$ | Element temperature | K | $\theta$ |
| $T_s$ | Temperature of solid | K | $\theta$ |
| $\Delta T$ | Temperature driving force | deg K | $\theta$ |
| $t$ | Time | s | $\mathbf{T}$ |
| $u_b$ | Velocity of rise of bubble | m/s | $\mathbf{LT^{-1}}$ |
| $u_c$ | Superficial velocity of fluid (open tube) | m/s | $\mathbf{LT^{-1}}$ |
| $u_i$ | Value of $u_c$ at infinite dilution | m/s | $\mathbf{LT^{-1}}$ |
| $u_{mf}$ | Minimum value of $u_c$ at which fluidisation occurs | m/s | $\mathbf{LT^{-1}}$ |
| $u_0$ | Free-falling velocity of particle in infinite fluid | m/s | $\mathbf{LT^{-1}}$ |
| $V$ | Voltage applied to element | W/A | $\mathbf{ML^2T^{-3}A^{-1}}$ |
| $V_B$ | Volume of bubble | m$^3$ | $\mathbf{L^3}$ |
| $W$ | Mass of solids in bed | kg | $\mathbf{M}$ |
| $y$ | Mol fraction of vapour in gas stream | — | — |
| $y_0$ | Mol fraction of vapour in inlet gas | — | — |
| $y^*$ | Mol fraction of vapour in equilibrium with solids | — | — |
| $Z$ | Total height of bed | m | $\mathbf{L}$ |
| $z$ | Height above bottom of bed | m | $\mathbf{L}$ |
| $\alpha$ | Ratio of gas velocities in bubble and emulsion phases | — | — |
| $\mu$ | Viscosity of fluid | Ns/m$^2$ | $\mathbf{ML^{-1}T^{-1}}$ |
| $\rho$ | Density of fluid | kg/m$^3$ | $\mathbf{ML^{-3}}$ |
| $\rho_c$ | Density of suspension | kg/m$^3$ | $\mathbf{ML^{-3}}$ |
| $\rho_s$ | Density of particle | kg/m$^3$ | $\mathbf{ML^{-3}}$ |
| $\psi$ | Particle stream function | — | — |
| $Ga$ | Galileo number $(\rho(\rho_s-\rho)gd^3/\mu^2)$ | — | — |
| $Nu$ | Nusselt number $(hd_t/k)$ | — | — |
| $Nu'$ | Nusselt number $(hd/k)$ | — | — |
| $Pr$ | Prandtl number $(c_p\mu/k)$ | — | — |
| $Re$ | Tube Reynolds number $(u_cd_t\rho/\mu)$ | — | — |
| $Re'$ | Particle Reynolds number $(u_cd\rho/\mu)$ | — | — |
| $Re_1$ | Bed Reynolds number $(u_c\rho/S(1-e)\mu)$ | — | — |
| $Re'_{mf}$ | Particle Reynolds number at minimum fluidising velocity $(u_{mf}\rho d/\mu)$ | — | — |
| $Re'_0$ | Particle Reynolds number $(u_0d\rho/\mu)$ | — | — |
| $Re^*$ | Particle Reynolds number $(u_cd\rho/(1-e)\mu)$ | — | — |
| $Sc$ | Schmidt number $(\mu/\rho D)$ | — | — |
| $Sh'$ | Sherwood number $(h_Dd/D)$ | — | — |

# CHAPTER 7

# Pneumatic and Hydraulic Conveying

## 7.1. INTRODUCTION

In a fluidised bed a gas or liquid is passed vertically upwards through a bed of particles in such a way that the bulk of the particles is retained in the system. Unless there is extensive channelling in the bed, the fluid velocity will be less than the terminal falling velocity of the particles. At higher velocities, the particles will be entrained in the stream of fluid and transport will occur. In such cases solid particles will normally be introduced continuously into the transport line. Fluidisation processes involve passing the fluid vertically upwards, but transport can be carried out in any desired direction.

Pneumatic conveying has been practised over a long period for the conveying of grain and the unloading of grain barges, but it is now being used extensively for the conveying of chemicals. It has the great advantage that it does not involve contamination of the material and pneumatic systems are, in fact, frequently used as dryers. Pneumatic transport systems can be operated under either reduced or positive pressure. In the former case the solids are readily introduced at the low pressure suction point, and the use of complicated solid feeders is obviated. However, positive pressure systems are usually to be preferred in the chemical industry where it may be necessary to transport solids over large distances.

Hydraulic conveying also has been used for a long while in, for example, the china clay industry where material is conveyed in open channels from the pits to the treatment plants. There are a number of large installations in which coal is carried over long distances in pipelines but, of course, the disadvantage of the process is that it gives a wet product. Where there are no existing transport facilities available, a hydraulic conveying system can be far more economical than the provision of road or rail facilities.

The fundamental principles underlying pneumatic and hydraulic conveying are similar, but in practice the two systems exhibit different characteristics largely because of the widely different physical properties of gases and liquids. Thus, for instance, in a hydraulic system the densities of the particles and the fluid will be similar, whereas the densities of the two phases will differ by orders of magnitude in a pneumatic system. It is therefore convenient to consider pneumatic and hydraulic conveying separately.

The principles of conveying vertically are essentially simpler than those relating to horizontal conveying. In the former case the weight of the particles is directly balanced by the upward drag of the fluid on the particles. In horizontal conveying, however, the mechanism of suspension of the particles is more complex, because the vertical force due to the weight of the particles has to be balanced by the secondary effects arising from the action of a horizontal drag force.

## 7.2. PNEUMATIC CONVEYING

### 7.2.1. Vertical Transport

In vertical conveying, the velocity of the gas relative to the particles is close to their terminal falling velocity, and the additional pressure drop due to their presence is approximately that determined by the weight of particles in the pipeline. At high gas velocities the particles are uniformly dispersed in the gas, but when the velocity is reduced, there is a tendency for alternate slugs of high and low solids content to form.

The conveying of fine particles in vertical pipes of diameters 25 mm, 50 mm, and 75 mm has been studied by Boothroyd[1]. He measured the pressure gradient in the pipeline, and found that the frictional pressure drop was less than that for air alone in the 25 mm pipe, but was greater in the larger pipes. This effect was attributed to the fact that the extent to which the fluid turbulence was affected by the presence of the particles was markedly influenced by pipe size.

### 7.2.2. Horizontal Transport

In a horizontal pipeline the distribution of the solids over the cross-section becomes progressively less uniform as the velocity is reduced. The following types of flow are commonly encountered in sequence at decreasing gas velocities and have been carefully observed in a pipeline of 25 mm diameter.

1. *Uniform suspended flow*
   The particles are evenly distributed over the cross-section over the whole length of pipe.

2. *Non-uniform suspended flow*
   The flow is similar to that described above but there is a tendency for particles to flow preferentially in the lower portion of the pipe. If there is an appreciable size distribution, the larger particles are found predominantly at the bottom.

3. *Slug flow*
   As the particles enter the conveying line, they tend to settle out before they are fully accelerated. They form dunes which are then swept bodily downstream giving an uneven longitudinal distribution of particles along the pipeline.

4. *Dune flow*
   The particles settle out as in slug flow but the dunes remain stationary with particles being conveyed above the dunes and also being swept from one dune to the next.

5. *Moving bed*
   Particles settle out near the feed point and form a continuous bed on the bottom of the pipe. The bed develops gradually throughout the length of the pipe and moves slowly forward. There is a velocity gradient in the vertical direction in the bed, and conveying continues in suspended form above the bed.

6. *Stationary bed*
   The behaviour is similar to that of a moving bed, except that there is virtually no movement of the bed particles. The bed can build up until it occupies about three-quarters of the cross-section. Further reduction in velocity quickly gives rise to a complete blockage.

### 7. *Plug flow*

Following slug flow the particles, instead of forming stationary dunes, gradually build up over the cross-section until they eventually cause a blockage. This type of flow is less common than dune flow.

The mechanism of suspension is related to the type of flow pattern obtained. Suspended types of flow are usually attributable to dispersion of the particles by the action of the turbulent eddies in the fluid. In turbulent flow, the vertical component of the eddy velocity will lie between one-seventh and one-fifth of the forward velocity of the fluid and, if this is more than the terminal falling velocity of the particles, they will tend to be supported in the fluid. In practice it is found that this mechanism is not as effective as might be thought because there is a tendency for the particles to damp out the eddy currents.

If the particles tend to form a bed, they will be affected by the lateral dispersive forces described by Bagnold[2,3]. A fluid in passing through a loose bed of particles exerts a dilating action on the system. This gives rise to a dispersion of the particles in a direction at right angles to the flow of fluid.

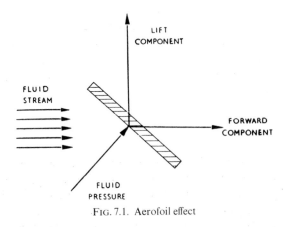

FIG. 7.1. Aerofoil effect

If a particle presents a face inclined at an angle to the direction of motion of the fluid, it may be subjected to an upward lift due to the "aerofoil" effect. In Fig. 7.1 a flat plate is shown at an angle to a stream of fluid flowing horizontally. The fluid pressure acts normally at the surface and thus produces forces with vertical and horizontal components as shown.

If the particle rotates in the fluid, it will be subjected to a lift on the Magnus Principle. Figure 7.2 shows a section through a cylinder rotating in a fluid stream. At its upper edge the cylinder and the fluid are both moving in the same direction, but at its lower edge they are moving in opposite directions. The fluid above the cylinder is therefore accelerated, and that below the cylinder is retarded. Thus the pressure is greater below the cylinder and an upward force is exerted.

The above processes, other than the action of the dispersive forces, can result in an upward or downward displacement of an individual particle, there being approximately equal chances of the net force acting upwards or downwards. However, because the gravitational force gives rise to a tendency for the concentration of particles to be greater at the bottom of the pipe, the overall effect is to lift the particles.

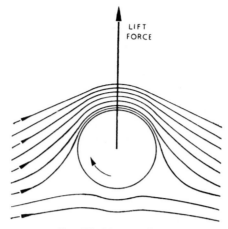

FIG. 7.2. Magnus effect

The principal characteristic of a particle which determines the dominant suspension mechanism is its terminal falling velocity. Particles with low falling velocities will be readily suspended by the action of the eddies, whereas the dispersive forces will be most important with particles of high falling velocities. In a particular case, of course, the fluid velocity will also be an important factor, full suspension of a given particle occurring more readily at high velocities.

*Energy for Horizontal Conveying*

The energy required for conveying can conveniently be considered in two parts: that required for the flow of the air alone, and the additional energy necessitated by the presence of the particles. It should be noted, however, that the fluid friction will itself be somewhat modified for the following reasons: the total cross-sectional area will not be available for the flow of fluid; the pattern of turbulence will be affected by the solids; and the pressure distribution through the pipeline will be different, and hence the gas density at a given point will be affected by the solids.

The presence of the solids is responsible for an increased pressure gradient for a number of reasons. If the particles are introduced from a hopper, they will have a lower forward velocity than the fluid and therefore have to be accelerated. Because the relative velocity is greatest near the feed point and progressively falls as the particles are accelerated, their velocity will initially increase rapidly and, as the particles approach their limiting velocities, the acceleration will become very small. The pressure drop due to acceleration is therefore greatest near the feed point. In pneumatic conveying, the air is expanding continuously along the line and therefore the solid velocity is also increasing. Secondly, work must be done against the action of the earth's gravitational field because the particles must be lifted from the bottom of the pipe each time they drop. Finally, particles will collide with one another and with the walls of the pipe, and therefore their velocities will fall and they will need to be accelerated again. Collisions between particles will be less frequent and result in less energy loss than impacts with the wall, because the relative velocity is much lower in the former case.

The transference of energy from the gas to the particles arises from the existence of a relative velocity. The particles will always be travelling at a lower velocity than the gas. The loss of energy by a particle will generally occur on collision and thus be a discontinuous process. The acceleration of the particle will be a gradual process occurring after each collision, the rate of transfer falling off as the particle approaches the gas velocity.

The accelerating force exerted by the fluid on the particle will be a function of the properties of the gas, the shape and size of the particle, and the relative velocity. It will also depend on the dispersion of the particles over the cross-section and the shielding of individual particles. The process is complex and therefore it is not possible to develop a precise analytical treatment, but it is obviously important to know the velocity of the particles.

*Determination of Solid Velocities*

The determination of the velocity of individual particles can be carried out in a number of ways. First, the particles in a given section of pipe can be isolated using two rapidly acting shutters or valves, and the quantity of particles trapped measured by removing the intervening section of pipe. This method was used by Segler[4], Clark[5], and Mitlin[6]. It is, however, a cumbersome method, very time-consuming, and dependent upon extremely good synchronisation of the shutters. Another method is to take two photographs of the particles in the pipeline at short time-intervals and to measure the distance the particles have travelled in the time. This method is restricted to very low concentrations of relatively large particles, and does not permit the measurement of more than a few particles[7]. A third method consists of measuring the time taken for a "tagged" particle to travel between two points. The particle can be either radioactive or magnetic[1]. The method is difficult to operate and gives results applicable only to an isolated particle.

A novel method was developed by McLeman[8] who injected a pulse of air into the conveying line. As a result there was a very short period during which the walls at any particular point were not subject to bombardment by particles, and the noise level was substantially reduced. By placing two transducers in contact with the wall of the pipe at a known distance apart and connecting each to a thyratron, it was possible to arrange for the first to start a frequency counter and for the second to stop the counter. A very accurate method of timing the air pulse was thus provided. It was found that the pulse retained its identity over a long distance, and this suggested that the velocities of all the particles tended to be the same. The method enabled extremely rapid and accurate measurements of solid velocity to be obtained. Over a 16 m distance the error was less than 1 per cent.

The importance of obtaining accurate measurements of solid velocity is associated with the fact that the drag exerted by the fluid on the particle is approximately proportional to the square of the relative velocity. As the solid velocity frequently approaches quite close to the air velocity, the necessity for very accurate values is apparent.

*Pressure Drops and Solid Velocities*

When solid particles are introduced into an air stream a large amount of energy is required to accelerate the particles, and the acceleration period occupies a considerable

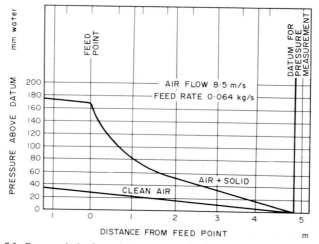

FIG. 7.3. Pressure in horizontal 25 mm conveying line for transport of cress seed

length of pipe. In order to obtain values of pressure gradients and solid velocities under conditions approaching equilibrium, measurements must be made at a considerable distance from the feed point. Much of the earlier experimental work suffered from the facts that conveying lines were much too short and that the pressure gradients were appreciably influenced by the acceleration of the particles. A typical curve obtained by Clark *et al.*[5], for the pressure distribution along a 25 mm diameter conveying line, is shown in Fig. 7.3. It will be noted that the pressure gradient gradually diminishes from a very high value in the neighbourhood of the feed point to an approximately constant value at distances greater than about 2 m. It has been checked experimentally that the solid velocity is increasing in

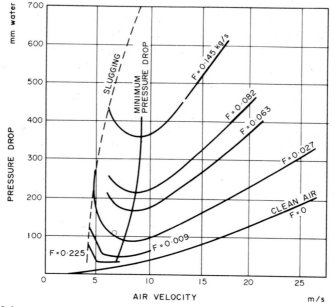

FIG. 7.4. Effect of air rate on pressure drop for transport of cress seed in 4·8 m horizontal length of 25 mm pipe

the region of decreasing pressure gradient and that the length of the acceleration period increases with the weight of the particles, as might be expected. When the pressure gradient is approximately constant, so is the solid velocity. A further factor which makes it necessary to obtain measurements over a long section of pipe is that the pressure gradient does in many cases exhibit a wave form[8]. This appears to be associated with the tendency for dune formation to occur within the pipe, and thus the measured value of the pressure gradient may be influenced by the exact location of the pressure tappings. It is therefore concluded that measurements should be made over a length of at least 15 m of pipe, and remote from the solids feed point.

If the pressure drop is plotted against air flow rate as in Fig. 7.4, it is seen that at a given feed rate the curve always passes through a minimum. At air velocities above the minimum of the curve, the solids are in suspended flow, but at lower velocities particles are deposited on the bottom of the pipe and there is a serious risk of blockage occurring.

TABLE 7.1. *Physical Properties of the Solids Used for Pneumatic Conveying*[8]

| Material | Shape | Particle size (mm) | | Density (kg/m³) | Free-falling velocity $u_0$ (m/s) |
| | | Range | Mean | | |
| --- | --- | --- | --- | --- | --- |
| Coal A | Rounded | 1·5–Dust | 0·75 | 1400 | 2·80 |
| Coal B | Rounded | 1·3–Dust | 0·63 | | 2·44 |
| Coal C | Rounded | 1·0–Dust | 0·50 | | 2·13 |
| Coal D | Rounded | 2·0–Dust | 1·0 | | 3·26 |
| Coal E | Rounded | 4·0–Dust | 2·0 | | 3·72 |
| Perspex A | Angular | 2·0–1·0 | 1·5 | 1185 | 3·73 |
| Perspex B | Angular | 5·0–2·5 | 3·8 | | 5·00 |
| Perspex C | Spherical | 1·0–0·5 | 0·75 | | 2·35 |
| Polystyrene | Spherical | 0·4–0·3 | 0·36 | 1080 | 1·62 |
| Lead | Spherical | 1·0–0·15 | 0·30 | 11080 | 8·17 |
| Brass | Porous, feathery filings | 0·6–0·2 | 0·40 | 8440 | 4·08 |
| Aluminium | Rounded | 0·4–0·1 | 0·23 | 2835 | 3·02 |
| Rape seed | Spherical | 2·0–1·8 | 1·91 | 1080 | 5·91 |
| Radish seed | Spherical | 2·8–2·3 | 2·5 | 1065 | 6·48 |
| Sand | Nearly spherical | 1·5–1·0 | 1·3 | 2610 | 4·66 |
| Manganese dioxide | Rounded | 1·0–0·25 | 0·75 | 4000 | 5·27 |

Accurate measurements of solid velocities and pressure gradients were made by McLeman[8] using a continuously operating system in which the solids were separated from the discharged air in a cyclone separator and introduced again to the feed hopper at the high pressure end of the system by means of a specially constructed rotary valve. A 25 mm diameter pipeline was used with two straight lengths of about 35 m joined by a semicircular bend. Experiments were carried out with air velocities up to 35 m/s. The solids used and their properties are listed in Table 7.1.

The velocities of solid particles $u_s$ (m/s) could be represented in terms of the air velocity $u_a$ (m/s), the free-falling velocity of the particles $u_0$ (m/s), and the density of the solid particles $\rho_s$ (kg/m³) by the following equation:

$$u_a - u_s = \frac{u_0}{0 \cdot 468 + 7 \cdot 25 \sqrt{u_0/\rho_s}} \qquad (7.1)$$

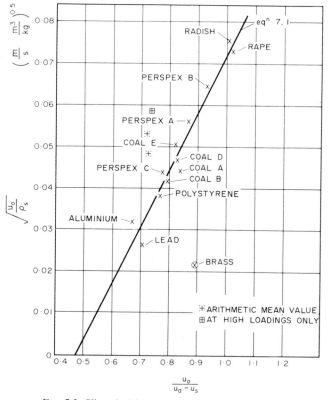

FIG. 7.5. Slip velocities $(u_a - u_s)$ for various materials

Deviations from equation 7.1 were noted only at high loadings with fine solids of wide size distribution. Experimental results are plotted in Fig. 7.5.

The additional pressure drop due to the presence of solids in the pipeline $-\Delta P_x$ could be expressed in terms of the solid velocity, the terminal falling velocity of the particles and the feed rate of solids $F$ (kg/s). Equation 7.2 represents the experimental results to within $\pm 10$ per cent:

$$\frac{-\Delta P_x}{-\Delta P_{air}} \frac{u_S^2}{F} = \frac{2805}{u_0} \tag{7.2}$$

In Fig. 7.6 $(-\Delta P_x / -\Delta P_{air}) u_S^2 / F$ is plotted against $1/u_0$ for a 21 m length of pipe. $-\Delta P_{air}$ is the pressure drop for the flow of air alone at the pressure existing within the pipe.

The above relations for solid velocity and pressure drop are applicable only in the absence of electrostatic charging of the particles. Many materials, including sand, become charged during transport and cause the deposition of a charged layer on the surface of the pipe. The charge remains on the earthed pipeline for long periods but can be removed by conveying certain materials, including coal and perspex, through the pipe. The charging is thought to be associated with attrition of the particles and therefore to be a relatively slow process, while the discharging is entirely an electrical phenomenon and therefore more rapid. The effect of electrostatic charging is to increase the frequency of collisions between the particles and the wall. As a result the velocity of the solids is substantially reduced and the excess pressure difference $-\Delta P_x$ may be increased by as much as tenfold. The results included in Fig. 7.6 are those obtained before an appreciable charge has built up.

The effect of pipe diameter on the pressure drop in a conveying system is seen by examining the results of Segler[4] who conveyed wheat grain in pipes up to 400 mm diameter. Taking a value of the solids velocity for wheat from the work of Gasterstädt[9], $(-\Delta P_x/-\Delta P_{air})u_s^2 u_0/F$ is plotted against pipe diameter in Fig. 7.7. It is seen that the value of the constant in equation 7.2 decreases with pipe diameter and that the results are

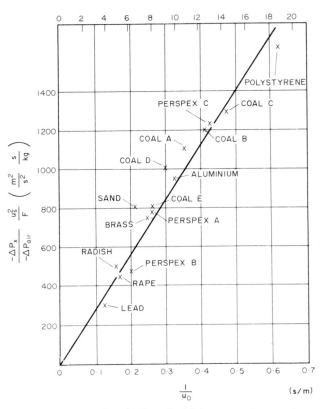

FIG. 7.6. Pressure drop for flow of solids in pneumatic conveying

consistent with those obtained by McLeman. Thus, for calculating the pressure gradient in a large diameter pipe, reference should be made to Fig. 7.7 for the value of the constant in equation 7.2.

The above correlations all apply in straight lengths of pipe with the solids suspended in the gas stream. Under conditions of slug or dune flow the pressure gradients will be considerably greater. It is generally found that the economic velocity corresponds approximately with the minimum of the curves in Fig. 7.4. If there are bends in the conveying line, the energy requirement for conveying will be increased because the particles are retarded and must be re-accelerated following the bend. Bends should be eliminated wherever possible because, in addition, the wear tends to be very high.

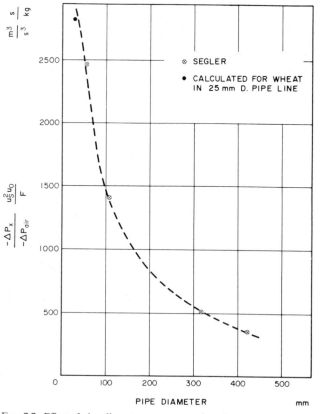

FIG. 7.7. Effect of pipe diameter on pressure drop for transport of wheat

## 7.3. HYDRAULIC CONVEYING

The main differences between hydraulic and pneumatic conveying arise from the large differences between the physical properties of a gas and a liquid. The flow velocities will be very much lower and the solids will be moving at velocities much closer to that of the fluid. The mechanism of suspension of the particles is similar in a horizontal system, but it is more common to operate under conditions where there is a tendency for the formation of a bed of solids. This applies particularly when transport is effected in an open channel where a blockage is not likely to be so serious.

A broad distinction can be drawn between suspensions of very fine particles which behave as homogeneous liquids at virtually all conditions of transport, and suspensions of coarser particles which tend to segregate during handling. The former will give similar behaviour irrespective of the direction of flow, whereas the latter may show marked differences in horizontal and vertical flow. The problems of hydraulic transport will therefore be considered under three headings:

(a) Transport of homogeneous non-settling suspensions.
(b) Transport of settling suspensions in horizontal flow.
(c) Transport of settling suspensions in vertical flow.

### 7.3.1. Homogeneous Flow

Suspensions of fine particles, particularly at high concentrations, exhibit such low rates of sedimentation that they can be considered as homogeneous fluids. However, they generally do not behave as Newtonian fluids, but show a dependence of apparent viscosity on rate of shear. The characteristics of a suspension will be dependent on the degree of flocculation of the particles and, normally, a flocculated suspension will show shear thinning properties with its apparent viscosity decreasing as the shear rate is increased. A deflocculated suspension will often show approximately Newtonian characteristics, but it may have a tendency for its apparent viscosity to increase with shear rate as the result of the build up of a "structure".

An extensive study of the flow behaviour of flocculated kaolin suspensions, formed by suspending kaolin in tap water, has been carried out by Heywood[10]. Measurements of pressure drop–flowrate relationships have been obtained for flow through both a laboratory capillary tube viscometer and for a 42 mm diameter conveying line arranged as part of a recirculating system. Fractional volumetric concentrations ranged from 0·086 to 0·234. In all cases the suspensions exhibited *shear thinning*, with the apparent viscosity decreasing with increase in rate of shear and it was possible to express the relation between shear stress and shear rate by the Ostwald–de Waele *power law* model, i.e.

$$R_y = k \left( \frac{du_x}{dy} \right)^n \tag{7.3}$$

where    $R_y$ is the shear stress in the fluid at a distance $y$ from the wall,
    $u_x$ is the point value of velocity at that position,
    $n$ is known as the *flow index*, and
    $k$ is known as the *consistency*.

The apparent viscosity $\mu_a$ of the fluid at any shear rate $du_x/dy$ is then:

$$\mu_a = k \left( \frac{du_x}{dy} \right)^{n-1} \tag{7.4}$$

Thus for $n$ values less than unity, the apparent viscosity decreases as shear rate is increased.

For a power law fluid, it can be shown (Volume 3, chapter 6) that, for laminar flow:

$$n = \frac{d[\ln R]}{d[\ln (8u/d_t)]} \tag{7.5}$$

where    $R$ is the wall shear stress $[=(-\Delta P d_t/4l)]$,
    $u$ is the mean velocity of flow, and
    $d_t$ is the pipe diameter.

Thus values of the flow index $n$ are most conveniently obtained from the slope of the plot of $\log R$ against $\log (8u/d_t)$.

Values of $n$ and $k$ for the flocculated kaolin suspensions are given in Table 7.2. It will be noted that $n$ falls only from 0·23 to 0·13 as the concentration is increased, but that $k$ increases by a factor of 50 over this range.

Results obtained for flow in the pipeline have been plotted (Fig. 7.8) as wall shear stress $R$ against wall shear rates $(du_x/dy)_{y=0}$ using logarithmic co-ordinates.

TABLE 7.2. *Power Law Parameters for Flocculated Kaolin Suspensions*

| Solids volume fraction $C$ | Flow index $n$ | Consistency $k$ $(Ns^n m^{-2})$ |
|---|---|---|
| 0·086 | 0·23 | 0·89 |
| 0·122 | 0·18 | 2·83 |
| 0·142 | 0·16 | 4·83 |
| 0·183 | 0·15 | 15·3 |
| 0·220 | 0·14 | 32·4 |
| 0·234 | 0·13 | 45·3 |

The shear rate at the wall is readily calculated from the expression for the laminar velocity profile for a power law fluid, given in Volume 3, chapter 6:

$$\frac{u_x}{u} = \frac{3n+1}{n+1}\left[1 - \left(1 - \frac{2y}{d_t}\right)^{(n+1)/n}\right] \tag{7.6}$$

Differentiating with respect to $y$ gives:

$$\frac{1}{u}\frac{du_x}{dy} = \frac{3n+1}{n}\cdot\frac{2}{d_t}\cdot\left(1 - \frac{2y}{d_t}\right)^{1/n}$$

At the wall ($y = 0$), the velocity gradient or shear rate is given by:

$$\left(\frac{du_x}{dy}\right)_{y=0} = \frac{3n+1}{n}\left(\frac{2u}{d_t}\right) \tag{7.7}$$

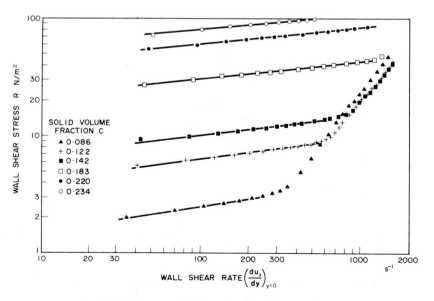

FIG. 7.8. Rheograms for flocculated kaolin suspensions

In Fig. 7.8, it will be noted that for each suspension there is a critical wall shear rate which marks the point of transition from streamline to turbulent flow. In the turbulent region all the curves tend to merge but it should be noted that the abscissa is no longer the true wall shear rate because equation 7.7 applies only to the laminar region.

It is interesting to note that the pressure drop for a constant volumetric rate of flow of suspension in the pipeline can under certain circumstances be substantially reduced by injecting air into the line. When the suspension is in laminar flow, the pressure drop over the pipeline increases only very slowly with increase in flowrate provided that the flow index $n$ is substantially less than unity. If air is injected into the line in such a way that it forms alternate slugs with the suspension, the linear velocity of the suspension will be increased, but the pressure gradient along the slug of suspension will increase only slightly. However, there will be a reduction in the area of pipe wall with which the suspension comes into contact and, on the assumption of no slip between the air and the suspension, the wetted surface will be reduced by an amount which can be directly calculated from the ratio of the air to the slurry flowrates. Reduction of pressure drop in the pipeline becomes

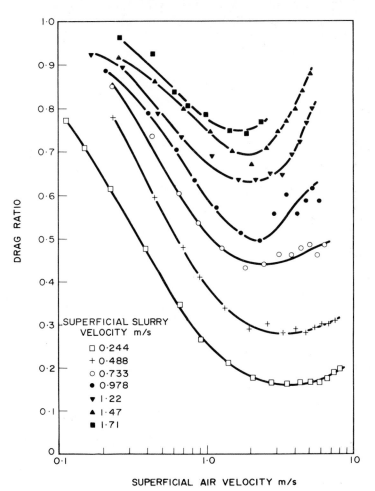

FIG. 7.9. Drag ratio data for a flocculated kaolin slurry

progressively more marked until the air rate has accelerated the slurry to the point at which its flow becomes turbulent. In Fig. 7.9, the drag ratio (pressure gradient with air injection/pressure gradient without air) is plotted against superficial air velocity for a number of different flowrates of flocculated kaolin suspension (volumetric concentration $C = 0.22$). It will be noted that the drag reduction effect is greatest at low slurry flowrates and that it is possible to effect up to a 6-fold reduction in pressure drop. Air injection has considerable possibilities for reduction of total pressures and pressure gradients in conveying systems. However, where it is possible to add a suitable deflocculating agent, even greater reductions in pressure gradient may be achieved as it is generally found that deflocculated suspensions offer much lower resistances to flow.

The results described above for a power law fluid, with flow index $n$ substantially less than unity, are broadly similar to those obtained with a suspension, such as a flocculated suspension of fine anthracite coal in water, which exhibits a finite yield stress.

Shear-thinning suspensions, whether their rheological behaviour can be described by a power law or by a model incorporating a yield stress, are very suitable for the large scale conveyance of coarse solids. Because the region of high shear is confined to the fluid close to the pipe wall, the suspension will show a high apparent viscosity over the greater part of the cross-section and a low effective viscosity near the walls. Thus large particles will tend to settle only very slowly as they will travel in the high viscosity region and will be well supported and will tend to have little contact with the walls. At the same time, the pressure gradient will remain low because of the low apparent viscosity of the fluid in contact with the pipe walls. There are interesting possibilities of reducing pressure gradients by air injection in such systems.

### 7.3.2. Horizontal Transport of Settling Suspensions

The most extensive experimental work on hydraulic conveying was carried out in France by Durand[11-13] and his co-workers who were particularly interested in conveying coal in pipelines over long distances. They were primarily concerned with obtaining information for design purposes, rather than investigating fundamentals. Work in the U.K. has been carried out by Worster and Denny[14], Turtle and Abbott[15], and Shook[16], and in Australia by Sinclair[17].

Durand investigated hydraulic conveying in pipes ranging from 40 mm to 560 mm in diameter, and used solid particles up to 25 mm in size at volumetric concentrations up to 22 per cent. Turtle and Abbott worked only with a 25 mm diameter pipeline, but explored a wider range of physical properties of solids using materials ranging from perspex of specific gravity 1·18 to zirconia of specific gravity 4·60. The following types of flow were obtained at successively lower velocities:

(a) Fully suspended homogeneous flow.
(b) Fully suspended heterogeneous flow with a marked concentration gradient over a vertical diameter.
(c) Suspended flow superimposed on a bed moving at a constant velocity at the bottom of the pipe.
(d) Suspended flow superimposed on a fixed bed, in which the top layers were frequently moving.

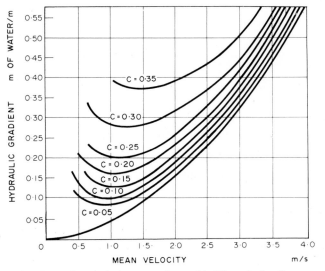

FIG. 7.10. Hydraulic gradient–velocity curves for sand in 25 mm hydraulic conveying line

(e) A stationary deposit, with particles moving over the surface in ripples, very few particles being suspended in the liquid.

(f) Isolated deposits on the bottom of the pipe which travelled along very slowly.

Pressure drops were measured over a test length of the 25 mm pipe and plotted as hydraulic gradient against the mean velocity of flow to give, for each material, curves of the form shown in Fig. 7.10 which refers to sand of 72–85 B.S. mesh size (*ca.* 200 $\mu$m). It will be noted that for each volumetric concentration of solids the curve exhibits a minimum point corresponding approximately with the velocity at which deposition of a bed commences. The economic operating velocity is also close to this velocity.

In the graph shown, the curves for different concentrations tend to become parallel to one another as the velocity is increased. This is a characteristic of a material in heterogeneous suspended flow. Similar curves are obtained with systems giving homogeneous suspended flow though the curves diverge at high velocities. For systems giving rise to a bed the curves tend to converge at high velocities. These characteristics have been predicted theoretically[15].

For materials tending to form a bed deposit, the data on pressure drop can be represented by the equation:

$$\frac{i - i_w}{C i_w} = 66 \frac{g d_t}{u^2} (s - 1)$$

(7.8)

where    $i$ is the hydraulic gradient in the pipeline (m water/m pipe),
   $i_w$ is the hydraulic gradient for water alone at the same rate of flow,
   $C$ is the fractional volumetric concentration of solids,
   $g$ is the acceleration due to gravity,
   $s$ is the specific gravity of the particles,
   $d_t$ is the pipe diameter, and
   $u$ is the mean velocity of flow.

It should be noted that $[(i-i_w)/i_w]$ is equivalent to $(-\Delta P_x/-\Delta P_{air})$ in equation 7.2. For materials forming a heterogeneous suspension:

$$\frac{i-i_w}{Ci_w} = 1100\frac{gd_t}{u^2}\frac{u_0}{u}(s-1) \tag{7.9}$$

where $u_0$ is the terminal falling velocity of the particles.

If the material forms a homogeneous suspension, the pressure drop is accurately determined by assuming it to behave as a single fluid.

Durand has investigated the effect of pipe diameter but has used a smaller range of solids. He suggests that for heterogeneous suspensions:

$$\frac{i-i_w}{Ci_w} = 121\left\{\frac{gd_t(s-1)}{u^2}\frac{u_0}{[gd(s-1)]^{1/2}}\right\}^{1\cdot5} \tag{7.10}$$

where $d$ is the particle diameter.

When the particle diameter is expressed in terms of the free-falling velocity and the properties of the fluid and particles, it is found that equations 7.9 and 7.10 are very similar, for the range of conditions normally used. The effect of pipe diameter is in some doubt, however.

The nature of the flow in a horizontal pipe was studied by Shook[16]. He examined the distribution of solids over the cross-section of a 25 mm diameter pipe and measured the velocity of the fluid relative to the solids. The distribution of solids was measured by fitting a flow divider in the form of a horizontal knife-edge to the end of the pipe. The distribution over the whole cross-section could be measured by varying the position of the divider and collecting separately the flows above and below the divider. The relative velocity between the water and particles was measured in two ways. First an average value was obtained by means of a salt injection method in which a pulse of salt solution was introduced into the pipeline, and the time taken for it to pass two sets of electrodes a known distance apart was recorded. This measurement gave the actual linear velocity of the water and, from a knowledge of the volumetric flowrate and the concentration of solids, the relative velocity could be calculated. The other method involved measuring the distribution of the velocity over the cross-section with a modified form of pitot tube.

Over the range of experiments carried out, it was found that there was an appreciable concentration gradient in a vertical direction. With coarse materials only, there was a tendency for the concentration at the bottom part of the pipe to be constant. This provided evidence that the fine materials were travelling as heterogeneous suspensions, but that the coarser materials readily formed a moving bed.

With fine materials (fine sand and perspex) the relative velocity between the liquid and solids was negligible. With coarser materials (coarse sand and gravel), the relative velocity was substantially constant over the cross-section; it tended to increase with concentration of solids and to decrease with mixture velocity. In all cases, the form of the velocity profile over the cross-section was disturbed; it was generally flattened and the position of the maximum velocity was raised.

The results were consistent with the hypothesis that the fine particles were supported by the turbulent eddies of the fluid, and the coarser particles of sand and gravel were acted upon by the Bagnold dispersive forces. An interesting effect was observed with perspex particles for which the concentration was a maximum at a short distance above the bottom of the pipe, leaving a zone virtually free of particles close to the bottom of the pipe.

It was found by Gliddon in his studies of vertical conveying that there was frequently a layer of water free of particles close to the walls. It is suggested that the perspex particles are supported largely by Magnus type forces and that the clear liquid layer occurs because the radial force on the particles is greatest close to the walls where the velocity gradients are a maximum.

In classifying materials according to mechanism of support, it should be remembered that the principal effect may vary as the water velocity is changed. Furthermore, the turbulent eddies are often not as significant as might be expected because they tend to be damped out by the particles themselves.

Shook and Daniel[18] have further extended studies on flow of suspensions in horizontal pipes under circumstances which give rise to a stable stationary deposit and where, consequently, Bagnold's dispersive forces are responsible for the suspension of the particles. They measured hydraulic gradients over a range of conditions and obtained the concentration of solids held in the pipeline by means of a gamma-ray device. A theoretical equation for the hydraulic gradient has been derived, and empirical constants in the equation were obtained using the experimental results. The final equation was very similar to that obtained previously by Abbott[15].

Many of the difficulties associated with the contamination of the solid particles by the conveying fluid can be obviated by enclosing the material to be conveyed in a solid capsule. The theoretical aspects of flow of capsules in pipes have been fully discussed by Hodgson and Charles[19].

### 7.3.3. Vertical Transport

Durand[20] has also studied vertical transport of sand and gravel of particle size ranging between 0·18 mm and 4·57 mm in a 150 mm diameter pipe, and Worster and Denny[21] conveyed coal and gravel in vertical pipes of diameters 75, 100, and 150 mm. They concluded that the pressure drop for the slurry was the same as for the water alone, if due allowance was made for the static head attributable to the solids in the pipe.

Gliddon[22] conveyed particles of specific gravities ranging from 1·19 to 4·56, of sizes 0·10 to 3·8 mm, in a 25 mm diameter pipe 12·8 m tall and in a 50 mm pipe 6·7 m tall. The particles used had a thirtyfold variation in terminal falling velocity. It was found that the larger particles had little effect on the frictional losses, provided the static head due to the solids was calculated on the assumption that the particles had a velocity relative to the liquid equal to their terminal falling velocities. Furthermore, it was shown photographically that at high velocities these particles travel in a central core, and thus the frictional forces at the wall will be unaffected by their presence. Very fine particles of sand give suspensions which behave as homogeneous fluids, and the hydraulic gradient due to friction is the same as for horizontal flow. When the settling velocity cannot be neglected in comparison with the liquid velocity, the hydraulic gradient is found to be given by:

$$\frac{i - i_w}{Cl_w} = 0 \cdot 0037 \left(\frac{gd_t}{u^2}\right)^{1/2} \frac{d_t}{d} s^2. \tag{7.11}$$

Cloete et al.[23] conveyed sand and glass ballotini particles at high concentrations through vertical pipes, 12·5 and 19 mm diameter, at velocities up to 3 m/s. They measured pressure gradients, and obtained values for the concentration of particles in the pipe, using a $\gamma$-ray system. At velocities up to 0·7 m/s the frictional head was similar to that for water, but at higher velocities it became increasingly greater than for water as the velocity was increased. The ratio of the solid velocity to the liquid velocity remained approximately constant at all flowrates, and appeared to be a characteristic of the solid.

### 7.3.4. Industrial Applications of Hydraulic Transport

An industrial application of hydraulic conveying is illustrated in Fig. 7.11 where the coal tailings underflow from a thickener is pumped 1·1 km through a lift of 25 m to the lagoon area at a rate of 15 kg/s.

FIG. 7.11. Discharge from the pipeline to lagoon for coal tailings disposal

The conveying system consists of two high-pressure water pumps arranged in conjunction with a thickener and two lock hoppers. Continuous pumping takes place by expelling slurry from the lock hoppers with clean high-pressure water so that the solids in the slurry are not in contact with the working parts of the pump during the operation.

Figure 7.12 illustrates the operation of the scheme. One lock hopper is filled with clean water and delivers slurry to the disposal pipeline at a dry solids rate of 7·9 kg/s and a slurry pipe velocity of 2·5 m/s. The other lock hopper is filled with slurry and expels the dirty water from the previous cycle to the thickener. The hoppers reverse their duties through automatic stop valves and synchronous timers ensure that the slurry flow is continuous.

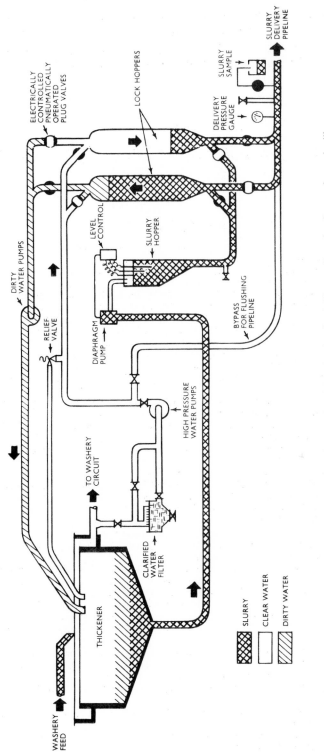

Fig. 7.12. Diagrammatic arrangement of the International Combustion Lock Hopper Pumping System for coal tailings

Slurry from the thickener is drawn into the hoppers through the diaphragm pump and slurry hopper. High-pressure water is admitted to the top of the vessels from one of the high-pressure water pumps and dirty water is withdrawn from the hoppers to the thickener by one of the Vacseal pumps (see Volume 1, chapter 6).

## 7.4. FURTHER READING

ZANDI, I (ed.): *Advances in Solid–Liquid Flow in Pipes and its Application* (Pergamon Press, 1971).
ZENZ, F. A. and OTHMER, D. F.: *Fluidization and Fluid-Particle Systems* (Reinhold, 1960).

## 7.5 REFERENCES

1. BOOTHROYD, R. G.: *Trans. Inst. Chem. Eng.* **44** (1966) T306. Pressure drop in duct flow of gaseous suspensions of fine particles.
2. BAGNOLD, R. A.: *Proc. Inst. Civ. Eng.* (iii) **4** (1955) 174. Some flume experiments on large grains but little denser than the transporting fluid, and their implications.
3. BAGNOLD, R. A.: *Phil. Trans.* **249** (1957) 235. The flow of cohesionless grains in fluids.
4. SEGLER, G.: *Z. Ver. deut. Ing.* **79** (1935) 558. Untersuchungen an Körnergebläsen und Grundlagen für ihre Berechnung. (Pneumatic Grain Conveying (1951), National Institute of Agricultural Engineering.)
5. CLARK, R. H., CHARLES, D. E., RICHARDSON, J. F., and NEWITT, D. M.: *Trans. Inst. Chem. Eng.* **30** (1952) 209. Pneumatic conveying. Part I. The pressure drop during horizontal conveyance.
6. MITLIN, L.: University of London, Ph.D. thesis (1954). A study of pneumatic conveying with special reference to solid velocity and pressure drop during transport.
7. JONES, C. and HERMGES, G.: *Brit. J. Appl. Phys.* **3** (1952) 283. The measurement of velocities for solid–fluid flow in a pipe.
8. RICHARDSON, J. F. and MCLEMAN, M.: *Trans. Inst. Chem. Eng.* **38** (1960) 257. Pneumatic conveying. Part II. Solids velocities and pressure gradients in a one-inch horizontal pipe.
9. GASTERSTÄDT, J.: *Forsch. Arb. Geb. Ing. Wes.* No. 265 (1924) 1–76. Die experimentelle Untersuchung des pneumatischen Fördervorganges.
10. HEYWOOD, N.: University of Wales, Ph.D. thesis (1976). Air injection into suspensions flowing in horizontal pipelines.
11. DURAND, R.: *Houille Blanche* **6** (1951) 384. Transport hydraulique des matériaux solides en conduite.
12. DURAND, R.: *Houille Blanche* **6** (1951) 609. Transport hydraulique de graviers et galets en conduite.
13. DURAND, R.: *Proc. of the Minnesota International Hydraulic Convention* (1953) 89. Basic relationships of transportation of solids in pipes—experimental research.
14. WORSTER, R. C. and DENNY, D. F.: *Proc. of the Third Annual Conference of the British Hydromechanics Research Association*, Part II (1954). Transport of solids in pipes.
15. NEWITT, D. M., RICHARDSON, J. F., ABBOTT, M., and TURTLE, R. B.: *Trans. Inst. Chem. Eng.* **33** (1955) 93. Hydraulic conveying of solids in horizontal pipes.
16. NEWITT, D. M., RICHARDSON, J. F., and SHOOK, C. A.: Third Congress of the European Federation of Chemical Engineering (1962). *The Interaction between Fluids and Particles*, 87. Hydraulic conveying of solids in horizontal pipes. Part II. Distribution of particles and slip velocities.
17. SINCLAIR, C. G.: Third Congress of the European Federation of Chemical Engineering (1962). *The Interaction between Fluids and Particles*, 78. The limit deposit-velocity of heterogeneous suspensions.
18. SHOOK, C. A. and DANIEL, S. M.: *Can. J. Chem. Eng.* **43** (1965) 56. Flow of suspensions of solids in pipelines. Part 1. Flow with a stable stationary deposit.
19. HODGSON, G. W. and CHARLES, M. E.: *Can. J. Chem. Eng.* **41** (1963) 43. The pipeline flow of capsules.
20. DURAND, R.: *Houille Blanche* **8** (1953) 124. Écoulements de mixture en conduites verticales—influence de la densité des matériaux sur les caractéristiques de refoulement en conduite horizontale.
21. WORSTER, R. C. and DENNY, D. F.: *Proc. Inst. Mech. Eng.* **169** (1955) 563. The hydraulic transport of solid material in pipes.
22. NEWITT, D. M., RICHARDSON, J. F., and GLIDDON, B. J.: *Trans. Inst. Chem. Eng.* **39** (1961) 93. Hydraulic conveying of solids in vertical pipes.
23. CLOETE, F. L. D., MILLER, A. I., and STREAT, M.: *Trans. Inst. Chem. Eng.* **45** (1967) T392. Dense phase flow of solids–water mixtures through vertical pipes.

## 7.6. NOMENCLATURE

| | | Units in SI System | Dimensions in **MLT** |
|---|---|---|---|
| $C$ | Fractional volumetric concentration of solids | — | — |
| $d$ | Particle diameter | m | **L** |
| $d_t$ | Pipe diameter | m | **L** |
| $F$ | Mass rate of feed of solids | kg/s | **MT**$^{-1}$ |
| $g$ | Acceleration due to gravity | m/s$^2$ | **LT**$^{-2}$ |
| $i$ | Hydraulic gradient in pipeline | — | — |
| $i_w$ | Hydraulic gradient for water alone | — | — |
| $k$ | Consistency | N s$^n$/m$^2$ | **ML**$^{-1}$**T**$^{n-2}$ |
| $\mu_a$ | Apparent viscosity | N s/m$^2$ | **ML**$^{-1}$**T**$^{-1}$ |
| $n$ | Flow index | — | — |
| $-\Delta P_x$ | Additional pressure drop attributable to solids | N/m$^2$ | **ML**$^{-1}$**T**$^{-2}$ |
| $-\Delta P_{\text{air}}$ | Pressure drop for flow of air alone at pressure in pipe | N/m$^2$ | **ML**$^{-1}$**T**$^{-2}$ |
| $R$ | Wall shear stress | N/m$^2$ | **ML**$^{-1}$**T**$^{-2}$ |
| $R_y$ | Shear stress at distance $y$ from wall | N/m$^2$ | **ML**$^{-1}$**T**$^{-2}$ |
| $s$ | Specific gravity of solids | — | — |
| $u_a$ | Air velocity | m/s | **LT**$^{-1}$ |
| $u$ | Mean velocity of mixture | m/s | **LT**$^{-1}$ |
| $u_S$ | Velocity of particles | m/s | **LT**$^{-1}$ |
| $u_0$ | Free-falling velocity of particles | m/s | **LT**$^{-1}$ |
| $u_x$ | Velocity at distance $y$ from wall | m/s | **LT**$^{-1}$ |
| $y$ | Distance from wall | m | **L** |
| $\rho_s$ | Density of particles | kg/m$^3$ | **ML**$^{-3}$ |

# CHAPTER 8

# *Gas Cleaning*

## 8.1. INTRODUCTION

The necessity for removing suspended dust and mist from a gas arises not only in the treatment of effluent gas from a works before it is discharged into the atmosphere, but also in processes where solids or liquids are carried over in the vapour or gas stream. Thus in an evaporator it is frequently necessary to eliminate droplets which become entrained in the vapour, and in a plant involving a fluidised solid the removal of fine particles is necessary, first to prevent loss of material, and secondly to prevent contamination of the gaseous product. Further, in all pneumatic conveying plants, some form of separator must be provided at the downstream end.

Whereas relatively large particles with settling velocities greater than about 0·3 m/s readily disengage themselves from a gas stream, fine particles tend to follow the same path as the gas and separation is therefore difficult. In practice, dust particles may have an average diameter of about one-hundredth of a millimetre and a settling velocity of about 0·003 m/s, so that a simple gravity settling vessel would be impracticable because of the long time required for settling and the large size of separator which would be required for a given throughput of gas.

The main problems involved in the removal of particles from a gas stream have been reviewed by Ashman[1] and more recently by Stairmand[2], who point out that the main reasons for removing particles from an effluent gas are:

(a) For reasons of the health of the operators in the plant and of the surrounding population. It is stated the chief danger arises from the inhaling of dust particles, and the most dangerous range of sizes is generally between about 0·5 and 3 μm.

(b) In order to eliminate explosion risks. A number of carbonaceous materials and finely powdered metals give rise to explosive mixtures with air, and flame can be propagated over large distances.

(c) In order to prevent wastage of valuable materials.

(d) Because the gas itself may be required for use in a further process, as, for example, blast furnace gas used for firing stoves.

The size range of commercial aerosols and the methods available for determination of particle size and for removing the particles from the gas are shown in Fig. 8.1, which is taken from Ashman's paper. It will be noted that the ranges over which the various items of equipment operate overlap to some extent, and the choice of equipment will depend not only on the particle size but also on such factors as the quantity of gas to be handled, the concentration of the dust or mist, and the physical properties of the particles. Tables 8.1 and 8.2[2] provide further information as an aid to the selection of suitable equipment.

Separation equipment may depend on one or more of the following principles and in some plant the relative importance of each is difficult to assess:

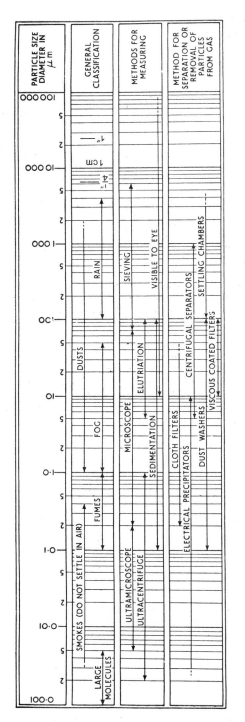

Fig. 8.1. Characteristics of aerosols and separators

(a) Gravitational settling.
(b) Centrifugal separation.
(c) Inertia or momentum processes.
(d) Filtration.
(e) Electrostatic precipitation.
(f) Washing with a liquid.
(g) Agglomeration of solid particles and coalescence of liquid droplets.

The classification will be used to describe commercially available equipment in the following section.

TABLE 8.1. *Summary of Dust Arrestor Performance*[2]

| Type of equipment | Field of application | Pressure loss |
|---|---|---|
| Settling chambers | Removal of coarse particles, larger than about 100–150 µm | Below 50 N/m² |
| Scroll collectors, Shutter collectors, Low-pressure-drop cyclones | Removal of fairly coarse dusts down to about 50–60 µm | Below 250 N/m² |
| High-efficiency cyclones | Removal of average dusts in the range 10–100 µm | 250–1000 N/m² |
| Wet washers (including spray towers, venturi scrubbers, etc.) | Removal of fine dusts down to about 5 µm (or down to sub-micron sizes for the high-pressure-drop type) | 250–600 N/m² or more |
| Bag filters | Removal of fine dusts and fumes, down to about 1 µm or less | 100–1000 N/m² |
| Electrostatic precipitators | Removal of fine dusts and fumes down to 1 µm or less | 50–250 N/m² |

TABLE 8.2. *Efficiency of Dust Collectors*\*[2]

| Dust collector | Efficiency at 5 µm (%) | Efficiency at 2 µm (%) | Efficiency at 1 µm (%) |
|---|---|---|---|
| Medium-efficiency cyclone | 27 | 14 | 8 |
| High-efficiency cyclone | 73 | 46 | 27 |
| Low-pressure-drop cellular cyclone | 42 | 21 | 13 |
| Tubular cyclone | 89 | 77 | 40 |
| Irrigated cyclone | 87 | 60 | 42 |
| Electrostatic precipitator | 99 | 95 | 86 |
| Irrigated electrostatic precipitator | 98 | 97 | 92 |
| Fabric filter | 99·8 | 99·5 | 99 |
| Spray tower | 94 | 87 | 55 |
| Wet impingement scrubber | 97 | 95 | 80 |
| Self-induced spray deduster | 93 | 75 | 40 |
| Disintegrator | 98 | 95 | 91 |
| Venturi scrubber | 99·8 | 99 | 97 |

\*For dust of density 2700 kg/m³.

## 8.2. GAS CLEANING EQUIPMENT

### 8.2.1. Gravity Separators

If the particles are large, they will settle out of the gas stream if the cross-sectional area is increased. The velocity will then fall so that the eddy currents which are maintaining the

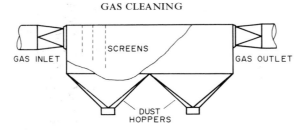

FIG. 8.2. Settling chamber

particles in suspension are suppressed. In most cases, however, it is necessary to introduce baffles or screens as shown in Fig. 8.2, or to force the gas over a series of trays as shown in Fig. 8.3 which depicts a separator used for removing dust from sulphur dioxide produced by the combustion of pyrites; this equipment is suitable when the concentration of particles is high, because it is easily cleaned by opening the doors at the side. Gravity separators are seldom used now as they are very bulky and will not remove particles smaller than 50–100 μm.

This principle of separation is little used since the equipment must be very large in order to reduce the gas velocity to a reasonably low value to allow the finer particles to settle. For example, a settling chamber to remove particles with a diameter of 20 μm and density 2000 kg/m$^3$ from a gas stream flowing at 10 m$^3$/s would have a volume of about 3000 m$^3$. Clearly, this very large volume is a severe limitation and this type of equipment is normally

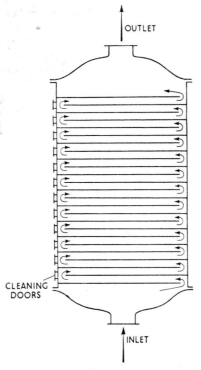

FIG. 8.3. Tray separator

restricted to small plants as a pre-separator to reduce the load on a more efficient secondary collector.

### 8.2.2. Centrifugal Separators

The rate of settling of particles in a gas stream can be greatly increased if centrifugal rather than gravitational forces are employed. In the cyclone separator (Figs. 8.4 and 8.5) the gas is introduced tangentially into a cylindrical vessel at a velocity of about 30 m/s and

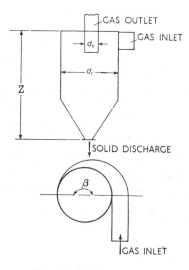

FIG. 8.4. Cyclone separator

the clean gas is taken off through a central outlet at the top. The solids are thrown outwards against the cylindrical wall of the vessel, and then move away from the gas inlet and are collected in the conical base of the plant. This separator is very effective unless the gas contains a large proportion of particles less than about 10 μm in diameter and is equally effective when used with either dust- or mist-laden gases. It is now the most commonly used general purpose separator.

The effects of the dimensions of the separator on its efficiency have been studied experimentally by Ter Linden[3] and Stairmand[4,5] who have studied the flow pattern in a glass cyclone separator with the aid of smoke injected into the gas stream. The gas is found to move spirally downwards, gradually approaching the central portion of the separator, and then to rise and leave through the central outlet at the top. The magnitude and direction of the gas velocity have been explored by means of small pitot tubes, and the pressure distribution has been measured. The tangential component of the velocity of the gas appears to predominate throughout the whole depth, except within a highly turbulent central core of diameter about 0·4 times that of the gas outlet pipe. The radial component of the velocity acts inwards, and the axial component is away from the gas inlet near the walls of the separator but is in the opposite direction in the central core. Pressure measurements indicate a relatively high pressure throughout, except for a region of

FIG. 8.5. Cyclone separator

reduced pressure corresponding to the central core. Any particle is therefore subjected to two opposing forces in the radial direction, the centrifugal force which tends to throw it to the walls and the drag of the fluid which tends to carry the particle away through the gas outlet. Both of these forces are a function of the radius of rotation and of the size of the particles, with the result that particles of different sizes tend to rotate at different radii. As the outward force on the particles increases with the tangential velocity and the inward force increases with the radial component, the separator should be designed so as to make the tangential velocity as high as possible and the radial velocity low. This is generally effected by introducing the gas stream with a high tangential velocity, with as little shock as possible, and by making the height of the separator large.

The radius at which a particle will rotate within the body of a cyclone corresponds to the position where the net radial force on the particle is zero. The two forces acting are the centrifugal force outwards and the frictional drag of the gas acting inwards.

Consider a spherical particle of diameter $d$ rotating at a radius $r$. Then the centrifugal force is:

$$\frac{mu_t^2}{r} = \frac{\pi d^3 \rho_s u_t^2}{6r} \tag{8.1}$$

where $m$ is the mass of the particle, and
$\quad u_t$ is the tangential component of the velocity of the gas.

It is assumed here that there is no slip between the gas and the particle in the tangential direction.

If the radial velocity is low, the inward radial force due to friction will, from Chapter 3, be equal to $3\pi\mu du_r$, where

$\mu$ is the viscosity of the gas, and

$u_r$ is the radial component of the velocity of the gas.

The radius $r$, at which the particle will rotate at equilibrium, is then given by:

$$\frac{\pi d^3 \rho_s u_t^2}{6r} = 3\pi\mu du_r$$

i.e.

$$\frac{u_t^2}{r} = \frac{18\mu}{d^2 \rho_s} u_r \qquad (8.2)$$

Now the free-falling velocity of the particle $u_0$ when the density of the particle is large compared with that of the gas is given from equation 3.17 as:

$$u_0 = \frac{d^2 g \rho_s}{18\mu} \qquad (8.3)$$

Substituting in equation 8.2:

$$\frac{u_t^2}{r} = \frac{u_r}{u_0} g$$

or

$$u_0 = \frac{u_r}{u_t^2} rg \qquad (8.4)$$

Thus the higher the terminal falling velocity of the particle, the greater is the radius at which it will rotate and the easier it is to separate. If it is assumed that a particle will be separated provided it tends to rotate outside the central core of diameter $0.4d_0$, the terminal falling velocity of the smallest particle which will be retained is found by substituting $r = 0.2d_0$ in equation 8.4, i.e.

$$u_0 = \frac{u_r}{u_t^2} 0.2d_0 g \qquad (8.5)$$

In order to calculate $u_0$, it is necessary to evaluate $u_r$ and $u_t$ for the region outside the central core. The radial velocity $u_r$ is found to be approximately constant at a given radius and to be given by the volumetric rate of flow of gas divided by the cylindrical area for flow at the radius $r$. Thus, if $G$ is the mass rate of flow of gas through the separator and $\rho$ is its density, the linear velocity in a radial direction at a distance $r$ from the centre is given by:

$$u_r = \frac{G}{2\pi r Z \rho} \qquad (8.6)$$

where $Z$ is the depth of the separator.

The tangential velocity is found experimentally to be inversely proportional to the square root of the radius at all depths. Then, if $u_t$ is the tangential component of the velocity at a radius $r$, and $u_{t0}$ is the corresponding value at the circumference of the separator:

$$u_t = u_{t0} \sqrt{\frac{d_t}{2r}} \qquad (8.7)$$

Further, it is found that $u_{t0}$ is approximately equal to the velocity with which the gas stream enters the cyclone separator. If these values for $u_r$ and $u_t$ are now substituted into equation 8.5, the terminal falling velocity of the smallest particle which the separator will retain is given by:

$$u_0 = \frac{G}{2\pi \times 0 \cdot 2 d_0 \rho Z} \frac{2 \times 0 \cdot 2 d_0}{d_t} \frac{1}{u_{t0}^2} 0 \cdot 2 d_0 g$$

$$= \frac{0 \cdot 2 G d_0 g}{\pi \rho Z d_t u_{t0}^2} \tag{8.8}$$

If the cross-sectional area of the inlet is $A_i$, $G = A_i \rho u_{t0}$ and:

$$u_0 = \frac{0 \cdot 2 A_i^2 d_0 \rho g}{\pi Z d_t G} \tag{8.9}$$

A small inlet and outlet therefore result in the separation of smaller particles, but as the pressure drop over the separator varies with the square of the inlet velocity and the square of the outlet velocity[4], the practical limit is set by the permissible drop in pressure. The depth and diameter of the body should be as large as possible because the former determines the radial component of the gas velocity and the latter controls the tangential component at any radius. In general, the larger the particles, the larger should be the diameter of the separator because the greater is the radius at which they rotate. The larger the diameter the greater too is the inlet velocity which can be used without causing turbulence within the separator. The factor which ultimately settles the maximum size is, of course, the cost. Because the separating power is directly related to the throughput of gas, the cyclone separator is not very flexible though its efficiency can be improved at low throughputs by restricting the area of the inlet with a damper and thereby increasing the velocity. Generally, however, it is better to use a number of cyclones in parallel and to keep the load on each approximately the same whatever the total throughput.

Because the vertical component of the velocity in the cyclone is downwards everywhere outside the central core, the particles will rotate at a constant distance from the centre and move continuously downwards until they settle on the conical base of the plant. Continuous removal of the solids is desirable so that the particles do not get entrained again in the gas stream due to the relatively low pressures in the central core. Entrainment is reduced to a minimum if the separator has a deep conical base of small angle.

Though the sizes of particles which will be retained or lost by the separator can be calculated, it is found in practice that some smaller particles are retained and some larger particles are lost. The small particles which are retained have in most cases collided with other particles and adhered to form agglomerates which behave as large particles. Relatively large particles are lost because of eddy motion within the cyclone separator and because they tend to bounce off the walls of the cylinder back into the central core of fluid. If the gas contains a fair proportion of large particles, it is desirable to remove them in a preliminary separator before the gas is fed into the cyclone.

The efficiency of the cyclone separator is greater for large than for small particles and increases with the throughput until the point is reached where excessive turbulence is created. Figure 8.6 shows the efficiency of collection plotted against particle size for an experimental separator for which the theoretical "cut" occurs at about 10 μm. It will be noted that an appreciable quantity of fine material is collected, largely as a result of

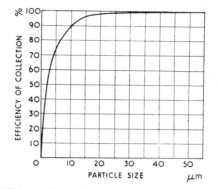

FIG. 8.6. Efficiency of a cyclone separator for various sizes of particles

agglomeration, and some of the coarse material is lost with the result that a sharp cut is not obtained.

The effect of the arrangement and size of the gas inlet and outlet has been investigated and it has been found that the inlet angle $\beta$ (Fig. 8.4) should be of the order of 180°. Further, the depth of the inlet pipe should be small, and a square section is generally preferable to a circular one because a greater area is then obtained for a given depth. The outlet pipe should extend downwards well below the inlet in order to prevent short-circuiting.

Various modifications can be made to improve the operation of the cyclone separator in special cases. If there is a large proportion of fine material present, a bag filter may be attached to the clean gas outlet. Alternatively, the smaller particles may be removed by means of a spray of water which is injected into the separator. In some cases, the removal of the solid material is facilitated by running a stream of water down the walls and this also reduces the risk of the particles becoming re-entrained in the gas stream. The chief difficulty lies in wetting the particles with the liquid.

Because the separation of the solid particles which have been thrown out to the walls is dependent on the flow of gas parallel to the axis rather than to the effect of gravity, the cyclone can be mounted in any desired direction. In many cases horizontal cyclone

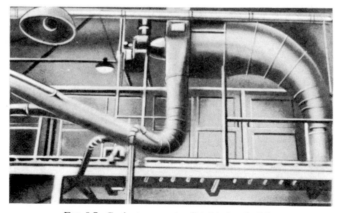

FIG. 8.7. Cyclone separator fitted in bend of duct

separators are used and occasionally the separator is fixed at the junction of two mutually perpendicular pipes and the axis is then a quadrant of a circle as shown in Fig. 8.7. The cyclone separator is usually mounted vertically, except where there is a shortage of headroom, because removal of the solids is more readily achieved especially if there are some large particles present.

A double cyclone separator is sometimes used when the range of the size of particles in the gas stream is large. It consists of two cyclone separators, one inside the other. The gas stream is introduced tangentially into the outer separator and the larger particles are deposited. The partially cleaned gas then passes into the inner separator through

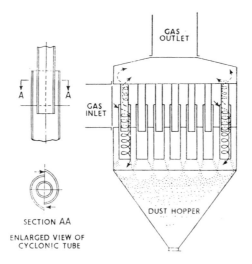

FIG. 8.8. Multi-tube cyclone separator

tangential openings and the finer particles are deposited there because the separating force in the inner separator is greater than that in the outer cylinder. In Fig. 8.8, a multi-cyclone is illustrated in which the gas is subjected to further action in a series of tubular units, the number of which can be varied with the throughput. This separator is therefore rather more flexible than the simple cyclone.

Szekely and Carr[6] have studied the heat transfer between the walls of a cyclone and a gas–solid suspension. They have shown that the mechanism of heat transfer is quite different from that occurring in a fluidised bed. There is a high rate of heat transfer directly from the wall to the particles, but the transfer direct to the gas is actually reduced. Overall, the heat transfer rate at the walls is slightly greater than that obtained in the absence of the particles.

### 8.2.3. Inertia or Momentum Separators

Momentum separators rely on the fact that the momentum of the particles is far greater than that of the gas, so that the particles do not follow the same path as the gas if the direction of motion is suddenly changed. An apparatus devised for the separation of particles from mine gas is shown in Fig. 8.9, from which it is seen that the direction of the gas is changed suddenly at the end of each baffle. Again the separator shown in Fig. 8.10 consists of a number of vessels—up to about thirty connected in series—in each of which the gas impinges on a central baffle. The dust drops to the bottom and the velocity of the gas must be arranged so that it is sufficient for effective separation to be made without the danger of re-entraining the particles at the bottom of each vessel.

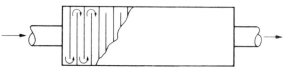

FIG. 8.9. Baffled separator

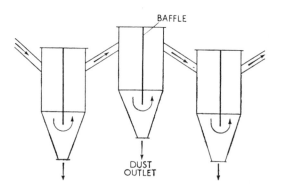

FIG. 8.10. Battery of momentum separators

As an alternative to the use of rigid baffles, the separator may be packed with a loose fibrous material. In this case the separation will be attributable partly to gravitational settling on to the packing, partly to inertia effects and partly to filtration. If the packing is moistened with a viscous liquid, the efficiency is improved because the film of liquid acts as an effective filter and prevents the particles being picked up again in the gas stream. The viscous filter (Fig. 8.11) consists of a series of corrugated plates, mounted in a frame and covered with a non-drying oil: these units are then arranged in banks to give the required area. They are readily cleaned and offer a low resistance to flow. Packs of slag wool offer a higher resistance but are very effective.

### 8.2.4. Fabric Filters

Fabric filters include all types of bag filters in which the filter medium is in the form of a woven or felted textile fabric which may be arranged as a tube or supported on a suitable framework. This type of filter is capable of removing particles of a size down to 1 μm or less by the use of glass fibre paper or pads[7]. In a normal fabric filter, particles smaller than the apertures in the fabric will be trapped by impingement on the fine "hairs" which span the apertures. Typically, the main strands of the material may have a diameter of 500 μm, spaced 100–200 μm apart. The individual textile fibres with a diameter of 5–10 μm criss-

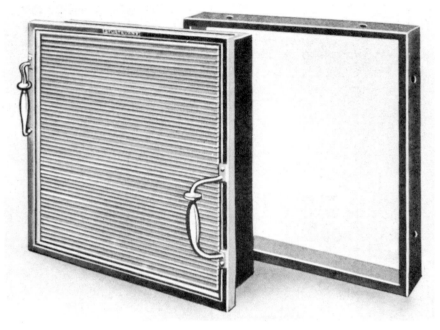

FIG. 8.11. Viscous filter

cross the aperture and form effective impingement targets capable of removing particles of sizes down to 1 µm.

In the course of operation, filtration efficiency will be low until a loose "floc" builds up on the fabric surface and it is this which provides the effective filter for the removal of fine particles. The cloth will require cleaning from time to time to avoid excessive build-up of solids which gives rise to a high pressure drop. The velocity at which the gases pass through the filter must be kept low, typically 0·005 to 0·03 m/s, in order to avoid compaction of the floc and consequently high pressure drops, or to avoid local breakdown of the filter bed which would allow large particles to pass the filter.

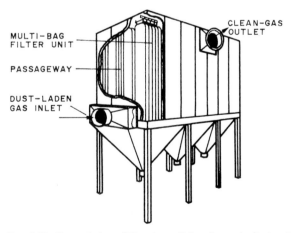

FIG. 8.12. General view of "bag-house" (low face-velocity type)

There are three main types of bag filter. The simplest, which is shown in Fig. 8.12, consists of a number of elements assembled together in a "bag-house". This is the cheapest type of unit and operates with a velocity of about 0·01 m/s across the bag surface.

A more sophisticated and robust version incorporates some form of automatic bag-shaking mechanism which may be operated by mechanical, vibratory or air-pulsed methods. A heavier fabric allows higher face velocities, up to 0·02 m/s, to be used and permits operation under more difficult conditions than the simpler bag-house type can handle.

The third type of bag filter is the reverse-jet filter[8], illustrated in Figs. 8.13 and 8.14, with face velocities of about 0·05 m/s and with the capability of dealing with high dust

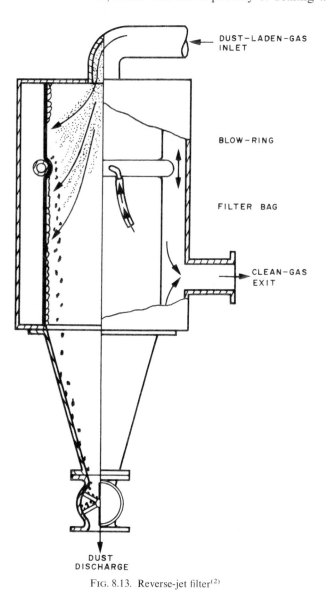

FIG. 8.13. Reverse-jet filter[2]

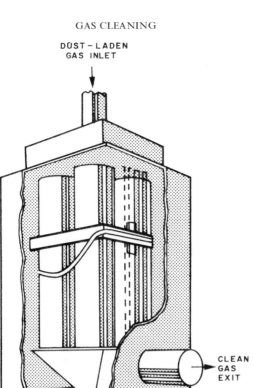

DUST – LADEN
GAS INLET

CLEAN
GAS
EXIT

FIG. 8.14. General view of reverse-jet filter (high face-velocity type)[2]

concentrations at high efficiencies; this type of filter can deal with difficult mixtures in an economic and compact unit. Use of the blow ring enables the cake to be dislodged in a cleaning cycle which takes only a few seconds.

### 8.2.5. Electrostatic Precipitators

When the gas contains very fine particles, an electrostatic precipitator is generally employed because its efficiency is highest when the particle size is very small. As the capital cost and running costs are relatively high it is usual to remove the coarser particles in a preliminary separator, such as a cyclone separator, and to use the electrostatic precipitator as an eliminator for the very fine material.

If the gas is passed between two electrodes charged to a potential difference of from 10 to 60 kV, it is subjected to the action of a corona discharge. Ions which are given off by the smaller electrode—on which the charge density is greater—attach themselves to the particles which are then carried to the larger electrode under the action of the electric field. The smaller electrode is known as the discharge electrode and the larger one, which is usually earthed, as the receiving electrode. Most industrial gases are sufficiently

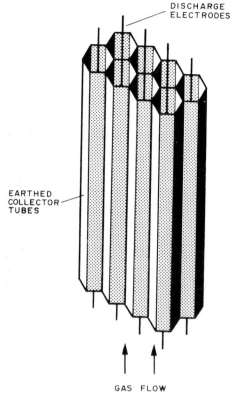

FIG. 8.15. General arrangement of wire-in-tube precipitator

conducting to be readily ionised, the most important conducting gases being carbon dioxide, carbon monoxide, sulphur dioxide, and water vapour, but if the conductivity is low, water vapour can be added. The electrodes may be in the form of plates, but the cylindrical type with the receiving electrode in the form of a metal pipe and the discharge electrode as a central wire is more efficient for large installations. The potential difference is usually determined by the tendency for arcing and, as the gap is gradually reduced as the solids or liquid droplets collect on the receiving electrode, continuous removal is desirable. This is usually achieved by rapping the electrodes, but in some cases, especially where the particles form sticky agglomerates, the electrodes are irrigated. With very sticky solids, oil vapour is sometimes injected into the gas stream so that a mobile sludge is precipitated on the electrode.

The two main types of precipitator are shown diagrammatically in Figs. 8.15 and 8.16[2].

The gas velocity over the electrodes usually varies between about 0·6 and 3 m/s, with an average contact time of about 2 s. The maximum velocity is determined by the maximum distance through which any particle must move in order to reach the receiving electrode, and by the attractive force acting on the particle. This force is given by the product of the charge on the particle and the strength of the electrical field, but calculation of the path of a particle is difficult because it gradually becomes charged as it enters the field and the force therefore increases during the period of charging; this rate of charging cannot be estimated with any degree of certainty. The particle moves towards the collecting electrode under the action of the accelerating force due to the electrical field and the retarding force of fluid friction, and the maximum rate of passage of gas is that which will just allow the most unfavourably located particle to reach the collecting electrode before the gas leaves the precipitator. Collection efficiencies of nearly 100 per cent can be obtained at low gas velocities but the economic limit is usually about 99 per cent.

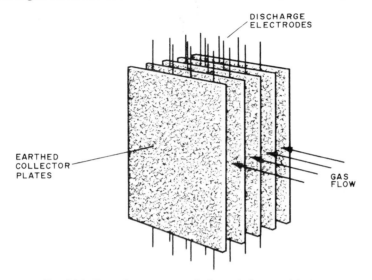

FIG. 8.16. General arrangement of wire-and-plate precipitator

Plate-type precipitators are now generally used only in small installations and the gas is often passed through an ioniser before it enters the dust collector cell. In this way the risk of arcing is reduced. Figure 8.17 shows an ionising unit in which the charged electrodes at about 13 kV are in the form of fine wires and the earthed electrodes are metal tubes. The charged particles then pass from the ioniser to the dust collector (Fig. 8.18) in which alternate plate electrodes are earthed and charged to about 6 kV. There is virtually no corona discharge here and the particles are simply collected on the plates.

Electrostatic precipitators are made in a very wide range of sizes and will handle up to about 50 m³ gas/s. Although they operate more satisfactorily at low temperatures, they can be used up to about 800 K. Pressure drops over the separator are low.

### 8.2.6. Liquid Washing

If the gas contains an appreciable proportion of fine particles, liquid washing provides an effective method of cleaning which gives a gas of high purity. In the spray column illustrated in Fig. 8.19, the gas flows upwards through a set of primary sprays to the main part of the column where it flows countercurrently to a water spray which is redistributed at intervals. In some cases packed columns are used for gas washing, but it is generally better to arrange the packing on a series of trays to facilitate cleaning. Figure 8.20 shows a venturi scrubber in which water is injected at the throat and the separation is

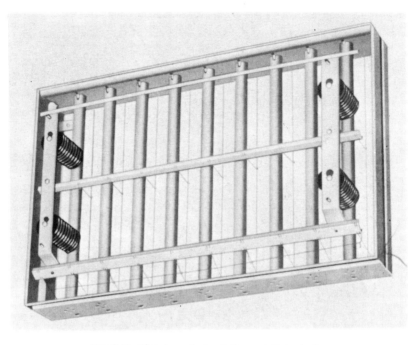

FIG. 8.17. Plate-type electrostatic precipitator ioniser

FIG. 8.18. Plate-type electrostatic precipitator collector

then carried out in a cyclone separator. In the cyclonic scrubber the gas is introduced tangentially into a cylindrical vessel and passes upwards through a water spray. Sometimes a venturi scrubber and a cyclonic scrubber are used together as shown in Fig. 8.21.

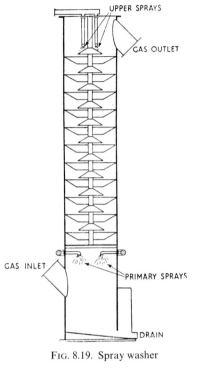

FIG. 8.19. Spray washer

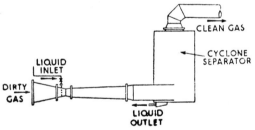

FIG. 8.20. Venturi washer with cyclone separator

The chief difficulty in operating a gas washer is wetting the particles because it is often necessary for the liquid to displace an adsorbed layer of gas. In some cases the gas is sprayed with a mixture of water and oil to facilitate wetting and it may then be bubbled through foam. This method of separation cannot be used, of course, where a dry gas is required.

### 8.2.7. Agglomeration and Coalescence

Separation of particles or droplets is often facilitated by first effecting an increase in the effective size of the individual particles by causing them to agglomerate or coalesce, and

FIG. 8.21. Venturi washer with cyclonic scrubber

then separating the enlarged particles. A number of methods are available. Thus, if the dust- or mist-laden gas is brought into contact with a supersaturated vapour, condensation occurs on the particles which act as nuclei. Again if the gas is brought into an ultrasonic field, the vibrational energy of the particles is increased so that they collide and agglomerate. Ultrasonic agglomeration[9,10] has been found very satisfactory in causing fine dust particles to form agglomerates, about 10 µm in diameter, which can then be removed in a cyclone separator. In the Calder–Fox scrubber[11], which has been developed largely for the removal of acid sprays, the gas is forced through constrictions in order to increase the turbulence and effect coalescence, and the enlarged droplets are then caused to impinge on a series of baffles.

The scrubber (Fig. 8.22) is fitted into an enlargement of the line carrying the gas, and is connected to the main portion of the duct by two conical sections. It consists of a number of perforated plates, divided into two groups, the agglomerator plates (Fig. 8.23) and the collector plates. There are two agglomerator plates, about 3 mm apart, which are drilled with a series of orifices at 9·5–12·5 mm centres. The first is the orifice plate which has 3 mm holes, and the other is the impact plate which has 6 mm holes which are staggered so that there is no straight run through the two plates and the direction of flow is abruptly changed. Although the main function of these plates is to effect coalescence of the drops, some liquid does separate from the gas at this stage and is run off from the separator. Most of the separation occurs at the three collector plates which are drilled with 2 mm holes (not staggered) and separated by 3 mm. These plates have the maximum number of holes consistent with adequate mechanical strength. The liquid which is separated here runs down over the surface of the plates as a thin film which acts as a very effective scrubbing agent.

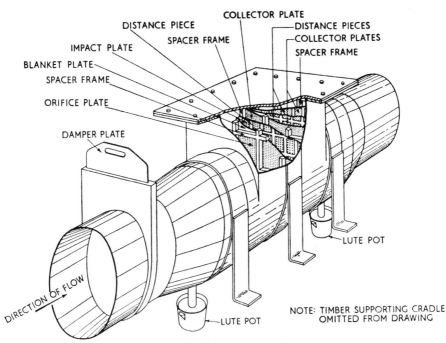

FIG. 8.22. The Calder–Fox scrubber

The final design of the Calder–Fox scrubber has been based on extensive preliminary investigations, during the course of which it was found that a better performance was obtained, for a given pressure drop, by using a high velocity through a short constriction rather than a lower velocity through a long constriction. In practice the velocity in the orifices is maintained at about 30 m/s by using a blanket plate when the throughput is

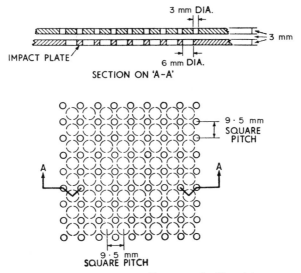

FIG. 8.23. Arrangement of impact and orifice plates

reduced. The minimum size of particle which can be removed is found to be directly proportional to the distance between the orifice and impact plates, and inversely proportional to the square root of the pressure drop. A limit is set to the maximum velocity at the impact plate, not only by the permissible pressure drop, but also by the tendency for re-entrainment: this is generally greatest with liquids of low viscosities.

The Calder–Fox scrubber is widely used for arresting sulphuric acid spray; scrubbers for this purpose are usually constructed of lead. When used in conjunction with hydrochloric acid plant, the orifices are formed with glass bars fixed in a corrosion resistant frame. Droplets down to about 2 μm in diameter can be removed from the gas stream.

## 8.3. FURTHER READING

DORMAN, R. G.: *Dust Control and Air Cleaning* (Pergamon Press, Oxford, 1974).
NONHEBEL, G. (Ed.): *Gas Purification Processes for Air Pollution Control*, 2nd edn. (Newnes–Butterworths, London, 1972).
WHITE, H. J.: *Industrial Electrostatic Precipitation* (Addison-Wesley Publ. Co. Inc. and Pergamon Press, London, 1963).

## 8.4. REFERENCES

1. ASHMAN, R.: *Proc. Inst. Mech. Eng.* **1B** (1952) 157. Control and recovery of dust and fume in industry.
2. STAIRMAND, C. J.: in *Gas Purification Processes for Air Pollution Control*, ed. G. NONHEBEL, 2nd edn. (Newnes–Butterworths, London, 1972).
3. TER LINDEN, A. J.: *Proc. Inst. Mech. Eng.* **160** (1949) 233. Cyclone dust collectors.
4. STAIRMAND, C. J.: *Engineering* **168** (1949) 409. Pressure drop in cyclone separators.
5. STAIRMAND, C. J.: *J. Inst. Fuel* **29** (1956) 58. The design and performance of modern gas cleaning equipment.
6. SZEKELY, J. and CARR, R.: *Chem. Eng. Sci.* **21** (1966) 1119. Heat transfer in a cyclone.
7. DORMAN, R. G. and MAGGS, F. A. P.: *Chem. Engnr.* (Oct. 1976) 671. Filtration of fine particles and vapours from gases.
8. HERSEY, H. J., Jr.: *Ind. Chem. Mfr.* **31** (1955) 138. Reverse-jet filters.
9. DALLAVALLE, J. M.: *Micromeritics*, 2nd edn. (Pitman, 1948).
10. WHITE, S. T.: *Heat Vent.* **45** (Sept. 1948) 59. Inaudible sound: a new tool for air cleaning.
11. LOWRIE-FAIRS, G.: *Trans. Inst. Chem. Eng.* **22** (1944) 110; *Proc. Chem. Eng. Gp.* **26** (1944) 110. Calder–Fox scrubbers and the factors influencing their performance.

## 8.5. NOMENCLATURE

|  |  | Units in SI System | Dimensions in **MLT** |
|---|---|---|---|
| $A_i$ | Cross-sectional area of gas inlet to cyclone separator | $m^2$ | $L^2$ |
| $d$ | Diameter of sphere or characteristic dimension of particle | m | L |
| $d_t$ | Diameter of cyclone body | m | L |
| $d_0$ | Diameter of outlet of cyclone separator | m | L |
| $G$ | Mass rate of flow of fluid | kg/s | $MT^{-1}$ |
| $g$ | Acceleration due to gravity | $m^2/s$ | $LT^{-2}$ |
| $m$ | Mass of particle | kg | M |
| $r$ | Radius at which particle rotates | m | L |
| $u_r$ | Radial component of velocity of gas in separator | m/s | $LT^{-1}$ |
| $u_t$ | Tangential component of velocity of gas in separator | m/s | $LT^{-1}$ |
| $u_{t0}$ | Tangential component of gas velocity at circumference | m/s | $LT^{-1}$ |
| $u_0$ | Free-falling velocity of particle | m/s | $LT^{-1}$ |
| $Z$ | Total height of separator | m | L |
| $\beta$ | Angle of entry into cyclone separator | — | — |
| $\mu$ | Viscosity of fluid | $N\,s/m^2$ | $ML^{-1}T^{-1}$ |
| $\rho$ | Density of fluid | $kg/m^3$ | $ML^{-3}$ |
| $\rho_s$ | Density of solid | $kg/m^3$ | $ML^{-3}$ |

# CHAPTER 9

## *Filtration*

### 9.1. INTRODUCTION

The separation of solids from a suspension in a liquid by means of a porous medium or screen which retains the solids and allows the liquid to pass is termed filtration.

In general, the pores of the medium will be larger than the particles which are to be removed, and the filter will work efficiently only after an initial deposit has been trapped in the medium. In the chemical laboratory, filtration is often carried out in a form of Buchner funnel, and the liquid is sucked through the thin layer of particles using a source of vacuum: in even simpler cases the suspension is poured into a conical funnel fitted with a filter paper. In the industrial equivalent of such an operation difficulties are involved in the mechanical handling of much larger quantities of suspension and solids. A thicker layer of solids has to form and, in order to achieve a high rate of passage of liquid through the solids, higher pressures will be needed, and it will be necessary to provide a far greater area. A typical filtration operation is illustrated in Fig. 9.1, which shows the filter medium, in this case a cloth, its support and the layer of solids, or filter cake, which has already formed.

The volumes of the suspensions to be handled will vary from the extremely large quantities involved in water purification and ore handling in the mining industry to relatively small quantities in the fine chemical industry where the variety of solids will be considerable. In most instances in the chemical industry it is the solids that are wanted and their physical size and properties are of paramount importance. Thus the main factors to be considered when selecting equipment and operating conditions are:

(a) The properties of the fluid, particularly its viscosity, density and corrosive properties.

(b) The nature of the solid—its particle size and shape, size distribution, and packing characteristics.

(c) The concentration of solids in suspension.

(d) The quantity of material to be handled, and its value.

(e) Whether the valuable product is the solid, the fluid, or both.

(f) Whether it is necessary to wash the filtered solids.

(g) Whether very slight contamination caused by contact of the suspension or filtrate with the various components of the equipment is detrimental to the product.

(h) Whether the feed liquor may be heated.

(i) Whether any form of pretreatment will be helpful.

Filtration is essentially a mechanical operation and is less demanding in energy than evaporation or drying where the high latent heat of the liquid, which is usually water, has to be provided. In the typical operation shown in Fig. 9.1 the cake gradually builds up on the medium and the resistance to flow progressively increases. During the initial period of flow particles are deposited in the surface layers of the cloth to form the true filtering

321

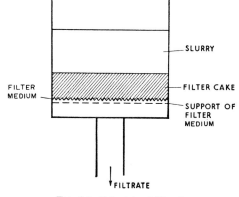

FIG. 9.1.  Principle of filtration

medium. This initial deposit may be formed from a special initial flow of precoat material which is discussed later. The most important factors on which the rate of filtration then depends will be:

(a) The drop in pressure from the feed to the far side of the filter medium.
(b) The area of the filtering surface.
(c) The viscosity of the filtrate.
(d) The resistance of the filter cake.
(e) The resistance of the filter medium and initial layers of cake.

The type of filtration described above is usually referred to as *cake filtration*; the proportion of solids in the suspension is large and most of the particles are collected in the filter cake which can subsequently be detached from the medium. Where the proportion of solids is very small, as for example in air or water filtration, the particles will often be considerably smaller than the pores of the filter medium and will penetrate a considerable depth before being captured; such a process is called *deep bed filtration*. In this chapter, attention will be focused on *cake filtration*, except in the short section on bed filters which provides an example of *deep bed filtration*—a subject which is treated in detail by Ives[1,2].

## 9.2. THE THEORY OF FILTRATION

In Chapter 4, equations were obtained for the calculation of the rate of flow of a fluid through a bed of granular material, and these will now be applied to the flow of filtrate through the filter cake. Some differences in general behaviour may be expected, however, because the cases so far considered relate to uniform fixed beds, whereas in filtration the bed is steadily growing in thickness. Thus if the filtration pressure is constant, the rate of flow will progressively diminish, whereas if the flowrate is to be maintained constant, the pressure must be gradually increased.

The mechanical details of the equipment, particularly of the flow channel and the support for the medium, influence the way the cake is built up and the ease with which it may be removed. A uniform structure is very desirable for good washing and cakes formed from particles of very mixed sizes and shapes present special problems. Although filter cakes are complex in their structure and cannot truly be regarded as composed of rigid non-deformable particles, the method of relating the flow parameters developed in

Chapter 4 is useful in describing the flow within the filter cake. For recent ideas on the theory of filtration and their importance in design, the reader is referred to a paper by Suttle[3]. It must be noted that there are two quite different methods of operating a batch filter: if the pressure is kept constant then the rate of flow will progressively diminish, whereas if the flowrate is to be kept constant then the pressure must be gradually increased. Because the particles forming the cake are small and the flow through the bed is slow, streamline conditions are almost invariably obtained, and therefore at any instant may be represented as (see equation 4.9):

$$u = \frac{1}{A}\frac{dV}{dt} = \frac{e^3}{5(1-e)^2 S^2}\frac{-\Delta P}{\mu l} \tag{9.1}$$

In this equation, $V$ is the volume of filtrate which has passed in time $t$, $A$ is the total cross-sectional area of the filter cake, $u$ is the superficial velocity of the filtrate, $l$ is the cake thickness, $S$ is the specific surface of the particles, $e$ is the voidage, $\mu$ is the viscosity of the filtrate, and $\Delta P$ is the applied pressure difference.

In deriving this equation it is assumed that the cake is uniform and that the voidage is constant throughout. In the deposition of a filter cake this is unlikely to apply and $e$ will depend on the nature of the support, including its geometry and surface structure, and on the rate of deposition. The initial stages in the formation of the cake are therefore of special importance for the following reasons:

(a) For any filtration pressure, the rate of flow is greatest at the beginning of the process since the resistance is then a minimum.

(b) High initial rates of filtration may result in plugging of the pores of the filter cloth and cause a very high resistance to flow.

(c) The orientation of the particles in the initial layers can appreciably influence the structure of the whole filter cake.

Filter cakes can be divided into two classes, incompressible cakes and compressible cakes. In the case of an incompressible cake, the resistance to flow of a given volume of cake is not appreciably affected either by the pressure difference across the cake or by the rate of deposition of material. On the other hand, with a compressible cake increase of the pressure difference or of the rate of flow causes the formation of a denser cake with a higher resistance. For incompressible cakes $e$ in equation 9.1 can be taken as constant and the quantity $e^3/[5(1-e)^2 S^2]$ is then a property of the particles forming the cake and should be constant for a given material.

Therefore,

$$\frac{1}{A}\frac{dV}{dt} = \frac{-\Delta P}{r\mu l} \tag{9.2}$$

where

$$r = \frac{5(1-e)^2 S^2}{e^3} \tag{9.3}$$

Equation 9.2 is the basic filtration equation and $r$ is termed the specific resistance. It is seen to depend on $e$ and $S$. For incompressible cakes it is taken as constant, but it will depend on rate of deposition, nature of particles, and on forces between the particles.

### 9.2.1. Relation between Thickness of Cake and Volume of Filtrate

In equation 9.2, the variables $l$ and $V$ are connected, and the relation between them can be obtained by making a material balance between the solids in the slurry and in the cake.

Mass of solids in filter cake $= (1-e)Al\rho_s$ (where $\rho_s$ is the density of the solids)
Mass of liquid retained in the filter cake $= eAl\rho$ (where $\rho$ is the density of the filtrate)
Then if $J$ is the mass fraction of solids in the original suspension:

$$(1-e)lA\rho_s = \frac{(V+eAl)\rho J}{1-J}$$

i.e.

$$(1-J)(1-e)Al\rho_s = JV\rho + AeJl\rho$$

so that

$$l = \frac{JV\rho}{A\{(1-J)(1-e)\rho_s - Je\rho\}} \qquad (9.4)$$

and

$$V = \frac{\{\rho_s(1-e)(1-J) - e\rho J\}Al}{\rho J} \qquad (9.5)$$

If $v$ is the volume of cake deposited by unit volume of filtrate then:

$$v = \frac{lA}{V} \quad \text{or} \quad l = \frac{vV}{A} \qquad (9.6)$$

and from equation 9.5:

$$v = \frac{J\rho}{(1-J)(1-e)\rho_s - Je\rho} \qquad (9.7)$$

Substituting for $l$ in equation 9.2:

$$\frac{1}{A}\frac{dV}{dt} = \frac{(-\Delta P}{r\mu}\frac{A}{vV}$$

or

$$\frac{dV}{dt} = \frac{A^2(-\Delta P)}{r\mu vV} \qquad (9.8)$$

Equation 9.8 can be regarded as the basic relation between $(-\Delta P)$, $V$, and $t$. Two important types of operation will now be considered: (i) where the pressure difference is maintained constant and (ii) where the rate of filtration is maintained constant.

*For a filtration at constant rate*

$$\frac{dV}{dt} = \frac{V}{t} = \text{constant}$$

so that

$$\frac{V}{t} = \frac{A^2(-\Delta P)}{r\mu V v} \tag{9.9}$$

i.e.

$$\frac{t}{V} = \frac{r\mu v}{A^2(-\Delta P)} V \tag{9.10}$$

and $(-\Delta P)$ is directly proportional to $V$.

*For a filtration at constant pressure difference*

$$\frac{V^2}{2} = \frac{A^2(-\Delta P)t}{r\mu v} \tag{9.11}$$

i.e.

$$\frac{t}{V} = \frac{r\mu v}{2A^2(-\Delta P)} V \tag{9.12}$$

Thus for a constant pressure filtration, there is a linear relation between $V^2$ and $t$ or between $t/V$ and $V$.

Filtration at constant pressure is more frequently adopted in practical conditions but it must be remembered that the pressure difference is normally gradually built up to its ultimate value; if this takes a time $t_1$ during which a volume $V_1$ of filtrate passes then on integration of equation 9.12:

$$\tfrac{1}{2}(V^2 - V_1^2) = \frac{A^2(-\Delta P)}{r\mu v}(t - t_1) \tag{9.13}$$

i.e.

$$\frac{t - t_1}{V - V_1} = \frac{r\mu v}{2A^2(-\Delta P)}(V - V_1) + \frac{r\mu v V_1}{A^2(-\Delta P)} \tag{9.14}$$

Thus there is a linear relation between $V^2$ and $t$ and between $(t - t_1)/(V - V_1)$ and $V - V_1$. Here $t - t_1$ represents the time of the constant pressure filtration and $V - V_1$ the corresponding volume of filtrate obtained.

Ruth[4-7] has made measurements on the flow in a filter cake and has concluded that the resistance is somewhat greater than that indicated by equation 9.1; he assumes that part of the pore space is rendered ineffective for the flow of filtrate because of the adsorption of ions on the surface of the particles. This is not borne out by Grace[8] or by Hoffing and Lockhart[9] who determined the relation between flowrate and pressure difference, both by means of permeability tests on a fixed bed and by filtration tests using suspensions of quartz and diatomaceous earth.

### 9.2.2. Flow of Liquid through the Cloth

Experimental work[10] on the flow of the liquid under streamline conditions has shown that the flowrate is directly proportional to the pressure difference. It is the resistance of the cloth plus initial layers of deposited particles that is important since the latter not only form the true medium but also tend to block the pores of the cloth thus increasing its resistance. Cloths may have to be discarded because of high resistance well before they are mechanically worn. No true analysis of the buildup of resistance is possible because the resistance will depend on the way in which the pressure is developed and small variations in support geometry can have an important influence. It is therefore usual to combine the resistance of the cloth with that of the first few layers of particles and suppose that this corresponds to a thickness $L$ of cake as deposited at a later stage. The resistance to flow through the cake and cloth combined is considered in the next section.

### 9.2.3. Flow of Filtrate through the Cloth and Cake Combined

Suppose the filter cloth and the initial layers of cake to be equivalent to a thickness $L$ of cake as deposited at a later stage in the process. Then if $-\Delta P$ is the pressure drop across the cake and cloth combined:

$$\frac{1}{A}\frac{dV}{dt} = \frac{(-\Delta P)}{r\mu(l+L)}$$

(9.15)

(compare equation 9.2)

i.e.

$$\frac{dV}{dt} = \frac{A(-\Delta P)}{r\mu\left(\dfrac{Vv}{A}+L\right)} = \frac{A^2(-\Delta P)}{vr\mu\left(V+\dfrac{LA}{v}\right)}$$

(9.16)

This equation can be integrated between the limits $t = 0$, $V = 0$ and $t = t_1$, $V = V_1$ for

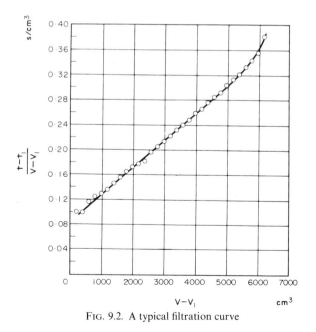

FIG. 9.2. A typical filtration curve

constant rate filtration, and $t = t_1$, $V = V_1$ and $t = t$, $V = V$ for a subsequent constant pressure filtration.

For the period of constant rate filtration:

$$\frac{V_1}{t_1} = \frac{A^2(-\Delta P)}{vr\mu\left(V_1 + \dfrac{LA}{v}\right)}$$

i.e.

$$\frac{t_1}{V_1} = \frac{r\mu v}{A^2(-\Delta P)}V_1 + \frac{r\mu L}{A(-\Delta P)}$$

i.e.

$$V_1^2 + \frac{LA}{v}V_1 = \frac{A^2(-\Delta P)}{r\mu v}t_1 \tag{9.17}$$

For a subsequent constant pressure filtration:

$$\tfrac{1}{2}(V^2 - V_1^2) + \frac{LA}{v}(V - V_1) = \frac{A^2(-\Delta P)}{r\mu v}(t - t_1) \tag{9.18}$$

i.e.

$$(V - V_1 + 2V_1)(V - V_1) + \frac{2LA}{v}(V - V_1) = \frac{2A^2(-\Delta P)}{r\mu v}(t - t_1)$$

i.e.

$$\frac{t - t_1}{V - V_1} = \frac{r\mu v}{2A^2(-\Delta P)}(V - V_1) + \frac{r\mu v V_1}{A^2(-\Delta P)} + \frac{r\mu L}{A(-\Delta P)} \tag{9.19}$$

Thus there is a linear relation between $(t - t_1)/(V - V_1)$ and $V - V_1$ (as shown in Fig. 9.2) and the slope is proportional to the specific resistance, as in the case of the flow of the filtrate through the filter cake alone (cf. equation 9.14), but the line does not go through the origin.

The intercept on the $(t - t_1)/(V - V_1)$ axis should enable $L$, the equivalent thickness of the cloth, to be calculated but reproducible results are not obtained because this resistance is critically dependent on the exact manner in which the operation is commenced. The time at which measurement of $V$ and $t$ is commenced does not affect the slope of the curve, only the intercept. It will be noted that a linear relation between $t$ and $V^2$ is no longer obtained when the cloth resistance is appreciable.

### 9.2.4. Compressible Filter Cakes

Nearly all filter cakes are compressible to some extent but in many cases the degree of compressibility is so small that the cake can, for practical purposes, be regarded as incompressible. The evidence for compressibility is that the specific resistance is a function of the pressure difference across the cake. Compressibility may be a reversible or an irreversible process. Most filter cakes are inelastic and the greater resistance offered to flow at high pressure differences is caused by the more compact packing of the particles forming the filter cake. Thus the specific resistance of the cake will correspond to that for the highest pressure difference to which the cake is subjected, even though this maximum pressure difference may be maintained for only a short time. It is therefore important that the filtration pressure should not be allowed to exceed the normal operating pressure at any stage. In elastic filter cakes the elasticity is attributable to compression of the particles themselves: they are rarely encountered, but some forms of carbon can give rise to elastic cakes.

Because of the frictional losses arising from the flow of filtrate through the cake, there will be a fluid pressure gradient. The particles will exert a retarding action on the fluid but the fluid will exert a compressing action on the particles and therefore tend to consolidate the cake. Since the force exerted by the fluid on each particle is transmitted to the particles deeper in the cake, the compressive force varies from a minimum of zero at the free surface of the cake to a maximum at the filter medium. The actual compressive pressure will depend on the structure of the cake and the nature of the contacts between the particles, but it can be expressed as a function of the difference between the pressure at the surface of the cake $P_1$ and that at a depth $z$ in the cake $P_z$ (see Fig. 9.3). Thus since the voidage $e_z$ at any point is a function of the compressive force at that point $e_z$ is a function of $(P_1 - P_z)$.

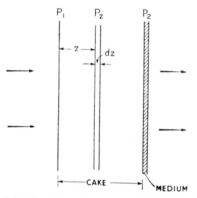

FIG. 9.3. Flow through a compressible filter cake

The voidage thus decreases from the free surface to the filter medium and the resistance shows a corresponding increase. If the filtration surface is horizontal and facing upwards, the effect of the fluid drag will be supplemented by gravity. Since the resistance varies throughout the depth of the cake, the filtration equation 9.2 must be expressed in differential form:

$$\frac{1}{A}\frac{dV}{dt} = \frac{1}{r_z\mu}\left(-\frac{dP_z}{dz}\right) \tag{9.20}$$

where $r_z$ is the point value of the specific resistance. This equation may be written as:

$$\frac{1}{A}\frac{dV}{dt} = \frac{e_z^3}{5(1-e_z)^2 S^2}\frac{1}{\mu}\left(-\frac{dP_z}{dz}\right) \tag{9.21}$$

The small element of cake thickness $dz$ must now be related to the small volume $dV$ of filtrate passing during its formation. As the cake is compressible, the volume of cake deposited by the flow of unit volume of filtrate will not be constant, but the mass of solids deposited $c$ will be almost independent of the conditions under which the cake is formed. Thus:

$$dz = dV \frac{c}{(1-e_z)\rho_s A} \qquad (9.22)$$

Thus, from equation 9.21:

$$\frac{1}{A}\frac{dV}{dt} = \frac{e_z^3}{5(1-e_z)^2 S^2} \frac{(1-e_z)\rho_s A}{c} \frac{1}{\mu}\left(-\frac{dP_z}{dz}\right)$$

$$\therefore \quad \frac{dV}{dt} = \frac{e_z^3 \rho_s}{5(1-e_z)S^2} \frac{A^2}{\mu c}\left(-\frac{dP_z}{dz}\right) \qquad (9.23)$$

$$= \frac{A^2}{\mu c \bar{r}_z}\left(-\frac{dP_z}{dz}\right) \qquad (9.24)$$

where

$$\bar{r}_z = \frac{5(1-e_z)S^2}{e_z^3 \rho_s} \qquad (9.25)$$

The specific resistance term $\bar{r}$ is based on the flow of filtrate through unit mass of cake deposited on unit area, whereas $r$ was based on the flow through a unit cube of cake. Comparison of equations 9.2 and 9.24 shows that for an incompressible cake:

$$c\bar{r} = vr$$

i.e.

$$\bar{r} = r\frac{v}{c} \qquad (9.26)$$

The dimensions of $\bar{r}$ are therefore $\mathbf{L}^{-2} \times \mathbf{L}^3 \mathbf{M}^{-1} = \mathbf{M}^{-1}\mathbf{L}$. For any instant in a constant pressure filtration, on integration through the whole depth of the cake:

$$\int_0^V \frac{dV}{dt}\,dV = \frac{A^2}{\mu c}\int_{P_1}^{P_2}\frac{(-dP_z)}{\bar{r}_z}$$

At any instant, $dV/dt$ is constant throughout the cake and therefore:

$$\frac{dV}{dt} = \frac{A^2}{V\mu c}\int_{P_1}^{P_2}\frac{(-dP_z)}{\bar{r}_z} \qquad (9.27)$$

Now $\bar{r}_z$ has been shown to be a function of the pressure difference $P_1 - P_z$, but is independent of the absolute value of the pressure. Suppose that:

$$\bar{r}_z = \bar{r}'(P_1 - P_z)^{n'} \qquad (9.28)$$

where $\bar{r}'$ is independent of $P_z$. Then:

$$\int_{P_1}^{P_2}\frac{(-dP_z)}{\bar{r}_z} = \frac{1}{\bar{r}'}\int_{P_2}^{P_1}\frac{dP}{(P_1 - P_z)^{n'}}$$

$$= \frac{1}{\bar{r}'}\frac{(P_1 - P_2)^{1-n'}}{1-n} \qquad (n' < 1)$$

$$= \frac{1}{\bar{r}'}\frac{(-\Delta P)^{1-n'}}{1-n} \qquad (9.29)$$

Thus:

$$\frac{\mathrm{d}V}{\mathrm{d}t} = \frac{A^2}{V\mu c\bar{r}'} \frac{(-\Delta P)}{(1-n')(-\Delta P)^{n'}}$$

$$= \frac{A^2(-\Delta P)}{V\mu c\bar{r}''(-\Delta P)^{n'}} \qquad (9.30)$$

where $\bar{r}'' = (1-n')\bar{r}'$

$$= \frac{A^2(-\Delta P)}{V\mu c\bar{r}} \qquad (9.31)$$

where $\bar{r}$ is the mean resistance as defined by equation 9.30, i.e.

$$\bar{r} = \bar{r}''(-\Delta P)^{n'} \qquad (9.32)$$

Heertjes[11] has studied the effect of pressure on the porosity of a filter cake and suggests that, as the pressure is increased above atmospheric, the porosity decreases in proportion to some power of the excess pressure.

Grace[12] has started the important task of relating the anticipated resistance to the physical properties of the feed slurry. Valleroy and Maloney[13] have examined the resistance of an incompressible bed of spherical particles when measured in a permeability cell, a vacuum filter, and a centrifuge, and emphasise the need for caution in applying laboratory data to units of different geometry.

Tiller and Huang[14] give further details of the problem of developing a usable design relationship for filter equipment. Recent work by Tiller[15] and Rushton[16] all show the difficulty in presenting practical conditions in a way which can be used analytically. It is very important to note that tests on slurries must be made with equipment that is geometrically similar to that proposed. This is a relatively new concept and means that specific resistance is very difficult to define in practice, since it will be determined by the nature of the filtering unit and the way in which the cake is initially formed and then built up.

## 9.3. FILTRATION PRACTICE

### 9.3.1. The Filter Medium

The function of the filter medium is generally to act as a support for the filter cake, while the initial layers of cake provide the true filter. The filter medium should be mechanically strong, resistant to the corrosive action of the fluid, and offer as little resistance as possible to the flow of filtrate. Woven materials are commonly used but granular materials and porous solids are useful for filtration of corrosive liquids in batch units. An important feature in the selection of a woven material is the ease of cake removal, since this is a key factor in the operation of modern automatic units. Ehlers[17] has discussed the selection of woven synthetic materials and Wrotnowski[18] that of non-woven materials. Further details of some of the newer materials are given in the literature[19].

## 9.3.2. Blocking Filtration

In the theory which has been presented previously it is assumed that there is a well-defined boundary between the filter cake and the filter cloth. The initial stages in the build-up of the filter cake are important, however, because they may have a large effect on the flow resistance and may seriously affect the useful life of the cloth.

The blocking of the pores of the filter medium by particles is an involved phenomenon, partly because of the very complex nature of the surface structure of the usual types of filter media and partly because the lines of movement of the particles are not well defined. At the commencement of filtration the manner in which the cake forms will lie between two extremes: the penetration of the pores by particles and the shielding of the entry to the pores by the particles forming bridges. Heertjes[11] considered a number of idealised cases in which suspensions of specified pore size distributions were filtered on a cloth with a regular pore distribution. First, it was assumed that an individual particle was capable on its own of blocking a single pore. Then, as filtration proceeded, successive pores would be blocked and the apparent value of the specific resistance of the filter cake would depend on the amount of solids deposited.

The pore and particle size distributions might, however, be such that more than one particle could enter a particular pore. In this case, the resistance of the pore would increase in stages as successive particles were trapped until the pore was completely blocked. In practice, however, it is much more likely that many of the pores will never become completely blocked and a cake of relatively low resistance will form over the entry to the partially blocked pore.

One of the most important variables affecting the tendency for blocking is the concentration of particles. The greater the concentration, the smaller will be the average distance between the particles, and the smaller will be the tendency for the particle to be drawn in to the streamlines directed towards the open pores. Instead, the particles in the concentrated suspension tend to distribute themselves fairly evenly over the filter surface and form bridges. As a result, suspensions of high concentration generally give rise to cakes of lower resistance than those formed from dilute suspensions.

## 9.3.3. Effect of Particle Sedimentation on Filtration

There are two important effects due to particle sedimentation which may affect the rate of filtration. First, if the sediment particles are all settling at approximately the same rate, as frequently occurs in a concentrated suspension in which the particle size distribution is not very wide, a more rapid build-up of particles will occur on an upward-facing surface and a correspondingly reduced rate of build-up will take place if the filter surface is facing downwards. Thus, there will be a tendency for accelerated filtration with downward-facing filter surfaces and reduced filtration rates for upward-facing surfaces. On the other hand, if the suspension is relatively dilute, so that the large particles are settling at a higher rate than the small ones, there will be a preferential deposition of large particles on an upward-facing surface during the initial stages of filtration, giving rise to a low resistance cake. Conversely for a downward-facing surface, fine particles will initially be deposited preferentially and the cake resistance will be correspondingly increased. It is thus seen that there can be complex interactions where sedimentation is occurring at an appreciable rate and that the orientation of the filter surface is an important factor.

### 9.3.4. Delayed Cake Filtration

In the filtration of a slurry, the resistance of the filter cake progressively increases and consequently, in a constant pressure operation, the rate of filtration falls. If the build-up of solids can be reduced, the effective cake thickness will be less and the rate of flow of filtrate will be increased.

In practice, it is sometimes possible to incorporate moving blades in the filter equipment so that the thickness of the cake is limited to the clearance between the filter medium and the blades. Filtrate then flows through the cake at an approximately constant rate but the solids are retained in suspension. Thus the solids concentration in the feed vessel will increase until the particles are in permanent physical contact with one another. At this stage the boundary between the slurry and the cake becomes ill-defined and a significant resistance to the flow of liquid develops within the slurry itself with a consequent reduction in the flowrate of filtrate.

By the use of this technique, a much higher rate of filtration can be achieved than is possible in a filter operated in a conventional manner. Furthermore, the resulting cake usually has a lower porosity because the blades effectively break down the bridges or arches which give rise to a structure in the filter cake; thus the final cake is significantly drier.

If the scrapers are in the form of rotating blades, the behaviour differs according to whether they are moving at low or at high speed. At low speeds, the cake thickness is reduced to the clearance depth each time the scraper blade passes, but cake then builds up again until the next passage of the scraper. If the blade is operated at high speed, there is little time for solids to build up between successive passages of the blade and the cake reaches an approximately constant thickness. Since particles tend to be swept across the surface of the cake by the moving slurry they will be trapped in the cake only if the drag force which the filtrate exerts on them is great enough. As the thickness of the cake increases the pressure gradient becomes less and there is a smaller force retaining particles in the cake surface. Thus the thickness of the cake tends to reach an equilibrium value, which can be considerably less than the clearance between the medium and the blades.

Experimental results for the effect of stirrer speed on the rate of filtration of a 10 per cent by weight suspension of clay are shown in Fig. 9.4 in which the filtrate volume collected per unit cross-section of filter is plotted against time, for several stirrer speeds[20].

The concentration of solids in the slurry in the feed vessel to the filter at any time can be calculated by noting that the volumetric rate of feed of slurry must be equal to the rate at which filtrate leaves the vessel. For a rate of flow of filtrate of $dV/dt$ out of the filter the rate of flow of slurry into the vessel must also be $dV/dt$ and the corresponding influx of solids is $(1-e_0)dV/dt$, where $(1-e_0)$ is the volume fraction of solids in the feed slurry. At any time $t$, the volume of solids in the vessel is $\mathbf{V}(1-e_V)$, where $\mathbf{V}$ is its volume and $(1-e_V)$ is the volume fraction of solids at that time. Thus a material balance on the solids gives:

$$(1-e_0)\frac{dV}{dt} = \frac{d}{dt}\{\mathbf{V}(1-e_V)\},\tag{9.33}$$

$$\frac{d(1-e_V)}{dt} = \frac{1}{\mathbf{V}}(1-e_0)\frac{dV}{dt}\tag{9.34}$$

For a constant filtration rate $dV/dt$, the fractional solids hold-up $(1-e_V)$ increases linearly

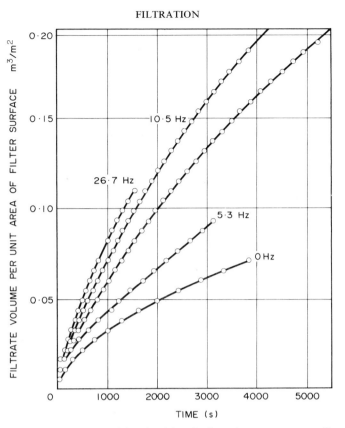

FIG. 9.4. Volume as function of time for delayed cake and constant pressure filtration

with time, until it reaches a limiting value when the resistance to flow of liquid within the slurry becomes significant. The filtration rate then drops rapidly to a near zero value.

### 9.3.5. Preliminary Treatment of Slurries before Filtration

If the slurry is dilute and the solid particles settle readily in the fluid, it may be desirable to effect a preliminary concentration in a thickener (see Chapter 5). The thickened suspension is then fed from the thickener to the filter and the quantity of material to be handled is thereby reduced.

The theoretical treatment has shown that the nature of the filter cake has a very pronounced effect on the rate of flow of filtrate and that it is, in general, desirable that the particles forming the filter cake should have as large a size as possible. More rapid filtration is therefore obtained if a suitable agent is added to the slurry to cause coagulation. If the solid material is formed in a chemical reaction by precipitation, the particle size can generally be controlled to a certain extent by the actual conditions of formation. For example, the particle size of the resultant precipitate can be controlled by varying the temperature and concentration, and sometimes the pH, of the reacting solutions. As indicated by Grace[8], a flocculated suspension gives rise to a more porous cake but the compressibility is greater. In many cases, crystal shape can be altered by adding traces of material which is selectively adsorbed on particular faces (see Chapter 14).

Filter aids are extensively used where the filter cake is relatively impermeable to the flow of filtrate. They are materials which pack to form beds of very high voidages and therefore they are capable of increasing the porosity of the filter cake if added to the slurry before filtration. Apart from economic considerations, there is an optimum quantity of filter aid which should be added in any given case. Whereas the presence of the filter aid reduces the specific resistance of the filter cake, it also results in the formation of a thicker cake. The actual quantity used will therefore depend on the nature of the material. The use of filter aids is normally restricted to operations in which the filtrate is valuable and the residue is a waste product. In some circumstances, however, the filter aid must be readily separable from the rest of the filter cake by physical or chemical means. Filter cakes incorporating filter aid are usually very compressible and care should therefore be taken to ensure that the good effect of the filter aid is not destroyed by employing too high a filtration pressure. Kieselguhr, which is a commonly used filter aid, has a voidage of about 0·85. Addition of relatively small quantities increases the voidage of most filter cakes; the resulting porosity normally lies between that of the filter aid and that of the filter solids. Sometimes the filter medium is "precoated" with filter aid, and a thin layer of the filter aid is removed with the cake at the end of each cycle.

In some cases the filtration time can be reduced by diluting the suspension in order to reduce the viscosity of the filtrate. This does, of course, increase the bulk to be filtered and is applicable only when the value of the filtrate is not affected by dilution.

### 9.3.6. Washing of the Filter Cake

When the wash liquid is miscible with the filtrate and has similar physical properties, the rate of washing at the same pressure difference will be about the same as the final rate of filtration. If the viscosity of the wash liquid is less, a somewhat greater rate will be obtained. However, channelling sometimes occurs, with the result that much of the cake is incompletely washed; the fluid passes preferentially through the channels, which are gradually enlarged by its continued passage. This does not occur during filtration because channels are self-sealing by virtue of deposition of solids from the slurry. Channelling is most marked with compressible filter cakes and can be minimised by using a smaller pressure difference for washing than for filtration.

Washing can be regarded as taking place in two stages. First, filtrate is displaced from the filter cake by wash liquid during the period of *displacement washing*; in this way up to 90 per cent of the filtrate may be removed. During the second stage, *diffusional washing*, solvent diffuses into the wash liquid from the less accessible voids and the following relation applies:

$$\frac{\text{volume of wash liquid passed}}{\text{cake thickness}} = \text{a constant} \times \log\frac{\text{initial concentration of solute}}{\text{concentration at particular time}} \tag{9.35}$$

Although an immiscible liquid is seldom used for washing, air is often used to effect partial drying of the filter cake. The rate of flow of air must normally be determined experimentally.

## 9.4. FILTRATION EQUIPMENT

The most suitable filter for any given operation is the one which will fulfil the requirements at the minimum overall cost. Since the cost of the equipment will be closely related to the filtering area, it is normally desirable to obtain a high overall rate of filtration. This involves the use of relatively high pressures but the maximum pressures are often limited by considerations of mechanical design. Although a higher throughput from a given filtering surface is obtained from a continuous filter than from a batch operated filter, it may sometimes be necessary to use a batch filter, particularly if the filter cake has a high resistance, because most continuous filters operate under reduced pressure and the maximum filtration pressure is therefore limited. Other features which are desirable in a filter include ease of discharge of the filter cake in a convenient physical form, and a method of observing the quality of the filtrate obtained from each section of the plant. These factors will be kept to the fore in the discussion of the types of equipment available. The most important types which will be referred to are filter presses, leaf filters, and continuous rotary filters. In addition, mention will be made of filters for special purposes, such as bag filters, and the disc type of filter which is used for the removal of small quantities of solids from a fluid.

The most important factors in filter selection are the specific resistance of the filter cake, the quantity to be filtered, and the solids concentration. For free-filtering materials, a rotary vacuum filter is generally the most satisfactory since it has a very high capacity for its size and does not require much manual attention. If the cake must be washed, the rotary drum is to be preferred to the rotary leaf. However, if a high degree of washing is required, it is usually desirable to repulp the filter cake and to filter a second time.

For large-scale filtration, there are three principal cases where a rotary vacuum filter will not be used. First, if the specific resistance is high, a positive pressure filter will be required; a filter press may well be suitable, particularly if the solid content is not so high that frequent dismantling of the press is necessary. Secondly, when efficient washing is required, a leaf filter is effective, because very thin cakes can be prepared and the risk of channelling during washing is reduced to a minimum. Finally, where only very small quantities of solids are present in the liquid, an edge filter may be employed.

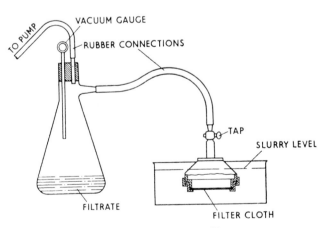

FIG. 9.5. Laboratory test filter

Whereas it may be possible to predict qualitatively the effect of the physical properties of the fluid and the solid on the filtration characteristics of a suspension, it is necessary in all cases to carry out a test on a sample before the large-scale plant can be designed. A simple vacuum filter which is used to obtain laboratory data is illustrated in Fig. 9.5; the area of the filter is 0·0065 m². The information on filtration rates and specific resistance obtained in this way can be directly applied to industrial fillers provided due account is taken of the compressibility of the filter cake. It cannot be stressed too strongly that data from any laboratory test cell must not be used without practical experience in the design of industrial units where the geometry of the flow channel is very different. The laying down of the cake influences the structure to such a marked extent.

A "compressibility–permeability" test cell has been developed by Ruth[7] and Grace[8] for testing the behaviour of slurries under various conditions of filtration.

### 9.4.1. Bed Filters

These filters provide an example of the application of the principles of *deep bed filtration* in which the particles penetrate into the interstices of the filter bed where they are trapped following impingement on the surfaces of the material of the bed.

For the purification of water supplies and for waste water treatment where the solid content is about 10 g/m³ or less, granular bed filters have largely replaced the former very slow sand filters[21]. They are formed from granular material of grain size 0·6–1·2 mm in beds of depth of 0·6–1·8 m. The very fine particles of solids are removed by mechanical action but the particles finally adhere as a result of surface electric forces or adsorption[22]. This operation has been analysed by Iwasaki[23] who gives the relation:

$$-\frac{\partial C}{\partial l} = \lambda C \tag{9.36}$$

On integration:

$$C/C_0 = e^{-\lambda l} \tag{9.37}$$

where    $C$ is the volume concentration of solids in suspension in the filter,
            $C_0$ is the value of $C$ at the surface of the filter,
             $l$ is the depth of the filter, and
             $\lambda$ is the filter coefficient.

If $u$ is the superficial flowrate of the slurry, the rate of flow of solids through the filter at depth $l$ is $uC$ per unit area. Thus the rate of accumulation of solids in a distance $dl = -u(\partial C/\partial l)dl$. If $\sigma$ is the volume of solids deposited per unit volume of filter at a depth $l$, the rate of accumulation can also be expressed as $(\partial \sigma/\partial t) dl$. Thus

$$-\frac{\partial C}{\partial l} = u\frac{\partial \sigma}{\partial t} \tag{9.38}$$

The problem is discussed further by Ives[22] and by Spielman and Friedlander[24]. The backwashing of these beds has presented problems and several techniques have been adopted. These include backflow of air followed by water, the flowrate of which may be high enough to give rise to fluidisation, the maximum hydrodynamic shear occurring at a voidage of about 0·7.

## 9.4.2. Bag Filters

Bag filters have now been almost entirely superseded for liquid filtration by other types of filter, but one of the few remaining types is the Taylor bag filter which has been widely used in the sugar industry. A number of long thin bags are attached to a horizontal feed tray and the liquid flows under the action of gravity so that the rate of filtration per unit area is very low; it is possible, however, to arrange a large filtering area in the plant, e.g. up to about 700 m². The filter is usually arranged in two sections so that each can be inspected separately without interrupting the operation.

Bag filters are still extensively used for the removal of dust particles from gases and can be operated either as pressure filters or as suction filters. Their use is discussed in Chapter 8.

## 9.4.3. The Filter Press

The filter press is made in two main forms, the plate and frame press and the recessed plate or chamber press.

*The Plate and Frame Press*

This type of filter consists of plates and frames arranged alternately and supported on a pair of rails as shown in Fig. 9.6. The plate has a ribbed surface (Fig. 9.7) and the edge stands slightly proud and is carefully machined. The hollow frame is separated from the plate by the filter cloth, and the press is closed either by means of a hand screw or hydraulically; the minimum pressure should be used in order to reduce wear on the cloths. A chamber is therefore formed between each pair of successive plates (Fig. 9.8). The slurry is introduced through a port in each frame and the filtrate passes through the cloth on each side so that two cakes are formed simultaneously in each chamber, and these join when the frame is full. The frames are usually square and may be between 100 mm and 1·5 m across and from 10 mm to 75 mm thick.

The slurry may be fed to the press through the continuous channel formed by the holes in the corners of the plates and frames, in which case it is necessary to cut corresponding holes in the cloths which themselves act as gaskets. Cutting of the cloth can be avoided by feeding through a channel at the side but rubber bushes must then be fitted so that a leak-tight joint is formed.

The filtrate runs down the ribbed surface of the plates and then is discharged through a cock into an open launder so that the filtrate from each plate can be inspected and any plate can be isolated if it is not giving a clear filtrate. In some cases the filtrate is removed through a closed channel but it is not then possible to observe the discharge from each plate separately.

In many filter presses, provision is made for steam heating so that the viscosity of the filtrate is reduced and a higher rate of filtration obtained. Materials, such as waxes, that solidify at normal temperatures can also be filtered in steam-heated presses. Steam heating also facilitates the production of a dry cake.

*Optimum time cycle.* The optimum thickness of cake to be formed in a filter press depends on the resistance offered by the filter cake and on the time taken to dismantle and

FIG. 9.6. Filter presses in a large chemical works

FIG. 9.7. Surface of filter plate

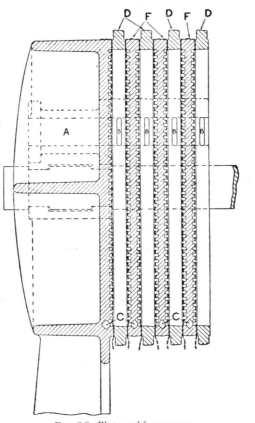

FIG. 9.8. Plate and frame press
A—inlet passage. B—feed ports. C—filtrate outlet. D—frames. F—plates

refit the press. Although the production of a thin filter cake results in a high average rate of filtration, it is necessary to dismantle the press more often and a greater time is therefore spent on this operation. For a filtration carried out entirely at constant pressure:

$$\frac{t}{V} = \frac{r\mu v}{2A^2(-\Delta P)} V + \frac{r\mu L}{A(-\Delta P)} \qquad (9.39)$$

(from equation 9.19)

$$= B_1 V + B_2 \qquad (9.40)$$

where $B_1$ and $B_2$ are constant.

Thus the time of filtration $t$ is given by:

$$t = B_1 V^2 + B_2 V \qquad (9.41)$$

The time of dismantling and assembling the press ($t'$, say) is substantially independent of the thickness of cake produced. The total time of a cycle in which a volume $V$ of filtrate is collected is then $t + t'$ and the overall rate of filtration is given by:

$$W = \frac{V}{B_1 V^2 + B_2 V + t'}$$

$W$ is a maximum when $dW/dV = 0$.

Differentiating $W$ with respect to $V$ and equating to zero:

$$B_1 V^2 + B_2 V + t' - V(2B_1 V + B_2) = 0$$

i.e.

$$t' = B_1 V^2 \qquad (9.42)$$

i.e.

$$V = \sqrt{\frac{t'}{B_1}} \qquad (9.43)$$

If the resistance of the filter medium is neglected, $t = B_1 V^2$ and the time during which filtration is carried out is exactly equal to the time the press is out of service. In practice, in order to obtain the maximum overall rate of filtration, the filtration time must always be somewhat greater in order to allow for the resistance of the cloth (represented by the term $B_2 V$). In general, the lower the specific resistance of the cake, the greater will be the economic thickness of the frame.

*Washing.* Two methods of washing can be employed, "simple" washing and "through" or "thorough" washing. With simple washing, the wash liquid is fed in through the same channel as the slurry but, as its velocity near the point of entry is high, erosion of the cake takes place. The channels which are thus formed gradually enlarge and uneven washing is usually obtained. It can be used only when the frame is not completely full.

In thorough washing, the wash liquid is introduced through a separate channel behind the filter cloth on alternate plates, known as washing plates (Fig. 9.9), and flows through the whole thickness of the cake, first in the opposite direction and then in the same direction as the filtrate. The area during washing is one-half of that during filtration and, in addition, the wash liquid has to flow through twice the thickness, so that the rate of

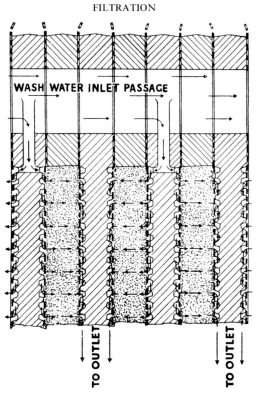

FIG. 9.9. Thorough washing

washing should therefore be about one-quarter of the final rate of filtration. The wash liquid is usually discharged through the same channel as the filtrate though sometimes a separate outlet is provided. Even with thorough washing some channelling occurs and several inlets are often provided so that the liquid is well distributed. If the cake is appreciably compressible, the minimum pressure should be used during washing, and in no case should the final filtration pressure be exceeded. After washing, the cake may be made easier to handle by removing excess liquid with compressed air.

For ease in identification, small buttons are embossed on the sides of the plates and frames, one of the non-washing plates, two on the frames and three on the washing plates (Fig. 9.10).

*The Chamber Press*

The chamber press (Fig. 9.11) is similar to the plate and frame type except that the use of frames is obviated by recessing the ribbed surface of the plates so that the individual filter chambers are formed between successive plates. In this type of press therefore the thickness of the cake cannot be varied.

The feed channel (Fig. 9.12) usually differs from that employed on the plate and frame press. All the chambers are connected by means of a comparatively large hole in the centre of each of the plates and the cloths are secured in position by means of screwed unions. Slurries containing relatively large solid particles can readily be handled in this type of

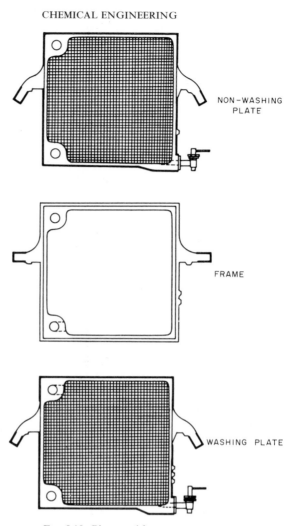

FIG. 9.10. Plates and frames

press without fear of blocking the feed channels. Developments in filter presses over the past 10 years have been towards the fabrication of larger units, made possible by mechanisation and the use of newer lighter materials for construction[25]. The plates of wood used in earlier times were limited in size because of limitations of pressures and large cast-iron plates presented difficulty in handling. Large plates are now frequently made of rubber mouldings or of polypropylene but here distortion may be a problem particularly if the temperature is high.

The second area of advance is in mechanisation which enables the opening and closing to be done automatically. Closing and opening may be effected by a ram driven hydraulically or by an electric motor. Plate transportation is effected by fitting triggers to two endless chains operating the plates, and labour costs have consequently been reduced very considerably. Improved designs have given better drainage which has led to improved washing; much shorter time cycles are now obtained and the cakes are thinner,

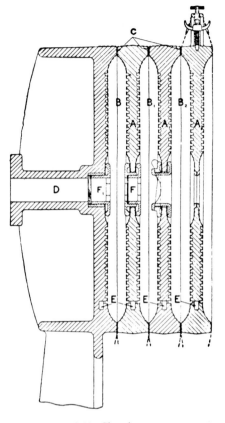

Fig. 9.11. Chamber press
A—plates. B—chambers.
C—machined rims of plates.
D—inlet. E—outlets.
F—screwed unions

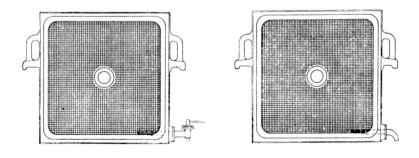

Fig. 9.12. Recessed plate

more uniform, and drier. These advantages have been rather more readily obtained with recessed plates where the cloth is subjected to less wear.

*Advantages of the Filter Press*

(a) Because of its basic simplicity it is versatile and may be used for a wide range of materials under varying operating conditions of cake thickness and pressure.
(b) Maintenance cost is low.
(c) It provides a large filtering area on a small floor space and few additional associated units are needed.
(d) Most joints are external and leakage is easily detected.
(e) High pressures are easily obtained.
(f) It is equally suitable whether the cake or the liquid is the main product.

*Disadvantages of the Filter Press*

(a) It is intermittent in operation and continual dismantling is apt to cause high wear on the cloths.
(b) Despite the improvements mentioned above it is fairly heavy on labour and is unsuitable for high throughputs.

### 9.4.4. Leaf Filters

The need to develop much larger capacity units was met by the introduction of the leaf filters, successively by Moore, Kelly, Sweetland and Vallez. The Moore filter operates under vacuum and is primarily conceived to give large areas using a simple form of construction. In the other three types the leaves are contained in a shell, thus enabling pressure operation to be obtained. The features of importance are cake uniformity and removal, ease of washing, and the fitting of the cloth.

*The Moore Filter*

The Moore filter consists of a number of leaves, sometimes of very large areas, supported on a rigid framework and with the filtrate discharge pipes connected to a common manifold. The battery of leaves is immersed in a tank of slurry and a vacuum is applied through the discharge manifold and filtration is continued until the required thickness of cake (3 to 10 mm) has been formed. The leaf assembly is then lifted out of the slurry tank and immersed in a second tank containing wash liquid. If the filter cake is of little value, a little fine mud may be introduced into the wash liquid to seal any cracks. After washing is complete, the cake is partially dried by sucking air through it, and then loosened from the filter cloth by the application of compressed air internally so that it is readily scraped off. Alternatively, the slurry is withdrawn from the tank at the end of the filtration and is replaced with wash liquid so that it is not necessary to move the leaf assembly.

The filter operates under reduced pressure, so that the pressure difference is limited and hot liquids cannot be filtered because they tend to boil. The Moore filter is now rarely used.

*The Kelly Filter*

In the Kelly filter, a number of vertical rectangular leaves as shown in Fig. 9.13 are arranged longitudinally in a horizontal cylindrical shell. The ends can be opened so that the battery of leaves can be slid out on a pair of rails. The output from each leaf can be observed and controlled independently.

The slurry is pumped into a cylindrical casing and the air is vented to the atmosphere, after which filtration is carried out until a cake of the required thickness is formed; if the slurry tends to sediment, it can be agitated by continuously circulating with a pump. The slurry is then withdrawn and replaced by wash liquid; during this operation, the pressure in the casing is maintained with compressed air in order to prevent the cake from cracking. The cake may then be partially dried with air and discharged after the leaves have been withdrawn from the shell; the application of compressed air to the inside of the leaves helps to loosen the cakes.

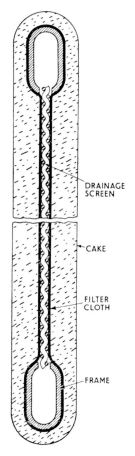

FIG. 9.13. Filter leaf from a Kelly filter

## The Sweetland Filter

The Sweetland filter employs circular leaves, all the same size, mounted transversely in a horizontal shell, which is split into two halves with the lower portion hinged so as to give easy access to the leaves (Fig. 9.14). The filtrate from each leaf enters a common manifold through which the liquid leaves. The operating cycle is similar to that for the Kelly filter, except that the cake is discharged without removing the leaves.

Many Sweetland filters are fitted with a perforated pipe from which the wash water can be sprayed on to the cakes. By this means the cakes are broken up and then re-formed filtering the new slurry that has been produced in the shell. This method of washing is particularly useful in cases where the cake is liable to cracking, which causes the wash water to by-pass the bulk of the material. It can also be used to sluice any adhering solids from the cloths.

The Sweetland overcomes many of the difficulties with the other leaf filters. The more uniform cake which can be discharged without removing the leaves is a distinct advantage and the washing is more effective; it will handle hot slurries. The cost and greater difficulty in fitting the leaves are drawbacks.

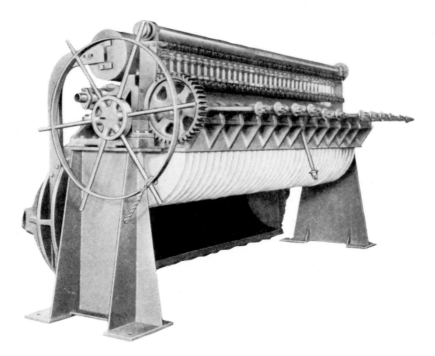

FIG. 9.14. Sweetland filter

## The Vallez Filter

This filter is similar to the Sweetland filter in that it employs transverse vertical leaves in a cylindrical casing. The circular leaves are mounted on a hollow shaft which is slowly rotated at between 0·01 and 0·03 Hz. The central portion of the leaf is in the form of a hub

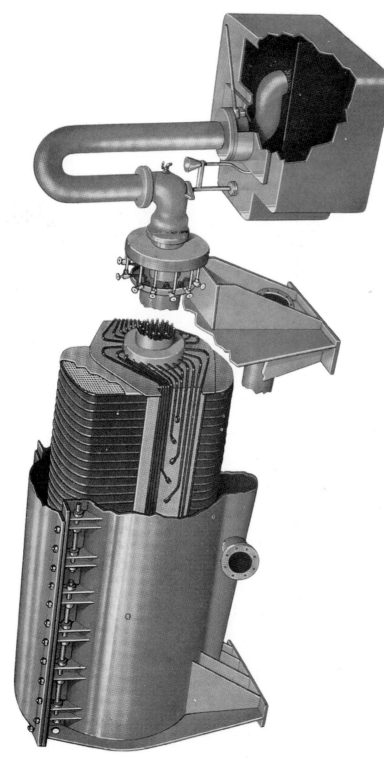

FIG. 9.15. The Rota Leaf filter

Fig. 9.16. Niagara horizontal tank filter used for beer filtration

which is wide enough to give the correct leaf spacing. Ports in the hubs are so placed that they correspond with similar ports on the shaft, which also functions as the filtrate outlet.

The slurry is introduced into the shell of the filter and filtration is continued until the correct thickness of cake has been formed. The cake is washed and air dried, as in the other forms of leaf filter, and the leaves are kept rotating during these operations. The cake is discharged by the application of compressed air, and falls to the bottom of the shell of the filter. The solid is then moved to the outlet, situated in the centre of the bottom of the press, by means of two screw conveyors.

The Vallez filter produces an even cake, and rapid handling is achieved since the press is not opened. The mechanical complications add to the capital cost and to the maintenance cost. The Rota Leaf filter illustrated in Fig. 9.15 operates on the same principle and with a single leaf area of 2·1 m² is available with a total filtration area of up to 117 m².

*The Niagara Filter*

The Niagara pressure leaf filter is made in a number of different forms. The horizontal filter, illustrated in Fig. 9.16, consists of a horizontal cylindrical tank in which is located a

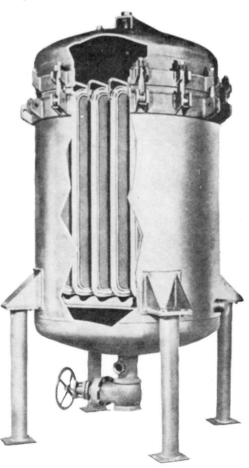

FIG. 9.17. Vertical Niagara filter

bank of transverse filter leaves mounted on a framework which can be slid on tracks from the shell for discharge of the cake. The cake is readily removed by tapping the leaves with a rubber mallet. The cake builds up evenly on the leaves and can therefore be effectively washed and dried *in situ*. It is used where large quantities of solid are to be obtained relatively free of moisture.

When large quantities of liquid with low concentrations of solid are to be filtered, the vertical filter, shown in Fig. 9.17, is frequently used. The filter cake is removed by slucing from the leaves and collecting the discharged slurry from the bottom of the vessel.

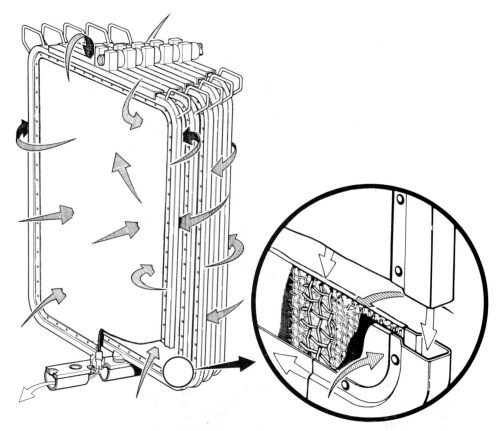

FIG. 9.18. Filter leaves for a Niagara filter

The construction of the filter leaves is shown in Fig. 9.18. For many applications an all-metal leaf is used, but for fine suspensions the leaf may be cloth covered.

One of the disadvantages of the normal leaf type of filter is that the contents of the shell of the filter must be blown back to the storage system before the filter can be opened. This difficulty is obviated in the Niagara horizontal plate filter in which the slurry is introduced under pressure from a central manifold to the individual plates. The flow path of the liquid can be seen in Fig. 9.19. In order to discharge the cake the plates are removed from the tank as a unit. This filter is particularly suitable for removing small quantities of solid from large amounts of liquid and for small batch filtrations.

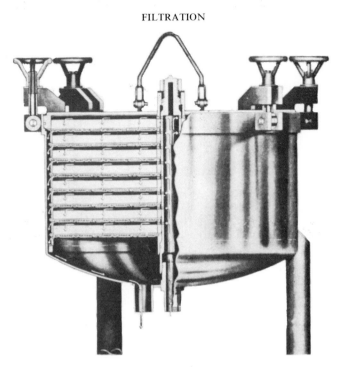

FIG. 9.19. Niagara horizontal plate filter

### 9.4.5. Continuous Rotary Filters

The two main types of continuous rotary filters are the rotating drum and the rotating disc filters. These are suction filters in which filtration, washing, partial drying, and discharge of the cake all take place automatically. Labour requirements are therefore low, and very economical operation can be obtained, especially if a thick suspension is fed to the filter. Rotary filters were developed for use in the mining industry, but are now widely used in the chemical industry where calcium carbonate and ammonium sulphate crystals, for example, are filtered on a large scale.

*Drum Filters*

The rotary drum filter which is shown in Fig. 9.20 consists of a cylinder (Fig. 9.21) mounted horizontally, with the outer surfaces formed of perforated plate or special drainage members over which the filter cloth is fixed. The cloth is sometimes separated from the drum by a coarse metal gauze, so that the effective area for filtration is a maximum, and is held in position with wires or with a light metal screen which also protects it from damage. In all but the earliest models of drum filter, the cylinder is divided into a number of sectors and a separate connection is made between each sector and a special rotary valve. The drum is immersed to the required depth in the slurry, which is agitated to prevent settling of the solids and vacuum is applied to those sectors of the drum which are submerged. A cake of the desired thickness, up to about 100 mm, is produced by adjusting the speed of rotation of the drum. Increase of speed results in the formation of a

Fig. 9.20.  30 m² ROVAC pipe type drum filter handling sludge on an arc furnace fume cleaning plant

thinner cake and consequently a higher rate of filtration. The capacity usually lies between 0·1 and 5 kg solid/m² s, and varies with the nature of the cake.

The filtration takes place at approximately constant pressure, except in the initial stages when the sector is being evacuated. This initial phase occupies about 3 per cent of the filtration time for the slow filtering materials and up to 20 per cent for the free filtering materials.

Each sector is immersed in turn in the slurry and the cake is then washed and partially dried by means of a current of air. Finally, pressure is applied under the cloth to aid the removal of the cake. The special valve consists of three main components as follows, the first two rotating with the drum and the third remaining stationary.

(a)  The valve seat which is a flat disc with ports connecting to the sectors of the filter.
(b)  The removable wear plate, which has holes corresponding with those on the valve seat and rotates in contact with it.
(c)  The valve head, shown in Fig. 9.22, has an annular slot which connects with the filtrate outlet, the wash liquid outlet, the vacuum supply for drying the cake, and the compressed air for loosening the cake, so that each sector of the filter is connected to these in turn. A movable bridge in the slot enables the fraction of the cycle occupied by filtering and washing to be varied. Commonly, one-third of the cycle is used for filtration, one-half for washing and air drying, and the remaining sixth for the removal of the cake.

Fig. 9.21.  Stainless steel drum for a filter during construction

Fig. 9.22.  Valve head, showing movable bridge

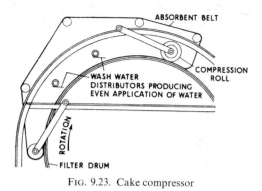

FIG. 9.23. Cake compressor

Washing is usually carried out by spraying the filter cake after it leaves the slurry, and the wash liquid then leaves the filter through the rotary valve and is collected separately from the filtrate. If the cake cracks badly as it leaves the slurry, it may be consolidated by means of a cake compressor which consists of an endless canvas belt (Fig. 9.23) forced against the drum by means of two heavy rollers and driven by friction with the cake so that no slip occurs. The wash liquid is applied to the top of the drum and percolates through the belt and any excess runs off and is collected. The cake compressor can also be used without the application of liquid in order to obtain a comparatively dry cake. When washing is very difficult, the unwashed cake may be discharged and then mixed with wash liquid to form a fresh slurry which is then refiltered.

The capacity of the vacuum pump will be determined very largely by the amount of air sucked through the filter cake during the washing and air-drying periods when, in most cases, there will be a simultaneous flow of both liquid and air. It is not possible to calculate with any accuracy the rate of passage of air under these circumstances and the capacity of the vacuum system must be determined as a result of test experiments. A typical layout of an installation is shown in Fig. 9.24, from which it is seen that the air and liquid are removed separately.

There are several methods of removing the filter cake from the drum but it is usual to employ an adjustable doctor knife. The discharge is facilitated by the application of compressed air or steam to the underside of the cloth as this loosens the cake and raises the

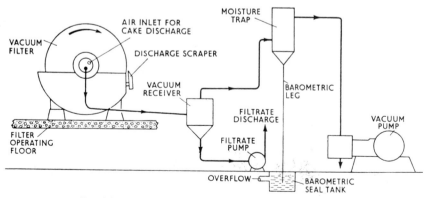

FIG. 9.24. Typical layout of rotary drum filter installation

FIG. 9.25. String discharge filtration showing the aligning and cleaning comb

filter cloth slightly, bringing it into close contact with the knife. Careful adjustment of the doctor knife is essential, for if it is set too close to the drum the cloth may be damaged.

When the cake is difficult to remove, or when it is desirable to deliver it intact, a string discharge is employed as shown in Fig. 9.25. A series of endless strings, in fibres, nylon or other materials, pass around the drum of the filter over a discharge roll and through the aligning and cleaning comb. In operation the drum rotates, and, under suction, a cake is formed on the filter cloth bridging over the strings. The strings leave the drum tangentially, and, with the release of vacuum, the filter cake is carried in a continuous sheet to the discharge roll. Filter cake is removed largely by flexure of the strings at the discharge roll.

An American company has developed a combined filtering and drying plant. A continuous belt, rather like a bed-spring in construction, passes round the underside of the drum filter and the filter cake is deposited on the belt to which it adheres. The belt leaves the filter towards the top and is then carried through a cabinet dryer. It then returns to the underside of the drum filter after the cake has been removed by agitation. The metal belt, which assists the drying of the material by virtue of its good heat-conducting properties, is formed into loops which are carried through the dryer on a slat conveyor. With finely divided materials the total loss of solid from the belt is as little as 1 or 2 per cent.

The slurry is agitated sufficiently to prevent settling of the particles, but with materials that settle very rapidly, for example slurries of coal dust or suspensions of crystals in their mother liquors, it is more satisfactory to apply the slurry from a trough at the top of the filter drum. Because the larger particles settle more rapidly they are deposited near the filter cloth and the smaller particles form the outer portions of the filter cake. A cake of relatively high porosity is thereby obtained and high filtration rates are achieved.

Recent developments in rotary filters include increased size, new materials of construction, and improved methods of cake removal and drying. Drums are now fabricated to give surfaces of 60–100 m$^2$ as compared with the 20 m$^2$ of the older cast iron drums. New construction materials such as stainless steel, titanium, epoxy resins and plastics, such as PVC, all give much improved corrosion resistance for many slurries and hence longer life. The replacement of the knife system by some form of belt has given better cake discharge and permitted the use of thinner filtering media such as synthetic fibres. The belt provides some support for the cake and materially assists the effect of compressed air for lifting off the cake. Drying can be improved by totally covering the filter with a hood. Improvements have also been made in techniques for reducing cake cracking.

*The precoat filter.* When the material forms a cake of high resistance, the process may be facilitated by the use of a precoat filter. A thick layer of a very free filtering material such as kieselguhr or asbestos is first built up on the filtering surface. Filtration is then carried out through this layer and the doctor blade is so arranged that it automatically progresses towards the drum. A very thin cake can then be formed and the whole of it removed together with a thin layer of the precoat material.

*Advantages of the rotary drum filter.* (a) The filter is entirely automatic in action and the manpower requirements are therefore reduced to a minimum.

(b) Cakes of varying thickness can be built up by altering the speed of rotation of the filter. The usual range of thicknesses is from 3 mm with fine solids to as much as 100 mm with very coarse solids.

(c) The filter has a very large capacity for its size, and is extensively used for the filtration of large quantities of free filtering material.

(d) If the cake consists of coarse solids, most of the liquid can be removed from it before discharging.

*Disadvantages.* (a) It is a vacuum filter and therefore the maximum available pressure difference is limited and difficulty is encountered in the filtration of hot liquids because of their tendency to boil.

(b) The filter cannot be used for materials that form relatively impermeable filter cakes or cakes that are difficult to remove from the cloth. The filtering properties of the slurry may be improved however by the use of filter aids.

(c) Good washing is not easily obtained, but double filtering will frequently assist.

(d) It is difficult to obtain a dry cake.

(e) The capital cost of the filter and vacuum equipment is high.

*The Rotary Disc Filter*

The rotary disc filter (Fig. 9.26) consists of a number of circular filter leaves mounted on

FIG. 9.26. Rotary disc filter showing all metal sectors

a heavy horizontal tubular shaft and spaced by means of hubs. The construction of the leaves can be seen and each sector of the disc is connected to a separate outlet and a continuous channel is formed by the outlets from the corresponding sectors of the other leaves when the discs are bolted together. These channels connect to a rotary valve, similar to that on the drum filter.

The operation of the rotary disc filter is very similar to that of the drum filter, though the discharge of the cake is more difficult. It can be operated so that several slurries are filtered simultaneously by arranging for the discs to dip into a number of different tanks. In this case it is not possible to separate the various filtrates, however. Its main advantage over the drum filter is that it gives a very much larger filtering area in the same floor space.

### 9.4.6. Horizontal Belt Filters

Another form of completely automatic continuous vacuum filter is the Landskrona band filter[26]. A permeable belt passes over two pulleys arranged at the same vertical height. The slurry is fed on to the upper side of the belt at one end, and the filtrate is collected underneath. The speed of the belt (about 0·05 m/s) and the distance over which the slurry is applied are so adjusted as to produce a cake of a suitable thickness. The cake is then subjected to the action of a spray of wash water and finally to air drying; the solid is discharged as the belt passes over the end pulley.

This filter is used in Sweden, where it was developed, for filtering phosphoric acid liquors from sludges containing calcium sulphate. It gives a very uniform filter cake free from cracks; washing is therefore good. The cost per unit area is greater than for the rotary filter.

FIG. 9.27. Prayon filter

A newer and interesting development is the Adpec[27] filter which has an intermittently moving belt. The slurry is fed to one end and the vacuum applied to the underside boxes and filtering occurs with the belt stationary. As the discharge roll at the other end moves inwards it trips the system so that the vacuum is relieved, the belt moved forward, thus avoiding the problem of pulling the belt continuously over the sections under vacuum.

### 9.4.7. Prayon Continuous Filter

This filter has been developed in Belgium and Luxemburg, particularly for the vacuum filtration of the highly corrosive liquids formed in the production of phosphoric acid and phosphates. It is made in five sizes providing filtering surfaces from about 2 up to 40 m$^2$ and the largest model is illustrated in Fig. 9.27. It consists of a number of horizontal cells held in a rotating frame, and each cell passes in turn under a slurry feed pipe and a series of wash liquor inlets. In the model illustrated, the connections are arranged so as to provide three countercurrent washes. The timing of the cycle is regulated to give the required times for filtration, washing and cake drying. The discharge of the cake is by gravity as the cell is automatically turned upside-down, and can be facilitated if necessary by washing or the application of back pressure. The cloth is cleaned by spraying with wash liquid and is allowed to drain before it reaches the feed point again.

The filter has a number of advantages including:
(a) A large filtration area, up to 85 per cent of which is effective at any instant.
(b) Low operating and maintenance costs and simple cloth replacement.
(c) Clean discharge of solids.
(d) Fabrication in a wide range of corrosion-resistant materials.
(e) Countercurrent washing.
(f) Operation at a pressure down to 15 kN/m$^2$.

### 9.4.8. Filter Cartridges

*The Metafilter*

The Metafilter, which employs a filter bed deposited on a base of rings mounted on a fluted rod, is extensively used for clarifying liquids containing small quantities of very fine suspended solids. The rings are accurately pressed from sheet metal of very uniform thickness and are made in a large number of corrosion-resistant metals, though stainless steels are usually employed. The standard rings are 22 mm in external diameter, 16 mm in internal diameter and 0·8 mm thick, and are scalloped on one side, as shown in Fig. 9.28, so that the edges of the discs are separated by a distance of between 0·025 and 0·25 mm according to requirements. The pack is formed by mounting the rings, all the same way up, on the drainage rod and tightening them together by a nut at one end against a boss at the other (Fig. 9.29). The packs are mounted in the body of the filter which is operated either by positive or by reduced pressure.

The bed is formed by feeding to the filter a dilute suspension of material, usually a form of kieselguhr, which is strained by the packs to form a bed about 3 mm thick. Kieselguhr is available in a number of grades and forms a bed of loose structure which is capable of trapping particles much smaller than the channels. During filtration the solids build up mainly on the surface and do not generally penetrate more than 0·5 mm into the bed. The filtrate passes between the discs and leaves through the fluted drainage rod, and operation is continued until the resistance becomes too high. The filter is then cleaned by back-flushing; this causes the filter cake to crack and peel away. In some cases the cleaning may be incomplete as a result of channelling. If for any reason the spaces between the rings become blocked, the rings may be quickly removed and washed.

The Metafilter is widely used for filtering domestic water, beer, organic solvents and oils. The filtration characteristics of clay-like materials can often be improved by the continuous introduction of a small quantity of filter aid to the slurry as it enters the filter. On the other hand, when the suspended solid is relatively coarse, the Metafilter will operate successfully as a strainer, without the use of a filter bed.

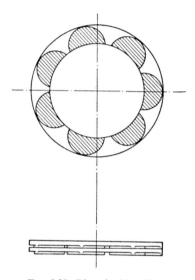

FIG. 9.28. Rings for Metafilter

The Metafilter is very robust and is economical in use because there is no filter cloth and the bed is easily replaced and hence labour charges are low. Mono pumps or diaphragm pumps are most commonly used for feeding the filter; these have been discussed in Volume 1, Chapter 6.

A different form of filter is available for lubricating oils (Fig. 9.30). In this, leaves or *pockets* are used in which a pair of filter papers, 100 mm in diameter and united at the edges, enclose a drainage support made up of two thin perforated metal discs, of which the inner surfaces are kept out of contact by burrs resulting from the piercing. A 16 mm diameter hole is provided through the centre of the pockets which are mounted on a drainage pipe. Filtration takes place over the whole surface of the papers, and these are so well supported that very high pressures can be applied, even 1400 kN/m², without causing the paper to break down. The filtration is very fine, so that fine suspended carbon can be cleared from crankcase oil, and very large areas of filter surface can be obtained in a small volume, as much as 0·2 m² of filter surface in a cylinder 100 mm diameter and 25 mm deep.

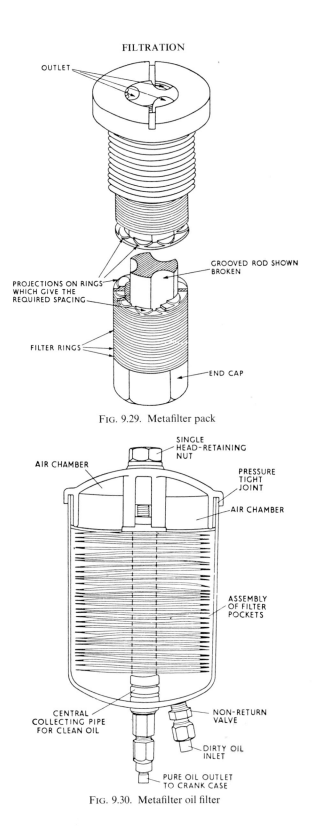

FIG. 9.29. Metafilter pack

FIG. 9.30. Metafilter oil filter

*The Streamline Filter*

The Streamline filter is geometrically similar to the Metafilter, but the cylindrical filter packs (Fig. 9.31) are formed from compressed discs about 50 mm in diameter made of woven synthetic materials such as Terylene or nylon. No filter bed is required and the liquid flows radially inwards through the small spaces between and through the individual discs. The filter pack is considerably cheaper than that in the Metafilter, but it needs replacing every few years and the same pack cannot be used for filtering a number of different liquids. The Streamline filter is extensively used as a petrol filter and also for oils.

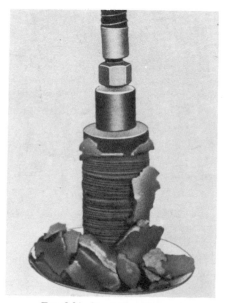

FIG. 9.31. Streamline filter pack

## 9.5. FILTRATION IN A CENTRIFUGE

When filtration is carried out in a centrifuge, the driving force is the centrifugal pressure due to the liquid, and this will not be affected by the presence of solid particles deposited on

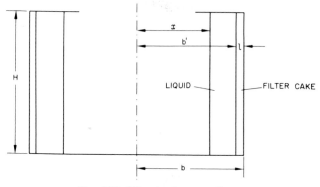

FIG. 9.32. Filtration in a centrifuge

the walls. This force has to overcome the friction caused by the flow of liquid through the filter cake, the cloth, and the supporting gauze and perforations. The resistance of the filter cake will increase as solids are deposited but the other resistances will remain approximately constant throughout the process. Consider a filtration in a basket of radius $b$, and suppose that the suspension is introduced at such a rate that the inner radius of the liquid surface remains constant (Fig. 9.32). At some time $t$ after the commencement of filtration, a filter cake of thickness $l$ will have been built up and the radius of the interface between the cake and the suspension will be $b'$.

If $dP'$ is the pressure difference across a small thickness $dl$ of cake, the velocity of flow of the filtrate will be given by:

$$u = \frac{1}{r\mu}\left(\frac{-dP'}{dl}\right) \qquad \text{(from equation 9.2)}$$

where $r$ is the specific resistance of the filter cake and $\mu$ is the viscosity of the filtrate.

If the centrifugal force is large compared with the gravitational force, the filtrate will flow in an approximately radial direction, and will be evenly distributed up the height of the bowl. The area available for flow will increase towards the walls of the basket. If $dV$ is the volume of filtrate flowing through the filter cake in time $dt$:

$$u = \frac{1}{2\pi b' H}\frac{dV}{dt}$$

Thus

$$\frac{1}{r\mu}\left(\frac{-dP'}{dl}\right) = \frac{1}{2\pi b' H}\frac{dV}{dt} \qquad (9.44)$$

$$-dP' = \frac{r\mu\,dl}{2\pi b' H}\frac{dV}{dt}$$

and thus the total pressure drop through the cake at time $t$ is given by:

$$-\Delta P' = \frac{ru}{2\pi H}\frac{dV}{dt}\int_0^l \frac{dl}{b'}$$

$$= \frac{r\mu}{2\pi H}\frac{dV}{dt}\int_{b'}^b \frac{db'}{b'}$$

$$= \frac{r\mu}{2\pi H}\frac{dV}{dt}\ln\frac{b}{b'} \qquad (9.45)$$

If the resistance of the cloth is negligible this is equal to the centrifugal pressure. More generally, if the cloth (considered together with the supporting wall of the basket) is equivalent in resistance to a cake of thickness $L$, situated at the walls of the basket, the pressure drop $-\Delta P''$ across the cloth is given by:

$$\frac{-\Delta P''}{r\mu L} = \frac{1}{2\pi Hb}\frac{dV}{dt}$$

$$-\Delta P'' = \frac{r\mu}{2\pi H}\frac{dV}{dt}\frac{L}{b} \qquad (9.46)$$

Thus the total pressure drop across the filter cake and the cloth $(-\Delta P)$ (say) is given by:

$$(-\Delta P) = (-\Delta P') + (-\Delta P'')$$

Thus:
$$-\Delta P = \frac{r\mu}{2\pi H}\frac{dV}{dt}\left\{\ln\frac{b}{b'} + \frac{L}{b}\right\} \tag{9.47}$$

Before this equation can be integrated it is necessary to establish the relation between $b'$ and $V$. If $v$ is the bulk volume of incompressible cake deposited by the passage of unit volume of filtrate:

$$v\,dV = -2\pi b'H\,db'$$

$$\frac{dV}{dt} = -\frac{2\pi Hb'\,db'}{v\,dt} \tag{9.48}$$

and substituting for $dV/dt$ in the previous equation gives:

$$-\Delta P = -\frac{r\mu}{2\pi H}\frac{2\pi Hb'\,db'}{v\,dt}\left\{\ln\frac{b}{b'} + \frac{L}{b}\right\} \tag{9.49}$$

$$\therefore\quad \frac{v(-\Delta P)}{r\mu}\,dt = \left\{\ln\frac{b'}{b} - \frac{L}{b}\right\}b'\,db'$$

This may be integrated between the limits $b' = b$ and $b' = b'$ as $t$ goes from 0 to $t$. $-\Delta P$ is constant because the inner radius $x$ of the liquid is maintained constant:

$$\frac{(-\Delta P)vt}{r\mu} = \int_{b}^{b'}\left\{\left(\ln\frac{b'}{b} - \frac{L}{b}\right)b'\right\}db'$$

$$= \frac{1}{4}(b^2 - b'^2) + \frac{L}{2b}(b^2 - b'^2) + \frac{1}{2}b'^2\ln\frac{b'}{b} \tag{9.50}$$

$$\therefore\quad (b^2 - b'^2)\left(1 + 2\frac{L}{b}\right) + 2b'^2\ln\frac{b'}{b} = \frac{4(-\Delta P)vt}{r\mu}$$

$$= \frac{2vt\rho\omega^2}{r\mu}(b^2 - x^2) \tag{9.51}$$

$$\text{since } -\Delta P = \frac{1}{2}\rho\omega^2(b^2 - x^2) \qquad \text{(equation 5.106)}$$

From this equation the time $t$ taken to build up the cake to a given thickness $b'$ may be calculated. The corresponding volume of cake is given by:

$$Vv = \pi(b^2 - b'^2)H \tag{9.52}$$

and the volume of filtrate:

$$V = \frac{\pi}{v}(b^2 - b'^2)H \tag{9.53}$$

Haruni and Storrow[28] have carried out an extensive investigation of the flow of liquid through a cake formed in a centrifuge. They have concluded that, although the results of tests on a filtration plant and a centrifuge are often difficult to compare because of the effects of the compressibility of the cake, it is frequently possible to predict the flowrate in a centrifuge to within 20 per cent. They have also shown that, when the thickness varies with height in the basket, the flowrate can be calculated on the assumption that the cake has a uniform thickness equal to the mean value; this gives a slightly high value in most cases.

## 9.6. FILTRATION CALCULATIONS

**Example 9.1**

When an aqueous slurry is filtered in a plate and frame press, fitted with two 50 mm thick frames each 150 mm square, at 450 kN/m² pressure the frames are filled in 3500 s. How long will it take to produce the same volume of filtrate as is obtained from a single cycle when using a centrifuge with a perforate basket, 300 mm diameter and 200 mm deep? The radius of the inner surface of the slurry is maintained constant at 75 mm and the speed of rotation is 65 Hz.

Assume that the filter cake is incompressible and that the resistance of the cloth is equivalent to 3 mm of cake in both cases.

**Solution**

*In the filter press*

$$V^2 + 2\frac{AL}{v}V = \frac{2(-\Delta P)A^2 t}{r\mu v} \quad (V = 0 \text{ when } t = 0)$$

(from equation 9.18)

Now

$$V = \frac{lA}{v}$$

(equation 9.6)

$$\therefore \quad \frac{l^2 A^2}{v^2} + \frac{2AL}{v}\frac{lA}{v} = \frac{2(-\Delta P)A^2 t}{r\mu v}$$

$$\therefore \quad l^2 + 2Ll = \frac{2(-\Delta P)vt}{r\mu}$$

*For one cycle*

$$l = 25 \text{ mm} = 0.025 \text{ m}; \quad L = 3 \text{ mm} = 0.003 \text{ m}$$

$$-\Delta P = (450 - 101.3) = 348.7 \text{ kN/m}^2 = 3.49 \times 10^5 \text{ N/m}^2$$

$$t = 3500 \text{ s}$$

$$\therefore \quad 0.025^2 + (2 \times 0.003 \times 0.025) = 2 \times 3.49 \times 10^5 \times 3500 \times \frac{v}{r\mu}$$

$$\therefore \quad \frac{r\mu}{v} = 3.15 \times 10^{12}$$

In the centrifuge

$$(b^2 - b'^2)\left(1 + 2\frac{L}{b}\right) + 2b'^2 \ln\frac{b'}{b} = \frac{2vt\rho\omega^2}{r\mu}(b^2 - x^2)$$

(equation 9.51)

$$b = 0.15 \text{ m}$$

$$H = 0.20 \text{ m}$$

$$\text{Volume of cake} = 2 \times 0.050 \times 0.15^2 = 0.00225 \, \text{m}^3$$

$$\therefore \quad \pi(b^2 - b'^2) \times 0.20 = 0.00225$$

$$\therefore \quad (b^2 - b'^2) = 0.00358$$

$$\therefore \quad b'^2 = 0.15^2 - 0.00358 = 0.0189 \, \text{m}^2$$

$$\therefore \quad b' = 0.138 \, \text{m}$$

$$x = 75 \, \text{mm} = 0.075 \, \text{m}$$

$$\omega = 65 \times 2\pi = 408.4 \, \text{radians/s}$$

The time taken to produce the same volume of filtrate or cake as in one cycle of the filter press is therefore given by:

$$(0.15^2 - 0.138^2)(1 + 2 \times 0.003/0.15) + 2(0.0189) \ln (0.138/0.15)$$

$$= \frac{2 \times t \times 1000 \times 408.4^2}{3.15 \times 10^{12}} (0.15^2 - 0.075^2)$$

$$0.00359 - 0.00315 = 1.787 \times 10^{-6} \, t$$

$$\therefore \quad t = \frac{4.4 \times 10^{-4}}{1.787 \times 10^{-6}}$$

$$= \underline{\underline{246 \, \text{s}}} \, (= 4 \, \text{min})$$

## Example 9.2

A slurry is filtered in a plate and frame press containing 12 frames, each 0.3 m square and 25 mm thick. During the first 200 s, the filtration pressure is slowly raised to the final value of 500 kN/m$^2$ and, during this period, the rate of filtration is maintained constant. After the initial period, filtration is carried out at constant pressure and the cakes are completely formed in a further 900 s. The cakes are then washed at 375 kN/m$^2$ for 600 s, using *thorough washing*. What is the volume of filtrate collected per cycle and how much wash water is used?

A sample of the slurry had previously been tested, using a vacuum leaf filter of 0.05 m$^2$ filtering surface and a vacuum equivalent to an absolute pressure of 30 kN/m$^2$. The volume of filtrate collected in the first 300 s was 250 cm$^3$ and, after a further 300 s, an additional 150 cm$^3$ was collected. Assume the cake to be incompressible and the cloth resistance to be the same in the leaf as in the filter press.

## Solution

In the leaf filter, filtration is at constant pressure from the start. Thus

$$V^2 + 2\frac{AL}{v}V = 2\frac{(-\Delta P)A^2}{r\mu v}t \qquad \text{(from equation 9.18)}$$

In the filter press, a volume $V_1$ of filtrate is obtained under constant rate conditions in time $t_1$, and filtration is then carried out at constant pressure.

Thus

$$V_1^2 + \frac{AL}{v} V_1 = \frac{(-\Delta P)A^2}{r\mu v} t_1 \qquad \text{(from equation 9.17)}$$

and

$$(V^2 - V_1^2) + 2\frac{AL}{v}(V - V_1) = 2\frac{(-\Delta P)A^2}{r\mu v}(t - t_1)$$

$$\text{(from equation 9.18)}$$

*For the leaf filter*

When $t = 300\,\text{s}$, $V = 250\,\text{cm}^3$ and when $t = 600\,\text{s}$, $V = 400\,\text{cm}^3$

$A = 0.05\,\text{m}^2$ and $-\Delta P = (101.3 - 30) = 71.3\,\text{kN/m}^2$

Thus

$$250^2 + 2(0.5\,L/v)\,250 = 2(71.3 \times 0.05^2/r\mu v)300$$

and

$$400^2 + 2(0.5\,L/v)400 = 2(71.3 \times 0.05^2/r\mu v)600$$

i.e.

$$62{,}500 + 25\frac{L}{v} = \frac{106.95}{r\mu v}$$

and

$$160{,}000 + 40\frac{L}{v} = \frac{213.9}{r\mu v}$$

Hence

$$L/v = 3500 \quad \text{and} \quad r\mu v = 7.13 \times 10^{-4}$$

*For the filter press*

The volume of filtrate $V_1$ collected during the constant rate period on the filter press is given by ($A = 2.16\,\text{m}^2$, $-\Delta P = (500 - 101.3) = 398.7\,\text{kN/m}^2$, $t = 200\,\text{s}$):

$$V_1^2 + 2.16 \times 3500\,V_1 = (398.7 \times 2.16^2/7.13 \times 10^{-4})200$$

i.e.

$$V_1^2 + 7560V_1 - 5.218 \times 10^8 = 0$$

$$\therefore \quad V_1 = -3780 + \sqrt{1.429 \times 10^7 + 5.218 \times 10^8} = 1.937 \times 10^4\,\text{cm}^3$$

For the constant pressure period

$$t - t_1 = 900\,s$$

The total volume of filtrate collected is therefore given by

$$(V^2 - 3.75 \times 10^8) + 15{,}120(V - 1.937 \times 10^4) = 5.218 \times 10^6 \times 900$$

i.e.

$$V^2 + 15{,}120\,V - 53{\cdot}64 \times 10^8 = 0$$

$$\therefore \quad V = -7560 + \sqrt{5{\cdot}715 \times 10^7 + 53{\cdot}64 \times 10^8}$$

$$= 6{\cdot}607 \times 10^4 \text{ cm}^3 \text{ or } \underline{0{\cdot}066 \text{ m}^3}$$

Final rate of filtration

$$= \frac{-\Delta P A^2}{r \mu v (V + AL/v)} \qquad\qquad \text{(from equation 9.16)}$$

$$= \frac{398{\cdot}7 \times 2{\cdot}16^2}{7{\cdot}13 \times 10^{-4}(6{\cdot}607 \times 10^4 + 2{\cdot}16 \times 3500)} = 35{\cdot}4 \text{ cm}^3/\text{s}$$

If the viscosity of the filtrate is the same as that of the wash-water,

Rate of washing at $500 \text{ kN/m}^2 = 35{\cdot}4 \text{ cm}^3/\text{s}$
Rate of washing at $375 \text{ kN/m}^2 = 35{\cdot}4(375 - 101{\cdot}3)/(500 - 101{\cdot}3) = 24{\cdot}3 \text{ cm}^3/\text{s}$

Thus the amount of wash-water passing in $600\,\text{s} = (600 \times 24{\cdot}3) = 1{\cdot}458 \times 10^4 \text{ cm}^3$ or $\underline{0{\cdot}0146 \text{ m}^3}$.

### Example 9.3

A slurry, containing $0{\cdot}2$ kg of solid (specific gravity $3{\cdot}0$) per kilogram of water, is fed to a rotary drum filter $0{\cdot}6$ m long and $0{\cdot}6$ m diameter. The drum rotates at one revolution in $350\,\text{s}$ and 20 per cent of the filtering surface is in contact with the slurry at any instant. If filtrate is produced at the rate of $0{\cdot}125 \text{ kg/s}$ and the cake has a voidage of $0{\cdot}5$, what thickness of cake is produced when filtering at a pressure of $35 \text{ kN/m}^2$?

The rotary filter breaks down and the operation has to be carried out temporarily in a plate and frame press with frames $0{\cdot}3$ m square. The press takes $100\,\text{s}$ to dismantle and $100\,\text{s}$ to reassemble and, in addition, $100\,\text{s}$ is required to remove the cake from each frame. If filtration is to be carried out at the same overall rate as before, with an operating pressure of $275 \text{ kN/m}^2$, what is the minimum number of frames that needs to be used and what is the thickness of each? Assume the cakes to be incompressible and neglect the resistances of the filter media.

### Solution

*Drum filter*

Area of filtering surface $= 0{\cdot}6 \times 0{\cdot}6\pi = 0{\cdot}36\pi \text{ m}^2$

Rate of filtration $\qquad = 0{\cdot}125 \text{ kg/s}$

$$= \frac{0{\cdot}125}{1000} = 1{\cdot}25 \times 10^{-4} \text{ m}^3/\text{s of filtrate}$$

Volumetric rate of deposition of solids (bulk)

$$= 1{\cdot}25 \times 10^{-4} \times 0{\cdot}2 \times \frac{1}{0{\cdot}5} \times \frac{1}{3{\cdot}0} = 1{\cdot}67 \times 10^{-5}\,\mathrm{m^3/s}$$

One revolution takes 350 s; therefore the given piece of filtering surface is immersed for $350 \times 0{\cdot}2 = 70\,\mathrm{s}$

Bulk volume of cake deposited per revolution $= 1{\cdot}67 \times 10^{-5} \times 350 = 5{\cdot}85 \times 10^{-3}\,\mathrm{m^3}$

Thickness of cake produced $= \dfrac{5{\cdot}85 \times 10^{-3}}{0{\cdot}36} = 5{\cdot}2 \times 10^{-3}$ or $\underline{\underline{5{\cdot}2\,\mathrm{mm}}}$

*Properties of filter cake*

Now

$$\frac{dV}{dt} = \frac{-\Delta P A}{r\mu l} = \frac{-\Delta P A^2}{r\mu V v} \qquad \text{(from equations 9.2 and 9.8)}$$

At constant pressure

$$V^2 = \frac{2}{r\mu v}(-\Delta P)A^2 t = K(-\Delta P)A^2 t \text{ (say)} \qquad \text{(from equation 9.11)}$$

Then expressing pressures, areas, times and volumes in $\mathrm{kN/m^2}$, $\mathrm{m^2}$, s and $\mathrm{m^3}$, respectively, for one revolution of the drum:

$$(1{\cdot}25 \times 10^{-4} \times 350)^2 = K(101{\cdot}3 - 35)(0{\cdot}36\pi)^2 \times 70$$

(since each element of area is immersed for one-fifth of a cycle), i.e.

$$K = 3{\cdot}22 \times 10^{-7}$$

*Filter press*

Use a filter press with $n$ frames of thickness $d$ m.

Total time, for one complete cycle of press $= t_f + 100n + 200\,\mathrm{s}$

where $t_f$ is the time during which filtration is occurring.

Overall rate of filtration $= \dfrac{V_f}{t_f + 100n + 200} = 1{\cdot}25 \times 10^{-4}\,\mathrm{m^3/s}$

where $V_f$ is the total volume of filtrate per cycle.

Now $V_f$ = Volume of frames/volume of cake deposited by unit volume of filtrate ($v$)
$= 0{\cdot}3^2 nd/(0{\cdot}2/(0{\cdot}5 \times 3{\cdot}0)) = 0{\cdot}675nd$
But $V_f^2 = 3{\cdot}22 \times 10^{-7}(275 - 101{\cdot}3)(2n \times 0{\cdot}3 \times 0{\cdot}3)^2 t_f$
$= (0{\cdot}675nd)^2$

i.e.

$$t_f = 2{\cdot}576 \times 10^5 d^2$$

Thus:

$$1\cdot25 \times 10^{-4} = 0\cdot675nd/(2\cdot576 \times 10^5 d^2 + 100n + 200)$$

i.e.

$$31\cdot45d^2 + 0\cdot0125n + 0\cdot0250 = 0\cdot675nd$$

giving:

$$n = \frac{0\cdot0250 + 31\cdot45d^2}{0\cdot675d - 0\cdot0125}$$

$n$ is a minimum when $dn/dd = 0$, i.e. when:

$$(0\cdot675d - 0\cdot0125) \times 62\cdot9d - (0\cdot0250 + 31\cdot45d^2) \times 0\cdot675 = 0$$
$$d^2 - 0\cdot0370d - 0\cdot000796 = 0$$
$$d = 0\cdot0185 \pm \sqrt{0\cdot000343 + 0\cdot000796}$$
$$= 0\cdot0522\,\text{m} \quad \text{or} \quad \underline{52\cdot2\,\text{mm}}$$

Hence

$$n = (0\cdot0250 + 31\cdot45 \times 0\cdot0522^2)/(0\cdot675 \times 0\cdot0522 - 0\cdot0125)$$

$$= 4\cdot87$$

Thus a minimum of 5 frames must be used.

The sizes of frames which will give exactly the required rate of filtration when five are used are given by

$$0\cdot0250 + 31\cdot45d^2 = 3\cdot375d - 0\cdot0625$$

i.e.

$$d^2 - 0\cdot107 + 0\cdot00278 = 0$$
$$\therefore \quad d = 0\cdot0535 \pm \sqrt{0\cdot00285 - 0\cdot00278}$$
$$= 0\cdot044 \quad \text{or} \quad 0\cdot063\,\text{m}$$

i.e. 5 frames of thickness 44 mm or 63 mm will give exactly the required filtration rate; intermediate sizes give higher rates.

Thus any frame thickness between 44 and 63 mm will be satisfactory. In practice, 2 in (50·4 mm) frames would be used.

## Example 9.4

A sludge is filtered in a plate and frame press fitted with 25 mm frames. For the first 600 s the slurry pump runs at maximum capacity. During this period the pressure rises to $415\,\text{kN/m}^2$ and a quarter of the total filtrate is obtained. The filtration takes a further 3600 s to complete at constant pressure and 900 s is required for emptying and resetting the press.

It is found that if the cloths are precoated with filter aid to a depth of 1·6 mm, the cloth resistance is reduced to a quarter of its former value. What will be the increase in the overall throughput of the press if the precoat can be applied in 180 s?

**Solution**

*Case 1*

$$\frac{\mathrm{d}V}{\mathrm{d}t} = \frac{A^2(-\Delta P)}{vr\mu\left(V + \frac{AL}{v}\right)} = \frac{a}{V+b} \qquad \text{(equation 9.16)}$$

For constant rate filtration

$$\frac{V_0}{t_0} = \frac{a}{V_0 + b}$$

i.e. $$V_0^2 + bV_0 = at_0$$

For constant pressure filtration $\frac{1}{2}(V^2 - V_0^2) + b(V - V_0) = a(t - t_0)$

$$t_0 = 600\,\text{s}, \quad t - t_0 = 3600\,\text{s}, \quad V_0 = \frac{V}{4}$$

$$\frac{V^2}{16} + b\frac{V}{4} = 600a$$

$$\frac{1}{2}\left(V^2 - \frac{V^2}{16}\right) + b\left(V - \frac{V}{4}\right) = 3600a$$

$$\therefore \quad 3600a = \tfrac{15}{32}V^2 + \tfrac{3}{4}bV = \tfrac{3}{8}V^2 + \tfrac{3}{2}bV$$

$$\therefore \quad b = \frac{V}{8}$$

and

$$a = \frac{1}{600}\left\{\frac{V^2}{16} + \frac{V^2}{32}\right\} = \frac{3}{19,200}V^2$$

Cycle time $= 900 + 4200 = 5100\,\text{s}$

Filtration rate per second $= V/5100 = 0{\cdot}000196V$

*Case 2*

$$\frac{V_1}{t_1} = \frac{a}{V_1 + \dfrac{b}{4}} = \frac{V_0}{t_0} = \frac{a}{V_0 + b}$$

$$\frac{1}{2}\left(\frac{49}{64}V^2 - V_1^2\right) + \frac{b}{4}\left(\frac{7}{8}V - V_1\right) = a(t - t_1)$$

$$\therefore \quad \frac{V}{2400} = \frac{\dfrac{3}{19,200}V^2}{V_1 + \dfrac{V}{32}}$$

$$\therefore \quad t_1 = \frac{t_0}{V_0}V_1 = \frac{600}{V/4} \times \frac{11}{32}V = \frac{3300}{4}\,\text{s} = 825\,\text{s}$$

Substituting:

$$\frac{1}{2}\left(\frac{49}{64}V^2 - \frac{121}{1024}V_1^2\right) + \frac{1}{4}\frac{V}{8}\left(\frac{7}{8}V - \frac{11}{32}V\right) = \frac{3}{19,200}V^2(t - t_1)$$

i.e.

$$\frac{49}{128} - \frac{121}{2048} + \frac{17}{1024} = \frac{3}{19,200}(t - t_1)$$

$$t - t_1 = \frac{19,200}{3} \times \frac{784 - 121 + 34}{2048}$$

$$= 2178 \text{ s}$$

$$\text{Cycle time} = 180 + 900 + 825 + 2178 = 4083 \text{ s}$$

$$\text{Filtration rate per second} = \frac{7}{8} \times \frac{V}{4083} = 0.000214\,V$$

$$\text{Increase} = \frac{(0.000214 - 0.000196)V}{0.000196\,V} = 9.1 \text{ per cent}$$

## 9.7. FURTHER READING

Ives, K. J. (ed.): *The Scientific Basis of Filtration* (Noordhoff, Leyden, 1975).
Purchas, D. B.: *Industrial Filtration of Liquids*, 2nd edn. (Leonard Hill, London, 1971).
Suttle, H. K. (ed.): *Process Engineering Technique Evaluation–Filtration* (Morgan-Grampian (Publishers) Ltd., London, 1969).

## 9.8. REFERENCES

1. Ives, K. J. (ed.): *The Scientific Basis of Filtration* (Noordhoff, Leyden, 1975).
2. Ives, K. J.: *Trans. I. Chem. E.* **48** (1970) T94. Advances in deep-bed filtration.
3. Suttle, H. K.: *Chem. Engineer*, No. 314 (Oct. 1976) 675. Development of industrial filtration.
4. Ruth, B. F., Montillon, G. H. and Montonna, R. E.: *Ind. Eng. Chem.* **25** (1933) 76 and 153. Studies in filtration. I. Critical analysis of filtration theory. II. Fundamental axiom of constant-pressure filtration.
5. Ruth, B. F.: *Ind. Eng. Chem.* **27** (1935) 708 and 806. Studies in filtration. III. Derivation of general filtration equations. IV. Nature of fluid flow through filter septa and its importance in the filtration equation.
6. Ruth, B. F. and Kempe, L.: *Trans. Am. Inst. Chem. Eng.* **33** (1937) 34. An extension of the testing methods and equations of batch filtration practice to the field of continuous filtration.
7. Ruth, B. F.: *Ind. Eng. Chem.* **38** (1946) 564. Correlating filtration theory with practice.
8. Grace, H. P.: *Chem. Eng. Prog.* **49** (1953) 303, 367, and 427. Resistance and compressibility of filter cakes.
9. Hoffing, E. H. and Lockhart, F. J.: *Chem. Eng. Prog.* **47** (1951) 3. Resistance to filtration.
10. Carman, P. C.: *Trans. Inst. Chem. Eng.* **16** (1938) 168. Fundamental principles of industrial filtration.
11. Heertjes, P. M.: *Chem. Eng. Sci.* **6** (1957) 190 and 269. Studies in filtration.
12. Grace, H. P.: *Chem. Eng. Prog.* **49**, No. 7 (1953) 367. Resistance and compressibility of filter cakes.
13. Valleroy, V. V. and Maloney, J. O.: *A.I.Ch.E.Jl.* **6** (1960) 382. Comparison of the specific resistances of cakes formed in filters and centrifuges.
14. Tiller, F. M. and Huang, C. J.: *Ind. Eng. Chem.* **53** (1961) 529. Filtration equipment. Theory.
15. Tiller, F. M. and Shirato, M.: *A.I.Ch.E.Jl.* **10** (1964) 61. The role of porosity in filtration: VI. New definition of filtration resistance.
16. Rushton, A. and Hameed, M. S.: *Filt. and Sepn.* **6** (1969) 136. The effect of concentration in rotary vacuum filtration.
17. Ehlers, S.: *Ind. Eng. Chem.* **53**, No. 8 (Aug. 1961) 552. The selection of filter fabrics re-examined.
18. Wrotnowski, A. C.: *Chem. Eng. Prog.* **58**, No. 12 (Dec. 1962) 61. Nonwoven filter media.

19. PURCHAS, D. B.: *Industrial Filtration of Liquids*, 2nd edn. (Leonard Hill, London, 1971).
20. TILLER, F. M. and CHENG, K. S.: *Filt. and Sepn.* **14** (1977) 13. Delayed cake filtration.
21. CLEASBY, J. L.: *Chem. Engr.* No. 314 (Oct. 1976) 663. Filtration with granular beds.
22. IVES, K. J.: *Proc. Int. Water Supply Assn. Eighth Congress*, Vienna. Vol. 1. (Intern. Water Supply Assn., London, 1969). Special Subject No. 7. Theory of Filtration.
23. IWASAKI, T.: *J. Am. Water Works Assn.* **29** (1937) 1591. Some notes on sand filtration.
24. SPIELMAN, L. A. and FRIEDLANDER, S. K.: *J. Colloid and Interface. Sci.* **46** (1974) 22. Role of the electrical double layer in particle deposition by convective diffusion.
25. CHERRY, G. B.: *Filt. and Sepn.* **11** (1974) 181. New developments in filter plates and filter presses.
26. PARRISH, P. and OGILVIE, H.: *Calcium Superphosphates and Compound Fertilisers. Their Chemistry and Manufacture* (Hutchison, 1939).
27. BOSLEY, R.: *Filt. and Sepn.* **11** (1974) 138. Vacuum filtration equipment innovations.
28. HARUNI, M. M. and STORROW, J. A.: *Ind. Eng. Chem.* **44** (1952) 2751; *Chem. Eng. Sci.* **1** (1952) 154; **2** (1953) 97, 108, 164, 203. Hydroextraction.

## 9.9. NOMENCLATURE

| | | Units in SI System | Dimensions in **MLT** |
|---|---|---|---|
| $A$ | Cross-sectional area of bed or filtration area | $m^2$ | $L^2$ |
| $B_1$ | Coefficient | $s/m^6$ | $L^{-6}T$ |
| $B_2$ | Coefficient | $s/m^3$ | $L^{-3}T$ |
| $b$ | Radius of basket of centrifuge | m | $L$ |
| $b'$ | Radius of inner surface of filter cake at height $y$ | m | $L$ |
| $C$ | Volume concentration of solids in the filter | — | — |
| $C_0$ | Value of $C$ at filter surface | — | — |
| $c$ | Mass of solids deposited by passage of unit volume of filtrate | $kg/m^3$ | $ML^{-3}$ |
| $e$ | Voidage of bed or filter cake | — | — |
| $e_0$ | Liquid fraction in feed slurry | — | — |
| $e_V$ | Liquid fraction in slurry in vessel | — | — |
| $e_z$ | Voidage at distance $z$ from surface | — | — |
| $H$ | Depth of basket | m | $L$ |
| $I$ | Coefficient | $(kg/m\,s^2)^{-n'}$ | $(ML^{-1}T^{-2})^{-n'}$ |
| $J$ | Mass ratio of solids to liquid in slurry | — | — |
| $L$ | Thickness of filter cake with same resistance as cloth | m | $L$ |
| $l$ | Thickness of filter cake or bed | m | $L$ |
| $n, n'$ | Indices | — | — |
| $P_1$ | Pressure at downstream face of cake | $N/m^2$ | $ML^{-1}T^{-2}$ |
| $P_2$ | Pressure at upstream face of cake | $N/m^2$ | $ML^{-1}T^{-2}$ |
| $P_z$ | Pressure at distance $z$ from surface | $N/m^2$ | $ML^{-1}T^{-2}$ |
| $-\Delta P$ | Total drop in pressure | $N/m^2$ | $ML^{-1}T^{-2}$ |
| $-\Delta P'$ | Drop in pressure across cake | $N/m^2$ | $ML^{-1}T^{-2}$ |
| $-\Delta P''$ | Pressure drop across cloth | $N/m^2$ | $ML^{-1}T^{-2}$ |
| $r$ | Specific resistance of filter cake | $1/m^2$ | $L^{-2}$ |
| $r_z$ | Specific resistance at distance $z$ from surface | $1/m^2$ | $L^{-2}$ |
| $\bar{r}$ | Specific resistance based on unit mass of cake | $m/kg$ | $M^{-1}L$ |
| $\bar{r}', \bar{r}''$ | Functions of $\bar{r}$ independent of $\Delta P$ | $m^{n'+1}s^{2n'}/kg^{n+1}$ | $M^{-(n'+1)}L^{n'+1}T^{2n'}$ |
| $\bar{r}_z$ | Value of $\bar{r}$ at distance $z$ from surface | $m/kg$ | $M^{-1}L$ |
| $S$ | Specific surface | $1/m$ | $L^{-1}$ |
| $t$ | Time | s | $T$ |
| $t'$ | Time of dismantling filter press | s | $T$ |
| $t_1$ | Time at beginning of operation | s | $T$ |
| $u$ | Mean velocity of flow calculated over the whole area | $m/s$ | $LT^{-1}$ |
| $V$ | Volume of liquid flowing in time $t$ | $m^3$ | $L^3$ |
| $V_1$ | Volume of liquid passing in time $t_1$ | $m^3$ | $L^3$ |
| $\mathbf{V}$ | Volume of vessel | $m^3$ | $L^3$ |
| $v$ | Volume of cake deposited by unit volume of filtrate | — | — |
| $W$ | Overall volumetric rate of filtration | $m^3/s$ | $L^3T^{-1}$ |

|   |   | Units in SI System | Dimensions in MLT |
|---|---|---|---|
| $x$ | Radius of inner surface of liquid at height $y$ | m | L |
| $y$ | Distance from bottom of basket | m | L |
| $z$ | Distance from surface of filter cake | m | L |
| $\lambda$ | Filter coefficient | l/m | $L^{-1}$ |
| $\mu$ | Viscosity of fluid | N s/m$^2$ | $ML^{-1}T^{-1}$ |
| $\rho$ | Density of fluid | kg/m$^3$ | $ML^{-3}$ |
| $\rho_s$ | Density of solids | kg/m$^3$ | $ML^{-3}$ |
| $\sigma$ | Volume of solids deposited per unit volume of filter | — | — |
| $\omega$ | Angular velocity | rad/s | $T^{-1}$ |

CHAPTER 10

# *Leaching*

## 10.1. INTRODUCTION

### 10.1.1. General Principles

Leaching refers to the extraction of a soluble constituent from a solid by means of a solvent. The process may be employed either for the production of a concentrated solution of a valuable solid material, or in order to free an insoluble solid, such as a pigment, from a soluble material with which it is contaminated. The method used for the extraction will be determined by the proportion of soluble constituent present, its distribution throughout the solid, the nature of the solid and the particle size.

If the solute is uniformly dispersed in the solid, the material close to the surface will first be dissolved, leaving a porous structure in the solid residue. The solvent will then have to penetrate this outer layer before it can reach further solute, and the process will become progressively more difficult and the extraction rate will fall. If the solute forms a very high proportion of the solid, this porous structure may break down almost immediately to give a fine deposit of insoluble residue, and access of solvent to the solute will not be impeded. Generally the process can be considered in three parts: first the change of phase of the solute as it dissolves in the solvent, secondly its diffusion through the solvent in the pores of the solid to the outside of the particle, and thirdly the transfer of the solute from the solution in contact with the particles to the main bulk of the solution. Any one of these three processes may be responsible for limiting the extraction rate, though the first process usually occurs so rapidly that it has a negligible effect on the overall rate.

In some cases the soluble material is distributed in small isolated pockets in a material which is impermeable to the solvent such as gold dispersed in rock, for example. In such cases the material is crushed so that all the soluble material is exposed to the solvent. If the solid has a cellular structure, the extraction rate will generally be comparatively low because the cell walls provide an additional resistance. In the extraction of sugar from beet, the cell walls perform the important function of impeding the extraction of undesirable constituents of relatively high molecular weight and the beet should therefore be prepared in long strips so that a relatively small proportion of the cells is ruptured. In the extraction of oil from seeds, the solute is itself liquid and may diffuse towards the solvent.

### 10.1.2. Factors Influencing the Rate of Extraction

The selection of the equipment for an extraction process will be influenced by the factors which are responsible for limiting the extraction rate. Thus, if the diffusion of the solute

375

through the porous structure of the residual solids is the controlling factor, the material should be of small size so that the distance the solute has to travel is small. On the other hand if diffusion of the solute from the surface of the particles to the bulk of the solution is sufficiently slow to control the process, a high degree of agitation of the fluid is called for.

There are four important factors to be considered as follows:

*Particle size.* The particle size influences the extraction rate in a number of ways. The smaller the size, the greater is the interfacial area between the solid and liquid and therefore the higher is the rate of transfer of material; further, the smaller is the distance the solute must diffuse within the solid as already indicated. On the other hand, the surface may not be so effectively used with a very fine material if circulation of the liquid is impeded, and separation of the particles from the liquid and drainage of the solid residue are made more difficult. It is generally desirable that the range of particle size should be small so that each particle requires approximately the same time for extraction and, in particular, the production of a large amount of fine material should be avoided as it may wedge in the interstices of the larger particles and impede the flow of the solvent.

*Solvent.* The liquid chosen should be a good selective solvent and its viscosity should be sufficiently low for it to circulate freely. Generally a relatively pure solvent will be used initially, but as the extraction proceeds the concentration of solute will increase and the rate of extraction will progressively decrease, first because the concentration gradient will be reduced, and secondly because the solution will generally become more viscous.

*Temperature.* In most cases, the solubility of the material which is being extracted will increase with temperature to give a higher rate of extraction. Further the diffusion coefficient will be expected to increase with rise in temperature and this will also improve the rate. In some cases, the upper limit of temperature is determined by secondary considerations, such as the necessity of preventing enzyme action during the extraction of sugar for example.

*Agitation of the fluid.* Agitation of the solvent is important because it increases the eddy diffusion and therefore increases the transfer of material from the surface of the particles to the bulk of the solution as indicated in the following section. Further, agitation of suspensions of fine particles prevents sedimentation and more effective use is made of the interfacial surface.

## 10.2. MASS TRANSFER IN LEACHING OPERATIONS

Mass transfer rates within the porous residue are difficult to assess because it is impossible to define the shape of the channels through which transfer must take place. It is possible, however, to obtain an approximate indication of the rate of transfer from the particles to the bulk of the liquid. Using the idea of a thin film being responsible for the resistance to transfer, the equation for mass transfer may be written as:

$$\frac{dM}{dt} = \frac{k'A(c_s - c)}{b} \tag{10.1}$$

where  $A$  is the area of the solid–liquid interface,
$b$  is the effective thickness of the liquid film surrounding the particles,

$c$ is the concentration of the solute in the bulk of the solution at time $t$,

$c_s$ is the concentration of the saturated solution in contact with the particles,

$M$ is the mass of solute transferred in time $t$, and

$k'$ is the diffusion coefficient. (This is approximately equal to the liquid phase diffusivity $D_L$, as discussed in Volume 1, Chapter 8. It will be taken here as constant.)

Considering a batch process in which $V$ is the total volume of solution, assumed to remain constant; then:

$$\mathrm{d}M = V\,\mathrm{d}c$$

and

$$\frac{\mathrm{d}c}{\mathrm{d}t} = \frac{k'A(c_s - c)}{bV}$$

The time $t$ taken for the concentration of the solution to rise from its initial value $c_0$ to a value $c$ is found by integration, on the assumption that both $b$ and $A$ remain constant. Rearranging:

$$\int \frac{\mathrm{d}c}{c_s - c} = \int \frac{k'A}{Vb}\,\mathrm{d}t$$

$$\therefore \quad \ln \frac{c_s - c_0}{c_s - c} = \frac{k'A}{Vb}t \tag{10.2}$$

If pure solvent is used initially, $c_0 = 0$, and:

$$1 - \frac{c}{c_s} = \mathrm{e}^{-(k'A/bV)t}$$

or

$$c = c_s(1 - \mathrm{e}^{-(k'A/bV)t}) \tag{10.3}$$

This shows that the solution approaches a saturated condition exponentially.

In most cases the interfacial area will tend to increase during the extraction, and when the soluble material forms a very high proportion of the total solid, complete disintegration of the particles may occur. Although this results in an increase in the interfacial area, the rate of extraction will probably be reduced because the free flow of the solvent will be impeded and the effective value of $b$ will be increased.

Work on the rate of dissolution of regular shaped solids in liquids has been carried out by Linton and Sherwood[1], to which reference has already been made in Volume 1, Chapter 8. They used benzoic acid, cinnamic acid, and $\beta$-naphthol as solutes, and water as solvent. For streamline flow their results were satisfactorily correlated on the assumption that transfer took place as a result of molecular diffusion alone. For turbulent flow through small tubes cast from each of the materials, the rate of mass transfer could be predicted from the pressure drop by using the "$j$-factor" for mass transfer. In their experiments with benzoic acid, they found unduly high rates of transfer because the area of the solids was increased as a result of pitting.

The effect of agitation, produced by a rotary stirrer, on mass transfer rates has been investigated by Hixson and Baum[2] who measured the rate of dissolution of pure salts in water. The degree of agitation was expressed by means of a dimensionless group $(Nd^2\rho/\mu)$ in which:

   $N$ is the number of revolutions of the stirrer per unit time,
   $d$ is the diameter of the vessel,
   $\rho$ is the density of the liquid, and
   $\mu$ is its viscosity.

This group will be referred to in Chapter 13, in which the power requirements for agitators are discussed.

For values of $(Nd^2\rho/\mu)$ less than 67,000, the results are correlated by the equation:

$$\frac{K_L d}{D_L} = 2{\cdot}7 \times 10^{-5} \left(\frac{Nd^2\rho}{\mu}\right)^{1{\cdot}4} \left(\frac{\mu}{\rho D_L}\right)^{0{\cdot}5} \tag{10.4}$$

and for higher values of $(Nd^2\rho/\mu)$:

$$\frac{K_L d}{D_L} = 0{\cdot}16 \left(\frac{Nd^2\rho}{\mu}\right)^{0{\cdot}62} \left(\frac{\mu}{\rho D_L}\right)^{0{\cdot}5} \tag{10.5}$$

where $K_L$ is the mass transfer coefficient, equal to $k'/b$ in equation 10.1.

Further experimental work was carried out on the rates of melting of a solid in a liquid, using a single component system, and Hixson and Baum were able to express their results for the heat transfer coefficient as:

$$\frac{hd}{k} = 0{\cdot}207 \left(\frac{Nd^2\rho}{\mu}\right)^{0{\cdot}63} \left(\frac{C_p\mu}{k}\right)^{0{\cdot}5} \tag{10.6}$$

for values of $(Nd^2\rho/\mu)$ greater than 67,000.

   Here $h$ is the heat transfer coefficient,
       $k$ is the thermal conductivity of the liquid, and
       $C_p$ is its specific heat.

Thus it will be seen from equations 10.5 and 10.6 that at high degrees of agitation the ratio of the heat and mass transfer coefficients is almost independent of the speed of the agitator and:

$$\frac{K_L}{h} = 0{\cdot}77 \left(\frac{D_L}{\rho C_p k}\right)^{0{\cdot}5} \tag{10.7}$$

Piret et al.[3] have attempted to reproduce the conditions in a porous solid using banks of capillary tubes, beds of glass beads and porous spheres, by measuring the rate of transfer of a salt as solute through water to the outside of the system. They have shown that the rate of mass transfer is that which would be predicted for an unsteady transfer process and have satisfactorily taken into account the shapes of the pores.

In a more recent theoretical study, Chorny and Krasuk[4] have analysed the diffusion process in extraction from simple regular solids, assuming constant diffusivity.

## 10.3. EQUIPMENT FOR LEACHING

### 10.3.1. Processes Involved

Three distinct processes are involved in normal leaching operations:
(a) Dissolving the soluble constituent.
(b) Separating the solution, so formed, from the insoluble solid residue.

(c) Washing the solid residue in order to free it of unwanted soluble matter or to obtain, as product, as much of the soluble material as possible.

Leaching has in the past been carried out mainly as a batch process but many continuous plants are now being developed. The type of equipment employed will depend on the nature of the solid—whether it is granular or cellular and whether it is coarse or fine. The normal distinction between coarse and fine solids is that the former have sufficiently large settling velocities for them to be readily separable from the liquid, whereas the latter can be maintained in suspension with the aid of only a small amount of agitation. Generally the solvent can be allowed to percolate through beds of the coarse materials, whereas the fine solids offer too high a resistance.

As already pointed out, the rate of extraction will, in general, be a function of the relative velocity between the liquid and the solid. In some plants the solid is stationary and the liquid flows through the bed of particles; in some continuous plants the solid and liquid move countercurrently.

### 10.3.2. Extraction from Cellular Materials

With seeds such as soya beans, containing only about 15 per cent of oil, solvent extraction is often used because mechanical methods are then not very efficient. Light petroleum fractions are generally used as solvents. Trichlorethylene has been used where fire risks are serious and acetone or ether where the material is very wet. A batch plant for the extraction of oil from seeds is illustrated in Fig. 10.1. This consists of a vertical cylindrical vessel divided into two sections by a slanting partition. The upper section is filled with the charge of seeds which is sprayed with fresh solvent from a distributor. The solvent percolates through the bed of solids and drains into the lower compartment where, together with any water extracted from the seeds, it is continuously boiled off by means of a steam coil. The vapours are taken to an external condenser and the mixed liquid is passed to a separating box from which the solvent is continuously fed back to the plant and the water is run to waste. By this means a concentrated solution of the oil is produced by the continued application of pure solvent to the seeds.

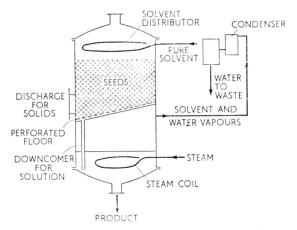

FIG. 10.1. Batch plant for extraction of oil from seeds

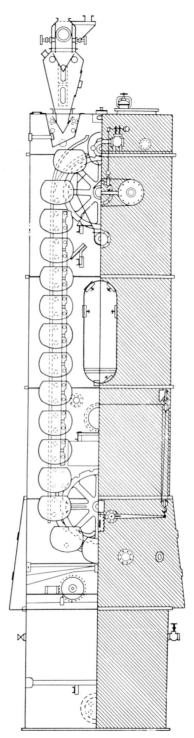

FIG. 10.2. Bollmann extractor

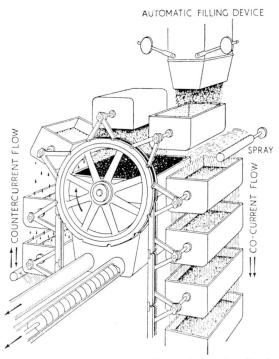

AUTOMATIC FILLING DEVICE

COUNTERCURRENT FLOW

SPRAY

CO-CURRENT FLOW

FIG. 10.3. Bollmann extractor—filling and emptying of baskets

The Bollman continuous moving bed extractor[5] (Fig. 10.2) consists of a series of perforated baskets, arranged as in a bucket elevator, contained in a vapour-tight vessel, and is widely used with seeds which do not disintegrate on extraction. Solid is fed into the top basket on the downward side and is discharged from the top basket on the upward side, as shown in Fig. 10.3. The solvent is sprayed on to the solid which is about to be discarded, and passes downwards through the baskets so that countercurrent flow is achieved. The solvent is then finally allowed to flow down through the remaining baskets in co-current flow. A typical extractor will move at about 0·3 mHz (1 revolution per hour), each basket containing about 350 kg of seeds. Generally about equal weights of seeds and solvent are used and the final solution, known as miscella, contains about 25 per cent of oil.

The Bonotto extractor[5] consists of a tall cylindrical vessel with a series of slowly rotating horizontal trays. The solid is fed continuously on to the top tray near its outside edge and a stationary scraper, attached to the shell of the plant, causes it to move towards the centre of the plate. It then falls through an opening on to the plate beneath, and another scraper moves the solids outwards on this plate which has a similar opening near its periphery. By this means the solid is moved across each plate, in opposite directions on alternate plates, until it reaches the bottom of the tower from which it is removed by means of a screw conveyor. The extracting liquid is introduced at the bottom and flows upwards so that continuous countercurrent flow is obtained, but a certain amount of mixing of solvent and solution takes place when the density of the solution rises as the concentration increases.

A more recently developed continuous extractor illustrated in Fig. 10.4 consists of a

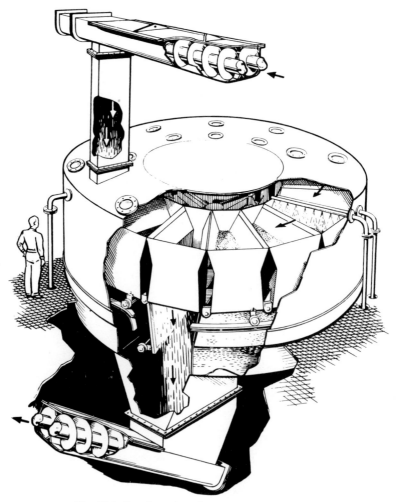

Fig. 10.4. Rosedowns' continuous solvent extractor

cylindrical tank in which 18 sector-shaped perforated cells slowly rotate. The material to be extracted is continuously fed in at a fixed point at such a rate that a cell is filled in the time taken to pass the feeder. It is then sprayed with a number of streams of solvent, each successive stream being more dilute than the previous one, and finally with pure solvent before it is discharged.

### 10.3.3. Leaching of Coarse Solids

The simple batch plant which is used for coarse solids consists of a cylindrical vessel in which the solids rest on a perforated support. The solvent is sprayed over the solids and, after extraction is complete, the residue is allowed to drain. If the solid contains such a high proportion of solute that it disintegrates, it is treated with solvent in a tank and the solution is decanted.

In the simple countercurrent system, the solid is contained in a number of tanks and the solvent flows through each in turn. The first vessel contains solid which is almost extracted

FIG. 10.5. Dorr classifier

and the last contains fresh solid. After a time, the first tank is disconnected and a fresh charge is introduced at the far end of the battery. The solvent may flow by gravity or be fed by positive pressure, and is generally heated before it enters each tank. The system is unsatisfactory in that it involves frequent interruption while the tanks are replaced, and countercurrent flow is not obtained within the units themselves.

A continuous unit in which countercurrent flow is obtained is the tray classifier, such as the Dorr classifier (Fig. 10.5). Solid is introduced near the bottom of a sloping tank and is gradually moved up by means of a rake. The solvent enters at the top and flows in the opposite direction to the solid and passes under a baffle before finally being discharged over a weir (Fig. 10.6). The classifier operates satisfactorily provided the solid does not disintegrate, and it can be arranged so that the solids are given ample time to drain before they are discharged. A number of these units can be connected in series to give countercurrent flow.

A plant has been successfully developed in Australia for the extraction of potassium sulphate from alums containing about 25 per cent soluble constituents. After roasting, the material is soft and porous with a size range from 12 mm to very fine particles, with 95 per cent greater than 100-mesh (0·15 mm), and is leached at 373 K with a solution that is

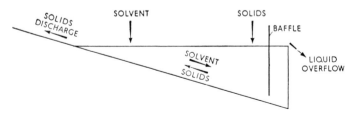

FIG. 10.6. Flow of solids and liquids in Dorr classifier

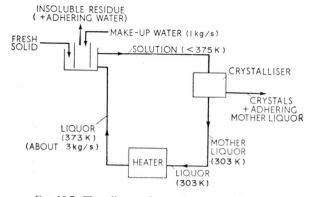

FIG. 10.7. Flow diagram for continuous leaching plant

saturated at 303 K, the flow being as shown in Fig. 10.7. The make-up water, which is used for washing the extracted solid, is required to replace that removed in the residue of spent solid, in association with the crystals, and by evaporation in the leaching tank and the crystalliser.

The leaching plant (Fig. 10.8) consists of an open tank, 3 m in diameter, into the outer portion of which the solid is continuously introduced from an annular hopper. Inside the tank a 1·8 m diameter vertical pipe rotates very slowly at the rate of about one revolution every 2400 s. It carries three ploughs stretching to the circumference of the tank, and these gradually take the solid through holes into the inside of the pipe. A hollow shaft, about 1 m in diameter, rotates in the centre of the tank at about one revolution in 200 s and carries a screw conveyor which lifts the solid and finally discharges it through an opening, so that it

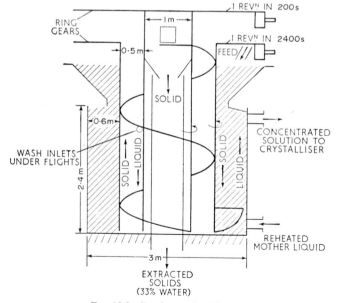

FIG. 10.8. Continuous leaching tank

falls down the shaft and is deflected into a waste pipe passing through the bottom of the tank. Leaching takes place in the outer portion of the tank where the reheated mother liquor rises through the descending solid. The make-up water is introduced under the flutes of the screw elevator, flows down over the solid and then joins the reheated mother liquor. Thus countercurrent extraction takes place in the outer part of the tank and countercurrent washing in the central portion. The plant described is said to operate efficiently and gives between 85 and 90 per cent extraction, as compared with only 50 per cent in the batch plant which it replaced.

### 10.3.4. Leaching of Fine Solids

Whereas coarse solids can be leached by causing the solvent to pass through a bed of the material, fine solids offer too high a resistance to flow. Particles less than about 200-mesh (0·075 mm) may be maintained in suspension with only a small amount of agitation, and as the total surface area is large an adequate extraction can be effected in a reasonable time. Because of the low settling velocity of the particles and their large surface, the subsequent separation and washing operations are more difficult for fine materials than for coarse solids.

Agitation can be achieved either by the use of a mechanical stirrer or by means of compressed air. If a paddle stirrer is used, precautions must be taken to prevent the whole of the liquid being swirled, with very little relative motion occurring between solids and liquid. The stirrer is often placed inside a central tube (Fig. 10.9) and the shape of the blades so arranged that the liquid is lifted upwards through the tube; the liquid then discharges at the top and flows downwards outside the tube, thus giving continuous circulation. Other types of stirrers are discussed in Chapter 13 in the context of liquid–liquid mixing.

An example of an agitated vessel in which compressed air is used is the Pachuca tank (Fig. 10.10). This is a cylindrical tank with a conical bottom, fitted with a central pipe connected to an air supply. Continuous circulation is then obtained with the central pipe acting as an air lift. Additional air jets are provided in the conical portion of the base and are used for dislodging any material that settles out.

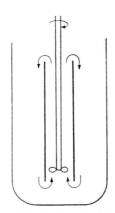

FIG. 10.9. Simple stirred tank

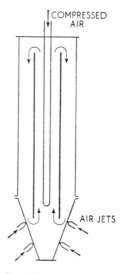

FIG. 10.10. Pachuca tank

The Dorr agitator (Fig. 10.11), which also uses compressed air for stirring, consists of a cylindrical flat-bottomed tank fitted with a central air lift inside a hollow shaft which slowly rotates. To the bottom of the shaft are fitted rakes which drag the solid material to the centre as it settles, so that it is picked up by the air lift. At the upper end of the shaft the air lift discharges into a perforated launder which distributes the suspension evenly over the surface of the liquid in the vessel. When the shaft is not rotating the rakes automatically fold up so as to prevent the plant from seizing up if it is shut down full of

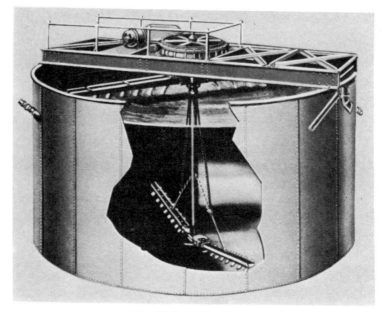

FIG. 10.11. Dorr agitator

slurry. This type of agitator can be used for batch or continuous operation. In the latter case the entry and delivery points are situated at opposite sides of the tank. The discharge pipe is often in the form of a flexible connection which can be arranged to take off the product from any desired depth. Many of these agitators are heated by steam coils.   If the soluble material dissolves very rapidly, extraction can be carried out in a thickener, such as the Dorr thickener described in Chapter 5. It gives only slight agitation and automatically separates the solid residue from the clear solution. Thickeners are also extensively used for separating the discharge from an agitator, and a number are frequently connected in series to give countercurrent washing of the residue.

## 10.4. COUNTERCURRENT WASHING

Where the residual solid after separation is still mixed with an appreciable amount of solution, it is generally desirable to pass it through a battery of washers, arranged to give countercurrent flow of the solids and the solvent as shown in Fig. 10.12. If the solids are relatively coarse a number of classifiers may be used, and with the more usual case of fine solids thickeners are generally employed. In each unit a liquid, referred to as the overflow, and a mixture of insoluble residue and solution, referred to as the underflow, are brought into contact so that intimate mixing is achieved and the solution leaving in the overflow has the same composition as that associated with the solids in the underflow. Each unit then represents an ideal stage. In some cases perfect mixing may not be achieved, and allowance must be made for the lower efficiency of the stage.

By means of a series of material balances, the compositions of all the streams in the system shown in Fig. 10.12 can be calculated on the assumption that the whole of the

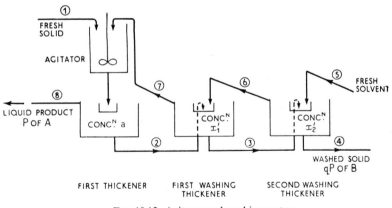

FIG. 10.12. Agitator and washing system

solute has been dissolved and that equilibrium is reached in each of the thickeners. In many cases it may only be necessary to determine the compositions of the streams entering and leaving, and the simplified methods given later may then be used.

In order to define such a system completely, the six quantities given as follows must be specified. The first four relate to the quantities and compositions of the materials used.

Quantity (e) specifies the manner in which each unit operates, and quantity (f) involves either the number of units or a specification of the duty required of the plant.

(a) The composition of the solvent fed to the system—in particular, the concentration of soluble material already present.

(b) The quantity of solvent used; alternatively the concentration of the solution to be produced may be specified, and the corresponding amount of solvent calculated from a material balance.

(c) The composition of the solid to be leached.

(d) The amount of solid fed to the system; alternatively, the amount of soluble or insoluble material required as product may be specified, and the necessary amount of solid feed then calculated from a material balance.

(e) The amount of liquid discharged with the solid in the underflow from each of the thickeners.

(f) The number and arrangement of the units; the purity of the product from the plant can then be calculated. Alternatively the required purity of the washed solid may be stated, and the number of units can then be calculated.

In the following example, a solid consisting of a soluble constituent **A** and an insoluble constituent **B** will be considered. Leaching is carried out with a pure solvent **S** and a solution is produced containing a mass $a$ of **A**, per unit mass of **S**; the total mass of **A** in solution is to be $P$. It will be assumed that the quantity of solvent removed in the underflow from each of the thickeners is the same, and is independent of the concentration of the solution in that thickener; suppose that unit mass of the insoluble material **B** removes a mass $s$ of solvent **S** in association with it. Perfect mixing in each thickener will be assumed and any adsorption of solute on the surface of the insoluble solid will be neglected. In a given thickener, therefore, the ratio of solute to solvent will be the same in the underflow as in the overflow.

The compositions of the various streams can be calculated in terms of three unknowns: $x_1'$ and $x_2'$, the ratios of solute to solvent in the first and second washing thickeners respectively, and $qP$, the amount of insoluble solid **B** in the underflow streams. An overall material balance is first made and then a balance is made on the agitator and its thickener combined, the first washing thickener and the second washing thickener. The procedure is as follows.

For the overall balance, streams 1 and 8 (Fig. 10.12) can be recorded immediately. Stream 4 is then obtained from the knowledge that the whole of **B** appears in the underflow, that the ratio **S/B** is $s$ and that the ratio **A/S** is $x_2'$. Stream 5 is obtained by difference. As pure solvent is fed to the system, a relation is obtained by equating the solute content of this stream to zero; this enables $q$ to be eliminated in terms of $x_2'$. For the agitator and separating thickener, streams 1 and 8 have already been determined, and stream 2 is obtained in the same way as stream 4 on the previous balance. Stream 7 is then obtained by difference. It is then possible to proceed in this manner from unit to unit because two streams are always common to consecutive units. Further equations are then obtained from the knowledge that the ratios **A/S** for the overflows from the two washing thickeners are equal to $x_1'$ and $x_2'$ respectively. Solution of these simultaneous equations for $x_1'$ and $x_2'$ gives each of the streams in terms of known quantities. It will be apparent that this procedure would be laborious if a larger number of units were involved, and therefore alternative methods are used to obtain the compositions of the end streams. The above method would also involve a number of trial and error solutions if it were necessary to calculate the number of units required to give a certain degree of washing. The application of the method will be readily understood from the following example.

**Example 10.1**

Caustic soda is manufactured by the lime-soda process of treating a solution of sodium carbonate in water ($0.25$ kg/s $Na_2CO_3$) with the theoretical requirement of lime, and after the reaction is complete the $CaCO_3$ sludge, containing by weight 1 part of $CaCO_3$ per 9 parts of water, is fed continuously to three thickeners in series and is washed countercurrently (Fig. 10.13). Calculate the necessary rate of feed of neutral water to the

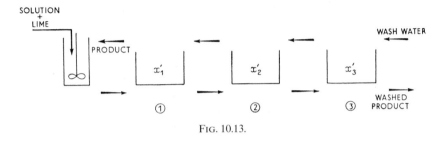

FIG. 10.13.

thickeners so that the calcium carbonate, on drying, contains only 1 per cent of sodium hydroxide. The solid discharged from each thickener contains 1 part by weight of calcium carbonate to 3 of water. The concentrated wash liquid is mixed with the contents of the agitator before being fed to the first thickener.

**Solution**

$$Na_2CO_3 + Ca(OH)_2 = 2NaOH + CaCO_3$$
$$106 \text{ kg} \qquad\qquad = 80 \text{ kg} \quad 100 \text{ kg}$$

Let $x'_1$, $x'_2$, $x'_3$ be the solute/solvent ratios in thickeners 1, 2, and 3. The quantities of $CaCO_3$, NaOH, and water in each of the streams will be calculated for every 100 kg of calcium carbonate.

| *Overall balance* | $CaCO_3$ | NaOH | Water |
|---|---|---|---|
| Feed from reactor | 100 | 80 | 900 |
| Feed as washwater | — | — | $W_f$ (say) |
| Product-underflow | 100 | $300x'_3$ | 300 |
| Product-overflow | — | $80-300x'_3$ | $600+W_f$ |

| *Thickener 1* | | | |
|---|---|---|---|
| Feed from reactor | 100 | 80 | 900 |
| Feed-overflow | — | $300(x'_1-x'_3)$ | $W_f$ |
| Product-underflow | 100 | $300x'_1$ | 300 |
| Product-overflow | — | $80-300x'_3$ | $600+W_f$ |

*Thickener 2*

| Feed-underflow | 100 | $300x'_1$ | 300 |
| Feed-overflow | — | $300(x'_2 - x'_3)$ | $W_f$ |
| Product-underflow | 100 | $300x'_2$ | 300 |
| Product-overflow | — | $300(x'_1 - x'_3)$ | $W_f$ |

*Thickener 3*

| Feed-underflow | 100 | $300x'_2$ | 300 |
| Feed-water | — | — | $W_f$ |
| Product-underflow | 100 | $300x'_3$ | 300 |
| Product-overflow | — | $300(x'_2 - x'_3)$ | $W_f$ |

Since the final underflow must contain only 1 per cent of NaOH:

$$\frac{300x'_3}{100} = 0{\cdot}01$$

If the equilibrium is achieved in each of the thickeners, the ratio of NaOH to water will be the same in the underflow and the overflow. Thus:

$$\frac{300(x'_2 - x'_3)}{W_f} = x'_3$$

$$\frac{300(x'_1 - x'_3)}{W_f} = x'_2$$

$$\frac{80 - 300x'_3}{600 + W_f} = x'_1$$

Solution of these four simultaneous equations gives

$$x'_3 = 0{\cdot}0033, \quad x'_2 = 0{\cdot}0142, \quad x'_1 = 0{\cdot}05, \quad W_f = 980$$

Thus the amount of water required for washing 100 kg/s $CaCO_3$ is 980 kg/s.

Solution fed to reactor contains 25 kg/s $Na_2CO_3$. This is equivalent to 23·6 kg/s $CaCO_3$.

Thus actual feed of water required = $(980 \times 23{\cdot}6/100)$
$$= 230 \text{ kg/s.}$$

## 10.5. CALCULATION OF THE NUMBER OF STAGES FOR COUNTERCURRENT WASHING

### 10.5.1. Batch Processes

The solid residue obtained from a batch leaching process may be washed by mixing it in a vat with liquid, allowing the mixture to settle, and then decanting the solution. This process can then be repeated until the solid is adequately washed. Suppose that, in each decantation operation, the ratio $R'$ of the amount of solvent decanted to that remaining in association with insoluble solid is a constant and independent of the concentration of the solution in the vat, then, after the first washing, the fraction of the soluble material remaining behind with the solid in the vat is $1/(R'+1)$. After the second washing, a fraction $1/(R'+1)$ of this remains behind, or $1/(R'+1)^2$ of the solute originally present is retained.

Similarly after $m$ washing operations, the fraction of the solute retained by the insoluble residue is $1/(R'+1)^m$.

### 10.5.2. Countercurrent Washing

If, in a battery of thickeners arranged in series for countercurrent washing, the amount of solvent removed with the insoluble solid in the underflow is constant, being independent of the concentration of the solution in the thickener, then the amount of solvent leaving each thickener in the underflow will then be the same, and therefore the amount of solvent in the overflow will also be the same. Hence the ratio of the solvent discharged in the overflow to that in the underflow is constant, and this will be taken as $R$. Thus:

$$R = \frac{\text{Amount of solvent discharged in the overflow}}{\text{Amount of solvent discharged in the underflow}} \tag{10.8}$$

If perfect mixing occurs in each of the thickeners and solute is not preferentially adsorbed on the surface of the solid, the concentration of the solution in the overflow will be the same as that in the underflow. If it is assumed that all the solute has been brought into solution in the agitators, then:

$$R = \frac{\text{Amount of solute discharged in the overflow}}{\text{Amount of solute discharged in the underflow}} \tag{10.9}$$

and

$$R = \frac{\text{Amount of solution discharged in the overflow}}{\text{Amount of solution discharged in the underflow}} \tag{10.10}$$

It will be noted that these relations apply only to the washing thickeners and not, in general, to the primary thickener in which the product from the agitators is first separated.

A system will now be considered consisting of $n$ washing thickeners arranged for countercurrent washing of a solid from a leaching plant, in which the whole of the soluble material is dissolved. The suspension is separated in a thickener and the underflow from this thickener is fed to the washing system as shown in Fig. 10.14.

The argument as follows is based on unit mass of insoluble solid.

Let $L_1, \ldots L_h, \ldots L_n$ be the amounts of solute in the overflows from washing thickeners 1 to $n$, respectively.

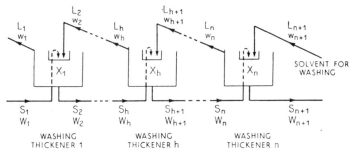

FIG. 10.14. Series of thickeners arranged for countercurrent washing

Let $w_1, \ldots w_h, \ldots w_n$ be the corresponding quantities of solution.

An amount $w_{n+1}$ of wash liquid, fed to the $n$th thickener, contains an amount $L_{n+1}$ of solute.

Let $S_2, \ldots S_{h+1}, \ldots S_{n+1}$ and $W_2, \ldots W_{h+1}, \ldots W_{n+1}$ be the amounts of solute and solution in the underflows from the thickeners.

Let $S_t$ and $W_1$ be the amounts of solute and solution with the solids which are fed to the system for washing. The solvent associated with these solids is taken as the same as that in the underflows from the washing thickeners.

Taking a solute balance on thickener $h$:

$$S_h - S_{h+1} = L_h - L_{h+1} \tag{10.11}$$

Taking a balance on solution:

$$W_h - W_{h+1} = w_h - w_{h+1} \tag{10.12}$$

Also

$$R = \frac{L_h}{S_{h+1}} = \frac{w_h}{W_{h+1}}$$

or

$$L_h = RS_{h+1} \tag{10.13}$$

and

$$w_h = RW_{h+1} \tag{10.14}$$

Taking a balance on solute for each of the thickeners in turn:

*Thickener $n$:*

$$S_n - S_{n+1} = L_n - L_{n+1} = RS_{n+1} - L_{n+1} = RS_{n+1} - L_{n+1}$$

*Thickener $n-1$:*

$$S_{n-1} - S_n = L_{n-1} - L_n = RS_n - RS_{n+1} = R^2 S_{n+1} - RL_{n+1}$$

.............................................................................................

*Thickener 2:*

$$S_2 - S_3 = L_2 - L_3 = RS_3 - RS_4 = R^{n-1}S_{n+1} - R^{n-2}L_{n+1}$$

*Thickener 1:*

$$S_1 - S_2 = L_1 - L_2 = RS_2 - RS_3 = R^n S_{n+1} - R^{n-1}L_{n+1}$$

Adding over the whole system:

$$S_1 - S_{n+1} = (R + R^2 + \ldots + R^n)S_{n+1} - (1 + R + \ldots + R^{n-1})L_{n+1}$$

or

$$S_1 = \frac{R^{n+1} - 1}{R - 1} S_{n+1} - \frac{R^n - 1}{R - 1} L_{n+1}$$

and

$$(R - 1)S_1 = (R^{n+1} - 1)S_{n+1} - (R^n - 1)L_{n+1} \tag{10.15}$$

Thus the amount of solute associated with the washed solid can be calculated in terms of the composition of the solid and of the wash liquid fed to the system. In many cases it will be stipulated that the amount of solute associated with the washed solid residue must not exceed a certain value; it is then possible to calculate directly the minimum number of thickeners necessary in order to achieve this.

If the liquid fed to the washing system is pure solvent, $L_{n+1}$ will be equal to zero and hence:

$$\frac{S_{n+1}}{S_1} = \frac{R-1}{R^{n+1}-1} \tag{10.16}$$

$(S_{n+1}/S_1)$ represents the fraction of the solute fed to the washing system which remains associated with the washed solids. If in a given case it is specified that this fraction should not exceed a value $f$, the minimum number of washing thickeners required is given by:

$$f = \frac{R-1}{R^{n+1}-1}$$

$$\therefore \quad R^{n+1} = 1 + (R-1)\frac{1}{f}$$

$$(n+1)\log R = \log\left\{1+(R-1)\frac{1}{f}\right\}$$

$$\therefore \quad n = \frac{\log\left\{1+(R-1)\dfrac{1}{f}\right\}}{\log R} - 1 \tag{10.17}$$

In general, $n$ will not work out as a whole number and the number of stages to be specified will be taken as the next higher number.

It is sometimes more convenient to work in terms of the total amount of *solution* entering and leaving each thickener. Since:

$$W_h - W_{h+1} = w_h - w_{h+1} \qquad\qquad \text{(equation 10.12)}$$

and

$$w_h = RW_{h+1} \qquad\qquad \text{(equation 10.14)}$$

using the same method as before, then:

$$(R-1)W_1 = (R^{n+1}-1)W_{n+1} - (R^n-1)w_{n+1} \tag{10.18}$$

### 10.5.3. Washing with Variable Underflow[6]

In the systems considered so far, the quantity of solvent, or of solution, removed in association with the insoluble solids has been assumed to be constant and independent of the concentration of solution in the thickener. Consideration will now be given to a similar countercurrent system in which the amount of solvent or solution in the underflow is a function of the concentration of the solution. This treatment is equally applicable to the washing thickeners alone or to the whole system involving agitator and thickeners.

The same notation will be employed as that used previously and shown in Fig. 10.14, that is:

$L$ and $w$ will denote solute and solution, respectively, in the liquid overflows and
$S$ and $W$ will denote solute and solution in the underflows.

The concentration of the solution in each thickener, defined as the ratio of solute to solution, will be denoted by the symbol $X$. Considering the overflow from thickener $h$:

$$X_h = \frac{L_h}{w_h} \tag{10.19}$$

and for the underflow:

$$X_h = \frac{S_{h+1}}{W_{h+1}} \tag{10.20}$$

It has already been stated (Section 10.4) that in order to define the system, it is necessary to specify the following quantities or other quantities from which they can be calculated by a material balance:

(a) The composition of the liquid used for washing, $X_{n+1}$.

(b) The quantity of wash liquid employed, $w_{n+1}$. Thus $L_{n+1}$ can be calculated from the relation $L_{n+1} = w_{n+1} X_{n+1}$ (equation 10.19).

(c) The composition of the solid to be washed, $S_1$ and $W_1$.

(d) The quantity of insoluble solid to be washed; this is taken as unity.

(e) The quantity of solution removed by the solid in the underflow from the thickeners; this will vary according to the concentration of the solution in the thickener, and it is therefore necessary to know the relation between $W_{h+1}$ and $X_h$. This must be determined experimentally under conditions similar to those under which the plant will operate. The data for $W_{h+1}$ should then be plotted against $X_{h+1}$. On the same graph it is convenient to plot values calculated for $S_{h+1}$ ($= W_{h+1} X_h$) (equation 10.20).

(f) The required purity of the washed solid, i.e. $S_{n+1}$. This automatically defines $X_n$ and $W_{n+1}$ whose values can be read off from the graph referred to under (e). Alternatively, the number of thickeners in the washing system may be given, and a calculation made of the purity of the product required. This problem is slightly more complicated and it will therefore be dealt with subsequently.

The solution of the problem depends on the application of material balances with respect to solute and to solution, first over the system as a whole and then over the first $h$ thickeners.

*Balance on the system as a whole.*

Solute:

$$L_{n+1} + S_1 = L_1 + S_{n+1}$$

Thus the solute in the liquid overflow from the system as a whole is given by:

$$L_1 = L_{n+1} + S_1 - S_{n+1} \tag{10.21}$$

Solution:

$$w_{n+1} + W_1 = w_1 + W_{n+1}$$

Thus the solution discharged in the liquid overflow is given by:

$$w_1 = w_{n+1} + W_1 - W_{n+1} \tag{10.22}$$

Then the concentration of the solution discharged from the system is obtained by substituting from equations 10.21 and 10.22 in 10.19:

$$X_1 = \frac{L_{n+1} + S_1 - S_{n+1}}{w_{n+1} + W_1 - W_{n+1}} \tag{10.23}$$

*Balance on the first h thickeners.*

Solute:

$$L_{h+1} + S_1 = L_1 + S_{h+1}$$

Thus the amount of solute in the liquid fed to thickener $h$

$$= L_{h+1} = L_1 + S_{h+1} - S_1$$

$$= L_{n+1} - S_{n+1} + S_{h+1} \text{ (from equation 10.21)} \tag{10.24}$$

Solution:

$$w_{h+1} + W_1 = w_1 + W_{h+1}$$

Thus the amount of solution fed to thickener $h$

$$= w_{h+1} = w_1 + W_{h+1} - W_1$$

$$= w_{n+1} - W_{n+1} + W_{h+1} \text{ (from equation 10.22)} \tag{10.25}$$

Thus the concentration of the solution fed to thickener $h$ is given by substituting from equations 10.24 and 10.25 in equation 10.19:

$$X_{h+1} = \frac{L_{n+1} - S_{n+1} + S_{h+1}}{w_{n+1} - W_{n+1} + W_{h+1}} \tag{10.26}$$

In equation 10.23, all the quantities except $X_1$ are known, and therefore $X_1$ can be calculated. (It should be noted here that if, instead of the quantity of wash liquid fed to the system, the concentration of the solution leaving the system had been given, equation 10.23 could have been used to calculate $w_{n+1}$.) Now that $X_1$ has been evaluated, the solution of the problem depends on the application of equation 10.26 in successive stages. The only unknown quantities in equation 10.26 are $X_{h+1}$, $S_{h+1}$ and $W_{h+1}$.

Applying equation 10.26 to the first stage ($h = 1$), then:

$$X_2 = \frac{L_{n+1} - S_{n+1} + S_2}{w_{n+1} - W_{n+1} + W_2} \tag{10.27}$$

Since $X_1$ is now known, the values of $S_2$ and $W_2$ can be read off from the graph, in which $S_{h+1}$ and $W_{h+1}$ are plotted against $X_h$. After substitution of these values in the equation, $X_2$ can be calculated. The next step is to apply equation 10.26 for $h = 2$. $X_2$ is now known so that $S_3$ and $W_3$ can be read off from the graph, and the value of $X_3$ can then be calculated. It is thus possible to apply equation 10.26 in this way in successive stages until the value obtained for $S_{h+1}$ is brought down to the specified value of $S_{n+1}$. The number of washing thickeners required to reduce the solute associated with the washed solid to a specified figure is thus readily calculated. In general, of course, it will not be possible to choose the number of thickeners so that $S_{h+1}$ is exactly equal to $S_{n+1}$.

It will be seen that the purity of the washed solid must be known before equations 10.23

and 10.26 can be applied. If in a given problem it is necessary to calculate the degree of washing obtained by the use of a certain number of washing thickeners, an initial assumption of the values of $S_{n+1}$ and $W_{n+1}$ must be made before the problem can be tackled. As a first step an average value for $R$ may be taken and $S_{n+1}$ calculated from equation 10.16. The method, as already given, should then be applied for the number of thickeners specified in the problem, and the calculated and assumed values of $S_{n+1}$ compared. If the calculated value is higher than the assumed value, the latter is too low. The calculated values of $S_{n+1}$ can then be plotted against the corresponding assumed values. The correct solution is then denoted by the point at which the two values agree.

## Example 10.2

A plant produces 100 kg/s ($= 864$ tonnes per day) of titanium dioxide pigment which must be 99·9 per cent pure when dried. The pigment is produced by precipitation and the material, as prepared, is contaminated with 1 kg of salt solution, containing 0·55 kg of salt, per kg of pigment. The material is washed countercurrently with water in a number of thickeners arranged in series. How many thickeners will be required if water is added at the rate of 200 kg/s and the solid discharged from each thickener removes 0·5 kg of solvent per kg of pigment?

What will be the required number of thickeners if the amount of solution removed in association with the pigment varies in the following way with the concentration of the solution in the thickener?

| Concentration of solution (kg solute/kg solution) | Amount of solution removed (kg solution/kg pigment) |
|---|---|
| 0 | 0·30 |
| 0·1 | 0·32 |
| 0·2 | 0·34 |
| 0·3 | 0·36 |
| 0·4 | 0·38 |
| 0·5 | 0·40 |

The concentrated wash liquor is mixed with the material fed to the first thickener.

## Solution

*Part* 1

Overall balance on plant in kg/s

| | $TiO_2$ | Salt | Water |
|---|---|---|---|
| Feed from reactor | 100 | 55 | 45 |
| Wash liquor added | — | — | 200 |
| Washed solid | 100 | 0·1 | 50 |
| Liquid product | — | 54·9 | 195 |

Solvent in underflow from final washing thickener $= 50$.

Solvent in overflow will be the same as that supplied for washing, i.e. 200.

$$\left(\frac{\text{Solvent discharged in overflow}}{\text{Solvent discharged in underflow}}\right) = 4 \text{ for the washing thickeners.}$$

Liquid product from plant contains 54·9 kg of salt in 195 kg of solvent.

This ratio will be the same in the underflow from the first thickener.

Thus the material fed to the washing thickeners consists of 100 kg $TiO_2$, 50 kg solvent and $(50 \times 54\cdot9/195) = 14$ kg salt.

The required number of thickeners for washing is given by:

$$\frac{(4-1)}{(4^{n+1}-1)} = \frac{0\cdot1}{14}$$

i.e.

$$4^{n+1} = 421$$

giving

$$4 < n+1 < 5$$

### Part 2

The same symbols will be used as in the text.

By inspection of the data, it is seen that $W_{h+1} = 0\cdot30 + 0\cdot2X_h$. Then

$$S_{h+1} = W_{h+1}X_h = 0\cdot30X_h + 0\cdot2X_h^2 = 5W_{h+1}^2 - 1\cdot5W_{h+1}$$

Consider the passage of unit quantity of $TiO_2$ through the plant:

$$L_{n+1} = 0, \quad w_{n+1} = 2, \quad X_{n+1} = 0$$

since 200 kg/s pure solvent is used.

$$S_{n+1} = 0\cdot001 \quad \text{and therefore} \quad W_{n+1} = 0\cdot3007$$

$$S_1 = 0\cdot55 \quad \text{and} \quad W_1 = 1\cdot00$$

Thus the concentration in the first thickener is given by equation 10.23:

$$X_1 = \frac{L_{n+1} + S_1 - S_{n+1}}{w_{n+1} + W_1 - W_{n+1}} = \frac{(0 + 0\cdot55 - 0\cdot001)}{(2 + 1 - 0\cdot3007)} = \frac{0\cdot549}{2\cdot6993} = 0\cdot203$$

From equation 10.26:

$$X_{h+1} = \frac{L_{n+1} - S_{n+1} + S_{h+1}}{w_{n+1} - W_{n+1} + W_{h+1}} = \frac{(0 - 0\cdot001 + S_{h+1})}{(2 - 0\cdot3007 + W_{h+1})} = \frac{-0\cdot001 + S_{h+1}}{1\cdot7 + W_{h+1}}$$

Since

$$X_1 = 0\cdot203, \quad W_2 = (0\cdot30 + 0\cdot2 \times 0\cdot203) = 0\cdot3406$$

and

$$S_2 = 0\cdot3406 \times 0\cdot203 = 0\cdot0691$$

Thus

$$X_2 = \frac{(0\cdot0691 - 0\cdot001)}{(1\cdot7 + 0\cdot3406)} = \frac{0\cdot0681}{2\cdot0406} = 0\cdot0334$$

Since

$$X_2 = 0{\cdot}0334, \quad W_3 = 0{\cdot}30 + 0{\cdot}2 \times 0{\cdot}0334 = 0{\cdot}30668$$

and

$$S_3 = 0{\cdot}3067 \times 0{\cdot}0334 = 0{\cdot}01025$$

Thus

$$X_3 = \frac{(0{\cdot}01025 - 0{\cdot}001)}{(1{\cdot}7 + 0{\cdot}3067)} = \frac{0{\cdot}00925}{2{\cdot}067} = 0{\cdot}00447$$

Since

$$X_3 = 0{\cdot}00447, \quad W_4 = 0{\cdot}30089 \quad \text{and} \quad S_4 = 0{\cdot}0013$$

Hence, by the same method,

$$X_4 = 0{\cdot}000150$$

Since

$$X_4 = 0{\cdot}000150, \quad W_5 = 0{\cdot}30003 \quad \text{and} \quad S_5 = 0{\cdot}000045$$

Thus $S_5$ is less than $S_{n+1}$ and therefore 4 thickeners are required.

## 10.6. NUMBER OF STAGES FOR COUNTERCURRENT WASHING BY GRAPHICAL METHODS

### 10.6.1. Introduction

It is sometimes convenient to use graphical constructions for the solution of countercurrent leaching or washing problems. This may be done by a method similar to the McCabe–Thiele method for distillation which is discussed in Chapter 11, with the

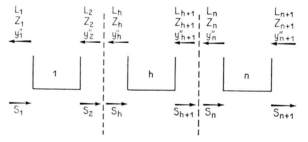

FIG. 10.15. Countercurrent washing system

overflow and underflow streams corresponding to the vapour and liquid respectively. The basis of this method will now be given, but a generally more convenient method involves the use of triangular diagrams which will be discussed later in some detail.

For the countercurrent washing system shown in Fig. 10.15, the ratio of solute to solvent in the overflow at any stage $y_h''$ can be related to the ratio of solute to insoluble solid in the underflow $S_h$ by means of a simple material balance. Using this notation:

$$y_h'' = \frac{L_h}{w_h - L_h} = \frac{L_h}{Z_h} \tag{10.28}$$

where $Z_h$ is the amount of solvent in the overflow per unit mass of insoluble solid in the underflow. If the solvent in the underflow is constant throughout the system, $Z_h$ will not be a function of concentration. Dropping the suffix of $Z$ therefore, and taking a balance on solute over thickeners $h$ to $n$ inclusive:

$$Z(y_h'' - y_{n+1}'') = S_h - S_{n+1}$$

or

$$y_h'' = \frac{S_h}{Z} - \frac{S_{n+1}}{Z} + y_{n+1}'' \qquad (10.29)$$

Thus a linear relation exists between $y_h''$ and $S_h$.

If the ratio of solvent in the overflow to solvent in the underflow from any thickener is equal to $R$:

$$\frac{L_h}{S_{h+1}} = \frac{w_h - L_h}{W_{h+1} - S_{h+1}} = \frac{Z y_h''}{S_{h+1}} = R$$

or

$$y_h'' = \frac{R}{Z} S_{h+1} \qquad (10.30)$$

This is the equation of a straight line of slope $R/Z$ which passes through the origin. Equations 10.29 and 10.30 can be represented on a $y'' - S$ diagram, as shown in Fig. 10.16. If pure solvent is used for washing, $y_{n+1}'' = 0$, and the intercept on the $S$-axis is $S_{n+1}$. As the two lines represent the relation between $y_h''$ and both $S_h$ and $S_{h+1}$, the change in composition of the underflow and overflow streams can be determined by a series of stepwise constructions, the number of steps required to change the composition of the overflow from $y_{n+1}''$ to $y_1''$ being the number of stages required.

For a variable underflow the relation between $y_h''$ and $S_{h+1}$ must be determined experimentally as the two curves are no longer straight lines, but the procedure is similar once they have been drawn. Further, it is assumed that each thickener represents an ideal stage and that the ratio of solute to solvent is the same in the overflow and the underflow. If each stage is only 80 per cent efficient, for example, equation 10.30 is no longer applicable, but the same method can be used except that each of the vertical steps will extend only 80 per cent of the way to the curve of $y_h''$ versus $S_{h+1}$.

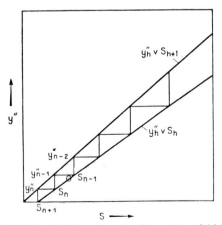

FIG. 10.16. Graphical construction for number of thickeners

Further use of graphical methods is discussed by Scheibel[7] though in the present text, attention will be confined to the use of right-angled triangular diagrams.

## 10.6.2. The Use of Right-angled Triangular Diagrams[8]

Suppose that a total mass $F$ of material is fed to a thickener and is separated into a mass $w'$ of overflow and a mass $W'$ of underflow: it will be assumed that the whole of the insoluble solid appears in the underflow. A material balance then gives:

$$F = W' + w' \qquad (10.31)$$

If $z$, $y$, and $x$ are the fractional compositions of $F$, $w'$, and $W'$, respectively, with respect to any one component in the mixture, the solute **A**, the insoluble solid **B**, or the solvent **S**, then, taking a material balance on any component:

$$Fz = W'x + w'y$$

or from equation 10.31,

$$(W' + w')z = W'x + w'y \qquad (10.32)$$

and

$$z = \frac{(W'x + w'y)}{(W' + w')} \qquad (10.33)$$

Thus for the solute and solvent respectively:

$$z_A = \frac{(W'x_A + w'y_A)}{(W' + w')}$$

$$z_S = \frac{(W'x_S + w'y_S)}{(W' + w')}$$

The composition of the three-component mixture can be represented on a right-angled triangular diagram. The proportion of solute **A** in the mixture is plotted as the abscissa, and the proportion of solvent as the ordinate; the proportion of insoluble solid is then obtained by difference. Let the point $a$ $(z_A, z_S)$ represent the composition of the material fed to the thickener, point $b$ $(x_A, x_S)$ the composition of the underflow and point $c$ $(y_A, y_S)$ the composition of the overflow as shown in Fig. 10.17.

The slope of the line $ab$ is given by:

$$\frac{z_S - x_S}{z_A - x_A} = \frac{\dfrac{W'x_S + w'y_S}{W' + w'} - x_S}{\dfrac{W'x_A + w'y_A}{W' + w'} - x_A} = \frac{w'y_S - w'x_S}{w'y_A - w'x_A} = \frac{(y_S - x_S)}{(y_A - x_A)} \qquad (10.34)$$

The slope of the line $bc$ is also $(y_S - x_S)/(y_A - x_A)$, however, so that $a$, $b$, and $c$ lie on the same straight line. Thus, if two streams are mixed, the composition of the mixture will be given by some point on the line joining the points representing the compositions of the constituent streams. Similarly, if one stream is subtracted from another, the composition of the resulting stream will lie at some point on the corresponding straight line produced.

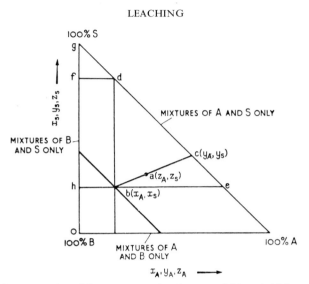

FIG. 10.17. Representation of three-component system on a right-angled triangular diagram

The location of the point will depend on the relative quantities in the two streams.
From equation 10.32,

$$(W' + w')z = W'x + w'y$$

$$\therefore \quad w'(z - y) = W'(x - z)$$

$$\therefore \quad \frac{w'}{W'} = \frac{x - z}{z - y} = \frac{x_A - z_A}{z_A - y_A} = \frac{x_S - z_S}{z_S - y_S} \qquad (10.35)$$

Thus the point representing the mixture divides the line $bc$ so that $ba/ac = w'/W'$; that is to say that $a$ is nearer to the point corresponding to the larger stream.

The proportion of insoluble solid in the underflow, for example, is given by the relation:

$$x_A + x_S + x_B = 1$$

$$\therefore \quad x_S = -x_A + (1 - x_B) \qquad (10.36)$$

Lines representing constant values of $x_B$ are therefore straight lines of slope $-1$: i.e. they are parallel to the hypotenuse of the right-angled triangle; the intercept on either axis is $(1 - x_B)$. Thus the hypotenuse represents mixtures containing no insoluble solid, and therefore the compositions of all possible overflows are represented by the hypotenuse. Further, it can be seen from the geometry of the diagram that:

$$fd = fg = x_A$$

and

$$Oh = x_S$$

thus

$$fh = bd = be$$

and

$$fh = 1 - x_A - x_S = x_B$$

The proportion of the third constituent, the insoluble solid **B**, is therefore given by the distance of the point from the hypotenuse, measured in a direction parallel to either of the main axes.

Points which lie within the triangle represent the compositions of real mixtures of the three components. Each vertex represents a pure component and each of the sides represents a two-component mixture. If two streams are mixed, the composition of the resultant stream is readily obtained and is represented by an addition point which must lie within the diagram. The composition of the material resulting from mixing a number of streams can be obtained by combining streams two at a time. If one stream is subtracted from another, a similar procedure is adopted to determine the composition of the remaining stream. The point so obtained is known as a difference point and must lie on the extension of the line joining the two given points, on the side nearer the one representing the mixed stream.

If an attempt is made to abstract from a stream more of a given component than is actually present, the composition of the resulting stream will be imaginary and will be represented by some point outside the triangle. The concept of an imaginary difference point is useful in its application to countercurrent flow processes.

*Effect of Saturation*

When the solute is initially present as a solid, the amount that can be dissolved in a given amount of solvent is limited by the solubility of the material. A saturated solution of the solute will be represented by some point, such as $A$, on the hypotenuse of the triangular diagram (Fig. 10.18). The line $OA$ represents the compositions of all possible mixtures of saturated solution with insoluble solid, since $x_S/x_A$ is constant at all points on this line. The part of the triangle above $OA$ therefore represents unsaturated solutions mixed with the insoluble solid **B**. If a mixture, represented by some point $N$, is separated into solid and liquid, it will yield a solid, of composition represented by $O$, i.e. pure component **B**, and an unsaturated solution $(N')$. Again the lower part of the diagram represents mixtures of insoluble solid, undissolved solute, and saturated solution. Thus, if a mixture $M$ is

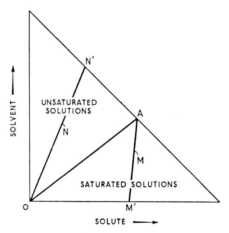

FIG. 10.18. Effect of saturation—solid state

separated into a liquid and a solid fraction, it will yield a liquid, indicated by point $A$ (i.e. a saturated solution), and a solid, consisting of **B** together with undissolved **A** ($M'$).

If the solute is initially in a liquid form and the solvent is completely miscible with it, the whole of the triangle will represent unsaturated conditions. If the solvent and solute are not completely miscible, the area can be divided into three distinct regions, as shown in Fig. 10.19. In region 1, the solvent is present as an unsaturated solution in the solute. In region 2, the liquid consists of two phases; a saturated solution of **A** in **S** and a saturated solution of **S** in **A**, in various proportions. In region 3, the liquid consists of an unsaturated solution of solute in solvent.

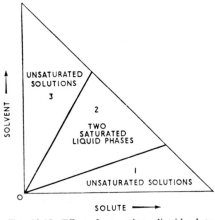

FIG. 10.19. Effect of saturation—liquid solute

In any plant, in which solvent is used to wash the adhering film of solution from the surface of the insoluble solid, the solution formed as a result of mixing in any of the units will consist of unsaturated solution. In an extracting plant, sufficient solvent will normally be added in order to dissolve the solute completely, so that, in practice, the solutions considered will rarely reach saturation.

*Representation of Underflow*

Considering the multistage industrial unit, in any equilibrium stage, the quantity of solution in the underflow may be a function of the concentration of the solution in the thickener, and the concentration of the overflow solution will be the same as that in the underflow. Let the curved line $EF$ (Fig. 10.20) represent the experimentally determined composition of the underflow for various concentrations. Any point $f$ on this line represents the composition of a mixture of pure **B** with a solution of composition $g$, and $Of/fg$ is the ratio of solution to solids in the underflow. If the amount of solution removed in the underflow is not affected by its concentration, the fractional composition of the underflow with respect to the insoluble material **B** ($x_B$) is a constant, and is represented by a straight line, through $E$, parallel to the hypotenuse. Let $EF'$ represent such a line. Point $E$ represents the composition of the underflow when the solution is infinitely weak, i.e. when

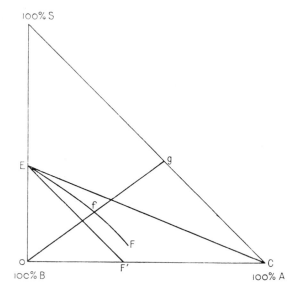

FIG. 10.20. Representation of underflow stream

it consists of pure solvent. If $K$ is the mass of solution removed in the underflow per unit mass of solids, the ordinate of $E$ is given by:

$$x_S = \frac{K}{K+1}$$

The equation of the line $EF'$ is therefore:

$$x_S = -x_A + \frac{K}{K+1}$$

If the ratio of solvent to insoluble solid is constant (and equal to $s$) in the underflow, however, the line $EF$ will be a straight line passing through the vertex $C$ of the triangle, which corresponds to pure solute $\mathbf{A}$. $E$ is then given by the coordinates:

$$x_A = 0, \quad x_S = \frac{s}{s+1}$$

### 10.6.3. Countercurrent Systems

Consider a countercurrent system consisting of $n$ thickeners, as shown in Fig. 10.21. The net flow to the right must necessarily be constant throughout the system, if no material enters or leaves at intermediate points.

$$\text{Net flow to the right} = F' = W_h' - w_h', \text{ etc.} \tag{10.37}$$

The point representing the stream $F'$ will be a difference point, since it will represent the composition of the stream which must be added to $w_h'$ to give $W_h'$. In general, in a countercurrent flow system of this sort, the net flow of all the constituents will not be in the same direction. Thus one or more of the fractional compositions in this difference stream

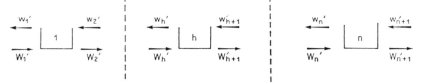

FIG. 10.21. Countercurrent extraction system

will be negative and the difference point will lie outside the triangle. A balance on the whole system, as shown in Fig. 10.21, gives:

$$W_1' - w_1' = W_{n+1}' - w_{n+1}'$$

and

$$W_1' x_1 - w_1' y_1 = W_{n+1}' x_{n+1} - w_{n+1}' y_{n+1}$$

for any of the three components.

The total net flow of material to the right at some intermediate point

$$= W_h' - w_h'$$

and the net flow of one of the constituents

$$= W_h' x_h - w_h' y_h$$

The fractional composition of the stream flowing to the right with respect to one of the components is given by:

$$x_d = \frac{W_h' x_h - w_h' y_h}{W_h' - w_h'} \tag{10.38}$$

If the direction of flow of this component is towards the right $x_d$ is positive, but if its direction of flow is to the left, $x_d$ is negative. For a countercurrent washing system, as shown in the diagram, the net flow of solvent at any stage will be to the left so that $x_{dS}$ is negative; the solute and insoluble residue will flow to the right, making $x_{dA}$ and $x_{dB}$ positive.

Consider a system as shown in Fig. 10.22, in which a dry solid is extracted with a pure solvent. The compositions of the solid and solvent, and their flowrates, are specified. It is desired to wash the residual solid so that it has not less than a certain degree of purity, and to calculate the number of thickeners required to achieve this. The compositions of the solid to be extracted, the washed solid and the solvent ($x_1, x_{n+1}$ and $y_{n+1}$ respectively) are therefore given. The composition of the concentrated solution leaving the system $y_1$ can then be calculated from a material balance over the whole plant.

The difference point, which represents the composition of the net stream of material flowing to the right at all stages, must lie on the straight line through points $x_1$ and $y_1$ produced (Fig. 10.23). Point $x_1$ has co-ordinates ($x_{A1}, x_{S1}$); $x_{A1}$ is the composition of the dry solid fed to the system; and $x_{S1}$ is zero because this solid has no solvent associated with it. Point $y_1$ represents the composition of the concentrated solution discharged from the plant, and will lie on the hypotenuse of the triangle. The difference point will lie on the line through points $x_{n+1}$ and $y_{n+1}$. Now $x_{n+1}$, the composition of the final product in the

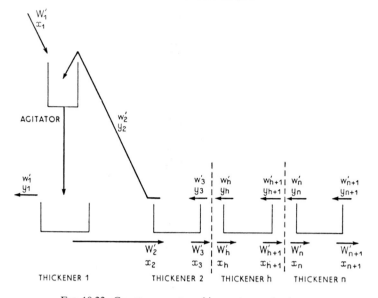

FIG. 10.22.  Countercurrent washing system and agitator

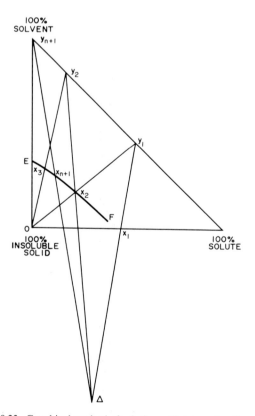

FIG. 10.23.  Graphical method of solution with triangular diagram

underflow, will lie on the line *EF* which represents the compositions of all possible underflows and which is constructed from the experimental data on the amount of solution discharged in the underflow for various concentrations in the thickeners. Because pure solvent is used for washing, $y_{n+1}$ will lie at the top vertex of the triangle. The difference point is then obtained as the point of intersection of these two straight lines. It will be denoted by $\Delta$.

As the difference point represents the difference between the underflow and the overflow at any point in the system, it is now possible to calculate the compositions of all the streams, by considering each thickener in turn.

For the agitator and thickener 1, the underflow, of composition $x_2$, will contain insoluble solid mixed with solution of the same concentration as that in the overflow $y_1$, on the assumption that equilibrium conditions are reached in the thickener. All such mixtures of solution and insoluble solid are represented by compositions on the line $Oy_1$. As this stream is an underflow, its composition must also be given by a point on the line *EF*. Thus $x_2$ is given by the point of intersection of *EF* and $Oy_1$. The composition $y_2$ of the overflow stream from thickener 2 must lie on the hypotenuse of the triangle and also on the line through points $\Delta$ and $x_2$. The composition $y_2$ is therefore determined. In this manner it is possible to find the compositions of all the streams in the system. The procedure is repeated until the amount of solute in the underflow has been reduced to a value not greater than $x_{n+1}$, and the number of thickeners is then readily counted. For the simple example illustrated, only two thickeners would be required.

This method of calculation can be applied to any system, provided that streams of material do not enter or leave at some intermediate point. If the washing system alone were considered (such as that shown in Fig. 10.14), the insoluble solid would be introduced, not as fresh solid free of solvent, but as the underflow from the thickener in which the mixture from the agitator is separated. Thus $x_1$ would lie on the line *EF* instead of on the *A* axis of the diagram.

### Example 10.3

Seeds, containing 20 per cent by weight of oil, are extracted in a countercurrent plant, and 90 per cent of the oil is recovered in a solution containing 50 per cent by weight of oil. If the seeds are extracted with fresh solvent and 1 kg of solution is removed in the underflow in association with every 2 kg of insoluble matter, how many ideal stages are required?

### Solution

This problem will be solved using the graphical method.
Since the seeds contain 20 per cent of oil:

$$x_{A1} = 0\cdot2 \quad \text{and} \quad x_{B1} = 0\cdot8$$

The final solution contains 50 per cent of oil. Thus

$$y_{A1} = 0\cdot5 \quad \text{and} \quad y_{S1} = 0\cdot5$$

The solvent which is used for extraction is pure and therefore

$$y_{S.n+1} = 1$$

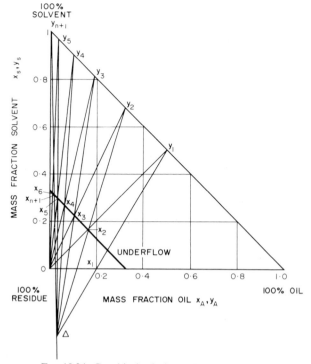

FIG. 10.24. Graphical solution to Example 10.3

Now 1 kg of insoluble solid in the washed product is associated with 0·5 kg of solution and 0·025 kg oil. Thus

$$x_{A.n+1} = 0\cdot0167, \quad x_{B.n+1} = 0\cdot6667 \quad \text{and} \quad x_{S.n+1} = 0\cdot3166$$

The mass fraction of insoluble material in the underflow is constant and equal to 0·667. The composition of the underflow is therefore represented, on the diagram, by a straight line parallel to the hypotenuse of the triangle with an intercept of 0·333 on the two main axes.

The difference point is now found by drawing in the two lines connecting $x_1$ and $y_1$ and $x_{n+1}$ and $y_{n+1}$.

The graphical construction described in the text is then used and it is seen from the diagram (Fig. 10.24) that $x_{n+1}$ lies in between $x_5$ and $x_6$.

Thus <u>5 thickeners</u> are adequate and produce the required degree of extraction.


### 10.6.4. Non-ideal Stages

If each stage in the extraction or washing system is not perfectly efficient, the ratio of solute to solvent in the overflow will be less than that in the underflow, and rather more units will be required than the number calculated by the method given. If the efficiency is independent of the concentration, allowance is simply made by dividing the theoretical number of ideal units by the efficiency. On the other hand, if the efficiency varies appreciably, account must be taken of this at each stage in the graphical construction. In

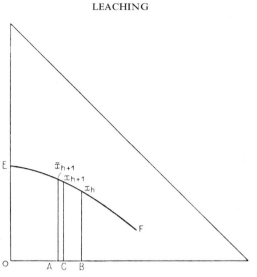

FIG. 10.25. Effect of stage efficiency

Fig. 10.25, $x_h$ represents the composition of the underflow fed to thickener $h$, and this composition would be changed to $\bar{x}_{h+1}$, say, in an ideal stage. The proportion of solute in the underflow is then reduced by an amount represented by $AB$. If the efficiency is less than unity, the change in the proportion of solute will be represented by $BC$, where $BC/AB$ is equal to the efficiency at the concentration considered, and the actual composition of the underflow is given by $x_{h+1}$. By this method it is possible to make allowance for a variable efficiency at each stage.

## 10.7. FURTHER READING

BACKHURST, J. R., HARKER, J. H., and PORTER, J. E.: *Problems in Heat and Mass Transfer* (Edward Arnold, London, 1974).

CHEN, NING HSING: *Chem. Eng., Albany* **71**, No. 24 (23 Nov. 1964) 125–8. Calculating theoretical stages in counter-current leaching.

HENLEY, E. J. and STAFFIN, H. K.: *Stagewise Process Design* (John Wiley, New York, 1963).

KARNOFSKY, G.: *Chem. Eng., Albany* **57** (Aug. 1950) 109. The Rotocel extractor.

KING, C. J.: *Separation Processes* (McGraw-Hill, New York, 1971).

MCCABE, W. L. and SMITH, J. C.: *Unit Operations of Chemical Engineering*, 3rd edn. (McGraw-Hill, New York, 1976).

MOLYNEUX, F.: *Ind. Chemist* **37,** No. 440 (Oct. 1961) 485–92. Prediction of "A" factor and efficiency in leaching calculations.

PAYNE, K. R.: *Ind. Chemist* **39,** No. 10 (Oct. 1963) 532–5. Isolation of alkaloids by batch solvent extraction.

PERRY, J. H. (Ed.) *Chemical Engineers Handbook*, 5th edn. (McGraw-Hill, New York, 1973).

SAWISTOWSKI, H. and SMITH, W.: *Mass Transfer Process Calculations* (Interscience, London, 1963).

TREYBAL, R. E.: *Mass Transfer Operations*, 2nd edn. (McGraw-Hill, New York, 1968).

## 10.8. REFERENCES

1. LINTON, W. H. and SHERWOOD, T. K.: *Chem. Eng. Prog.* **46** (1950) 258. Mass transfer from solid shapes to water in streamline and turbulent flow.

2. HIXSON, A. W. and BAUM, S. J.: *Ind. Eng. Chem.* **33** (1941) 478, 1433. Agitation: mass transfer coefficients in liquid–solid agitated systems. Agitation: heat and mass transfer coefficients in liquid–solid systems.

3. PIRET, E. L., EBEL, R. A., KIANG, C. T., and ARMSTRONG, W. P.: *Chem. Eng. Prog.* **47** (1951) 405 and 628. Diffusion rates in extraction of porous solids—1. Single phase extractions; 2. Two-phase extractions.
4. CHORNY, R. C. and KRASUK, J. H.: *Ind. Eng. Chem. Process Design and Development* **5**, No. 2 (Apr. 1966) 206–8. Extraction for different geometries. Constant diffusivity.
5. GOSS, W. H.: *J. Am. Oil Chem. Soc.* **23** (1946) 348. Solvent extraction of oilseeds.
6. RUTH, B. F.: *Chem. Eng. Prog.* **44** (1948) 71. Semigraphical methods of solving leaching and extraction problems.
7. SCHEIBEL, E. G.: *Chem. Eng. Prog.* **49** (1953) 354. Calculation of leaching operations.
8. ELGIN, J. C.: *Trans. Am. Inst. Chem. Eng.* **32** (1936) 451. Graphical calculation of leaching operations.

## 10.9. NOMENCLATURE

| | | Units in SI System | Dimensions in $\mathbf{MLT\theta}$ |
|---|---|---|---|
| $A$ | Area of solid–liquid interface | m$^2$ | $\mathbf{L^2}$ |
| $a$ | Mass of solute per unit mass of solvent in final overflow | kg/kg | — |
| $b$ | Thickness of liquid film | m | $\mathbf{L}$ |
| $C_p$ | Specific heat of solution | J/kg K | $\mathbf{L^2T^{-2}\theta^{-1}}$ |
| $c$ | Concentration of solute in solvent | kg/m$^3$ | $\mathbf{ML^{-3}}$ |
| $c_o$ | Initial concentration of solute in solvent | kg/m$^3$ | $\mathbf{ML^{-3}}$ |
| $c_g$ | Concentration of solute in solvent in contact with solid | kg/m$^3$ | $\mathbf{ML^{-3}}$ |
| $D_L$ | Liquid phase diffusivity | m$^2$/s | $\mathbf{L^2T^{-1}}$ |
| $d$ | Diameter of vessel | m | $\mathbf{L}$ |
| $F$ | Total mass of material fed to thickener | kg | $\mathbf{M}$ |
| $F'$ | Difference between underflow and overflow | kg | $\mathbf{M}$ |
| $f$ | Fraction of solute remaining with solids after washing | — | — |
| $h$ | Heat transfer coefficient | W/m$^2$ K | $\mathbf{MT^{-3}\theta^{-1}}$ |
| $K_L$ | Mass transfer coefficient | m/s | $\mathbf{LT^{-1}}$ |
| $K$ | Solution per unit mass of insoluble solid in underflow | kg/kg | — |
| $k$ | Thermal conductivity | W/m K | $\mathbf{MLT^{-3}\theta^{-1}}$ |
| $k'$ | A diffusion constant | m$^2$/s | $\mathbf{L^2T^{-1}}$ |
| $L$ | Mass of solute in overflow per unit mass of insoluble | kg/kg | — |
| $M$ | Mass of solute transferred in time $t$ | kg | $\mathbf{M}$ |
| $m$ | Number of batch washing thickeners | — | — |
| $N$ | Number of revolutions of stirrer in unit time | Hz | $\mathbf{T^{-1}}$ |
| $n$ | Number of countercurrent washing thickeners | — | — |
| $P$ | Mass of solute in overflow | kg | $\mathbf{M}$ |
| $q$ | Insoluble in underflow per unit mass of solute in overflow | kg/kg | — |
| $R$ | Solvent ratio overflow:underflow | kg/kg | — |
| $R'$ | Ratio of solvent decanted to solvent retained | kg/kg | — |
| $S$ | Solute in underflow per unit mass insoluble | kg/kg | — |
| $s$ | Solvent in underflow per unit mass insoluble | kg/kg | — |
| $t$ | Time | s | $\mathbf{T}$ |
| $V$ | Volume of solvent used for extraction | m$^3$ | $\mathbf{L^3}$ |
| $W$ | Mass of solution in underflow per unit mass of insoluble | kg/kg | — |
| $W'$ | Total mass of underflow | kg | $\mathbf{M}$ |
| $w$ | Mass of solution in overflow per unit mass of insoluble | kg/kg | — |
| $w'$ | Total mass of overflow | kg | $\mathbf{M}$ |
| $X$ | Mass of solute per unit mass of solution | kg/kg | — |
| $x$ | Fractional composition of underflow | — | — |
| $x_d$ | Fractional composition of difference stream | — | — |
| $x'$ | Mass of solute per unit mass of solvent | kg/kg | — |
| $y$ | Fractional composition of overflow | — | — |
| $y''$ | Ratio of solute to solvent in overflow | kg/kg | — |
| $Z$ | Solvent in overflow per unit mass insoluble | kg/kg | — |
| $z$ | Fractional composition of feed | — | — |
| $\mu$ | Viscosity of solution or liquid | Ns/m$^2$ | $\mathbf{ML^{-1}T^{-1}}$ |
| $\rho$ | Density of solution or liquid | kg/m$^3$ | $\mathbf{ML^{-3}}$ |

Suffixes

$A, B, S$ refer to solute, insoluble solid, solvent respectively

$1, \ldots h, \ldots n$ refer to liquid overflow or underflow feed to units $1, \ldots h, \ldots n$

CHAPTER 11

# *Distillation*

## 11.1. INTRODUCTION

The separation of liquid mixtures into their several components is one of the major processes of the chemical and petroleum industries, and distillation is the most widely used method of achieving this end; it is the key operation of the oil refinery. Throughout the chemical industry the demand for purer products, coupled with a relentless pursuit of greater efficiency, has necessitated continued research into the techniques of distillation. On the engineering side, distillation columns have to be designed with a larger range in capacity than any other types of chemical engineering equipment, with single columns from 0·3 to 10 m in diameter and from 3 m to upwards of 75 m in height. The demand on designers is not only to achieve the desired product quality at minimum cost, but also to provide constant purity of product even though there may be some variation in feed composition. A distillation unit must never be considered without its associated control unit, and usually it will operate in association with several other separate units.

The vertical cylindrical column provides in a compact form, with the minimum of ground utilisation, a large number of separate stages of vaporisation and condensation. In this chapter the basic problems of design will be considered and it will be seen that not only the physical and chemical properties, but also the fluid dynamics inside the unit, determine the number of stages required and the overall layout.

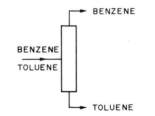

FIG. 11.1. Separation of binary mixture.

The separation of benzene from a mixture with toluene requires only a simple single unit as indicated in Fig. 11.1; and virtually pure products may be obtained. A more complex arrangement is shown in Fig. 11.2 where the columns for the purification of crude styrene formed by the dehydrogenation of ethyl benzene are shown. Here it is seen that several columns are required and that it is necessary to recycle some of the streams to the reactor.

In this chapter consideration is given to the theory of the process, methods of distillation and calculation of the number of stages required for both binary and multicomponent systems. Design methods are included for plate and packed columns using a variety of column internals.

411

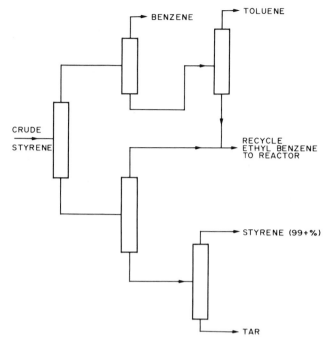

FIG. 11.2. Multicomponent separation.

## 11.2. VAPOUR–LIQUID EQUILIBRIUM

The composition of the vapour in equilibrium with a liquid of given composition is determined experimentally using an equilibrium still. The results are conveniently shown on a temperature–composition diagram (Fig. 11.3a, b, c). In the normal case (Fig. 11.3a), the curve $ABC$ shows the composition of the liquid which boils at any given temperature, and the curve $ADC$ the corresponding composition of the vapour at that temperature. Thus, a liquid of composition $x_1$ will boil at temperature $T_1$, and the vapour in equilibrium is indicated by point $D$ of composition $y_1$. It is seen that for any liquid composition $x$ the vapour formed will be richer in the more volatile component (where $x$ is the mol fraction of the more volatile component in the liquid, and $y$ in the vapour). Examples of mixtures giving this type of curve are benzene–toluene, $n$-heptane–toluene, and carbon disulphide–carbon tetrachloride.

In Fig. 11.3b and c, there is a critical composition $x_g$ where the vapour has the same composition as the liquid, so that no change occurs on boiling. Such critical mixtures are called azeotropes, and special methods are necessary to effect separation. For compositions other than $x_g$, the vapour formed will have a different composition from that of the liquid. It is important to note that these diagrams are for constant pressure conditions, and that the composition of the vapour in equilibrium with a given liquid will change with pressure.

For distillation purposes it is more convenient to plot $y$ against $x$ at a constant pressure, since the majority of industrial distillations take place at substantially constant pressure. This is shown in Fig. 11.4 where it must be remembered that the temperature varies along each of the curves.

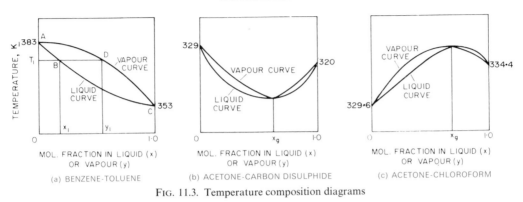

FIG. 11.3. Temperature composition diagrams

## 11.2.1. Partial Vaporisation and Partial Condensation

Suppose a mixture of benzene and toluene to be heated in a vessel, closed in such a way that the pressure remains atmospheric but no material can escape. If the mol fraction of the more volatile component (benzene) in the liquid is plotted as abscissa, and the temperature at which the mixture boils as ordinate, then the boiling curve is obtained as shown by *ABCJ* (Fig. 11.5). The corresponding dew point curve *ADEJ* shows the temperature at which a vapour of composition $y$ starts to condense.

If a mixture of composition $x_2$ is at a temperature $T_3$ below its boiling point ($T_2$), as shown by point $G$ on the diagram, then on heating at constant pressure the following changes will occur:

(a) When the temperature reaches $T_2$, the liquid will boil, as shown by point $B$, and some vapour of composition $y_2$, shown by point $E$, is formed.

(b) On further heating the composition of the liquid will change, because of the loss of the more volatile component to the vapour. The boiling point will therefore rise to some temperature $T'$. At this temperature the liquid will have a composition represented by point $L$, and the vapour a composition represented by point $N$. Since no material is lost from the system, there will be a change in the proportion of liquid to vapour, the ratio being:

$$\frac{\text{Liquid}}{\text{Vapour}} = \frac{MN}{ML}$$

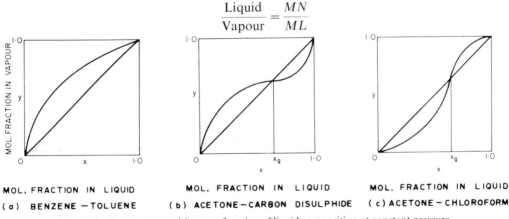

FIG. 11.4. Vapour composition as a function of liquid composition at constant pressure

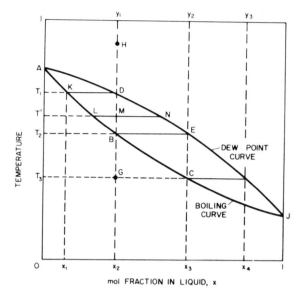

FIG. 11.5. Effect of partial vaporisation and condensation at the boiling point

(c) On further heating to a temperature $T_1$, all of the liquid is vaporised to give vapour $D$ of the same composition $y_1$ as the original liquid.

It is seen that partial vaporisation of the liquid gives a vapour richer in the more volatile component than the liquid. If the vapour initially formed, as for instance at point $E$, is at once removed by condensation, then a liquid of composition $x_3$ is obtained, represented by point $C$. The step $BEC$ may be regarded as representing an ideal stage, since the liquid passes from composition $x_2$ to a liquid of composition $x_3$, which is a greater enrichment in the more volatile component than can be obtained by any other degree of vaporisation.

Starting with superheated vapour represented by point $H$, on cooling to $D$ condensation will commence, and the first drop of liquid will have a composition $K$. Further cooling to $T'$ will give liquid $L$ and vapour $N$. Thus, partial condensation brings about enrichment of the vapour in the more volatile component in the same manner as partial vaporisation. The industrial distillation column is in essence a series of units in which these two processes of partial vaporisation and partial condensation are effected simul taneously.

### 11.2.2. Partial Pressures, and Dalton's, Raoult's, and Henry's Laws

The partial pressure $P_A$ of component $\mathbf{A}$ in a mixture of vapours is the pressure that would be exerted by component $\mathbf{A}$ at the same temperature, if present in the same volumetric concentration as in the mixture.

By Dalton's Law of Partial Pressures, $P = \Sigma P_A$, i.e. the total pressure is equal to the summation of the partial pressures. Then, in an ideal mixture, the partial pressure is proportional to the mol fraction of the constituent in the vapour phase, and:

$$P_A = y_A P \qquad (11.1)$$

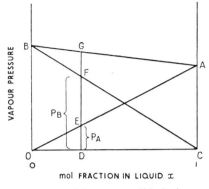

FIG. 11.6. Partial pressures of ideal mixtures

A simple relation for the partial pressure developed by a liquid solute **A** in a solvent liquid **B** is given by Henry's Law in the form:

$$P_A = \mathcal{H}' x_A \tag{11.2}$$

According to this law, the partial pressure is directly proportional to the mol fraction $x_A$ of the solute in the solvent. It is found that this holds only for dilute solutions.

The partial pressure is related to the concentration in the liquid phase by Raoult's Law which can be written:

$$P_A = P^\circ_A x_A \tag{11.3}$$

where $P^\circ_A$ is the vapour pressure of pure **A** at the same temperature. This relation is usually found to be true only for high values of $x_A$, or correspondingly low values of $x_B$, but mixtures of organic isomers and some hydrocarbons follow the law closely. If **A** is regarded as the solute and **B** as the solvent, then Henry's Law applies when $x_A$ is small, and Raoult's Law when $x_B$ is small.

If the mixture follows Raoult's Law, then the vapour pressure of a mixture can be obtained graphically from a knowledge of the vapour pressure of the two components. Thus, in Fig. 11.6, $OA$ represents the partial pressure $P_A$ of **A** in a mixture, and $CB$ the partial pressure of **B**, the total pressure being shown by the line $BA$. Thus, in a mixture of composition $D$, the partial pressure $P_A$ is given by $DE$, $P_B$ by $DF$, and the total pressure $P$ by $DG$.

Figure 11.7 shows the partial pressure of one component **A** plotted against the mol fraction for a mixture that is not ideal. It will be found that over the range $OC$ the mixture follows Henry's Law, and over $BA$ it follows Raoult's Law. Although most mixtures show wide divergences from ideality, one of the laws is usually followed at very high and very low concentrations.

If it is known that the mixture follows Raoult's Law, then the values of $y_A$ for various values of $x_A$ may be calculated from a knowledge of the vapour pressures of the two components at various temperatures. Thus:

$$P_A = P^\circ_A x_A$$

and:

$$P_A = P y_A$$

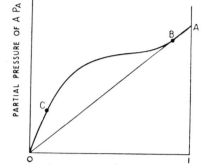

FIG. 11.7. Partial pressures of non-ideal mixtures

so that:

$$y_A = \frac{P_A^\circ x_A}{P}, \quad \text{and} \quad y_B = \frac{P_B^\circ x_B}{P} \tag{11.4}$$

But:

$$y_A + y_B = 1$$

$$\frac{P_A^\circ x_A}{P} + \frac{P_B^\circ (1 - x_A)}{P} = 1$$

giving:

$$x_A = \frac{P - P_B^\circ}{P_A^\circ - P_B^\circ} \tag{11.5}$$

### Example 11.1

For a mixture of $n$-heptane (**A**), and toluene (**B**), at 373 K and $101 \cdot 3 \, \text{kN/m}^2$

$$P_A^\circ = 106 \, \text{kN/m}^2$$

$$P_B^\circ = 73 \cdot 7 \, \text{kN/m}^2$$

$$x_A = \frac{101 \cdot 3 - 73 \cdot 7}{106 - 73 \cdot 7} = \underline{\underline{0 \cdot 856,}}$$

and

$$y_A = \frac{106 \times 0 \cdot 856}{101 \cdot 3} = \underline{\underline{0 \cdot 896.}}$$

Equilibrium data usually have to be found by tedious laboratory methods. Some proposals have been made which enable the complete diagram to be deduced with reasonable accuracy from a relatively small number of experimental values. Some of these methods are discussed by Robinson and Gilliland[1] and by Thornton and Garner[2].

### 11.2.3. Relative Volatility ($\alpha$)

The relationship between the composition of the vapour $y_A$ and of the liquid $x_A$ in equilibrium can also be expressed in another way, which is particularly useful in distillation calculations. If the ratio of the partial pressure to the mol fraction in the liquid is defined as the volatility, then:

$$\text{Volatility of } \mathbf{A} = \frac{P_A}{x_A} \quad \text{and} \quad \text{Volatility of } \mathbf{B} = \frac{P_B}{x_B}$$

The ratio of these two volatilities is known as the relative volatility $\alpha$ given by:

$$\alpha = \frac{P_A x_B}{x_A P_B}$$

Substituting $P y_A$ for $P_A$, and $P y_B$ for $P_B$:

$$\alpha = \frac{y_A x_B}{y_B x_A} \tag{11.6}$$

or

$$\frac{y_A}{y_B} = \alpha \frac{x_A}{x_B} \tag{11.7}$$

This gives a valuable relation between the ratio of $\mathbf{A}$ and $\mathbf{B}$ in the vapour and that in the liquid.

Since with a binary mixture $y_B = 1 - y_A$, and $x_B = 1 - x_A$:

$$\alpha = \frac{y_A}{1 - y_A} \frac{1 - x_A}{x_A}$$

or

$$y_A = \frac{\alpha x_A}{1 + (\alpha - 1)x_A} \tag{11.8}$$

and

$$x_A = \frac{y_A}{\alpha - (\alpha - 1)y_A} \tag{11.9}$$

This relation enables the composition of the vapour to be calculated for any desired value of $x$, if $\alpha$ is known. For separation to be achieved, $\alpha$ must not equal 1, and considering the more volatile component as $\alpha$ increases above unity, $y$ increases and the separation becomes much easier. Equation 11.7 is useful in calculation of plate enrichment and finds wide application and multicomponent distillation.

From the definition of the volatility of a component, it is seen that for an ideal system the volatility is numerically equal to the vapour pressure of the pure component. Thus the relative volatility $\alpha$ may be expressed as:

$$\alpha = \frac{P_A^\circ}{P_B^\circ} \tag{11.10}$$

This also follows by applying equation 11.1 from which $P_A/P_B = y_A/y_B$, so that:

$$\alpha = \frac{P_A x_B}{P_B x_A} = \frac{P_A^\circ x_A x_B}{P_B^\circ x_B x_A} = \frac{P_A^\circ}{P_B^\circ}$$

Whilst $\alpha$ does vary somewhat with temperature, it remains remarkably steady for many systems, and a few values to illustrate this point are given in Table 11.1.

TABLE 11.1. *Relative Volatility of Mixtures of Benzene and Toluene*

| Temperature (K) | 353 | 363 | 373 | 383 |
|---|---|---|---|---|
| $\alpha$ | 2·62 | 2·44 | 2·40 | 2·39 |

It will be seen that $\alpha$ rises as the temperature falls so that it is sometimes worthwhile to reduce the boiling point by operating at reduced pressure. When equation 11.9 is used to construct the equilibrium curve an average value of $\alpha$ must be taken over the whole column. This is valid if the relative volatilities at the top and bottom of the column are less than 15 per cent apart[3]. If they differ by more than this figure, the equilibrium curve must be constructed incrementally by calculating the relative volatility at several points along the column.

Another frequently used relationship for vapour–liquid equilibrium is the simple equation:

$$y_A = Kx_A \qquad (11.11)$$

For many systems $K$ is constant over an appreciable temperature range and equation 11.11 may be used to determine the vapour composition at any stage. The method is particularly suited to multicomponent systems and is discussed further in Section 11.7.1.

## 11.3. THE METHODS OF DISTILLATION— TWO COMPONENT MIXTURES

From curve *a* of Fig. 11.4 it is seen that, for a binary mixture with a normal $y$–$x$ curve, the vapour is always richer in the more volatile component than the liquid from which it is formed. There are three main methods used in distillation practice which all rely on this basic fact; they are:

(a) differential distillation,
(b) flash or equilibrium distillation, and
(c) rectification.

Of these rectification is much the most important, and differs from the other two methods in that part of the vapour is condensed and returned as liquid to the still, whereas, in the other methods, all the vapour is either removed as such, or is condensed as product.

### 11.3.1. Differential Distillation

The simplest example of batch distillation is a single stage, differential distillation, starting with a still pot, initially full, heated at a constant rate. In this process the vapour formed on boiling the liquid is removed at once from the system. Since this vapour is richer in the more volatile component than the liquid, it follows that the liquid remaining becomes steadily weaker in this component, with the result that the composition of the product progressively alters. Thus, whilst the vapour formed over a short period is in equilibrium with the liquid, the total vapour formed is not in equilibrium with the residual liquid. At the end of the process the liquid which has not been vaporised is removed as the bottom product. The analysis of this process was first proposed by Rayleigh[4].

Let $S$ be the number of mols of material in the still and $x$ be the mol fraction of component **A**. Suppose an amount $dS$, containing a mol fraction $y$ of **A**, be vaporised.

Then a material balance on component **A** gives:

$$y \, dS = d(Sx)$$

$$= S \, dx + x \, dS$$

$$\int_{S_0}^{S} \frac{dS}{S} = \int_{x_0}^{x} \frac{dx}{y - x}$$

and

$$\ln \frac{S}{S_0} = \int_{x_0}^{x} \frac{dx}{y - x} \tag{11.12}$$

The integral on the right-hand side can be solved graphically if the equilibrium relationship between $y$ and $x$ is available. In some cases a direct integration is possible. Thus, if over the range concerned the equilibrium relationship is a straight line of the form $y = mx + c$:

$$\ln \frac{S}{S_0} = \frac{1}{m-1} \ln \frac{(m-1)x + c}{(m-1)x_0 + c}$$

or

$$\frac{S}{S_0} = \left( \frac{y - x}{y_0 - x_0} \right)^{1/(m-1)}$$

i.e.

$$\frac{y - x}{y_0 - x_0} = \left( \frac{S}{S_0} \right)^{m-1} \tag{11.13}$$

From this equation the amount of liquid to be distilled in order to obtain a liquid of given concentration in the still may be calculated and from this the average composition of the distillate can be found by a mass balance.

Alternatively, if the relative volatility may be assumed constant over the range concerned, then $y = ax/(1 + (\alpha - 1)x)$ (equation 11.8) can be substituted in equation 11.12. This leads to the solution:

$$\ln \frac{S}{S_0} = \frac{1}{\alpha - 1} \ln \left[ \frac{x(1 - x_0)}{x_0(1 - x)} \right] + \ln \frac{1 - x_0}{1 - x} \tag{11.14}$$

As this process consists of only a single stage, a complete separation is impossible unless the relative volatility is infinite. Application is restricted to conditions where a preliminary separation is to be followed by a more rigorous distillation, where high purities are not required, or where the mixture is very easily separated.

## 11.3.2. Flash or Equilibrium Distillation

This method is frequently carried out as a continuous process. The feed is passed into the still, where part is vaporised, the vapour remaining in contact with the liquid. The mixture of vapour and liquid leaves the still and is separated so that the vapour is in equilibrium with the liquid. This system is widely used in the pipe stills of the petroleum industry where, for instance, a crude oil enters a pipe still at 440 K and about 900 kN/m$^2$,

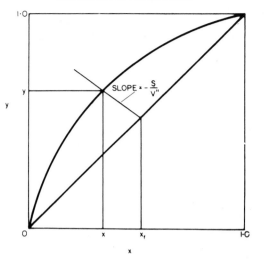

FIG. 11.8. Diagram for solution of flash distillation problems

and leaves at 520 K and a pressure of only 400 kN/m², some 15 per cent being vaporised. An analysis of the process can be given as below, for a binary mixture of **A** and **B**.

Let  $V''$  be the number of mols of vapour formed, and $y$ the mol fraction of **A**,
    $S$  be the number of mols of liquid, and $x$ the mol fraction of **A**, and
    $F''$  be the number of mols of feed of mol fraction $x_f$ of **A**.

Then a material balance on the more volatile component **A** gives:

$$F''x_f = V''y + Sx$$

But

$$F'' = V'' + S$$

$$\therefore \quad \frac{V''}{F''} = \frac{x_f - x}{y - x}$$

or

$$y = \frac{F''}{V''}x_f - \left(\frac{F''}{V''} - 1\right)x \tag{11.15}$$

Equation 11.15 represents a straight line of slope:

$$-\frac{F'' - V''}{V''} = -\frac{S}{V''},$$

passing through the point $(x_f, x_f)$. The values of $x$ and $y$ required must satisfy not only this equation but also the appropriate equilibrium data. Thus these values may be determined graphically using an $x$–$y$ diagram as shown in Fig. 11.8.

### 11.3.3. Rectification

In the two processes outlined previously, the vapour leaving the still at any time is in equilibrium with the liquid remaining, and there will normally be only a small increase in concentration of the more volatile component. The essential merit of rectification is that it enables a vapour to be obtained that is substantially richer than the liquid left in the still. This is achieved by an arrangement known as a fractionating column which enables successive vaporisation and condensation to be accomplished in one unit. Detailed consideration of the process is given in Section 11.4.

### 11.3.4. Batch Distillation

In batch distillation, which is considered in detail in Section 11.6, the more volatile component is evaporated from the still which therefore becomes progressively richer in the less volatile constituent. Distillation is continued, either until the residue of the still contains a material with an acceptably low content of the volatile material, or until the distillate is no longer sufficiently pure in respect of the volatile content.

## 11.4. THE FRACTIONATING COLUMN

### 11.4.1. The Fractionating Process

The operation of a fractionating column may be followed by reference to Fig. 11.9. The column consists of a cylindrical structure divided into sections by a series of perforated trays which permit the upward flow of vapour. The liquid reflux flows across each tray,

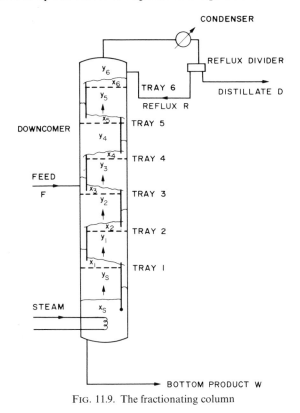

FIG. 11.9. The fractionating column

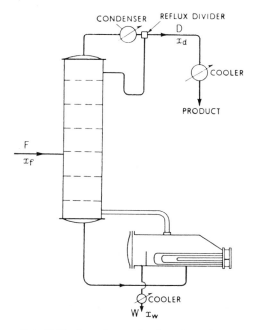

FIG. 11.10. A continuous fractionating column

over a weir and downcomer to the tray below. The vapour rising from the top tray passes
to a condenser and then to some form of reflux divider where part is withdrawn as product
$D$ and the remainder returned to the top tray as reflux. This reflux stream is frequently
passed from the condenser through a reflux drum and is then pumped to the column at a
rate determined by a suitable control device.

The liquid in the base of the column is frequently heated, either by condensing steam or
by a hot oil stream, and the vapour rises through the perforations to tray 1. A more
commonly used arrangement with an external reboiler is shown in Fig. 11.10. Here the
liquid from the still passes into the reboiler where it flows over the tubes and weir and
leaves as the bottom product; the more volatile material returns as vapour to the still.
Vapour of composition $y_s$ enters tray 1 where it is partially condensed and then
revaporised to give vapour of composition $y_1$. This operation of partial condensation of
the rising vapour and partial vaporisation of the reflux liquid is repeated on each tray.
Vapour of composition $y_6$ from the top tray is condensed to give the top product $D$ and the
reflux $R$ both of the same composition $y_6$. The feed stream is introduced on some
intermediate tray where the liquid has approximately the same composition as the feed.
The part of the column above the feed point is known as the rectifying section; the lower
portion is known as the stripping section. On an ideal tray the vapour rising from it will be
in equilibrium with the liquid leaving, though in practice a smaller degree of enrichment
will occur.

In analysing the operation on each tray it is important to note that the vapour rising to
it, and the reflux flowing down to it, are not in equilibrium, and adequate rates of mass and
heat transfer are essential for the proper functioning of the tray.

The tray described above is known as a sieve tray and has perforations of up to about
12 mm diameter, but there are several alternative arrangements for promoting mass

transfer on the tray; valve units, bubble caps and other devices are described in Section 11.10.1. In all cases the aim is to obtain good mixing of vapour and liquid with low drop in pressure across the tray.

On each tray the system tends to reach equilibrium because:

(a) Some of the less volatile component condenses from the rising vapour into the liquid thus increasing the concentration of the more volatile component (M.V.C.) in the vapour.

(b) Some of the M.V.C. is vaporised from the liquid on the tray thus decreasing the concentration of the M.V.C. in the liquid.

The number of molecules passing in each direction from vapour to liquid and in reverse will be approximately the same since the heat given out by one molecule of the vapour on condensing is approximately equal to the heat required to vaporise one molecule of the liquid. The problem is thus one of equimolecular counterdiffusion, described in Volume 1, Chapter 8.

In the arrangement discussed above, the feed is introduced continuously to the column and two product streams are obtained, one at the top much richer than the feed in the M.V.C. and the second from the base of the column weaker in the M.V.C. For the separation of small quantities of mixtures a batch still may be used (commonly in the fine organic chemical industry). Here the column rises directly from a large drum which acts as the still and reboiler and holds the charge of feed. The trays in the column now form a rectifying column and distillation is continued until it is no longer possible to obtain the desired product quality from the column. The concentration of the M.V.C. will steadily fall in the liquid remaining in the still so that enrichment to the desired level of the M.V.C. will not be possible. This problem is discussed in more detail later in Section 11.6.

A complete unit will normally consist of a feed tank, a feed heater, a column with boiler, a condenser, an arrangement for returning part of the condensed liquid as reflux, and coolers to cool the two products before running them to storage. The reflux liquor may be allowed to flow back by gravity to the top plate of the column or, as in the larger units of the petroleum industry, it is run back to a drum from which it is pumped to the top of the column. The control of the reflux on very small units is conveniently effected by hand-operated valves, and with the larger units by adjusting the delivery from a pump. In many cases the reflux is divided by means of an electromagnetically operated device which diverts the top product either to the product line or to the reflux line for controlled time intervals.

## 11.4.2. Number of Plates Required in a Distillation Column

To develop a method for the design of distillation units to give the desired fractionation it is necessary in the first instance to develop an analytical approach which will enable the necessary number of trays to be calculated. First the heat and material flows over the trays, the condenser, and the reboiler must be established. Thermodynamic data are required to establish how much mass transfer is needed to establish equilibrium between the streams clearing each tray. The required diameter of the column will be dictated by the necessity to accommodate the desired flowrates, to operate within the available drop in pressure while at the same time obtaining the desired degree of mixing of the streams on each tray.

Four streams are involved in the transfer of heat and material across a plate, as shown in Fig. 11.11.

Plate $n$ receives liquid $L_{n+1}$ from plate $n+1$ above,
and vapour $V_{n-1}$ from plate $n-1$ below.
Plate $n$ supplies liquid $L_n$ to plate $n-1$,
and vapour $V_n$ to plate $n+1$.

The action of the plate is to bring about mixing so that the vapour $V_n$, of composition $y_n$, reaches equilibrium with the liquid $L_n$, of composition $x_n$. The streams $L_{n+1}$ and $V_{n-1}$ cannot be in equilibrium and, during the interchange process on the plate, some of the

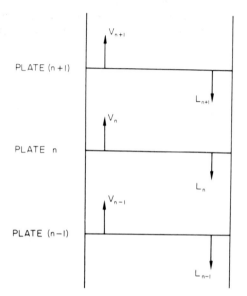

FIG. 11.11. Material balance over a plate

more volatile component is vaporised from the liquid $L_{n+1}$, decreasing its concentration to $x_n$, and some of the less volatile component is condensed from $V_{n-1}$, increasing the vapour concentration to $y_n$. The heat to vaporise the more volatile component from the liquid is supplied by partial condensation of the vapour $V_{n-1}$. Thus the resulting effect is that the more volatile component is passed from the liquid running down the column to the vapour rising up, whilst the less volatile component is transferred in the opposite direction.

*Heat Balance over a Plate*

A heat balance across plate $n$ can be written:

$$L_{n+1} H_{n+1}^L + V_{n-1} H_{n-1}^V = V_n H_n^V + L_n H_n^L + \text{losses} + \text{heat of mixing} \qquad (11.16)$$

where   $H_n^L$ is the enthalpy per mol of the liquid on plate $n$, and
$H_n^V$ is the enthalpy per mol of the vapour rising from plate $n$.

This equation is difficult to handle for the majority of mixtures, and some simplifying assumptions are usually made. Thus, with good lagging the heat losses will be small and can be neglected, and for an ideal system the heat of mixing is zero. For such mixtures, the molar heat of vaporisation can be taken as constant and independent of the composition. Thus, one mol of vapour $V_{n-1}$ on condensing releases sufficient heat to liberate one mol of vapour $V_n$. It follows that $V_n = V_{n-1}$, so that the molar vapour flow is constant up the column unless material enters or is withdrawn from the section. The temperature change from one plate to the next will be small, and $H_n^L$ can be taken as equal to $H_{n+1}^L$. Applying these simplifications to equation 11.16, it is seen that $L_n$ equals $L_{n+1}$, so that the mols of liquid reflux are also constant in this section of the column. Thus $V_n$ and $L_n$ are constant over the rectifying section, and $V_m$ and $L_m$ over the stripping section.

For these conditions there are two basic methods for determining the number of plates required; the first is due to Sorel[5] and later modified by Lewis[6], and the second is due to McCabe and Thiele[7]. The Lewis method is used here first for binary systems and it is further employed in Section 11.7.4 for calculations involving multicomponent mixtures; it is also the basis of modern computerised methods. The McCabe method is particularly important since it introduces the idea of the operating line which is a common important concept in multistage chemical engineering operations. The best assessment of these methods and their various applications is given by Underwood[8].

When the molar heat of vaporisation varies appreciably and the heat of mixing is no longer negligible these methods have to be modified, and alternative techniques are frequently used as discussed in Section 11.5.

*Calculation of Number of Plates Using the Lewis–Sorel Method*

Suppose a unit is operating as shown in Fig. 11.12, so that a binary feed $F$ is distilled to give a top product $D$ and a bottom product $W$, with $x_f$, $x_d$, and $x_w$ as the corresponding mol fractions of the more volatile component. The vapour $V_t$ rising from the top plate is condensed, and part is run back as liquid at its boiling point to the column as reflux, the remainder being withdrawn as product.

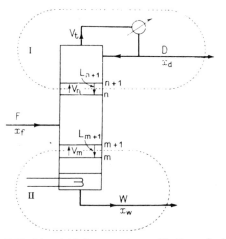

FIG. 11.12. Material balances at top and bottom of column

A material balance above plate $n$ (indicated by the loop I in Fig. 11.12) gives:

$$V_n = L_{n+1} + D \tag{11.17}$$

Expressing this balance on the more volatile component:

$$y_n V_n = L_{n+1} x_{n+1} + D x_d$$

Thus:

$$y_n = \frac{L_{n+1}}{V_n} x_{n+1} + \frac{D}{V_n} x_d \tag{11.18}$$

This equation gives a relation between the composition of the vapour rising to the plate and the composition of the liquid on any plate above the feed plate. Since the mols of liquid overflow are constant, $L_n$ may be replaced by $L_{n+1}$ and:

$$y_n = \frac{L_n}{V_n} x_{n+1} + \frac{D}{V_n} x_d \tag{11.19}$$

Similarly, taking a material balance for the total streams and for the more volatile component from the bottom to above plate $m$ (indicated by the loop II in Fig. 11.12), and noting that $L_m = L_{m+1}$:

$$L_m = V_m + W$$

and

$$y_m V_m = L_m x_{m+1} - W x_w$$

$$\therefore \quad y_m = \frac{L_m}{V_m} x_{m+1} - \frac{W}{V_m} x_w \tag{11.20}$$

This is similar to equation 11.19, and gives the corresponding relation between the compositions of the vapour rising to a plate and the liquid on the plate, for the section below the feed plate. These two equations are the equations of the operating lines.

In order to calculate the change in composition from one plate to the next, the equilibrium data are used to find the composition of the vapour above the liquid, and the enrichment line to calculate the composition of the liquid on the next plate. This method can then be repeated up the column, using equation 11.20 for sections below the feed point, and equation 11.19 for sections above the feed point.

**Example 11.2**

A mixture of benzene and toluene containing 40 mol per cent of benzene is to be separated to give a product of 90 mol per cent of benzene at the top, and a bottom product with not more than 10 mol per cent of benzene. The feed is heated so that it enters the column at its boiling point, and the vapour leaving the column is condensed but not cooled, and provides reflux and product. It is proposed to operate the unit with a reflux ratio of 3 kmol/kmol product. It is required to find the number of theoretical plates needed and the position of entry for the feed. The equilibrium diagram for operating at $101 \cdot 3 \, \mathrm{kN/m^2}$ pressure is shown in Fig. 11.13.

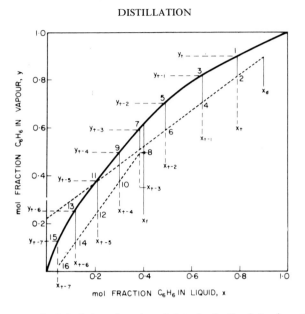

FIG. 11.13. Calculation of number of plates by the Sorel–Lewis method

**Solution**

Considering 100 kmol of feed a total material balance gives:

$$100 = D + W$$

A balance on the M.V.C. (benzene) gives:

$$100 \times 0.4 = 0.9D + 0.1W$$

Thus:

$$40 = 0.9(100 - W) + 0.1W$$

whence:

$$W = 62.5 \quad \text{and} \quad D = 37.5 \, \text{kmol}$$

Using the notation in Fig. 11.12:

$$L_n = 3D = 112.5$$

and

$$V_n = L_n + D = 150$$

Thus, the top operating line from equation 11.19 is:

$$y_n = \frac{112.5}{150} x_{n+1} + \frac{37.5 \times 0.9}{150}$$

or

$$y_n = 0.75 x_{n+1} + 0.225 \tag{11.21}$$

Since the feed is all liquid at its boiling point, it will all run down as increased reflux to the plate below. Thus:

$$L_m = L_n + F$$

$$= 112.5 + 100 = 212.5$$

Again:

$$V_m = L_m - W$$

$$= 212 \cdot 5 - 62 \cdot 5 = 150 = V_n$$

$$\therefore \quad y_m = \frac{212 \cdot 5}{150} x_{m+1} - \frac{62 \cdot 5}{150} \times 0 \cdot 1 \qquad \text{(equation 11.20)}$$

i.e.

$$y_m = 1 \cdot 415 x_{m+1} - 0 \cdot 042 \qquad (11.22)$$

With the two equations 11.21 and 11.22 and the equilibrium curve, the composition on the various plates can be calculated either by working from the still up to the condenser, or in the reverse direction. Since all the vapour from the column is condensed, the composition of the vapour $y_t$ from the top plate must equal that of the product $x_d$, and that of the liquid returned as reflux $x_r$. The composition $x_t$ of the liquid on the top plate is found from the equilibrium curve and since it is in equilibrium with vapour of composition $y_t = 0 \cdot 90$, $x_t = 0 \cdot 79$.

The value of $y_{t-1}$ is obtained from equation 11.21 as:

$$y_{t-1} = 0 \cdot 75 \times 0 \cdot 79 + 0 \cdot 225 = 0 \cdot 593 + 0 \cdot 225 = 0 \cdot 818$$

$x_{t-1}$ is obtained from equilibrium curve as $0 \cdot 644$

$$y_{t-2} = 0 \cdot 75 \times 0 \cdot 644 + 0 \cdot 225 = 0 \cdot 483 + 0 \cdot 225 = 0 \cdot 708$$

$x_{t-2}$ from equilibrium curve $= 0 \cdot 492$

$$y_{t-3} = 0 \cdot 75 \times 0 \cdot 492 + 0 \cdot 225 = 0 \cdot 369 + 0 \cdot 225 = 0 \cdot 594$$

$x_{t-3}$ from equilibrium curve $= 0 \cdot 382$

This last value of composition is sufficiently near to that of the feed for the feed to be introduced on plate $t-3$. For the lower part of the column, the operating line equation 11.22 will be used.

$$y_{t-4} = 1 \cdot 415 \times 0 \cdot 382 - 0 \cdot 042 = 0 \cdot 540 - 0 \cdot 042 = 0 \cdot 498$$

$x_{t-4}$ from equilibrium curve $= 0 \cdot 298$

$$y_{t-5} = 1 \cdot 415 \times 0 \cdot 298 - 0 \cdot 042 = 0 \cdot 421 - 0 \cdot 042 = 0 \cdot 379$$

$x_{t-5}$ from equilibrium curve $= 0 \cdot 208$

$$y_{t-6} = 1 \cdot 415 \times 0 \cdot 208 - 0 \cdot 042 = 0 \cdot 294 - 0 \cdot 042 = 0 \cdot 252$$

$x_{t-6}$ from equilibrium curve $= 0 \cdot 120$

$$y_{t-7} = 1 \cdot 415 \times 0 \cdot 120 - 0 \cdot 042 = 0 \cdot 169 - 0 \cdot 042 = 0 \cdot 127$$

$x_{t-7}$ from equilibrium curve $= 0 \cdot 048$.

This liquid $x_{t-7}$ is slightly weaker than the minimum required and may be withdrawn as the bottom product. Thus, $x_{t-7}$ will correspond to the reboiler, and there will be seven plates in the column.

## The Method of McCabe and Thiele[7]

The simplifying assumptions of constant molal heat of vaporisation, of no heat losses, and of no heat on mixing, led to a constant molal vapour flow and a constant molar reflux flow in any section of the column, i.e. $V_n = V_{n+1}$, $L_n = L_{n+1}$, etc. Using these simplifications, the two enrichment equations were obtained:

$$y_n = \frac{L_n}{V_n} x_{n+1} + \frac{D}{V_n} x_d \qquad \text{(equation 11.19)}$$

and

$$y_m = \frac{L_m}{V_m} x_{m+1} - \frac{W}{V_m} x_w. \qquad \text{(equation 11.20)}$$

These equations were used in the Lewis–Sorel method to calculate the relation between the composition of the liquid on a plate and the composition of the vapour rising to that plate. McCabe and Thiele pointed out that, since these equations represent straight lines connecting $y_n$ with $x_{n+1}$ and $y_m$ with $x_{m+1}$, they can be drawn on the same diagram as the equilibrium curve to give a simple graphical solution for the number of stages required. Thus, the line of equation 11.19 will pass through the points 2, 4, 6, etc., shown in Fig. 11.13, and similarly the line of equation 11.20 will pass through points 8, 10, 12, 14.

In equation 11.19, if $x_{n+1} = x_d$, then:

$$y_n = \frac{L_n}{V_n} x_d + \frac{D}{V_n} x_d = x_d \qquad (11.23)$$

so that this equation represents a line passing through the point $y_n = x_{n+1} = x_d$. Further, if $x_{n+1}$ is put equal to zero, then $y_n = Dx_d/V_n$, giving a second easily determined point. The top operating line is therefore drawn through two points of coordinates $(x_d, x_d)$ and $(0, (Dx_d/V_n))$.

For the bottom operating line, equation 11.20, if $x_{m+1} = x_w$, then:

$$y_m = \frac{L_m}{V_m} x_w - \frac{W}{V_m} x_w$$

Since $V_m = L_m - W$, it follows that $y_m = x_w$. Thus the bottom operating line passes through the point $C$, i.e. $(x_w, x_w)$, and has a slope $L_m/V_m$. When the two operating lines have been drawn in, the number of stages required may be found by drawing steps between the operating line and the equilibrium curve starting from point $A$.

This method is one of the most important concepts in chemical engineering and is an invaluable tool for the solution of distillation problems. The assumption of constant molal overflow is not limiting since in very few systems do the molal heats of vaporisation differ by more than 10 per cent. The method does have limitations, however, and should not be employed when the relative volatility is less than 1·3 or greater than 5, when the reflux ratio is less than 1·1 times the minimum, or when more than twenty-five theoretical trays are required[3]. In these circumstances, the Ponchon–Savarit method described in Section 11.5 should be used.

## Example 11.3. The McCabe–Thiele Method

Example 11.2 will now be worked using this method. Thus, with a feed composition $x_f$ of 0·4, the top composition $x_d$ is to have a value of 0·9 and the bottom composition $x_w$ of 0·10. The reflux ratio $L_n/D = 3$.

**Solution** (Fig. 11.14)

1. By a material balance for a feed of 100 kmol:

$$V_n = V_m = 150; \quad L_n = 112·5; \quad L_m = 212·5; \quad D = 37·5; \quad W = 62·5.$$

2. Draw the equilibrium curve and the diagonal line.
3. The equation of the top operating line is:

$$y_n = 0·75x_{n+1} + 0·225 \qquad \text{(equation 11.21)}$$

Thus line $AB$ is drawn through the two points $A$ (0·9, 0·9) and $B$ (0, 0·225).

4. The equation of the bottom operating line is:

$$y_m = 1·415x_{m+1} - 0·042 \qquad \text{(equation 11.22)}$$

This equation is represented by line $CD$ drawn through $C$ (0·1, 0·1) at a slope of 1·415.

5. Starting at point $A$, draw the horizontal line to cut the equilibrium curve on point 1. Drop the vertical through 1 to the operating line at point 2. Proceed in this way to obtain points 3–6.

6. Draw horizontal line through 6 to cut equilibrium curve on point 7, and vertical through 7 to lower enrichment line at point 8. Proceed to get points 9–16.

7. Count the number of stages, i.e., points 2, 4, 6, 8, 10, 12, 14, which gives the number of plates as 7.

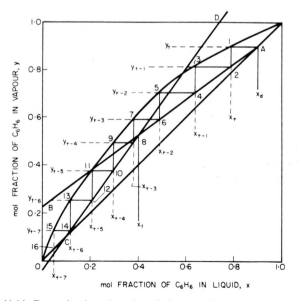

FIG. 11.14. Determination of number of plates by the McCabe–Thiele method

*Enrichment in Still and Condenser*

Point 16 in Fig. 11.14 represents the concentration of the liquor in the still. The concentration of the vapour is represented by point 15, so that the enrichment represented by the increment 16–15 is achieved in the boiler or still body. Again, the concentration on the top plate is given by point 2, but the vapour from this plate has a concentration given by point 1, and the condenser by completely condensing the vapour gives a product of equal concentration, represented by point $A$. The still and condenser together, therefore, provide enrichment $(16-15)+(1-A)$, which is equivalent to one ideal stage. Thus, the actual number of theoretical plates required is one less than the number of stages shown on the diagram. From a liquid in the still (point 16) to the product (point $A$) there are eight steps, but the column need only contain seven theoretical plates.

*The Intersection of the Operating Lines*

It will have been seen from the example shown in Fig. 11.14 in which the feed enters as liquid at its boiling point that the two operating lines intersect at a point having an $X$-coordinate of $x_f$. The locus of the point of intersection of the operating lines is of considerable importance since, as will be seen, it is dependent on the temperature and physical condition of the feed.

If the two operating lines intersect at a point with coordinates $(x_q, y_q)$, then from equations 11.19 and 11.20:

$$V_n y_q = L_n x_q + D x_d \tag{11.24}$$

and

$$V_m y_q = L_m x_q - W x_w \tag{11.25}$$

whence

$$y_q(V_m - V_n) = (L_m - L_n)x_q - (Dx_d + Wx_w) \tag{11.26}$$

A material balance over the feed plate gives:

$$F + L_n + V_m = L_m + V_n$$

or

$$V_m - V_n = L_m - L_n - F \tag{11.27}$$

To obtain a relation between $L_n$ and $L_m$, it is necessary to take an enthalpy balance over the feed plate, and to consider what happens when the feed enters the column. If the feed is all in the form of liquid at its boiling point, the reflux $L_m$ overflowing to the plate below will be $L_n + F$. If however the feed is a liquid at a temperature $T_f$, less than the boiling point, some vapour rising from the plate below will condense to provide sufficient heat to bring the feed liquor to the boiling point.

Let $H_f$ be the enthalpy per mol of feed, and $H_{fs}$ the enthalpy of one mol of feed at its boiling point. Then the heat to be supplied to bring feed to the boiling point is $F(H_{fs} - H_f)$, and the number of mols of vapour to be condensed to provide this heat is $F(H_{fs} - H_f)/\lambda$, where $\lambda$ is the molar latent heat of the vapour.

Then, reflux liquor

$$L_m = L_n + F + \frac{F(H_{fs} - H_f)}{\lambda}$$

$$= L_n + F\left(\frac{\lambda + H_{fs} - H_f}{\lambda}\right)$$

$$= L_n + qF \qquad (11.28)$$

where

$$q = \frac{\text{heat to vaporise 1 mol of feed}}{\text{molar latent heat of the feed}}$$

Thus, from equation 11.27:

$$V_m - V_n = qF - F \qquad (11.29)$$

A material balance of the more volatile component over the whole column gives:

$$Fx_f = Dx_d + Wx_w$$

Thus, from equation 11.26:

$$F(q-1)y_q = qFx_q - Fx_f$$

or

$$y_q = \frac{q}{q-1}x_q - \frac{x_f}{q-1} \qquad (11.30)$$

This equation is commonly known as the equation of the $q$-line. If $x_q = x_f$, then $y_q = x_f$. Thus, the point of intersection of the two operating lines lies on the straight line of slope

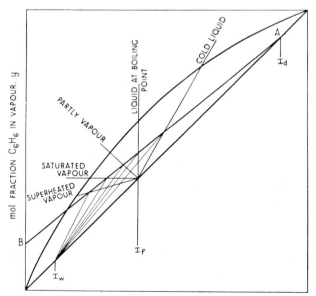

FIG. 11.15. Effect of the condition of the feed on the intersection of the operating lines for a fixed reflux ratio

$q/(q-1)$ passing through the point $(x_f, x_f)$. When $y_q = 0, x_q = x_f/q$. The line can thus be drawn through two easily determined points. From the definition of $q$, it follows that the slope of the $q$-line is governed by the nature of the feed.

| | | |
|---|---|---|
| (a) Cold feed as liquor | $q > 1$ | $q$ line ╱ |
| (b) Feed at boiling point | $q = 1$ | $q$ line │ |
| (c) Feed partly vapour | $0 < q < 1$ | $q$ line ╲ |
| (d) Feed saturated vapour | $q = 0$ | $q$ line — |
| (e) Feed superheated vapour | $q < 0$ | $q$ line ╱ |

These various conditions are indicated in Fig. 11.15.

Alteration in the slope of the $q$-line will alter the liquid concentration at which the two operating lines cut each other for a given reflux ratio. This will mean a slight alteration in the number of plates required for the given separation. Whilst the change in the number of plates will usually be rather small, if the feed is cold, there will be an increase in reflux flow below the feed plate, and hence an increased heat consumption from the boiler per mol of distillate.

### 11.4.3. The Importance of the Reflux Ratio

*Influence on the Number of Plates Required*

The ratio $L_n/D$ of the top overflow to the quantity of product is denoted by $R$ and this enables the equation of the operating line to be expressed in another way, which is often more convenient. Thus, introducing $R$ in equation 11.19:

$$y_n = \frac{L_n}{L_n + D} x_{n+1} + \frac{D}{L_n + D} x_d \tag{11.31}$$

$$= \frac{R}{R+1} x_{n+1} + \frac{x_d}{R+1} \tag{11.32}$$

Any change in $R$ will therefore modify the slope of the operating line and, as can be seen from Fig. 11.14, will alter the number of plates required for a given separation. If $R$ is known, the top line is most easily drawn by joining point $A$ $(x_d, x_d)$ to $B(0, x_d/(R+1))$ (Fig. 11.16). This method avoids the calculation of the actual flow rates $L_n, V_n$, when only the number of plates is to be estimated.

If no product is withdrawn from the still ($D = 0$), the column is said to operate under conditions of total reflux and, as seen from equation 10.29, the top operating line has its maximum slope of unity, and coincides with the line $x = y$. If the reflux ratio is reduced, the slope of the operating line is reduced and more stages are required to pass from $x_f$ to $x_d$, as shown by the line $AK$ in Fig. 11.16. Further reduction in $R$ will eventually bring the operating line to $AE$, when an infinite number of stages is needed to pass from $x_d$ to $x_f$. This arises from the fact that under these conditions the steps become very close together at liquid compositions near $x_f$, and no enrichment occurs from the feed plate to the plate above. These conditions are known as minimum reflux, and the reflux ratio is denoted by $R_m$. Any small increase in $R$ beyond $R_m$ will give a workable system, though a large number of plates will be required. It is important to note that any line such as $AG$, which is equivalent to a smaller value of $R$ than $R_m$, represents an impossible condition, since it is impossible to pass beyond point $G$ towards $x_f$. Two important deductions may be made;

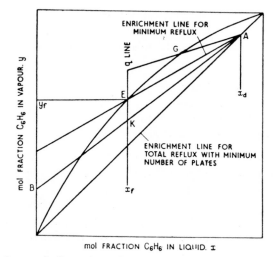

FIG. 11.16. Influence of reflux ratio on the number of plates required for a given separation

first that the minimum number of plates is required for a given separation at conditions of total reflux, and secondly that there is a minimum reflux ratio below which it is impossible to obtain the desired enrichment, however many plates are used.

*Calculation of Minimum Reflux Ratio $R_m$*

Figure 11.16 represents conditions where the $q$-line is vertical, and the point $E$ lies on the equilibrium curve and has co-ordinates $(x_f, y_f)$. Then the slope of the line $AE$ is given by:

$$\frac{R_m}{R_m+1} = \frac{x_d - y_f}{x_d - x_f}$$

or

$$R_m = \frac{x_d - y_f}{y_f - x_f} \tag{11.33}$$

If the $q$-line is horizontal as in Fig. 11.17 the enrichment line for minimum reflux is given by $AC$, where $C$ has co-ordinates $(x_c, y_c)$. Then:

$$\frac{R_m}{R_m+1} = \frac{x_d - y_c}{x_d - x_c}$$

or, since $y_c = x_f$:

$$R_m = \frac{x_d - y_c}{y_c - x_c} = \frac{x_d - x_f}{x_f - x_c} \tag{11.34}$$

*Underwood and Fenske equations.* For ideal mixtures, or where over the concentration range concerned the relative volatility can be taken as constant, $R_m$ can be obtained analytically from the physical properties of the system. Thus, if $x_{nA}$ and $x_{nB}$ are the mol fractions of two components **A** and **B** in the liquid on any plate $n$, then a material

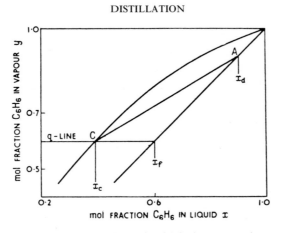

FIG. 11.17. Minimum reflux ratio with feed as saturated vapour

balance over the top portion of the column above plate $n$ gives:

$$V_n y_{nA} = L_n x_{(n+1)A} + D x_{dA} \tag{11.35}$$

and

$$V_n y_{nB} = L_n x_{(n+1)B} + D x_{dB} \tag{11.36}$$

Under conditions of minimum reflux, a column has to have an infinite number of plates, or alternatively the composition on plate $n$ is equal to that on plate $n+1$. Dividing equation 11.35 by equation 11.36 and using the relations $x_{(n+1)A} = x_{nA}$ and $x_{(n+1)B} = x_{nB}$:

$$\frac{\alpha x_{nA}}{x_{nB}} = \frac{y_{nA}}{y_{nB}} = \frac{L_n x_{nA} + D x_{dA}}{L_n x_{nB} + D x_{dB}}$$

$$\therefore \quad R_m = \left(\frac{L_n}{D}\right)_{\min} = \frac{1}{\alpha - 1}\left[\frac{x_{dA}}{x_{nA}} - \alpha \frac{x_{dB}}{x_{nB}}\right] \tag{11.37}$$

In this analysis, $\alpha$ is taken as the volatility of **A** relative to **B**. There is therefore, in general, a different value of $R_m$ for each plate. In order to produce any separation of the feed, the minimum relevant value of $R_m$ is that for the feed plate, so that the minimum reflux ratio for the desired separation is:

$$R_m = \frac{1}{\alpha - 1}\left[\frac{x_{dA}}{x_{fA}} - \alpha \frac{x_{dB}}{x_{fB}}\right] \tag{11.38}$$

For a binary system this becomes:

$$R_m = \frac{1}{\alpha - 1}\left[\frac{x_{dA}}{x_{fA}} - \alpha \frac{(1 - x_{dA})}{(1 - x_{fA})}\right] \tag{11.39}$$

This relation could be obtained by putting $y = \alpha x/(1 + (\alpha - 1)x)$ (equation 11.8) in equation 11.33:

$$R_m = \frac{x_d - \dfrac{\alpha x_f}{1 + (\alpha - 1)x_f}}{\dfrac{\alpha x_f}{1 + (\alpha - 1)x_f} - x_f} = \frac{1}{\alpha - 1}\left[\frac{x_d}{x_f} - \frac{\alpha(1 - x_d)}{(1 - x_f)}\right] \tag{11.40}$$

*The Number of Plates at Total Reflux. Fenske's Method*[9]

For conditions in which the relative volatility is constant, Fenske derived an equation for calculating the required number of plates for a desired separation. Since no product is withdrawn from the still, the equations of the two operating lines become:

$$y_n = x_{n+1} \quad \text{and} \quad y_m = x_{m+1} \tag{11.41}$$

Consider two components **A** and **B**, the concentrations of which in the still are $x_{sA}$ and $x_{sB}$. Then the composition on the first plate is obtained by:

$$\left(\frac{x_A}{x_B}\right)_1 = \left(\frac{y_A}{y_B}\right)_s = \alpha_s\left(\frac{x_A}{x_B}\right)_s$$

where the subscript outside the bracket indicates the plate, and $s$ the still. For plate 2:

$$\left(\frac{x_A}{x_B}\right)_2 = \left(\frac{y_A}{y_B}\right)_1 = \alpha_1\left(\frac{x_A}{x_B}\right)_1 = \alpha_1\alpha_s\left(\frac{x_A}{x_B}\right)_s$$

And for plate $n$:

$$\left(\frac{x_A}{x_B}\right)_n = \left(\frac{y_A}{y_B}\right)_{n-1} = \alpha_1\alpha_2\alpha_3\ldots\alpha_{n-1}\alpha_s\left(\frac{x_A}{x_B}\right)_s$$

If an average value of $\alpha$ is used:

$$\left(\frac{x_A}{x_B}\right)_n = \alpha_{av}^n\left(\frac{x_A}{x_B}\right)_s$$

In most cases total condensation occurs in the condenser, so that:

$$\left(\frac{x_A}{x_B}\right)_d = \left(\frac{y_A}{y_B}\right)_n = \alpha_n\left(\frac{x_A}{x_B}\right)_n = \alpha_{av}^{n+1}\left(\frac{x_A}{x_B}\right)_s$$

$$n+1 = \frac{\log\left\{\left(\frac{x_A}{x_B}\right)_d\left(\frac{x_B}{x_A}\right)_s\right\}}{\log\alpha_{av}} \tag{11.42}$$

and $n$ is the required number of theoretical plates in the column.

It is important to note that, in this derivation, only the relative volatilities of two components have been used; the same relation may be applied to two components of a multicomponent mixture, as will be seen in Section 11.7.6.

**Example 11.4**

For the separation of a mixture of benzene and toluene, as in Example 11.2, $x_d = 0.9$, $x_w = 0.1$, and $x_f = 0.4$, using an average value of 2·4 for the volatility of benzene relative to toluene, the number of plates required at total reflux is given by:

$$n+1 = \frac{\log\left\{\left(\frac{0.9}{0.1}\right)\left(\frac{0.9}{0.1}\right)\right\}}{\log 2.4} = 5.0$$

Thus the number of theoretical plates in the column is <u>four</u>, and this is independent of the composition of the feed.

If the feed is liquid at its boiling point, then the minimum reflux ratio $R_m$ is given by equation 11.40:

$$R_m = \frac{1}{\alpha - 1}\left[\frac{x_d}{x_f} - \alpha\frac{(1 - x_d)}{(1 - x_f)}\right]$$

$$= \frac{1}{2\cdot4 - 1}\left[\frac{0\cdot9}{0\cdot4} - \frac{2\cdot4 \times 0\cdot1}{0\cdot6}\right]$$

$$= \underline{\underline{1\cdot32}}$$

Using the graphical construction shown in Fig. 11.17, since $y_f$ is 0·61, the value of $R_m$ is obtained as:

$$R_m = \frac{x_d - y_f}{y_f - x_f} = \frac{0\cdot9 - 0\cdot61}{0\cdot61 - 0\cdot4} = \underline{\underline{1\cdot38}}$$

*Selection of Economic Reflux Ratio*

The cost of a distillation unit may be considered as made up of the capital cost of the column, determined largely by the number and diameter of the plates, and the operating costs, determined by the steam and cooling water requirements. The depreciation charges can be taken as a percentage of the capital cost, and the two together taken as the overall charges. The steam required will be proportional to $V_m$, which can be taken as $V_n$ where the feed is liquid at its boiling point. From a material balance over the top portion of the column, $V_n = D(R + 1)$, so that the steam required per mol of product is proportional to $R + 1$; it will be a minimum when $R$ equals $R_m$, and will steadily rise as $R$ is increased. The relationship between the number of plates $n$ and the reflux ratio $R$, as derived by Gilliland[10], is discussed later in Section 11.7.7.

The reduction in the required number of plates as $R$ is increased beyond $R_m$ will tend to reduce the cost of the column. For a column separating a benzene–toluene mixture, where $x_f = 0\cdot79$, $x_d = 0\cdot99$ and $x_w = 0\cdot01$ the numbers of theoretical plates as given by the McCabe–Thiele method are given below for various values of $R$. The minimum reflux ratio for this case is 0·81.

| Reflux Ratio $R$ | 0·81 | 0·9 | 1·0 | 1·1 | 1·2 |
|---|---|---|---|---|---|
| No. of Plates | $\infty$ | 25 | 22 | 19 | 18 |

Thus, an increase in $R$, at values near $R_m$, gives a marked reduction in the number of plates, but at higher values of $R$, further increases have little effect on the number. Increasing the reflux ratio from $R_m$ therefore affects the capital and operating cost of a column in the following ways:

(a) The operating costs rise and are approximately proportional to $(R + 1)$.
(b) The capital cost initially falls since the number of plates falls off rapidly at this stage.
(c) The capital cost rises at high values of $R$, since there is then only a very small reduction in the number of plates, but the diameter, and hence area, continually increases because the vapour load becomes greater. The associated condenser and reboiler will also be larger and more expensive.

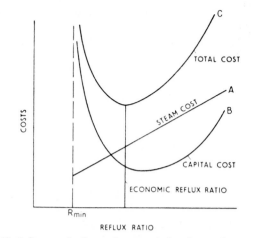

FIG. 11.18.  Influence of reflux ratio on capital and operating costs of still

The total charges are found by adding the fixed and operating charges as shown in Fig. 11.18, where curve $A$ shows the steam costs and $B$ the fixed costs. The final picture is shown by curve $C$ which has a minimum value corresponding to the economic reflux ratio. There is no simple relation between $R_m$ and the optimum value, but practical values generally lie between 1·1 and 1·5 times the minimum, though much higher values are sometimes employed, particularly in the case of vacuum distillation. It is also useful to note that, for a fixed degree of enrichment from the feed to the top product, the number of trays required increases rapidly as the difficulty of separation increases, i.e. as the relative volatility approaches unity. A demand for a higher purity of product necessitates a very considerable increase in the number of trays, particularly when $\alpha$ is near unity; in these circumstances only a limited improvement in product purity may be obtained by increasing the reflux

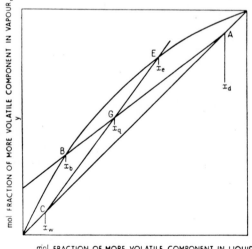

FIG. 11.19.  Location of feed point

ratio. The engineer must be careful to consider the increase in cost of plant resulting from specification of a higher degree of purity of production; at the same time he should assess the highest degree of purity that may be obtained with the proposed plant.

### 11.4.4. Location of Feed Point in a Continuous Still

From Fig. 11.19 it will be seen that, when stepping off plates down the top operating line $AB$, the bottom operating line $CE$ cannot be used until the value of $x_n$ on any plate is less than $x_e$. Again it is essential to pass to the lower line $CE$ by the time $x_n = x_b$. The best conditions will be those where the minimum number of plates is used. From the geometry of the figure, the largest steps in the enriching section occur down to the point of intersection of the operating lines at $x = x_q$. Below this value of $x$, the steps are larger on the lower operating line. Thus, although the column will operate for a feed composition between $x_e$ and $x_b$, the minimum number of plates will be required if $x_f = x_q$. For a binary mixture at its boiling point, this is equivalent to making $x_f$ equal the composition of the liquid on the feed plate.

### 11.4.5. Multiple Feeds and Sidestreams

For the purpose of this section, a sidestream will be defined as any product stream other than the overhead product and the residue ($S'$, $S''$, or $S'''$ in Fig. 11.20). Likewise, $F_1$ and $F_2$

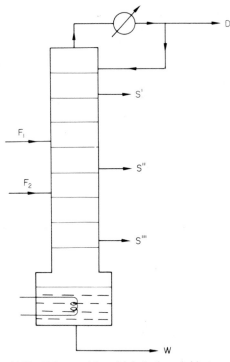

FIG. 11.20. Column with multiple feeds and sidestreams

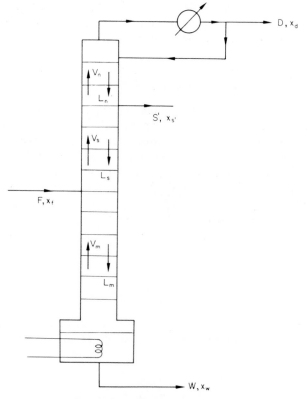

FIG. 11.21. Column with sidestream

constitute separate feed streams to the column. Sidestreams are most often removed with multicomponent systems, but they can be used with binary mixtures. For the purposes of illustration, a binary system will be considered, with one sidestream, as shown in Fig. 11.21. $S'$ represents the rate of removal of the sidestream and $x_{s'}$ its composition.

Assuming constant molar overflow, then for that part of the column above the sidestream the operating line is given by:

$$y_n = \frac{L_n}{V_n} x_{n+1} + \frac{Dx_d}{V_n} \qquad \text{(equation 11.19)}$$

as before. Balances for that part of the tower above a plate between the feed plate and the sidestream give:

$$V_s = L_s + S' + D \qquad (11.43)$$

and

$$V_s y_n = L_s x_{n+1} + S' x_{s'} + Dx_d \qquad (11.44)$$

$$\therefore \quad y_n = \frac{L_s}{V_s} x_{n+1} + \frac{S' x_{s'} + Dx_d}{V_s} \qquad (11.45)$$

Since the sidestream is normally removed as a liquid, $L_s = L_n - S'$ and $V_s = V_n$.

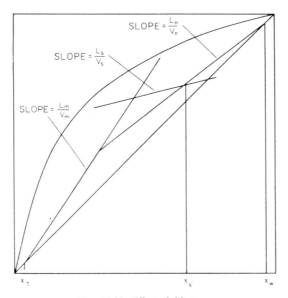

FIG. 11.22. Effect of sidestream

The line represented by equation 11.19 has a slope $L_n/V_n$ and passes through the point $(x_d, x_d)$. Equation 11.45 represents a line of slope $L_s/V_s$, which passes through the point $y = x = (S'x_{s'} + Dx_d)/(S' + D)$, which is the average molar composition of the overhead product and sidestream. Since $x_{s'} < x_d$, and $L_s < L_n$, this additional operating line cuts the $y = x$ line at a lower value than the upper operating line but has a smaller slope, as shown in Fig. 11.22. The two lines will intersect at $x = x_{s'}$. Plates are stepped off as before between the appropriate operating line and the equilibrium curve. It can be seen that the removal of a sidestream increases the number of plates required, owing to the decrease in liquid rate below the sidestream.

The effect of any additional sidestream or feed is to introduce an additional operating line for each component. In all other respects the method of calculation is identical with that for the straight separation of a binary mixture outlined earlier.

The Ponchon–Savarit method, using an enthalpy–composition diagram, can also be used to handle sidestreams and multiple feeds, but only for binary systems. This is dealt with in the next section.

## 11.5. CONDITIONS FOR VARYING OVERFLOW NON-IDEAL BINARY SYSTEMS

### 11.5.1. The Heat Balance

In previous sections the case of constant molar latent heat has been considered with no heat of mixing, and hence constant molar rate of reflux in the column. These simplifying assumptions are extremely useful in that they enable a simple geometrical method to be used for finding the change in concentration on the plates and, whilst they are rarely entirely true in industrial conditions, they often provide a convenient start for design

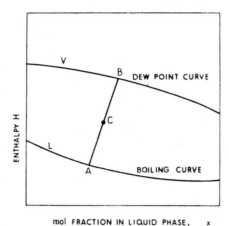

FIG. 11.23. Enthalpy–composition diagram, showing enthalpies of liquid and vapour

purposes. For a non-ideal system, where the molar latent heat is no longer constant and there is a substantial heat of mixing, the calculations become much more tedious. For binary mixtures of this kind a graphical model has been developed by Ruhemann[11], Ponchon[12], and Savarit[13], based on the use of an enthalpy–composition chart. A typical enthalpy–composition or $H - x$ chart is indicated in Fig. 11.23, where the upper curve $V$ is the dew-point curve, and the lower curve $L$ the boiling-point curve. The use of this diagram is based on the following geometrical properties, as illustrated in Fig. 11.24. A quantity of mixture in any physical state is known as a "phase" and is denoted by mass, composition and enthalpy. The phase is indicated upon the diagram by a point which shows enthalpy and composition, but which does not show the mass. If $m$ is the mass, $x$ the composition and $H$ the enthalpy of the phase, then the addition of two phases $A$ and $B$ to

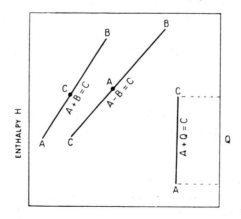

FIG. 11.24. Combination and separation of mixture on enthalpy–composition diagram

give phase $C$ is governed by:

$$m_A + m_B = m_C \tag{11.46}$$

$$m_A x_A + m_B x_B = m_C x_C \tag{11.47}$$

$$m_A H_A + m_B H_B = m_C H_C \tag{11.48}$$

Similarly, if an amount $Q$ of heat is added to a mass $m_A$ of a phase, the increase in enthalpy from $H_A$ to $H_C$ will be given by:

$$H_A + \frac{Q}{m_A} = H_C \tag{11.49}$$

Thus, the addition of two phases $A$ and $B$ is shown on the diagram by point $C$ on the straight line joining the two phases, whilst the difference $A - B$ is found by a point $C$ on the extension of the line $AB$. If, as shown in Fig. 11.23, a phase represented by $C$ in the region between the dew-point and boiling-point curves is considered, then this phase will divide into two phases $A$ and $B$ at the ends of a tie line through the point $C$, so that:

$$\frac{m_A}{m_B} = \frac{CB}{CA} \tag{11.50}$$

The $H-x$ chart, therefore, enables the effect of adding two phases with or without the addition of heat to be determined geometrically. The diagram may be drawn for unit mass or for one mol of material, though as a constant molar reflux does not now apply, it is more convenient to use unit mass as the basis. Thus, working with unit mass of product, the mass of the individual streams as proportions of the product will be calculated.

Figure 11.25 represents a continuous distillation unit operating with feed $F$ of composition $x_f$, and giving a top product $D$ of composition $x_d$ and a bottom product $W$ of composition $x_w$. In this analysis, the quantities in the streams $V$ of rising vapour and $L$ of reflux are given in mass units, such as kg/s, and the composition of the streams as mass fractions, $x$ referring to the liquid and $y$ to the vapour streams.

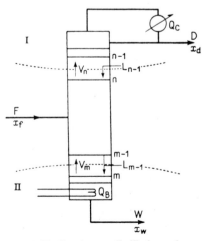

FIG. 11.25. Continuous distillation column

The plates are numbered from the top downwards, subscript $n$ indicating the rectifying and $m$ the stripping section.

$H^V$ and $H^L$ represent the enthalpy of a vapour and liquid stream respectively.

$Q_C$ is the heat removed in the condenser. (In this case no cooling of product considered.)

$Q_B$ is the heat put into the boiler.

Then the following relationships are obtained by taking material and heat balances:

$$V_n = L_{n-1} + D$$

or

$$V_n - L_{n-1} = D \tag{11.51}$$

$$V_n y_n = L_{n-1} x_{n-1} + D x_d$$

or

$$V_n y_n - L_{n-1} x_{n-1} = D x_d \tag{11.52}$$

$$V_n H_n^V = L_{n-1} H_{n-1}^L + D H_d^L + Q_C$$

or

$$V_n H_n^V - L_{n-1} H_{n-1}^L = D H_d^L + Q_C \tag{11.53}$$

Putting $H_d' = H_d^L + Q_C/D$, equation 11.53 may be written as:

$$V_n H_n^V = L_{n-1} H_{n-1}^L + D H_d'$$

or

$$V_n H_n^V - L_{n-1} H_{n-1}^L = D H_d' \tag{11.54}$$

Then from equations 11.51 and 11.52:

$$\frac{L_{n-1}}{D} = \frac{x_d - y_n}{y_n - x_{n-1}} \tag{11.55}$$

and from equations 11.51 and 11.54:

$$\frac{L_{n-1}}{D} = \frac{H_d' - H_n^V}{H_n^V - H_{n-1}^L} \tag{11.56}$$

whence

$$\frac{H_d' - H_n^V}{H_n^V - H_{n-1}^L} = \frac{x_d - y_n}{y_n - y_{n-1}} \tag{11.57}$$

or

$$y_n = \left[ \frac{H_d' - H_n^V}{H_d' - H_{n-1}^L} \right] x_{n-1} + \left[ \frac{H_n^V - H_{n-1}^L}{H_d' - H_{n-1}^L} \right] x_d \tag{11.58}$$

This equation (11.58) is that of any operating line relating the composition of the vapour $y_n$ rising from a plate to the composition of the liquid reflux entering the plate; or alternatively it represents the relation between the composition of the vapour and liquid streams between any two plates. From equation 11.57, it will be seen that all such operating lines pass through a common pole $N$ of coordinates $x_d$ and $H_d'$.

Alternatively, noting that the right-hand side of equations 11.51, 11.52, 11.53 are independent of conditions below the feed plate, a stream $N$ may be defined with mass equal to the difference between the vapour and liquid streams between two plates, of composition $x_d$ and of enthalpy $H_d'$. Then the three quantities $V_n$, $L_{n-1}$, and $N$ will be on a straight line passing through $N$, as shown in Fig. 11.26.

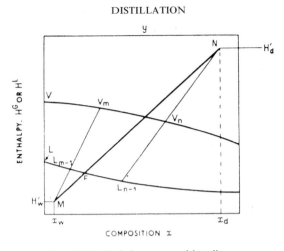

FIG. 11.26. Enthalpy–composition diagram

Below the feed plate a similar series of equations for material and heat balances may be written:

$$V_m + W = L_{m-1}$$

or:

$$-V_m + L_{m-1} = W \tag{11.59}$$

$$V_m y_m + W x_w = L_{m-1} x_{m-1}$$

or:

$$-V_m y_m + L_{m-1} x_{m-1} = W x_w \tag{11.60}$$

$$V_m H_m^V + W H_w^L = L_{m-1} H_{m-1}^L + Q_B$$

or:

$$-V_m H_m^V + L_{m-1} H_{m-1}^L = W H_w^L - Q_B \tag{11.61}$$

And putting:

$$H_w' = H_w^L - \frac{Q_B}{W} \tag{11.62}$$

$$-V H_m^V + L_{m-1} H_{m-1}^L = W H_w' \tag{11.63}$$

Then:

$$\frac{L_{m-1}}{W} = \frac{-x_w + y_m}{y_m - x_{m-1}} \tag{11.64}$$

and:

$$\frac{L_{m-1}}{W} = \frac{-H_w' + H_m^V}{H_m^V - H_{m-1}^L} \tag{11.65}$$

whence:

$$\frac{-H_w' + H_m^V}{H_m^V - H_{m-1}^L} = \frac{-x_w + y_m}{y_m - x_{m-1}} \tag{11.66}$$

This last equation represents any operating line below the feed plate, and shows that all such lines pass through a common pole $M$ of coordinates $x_w$ and $H_w'$. As with the rectifying section, a stream $M$ may be defined by mass $L_{m-1} - V_m$, its composition $x_w$ and enthalpy $H'_w$. Then:

$$F = M + N \tag{11.67}$$

and:

$$Fx_f = Mx_w + Nx_d \tag{11.68}$$

It thus follows that phases $F$, $M$, and $N$ will be on a straight line on the $H$–$x$ chart, as indicated in Figs. 11.26 and 11.27.

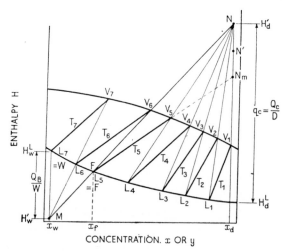

FIG. 11.27.  Determination of number of plates using enthalpy–composition diagram

### 11.5.2.  Determination of the Number of Plates on $H$–$x$ Diagram

The method adopted for determining the necessary number of plates for a desired separation is indicated in Fig. 11.27. The position of the feed $(F, x_f)$ is shown at $F$ on the boiling line and the pole $N$ is located as $(x_d, H'_d)$, where:

$$H'_d = H^L_d + \frac{Q_c}{D} \tag{11.69}$$

Pole $M$ is located as on the extension of $NF$ cutting the ordinate at $x_w$ in $M$.

The condition of the vapour leaving the top plate is shown at $V_1$ on the dew-point curve with abscissa $x_d$. The condition of the liquid on the top plate is then found by drawing the tie line $T_1$ from $V_1$ to $L_1$ on the boiling curve. The condition $V_2$ of the vapour on the second plate is found (equation 11.51) by drawing $L_1 N$ to cut the dew-point curve on $V_2$. $L_2$ is then found on the tie line $T_2$. The conditions of vapour and liquid $V_3$, $V_4$, $V_5$ and $L_3$, $L_4$ are found in the same way. Tie line $T_5$ gives $L_5$, which has the same composition as the feed. $V_6$ is then found using the line $MFV_6$, as this represents the vapour on the top plate of the stripping section. $L_6$, $L_7$ and $V_7$ are then found by similar construction. $L_7$ has the required composition $x_w$ of the bottoms.

Alternatively calculations may start with the feed condition and proceed up and down the column.

### 11.5.3. Minimum Reflux Ratio

The pole $N$ has coordinates $[x_d, H_d^L + Q_C/D]$. $Q_C/D$ is the heat removed in the condenser per unit mass of product (as liquid at its boiling point) and is represented as shown in Fig. 11.27. The number of plates in the rectifying section is determined, for a given feed $x_f$ and product $x_d$, by the height of this pole $N$. As $N$ is lowered to say $N'$ the heat $q_c$ falls, but the number of plates required increases. When $N$ lies at $N_m$ on the isothermal through $F$, $q_c$ is a minimum but the number of plates required becomes infinite. Since the tie lines have different slopes, it follows that there is a minimum reflux for each plate, and the tie line cutting the vertical axis at the highest value of $H$ will give the minimum practical reflux. This, as indicated above, will frequently correspond to the tie line through $F$.

From equations 11.57 and 11.69 and writing $Q_C/D = q_c$:

$$\frac{H_d^L + q_c - H_n^V}{H_n^V - H_{n-1}^L} = \frac{x_d - y_n}{y_n - x_{n-1}} \tag{11.70}$$

or:

$$q_c = (H_n^V - H_{n-1}^L)\left(\frac{x_d - y_n}{y_n - x_{n-1}}\right) + H_n^V - H_d^L \tag{11.71}$$

and:

$$(q_c)_{min} = (H_f^V - H_{f-1}^L)\left(\frac{x_d - y_f}{y_f - x_{f-1}}\right) + H_f^V - H_d^L \tag{11.72}$$

The advantage of the $H-x$ chart lies in the fact that the heat quantities required for the distillation are clearly indicated. Thus, the higher the reflux ratio the more heat must be removed per mol of product, and point $N$ rises. This shows at once that both $q_c$ and $Q_B$ are increased. The use of this method will be illustrated by considering the separation of ammonia from an ammonia–water mixture, as occurs in the ammonia absorption unit for refrigeration.

### Example 11.5

It is required to separate 1 kg/s of a solution of ammonia in water, containing 30 per cent by weight of ammonia, to give a top product of 99·5 per cent purity and a weak solution containing 10 per cent by weight of ammonia.

Calculate the heat required in the boiler and the heat to be rejected in the condenser, assuming a reflux 8 per cent in excess of the minimum and a column pressure of 1013 kN/m². The plates may be assumed to have an ideal efficiency of 60 per cent.

### Solution

Taking a material balance for the whole throughput and for the ammonia gives:

$$D + W = 1 \cdot 0$$

$$0 \cdot 995D + 0 \cdot 1W = 1 \cdot 0 \times 0 \cdot 3$$

Thus
$$D = 0 \cdot 22 \text{ kg/s}$$

and
$$W = 0 \cdot 78 \text{ kg/s}$$

The enthalpy–composition chart for this system is shown in Fig. 11.28. It is assumed that the feed $F$ and the bottom product $W$ are liquids at their boiling points.

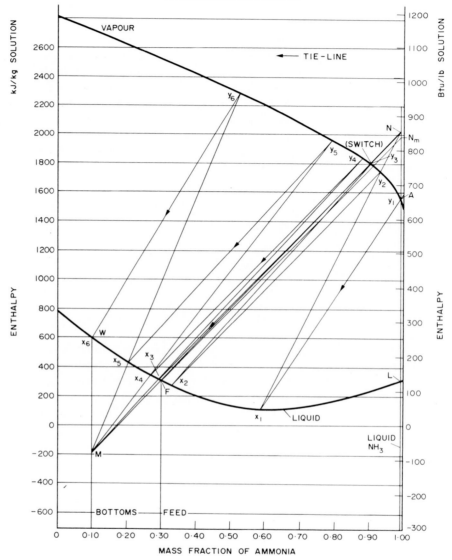

FIG. 11.28. Enthalpy–composition diagram for ammonia–water at $1.013 \, MN/m^2$ pressure

## Location of the Poles N and M

$N_m$ for minimum reflux will be found by drawing a tie-line through $F$, representing the feed, to cut the line $x = 0.995$ at $N_m$.

$$\text{The minimum reflux ratio } R_m = \frac{\text{length } N_m A}{\text{length } AL}$$

$$= \frac{1952 - 1547}{1547 - 295} = 0.323$$

Since the actual reflux is 8 per cent above the minimum:

$$NA = 1{\cdot}08N_m A$$

$$= 1{\cdot}08 \times 405 = 437$$

Point $N$ therefore has an ordinate $437 + 1547 = 1984$ and abscissa $0{\cdot}995$.
Point $M$ is found by drawing $NF$ to cut the line $x = 0{\cdot}10$, through $W$, at $M$.
Then the number of theoretical plates is found, as on the diagram, to be $5+$.
The number of plates to be provided $= (5/0{\cdot}6) = 8{\cdot}33$, say 9.

The feed is introduced just below the third ideal plate from the top, or just below the fifth actual plate.

Heat input at the boiler per unit mass of bottom product

$$= \frac{Q_B}{W} = 582 - (-209) = 791$$

$$\text{Heat input to boiler} = (791 \times 0{\cdot}78) = \underline{617\,\text{kW}}$$

$$\text{Condenser duty} = \text{length } NL \times D$$
$$= (1984 - 296) \times 0{\cdot}22 = \underline{372\,\text{kW}}$$

### 11.5.4. Multiple Feeds and Sidestreams

The enthalpy–composition approach can also be used to handle multiple feeds and sidestreams for binary systems. For the condition of constant molar overflow, each

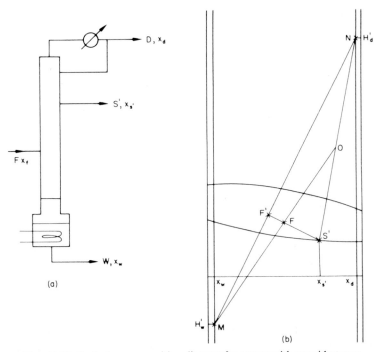

FIG. 11.29. Enthalpy–composition diagram for system with one sidestream

additional sidestream or feed adds a further operating line and pole point to the system.

Taking the same system as used in Fig. 11.21, with one sidestream only, the procedure is as shown in Fig. 11.29.

The upper pole point $N$ is located as before. The effect of removing a sidestream $S'$ from the system is to produce an effective feed $F'$, where $F' = F - S'$ and where $F'S'/F'F = F/S'$. Thus, once $S'$ and $F$ have been located in the diagram, the position of $F'$ may also be determined. The position of the lower pole point $M$, which must lie on the intersection of $x = x_w$ and the straight line drawn through $NF'$, can then be found. $N$ relates to the section of the column above the sidestream and $M$ to that part below the feed plate. A third pole point must be determined to handle that part of the column between the feed and the sidestream.

The pole point for the intermediate section must be on the limiting operating line for the upper part of the column, i.e. $NS'$. It must also lie on the limiting operating line for the lower part of the column, i.e. $MF$ or its extension. Thus the intersection of $NS'$ and $MF$ extended gives the position of the intermediate pole point $O$.

The number of stages required is determined in the same manner as before, using the upper pole point $N$ for that part of the column between the sidestream and the top, the intermediate pole point $O$ between the feed and the sidestream, and the lower pole point $M$ between the feed and the bottom.

For the case of multiple feeds, the procedure is similar and can be followed by reference to Fig. 11.30.

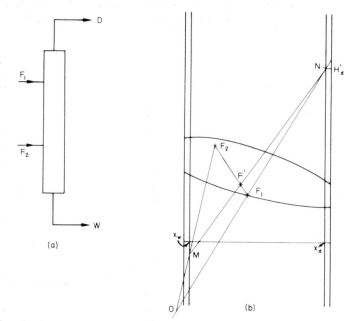

FIG. 11.30. Enthalpy–composition diagram for system with two feeds

## Example 11.6

A mixture containing equal parts by weight of carbon tetrachloride and toluene is to be fractionated to give an overhead product containing 95 wt per cent carbon tetrachloride, a bottom one of 5 wt per cent carbon tetrachloride, and a sidestream containing 80 wt per cent carbon tetrachloride. Both the feed and sidestream can be regarded as liquids at their boiling points.

The rate of withdrawal of the sidestream is 10 per cent of the column feed rate and the external reflux ratio is 2·5. Using the enthalpy composition method, determine the number of theoretical stages required, and the amounts of bottom product and distillate as percentages of the feed rate.

It may be assumed that the enthalpies of liquid and vapour are linear functions of composition. Enthalpy and equilibrium data are provided.

**Solution**

Basis: 100 kg feed.
Overall material balances give:

$$F = D + W + S'$$

i.e.,

$$100 = D + W + 10$$

$$Fx_f = Dx_d + Wx_w + S'x_{s'}$$

i.e.,

$$50 = 0.95D + 0.05W + 8$$

$$\therefore \quad D = 41.7 \text{ per cent}; \quad W = 48.3 \text{ per cent}$$

From the enthalpy data and the reflux ratio the upper pole point $M$ can be located (Fig. 11.31). Locate $F$ and $S'$ on the liquid line, and the position of the effective feed, such that $F'S'/F'F = 10$. Join $NF'$ and extend to cut $x = x_w$ at $M$, the lower pole point. Join $MF$ and extend to cut $NS'$ at $O$, the intermediate pole point. The number of stages required is then determined as shown in Fig. 11.31b.

Therefore 13 theoretical stages are required.

## 11.6. BATCH DISTILLATION

### 11.6.1. The Process

In the previous sections conditions have been considered in which there has been a continuous feed to the still and a continuous withdrawal of products from the top and bottom. In many instances chemical processes are carried out in batches, and it is more convenient to distil each batch separately. In these cases the whole of a batch is run into the boiler of the still and, on heating, the vapour is passed into a fractionation column, as indicated in Fig. 11.32. As with continuous distillation, the composition of the top product will depend on the still composition, the number of plates in the column and on the reflux ratio used. When the still is operating, since the top product will be relatively rich in the more volatile component, the liquid remaining in the still will become steadily weaker in this component. As a result, the purity of the top product will steadily fall. Thus, the still may be charged with $S_1$ mols of a mixture containing a mol fraction $x_{s1}$ of the more volatile component. Initially, with a reflux ratio $R_1$, the top product has a composition $x_{d1}$.

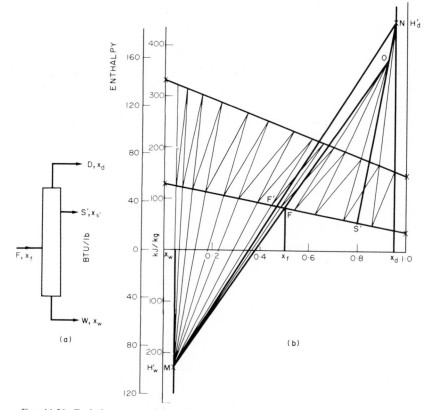

FIG. 11.31. Enthalpy–composition diagram for carbon tetrachloride–toluene separation (one sidestream)

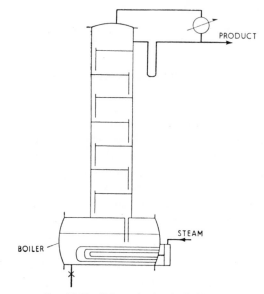

FIG. 11.32. Column for batch distillation

If after a certain interval of time the composition of the top product starts to fall, then, if the reflux ratio is increased to a new value $R_2$, it will be possible to obtain the same composition at the top as before, although the composition in the still is weakened to $x_{s2}$. This method of operating a batch still requires a continuous increase in the reflux ratio to maintain a constant quality of the top product.

An alternative method of procedure is to work with a constant reflux ratio and allow the composition of the top product to fall. For example, if a product of composition 0·9 with respect to the more volatile component is required, the composition initially obtained may be 0·95, and distillation is allowed to continue until the composition has fallen to some value below 0·9, say 0·82. The total product obtained will then have the required composition, provided the amounts of a given purity are correctly chosen.

One of the added merits of batch distillation lies in the fact that more than one product may be obtained. Thus, a binary mixture of alcohol and water may be distilled to obtain initially a high quality alcohol. As the composition in the still weakens with respect to alcohol, a second product may be removed from the top with a reduced concentration of alcohol. In this way it is possible to obtain not only two different quality products, but to reduce the alcohol in the still to a minimum value. This method of operation is particularly useful for handling small quantities of multi-component organic mixtures, since it is possible to obtain the different components at reasonable degrees of purity, one after the other. To obtain the maximum recovery of a valuable component, the charge remaining in the still after the first distillation may be added to the next batch.

### 11.6.2. Operation at Constant Product Composition

Suppose a column with four ideal plates be used to separate a mixture of ethyl alcohol and water. Initially there are in the still $S_1$ mols of liquor of mol fraction $x_{s1}$ with respect to the more volatile component (alcohol). The top product is to contain a mol fraction $x_d$, and this necessitates a reflux ratio $R_1$. Suppose the distillation to be continued till there are $S_2$ mols in the still, of mol fraction $x_{s2}$. Then, for the same number of plates the reflux ratio will have been increased to $R_2$. If the amount of product obtained is $D_b$ mols, by a material balance:

$$S_1 x_{s1} - S_2 x_{s2} = D_b x_d \tag{11.73}$$

and

$$S_1 - S_2 = D_b \tag{11.74}$$

$$\therefore \quad S_1 x_{s1} - (S_1 - D_b) x_{s2} = D_b x_d$$

$$\therefore \quad S_1 x_{s1} - S_1 x_{s2} = D_b x_d - D_b x_{s2}$$

$$\therefore \quad D_b = S_1 \left[ \frac{x_{s1} - x_{s2}}{x_d - x_{s2}} \right] = \frac{a}{b} S_1 \tag{11.75}$$

where $a$ and $b$ are as shown in Fig. 11.33. If $\phi$ is the intercept on the Y-axis for any operating line (equation 11.32), then:

$$\frac{x_d}{R+1} = \phi, \quad \text{or} \quad R = \frac{x_d}{\phi} - 1 \tag{11.76}$$

These equations enable the final reflux ratio to be found for any desired end concentration in the still, and also give the total quantity of distillate obtained. What is

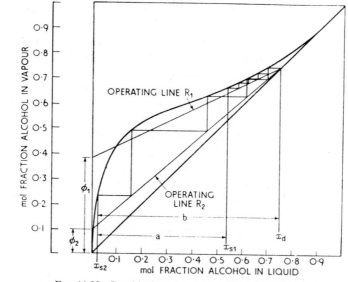

FIG. 11.33. Graphical representation of batch distillation

important, in comparing the operation at constant reflux ratio with that at constant product composition, is the difference in the total amount of steam used in the distillation, for a given quantity $D_b$ of product.

If the reflux ratio $R$ is assumed to be adjusted continuously to keep the top product at constant quality, then at any moment the reflux ratio is given by $R = dL_b/dD_b$. During the course of the distillation, the total reflux liquor flowing down the column is given by:

$$\int_0^{L_b} dL_b = \int_{R=R_1}^{R=R_2} R \, dD_b \tag{11.77}$$

To provide reflux $dL_b$ requires the removal of heat $\lambda \, dL_b$ in the condenser, where $\lambda$ is the latent heat per mol. Thus, the heat to be supplied in the boiler $Q_R$ to provide this reflux during the total distillation is given by:

$$Q_R = \lambda \int_0^{L_b} dL_b = \lambda \int_{R=R_1}^{R=R_2} R \, dD_b \tag{11.78}$$

This equation can be integrated graphically if the relation between $R$ and $D_b$ is first found. For any desired value of $R$, $x_s$ may be obtained by drawing the operating line, and marking off the steps corresponding to the given number of stages. The amount of product $D_b$ is then obtained from equation 11.75 and, if the corresponding values of $R$ and $D_b$ are plotted, graphical integration will give the value of $\int R \, dD_b$.

The minimum reflux ratio $R_m$ may be found for any given still concentration $x_s$ from equation 11.40.

### 11.6.3. Operation at Constant Reflux Ratio $R$

If the same column as before is operated at a constant reflux ratio $R$, the concentration of the M.V.C. in the top product will continuously fall. Over a small interval of time $dt$, the top product composition will fall from $x_d$ to $x_d - dx_d$. If in this time the amount of product obtained is $dD_b$, a material balance on the more volatile component gives:

$$\text{More volatile component removed in product} = dD_b \left[ x_d - \frac{dx_d}{2} \right]$$

$$= x_d \, dD_b \qquad (11.79)$$

and

$$x_d \, dD_b = -d(Sx_s)$$

But $dD_b = -dS$,

$$\therefore \quad -x_d \, dS = -S \, dx_s - x_s \, dS$$

$$\therefore \quad S \, dx_s = dS(x_d - x_s)$$

and

$$\int_{S_1}^{S_2} \frac{dS}{S} = \int_{x_{s1}}^{x_{s2}} \frac{dx_s}{x_d - x_s}$$

$$\ln \frac{S_1}{S_2} = \int_{x_{s2}}^{x_{s1}} \frac{dx_s}{x_d - x_s} \qquad (11.80)$$

The right-hand side of this equation can be integrated by plotting $1/(x_d - x_s)$ vs. $x_s$. This will enable the ratio of the initial to final quantity in the still to be found for any desired change in $x_s$, and hence the amount of distillate $D_b$. The heat to be supplied to provide the reflux will now be $Q_R = \lambda R D_b$ and hence the reboil heat required per mol of product can be compared with that of the first method.

## Example 11.7

A mixture of ethyl alcohol and water with 0·55 mol fraction of alcohol is distilled to give

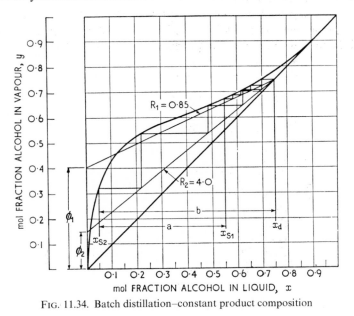

FIG. 11.34. Batch distillation–constant product composition

a top product of 0·75 mol fraction of alcohol. The column has four ideal plates and the distillation is stopped when the reflux ratio has to be increased beyond 4·0.

It is required to find the amount of distillate obtained, and the heat required per kmol of product.

**Solution**

For various values of $R$ the corresponding values of the intercept $\phi$ and the concentration in the still $x_s$ are given in the table. Values of $x_s$ are found as indicated in Fig. 11.34 for the two values of $R$ of 0·85 and 4. Then the amount of product is found from equation 11.75. Thus, for $R$ of 4:

$$D_b = 100\left[\frac{0·55 - 0·05}{0·75 - 0·05}\right] = 100\left(\frac{0·5}{0·7}\right) = \underline{\underline{71·4 \text{ kmol}}}$$

Values of $D_b$ found in this way are also given in the table.

| $R$ | $\phi$ | $x_s$ | $D_b$ |
|------|--------|--------|--------|
| 0·85 | 0·405 | 0·55 | 0 |
| 1·0 | 0·375 | 0·50 | 20·0 |
| 1·5 | 0·3 | 0·37 | 47·4 |
| 2·0 | 0·25 | 0·20 | 63·8 |
| 3·0 | 0·187 | 0·075 | 70·5 |
| 4·0 | 0·15 | 0·05 | 71·4 |

The relation between $D_b$ and $R$ is shown in Fig. 11.35 and the $\int_{R=0·85}^{R=4·0} R \, dD_b$ is given by area $OABC = \underline{96 \text{ kmol}}$.

Taking an average latent heat per kmol for the alcohol–water mixtures as 4000 kJ the heat to be supplied to provide the reflux $Q_R$ is $(96 \times 4000)/1000$ or approximately 380 MJ.

The heat to be supplied to provide the reflux per kmol of product is then $380/71·4$ or 5·32 MJ and the total heat per kmol of product is $5·32 + 4·0 = \underline{\underline{9·32 \text{ MJ}}}$.

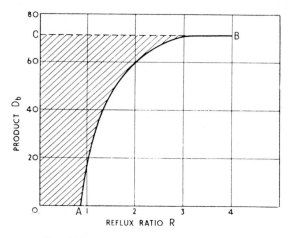

FIG. 11.35. Graphical integration for Example 11.7

## Example 11.8

If the same batch as above is distilled with a constant reflux ratio $R$ of 2·1, what will be the heat required and the average composition of the distillate if the distillation is stopped when the composition in the still has fallen to 0·105 mol fraction of alcohol?

## Solution

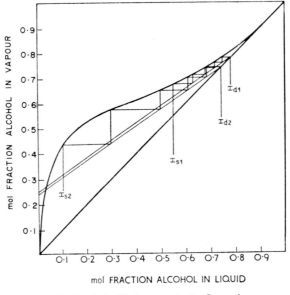

FIG. 11.36. Batch distillation–constant reflux ratio

The initial composition of the top product will be 0·78, as shown in Fig. 11.36, and the final composition will be 0·74. Values of $x_d$, $x_s$, $x_d - x_s$ and of $1/(x_d - x_s)$ for various values of $x_s$ and a constant reflux ratio are given in the following table.

*Ethyl Alcohol–Water Distillation at Constant R of 2·1 with Column of 4 Plates*

| Values of $x_s$ | $x_d$ | $x_d - x_s$ | $\dfrac{1}{x_d - x_s}$ |
|---|---|---|---|
| 0·550 | 0·780 | 0·230 | 4·35 |
| 0·500 | 0·775 | 0·275 | 3·65 |
| 0·425 | 0·770 | 0·345 | 2·90 |
| 0·310 | 0·760 | 0·450 | 2·22 |
| 0·225 | 0·750 | 0·525 | 1·91 |
| 0·105 | 0·740 | 0·635 | 1·58 |

Values of $x_s$ and $\dfrac{1}{x_d - x_s}$ are plotted in Fig. 11.37 from which $\displaystyle\int_{\cdot105}^{\cdot55} \dfrac{dx_s}{x_d - x_s}$ is found to be 1·1.

From equation 11.80, $\ln(S_1/S_2) = 1\cdot1$ and $(S_1/S_2) = 3\cdot0$.

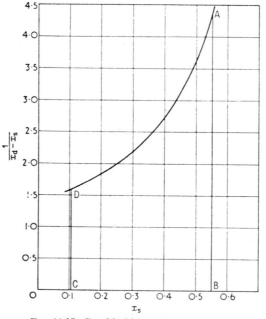

FIG. 11.37. Graphical integration for Example 11.8

Product obtained $D_b = S_1 - S_2 = 100 - 100/3 = 66.7$ kmol.

Amount of alcohol in product $= x_1 S_1 - x_2 S_2$
$$= 0.55 \times 100 - 0.105 \times 33.3$$
$$= 55 - 3.5 = 51.5 \text{ kmol}$$

$\therefore$   Average composition of product is $(51.5/66.7) = \underline{0.77 \text{ mol fraction alcohol.}}$

The heat required to provide the reflux is $4000 \times 2.1 \times 66.7 = 560,380$ kJ.

Heat required to provide reflux per kmol of product $= 560,380/66.7 = \underline{\underline{8400 \text{ kJ.}}}$

   Thus in the previous example the total heat required per kmol of product is $5320 + 4000 = 9320$ kJ and at constant reflux ratio it is $8400 + 4000 = 12,400$ kJ, but the average quality of product is 0.77 for the second case and only 0.75 for the first.

   A discussion on the relative merits of batch and continuous distillation is given by Ellis[14]. He shows that when a large number of plates is used and the reflux ratio approaches the minimum value, then continuous distillation has the lowest reflux requirement and hence operating costs. If a smaller number of plates is used and high purity product is not desired, then the advantages will probably be with batch distillation.

## 11.7. MULTICOMPONENT MIXTURES

### 11.7.1. Equilibrium Data

   For a binary mixture under constant pressure conditions the vapour–liquid equilibrium curve for either component is unique so that, if the concentration of either component is known in the liquid phase, the compositions of the liquid and of the vapour are fixed. It is on the basis of this single equilibrium curve that the McCabe–Thiele method was developed for the rapid determination of the number of theoretical plates required for a given separation. With a ternary system the conditions of equilibrium are more complex, for at constant pressure the mol fraction of two of the components in the liquid phase must be given before the composition of the vapour in equilibrium can be determined, even for an ideal system. Thus, the mol fraction $y_A$ in the vapour depends not only on $x_A$ in the liquid, but also on the relative proportions of the other two components.

To obtain experimentally the equilibrium relationships for a multicomponent mixture requires considerable data, and one of two methods of simplification is usually adopted. For many systems, particularly those consisting of chemically similar substances, the relative volatilities of the components remain constant over a wide range of temperature and composition. This is illustrated in Table 11.2 for mixtures of phenol, ortho and meta-cresols, and xylenols, where the volatilities are shown relative to ortho-cresol.

TABLE 11.2. *Volatilities Relative to o-Cresol*

|          | Temperatures (K) | | |
| --- | --- | --- | --- |
|          | 353   | 393   | 453   |
| Phenol   | 1·25  | 1·25  | 1·25  |
| o-Cresol | 1     | 1     | 1     |
| m-Cresol | 0·57  | 0·62  | 0·70  |
| Xylenols | 0·30  | 0·38  | 0·42  |

An alternative method, particularly useful for multicomponent mixtures of hydro-carbons in the petroleum industry, is to use the simple relation $y_A = Kx_A$. These $K$ values have been measured for a wide range of hydrocarbons at various pressures, and some values are shown in Fig. 11.38.

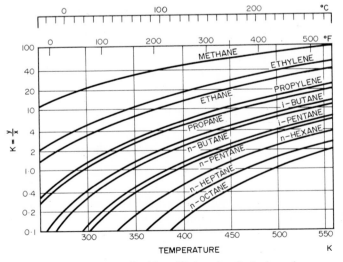

FIG. 11.38. Vapour–liquid equilibrium data for hydrocarbons

Some progress has been made in presenting methods for calculating ternary data from known data for the binary mixtures, though as yet no entirely satisfactory method is available.

### 11.7.2. The Feed and Product Compositions

With a binary system, if the feed composition $x_f$ and the top product composition $x_d$ are known for one component, then the composition of the bottoms $x_w$ can have any desired value, and a material balance will determine the amounts of the top and bottom products $D$ and $W$. This freedom of selecting the compositions does not apply for mixtures with three or more components. Gilliland and Reed[15] have determined the number of degrees of freedom for the continuous distillation of a multicomponent mixture. For the common case in which the feed composition, nature of the feed, and operating pressure are given, there remain only four variables that may be selected. If the reflux ratio $R$ is fixed and the number of plates above and below the feed plate are chosen to give the best use of the plates, then only two variables remain. The complete composition of neither the top nor bottom product can then be fixed at will. This means that some degree of trial and error is unavoidable in calculating the number of plates required for any desired separation. Thus, if a trial composition is taken, and it is found that for the given bottom composition the desired top composition is not obtained with the selected reflux ratio, then an adjustment must be made in the bottom composition. An exact fit in a calculation of this kind is not essential since the equilibrium data and the plate efficiency will be known with only limited accuracy. This problem is frequently simplified if a sharp cut is to be made between the components, so that all of the more volatile components appear in the top and all of the less volatile in the bottom product.

### 11.7.3. Light and Heavy Key Components

In the fractionation of multicomponent mixtures it frequently happens that the essential requirement is the separation between two components. Such components are called the key components and by concentrating attention on these it is possible to simplify the handling of complex mixtures. Suppose a four-component mixture **A–B–C–D**, in which **A** is the most volatile and **D** the least volatile, to be separated as shown in Table 11.3.

TABLE 11.3. *Separation of Multicomponent Mixture*

| Feed | Top product | Bottoms |
|:---:|:---:|:---:|
| A | A | |
| B | B | B |
| C | C | C |
| D | | D |

Then **B** is the lightest component appearing in the bottoms and is termed the light key component, and **C** is the heaviest component appearing in the distillate and is called the heavy key component. The main purpose of the fractionation is the separation of **B** from **C**. It should be noted that, if the top product is to consist of only **A** and **B** and the bottom of **B**, **C** and **D**, then the light key is **B** and the heavy key is **A**.

## 11.7.4. The Calculation of the Number of Plates Required for a Given Separation

One of the most successful methods for calculating the number of plates necessary for a given separation is due to Lewis and Matheson[16]. This is based on the Sorel–Lewis method, previously described for binary mixtures. If the composition of the liquid on any plate is known, then the composition of the vapour in equilibrium is calculated from a knowledge of the vapour pressures or relative volatilities of the individual components. The composition of the liquid on the plate above is then found by using an operating equation, as for binary mixtures, but in this case there will be a separate equation for each component.

Suppose a mixture of components $A$, $B$, $C$, $D$, etc., to have mol fractions $x_A$, $x_B$, $x_C$, $x_D$, etc., in the liquid and $y_A$, $y_B$, $y_C$, $y_D$, etc., in the vapour. Then

$$y_A + y_B + y_C + y_D + \ldots = 1 \tag{11.81}$$

and

$$\frac{y_A}{y_B} + \frac{y_B}{y_B} + \frac{y_C}{y_B} + \frac{y_D}{y_B} + \ldots = \frac{1}{y_B}$$

$$\alpha_{AB}\frac{x_A}{x_B} + \alpha_{BB}\frac{x_B}{x_B} + \alpha_{CB}\frac{x_C}{x_B} + \alpha_{DB}\frac{x_D}{x_B} + \ldots = \frac{1}{y_B}$$

$$\Sigma(\alpha_{AB}x_A) = \frac{x_B}{y_B}$$

$$y_B = \frac{x_B}{\Sigma(\alpha_{AB}x_A)} \tag{11.82}$$

And, similarly,

$$y_A = \frac{x_A\alpha_{AB}}{\Sigma(\alpha_{AB}x_A)}; \quad y_C = \frac{x_C\alpha_{CB}}{\Sigma(\alpha_{AB}x_A)}; \quad y_D = \frac{x_D\alpha_{DB}}{\Sigma(\alpha_{AB}x_A)}$$

Thus, the composition of the vapour is conveniently found from that of the liquid by use of the relative volatilities of the components.

## Example 11.9

A mixture of ortho, meta, and para mononitrotoluenes containing 60, 4, and 36 mol per cent respectively of the three isomers is to be continuously distilled to give a top product of 98 mol per cent ortho, and the bottom is to contain 12·5 mol per cent ortho. The mixture is to be distilled at a temperature of 410 K requiring a pressure in the boiler of about 6·0 kN/m². If a reflux ratio of 5 is used, how many plates will be required and what will be the approximate compositions of the product streams?

## Solution

The volatility of ortho relative to the para isomer can be taken as 1·70 and of the meta as 1·16 over the temperature range of 380 to 415 K. As a first estimate, suppose the distillate to contain 0·6 mol per cent meta and 1·4 mol per cent para. Then a material balance will give the composition of the bottoms.

Consider 100 kmol of feed.

Let $D$ and $W$ be the kmol of product and bottoms, and $x_{do}$ and $x_{wo}$ the mol fraction of the ortho in the distillate and bottoms. Total material balance gives:

$$100 = D + W$$

Ortho balance gives:

$$60 = Dx_{do} + Wx_{wo}$$

whence

$$60 = (100 - W)0·98 + 0·125W$$

from which

$$D = 55·56 \quad \text{and} \quad W = 44·44.$$

The material balance to give the composition and amount of the streams can then be obtained as:

| Component | Feed | | Distillate | | Bottoms | |
|---|---|---|---|---|---|---|
|  | kmol | mol% | kmol | mol% | kmol | mol% |
| Ortho $o$ | 60 | 60 | 54·44 | 98·0 | 5·56 | 12·5 |
| Meta $m$ | 4 | 4 | 0·33 | 0·6 | 3·67 | 8·3 |
| Para $p$ | 36 | 36 | 0·79 | 1·4 | 35·21 | 79·2 |
|  | 100 | 100 | 55·56 | 100 | 44·44 | 100 |

*Equations of operating lines.* The liquid and vapour streams in the column can be obtained as follows:

Above feed-point:

$$\text{Liquid downflow } L_n = 5D = 277·8$$
$$\text{Vapour up } V_n = 6D = 333·4$$

Below feed-point, taking all the feed as liquid at its boiling point:

$$\text{Liquid downflow} \quad L_m = L_n + F = 277·8 + 100 = 377·8$$
$$\text{Vapour up} \quad V_m = L_m - W = 377·8 - 44·44 = 333·4$$

The equations for the operating lines below the feed plate can then be written as:

$$y_m = \frac{L_m}{V_m}x_{m+1} - \frac{W}{V_m}x_w \qquad \text{(equation 11.20)}$$

Ortho

$$y_{mo} = \frac{377·8}{333·4}x_{m+1} - \frac{44·44}{333·4}x_w$$

$$= 1·133x_{m+1} - 0·0166$$

Meta

$$y_{mm} = 1·133x_{m+1} - 0·011$$

Para

$$y_{mp} = 1·133x_{m+1} - 0·105$$

$$(11.83)$$

and above the feed plate:

$$y_n = \frac{L_n}{V_n} x_{n+1} + \frac{D}{V_n} x_d \qquad \text{(equation 11.19)}$$

Ortho

$$\left. \begin{aligned} y_{no} &= \frac{277\cdot8}{333\cdot4} x_{n+1} + \frac{55\cdot56}{333\cdot4} \times 0\cdot98 \\ &= 0\cdot833 x_{n+1} + 0\cdot0163 \end{aligned} \right\}$$

Meta

$$y_{nm} = 0\cdot833 x_{n+1} + 0\cdot001$$

Para

$$y_{np} = 0\cdot833 x_{n+1} + 0\cdot002$$

(11.84)

*Composition of liquid on first plate.* The temperature of distillation has been fixed by considerations of safety at 410 K and, from a knowledge of the vapour pressures of the three components, the pressure in the still is found to be about $6 \text{ kN/m}^2$. The composition of the vapour in the still is found from the relation $y_{so} = \alpha_o x_{so}/\Sigma \alpha x_s$.

The liquid composition on the first plate is then found from equation 11.83; thus for ortho:

$$0\cdot191 = 1\cdot133 x_1 - 0\cdot0166$$

$$x_1 = 0\cdot183$$

The values of the compositions as found in this way are shown in the table below. The liquid on plate 7 has a composition with the ratio of the concentrations of ortho and para about that in the feed, and the feed will therefore be introduced on this plate. Above this plate the same method is used but the operating equations are equations 11.84. The vapour from the sixteenth plate has the required concentration of the ortho isomer, and the values for the meta and para are sufficiently near to take this as showing that sixteen ideal plates will be required.

Using the relation $y_{mo} = \alpha_o x_{mo}/\Sigma \alpha x_m$:

| Plate compositions below the feed plate | | | | | | | |
|---|---|---|---|---|---|---|---|
| Component | $x_s$ | $\alpha x_s$ | $y_s$ | $x_1$ | $\alpha x_1$ | $y_1$ | $x_2$ |
| o | 0·125 | 0·211 | 0·191 | 0·183 | 0·308 | 0·270 | 0·253 |
| m | 0·083 | 0·096 | 0·088 | 0·088 | 0·102 | 0·090 | 0·089 |
| p | 0·792 | 0·792 | 0·721 | 0·729 | 0·729 | 0·640 | 0·658 |
|   | 1 | 1·099 | 1 | 1 | 1·139 | 1 | 1 |

|   | $\alpha x_2$ | $y_2$ | $x_3$ | $\alpha x_3$ | $y_3$ | $x_4$ | $\alpha x_4$ |
|---|---|---|---|---|---|---|---|
| o | 0·430 | 0·357 | 0·330 | 0·561 | 0·450 | 0·411 | 0·698 |
| m | 0·103 | 0·086 | 0·086 | 0·100 | 0·080 | 0·080 | 0·093 |
| p | 0·658 | 0·557 | 0·584 | 0·584 | 0·470 | 0·509 | 0·509 |
|   | 1·191 | 1 | 1 | 1·245 | 1 | 1 | 1·300 |

#### Plate compositions below the feed plate (*cont.*)

| Component | $y_4$ | $x_5$ | $\alpha x_5$ | $y_5$ | $x_6$ | $\alpha x_6$ | $y_6$ |
|---|---|---|---|---|---|---|---|
| o | 0.537 | 0.488 | 0.830 | 0.613 | 0.556 | 0.944 | 0.674 |
| m | 0.071 | 0.072 | 0.083 | 0.061 | 0.063 | 0.073 | 0.052 |
| p | 0.392 | 0.440 | 0.440 | 0.326 | 0.381 | 0.381 | 0.274 |
| | 1 | 1 | 1.353 | 1 | 1 | 1.398 | 1 |

| | $x_7$ |
|---|---|
| o | 0.609 |
| m | 0.055 |
| p | 0.336 |
| | 1 |

#### Plate compositions above the feed plate

| Component | $x_7$ | $\alpha x_7$ | $y_7$ | $x_8$ | $\alpha x_8$ | $y_8$ | $x_9$ |
|---|---|---|---|---|---|---|---|
| o | 0.609 | 1.035 | 0.721 | 0.669 | 1.136 | 0.770 | 0.728 |
| m | 0.055 | 0.064 | 0.044 | 0.051 | 0.059 | 0.040 | 0.047 |
| p | 0.336 | 0.336 | 0.235 | 0.280 | 0.280 | 0.190 | 0.225 |
| | 1 | 1.435 | 1 | 1 | 1.475 | 1 | 1 |

| | $\alpha x_9$ | $y_9$ | $x_{10}$ | $\alpha x_{10}$ | $y_{10}$ | $x_{11}$ | $\alpha x_{11}$ |
|---|---|---|---|---|---|---|---|
| o | 1.238 | 0.816 | 0.782 | 1.330 | 0.856 | 0.832 | 1.415 |
| m | 0.054 | 0.035 | 0.041 | 0.047 | 0.030 | 0.035 | 0.040 |
| p | 0.225 | 0.149 | 0.177 | 0.177 | 0.144 | 0.133 | 0.133 |
| | 1.517 | 1 | 1 | 1.554 | 1 | 1 | 1.588 |

| | $y_{11}$ | $x_{12}$ | $\alpha x_{12}$ | $y_{12}$ | $x_{13}$ | $\alpha x_{13}$ | $y_{13}$ |
|---|---|---|---|---|---|---|---|
| o | 0.891 | 0.874 | 1.485 | 0.920 | 0.907 | 1.542 | 0.940 |
| m | 0.025 | 0.029 | 0.033 | 0.020 | 0.023 | 0.027 | 0.017 |
| p | 0.084 | 0.097 | 0.097 | 0.060 | 0.070 | 0.070 | 0.043 |
| | 1 | 1 | 1.615 | 1 | 1 | 1.639 | 1 |

| | $x_{14}$ | $\alpha x_{14}$ | $y_{14}$ | $x_{15}$ | $\alpha x_{15}$ | $y_{15}$ | $x_{16}$ |
|---|---|---|---|---|---|---|---|
| o | 0.932 | 1.585 | 0.957 | 0.953 | 1.620 | 0.970 | 0.968 |
| m | 0.019 | 0.022 | 0.013 | 0.014 | 0.016 | 0.010 | 0.010 |
| p | 0.049 | 0.049 | 0.030 | 0.033 | 0.033 | 0.020 | 0.022 |
| | 1 | 1.656 | 1 | 1 | 1.669 | 1 | 1 |

| | $\alpha x_{16}$ | $y_{16}$ |
|---|---|---|
| o | 1.632 | 0.980 |
| m | 0.012 | 0.007 |
| p | 0.022 | 0.013 |
| | 1.666 | 1 |

### 11.7.5. Minimum Reflux Ratio

In the distillation of binary mixtures, the minimum reflux ratio is given by the operating line which joins the product-composition to the point where the $q$ line cuts the equilibrium

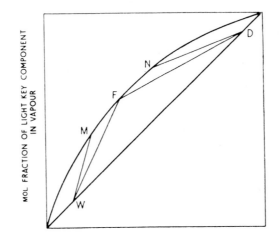

MOL FRACTION OF LIGHT KEY COMPONENT IN
LIQUID

FIG. 11.39. McCabe–Thiele diagram for two key components

curve. Thus, if Fig. 11.39 represents the McCabe–Thiele diagram for a binary mixture of the two key components of a multicomponent mixture, then $DF$ and $WF$ give the minimum reflux ratios for the rectifying and stripping sections respectively. Moving along these operating lines, the change in composition on adjacent plates becomes less and less until it becomes negligible at the feed plate. At $F$ the compositions are said to be "pinched".

In a multicomponent mixture, the pinch does not necessarily occur at the position corresponding to the feed plate. In general, there will be a relatively few plates above the feed plate in which the concentrations of components heavier than the heavy key are reduced to negligible proportions and then a true pinched condition occurs with only the heavy key and more volatile components present. Similarly, there will be a region in which the concentrations of materials lighter than the light key are reduced to very low values and thus there will be a second pinch below the feed plate. In multicomponent distillation there are thus two pinched-in regions. In locating these pinched-in regions it may be noted that:

(a) If there are no components lighter than the light key, then all of the components appear in the bottoms and the pinch in the stripping section will be near the feed plate.
(b) If there are no components heavier than the heavy key, then all of the components will appear in the top and the upper pinch is also at the feed plate.

If both of these conditions are true, then the two pinches coincide at the feed plate, as for a binary system. For the general case, a number of proposals have been made for locating the pinched regions and hence the minimum reflux ratio $R_m$. Of these, the methods of Colburn and Underwood will be mentioned. It may be noted that between the feed plate and the enriching pinch, the concentrations of components heavier than the heavy key fall off rapidly, so that the upper pinch can be regarded as containing the heaviest key and lighter components. Similarly the lower pinch has the lightest key and heavier components.

*Colburn's Method*[17] *for Minimum Reflux* $R_m$

Let **A** and **B** be the light and heavy key components of a multicomponent mixture. Applying the method given earlier (equation 11.37) for binary mixtures, the minimum reflux ratio $R_m$ is obtained from the relationship

$$R_m = \frac{1}{\alpha_{AB} - 1} \left\{ \frac{x_{dA}}{x_{nA}} - \alpha_{AB} \frac{x_{dB}}{x_{nB}} \right\} \tag{11.85}$$

Here $x_{dA}$ and $x_{nA}$ are the top and pinch compositions of the light key component, $x_{dB}$ and $x_{nB}$ are the top and pinch compositions of the heavy key component, and $\alpha_{AB}$ is the volatility of the light key relative to the heavy key component.

The difficulty in using this equation is that only the values of $x_{nA}$ and $x_{nB}$ are known in special cases where the pinch coincides with the feed composition. Colburn has suggested that an approximate value for $x_{nA}$ is given by:

$$x_{nA}\,(\text{approx.}) = \frac{r_f}{(1 + r_f)(1 + \Sigma \alpha x_{fh})} \tag{11.86}$$

and

$$x_{nB}\,(\text{approx.}) = \frac{x_{nA}}{r_f} \tag{11.87}$$

where    $r_f$ is the estimated ratio of the key components on the feed plate. For an all liquid feed at its boiling point, $r_f$ equals the ratio of the key components in the feed. Otherwise $r_f$ is the ratio of the key components in the liquid part of the feed.
$x_{fh}$ is the mol fraction of each component in the liquid portion of feed heavier than the heavy key, and
$\alpha$ is the volatility of the component relative to the heavy key.

Using this approximate value for $R_m$, equation 11.85 can be rearranged to give the concentrations of all the light components in the upper pinch as:

$$x_n = \frac{x_d}{(\alpha - 1)R_m + \alpha(x_{dB}/x_{nB})} \tag{11.88}$$

The concentration of the heavy key in the upper pinch is then obtained by difference, after obtaining the values for all the light components. The second term in the denominator is usually negligible, as the concentration of the heavy key in the top product is small.

A similar condition occurs in the stripping section, and the concentration of all components heavier than the light key is given by:

$$x_m = \frac{\alpha_{AB} x_w}{(\alpha_{AB} - \alpha)(L_m/W) + \alpha(x_{wA}/x_{mA})} \tag{11.89}$$

where  $x_m$ and $x_w$ are the compositions of a given heavy component at the pinch and in the bottoms,
$x_{mA}$ and $x_{wA}$ are the compositions of the light key component at the pinch and in the bottoms,
$L_m/W$ is the molar ratio of the liquid in the stripping section to the bottom product,
$\alpha_{AB}$ is the volatility of the light key relative to the heavy key, and
$\alpha$ is the volatility of the component relative to the heavy key.

Again, the second term in the denominator may usually be neglected.

The essence of Colburn's method is that he has provided an empirical relation between the compositions at the two pinches for the condition of minimum reflux. This enables the assumed value of $R_m$ to be checked. This relation may be written as:

$$\frac{r_m}{r_n} = \psi = \frac{1}{(1 - \Sigma b_m \alpha x_m)(1 - \Sigma b_n x_n)} \qquad (11.90)$$

where $r_m$ is the ratio of the light key to the heavy key in the stripping pinch,

$r_n$ is the ratio of the light key to the heavy key in the upper pinch,

$\Sigma b_m \alpha x_m$ is the summation of $b_m \alpha x_m$ for all components heavier than the heavy key in the lower pinch,

$\Sigma b_n x_n$ is the summation of $b_n x_n$ for all components lighter than the light key in the upper pinch, and

$b_m, b_n$ are the factors shown in Fig. 11.40.

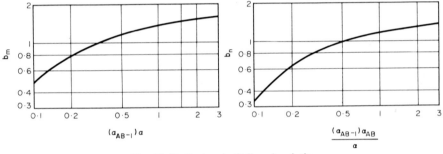

FIG. 11.40. Factors in Colburn's solution

## Example 11.10

A mixture of $n$-$C_4$ to $n$-$C_7$ hydrocarbons is to be distilled to give top and bottom products as shown in the table below. The distillation is effected at $792 \text{ kN/m}^2$ and the feed is at 372 K. The equilibrium values ($K$) are shown in Fig. 11.38. It is required to find the minimum reflux ratio. No cooling occurs in the condenser.

| Component | Feed | Distillate | | Bottoms | |
|---|---|---|---|---|---|
| | $F$ | $D$ | $x_d$ | $W$ | $x_w$ |
| $n$-$C_4$(light key) | 40 | 39 | 0·975 | 1 | 0·017 |
| $C_5$(heavy key) | 23 | 1 | 0·025 | 22 | 0·367 |
| $C_6$ | 17 | | | 17 | 0·283 |
| $C_7$ | 20 | | | 20 | 0·333 |
| | $F = 100$ | $D = 40$ | 1·000 | $W = 60$ | 1·000 |

**Solution**

1. *Estimation of top temperature $T_d$.* By dew-point calculation $\Sigma x_d = \Sigma(x_d/K)$.

| Component | $x_d$ | $T_d = 344\,\text{K}$ | | $T_d = 343\,\text{K}$ | |
|---|---|---|---|---|---|
| | | $K$ | $x_d/K$ | $K$ | $x_d/K$ |
| $n\text{-}C_4$ | 0·975 | 1·05 | 0·929 | 1·04 | 0·938 |
| $C_5$ | 0·025 | 0·41 | 0·061 | 0·405 | 0·062 |
| | | | 0·990 | | 1·000 |

Hence the top temperature $T_d = 343\,\text{K}$.

2. *Estimation of still temperature $T_s$.* $\Sigma x_w = \Sigma K x_w$.

| Component | $x_w$ | $T_s = 419\,\text{K}$ | | $T_s = 416\,\text{K}$ | |
|---|---|---|---|---|---|
| | | $K$ | $K x_w$ | $K$ | $K x_w$ |
| $n\text{-}C_4$ | 0·017 | 3·05 | 0·052 | 2·93 | 0·050 |
| $C_5$ | 0·367 | 1·6 | 0·586 | 1·54 | 0·565 |
| $C_6$ | 0·283 | 0·87 | 0·246 | 0·82 | 0·232 |
| $C_7$ | 0·333 | 0·49 | 0·163 | 0·46 | 0·153 |
| | 1·000 | | 1·047 | | 1·000 |

Hence the still temperature $T_s = 416\,\text{K}$.

3. *Calculation of feed condition.* To determine the nature of the feed, its boiling point $T_B$ must be found, i.e. where $\Sigma K x_f = 1$.

| Component | $x_f$ | $T_B = 377\,\text{K}$ | | $T_B = 376\,\text{K}$ | |
|---|---|---|---|---|---|
| | | $K$ | $K x_f$ | $K$ | $K x_f$ |
| $n\text{-}C_4$ | 0·40 | 1·80 | 0·720 | 1·78 | 0·712 |
| $C_5$ | 0·23 | 0·81 | 0·186 | 0·79 | 0·182 |
| $C_6$ | 0·17 | 0·39 | 0·066 | 0·38 | 0·065 |
| $C_7$ | 0·20 | 0·19 | 0·038 | 0·185 | 0·037 |
| | 1·00 | | 1·010 | | 0·996 |

Hence $T_B$ is approximately 376 K, and since the feed is at 372 K it may be assumed to be all liquid at its boiling point.

4. *Calculation of pinch temperatures.* The temperature of the pinches are taken in the first place at one-third and two-thirds of the difference between the still and top temperatures.

Thus the upper pinch temperature $T_n = 343 + 0.33(416 - 343) = 367$ K
and the lower pinch temperature $T_m = 343 + 0.67(416 - 343) = 391$ K

5. *Calculation of approximate minimum reflux ratio.* The calculations can be set down in the following way:

| Component | $T_n = 367$ K $\alpha$ | $T_m = 391$ K $\alpha$ | $x'_{fh}$ | $\alpha x_{fh}$ |
|---|---|---|---|---|
| $n\text{-}C_4$ | 2·38 | 2·00 | | |
| $C_5$ | 1·00 | 1·00 | | |
| $C_6$ | 0·455 | 0·464 | 0·17 | 0·077 |
| $C_7$ | 0·220 | 0·254 | 0·20 | 0·044 |
| | | | | 0·121 |

Then

$$r_f = \frac{x_{f4}}{x_{f5}} = \frac{0.40}{0.23} = 1.740$$

From equation 11.86:

$$X_{n4} = \frac{1.740}{(1 + 1.740)(1 + 0.121)} = 0.565$$

and

$$X_{n5} = \frac{0.565}{1.740} = 0.325$$

$$R_m = \frac{1}{(2.38 - 1)} \frac{0.975}{0.563} - 2.38\left(\frac{0.025}{0.325}\right)$$

$$= 1.12.$$

6. *The streams in the column*

$$L_n = DR_m = 40 \times 1.12 = 44.8 \text{ kmol}$$
$$V_n = L_n + D = 44.8 + 40 = 84.8 \text{ kmol}$$
$$L_m = L_n + F = 44.8 + 100 = 144.8 \text{ kmol}$$
$$V_m = L_m - W = 144.8 - 60 = 84.4 \text{ kmol}$$
$$L_m/W = 144.8/60 = 2.41$$

7. *Check on minimum reflux ratio*

$$X_n = \frac{x_d}{(\alpha - 1)R_m} \quad \text{for components lighter than heavy key}$$

For $n\text{-}C_4$

$$X_n = \frac{0.975}{(2.38 - 1)1.12} = 0.630$$

and $n$-$C_5$

$$x_n = 1 - 0.630 = 0.370$$

Temperature check for upper pinch. $\Sigma K x_n = 1.0$

$$\Sigma K x_n = (1.62 \times 0.630) + (0.68 \times 0.370) = 1.273$$

Upper pinch temperature of 367 K is incorrect. Try $T_n = 355$ K.

| Component | $K_{355}$ | $\alpha$ | $x_n$ | $K x_n$ |
|-----------|-----------|----------|-------|---------|
| $n$-$C_4$ | 1·35 | 2·55 | 0·562 | 0·759 |
| $C_5$ | 0·53 | 1 | 0·438 | 0·233 |
| | | | 1 | 0·992 |

This is sufficiently near, so take the upper pinch temperature as 355 K. Then

$$r_n = 0.562/0.438 = 1.282$$

Since there is no component lighter than light key $\Sigma b_n x_n = 0$. In the lower pinch

$$x_m = \frac{\alpha_{AB} x_w}{(\alpha_{AB} - \alpha) L_m / W}$$

from equation 11.89.

| Component | $K_{391}$ | $\alpha$ | $\alpha_{AB} - \alpha$ | $x_m$ | $K x_m$ |
|-----------|-----------|----------|------------------------|-------|---------|
| $n$-$C_4^*$ | 2·2 | 2 | | 0·384 | 0·845 |
| $C_5$ | 1·1 | 1 | 1·000 | 0·305 | 0·335 |
| $C_6$ | 0·51 | 0·464 | 1·536 | 0·153 | 0·078 |
| $C_7$ | 0·28 | 0·254 | 1·746 | 0·158 | 0·044 |
| | | | | 1 | 1·302 |

*$x_m$ by difference.

As the temperatures do not check $T_m \neq 391$ K. Try $T_m = 372$ K.

| Component | $K_{372}$ | $\alpha$ | $\alpha_{AB} - \alpha$ | $x_m$ | $K x_m$ | $b_m$ | $b_m \alpha_m x_m$ |
|-----------|-----------|----------|------------------------|-------|---------|-------|---------------------|
| $n$-$C_4^*$ | 1·70 | 2·30 | — | 0·428 | 0·729 | | |
| $C_5$ | 0·74 | 1 | 1·30 | 0·270 | 0·200 | | |
| $C_6$ | 0·35 | 0·47 | 1·83 | 0·148 | 0·052 | 1·22 | 0·085 |
| $C_7$ | 0·17 | 0·23 | 2·07 | 0·154 | 0·026 | 0·91 | 0·032 |
| | | | | 1 | 1·007 | | 0·117 |

*$x_m$ by difference.

This is sufficiently near, so $T_m = 372$ K. Then

$$r_m = 0.428/0.270 = 1.586$$
$$\alpha r_m / r_n = 1.586/1.282 = 1.235$$
$$\psi = 1/(1 - 0.117) = 1/0.883$$
$$= 1.132$$

Hence $R_m$ is not quite equal to 1·12.

8. *Second approximation to reflux ratio*

Now try $R_m = 1.08$.

$$L_n = 40 \times 1.08 = 43.2$$
$$V_n = 40 + 43.2 = 83.2 = V_m$$
$$L_m = 143.2$$

Take $T_n = 355\,\mathrm{K}$ as before.

| Component | $K_{355}$ | $\alpha$ | $x_n$ | $Kx_n$ | |
|-----------|-----------|----------|-------|--------|---|
| $n\text{-}C_4$ | 1·35 | 2·55 | 0·582 | 0·785 | |
| $C_5$ | 0·53 | 1 | 0·418 | 0·221 | |
| | | | 1 | 1·006 | Checks |

$$\therefore \quad r_n = 0.582/0.418 = 1.393$$

Take $T_m = 372\,\mathrm{K}$ as before.

| Component | $K_{372}$ | $\alpha$ | $\alpha_{AB} - \alpha$ | $x_m$ | $Kx_m$ | $b_m$ | $b_m \alpha_m x_m$ |
|-----------|-----------|----------|------------------------|-------|--------|-------|---------------------|
| $n\text{-}C_4^*$ | 1·70 | 2·30 | — | 0·424 | 0·721 | | |
| $C_5$ | 0·74 | 1 | 1·30 | 0·272 | 0·201 | | |
| $C_6$ | 0·35 | 0·47 | 1·83 | 0·149 | 0·052 | 1·22 | 0·085 |
| $C_7$ | 0·17 | 0·23 | 2·07 | 0·155 | 0·026 | 0·91 | 0·034 |
| | | | | 1 | 1 | | 0·119 |

*$x_m$ by difference.

$$\therefore \quad r_m = 0.424/0.272 = 1.558 \qquad \alpha r_m/r_n = 1.558/1.393 = 1.12$$

and

$$\psi = 1/(1 - 0.119) = 1.13$$
$$\therefore \quad R_m = 1.08 \text{ is near enough.}$$

Since there are no components lighter than the light key, the lower pinch should be expected to be near the feed plate as it is, but the general method of taking the pinch temperature as one-third and two-thirds up the column was used above.

The small change in $R$ from 1·12 to 1·08 gives a change in $r_n$ but very little change in $r_m$. It is seen that the first estimation for $R_m$ of 1·12 based on equation 11.86 for locating the upper pinch composition is nearly correct but that it gives the wrong pinch composition.

*Minimum Reflux Ratio $R_m$, using Underwood's Method*

For conditions where the relative volatilities remain constant, Underwood[18] has developed the following two equations from which $R_m$ may be calculated:

$$\frac{\alpha_A x_{fA}}{\alpha_A - \theta} + \frac{\alpha_B x_{fB}}{\alpha_B - \theta} + \frac{\alpha_C x_{fC}}{\alpha_C - \theta} + \ldots = 1 - q \qquad (11.91)$$

and

$$\frac{\alpha_A x_{dA}}{\alpha_A - \theta} + \frac{\alpha_B x_{dB}}{\alpha_B - \theta} + \frac{\alpha_C x_{dC}}{\alpha_C - \theta} + \ldots = R_m + 1 \tag{11.92}$$

where $x_{fA}$, $x_{fB}$, $x_{fC}$, $x_{dA}$, $x_{dB}$, $x_{dC}$, etc., are the mol fractions of components **A**, **B**, **C**, etc., in the feed and distillate, **A** being the light and **B** the heavy key,

$q$ is the ratio of the heat required to vaporise 1 mol of the feed to the molar latent heat of the feed, as in equation 11.28,

$\alpha_A$, $\alpha_B$, $\alpha_C$, etc., are the volatilities with respect to the least volatile component, and

$\theta$ is the root of equation 11.91, which lies between the values of $\alpha_A$ and $\alpha_B$.

If one component in the system has a relative volatility falling between those of the light and heavy keys, it is necessary to solve for two values of $\theta$.

## Example 11.11

Suppose a mixture of hexane, heptane, and octane to be separated to give products as shown in the table. What will be the value of the minimum reflux ratio, if the feed is liquid at its boiling point?

| Component | Feed | | Product | | Bottoms | | Relative volatility |
|---|---|---|---|---|---|---|---|
| | $F$ mol | $x_f$ | $D$ kmol | $x_d$ | $W$ kmol | $x_w$ | |
| Hexane | 40 | 0·40 | 40 | 0·534 | 0 | 0 | 2·70 |
| Heptane | 35 | 0·35 | 34 | 0·453 | 1 | 0·04 | 2·22 |
| Octane | 25 | 0·25 | 1 | 0·013 | 24 | 0·96 | 1·0 |

## Solution

From equation 11.91 light key (**A**) is heptane and heavy key (**B**) is octane. With $q = 1$,

$$\frac{2·70 \times 0·40}{2·70 - \theta} + \frac{2·22 \times 0·35}{2·22 - \theta} + \frac{1 \times 0·25}{1 - \theta} = 0$$

The required value of $\theta$ must satisfy the relation $\alpha_B < \theta < \alpha_A$, i.e. $1·0 < \theta < 2·22$. Try $\theta = 1·15$: then

$$\Sigma \frac{\alpha x_f}{\alpha - \theta} = -0·243$$

Try $\theta = 1·17$: then

$$\Sigma \frac{\alpha x_f}{\alpha - \theta} = -0·024$$

This is near enough, so that from equation 11.92

$$\frac{2·70 \times 0·534}{2·70 - 1·17} + \frac{2·22 \times 0·453}{2·22 - 1·17} + \frac{1·00 \times 0·013}{1·00 - 1·17} = 1·827$$

$$\therefore \quad R_m = \underline{\underline{0·827}}$$

## 11.7.6. Number of Plates at Total Reflux

The number of plates required for the desired separation under conditions of total reflux can be found by applying Fenske's equation (11.42) to the two key components. Thus:

$$n+1 = \frac{\log\left\{\left(\frac{x_A}{x_B}\right)_d \left(\frac{x_B}{x_A}\right)_s\right\}}{\log(\alpha_{AB})_{\text{av.}}} \tag{11.93}$$

### Example 11.12

For the separation of hexane, heptane, and octane as in the previous example, determine the number of theoretical plates required.

### Solution

$$n+1 = \frac{-\log\left[\frac{0.0453}{0.013} \times \frac{0.960}{0.040}\right]}{-\log 2.22} = \frac{\log 835}{\log 2.22}$$

$$= 8.5.$$

Thus, the minimum number of plates $= \underline{\underline{7.5, \text{ say } 8.}}$

## 11.7.7. Relation between Reflux Ratio and Number of Plates

Gilliland[10] has given an empirical relation between the reflux ratio $R$ and the number of plates $n$, in which only the minimum reflux ratio $R_m$ and the number of plates at total reflux $n_m$ are required. This is shown in Fig. 11.41, where $(R - R_m)/(R + 1)$ is plotted against the group $[(n+1) - (n_m + 1)]/(n+2)$.

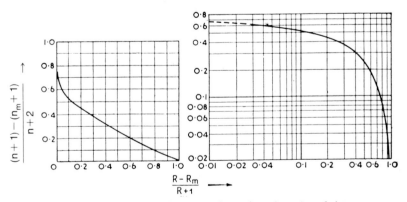

FIG. 11.41. Relation between reflux ratio and number of plates

**Example 11.13**

Using the example above investigate the change in $n$ with $R$ using Fig. 11.41. (Note: $R_m = 0.83$ and $(n_m + 1) = 8.5$.)

| Value of $R$ | $\dfrac{R - R_m}{R + 1}$ | $\dfrac{(n+1) - (n_m + 1)}{n + 2}$ | $n$ |
|---|---|---|---|
| 1 | $\dfrac{0.17}{2} = 0.085$ | 0.55 | 19 |
| 2 | $\dfrac{1.17}{3} = 0.390$ | 0.32 | 11.8 |
| 5 | $\dfrac{4.17}{6} = 0.695$ | 0.15 | 9.1 |
| 10 | $\dfrac{9.17}{11} = 0.833$ | 0.08 | 8.2 |

This illustration shows that the number of plates falls off rapidly at first, but slowly later, and a value of $R$ of nearly two is probably the most economic.

For calculations of the type illustrated by Example 11.13 a convenient nomograph has recently been produced by Zanker[19] to relate $R$, $R_m$, $n$, and $n_m$ so that any variable may be quickly found if the other three are known.

## 11.8. AZEOTROPIC AND EXTRACTIVE DISTILLATION

In the processes so far considered, the vapour becomes steadily richer in the more volatile component on successive plates. There are two types of mixtures where this steady increase in the concentration of the more volatile component either does not take place, or else takes place so slowly that an uneconomical number of plates is required. If, for instance, a mixture of ethyl alcohol and water is distilled, the concentration of the alcohol steadily increases until it reaches 96 per cent by weight, when the composition of the vapour equals that of the liquid, and no further enrichment occurs. This mixture is called an azeotrope, and cannot be separated by straightforward distillation. Such a condition is shown in the $y$–$x$ curves of Fig. 11.4 where it was seen that the equilibrium curve crosses the diagonal, indicating the existence of an azeotrope. A large number of azeotropic mixtures have been found, some of which are of great industrial importance, e.g. water–nitric acid, water–hydrochloric acid, and water–many alcohols.

The second type of problem occurs where the relative volatility of a binary mixture is very low, in which case continuous distillation of the mixture to give nearly pure products will require high reflux ratios with correspondingly high heat requirements; in addition, it will necessitate a tower of large cross-section containing many trays. An example of the second type of problem is the separation of $n$-heptane from methyl cyclohexane. Here the relative volatility is only 1.08 and a large number of plates is required to achieve separation.

The principle of azeotropic and of extractive distillation lies in the addition of a new substance to the mixture so as to increase the relative volatility of the two key components, and thus make separation relatively easy. Benedict and Rubin[20] have defined these two processes in the following way. In azeotropic distillation the substance added forms an azeotrope with one or more of the components in the mixture, and as a result is present on most of the plates of the column in appreciable concentration. With extractive distillation the substance added is relatively non-volatile compared with the components to be separated, and it is therefore fed continuously near the top of the column. This extractive agent runs down the column as reflux and is present in appreciable concentration on all the plates.

The third component added to the binary mixture is sometimes known as the *entrainer* or the *solvent*.

### 11.8.1. Azeotropic Distillation (Ethanol–Water)

Young[21], about 1902, found that if benzene were added to the ethanol–water azeotrope, then a ternary azeotrope was formed with a boiling point of 338·0 K, i.e. less than that of the binary azeotrope (351·3 K). The industrial production of ethanol from the azeotrope using this principle has been described by Guinot and Clark[22]; and the general arrangement of the plant is as indicated in Fig. 11.42. This requires the use of three atmospheric pressure fractionating columns, and a continuous two-phase liquid separator or decanter.

The azeotrope in the ethanol–water binary system has a composition of 89 mol per cent of ethanol[23]. Starting with a mixture containing a lower proportion of ethanol, it is not possible to obtain a product richer in ethanol than this by normal binary distillation. Near

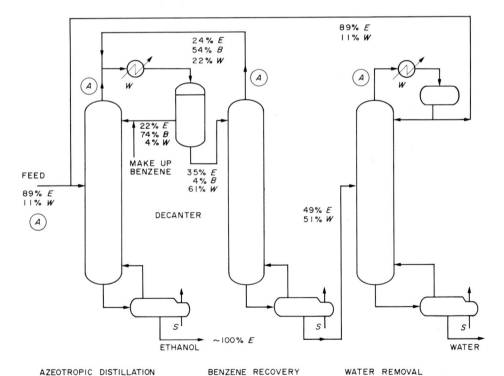

FIG. 11.42. Azeotropic distillation for separation of ethanol from water using benzene as entrainer. Compositions are given in mole per cent. $E$ = Ethanol, $B$ = Benzene, $W$ = Water, $S$ = Steam

azeotropic conditions exist at points marked *A* in Fig. 11.42. The addition of the relatively non-polar benzene entrainer serves to volatilise water (a highly polar molecule) to a greater extent than it volatilises ethanol (a moderately polar molecule) and a virtually pure ethanol product may be obtained. Equilibrium conditions for this system have been discussed by King[23] and Norman[24] and the latter shows how the number of plates required may be determined.

The first tower in Fig. 11.42 forms the ternary azeotrope as an overhead vapour. Nearly pure ethanol issues from the bottom. The ternary azeotrope is condensed and splits into two liquid phases in the decanter. The benzene-rich phase from the decanter serves as reflux, while the water–ethanol-rich phase passes to two towers, one for benzene recovery and the other for water removal. The azeotropic overheads from these successive towers are returned to appropriate points of the primary tower.

Figure 11.43 shows a composition profile for the azeotropic distillation column in the process shown in Fig. 11.42; this is taken from a solution presented by Robinson and Gilliland[1].

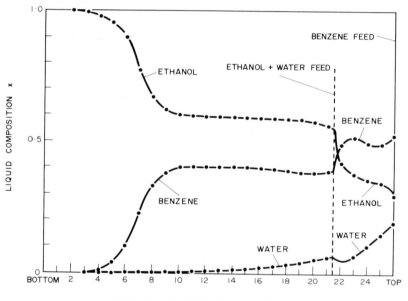

FIG. 11.43. Composition profile for azeotropic distillation of ethanol and water, with benzene as entrainer

### 11.8.2. Extractive Distillation

Extractive distillation is a method of rectification similar in purpose to azeotropic distillation. To a binary mixture which is difficult or impossible to separate by ordinary means a third component, termed a *solvent*, is added which alters the relative volatility of the original constituents, thus permitting the separation. The added solvent is, however, of low volatility and is itself not appreciably vaporised in the fractionator.

For a non-ideal binary mixture the partial pressure may be expressed in the form:

$$P_A = \gamma_A P_A^\circ x_A \tag{11.94}$$

$$P_B = \gamma_B P_B^\circ x_B \tag{11.95}$$

where $\gamma_A$ and $\gamma_B$ are the activity coefficients for the two components. The relative volatility $\alpha$ may thus be written:

$$\alpha = \frac{P_A}{P_B}\frac{x_B}{x_A}$$

$$= \frac{\gamma_A}{\gamma_B}\frac{P_A^\circ}{P_B^\circ} \tag{11.96}$$

The solvent added to the mixture in extractive distillation differentially affects the activities of the two components, and hence $\alpha$.

Such a process depends upon the difference in departure from ideality between the solvent and the components of the binary mixture to be separated. In the example given below both toluene and iso-octane separately form non-ideal liquid solutions with phenol, but the extent of the non-ideality with iso-octane is greater than that with toluene. When all three substances are present, therefore, the toluene and iso-octane themselves behave as a non-ideal mixture, and their relative volatility becomes high.

An example of extractive distillation[25] is the separation of toluene (b.p. = 384 K) from paraffin hydrocarbons of approximately the same molecular weight; this is either very difficult or impossible, owing to low relative volatility or azeotropic formation yet such a separation is necessary in the recovery of toluene from certain petroleum hydrocarbon mixtures. Using iso-octane (b.p. = 372·5 K) as an example of a paraffin hydrocarbon, Fig. 11.44a shows that iso-octane in this mixture is the more volatile, but the separation is obviously difficult. In the presence of phenol (b.p. = 454·6 K), however, the relative volatility of iso-octane increases, so that, with as much as 83 mol per cent phenol in the

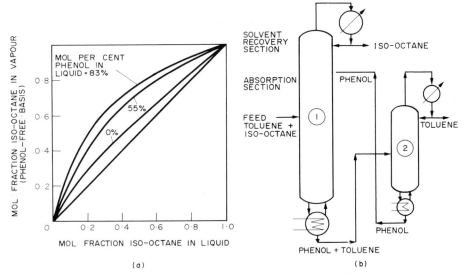

FIG. 11.44. Extractive distillation of toluene–iso-octane with phenol

liquid, the separation from toluene is relatively easy. A flowsheet for accomplishing this is shown in Fig. 11.44b, where the binary mixture is introduced more or less centrally into the extractive distillation tower (1), and phenol as the solvent is introduced near the top so as to be present in high concentration upon most of the trays in the tower. Under these conditions iso-octane is readily distilled as an overhead product, while toluene and phenol are removed as a residue. Although phenol is relatively high-boiling, its vapour pressure is nevertheless sufficient for some to appear in the overhead product. The solvent-recovery section of the tower, which may be relatively short, serves to separate the phenol from the iso-octane. The residue from the tower must be rectified in the auxiliary tower (2) to separate toluene from the phenol which is recycled, but this is a relatively easy separation. In practice, the paraffin hydrocarbon is a mixture rather than pure iso-octane, but the principle of the operation remains the same.

The solvent to be used will be selected on the basis of selectivity, volatility, ease of separation from the top and bottom products, and the cost. The selectivity is most easily assessed by determining the effect on the relative volatility of the two key components of addition of the solvent. The more volatile the solvent, the greater the percentage of solvent in the vapour, and the poorer the separation for a given heat consumption in the boiler. It is important to note that the solvent must not form an azeotrope with any of the components. Some of the problems of selecting the solvent are discussed by Scheibel[26] who points out that use may be made of the fact that, when two compounds show deviations from Raoult's Law, then one of these compounds shows the same type of deviation with any member of the homologous series of the other component. Thus the azeotropic mixture acetone (b.p. 329·6 K)–methanol (b.p. 337·9 K) has 20 mol per cent acetone and boils at 328·9 K, i.e. less than the boiling point of either component. Then any member of the series ethanol (357·5), propanol (370·4), water (373·2), butanol (391·0) may be used as an extractive agent, or in the series of ketones, methyl n-propyl ketone (375), methyl iso-butyl ketone (389·2). The advantage of using a solvent from the alcohol series is that the more volatile acetone will be taken overhead, though water would have the advantage of cheapness. Pratt[27] has given details of a method of calculation for extractive distillation using the system acetonitrile–trichlorethylene–water as an example.

Extractive distillation is usually a more desirable process than azeotropic distillation since no large quantities of solvent must be vaporised. Furthermore, a greater choice of added component is possible since the process is not dependent upon the accident of azeotrope formation. It cannot be conveniently carried out in batch operations, however.

Azeotropic and extractive-distillation equipment can be designed using the general methods for multicomponent distillation, and detailed discussion is available elsewhere[1,20,28,29].

## 11.9. STEAM DISTILLATION

Where the material to be distilled has a high boiling point, and particularly where decomposition might occur if direct distillation were employed, the process of steam distillation can be used. Steam is passed directly into the liquid in the still; the solubility of the steam in the liquid must be very low. Steam distillation is perhaps the most common example of differential distillation.

Two cases are possible. The steam may be superheated and provide sufficient heat to vaporise the material concerned, without itself condensing. Alternatively, some of the steam may condense, producing a liquid water phase. In either case, assuming the gas laws to apply, the composition of the vapour produced can be obtained from the following relationships:

$$\frac{m_A}{M_A} \bigg/ \frac{m_B}{M_B} = \frac{P_A}{P_B} = \frac{y_A}{y_B} = \frac{P_A}{P - P_A} \tag{11.97}$$

where the subscript $A$ refers to the component being recovered, and $B$ to steam, and

$m$ = mass,
$M$ = molecular weight,
$P_A, P_B$ = partial pressure of **A**, **B**, and
$P$ = total pressure.

If there is no liquid phase present, then from the phase rule there will be two degrees of freedom, and both the total pressure and the operating temperature can be fixed independently, and $P_B = P - P_A$ (which must not exceed the vapour pressure of pure water, if no liquid phase is to appear).

With a liquid water phase present, there will be only one degree of freedom, and setting the temperature or pressure fixes the system, the water and the other component each exerting a partial pressure equal to its vapour pressure at the boiling point of the mixture. In this case, the distillation temperature will always be less than that of boiling water at the total pressure in question. Consequently, a high boiling organic material can be steam-distilled at temperatures below 373 K at atmospheric pressure. By using reduced operating pressures, the distillation temperature may be reduced still further, with consequent economy of steam.

A convenient method of calculating the temperature and composition of the vapour, for the case where the liquid water phase is present, is by use of the diagram shown in Fig. 11.45, due to Hausbrand[30]. The parameter $(P - P_B)$ is plotted for total pressures of 101·3,

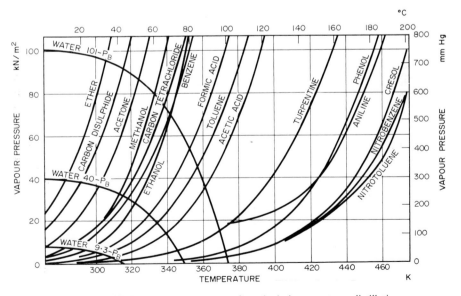

FIG. 11.45. Vapour pressure curves for calculations on steam distillation

40 and 9·3 kN/m², and the vapour pressures of a number of other materials are plotted directly against temperature. The intersection of the two appropriate curves gives the temperature of distillation and the molar ratio of water to organic material will be given by $(P - P_A)/P_A$. Thus, if nitrobenzene is distilled at atmospheric pressure with live saturated steam, the boiling point will be about 372 K and the mass-ratio of water to nitrobenzene in the vapour will be:

$$\frac{(101\cdot3 - 2\cdot7)}{2\cdot7} \times \frac{18}{123} = 5\cdot34$$

Where there is no liquid water phase present the steam consumption will be high, unless the steam is very highly superheated. With a water phase present, the boiling point of the mixture will be low, and consequently $P_A$ will have a low value. Thus, on a molar basis, the steam consumption will again be high, but owing to the relatively low molecular weight of steam, the consumption may not be excessive. Steam economy can be effected by using indirect heating of the still, having no liquid water phase present, or by operating under reduced pressure.

In any operation of this kind it is essential that the separation of the material being distilled from the water should be a relatively simple process.

When determining the number of stages required to effect a steam distillation, the steam flow must be included in the operating line equation for the lower part of the column. Using indirect heating and assuming constant molar overflow, the lower operating line for the organic material is:

$$y_m = \frac{L_m}{V_m} x_{m+1} - \frac{W}{V_m} x_w \qquad\qquad \text{(equation 11.20)}$$

This has a slope $L_m/V_m$ and cuts the $y = x$ line at $x = x_w$.

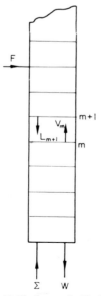

FIG. 11.46. Steam distillation

If $\Sigma$kmol of live steam are used as shown in Fig. 11.46, then considering that part of the column below the plate $(m+1)$:

$$V_m = L_{m+1} + \Sigma - W \qquad (11.98)$$

and

$$V_m y_m = L_{m+1} x_{m+1} - W x_w \qquad (11.99)$$

Assuming constant molar overflow, $L_m = L_{m+1} = W$ and $V_m = \Sigma$, giving again:

$$y_m = \frac{L_m}{V_m} x_{m+1} - \frac{W}{V_m} x_w \qquad (11.100)$$

This also has a slope of $L_m/V_m$ but cuts the $y = x$ line at $x = Wx_w/(W-\Sigma)$ (Fig. 11.47). When $x = x_w$, $y = 0$, corresponding to the composition of the vapour (steam) rising to the bottom plate. For a given external reflux ratio and feed condition, $L_m/V_m$ will be the same whether direct or indirect steam is used and the lower operating line must cut the $y = x$ line at the same value in each case, but, of course, $x_m$ for the indirect steam will be higher than $x_w$ for direct steam.

When stepping off the theoretical stages, the bottom step must start at $y = 0$, $x = x_w$. Hence the use of direct steam, although eliminating the still, dilutes the bottom material, and so increases the number of stages required in the lower part of the column.

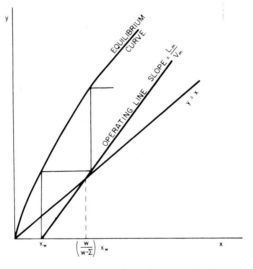

FIG. 11.47. Operating lines for steam distillation

## 11.10. PLATE COLUMNS

The process of distillation is carried out in many types of column, and it is convenient to consider them in relation to their internal arrangements. In plate or tray columns the operation is carried out in a stagewise manner, whereas with packed columns the process of mass transfer is continuous. Plate columns form the most important category and their features and performance will be taken first; packed columns will be considered later in Section 11.11.

The number of theoretical stages required to effect a required separation, and the corresponding rates for the liquid and vapour phases, can be determined by the procedures described earlier. To translate these quantities to form an actual design the following factors must be considered:

(a) The type of plate or tray.
(b) The vapour velocity, which is the major factor in determining the diameter of the column.
(c) The plate spacing, which is the major factor fixing the height of the column when the number of stages is known.

The detailed design methods will be discussed later, but it is first necessary to consider the range of trays available and some of their important features.

### 11.10.1. Types of Trays

The main requirement of a tray is that it should provide intimate mixing between the liquid and vapour streams, that it should be suitable for handling the desired rates of vapour and liquid without excessive entrainment or flooding, that it should be stable in operation, and that it should be reasonably easy to erect and maintain. In many cases, particularly with vacuum distillation, it is essential to arrange for the drop in pressure over the tray to be a minimum.

The arrangements for the liquid flow over the tray depend largely on the ratio of liquid to vapour flow. Three layouts are shown in Fig. 11.48, of which the cross-flow is much the most frequently used.

(a) *Cross-flow.* Normal, with good length of liquid path giving good opportunity for mass transfer.
(b) *Reverse.* Downcomers much reduced in area, and very long liquid path. Suitable for low liquid vapour ratios.
(c) *Double-pass.* Liquid flow splits into two directions, so that this system will handle high liquid vapour ratios.

The liquid reflux flows across each tray and enters the downcomer via a weir, the height of which largely determines the amount of liquid on the tray. The downcomer is taken beneath the liquid surface on the tray below, thus forming a vapour seal. The vapour flows upwards through risers into caps, or through simple perforations in the tray.

*The bubble-cap tray.* This is probably the most widely used tray because of its range of operation, but it is now tending to be superseded by other types. The general construction

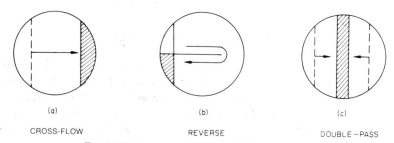

(a)        (b)        (c)

CROSS-FLOW      REVERSE      DOUBLE–PASS

FIG. 11.48. Arrangements for liquid flow over a tray

FIG. 11.49. 5·6 m diameter bubble tray (stainless steel) fitted with 1118 100 mm diameter caps

is shown in Figs. 11.49 and 11.50. The individual caps are mounted on risers and have rectangular or triangular slots cut around their sides. The caps are held in position by some form of spider, and the areas of the riser and the annular space around the riser should be kept about equal. With small trays, the reflux passes to the tray below over two or three circular weirs, and with the larger trays through segmental downcomers.

*Sieve or perforated trays.* These are much simpler in construction, requiring the forming of small holes in the tray. The liquid flows across the tray and down the segmental downcomer. Figure 11.51 indicates the general form of tray layout.

*Valve trays.* These can be regarded as a cross between bubble-cap and sieve trays. The construction is similar to that of cap types, but there are no risers and no slots. It is important to note that with most types of valve tray the extent of the opening may be varied by the vapour flow, so that the trays can operate over a wide range of flows. Because of their flexibility and price, they are tending to replace bubble-cap trays. Figure 11.52 shows a typical tray.

These three tray types have a common feature in that they all have separate downcomers for the passage of liquid from each tray to the one below. There is another class of tray which has no separate downcomers while still employing a tray type of construction giving a hydrodynamic performance between that of a packed and a plate column. Two examples of this type of device are the Kittel plate and a Turbogrid tray[31]. Design data for these trays are sparse in the literature and the manufacturer's advice should be sought on possible applications.

## 11.10.2. Factors Determining Column Performance

The performance of a column can be judged in relation to two separate but related criteria. First, if in a column the vapour leaving a tray is in equilibrium with the liquid leaving, this gives a theoretical tray and provides a standard of performance. Secondly, the relative performance of, say, two columns of the same diameter must be considered in relation to their capacity for liquid and vapour flow. The mass transfer criteria have

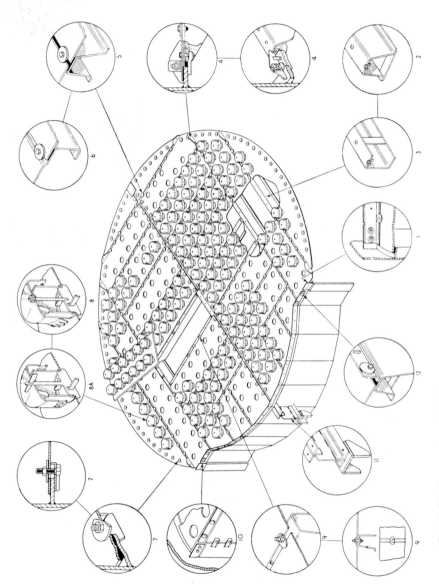

Fig. 11.50. Construction of bubble-cap plate, showing engineering features of Glitsch "Truss-type" bubble trays

Glitsch trays are provided with adjustable weir seal plates at each end to ensure a liquid-tight seal between the end of the weir and the tower wall.

## 2 Trapezoidal Minor Truss

Designed to provide a flat surface on top with accessibility to the lower side, this shape allows maximum rigidity and strength at a minimum of material and labour cost. The neutral axis of the section is slightly above the centre line in order to reduce stresses at the top or compression side of the structure. Tension braces maintain the section modulus of the truss in the event of minor surges or explosions.

## 3 Channel Minor Truss

Like the trapezoidal truss, this type of minor truss is designed to afford a flat surface on top with easy accessibility to the lower side, with simplicity, economy and ruggedness as its important advantages. Attention is directed to the nuts which are welded to the underside of this type of truss as well as the trapezoidal truss.

## 4 Top-type Truss Hangers

Built-up truss hangers are welded to the ends of the minor trusses to become integral parts of the trusses. The top-type hanger is provided with a clamping device for attaching the minor trusses to the tray support rings, thereby eliminating the necessity of individual brackets which otherwise would have to be welded to the tower wall. The truss hangers are also provided with individual clip angles to ensure that the outer edges of the tray floors do not tend to lift when load is applied to the tray.

## 5 Thermal Expansion Joint between Tray Floors and Supporting Trusses

Two important advantages are obtained from the use of frictional washers for holding down tray floor sections. Firstly, they provide for thermal expansion and contraction of the tray floor without risk of warping or buckling the relatively thin alloy sheet, the washers permitting independent lateral movement of all sections.

Secondly, they eliminate the need for matching holes among component parts of the tray, with a consequent saving in fabrication and erection costs.

The washers are bolted to nuts welded to the underside of the trusses (see diagrams 2 and 3).

## 6 An Integral Minor Supporting Truss

This method provides a minor truss section which is formed integrally with the tray floor.

## 7 Peripheral Tray Clamp

The clamps illustrated form a simple and effective means of holding the periphery of the tray to the tray support ring, and they may be installed from compensate for "out-of-roundness" of the tower, and the clamps serve to prevent bending stress on the tray floors. The use of these clamps allows for expansion and contraction due to temperature. They also eliminate the need for drilling the tray support ring and for matching holes. The round frictional washer, plus if required an asbestos gasket, is provided as a means of sealing the slotted holes.

## 8 Removable Cap and Riser Assembly

This is one of the many unique Glitsch cap and riser assemblies. It incorporates a removable cap and riser, together with a "snap-in" frog hold-down device to permit speedy, simplified installation from the upper side of the tray. The cap is complete with internal riser lugs which add rigidity and ruggedness to the assembly. An extruded riser opening provides an excellent method of centring the riser and adds considerable rigidity to the tray floor.

## 8A

The assembly is available either with a bolt and nut or with a wedge type hold-down, and both types are suitable for either round or rectangular riser openings.

## 9 Hold-down Device for Manway Section

This Glitsch device provides accessibility from either side of the tray. The assembly consists of two manway lock clamps—one to be used on each side of the truss member, or the manway batten strip, in conjunction with a stud bolt. To free the manway section requires only loosening of the nut from either side and turning the entire assembly.

## 10 Downcomer Clamping Bars

Downcomer clamping bars are designed to effect a liquid and vapour tight seal at the outer edges of downcomers, and are usually welded to the tower wall by the tower manufacturers.

## 11 Fabricated Major Beams

On larger trays, usually greater than 3 m diameter, where designs indicate heavy loads or explosion service, one or more major beams is provided. These beams are designed to meet the permissible maximum deflection allowances for loaded trays and to keep stresses within the allowable limit. Beams are attached to the tower wall by means of foot rests and vertical bolting plates, which can be welded to the tower wall either in the shop or in the field. When operating conditions permit, it is possible to hang these major supporting beams from the tower rings.

## 12 Adjustable Weirs

Provision for the alteration of weir levels is another useful feature frequently embodied in Glitsch trays in order to cater for possible process changes. In these cases weir plates are provided with slots to permit vertical adjustment.

FIG. 11.51.  A perforated plate tray for a 3·2 m column

FIG. 11.52.  Valve tray

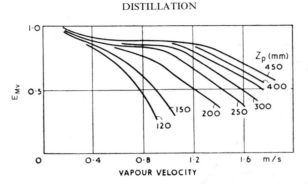

FIG. 11.53. Plate efficiency $E_{Mv}$ as a function of vapour velocity for various plate spacings

already been discussed, and it is helpful now to consider many of the features influencing the performance, particularly in relation to capacity. These main features are:

(a) Liquid and vapour velocities.
(b) Physical properties of the liquid and vapour.
(c) Extent of entrainment of liquid by rising vapour streams.
(d) The hydraulics of the flow of liquid and vapour across and through the tray.

It has been found by a number of workers[33–35] that the vapour velocity is a prime factor in determining the diameter of a column. Kirschbaum[32] using an equimolar mixture of ethanol and water on a 400 mm diameter plate containing 15 bubble caps, obtained results which may be summarised as follows:

(a) For all plate spacings the efficiency $E_{Mv}$ (defined later as equation 11.126, as the ratio of the actual change in liquid composition on a plate to that which would be obtained if the liquid left in equilibrium with the vapour) decreases as the velocity is increased (Fig. 11.53). This is mainly due to the reduction in contact time between the phases.
(b) The decrease in efficiency is much less with high plate spacings $(Z_p)$ so that with $Z_p$ greater than 400 mm $E_{Mv}$ remains constant over wide ranges of velocity.
(c) The capacity limit may be that of the downcomers to carry the reflux, rather than that of the caps to handle the vapour.

By working at various pressures, Kirschbaum was able to assess the influence of vapour density as shown in Fig. 11.54. First, at very low vapour velocities, $E_{Mv}$ is not much influenced by pressure. Then, for a given value of $E_{Mv}$, much greater velocities are possible at low pressures. This is seen by plotting $E_{Mv}$ against the mass velocity at different plate

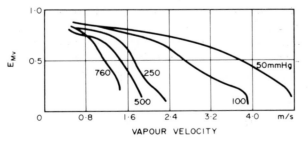

FIG. 11.54. Plate efficiency $E_{Mv}$ as a function of vapour velocity for various pressures. Plate spacing 200 mm

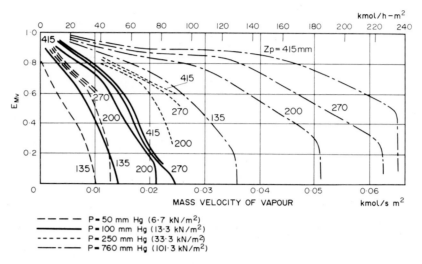

FIG. 11.55. Plate efficiency $E_{Mv}$ as a function of mass velocity for various plate spacings (mm) and pressures

spacings and at different pressures. For a given increase in mass velocity, the decrease in $E_{Mv}$ is more pronounced the lower the working pressure and the smaller the plate spacing (Fig. 11.55). This is shown by the data in Table 11.2.

TABLE 11.2. *Effect of Throughput on Change in Working Pressure or Change in Plate Efficiency*

| Plate spacing 200 mm | | | |
|---|---|---|---|
| Plate efficiency ($E_{Mv}$) | Total pressure (kN/m²) | Vapour density (kg/m³) | Throughput (kmol/m² s) |
| 0·8 | 101·3 | 1·254 | 0·025 |
| | 6·7 | 0·10 | 0·0039 |
| 0·7 | 101·3 | 1·254 | 0·033 |
| | 6·7 | 0·10 | 0·0069 |

Thus, reducing $E_{Mv}$ from 0·8 to 0·7 at 101·3 kN/m² pressure increases the throughput by 0·008 kmol/m² s or 33 per cent, but at 6·7 kN/m² pressure by 0·003 kmol/m² s or 55 per cent. Raising the pressure from 6·7 to 101·3 kN/m², with $E_{Mv}$ at 0·8, increases the throughput by 0·0211 kmol/m² s or 268 per cent, whilst if the column is operated with $E_{Mv}$ equal to 0·7 the increase is 0·0261 kmol/m² s or 215 per cent.

The effects of liquid viscosity have been studied by Drickamer and Bradford[36] and O'Connell[37] and are discussed later. Surface tension influences operation with sieve trays, both in relation to foaming and the stability of bubbles.

### Operating Ranges for Trays

It is instructive to note that for a given tray layout there are certain limits of flow of vapour and liquid within which stable operation is obtained. The range is indicated in the

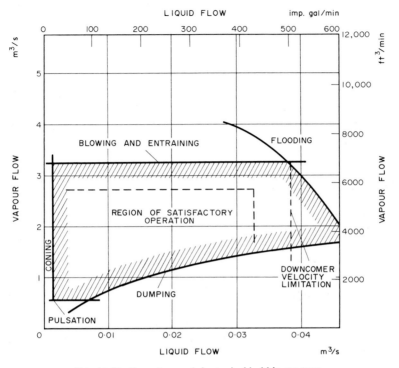

FIG. 11.56. Capacity graph for typical bubble-cap tray

general diagram shown in Fig. 11.56, which relates to a bubble-cap plate. The region of satisfactory operation is bounded by areas where undesirable phenomena occur. Coning occurs at low liquid rates, where the vapour forces the liquid back from the slots and passes out as a continuous stream, with a consequent loss in efficiency. Low vapour rates result in pulsating vapour flow or dumping. With low liquid rates, vapour passes through the slots intermittently, but with higher liquid rates some slots dump liquid rather than passing vapour. Both pulsating vapour flow and dumping, which can be referred to jointly as weeping, result in poor efficiency. At very high vapour rates, the vapour bubbles carry liquid as spray or droplets to the plate above, giving excessive entrainment. With high liquid rates, a point is reached where the drop in pressure across the plate equals the liquid head in the downcomer. Beyond this point, the liquid builds up and floods the tray.

*Entrainment*

The extent of entrainment of the liquid by the vapour rising over a plate has been studied by many workers. The entrainment has been found to vary with the vapour velocity in the slot or perforation, and the spacing used. Strang[38], using an air–water system, found that entrainment was small until a critical vapour velocity was reached, above which it increased rapidly. Similar results from Peavy and Baker[39] and Colburn[40] have shown the effect on tray efficiency, which is not seriously affected until the entrainment exceeds 0·1 kmol of liquid per kmol of vapour. The entrainment on sieve trays is discussed in Section 11.10.4.

*Hydraulics of the Flow of Reflux and Vapour through a Column*

The design of the tray fittings and the downcomers will influence the column performance. It is convenient to consider this separately for bubble cap trays and sieve trays. What is required is the diameter of the tower, the tray spacing and the detailed design of the tray.

### 11.10.3. Bubble Cap Trays. General Design Method

In considering the design of a column for a given separation, the number of stages required and the flowrates of the liquid and vapour streams must first be determined using the general methods already outlined. In the mechanical design of the column, tower diameter, tray spacing, and the detailed layout of each tray will be investigated. In the first place, a diameter is established on the criterion of freedom from liquid entrainment in the vapour stream, and then the weirs and the downcomers are designed to handle the required liquid flow. It is then possible to consider the tray geometry in more detail, and, finally, to examine the general operating conditions for the tray and to establish its optimum range of operation. The method outlined here is based on the early work of Carey *et al.*[33] and Souders and Brown[34], and on the more recent work of Bolles[41], Fair and Matthews[42], and Lockhart and Leggett[43]. The specialised texts of Robinson and Gilliland[1], Hengstebeck[44] and Smith[29], the Final Report of the ABCM/BCPMA Distillation Panel[45] and the *Bubble Tray Design Manual* of the American Institute of Chemical Engineers[46] give extensive coverage of this important area of chemical engineering work. The methods given for the design of columns for distillation are also applicable for gas absorption columns.

Because small diameter towers are frequently installed in buildings, the tray spacing is made as small as possible (150–300 mm). However, for the larger columns used in the petroleum and petrochemical industries the spacing is commonly 600 mm or more; this facilitates maintenance and enables high flowrates to be used. In the large columns separate manholes are not necessary for each tray since, by arranging for sections of a tray to be removable, it is relatively easy to climb up and down the column.

Bubble-cap trays are now rarely used for new installations on account of their high cost and their high pressure drop; in addition, difficulties arise in large columns because of the large hydraulic gradients which are set up across the trays. They are capable of dealing with very low liquid rates and are therefore useful for operation at low reflux ratios. There are still many bubble-cap columns in use and the following summary of design considerations is given in particular to enable existing equipment to be assessed for new studies.

*Tray Spacing*

The Souders and Brown equation[34] enables the column diameter to be found based on considerations of tray spacing:

$$G' = B\sqrt{\rho_V(\rho_L - \rho_V)} \tag{11.101}$$

The constant $B^{(34)}$ is a function of tray spacing and the liquid seal on the tray. More recently[42] it has been suggested that equation 11.101 was too conservative and two new parameters, $F_{lv}$ and $C_{sb}$, defined below, have been suggested.

$$F_{lv} = \frac{L'}{G'}\sqrt{\frac{\rho_v}{\rho_L}} \quad \text{and} \quad C_{sb} = u_n\sqrt{\frac{\rho_v}{\rho_L - \rho_v}}$$

$L'$ and $G'$ are the mass flowrates of liquid and vapour, respectively, per unit total area, and $u_n$ is the velocity of the vapour based on the net area above the tray (Area of column less area of one downcomer). The parameters are shown in Fig. 11.57 for various tray spacings for liquid with a surface tension of $0 \cdot 02 \, \text{N/m} \, (= 0 \cdot 02 \, \text{J/m}^2)$.

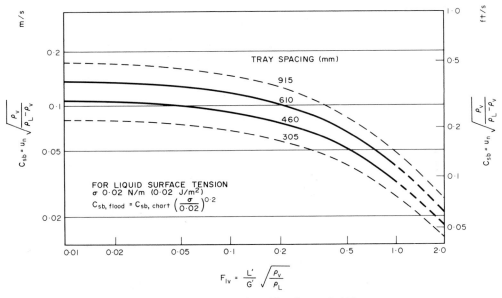

FIG. 11.57. Generalised correlation of flooding on bubble-cap trays

Figure 11.57 is used by calculating the value of $L'/G' \sqrt{\rho_v/\rho_L}$ and guessing a tray spacing so that the chart value $C_{sb}$ may be found; when multiplied by $(\sigma/0 \cdot 02)^{0 \cdot 2}$ this gives the value of $C_{sb}$ at flooding from which the vapour velocity based on the net area can be calculated under flooding conditions; this enables the column diameter to be calculated on the assumption that a velocity equal to some fixed fraction of the flooding velocity is to be used. Consideration of the tray hydraulics will ascertain whether the initial tray spacing was chosen correctly. If not, the procedure must be repeated for a new spacing.

*Tray Layout Based on Tray Hydraulics*

It is necessary to establish the dynamics of the flow of the liquid across the tray, and to determine the drop in pressure over the tray. This involves selecting a suitable cap and fixing the length and height of the exit weir. Proposals from Bolles[47] and Lockhart and Leggett[43] help in forming a picture of the relative importance of several parameters.

Details of four types of tray which can cope with different ratios of liquid to vapour are given in Table 11.4. The first three have one pass of liquid and the fourth two passes. The low riser type is used where the vapour load is low.

TABLE 11.4. *General Types of Tray*

|  | Low riser | Medium riser | High riser | Two-pass |
|---|---|---|---|---|
| Riser area as % of tower area | 10 | 12·5 | 15 | 10 |
| Downcomer area as % of tower area | 20 | 14 | 8 | 17·5 |
| Weir length as % of tower diameter | 87 | 79·7 | 68·5 | 84·2 |
| Weir height (mm) | 64 | 64 | 64 | 24 |

It is desirable that the height of the weir plus the crest over the weir $(h_w + h_{ow})$ should be between 65 and 125 mm. The liquid flow should not exceed 25 cm$^3$/s per mm of weir length; for higher flows two-pass trays should be used. The froth in the downcomer has a low specific gravity, and velocities should not exceed 0·2 m/s with 0·75 m tray spacing. On the whole, inlet weirs have not been favoured although they may give a more even flow over the tray from the downcomer.

*Bubble Caps*

Typical bubble-cap trays are shown in Fig. 11.50 and in Fig. 11.58a and b. Vapour from one tray enters the tray above through risers (1) and then passes through the cap and out through the slots to give a stream of bubbles which coalesce. Since the main aim is to promote intimate mixing between the vapour and the liquid, narrow slots of 1·5–5 mm

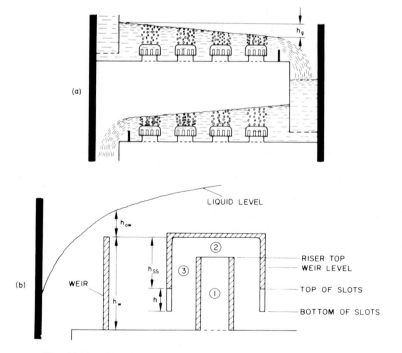

FIG. 11.58. (*a*) Typical bubble-cap tray. (*b*) Enlarged view of bubble-cap

are generally used. To minimise friction losses in the cap, the areas of the risers (1), reversal section (2) and the annulus (3) are made approximately equal. Caps made from pressed steel or alloys are generally preferred to those from cast-iron, since they are lighter and simpler to handle. The caps are usually spaced on equilateral triangular centres, the minimum distance between the caps being 25–40 mm. A clearance of about 75 mm should be allowed between the last row of caps and the exit weir, and in no case should a cap be brought nearer the wall than 40 mm. The bottom, or skirt, of the cap is frequently placed 10–25 mm above the plate, but a greater clearance is necessary both for high liquid rates and for liquids containing suspended solids. The clearance is reduced to a minimum of zero for high vacuum towers where the pressure drop over a tray must be kept to a very low value. The distance between the top of the slot and the top of the exit weir is known as the static submergence ($h_{ss}$), and this determines to a great extent the depth of liquid through which the bubbles must pass. The slot velocity is frequently 3 m/s or more. It is important to realise that a foam forms on the tray and a shallow layer of clear liquid collects at the bottom with an aerated mass above. In normal operation, the vapour rate is such that the slots are fully open and the streams from neighbouring caps intermingle with each other. Full details of bubble-cap geometry are given by Bolles[41].

*Drop in Pressure through the Cap*

The drop in pressure as the vapour passes through the cap arises from the loss at the entrance to the risers $h_r$, the loss on reversal in the annulus due to friction $h_{rc}$ and the loss in passing through the slots $h_s$.

$h_r$ and $h_{rc}$ are combined in equation 11.102; the dry-cap coefficient $K_c$ is obtained from Fig. 11.59:

$$h_r + h_{rc} = K_c \frac{u_r^2}{2g} \frac{\rho_v}{\rho_L} \tag{11.102}$$

If $h_s$ is the depth of the slot that is open during the passage of the vapour, then:

$$h_s = 3.26 \times 10^4 \left(\frac{\rho_v}{\rho_L - \rho_v}\right)^{1/3} \left(\frac{v_s}{w}\right)^{2/3} \text{mm} \tag{11.103}$$

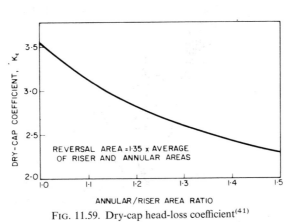

FIG. 11.59. Dry-cap head-loss coefficient[41]

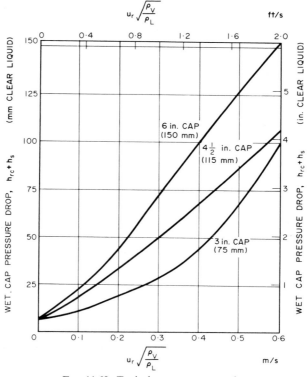

FIG. 11.60. Typical wet-cap pressure drops

The total loss in head through the cap is then $h_r + h_{rc} + h_s$. Some practical values of this total pressure drop through the wetted-cap are shown in Fig. 11.60, which includes work by Dauphine[48] and Rhys and Minich[49]. The velocity $u_r$ through the riser is in m/s.

### Drop in Pressure over Tray

The vapour in passing through a tray suffers loss in pressure in the cap, then through the slots and finally in passing up through the aerated liquid. This gives the drop in pressure (expressed as head of liquid) for the vapour as:

$$h_p = h_{rc} + h_s + h_{al} \tag{11.104}$$

where $h_{al}$ is the loss in passing through the aerated liquid above the top of slots. The aerated liquid has a density less than that of the clear liquid so that, taking the point where the liquid height on the plate is the mean of the extreme values;

$$h_{al} = \beta h_{ds} = \beta[h_{ss} + h_{ow} + \tfrac{1}{2}h_g] \tag{11.105}$$

where $h_{ds}$ is the dynamic slot seal and $\beta$ is the aeration factor. The aeration factor has been evaluated for various vapour velocities, but a value of 0·8 for low vapour velocities over the tray (say 0·6–1·5 m/s) and a value of 0·6 for high vapour velocities are reasonable approximations. The two remaining terms to consider are the height of the crest over the weir $h_{ow}$ and the liquid gradient term $h_g$.

*Crest over weir.* The liquor passing through the downcomer flows over an inlet weir, across the plate, and then over the exit weir. The height of the crest forming over the exit weir $h_{ow}$ can be found by modification of the Francis weir formula. In this case the liquid approaches the weir with an appreciable velocity, and the height of the crest will vary with the liquid flow. Bolles[47] has given a graphical solution for finding $h_{ow}$ from the equation:

$$h_{ow} = 68,175F_w \left[ \frac{Q'}{w} \right]^{2/3} \tag{11.106}$$

where  $h_{ow}$ = height of liquid over weir in mm,
$\quad\quad Q'$ = liquor rate in $m^3/s$,
$\quad\quad w$ = weir length in mm, and
$\quad\quad F_w$ = factor as given by Bolles which varies from 1·0 to 1·2.

*Hydraulic gradient.* The liquid flowing under the downcomer will rise up to a height greater than that over the exit weir, thus giving a hydraulic gradient to force the liquid across the tray. This difference of head $h_g$ cannot readily be expressed in terms of the dimensions of the system because the liquid meets so many obstacles in the form of caps. Bolles[41] has presented a complex expression for calculating the hydraulic gradient, together with design charts which are relatively easy to use. Other work on liquid gradients has been done by Kemp and Pyle[50], Davies[51] and Good *et al.*[52]. Lockhart and Leggett[43] have pointed out that the experimental data have been obtained with simulated conditions, using trays with long calming sections between the last row of caps and the exit weir. In this way the liquid is fully de-aerated before passing over the weir, whereas in practice this is not the case. They give a chart (Fig. 11.61) in which the values of the liquid gradient are much larger; these values are recommended for calculating the height to which the liquid will rise in the downcomer.

*Tray Spacing*

The necessary tray spacing to avoid entrainment of the liquid by the vapour has been discussed above. To enable the downcomer to handle the flow, the tray spacing must be such that the liquid does not rise up the downcomer to a height much above half-way between two trays. The head of liquid in the downcomer $Z'_p$ is made up of the head of liquid on the tray $(h_w + h_{ow})$, plus the head lost by the vapour $h_p$, plus the head lost at the bottom of the downcomer $h_d$, plus the liquid gradient term $h_g$. Thus:

$$Z'_p = h_p + h_w + h_{ow} + h_d + h_g \tag{11.107}$$

$$= h_{rc} + h_s + h_{al} + h_w + h_{ow} + h_d + h_g$$

$$= h_{rc} + h_s + \beta[h_{ss} + h_{ow} + \tfrac{1}{2}h_g] + h_w + h_{ow} + h_d + h_g \tag{11.108}$$

An approximate relation has been given by Bolles[47] and Kirkbride[53] as:

$$Z'_p = 2(h_w + h_{ow}) + h_d + h_{rc} \tag{11.109}$$

Then, for safe operation, the tray spacing is given by:

$$Z_p = 2Z'_p \tag{11.110}$$

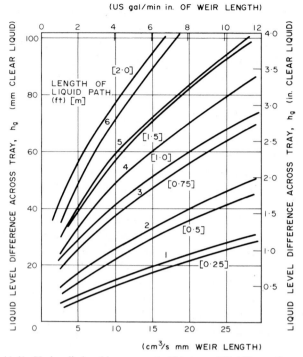

FIG. 11.61. Hydraulic head loss across bubble-trays (75–150 mm dia. caps)

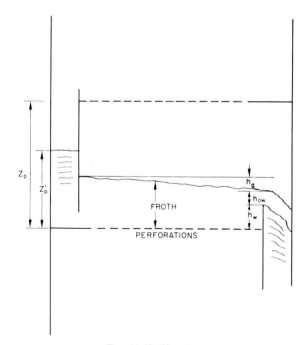

FIG. 11.62. Sieve tray

Thus for the assumed tray spacing in Fig. 11.57, the various head losses are calculated and $Z'_p$ found. The tray spacing required is $2Z'_p$ and this value must then be compared with the original assumed value.

### 11.10.4. Sieve Trays

Studies on sieve trays have shown that they can offer several advantages over the bubble-cap tray, and their simpler and cheaper construction has led to their increasing use. The general form of the flow on a sieve tray is indicated in Fig. 11.62, and it is seen that the tray forms a typical cross-flow system with perforations in the tray taking the place of the more complex bubble caps. The hydraulic flow conditions will be examined for such a tray in the same manner as for the bubble-cap tray by considering entrainment, flooding, pressure loss, and so on. It is important to notice the key differences in operation between these two types of tray. With the sieve tray the vapour passes vertically through the holes into the liquid on the tray, whereas with the bubble cap the vapour issues in an approximately horizontal direction from the slots. With the sieve plate the vapour velocity through the perforations must be not less than a certain minimum value in order to prevent the weeping of the liquid stream down through the holes. At the other extreme, a very high vapour velocity will lead to excessive entrainment and loss of tray efficiency. Thus there will be an operating region of the same general shape as with bubble caps, as shown in Fig. 11.63. The arrangements for perforations, weirs and the hydraulic flow will

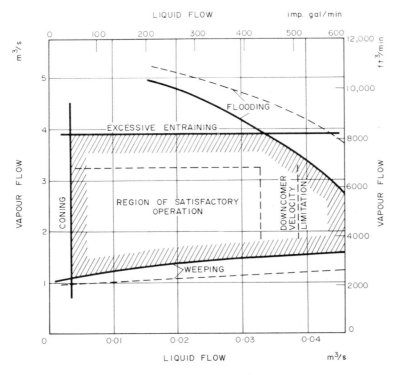

FIG. 11.63. Capacity graph for typical sieve tray

be considered using the approach of Mayfield et al.[54], Leibson et al.[55], Bain and van Winkle[56] and Fair[57].


### Tray Diameter

The diameter of a sieve tray is found by a method similar to that used for bubble-cap trays. The relevant flooding correlation is presented as Fig. 11.64 where, using Fair's[57] approach, the Souders and Brown[34] parameter $C_{sb}$ is plotted against the flow parameter $F_{lv}$. It is important to make corrections for values of hole/active area ratio as follows:

| Hole/active area ratio | Multiply $C_{sb}$ by |
| --- | --- |
| 0·10 | 1·0 |
| 0·08 | 0·9 |
| 0·06 | 0·8 |

The velocity $u_n$ is based on the net area and velocities up to 0·8 times the flooding velocity may be used.

The fraction of the total plate area represented by the perforations would be expected to have some influence on the maximum velocity for the vapour. Thus, if the spacing between the holes is reduced too much, the liquid will be held above the tray and the foaming of the aerated mass will increase. Eld[58] has suggested a rather different equation to give the

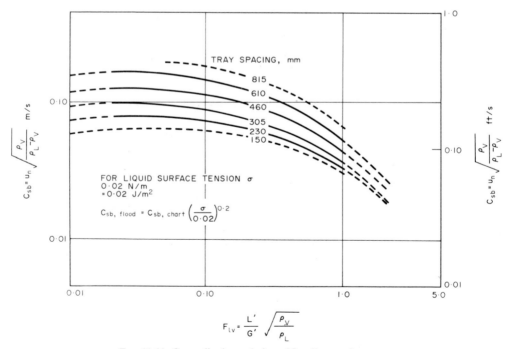

FIG. 11.64. Generalised correlation of flooding on sieve trays

maximum vapour velocity $u_e$; this involves the cube root of the density difference rather than the square root:

$$u_e = C_e \left( \frac{\rho_L - \rho_v}{\rho_v} \right)^{1/3} \qquad (11.111)$$

Some values of $C_e$ are given in Table 11.5.

TABLE 11.5. *Values of $C_e$ (m/s) for Use in Equation 11.111*

| Tray spacing | | % perforations on tray | | |
|---|---|---|---|---|
| (in) | (mm) | 5 | 10 | 15 |
| 12 | 305 | 2·3 | 1·7 | 1·3 |
| 18 | 460 | 3·5 | 2·6 | 2·0 |
| 24 | 610 | 4·6 | 3·4 | 2·7 |
| 36 | 915 | 7·0 | 5·1 | 3·9 |

If the system foams badly, these values should be reduced.

Gerster[59] has shown that all data for flooding on bubble-cap, sieve, and valve trays can be correlated together if, instead of tray spacing, a parameter $h_w + h_{ow}$ is used.

*Entrainment*

Fair[57] produced a correlation relating the fractional entrainment to the parameter $F_{lv}$ with the ratio of the vapour velocity to that at flooding as parameter, as shown in Fig. 11.65. The fractional entrainment is defined as (liquid entrainment)/(liquid flowrate + liquid entrainment) and, for most operations, its value should not exceed 0·10.

*Perforations*

The holes range from 3 mm to 25 mm, though 5–10 mm are the most widely used. The holes are located on triangular centres with a pitch to diameter ratio of 2 to 4·5. A space of about 100 mm must be left between the base of the downcomer and the first row of holes, and between the last row and the exit weir. The active area of the tray is the total area of cross-section less the area of both downcomers. For normal sieve trays 5–15 per cent of the active area may be accounted for by the perforations.

*Weirs*

The length and height of the weir is found as for bubble-cap trays. The liquid flow is normally kept to less than $0·025 \text{ m}^3/\text{s}$ per metre of weir length. To give good contact it is necessary to have a minimum height $(h_w + h_{ow})$ of 50 mm. The weir crest $h_{ow}$ (mm) can be obtained from equation 11.106 as for bubble caps, or alternatively by:

$$h_{ow} = [1·7 \times 10^4 \, (\text{m}^3/\text{s})/l_w]^{2/3} \text{ mm} \qquad (11.112)$$

where $l_w$ is the length of the outlet weir in metres.

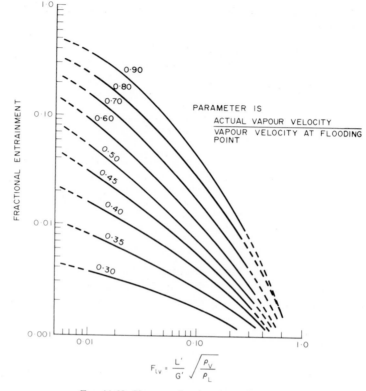

FIG. 11.65. Sieve tray fractional entrainment

$$F_{lv} = \frac{L'}{G'} \sqrt{\frac{\rho_V}{\rho_L}}$$

*Downcomers*

The problem is the same as with bubble caps; a sufficient path length must be provided for the froth to decay before it reaches the exit weir. To ensure that the entrained bubbles escape, the liquid velocity is kept below 0·3 m/s. The loss in head at the bottom of the downcomer $h_d$ is found as for bubble-cap trays.

*Pressure Drop over the Trays*

The drop in pressure for the vapour arises from the loss in passing through the holes, $h_o$, and that arising from the passage through the aerated liquid, taken as $\beta(h_w + h_{ow})$. The aeration factor $\beta$ has been given by Mayfield *et al.*[54] for various values of clear liquid depth; and values of 0·8 and 0·5 for values of $(h_w + h_{ow})$ of 50 mm and 115 mm, respectively, give reasonable accuracy. The drop in pressure through the perforations is given by the orifice equation:

$$u_o = C_o \left( 2g \frac{\rho_L}{\rho_v} h_o \right)^{1/2} \tag{11.113}$$

or

$$h_o = \frac{1}{2g} \frac{\rho_v}{\rho_L} \left( \frac{u_o}{C_o} \right)^2 \tag{11.114}$$

where   $C_o$ is the orifice coefficient given in Fig. 11.66[55],

    $h_o$ is the drop in pressure, expressed as head of liquid, and

    $u_o$ is the vapour velocity through the perforations.

It will be noticed that $C_o$ depends on the ratio of the tray thickness to the hole diameter. The head lost over the tray $h_p$ is then given by:

$$h_p = h_o + (h_w + h_{ow}) \tag{11.115}$$

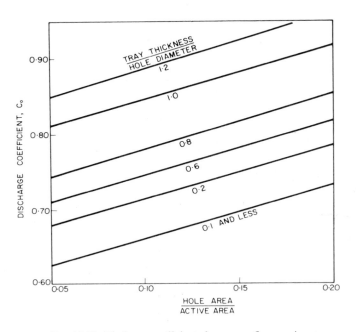

FIG. 11.66. Discharge coefficients for vapour flow on sieve trays

## Weeping

It is important to see that the vapour velocity is such that the liquid will not weep through the perforations. Some indication of these limiting conditions are shown in Fig. 11.67 taken from the work of Leibson et al.[55] and of Mayfield et al.[54]. Alternatively, one can approach this problem by arranging that the head lost at perforations $h_o$, together with that arising from surface tension $h_\sigma$, is greater than that of the clear liquid, i.e.:

$$h_o + h_\sigma > h_w + h_{ow} \tag{11.116}$$

## Liquid Gradient

The loss in head for the sieve tray is much less than the corresponding value with bubble caps at the same flow rates. The main loss occurs at the inlet to the tray where the maximum height of clear liquid is built up.

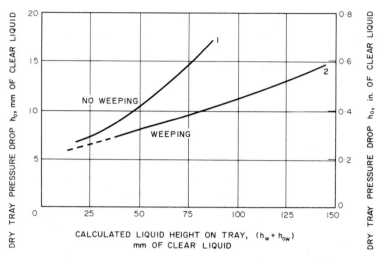

FIG. 11.67. Weeping correlation for sieve trays with 4·8 mm perforations

*Height of Liquid in Downcomer*

The height the liquid rises up the downcomer $Z'_p$ is then given in the same form as for bubble-cap trays:

$$Z'_p = h_p + h_w + h_{ow} + h_g + h_d \tag{11.117}$$

$$= h_o + \beta(h_w + h_{ow}) + h_w + h_{ow} + h_g + h_d \tag{11.118}$$

With present information this equation is sufficiently nearly given by:

$$Z'_p = h_o + 2(h_w + h_{ow}) + h_d \tag{11.119}$$

The tray spacing $Z_p$ is then given by:

$$Z_p = 2Z'_p \tag{11.120}$$

from which the original tray spacing may be checked.

### 11.10.5. Valve Trays

The valve tray, which may be regarded as intermediate between the bubble cap and the sieve tray, offers advantages over both. The feature of the tray is that liftable caps act as variable orifices which adjust themselves to changes in vapour flow. The valves are metal discs of up to about 38 mm diameter or metal strips which are raised above the openings in the tray deck as vapour passes through the trays. The caps are restrained by legs or spiders which limit the vertical movement and some types are capable of forming a total liquid seal when the vapour flow is insufficient to lift the cap.

Advantages claimed for valve trays include:

(a) Operation at the same capacity and efficiency as sieve trays.

(b) A low pressure drop which is fairly constant over a large portion of the operating range.

(c)  A high turndown ratio (i.e. can be operated at a small fraction of design capacity).

(d)  A relatively simple construction which leads to a cost of only 20 per cent higher than a comparable sieve tray.

Valve trays, because of their proprietary nature, are usually designed by the manufacturer, though it is possible to obtain an estimate of design and performance from published literature[60].

## Tray Diameter

The required diameter of the tray will depend on the way the liquid flows across the tray to the overflow weir. In large diameter columns it is necessary to split the liquid flow into a number of "passes" to reduce the hydraulic gradient across the tray. An estimate of the diameter of single, two-pass, and four-pass trays may be obtained from Fig. 11.68. The diameter of two-pass and four-pass trays should not be less than 1·5 m and 3·0 m respectively.

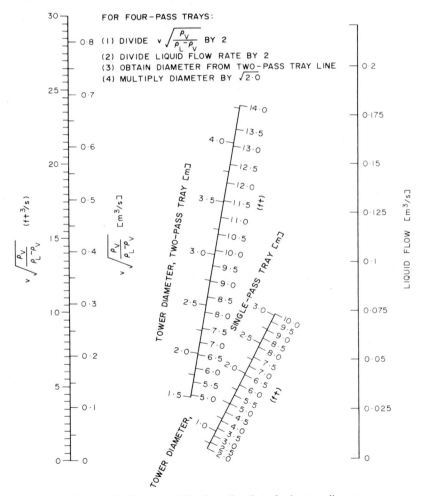

FIG. 11.68. Nomograph for the estimation of valve tray diameter

## Downcomer Area

The downcomer area is a function of the liquid flowrate, the degree of approach to flooding conditions and the downcomer design velocity. It should not be less than 11 per cent of the active area of the tray and is based on a liquid velocity (m/s) which is the smallest of the following values: (a) $0.17$ m/s; (b) value from Fig. 11.69; (c) values obtained from equations 11.121 and 11.122.

$$u_{dc} = 0.007\sqrt{(\rho_L - \rho_V)} \tag{11.121}$$

$$u_{dc} = 0.008\sqrt{Z_p(\rho_L - \rho_V)} \tag{11.122}$$

where $Z_p$ = tray spacing (m), and
$\rho_L - \rho_V$ = density difference, expressed in kg/m$^3$.

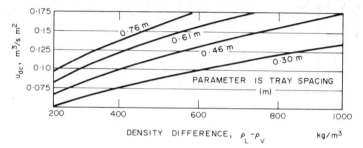

FIG. 11.69. Ideal downcomer velocity as a function of tray spacing and density difference

## Number of Valve Units

A knowledge of the number of valve units on each tray is necessary in order to calculate the vapour velocity through the orifices for use in pressure drop calculations. Only a detailed tray layout can give the number accurately though a range of 120–175 units/m$^2$ serves as a guide with a value of 150 units/m$^2$ as being a normal maximum. The hole area may then be estimated from the details of the valves selected. For Glitsch trays (Fig. 11.52) a hole area of 1 m$^2$ corresponds to 845 valve units.

## Pressure Drop

The pressure drop across a valve tray is a function of the liquid and vapour flowrates, the hole area, the type of valve, the material of construction and the dimensions of the weir.

The pressure drop for the dry tray is first calculated from equations 11.123 and 11.124 and the larger of the two values is then used.

$$-\Delta P_{\text{dry}} = 0.00135\, l_m \frac{\rho_m}{\rho_L} + K_1 u_o^2 \frac{\rho_v}{\rho_L} \tag{11.123}$$

or

$$-\Delta P_{\text{dry}} = K_2 u_o^2 \frac{\rho_v}{\rho_L} \tag{11.124}$$

where   $u_o$ = velocity through the holes (m/s),
   $l_m$ = deck thickness (mm),
   $-\Delta P_{dry}$ = dry tray pressure drop (m of liquid),
   $K_1, K_2$ = coefficients from Table 11.6, and
   $\rho_m$ = metal density (kg/m³).

TABLE 11.6. *Coefficients for Equations 11.123 and 11.124*

| Type of unit | $K_1$ | $K_2$ Deck thickness | | | |
|---|---|---|---|---|---|
| | | 14 swg(1·88 mm) | 12 swg(2·65 mm) | 10 swg(3·3 mm) | ¼ in.(6·35 mm) |
| Glitsch V-1 (sharp edge) | 0·0546 | 0·287 | 0·251 | 0·224 | 0·158 |
| Glitsch V-4 (venturi shaped orifice) | 0·0273 | 0·136 | 0·106 | 0·104 | — |

Data for $l_m$ and $\rho_m$ for commonly used trays may be obtained from Table 11.7.

TABLE 11.7. *Data for Materials of Construction*

| Thickness $l_m$ | | | | Density of valve materials $\rho_m$ | |
|---|---|---|---|---|---|
| swg | in. | mm | | lb/ft³ | kg/m³ |
| 20 | 0·037 | 0·94 | Mild steel | 480 | 7,700 |
| 18 | 0·050 | 1·27 | Stainless steel | 510 | 8,180 |
| 16 | 0·060 | 1·52 | Nickel | 553 | 8,850 |
| 14 | 0·074 | 1·88 | Monel | 550 | 8,820 |
| 12 | 0·104 | 2·64 | Titanium | 283 | 4,540 |
| 10 | 0·134 | 3·40 | Hastelloy | 560 | 8,980 |
| ¼ in. | 0·250 | 6·35 | Aluminium | 168 | 2,695 |
| | | | Copper | 560 | 8,980 |
| | | | Lead | 708 | 11,350 |

The total tray pressure drop is then obtained from equation 11.125:

$$-\Delta P_T = -\Delta P_{dry} + 0.554 \left(\frac{Q'}{L_W}\right)^{0.67} + 0.4\, h_w \qquad (11.125)$$

where   $h_w$ = weir height (m),
   $Q'$ = liquid flowrate (m³/s),
   $L_W$ = weir height (m), and
   $-\Delta P_T$ = total tray pressure drop (m of liquid).

## 11.10.6. Plate Efficiency

The number of ideal stages required for a desired separation may be calculated by one of the methods previously discussed but in practice it will normally be found that more trays are required than ideal stages. Then the ratio $n/n_p$ of the number of ideal stages $n$ to the number of actual trays $n_p$ represents the overall efficiency $E$ of the column. This overall efficiency may vary from 30 to 100 per cent[61]. The main reason for loss in efficiency is that

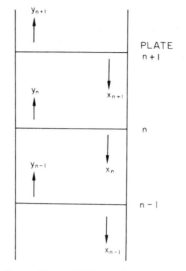

FIG. 11.70. Compositions of liquid and vapour streams from plates

the kinetics for the rate of approach to equilibrium, and the flow pattern on the plate, may be unfavourable, so that equilibrium between the vapour and liquid is not attained. Some empirical equations have been developed from which values of efficiency may be calculated, and this approach is of considerable value in giving a general picture of the problem. The proportion of liquid and vapour, and the physical properties of the mixtures on the trays, will vary up the column, and to obtain a sharper picture of the meaning of efficiency conditions on individual trays must be examined, as suggested by Murphree[62]. For a single ideal tray the vapour leaving is in equilibrium with the liquid leaving, and the ratio of the actual change in composition achieved to that which would occur if equilibrium between $y_n$ and $x_n$ were achieved is known as the Murphree plate efficiency $E_M$. Using the notation shown in Fig. 11.70, the plate efficiency expressed in vapour terms is given by:

$$E_{Mv} = \frac{y_n - y_{n-1}}{y_e - y_{n-1}} \tag{11.126}$$

where $y_e$ is the composition of the vapour that would be in equilibrium with the liquid of composition $x_n$ actually leaving the plate. This equation gives the efficiency in vapour terms, but if the concentrations in the liquid streams are used then the plate efficiency $E_{Ml}$ is given by:

$$E_{Ml} = \frac{x_{n+1} - x_n}{x_{n+1} - x_e} \tag{11.127}$$

where $x_e$ is the composition of the liquid that would be in equilibrium with the composition $y_n$ of the vapour actually leaving the plate.

The ratio $E_{Mv}$ is shown graphically in Fig. 11.71 where for any operating line $AB$ the enrichment that would be achieved by an ideal plate is $BC$, and that achieved with an actual plate is $BD$. Then the ratio $BD/BC$ represents the plate efficiency. The efficiency may vary from point to point on a tray. Local values of the Murphree efficiency will be designated $E_{mv}$ and $E_{ml}$.

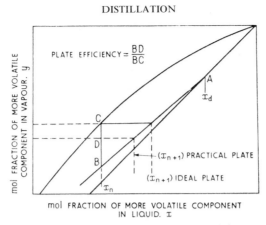

FIG. 11.71. Graphical representation of plate efficiency $E_{Mr}$

*Empirical Expressions for Plate Efficiency*

The efficiency of the individual plates will be expected to depend on the physical properties of the mixture, the geometrical arrangements of the trays, and the flowrates of the two phases. A simple empirical relationship for the overall efficiency, $E$, of columns handling petroleum hydrocarbons was given by Drickamer and Bradford[36] who related efficiency for the column to the average viscosity of the feed:

$$E = 0 \cdot 17 - 0 \cdot 616 \log_{10} \Sigma(x_f \mu_L) \qquad (11.128)$$

where $x_f$ is the mol fraction of the component in the feed, and
$\mu_L$ is the viscosity in mN s/m² at the mean tower temperature.
Further work, mainly with larger towers 3 m in diameter, suggested that higher efficiencies were obtained with larger diameters because of the longer liquid path. Thus, compared with a 0·9 m diameter tray, one of 3 m diameter might give up to 25 per cent greater efficiency.

### Example 11.14

Use equation 11.128 for the following data on the separation of a stream of $C_3$ to $C_6$ hydrocarbons to determine the plate efficiency.

| Component | Mol fraction in feed $x_f$ | $\mu_L$ (mN s/m²) | $\mu_l x_f$ |
|---|---|---|---|
| $C_3$ | 0·2 | 0·048 | 0·0096 |
| $C_4$ | 0·3 | 0·112 | 0·0336 |
| $C_5$ | 0·2 | 0·145 | 0·0290 |
| $C_6$ | 0·3 | 0·188 | 0·0564 |
| | | | 0·1286 |

Then $E = 0 \cdot 17 - 0 \cdot 616 \log 0 \cdot 1286 = \underline{\underline{0 \cdot 72}}$.

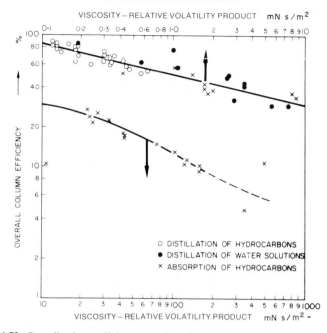

FIG. 11.72. Overall column efficiency $E$ as function of viscosity–relative volatility product

O'Connell[37] found that a rather better relation could be obtained by plotting the overall efficiency in relation to the product of the viscosity and the relative volatility of the key components. This relation has also been presented by Lockhart and Leggett[43] as shown in Fig. 11.72. Thus, in Example 11.14, taking $C_3$ and $C_5$ as key components, the relative volatility $\alpha$ is about 1·76 and the mean viscosity about 0·4 mN s/m², giving a product of 0·246. Then from Fig. 11.72, $E$ is found as 70 per cent.

Chu[63] has given a more complex correlation for overall efficiency $E$ by including the relative flow rates $L$ and $V$ of the phases and the effective submergence of the liquid $h_L$. Thus:

$$\log_{10} E = 1\cdot67 + 0\cdot30\log_{10}(L/V) - 0\cdot25\log_{10}(\mu_L \alpha) + 0\cdot30\,h_L \qquad (11.129)$$

where $L, V$ are the liquid and vapour flowrates in kmol/s,
$\mu_L$ is the viscosity of the liquid feed in mN s/m²,
$\alpha$ is the relative volatility of the key components, and
$h_L$ is the effective submergence in m, taken as the distance from the top of the slot to the weir lip plus half the slot height.

### Expressions for Plate Efficiency Related to Mass Transfer

By assuming that the vapour issuing from slots is in the form of spherical bubbles, Chu[63], Geddes[64], and Bakowski[65] have derived methods for expressing the efficiency $E$ in terms of transfer coefficients ($k_g$, $k_l$) and tray parameters such as the slot dimensions. These methods have proved very difficult to use because of the unreliability of data for calculating transfer coefficients, and the greater problem of calculating the interfacial areas. Probably the most successful analysis for determining efficiency in terms of mass transfer functions has been obtained by the American Institute of Chemical Engineers Research Project[46], and an outline of this work will be given here. Whilst this is a complex analysis containing parameters which are only roughly available, the method does outline some of the important factors involved and shows that some parameters are of little importance.

*Plate efficiency in terms of transfer units.* The process of mass transfer across a phase boundary has already been discussed in Volume 1, Chapter 8. A resistance to mass transfer exists within the fluid on each side of the interface, and the overall transfer rate of a component in a mixture depends on the sum of these resistances and the total driving force.

The concept of a transfer unit for a countercurrent mass transfer process was introduced and this will be developed further for distillation in packed columns in Section 11.11. The number of transfer units is defined as the integrated value of the ratio of the change in composition to the driving force. Thus, considering the vapour phase, the number of overall gas transfer units $\mathbf{N}_{OG}$ is given by:

$$\mathbf{N}_{OG} = \int \frac{dy}{y_e - y} \tag{11.130}$$

For the liquid phase, the corresponding number of overall liquid transfer units $\mathbf{N}_{OL}$ is given by:

$$\mathbf{N}_{OL} = \int \frac{dx}{x - x_e} \tag{11.131}$$

Equations 11.130 and 11.131 are derived later (equations 11.159 and 11.160).

The relation between $\mathbf{N}_{OG}$ and $\mathbf{N}_{OL}$ (see equation 11.165) is given by:

$$\frac{\mathbf{N}_{OL}}{\mathbf{N}_{OG}} = \frac{mG'}{L'} = \varepsilon \tag{11.132}$$

where    $m$ is the slope of the vapour–liquid equilibrium line ($y_e$ versus $x$), and
         $G'$ and $L'$ are the molar rates of flow of vapour and liquid, respectively, per unit cross-section of column.

The equation for transfer units may be applied to the mass transfer over a tray, and thus relate the local Murphree efficiency $E_{mv}$ to the overall transfer units $\mathbf{N}_{OG}$. With the notation in Fig. 11.70, the vapour $y_{n-1}$ rises from plate $n-1$, crosses the liquid on plate $n$ and leaves with composition $y_n$. The liquid flowing from plate $n+1$ through the downcomer crosses tray $n$ and leaves with composition $x_n$. It is supposed for this argument that there is no change in the composition of the liquid in a vertical plane through the liquid. Then applying the mass transfer equation for the flow of vapour on a vertical path and over a small element of plate area:

$$\mathbf{N}_{OG} = \int \frac{dy}{y_e - y} = -\ln \frac{y_e - y_n}{y_e - y_{n-1}} \tag{11.133}$$

or

$$\exp(-\mathbf{N}_{OG}) = \frac{y_e - y_n}{y_e - y_{n-1}}$$

whence

$$1 - \exp(-N_{OG}) = \frac{y_n - y_{n-1}}{y_e - y_{n-1}} = E_{mv} \qquad (11.134)$$

This analysis refers to a small area for vertical flow, and $E_{mv}$ is therefore the *point* or *local* Murphree efficiency. The relation between this point efficiency and the tray efficiency will depend upon the nature of the liquid mixing on the tray. If there is complete mixing of the liquid, $x = x_n$ for the liquid, and $y_e$ and $y$ will also be constant over a horizontal plane. Then the tray efficiency $E_{Mv} = E_{mv}$. With no mixing of the liquid, the liquid can be considered to be in plug flow. Then if $y_e = mx + b$ and $E_{mv}$ is taken as constant over tray, it may be shown[46] that:

$$E_{mv} = \frac{1}{\varepsilon}[\exp(\varepsilon E_{mv}) - 1] \qquad (11.135)$$

where

$$\varepsilon = \frac{mG'}{L'}$$

For intermediate cases where partial mixing of liquid occurs, the A.I.Ch.E. Manual[46] should be consulted.

*Plate efficiency in terms of liquid concentrations.* With the same concept for tray layout as in Fig. 11.70, relations for $E_{ml}$ and $E_{Ml}$ may be derived. Assuming that the vapour concentration does not change in a horizontal plane, a similar analysis to that above gives:

$$E_{ml} = 1 - \exp(-N_{OL}) \qquad (11.136)$$

The efficiencies $E_{mv}$ and $E_{ml}$ may be related by using the relation between $N_{OG}$ and $N_{OL}$ (equation 11.132). Thus:

$$\ln(1 - E_{ml}) = \varepsilon \ln(1 - E_{mv}) \qquad (11.137)$$

### Effect of Entrainment on Efficiency

For conditions where the entrainment may be assumed constant across a tray, Colburn[40] has suggested that the following expression gives, for entrainment $e'$ (mols/unit time, unit area), a correction to $E_{Mv}$, so that the new value of efficiency $E_a$ is given by:

$$E_a = \frac{E_{Mv}}{1 + (e' E_{Mv})/L'} \qquad (11.138)$$

### Experimental Work from A.I.Ch.E. Programme

Having now seen how the tray efficiencies can be related to the values of $N_{OG}$ and $N_{OL}$, experimentally determined results are now required for expressing the mass transfer in terms of degree of mixing, entrainment, geometrical arrangements on the trays and the operating conditions including mass flows. These are provided from the research programme, which gives expressions for the number of film transfer units $N_G$ and $N_L$ (see Section 11.11.3).

*Gas phase transfer.* The value of $N_G$ is expressed in terms of weir height $h_w$, gas flow expressed as $\bar{F}$, liquid flow $L_p$ and the Schmidt number $Sc_v$ for the gas phase. The two key relations are:

$$N_G = [0.776 + 0.0046\,h_w - 0.024\,\bar{F} + 105\,L_p]Sc_v^{-0.5} \tag{11.139}$$

and

$$N_G = -\ln(1 - E_{mv}) \tag{11.140}$$

Equation 11.140 gives the point efficiency for cases where all the resistance occurs in the gas phase. In these equations:

$h_w$ is the exit weir height (mm),
$\bar{F} = u\sqrt{\rho_v}$ where $u$ is the vapour rate (m/s) based on the bubbling area,
  and $\rho_v$ is gas density (kg/m$^3$),
$L_p$ is the liquid flow (m$^3$/s per m liquid flow path),
$\mu_v$ is the gas viscosity (N s/m$^2$),
$D_v$ is the gas diffusivity (m$^2$/s), and
$Sc_v$ is the Schmidt number $\mu_v/\rho_v D_v$.

*Liquid phase transfer.* Here the value of $N_L$ is expressed in terms of the $\bar{F}$-factor for gas flow, the time of contact $t_L$ (s), and the liquid diffusivity $D_L$ (m$^2$/s); the experimental work gives:

$$N_L = [4.13 \times 10^8 D_L]^{0.5}[0.022\,\bar{F} + 0.15]t_L \tag{11.141}$$

The residence time $t_L$ in seconds is expressed by the relation:

$$t_L = Z_c Z_L / L_p \tag{11.142}$$

$Z_c$ is the hold-up of liquid on the tray (m$^3$/m$^2$ of effective cross-section) and is given by:

$$Z_c = 0.043 + 0.191\,h_w - 0.013\,\bar{F} + 2.5\,L_p \tag{11.143}$$

$Z_L$ is the distance between the weirs in metres.

*Consideration of relationships for $N_G$ and $N_L$.* From the knowledge of $N_G$ and $N_L$, the value of $N_{OG}$ is obtained from equation 11.144 which is derived in the same way as equation 11.164.

$$\frac{1}{N_{OG}} = \frac{1}{N_G} + \frac{mG'}{L'}\frac{1}{N_L} \tag{11.144}$$

Then the point efficiency $E_{mv}$ is obtained from:

$$E_{mv} = 1 - \exp(-N_{OG}) \qquad \text{(equation 11.134)}$$

Whilst these expressions are difficult to use and involve some inconsistent assumptions about the liquid and vapour flow, they do bring out some useful features in relation to the tray efficiency. Thus $N_G$ varies linearly with $h_w$, $\bar{F}$, and $L_p$, but the important relation between $N_G$ and $E_{mv}$ is complex. The Manual[46] gives figures as a guide.

## 11.11. PACKED COLUMNS FOR DISTILLATION

In the bubble cap and perforated plate columns so far considered, a large interfacial area between the rising vapour and the reflux has been obtained by causing the vapour to bubble through the liquid. An alternative arrangement, which also provides the necessary large interfacial area for diffusion, is the packed column, in which the cylindrical shell of the column is filled with some form of packing. A common arrangement for distillation is as indicated in Fig. 11.73, where the packing may consist of rings, saddles, or other shaped particles, all of which are designed to provide a high interfacial area for transfer. These have already been referred to in Chapter 4. In these columns the vapour flows steadily up and the reflux steadily down, giving a true countercurrent system in contrast to the conditions in bubble cap columns, where the process of enrichment is stagewise.

### 11.11.1. Packings

A selection of suitable packing material is based on the same arguments as for absorption towers (Chapter 12), but for industrial units the most usual packings are rings, and the material of construction is determined by the corrosive nature of the fluids. It is important to realise that in a distillation system operating at high reflux ratios the weight of reflux is approximately equal to the weight of vapour, but that at low reflux ratios the weight of the liquid is only a small fraction of that of the vapour. Since the vapour has a much lower density than the liquid, the process is really one in which a small quantity of liquid passes through the vapour, and the establishment of good distribution of the liquid is more difficult than in absorption towers, where the two streams are more nearly balanced.

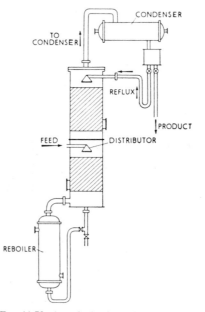

FIG. 11.73. A packed column for distillation

In Chapter 4 the characteristics of packings and their influence on column hydraulics was considered and in Chapter 12 the mass transfer aspects are covered. The choice of a packing for a particular application has therefore been fully discussed though it should be noted that, in the case of vacuum distillation for instance, pressure-drop considerations may be of overriding importance and there may be problems associated with the wetting of packing, which he termed the *height equivalent of a theoretical plate* (H.E.T.P.). As normal practice to increase the calculated height of packing by 40 per cent to allow for liquid maldistribution and wetting problems.

### 11.11.2. Calculation of Enrichment in Packed Columns

With the plate columns previously discussed, the vapour leaving a plate will be richer in the more volatile component than the vapour entering the plate, by one equilibrium step. Peters[66] supposed that this same enrichment of the vapour will occur in a certain height of packing, which he termed the *height equivalent of a theoretical plate* (H.E.T.P.). As all sections of the packing are physically the same, it is assumed that one equilibrium stage is represented by a given height of packing. Thus the required height of packing for any desired separation is given by H.E.T.P. × (No. of ideal stages required).

TABLE 11.8. *Constants for Use in Equation 11.145*

| Type of packing | Size (mm) | $C_1$ ($\times 10^{-5}$) | $C_2$ | $C_3$ |
|---|---|---|---|---|
| Rings | 6 | | | 1·24 |
| | 9 | 0·77 | −0·37 | 1·24 |
| | 12·5 | 7·43 | −0·24 | 1·24 |
| | 25 | 1·26 | −0·10 | 1·24 |
| | 50 | 1·80 | 0 | 1·24 |
| Saddles | 12·5 | 0·75 | −0·45 | 1·11 |
| | 25 | 0·80 | −0·14 | 1·11 |
| Raschig | 6 | 0·28 | 0·25 | 0·30 |
| rings of | 9 | 0·29 | 0·50 | 0·30 |
| protruded | 19 | 0·45 | 0·30 | 0·30 |
| metal | 25 | 0·92 | 0·12 | 0·30 |

This is a simple method of representation which has been widely used as a method of design. Despite this fact, there have been few developments in the theory. Murch[67] has given the following relationships for the H.E.T.P. from an analysis of the results of a number of workers. He has considered columns from 50 to 750 mm diameter and packed over heights of 0·9–3·0 m with rings, saddles, and other packings. Most of the results were for conditions of total reflux, with a vapour rate of 0·18–2·5 kg/m² s (25–80 per cent of flooding).

$$\text{H.E.T.P.} = C_1 G'^{C_2} d_c^{C_3} Z^{1/3} \frac{\alpha \mu_L}{\rho_L} \tag{11.145}$$

where the values of $C_1$, $C_2$, $C_3$ varied with packings as given in Table 11.8.

It should also be noted that the mixtures considered were mainly hydrocarbons with values of relative volatiles not much above three. In this equation,

$G'$ = mass velocity of vapour in $kg/m^2 s$ of tower area,
$d_c$ = column diameter in m,
$Z$ = packed height in m,
$\alpha$ = relative volatility,
$\mu_L$ = liquid viscosity in $N s/m^2$, and
$\rho_L$ = liquid density in $kg/m^3$.

Ellis[14] has presented the following general equation which is in consistent units for the H.E.T.P. $(Z_t)$ of packed columns using 25 and 50 mm. Raschig rings:

$$Z_t = 18d_r + 12m\left[\frac{G'}{L'} - 1\right]$$
(11.146)

where    $d_r$ is the diameter of the rings,
         $m$ is the average slope of equilibrium curve,
         $G'$ is the vapour flowrate, and
         $L'$ is the liquor flowrate.

In industrial practice, the H.E.T.P. concept is used to convert empirically the number of theoretical stages to packing height. As most data in the literature have been derived from small-scale operations, they do not provide a good guide to the values which will be obtained on full-scale plant. However, the values given in Table 11.9 may be used as a guide.

TABLE 11.9. Values of H.E.T.P.[3] for Full-scale Plant

| Type of packing/application | H.E.T.P. (m) |
|---|---|
| 25 mm diam. packing | 0·46 |
| 38 mm diam. packing | 0·66 |
| 50 mm diam. packing | 0·9 |
| Absorption duty | 1·5–1·8 |
| Small diameter columns | |
| ( <0·6 m diam.) | column diameter |
| Vacuum columns | values as above + 0·1 m |

For a particular type of packing, the ratio H.E.T.P./pressure drop is fairly constant for all sizes so that there is no advantage in attempting to improve the H.E.T.P. by using a smaller packing, since the disadvantages of the higher pressure drop will offset the savings made by reduction of packed height.

Further data on H.E.T.P. for packing smaller than 38 mm is presented in Fig. 11.74, where it will be seen that some of the newer packings, such as Pall rings and Mini rings, give a relatively constant value of H.E.T.P. over a wide range of gas rates[68].

## 11.11.3. The Method of Transfer Units (see also Chapter 12 and Volume 1, Chapter 8)

Peters' proposals are really the application of the stagewise mechanism for the plate column to the packed tower, where however the process is one of continuous

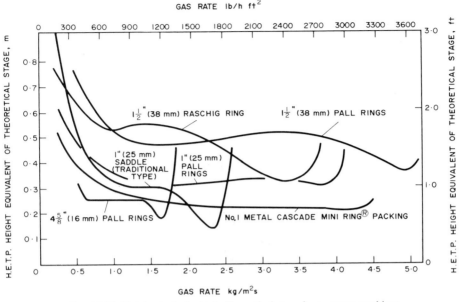

FIG. 11.74. Height equivalent of a theoretical stage for common packings

countercurrent mass transfer. The degree of separation is represented by the rate of change of composition of a component with height of packing, i.e. $dy/dZ$. This rate of change of composition will be dependent upon the equipment, the operating conditions, and on the diffusional potential across the two films. Over the vapour film this driving force is measured by $y_i - y$ or $(\Delta y)_f$ where $y_i$ is the mol fraction of the diffusing component at the interface and $y$ the value for the vapour.

The performance of the column can thus be represented by $J$, the change in composition with height for unit driving force:

$$J = \frac{dy/dZ}{(\Delta y)_f} \tag{11.147}$$

A relation between $dy/dZ$ and $(\Delta y)_f$ can be obtained on the basis of the two-film theory for diffusion (Volume 1, Chapter 8). For the vapour film, from Fick's Law (from Volume 1, Chapter 8):

$$N_A = -D_v \frac{dC}{dz} \tag{11.148}$$

where $N_A$ is the molar rate of transfer per unit area of interface of component $\mathbf{A}$,
$D_v$ is the vapour diffusion coefficient,
$C$ is the concentration in mols per unit volume, and
$z$ is the distance in direction of diffusion.
For an ideal gas, this gives:

$$N_A = -\frac{D_v}{RT}\frac{dP_A}{dz} = -\frac{D_v P}{RT}\frac{dy}{dz} \tag{11.149}$$

where $y$ is the mol fraction of $\mathbf{A}$.

The negative sign occurs because $z$ is taken in the direction of diffusion from the interface, and $y$ decreases in this direction.

For equimolecular counterdiffusion, integration gives:

$$N_A = -\frac{D_v P}{\mathbf{R}Tz}(\Delta y)_f \tag{11.150}$$

or

$$N_A = k_g'(\Delta y)_f \tag{11.151}$$

where

$$k_g' = -\frac{D_v P}{\mathbf{R}Tz} \tag{11.152}$$

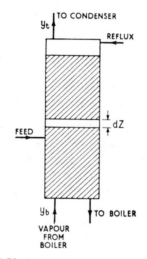

FIG. 11.75. Arrangement for a packed column

Consider a column as indicated in Fig. 11.75, in which the concentration of the more volatile component increases from $y_b$ to $y_t$. Over a small height of the column $dZ$ the rate of transfer can be written as:

$$k_g' a A(y_i - y)\,dZ \tag{11.153}$$

where    $a$ is the active interfacial area for transfer per unit volume of the column, and
$A$ is the cross-sectional area of the column.

But the mols of component **A** diffusing $=$ total mols of vapour $\times$ change in
mol fraction
$= G'A\,dy$

where    $G'$ is the vapour rate in the column in mol/unit time-unit cross-section,
and    $k_g'$ is known as the gas-film transfer coefficient, and is measured as mols/unit time-unit area-unit mol fraction difference.

$$\therefore \quad G'A\,dy = k_g' a A(y_i - y)\,dZ \tag{11.154}$$

or

$$J = \frac{dy/dZ}{y_i - y} = \frac{k'_g a}{G'}$$                    (from equation 11.147)

and

$$\int_{y_b}^{y_t} \frac{dy}{y_i - y} = \frac{k'_g a}{G'}$$                                        (11.155)

where $k'_g a$ is taken as constant over the column.

The group of the left-hand side of this equation represents the integrated ratio of the change in composition to the driving force tending to bring this about. This group has been defined by Chilton and Colburn[69] as the number of transfer units $\mathbf{N}_G$. The quantity $G'/(k'_g a)$, which is the reciprocal of the efficiency $J$ and has the dimensions of length, they define as the height of a transfer unit (H.T.U.), $\mathbf{H}_G$. Equation 11.155 can be written as:

$$Z = \mathbf{H}_G \mathbf{N}_G$$                                            (11.156)

The concentrations $y_i$, $y$ refer to conditions on either side of the gas film, and hence $\mathbf{N}_G$ is the number of gas-film transfer units, and $\mathbf{H}_G$ the height of a gas-film transfer unit.

For packed columns $(k'_g a)/G'$ represents a useful value for the efficiency, and the performance of a packed column is commonly represented by the simple term $\mathbf{H}_G$, a low value of $\mathbf{H}_G$ corresponding to an efficient column. If $\mathbf{H}_G$ is known, the necessary height of a column is found from equation 11.156, since $\mathbf{N}_G$ is determined from the change in concentration required and the shape of the equilibrium curve. Application of this technique is discussed further in Chapter 12.

The same number of mols pass through the liquid film and a similar series of equations can be obtained in terms of concentrations across the liquid film, i.e.:

$$G'\,dy = L'\,dx$$                                            (11.157)

where     $L'$ is the molar flow rate of liquid/unit area, and
              $x$ is the mol fraction of the more volatile component in the liquid.
If $k'_l a$ is the mass transfer coefficient for the liquid phase in mols/unit time-unit volume-unit mol fraction driving force, then:

$$AL'\,dx = k'_l a A (x - x_i)\,dZ$$

$$\therefore \quad \int_{x_b}^{x_t} \frac{dx}{x - x_i} = \frac{k'_l a}{L'}$$                                  (11.158)

or

$$\mathbf{N}_L = \frac{1}{\mathbf{H}_L} Z$$

where   $\mathbf{N}_L$ is the number of liquid film transfer units and
             $\mathbf{H}_L$ is the height of a liquid film transfer unit, which for distillation applications is
             presented in Table 11.10 as a function of type and size of packing[70].

*Overall Transfer Coefficients and Transfer Units*

The driving force over the gas film is taken as $y_i - y$ and over the liquid film as $x - x_i$. If $y_e$ is the concentration in the gas phase in equilibrium with concentration $x$ in the liquid

TABLE 11.10. *Values of $H_L$ for distillation*

| Packing size (in.) (mm) | | 0·5 (12) | 0·75 (18) | 1·0 (25) | 1·5 (40) | 2·0 (50) |
|---|---|---|---|---|---|---|
| Raschig type | ft | 0·24 | 0·30 | 0·34 | 0·47 | 0·58 |
|  | (m) | 0·073 | 0·092 | 0·104 | 0·143 | 0·177 |
| Intalox | ft | 0·20 | 0·26 | 0·29 | 0·40 | — |
|  | (m) | 0·061 | 0·079 | 0·089 | 0·122 | — |
| Pall rings | ft | — | — | — | 0·40 | 0·49 |
|  | (m) | — | — | — | 0·122 | 0·150 |

phase, then $y_e - y$ is taken as the overall driving force expressed in terms of $y$. Similarly $x - x_e$ is taken as the overall driving force in terms of $x$, where $x_e$ is the concentration in the liquid in equilibrium with a concentration $y$ in the vapour.

Then the overall driving forces $(\Delta y)_o$ and $(\Delta x)_o$ can be written as:

$$(\Delta y)_o = y_e - y = (y_e - y_i) + (y_i - y)$$

$$(\Delta x)_o = x - x_e = (x - x_i) + (x_i - x_e)$$

By analogy with the derivation for film coefficients, a series of overall transfer coefficients and overall transfer units based on these overall driving forces may be defined.

Thus, the number of overall gas transfer units:

$$\mathbf{N}_{OG} = \int \frac{\mathrm{d}y}{y_e - y} \tag{11.159}$$

and the number of overall liquid transfer units:

$$\mathbf{N}_{OL} = \int \frac{\mathrm{d}x}{x - x_e} \tag{11.160}$$

The heights of the overall transfer units are:

$$\mathbf{H}_{OG} = \frac{G'}{K'_g a} \tag{11.161}$$

and

$$\mathbf{H}_{OL} = \frac{L'}{K'_l a} \tag{11.162}$$

where $K'_g a$ and $K'_l a$ are overall transfer coefficients, based on gas or liquid concentrations in mol/unit time-unit volume-unit mol fraction driving force.

*Relation between Overall and Film Transfer Units*

From the above definitions, the following equation may be written:

$$\frac{\mathrm{d}Z}{\mathbf{H}_{OG}} = \frac{\mathrm{d}y}{y_e - y} \qquad \frac{\mathrm{d}Z}{\mathbf{H}_G} = \frac{\mathrm{d}y}{y_i - y}$$

$$\frac{\mathrm{d}Z}{\mathbf{H}_{OL}} = \frac{\mathrm{d}x}{x - x_e} \qquad \frac{\mathrm{d}Z}{\mathbf{H}_L} = \frac{\mathrm{d}x}{x - x_i}$$

$$\therefore \quad \frac{\mathbf{H}_{OG}}{y_e - y} = \frac{\mathbf{H}_G}{y_i - y}$$

and

$$\mathbf{H}_{OG} = \mathbf{H}_G \left[ \frac{y_e - y_i + y_i - y}{y_i - y} \right] = \mathbf{H}_G \left[ 1 + \frac{y_e - y_i}{y_i - y} \right]$$

If the equilibrium curve is straight over the range $y = y_e$ to $y = y_i$, then assuming equilibrium at the interface:

$$y_e - y_i = m(x - x_i)$$

and

$$\mathbf{H}_{OG} = \mathbf{H}_G + m \left( \frac{x - x_i}{y_i - y} \right) \mathbf{H}_G$$

But

$$\frac{x - x_i}{y_i - y} = \frac{dx}{dy} \frac{\mathbf{H}_L}{\mathbf{H}_G} = \frac{G'}{L'} \frac{\mathbf{H}_L}{\mathbf{H}_G}$$

since $G' \, dy = L' dx$.

$$\therefore \ \mathbf{H}_{OG} = \mathbf{H}_G + \frac{mG'}{L'} \mathbf{H}_L \tag{11.163}$$

Similarly:

$$\mathbf{H}_{OL} = \mathbf{H}_L + \frac{L'}{mG'} \mathbf{H}_G \tag{11.164}$$

Dividing equation 11.163 by equation 11.164:

$$\frac{\mathbf{H}_{OG}}{\mathbf{H}_{OL}} = \frac{mG'}{L'} = \frac{\mathbf{N}_{OL}}{\mathbf{N}_{OG}} \tag{11.165}$$

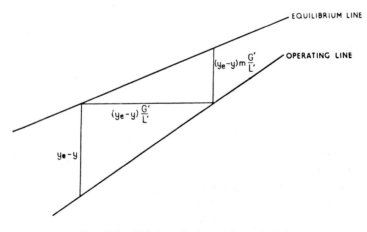

FIG. 11.76. Height equivalent of theoretical plate

This form of relationship can be written in terms of transfer coefficients (see Chapter 12):

$$\frac{1}{K_g'a} = \frac{1}{k_g'a} + \frac{m}{k_l'a} \qquad (11.166)$$

### Relation of H.T.U. to H.E.T.P.

In a theoretical plate, the mol fraction of the M.V.C. in the vapour will increase from $y$ to $y_e$, so that the total mass transfer is $G'(y_e - y)$. If the equilibrium curve can be considered straight over the height $(Z_t)$ of column equivalent to a theoretical plate, the logarithmic mean driving force may be used. Thus, referring to Fig. 11.76:

$$G'(y_e - y) = K_g'aZ_t(y_e - y)\frac{(mG'/L) - 1}{\ln(mG'/L)}$$

$$\therefore \quad Z_t = \frac{G'}{K_g'a}\frac{\ln(mG'/L)}{(mG'/L) - 1} = \mathbf{H}_{OG}\frac{\ln(mG'/L)}{(mG'/L) - 1}$$

$$= \mathbf{H}_{OG}\frac{\ln\{1 - [1 - (mG'/L)]\}}{(mG'/L) - 1}$$

$$= \mathbf{H}_{OG}\{1 + \tfrac{1}{2}[1 - (mG'/L)] + \tfrac{1}{3}[1 - (mG'/L)]^2 + \ldots\} \quad (11.167)$$

If the operating and equilibrium lines are parallel, $mG'/L = 1$, and:

$$Z_t = \mathbf{H}_{OG} \qquad (11.168)$$

Thus, the ratio of the height equivalent to a theoretical plate to the height of the transfer unit $(Z_t/\mathbf{H}_{OG})$ may be greater or less than unity, according to whether the slope of the operating line is greater or less than that of the equilibrium curve.

### Determination of Transfer Units (Experimental)

There have been a number of reports[71-73] on the influence of flow parameters and physical properties on the value of the height of a transfer unit. Most of the work has been carried out in small laboratory columns and great care must be exercised if these data are applied to large diameter units. Some general indication of the values of $\mathbf{H}_{OG}$ are given in Table 11.10 below, which gives values obtained by Furnas and Taylor[71] for experiments with ethanol–water mixtures at atmospheric pressure in a column 305 mm diameter operating at total reflux.

### Values of H.T.U. in Terms of Flow Rates and Physical Properties

Wetted-wall columns. The proposals that have been made for calculating transfer coefficients from physical data of the system and the operating liquid and vapour rates are really all related to conditions existing in a simpler unit in the form of a wetted-wall column. In the wetted-wall column (see Chapter 12), vapour rising from the boiler passes up the column which is lagged to prevent heat loss. The liquid flows down the walls, and it thus provides the simplest form of equipment giving countercurrent flow. The mass

TABLE 11.11. *Values of the Height of the Transfer Unit*
$$\mathbf{H}_{OG}$$

| Packing | Depth of packing (m) | Liquid rate (kg/m² s) | $\mathbf{H}_{OG}$ (m) |
|---|---|---|---|
| 50 mm Raschig rings | 3·0 | 1·06 | 0·670 |
| 25 mm Raschig rings | 3·0 | 1·02 | 0·366 |
| 25 mm Berl saddles | 2·75 | 0·195 | 0·427 |
| 25 mm Berl saddles | 2·75 | 1·25 | 0·335 |
| 12·5 mm Berl saddles | 3·0 | 0·25 | 0·457 |
| 12·5 mm Berl saddles | 3·0 | 1·196 | 0·274 |
| 9·5 mm Raschig rings | 2·44 | 0·416 | 0·396 |
| 9·5 mm Raschig rings | 2·44 | 0·780 | 0·305 |

transfer in the unit can be expressed by means of the $j$-factor of Chilton and Colburn which has already been discussed in Volume 1, Chapter 8:

$$j_d = \frac{k'_g}{G'}\left(\frac{\mu}{\rho D}\right)_v^{2/3} = 0·023\left(\frac{d_c u\rho}{\mu}\right)_v^{-0·17} \tag{11.169}$$

where the properties refer to the vapour. This type of unit has been studied by Gilliland and Sherwood[74], Chari and Storrow[75], Surowiec and Furnas[76] and others.

For a wetted wall column,

$$\frac{\text{area of interface}}{\text{volume of column}} = (4/d_c) = a$$

Then

$$\mathbf{H}_G = \frac{G'}{k'_g a} = 10·9\, d_c\, Re_v^{0·17}\, Sc_v^{2/3} \tag{11.170}$$

where the linear characteristic length is taken as the diameter of the column $d_c$. $Re_v$ and $Sc_v$ are the Reynolds and Schmidt numbers with respect to the vapour. Surowiec and Furnas were able to express their results, obtained with alcohol and water, in this form. For transfer through the liquid film, they derived an expression based on the analysis of heat transfer from a tube to a liquid flowing under viscous conditions down the inside of the tube.

They presented their equation as:

$$\mathbf{H}_L = B'Z\frac{M}{M_m}Re_l^{8/9}\, Sc_l^{10/9}\left(\frac{D_L^2}{gZ^3}\right)^{2/9} \tag{11.171}$$

where    $B'$ is a constant,
$M_m$ is the mean molecular weight of the liquid,
$M$ is the point value of the molecular weight,
$Z$ is the height of the tube, and
$Re_l$ and $Sc_l$ are the Reynolds and Schmidt numbers with respect to the liquid.

Colburn, however, has suggested that for mass transfer, the transfer in the liquid phase is from a vapour–liquid interface where the liquid velocity is a maximum to the wall where the liquid velocity is zero. With a liquid flowing inside the tube the heat transfer is from a layer of zero velocity at the wall to the fluid at the centre of the tube which is moving with a maximum velocity. Hatta[77] based his analysis on the more closely related process of diffusion of a gas into a liquid, and obtained the expression:

$$\mathbf{H}_L = B''Z \frac{M}{M_m} Re_l^{2/3} Sc_l^{5/6} \left(\frac{D_L^2}{gZ^3}\right)^{1/6} \tag{11.172}$$

It will be seen that, despite the difference in the arguments, the two equations are really of a similar nature.

*Packed columns.* The application of the ideas indicated above for wetted wall columns to the more complex case of packed columns requires the assumptions: (a) that the mechanism is unchanged and (b) that the expressions are valid over the much wider ranges of flow used in packed columns. This has been attempted by Smith[73] and Pratt[78]. Pratt started from the basic equation:

$$j_d = \frac{k_g'}{G}\left(\frac{\mu}{\rho D}\right)_v^{2/3} = \text{const } Re^{-0.2} \tag{11.173}$$

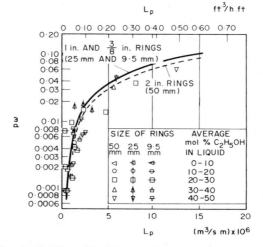

FIG. 11.77. Effect of liquid rate on degree of wetting of packing

He suggested, from the examination of the available data, that the importance of the degree of wetting can be taken into account by writing this in the form:

$$\frac{k_g}{G'} e \left(\frac{\mu}{\rho D}\right)_v^{2/3} = p\omega \left(\frac{d_e G'}{\mu e}\right)^{-0.25} \tag{11.174}$$

where    $G'$ is the mass velocity (mass rate per unit area),

$d_e$ is the hydraulic mean diameter for the packing,

$p$ is a constant,

$\omega$ is the fraction of the packing wetted, and

$e$ is the fractional voidage of the packing.

He gives several plots of $p\omega$ vs $L_p$, as indicated in Fig. 11.77, where $L_p$ is the liquid rate based on the periphery of the packing, in m³/s m. The periphery is taken as equal to $a^{-1}$(m³/m²), although this is only correct for geometrical systems such as stacked rings.

## 11.12. FURTHER READING

BACKHURST, J. R. and HARKER, J. H.: *Process Plant Design* (Heinemann Educational Books, London, 1973).
HOFFMAN, E. J.: *Azeotropic and Extractive Distillation* (Interscience Publishers, Inc., New York, 1964).
KING, J. K.: *Separation Processes* (McGraw-Hill Book Co., New York, 1971).
SAWISTOWSKI, H. and SMITH, W.: *Mass Transfer Process Calculations* (Wiley, Chichester, 1963).
SHERWOOD, T. K., PIGFORD, R. L., and WILKE, C. R.: *Mass Transfer* (McGraw-Hill Book Co., New York, 1974).
SMITH, B. D.: *Design of Equilibrium Stage Processes* (McGraw-Hill Book Co., New York, 1963).
TREYBAL, R. E.: *Mass Transfer Operations*, 2nd edn. (McGraw-Hill Book Co., New York, 1968).

## 11.13. REFERENCES

1. ROBINSON, C. S. and GILLILAND, E. R.: *Elements of Fractional Distillation*, 4th edn. (McGraw-Hill, 1950).
2. THORNTON, J. D. and GARNER, F. H.: *J. Appl. Chem. Suppl.* **1** (1951) 61. Vapour–liquid equilibria in hydrocarbon–non-hydrocarbon systems, 1: The system benzene–cyclohexane–furfuraldehyde.
3. FRANK, O.: *Chem. Eng. Albany* **84** (14 Mar. 1977) 111. Distillation design.
4. RAYLEIGH, LORD: *Phil. Mag.* (vi) **4**, No. 23 (1902) 521. On the distillation of binary mixtures.
5. SOREL, E.: *Distillation et Rectification Industrielle* (G. Carré et C. Naud, 1899).
6. LEWIS, W. K.: *Ind. Eng. Chem.* **1** (1909) 522. The theory of fractional distillation.
7. McCABE, W. L. and THIELE, E. W.: *Ind. Eng. Chem.* **17** (1925) 605. Graphical design of fractionating columns.
8. UNDERWOOD, A. J. V.: *Trans. Inst. Chem. Eng.* **10** (1932) 112. The theory and practice of testing stills.
9. FENSKE, M. R.: *Ind. Eng. Chem.* **24** (1932) 482. Fractionation of straight-run Pennsylvania gasoline.
10. GILLILAND, E. R.: *Ind. Eng. Chem.* **32** (1940) 1220. Multicomponent rectification. Estimation of the number of theoretical plates as a function of the reflux ratio.
11. RUHEMANN, M.: *Trans. Inst. Chem. Eng.* **25** (1947) 143. The ammonia absorption machine. *Ibid.* 152. A study of the generator and rectifier of an ammonia absorption machine.
12. PONCHON, M.: *Technique Moderne* **13** (1921) 20 and 55. Etude graphique de la distillation fractionnée industrielle.
13. SAVARIT, P.: *Arts et Métiers* **75** (1922) 65. Eléments de distillation.
14. ELLIS, S. R. M.: *Birmingham University Chemical Engineer* **5**, No. 1 (1953) 21. H.E.T.P. values in ring packed columns.
15. GILLILAND, E. R. and REED, C. E.: *Ind. Eng. Chem.* **34** (1942) 551. Degrees of freedom in multicomponent absorption and rectification columns.
16. LEWIS, W. K. and MATHESON, G. L.: *Ind. Eng. Chem.* **24** (1932) 494. Studies in distillation design of rectifying columns for natural and refining gasoline.
17. COLBURN, A. P.: *Trans. Am. Inst. Chem. Eng.* **37** (1941) 805. The calculation of minimum reflux ratio in the distillation of multicomponent mixtures.
18. UNDERWOOD, A. J. V.: *J. Inst. Petroleum* **32** (1946) 614. Fractional distillation of multi-component mixtures—calculation of minimum reflux ratio.
19. ZANKER, A.: *Hydrocarbon Processing* **56**, No. 5 (1977) 263. Nomograph replaces Gilliland Plot.
20. BENEDICT, M. and RUBIN, L. C.: *Trans. Am. Inst. Chem. Eng.* **41** (1945) 353. Extractive and azeotropic distillation.
21. YOUNG, S.: *Fractional Distillation* (Macmillan, 1903).
22. GUINOT, H. and CLARK, F. W.: *Trans. Inst. Chem. Eng.* **16** (1938) 189. Azeotropic distillation industry.
23. KING, J. K.: *Separation Processes* (McGraw-Hill Book Co., New York, 1971).
24. NORMAN, W. S.: *Trans. Inst. Chem. Eng.* **23** (1945) 66. The dehydration of ethanol by azeotropic distillation. *Ibid.* 89. Design calculations for azeotropic dehydration columns.
25. TREYBAL, R. E.: *Mass Transfer Operations*, 2nd edn. (McGraw-Hill Book Co., New York, 1968).
26. SCHEIBEL, E. G.: *Chem. Eng. Prog.* **44** (1948) 927. Principles of extractive distillation.
27. PRATT, H. R. C.: *Trans. Inst. Chem. Eng.* **25** (1947) 43. Continuous purification and azeotropic dehydration of acetonitrile produced by the catalytic acetic acid–ammonia reaction.
28. HOFFMAN, E. J.: *Azeotropic and Extractive Distillation* (Interscience Publishers Inc., New York, 1964).
29. SMITH, B. D.: *Design of Equilibrium Stage Processes* (McGraw-Hill, 1963).
30. HAUSBRAND, E.: *Principles and Practice of Industrial Distillation*, 6th edn., translated by TRIPP, E. H. (Wiley, 1926).
31. Engineering Staff, Shell Development Company, Emeryville, California: *Chem. Eng. Prog.* **50** (1954) 57. Turbogrid distillation trays.
32. KIRSCHBAUM, E.: *Distillation and Rectification* (Chemical Publishing Co., 1948).

33. CAREY, J. S., GRISWOLD, J., LEWIS, W. K., and McADAMS, W. H.: *Trans. Am. Inst. Chem. Eng.* **30** (1934) 504. Plate efficiencies in rectification of binary mixtures.
34. SOUDERS, M. and BROWN, G. G.: *Ind. Eng. Chem.* **26** (1934) 98. Design of fractionating columns.
35. CAREY, J. S.: *Chem. Met. Eng.* **46** (1939) 314. Plate-type distillation columns.
36. DRICKAMER, H. G. and BRADFORD, J. R.: *Trans. Am. Inst. Chem. Eng.* **39** (1943) 319. Overall plate efficiency of commercial hydrocarbon fractionating columns as a function of viscosity.
37. O'CONNELL, H. E.: *Trans. Am. Inst. Chem. Eng.* **42** (1946) 741. Plate efficiency of fractionating columns and absorbers.
38. STRANG, L. C.: *Trans. Inst. Chem. Eng.* **12** (1934) 169. Entertainment in a bubble-cap fractionating column.
39. PEAVY, C. C. and BAKER, E. M.: *Ind. Eng. Chem.* **29** (1937) 1056. Efficiency and capacity of a bubble-plate fractionating column.
40. COLBURN, A. P.: *Ind. Eng. Chem.* **28** (1936) 526. Effect of entertainment on plate efficiency in distillation.
41. BOLLES, W. L.: *Pet. Processing* **11** (1956) Optimum bubble-cap tray design. (Feb.) 65. Part I. Tray dynamics. (March) 82. Part II. Design standards. (April) 72. Part III. Design technique (May) 109. Part IV. Design example.
42. FAIR, J. R. and MATTHEWS, R. L.: *Pet. Refiner* **37,** No. 4 (1958) 153–8. Better estimate of entrainment from bubble-cap trays.
43. KOBE, J. A. and McKETTA, J. J.: *Advances in Petroleum Chemistry and Refining*, Vol. 1 (Interscience, 1958). Chapter 6. New Fractionating-tray Designs, by LOCKHART, F. J. and LEGGETT, C. W.
44. HENGSTEBECK, R. J.: *Distillation Principles and Design Procedures* (Reinhold, 1961).
45. Final Report by the ABCM/BCPMA Distillation Panel: *Distillation* (Chemical Industries Association Limited, London, 1968).
46. *Bubble Tray Design Manual* (American Institute of Chemical Engineers, New York, 1958).
47. BOLLES, W. L.: *Pet. Refiner* **25** (1946) 613. Rapid graphical method of estimating tower diameter and tray spacing of bubble-plate fractionators.
48. DAUPHINE, T. C.: Massachusetts Institute of Technology, Sc.D. thesis (1939). Pressure drops in bubble-trays.
49. RHYS, C. O. and MINICH, H. L.: *Regional Meeting A.I.Ch.E., Los Angeles* (6–9 Mar. 1949). Fractionator tray performance and design.
50. KEMP, H. S. and PYLE, C.: *Chem. Eng. Prog.* **45** (1949) 435. Hydraulic gradient across various bubble-cap plates.
51. DAVIES, J. A.: *Ind. Eng. Chem.* **39** (1947) 774. Bubble tray hydraulics.
52. GOOD, A. J., HUTCHINSON, M. H., and ROUSSEAU, W. C.: *Ind. Eng. Chem.* **34** (1942) 1445. Liquid capacity of bubble cap plates.
53. KIRKBRIDE, C. G.: *Pet. Refiner* **23** (1944) 321. Process design procedure for multi-component fractionators.
54. MAYFIELD, F. D., CHURCH, W. L., GREEN, A. C., LEE, D. C., and RASMUSSEN, R. W.: *Ind. Eng. Chem.* **44** (1952) 2238. Perforated-plate distillation columns.
55. LEIBSON, I., KELLEY, R. E., and BULLINGTON, L. A.: *Pet. Refiner*, **36,** No. 2 (1957) 127; No. 3, 288. How to design perforated trays.
56. BAIN, J. L. and VAN WINKLE, M.: *A.I.Ch.E.Jl.* **7** (1961) 363. A study of entrainment, perforated plate column—air–water system.
57. FAIR, J. R.: *Petrol/Chem. Engr*, **33,** No. 9 (1961) 211. How to predict sieve-tray entrainment and flooding.
58. ELD, A. C.: *Pet. Refiner* **32,** No. 5 (1953) 157. A new approach to tray design.
59. GERSTER, J. A.: *Recent Advances in Distillation*. Davis-Swindin Memorial Lecture, University of Loughborough (1964).
60. *Ballast Tray Manual*. Bulletin No. 4900 (revised) (Fritz Glitsch and Sons Inc., Dallas, Texas, 1970).
61. PERRY, J. H.: *Chemical Engineers' Handbook*, 5th edn. (McGraw-Hill, 1973). Revised by R. PERRY and C. CHILTON.
62. MURPHREE, E. V.: *Ind. Eng. Chem.* **17** (1925) 747. Rectifying column calculations—with particular reference to N component mixtures.
63. CHU, J. C., DONOVAN, J. R., BOSEWELL, B. C. and FURMEISTER, L. C.: *J. Appl. Chem.* **1** (1951) 529. Plate efficiency correlation in distilling columns and gas absorbers.
64. GEDDES, R. L.: *Trans. Am. Inst. Chem. Eng.* **42** (1946) 79. Local efficiencies of bubble plate fractionators.
65. BAKOWSKI, S.: *Chem. Eng. Sci.* **1** (1951/2) 266. A new method for predicting the plate efficiency of bubble-cap columns.
66. PETERS, W. A.: *Ind. Eng. Chem.* **14** (1922) 476. The efficiency and capacity of fractionating columns.
67. MURCH, D. P.: *Ind. Eng. Chem.* **45** (1953) 2616. Height of equivalent theoretical plate in packed fractionation columns.
68. EASTHAM, I.: Private communication (1977).
69. CHILTON, T. H. and COLBURN, A. P.: *Ind. Eng. Chem.* **27** (1935) 255, 904. Distillation and absorption in packed columns.
70. Tower Packings (Hydronyl Ltd., Stoke-on-Trent, England).

71. Furnas, C. C. and Taylor, M. L.: *Trans. Am. Inst. Chem. Eng.* **36** (1940) 135. Distillation in packed columns.
72. Duncan, D. W., Koffolt, J. H., and Withrow, J. R.: *Trans. Am. Inst. Chem. Eng.* **38** (1942) 259. The effect of operating variables on the performance of a packed column still.
73. Sawistowski, H. and Smith, W.: *Ind. Eng. Chem.* **51** (1959) 915. Performance of packed distillation columns.
74. Gilliland, E. R. and Sherwood, T. K.: *Ind. Eng. Chem.* **26** (1934) 516. Diffusion of vapours into air streams.
75. Chari, K. S. and Storrow, J. A.: *J. Appl. Chem.* **1** (1951) 45. Film resistances in rectification.
76. Surowiec, A. J. and Furnas, C. C.: *Trans. Am. Inst. Chem. Eng.* **38** (1942) 53. Distillation in a wetted-wall tower.
77. Hatta, S.: *J. Soc. Chem. Ind. Japan* **37** (1934) 275. On the theory of absorption of gases by liquids flowing as a thin layer.
78. Pratt, H. R. C.: *Trans. Inst. Chem. Eng.* **29** (1951) 195. The performance of packed absorption and distillation columns with particular reference to wetting.

# 11.14. NOMENCLATURE

| | | Units in SI System | Dimensions in $\mathbf{MLT}\theta$ |
|---|---|---|---|
| $A$ | Cross-sectional area of column | m$^2$ | $\mathbf{L}^2$ |
| $a$ | Interfacial surface per unit volume of column | m$^2$/m$^3$ | $\mathbf{L}^{-1}$ |
| $B$ | Constant in equation 11.101 | kg/m$^2$ s | $\mathbf{ML}^{-2}\mathbf{T}^{-1}$ |
| $B'$ | Constant in equation 11.171 | — | — |
| $B''$ | Constant in equation 11.172 | — | — |
| $b_m, b_n$ | Factors in equation 11.90 | — | — |
| $C$ | Concentration in mols/unit volume | kmol/m$^3$ | $\mathbf{ML}^{-3}$ |
| $C_D$ | Coefficient of discharge for slot | — | — |
| $C_e$ | Entrainment coefficient (equation 11.111) | m/s | $\mathbf{LT}^{-1}$ |
| $C_o$ | Orifice coefficient for sieve plate | — | — |
| $C_{sb}$ | Parameter $u_n \sqrt{\rho_v/(\rho_L - \rho_v)}$ | m/s | $\mathbf{LT}^{-1}$ |
| $c$ | Constant in vapour–liquid equilibrium relations | — | — |
| $D$ | Mols or mass of product per unit time | kg/s *or* kmol/s | $\mathbf{MT}^{-1}$ |
| $D_b$ | Mols of product in batch distillation | kmol | $\mathbf{M}$ |
| $D_L$ | Diffusivity in the liquid phase | m$^2$/s | $\mathbf{L}^2\mathbf{T}^{-1}$ |
| $D_v, D$ | Diffusivity in the vapour phase | m$^2$/s | $\mathbf{L}^2\mathbf{T}^{-1}$ |
| $d$ | Bubble diameter | m | $\mathbf{L}$ |
| $d_c$ | Column diameter | m | $\mathbf{L}$ |
| $d_r$ | Diameter of ring | m | $\mathbf{L}$ |
| $E$ | Average overall plate efficiency, $n/n_p$ | — | — |
| $E_a$ | Plate efficiency allowing for entrainment $e'$ | — | — |
| $E_m$ | Local Murphree plate efficiency | — | — |
| $E_M$ | Average Murphree plate efficiency | — | — |
| $e$ | Fractional voidage of packing | — | — |
| $e'$ | Entrainment (mols per unit time and unit cross-section) | kmol/m$^2$ s | $\mathbf{ML}^{-2}\mathbf{T}^{-1}$ |
| $F$ | Mols or mass of feed per unit time | kmol/s | $\mathbf{MT}^{-1}$ |
| $F_{lv}$ | Parameter $L'/G' \sqrt{\rho_v/\rho_L}$ | — | — |
| $F_w$ | Bolles' factor for flow over weir (equation 11.106) | s$^{2/3}$/m$^{1/3}$ | $\mathbf{L}^{-1/3}\mathbf{T}^{2/3}$ |
| $F''$ | Mols of feed | kmol | $\mathbf{M}$ |
| $\bar{F}$ | Parameter $u_v \sqrt{\rho_v}$ used in equation 11.139 | kg$^{1/2}$/m$^{1/2}$ s | $\mathbf{M}^{1/2}\mathbf{L}^{-1/2}\mathbf{T}^{-1}$ |
| $G'$ | Mols of vapour per unit time and unit cross-section | kmol/m$^2$ s | $\mathbf{ML}^{-2}\mathbf{T}^{-1}$ |
| $g$ | Acceleration due to gravity | m/s$^2$ | $\mathbf{LT}^{-2}$ |
| $H$ | Enthalpy per mol or unit mass | J/kmol, J/kg | $\mathbf{L}^2\mathbf{T}^{-2}$ |
| $H^L$ | Enthalpy per mol or unit mass of liquid | J/kmol, J/kg | $\mathbf{L}^2\mathbf{T}^{-2}$ |
| $H^V$ | Enthalpy per mol or unit mass of vapour | J/kmol, J/kg | $\mathbf{T}^2\mathbf{T}^{-2}$ |
| $H'_d$ | $H_d^L + (Q_C/D)$ | J/kmol, J/kg | $\mathbf{L}^2\mathbf{T}^{-2}$ |
| $H'_w$ | $H_w^L - (Q_B/W)$ | J/kmol, J/kg | $\mathbf{L}^2\mathbf{T}^{-2}$ |
| $\mathbf{H}$ | Height of transfer unit | m | $\mathbf{L}$ |
| $\mathbf{H}_G$ | Height of transfer unit—gas film | m | $\mathbf{L}$ |
| $\mathbf{H}_{OG}$ | Height of transfer unit—overall (gas concentrations) | m | $\mathbf{L}$ |
| $\mathbf{H}_L$ | Height of transfer unit—liquid film | m | $\mathbf{L}$ |

|  |  | Units in SI System | Dimensions in **MLT**$\theta$ |
|---|---|---|---|
| $\mathbf{H}_{OL}$ | Height of transfer unit—overall (liquid concentration) | m | **L** |
| $\mathscr{H}'$ | Henry's constant $(P_A/x_A)$ | N/m$^2$ | **ML$^{-1}$T$^{-2}$** |
| $h$ | Height of slot | m | **L** |
| $h_{al}$ | Head loss as vapour flows through aerated liquid (equation 11.105) | m | **L** |
| $h_d$ | Loss in head at bottom of downcomer | m | **L** |
| $h_{ds}$ | Dynamic slot seal | m | **L** |
| $h_g$ | Drop in head of liquid across plate | m | **L** |
| $h_g'$ | Value of $h_g$ uncorrected for vapour flow | m | **L** |
| $h_L$ | Effective submergence | m | **L** |
| $h_o$ | Head loss across orifice | m | **L** |
| $h_p$ | Loss in head of vapour passing through plate | m | **L** |
| $h_r$ | Loss in head due to contraction at cap riser | m | **L** |
| $h_{rc}$ | Loss in head due to riser and reversal inside cap | m | **L** |
| $h_s$ | Loss in head through slot (i.e. slot opening) | m | **L** |
| $h_{ss}$ | Static submergence (top of slot to weir) | m | **L** |
| $h_w$ | Height of exit weir | m | **L** |
| $h_{ow}$ | Height of crest over weir | m | **L** |
| $h_\sigma$ | Head loss due to surface tension effects | m | **L** |
| $J$ | Rate of change in composition with height for unit driving force | l/m | **L$^{-1}$** |
| $j_d$ | $j$-factor for mass transfer | — | — |
| $K$ | Equilibrium constant $(y/x)$ or coefficient | — | — |
| $K_c$ | Dry cap coefficient (equation 11.102) | — | — |
| $K_g$ | Overall mass transfer coefficient | m/s | **LT$^{-1}$** |
| $K_g'$ | Overall mass transfer coefficient (mols/unit time-unit area-unit mol fraction driving force) | kmol/m$^2$ s | **ML$^{-2}$T$^{-1}$** |
| $K_l$ | Overall mass transfer coefficient | m/s | **LT$^{-1}$** |
| $K_l'$ | Overall mass transfer coefficient | kmol/m$^2$ s | **ML$^{-2}$T$^{-1}$** |
| $k_g, k_g'$ | Film coefficients corresponding to $K_g, K_g'$, above | m/s, kmol/m$^2$ s | **LT$^{-1}$, ML$^{-2}$T$^{-1}$** |
| $k_l, k_l'$ | Film coefficients corresponding to $K_L, K_L'$ | m/s, kmol/m$^2$ s | **LT$^{-1}$, ML$^{-2}$T$^{-1}$** |
| $L$ | Liquid flow in mass or mols/unit time | kg/s, kmol/s | **MT$^{-1}$** |
| $L'$ | Liquid flow, mols/unit time-unit area | kmol/m$^2$ s | **ML$^{-2}$T$^{-1}$** |
| $L_b$ | Mols of liquid in batch distillation | kmol | **M** |
| $L_p$ | Liquid rate per unit periphery | m$^3$/m s | **L$^2$T$^{-1}$** |
| $l_m$ | Metal thickness | m | **L** |
| $M$ | Molecular weight | kg/kmol | — |
|  | or difference stream below feed plate | kmol/s | **MT$^{-1}$** |
| $M_m$ | Mean molecular weight | kg/kmol | — |
| $m$ | Mass of material | kg | **M** |
|  | or gradient of equilibrium curve | — | — |
| $N$ | Molar rate of transfer per unit area of interface | kmol/m$^2$ s | **ML$^{-1}$T$^{-2}$** |
|  | or difference stream above feed plate | kmol/s | **MT$^{-1}$** |
| $\mathbf{N}$ | Number of transfer units | — | — |
| $\mathbf{N}_G, \mathbf{N}_{OG}$ | Number of gas film and overall gas transfer units | — | — |
| $\mathbf{N}_L, \mathbf{N}_{OL}$ | Number of liquid film and overall liquid transfer units | — | — |
| $n$ | Number of (theoretical) plates | — | — |
| $n_m$ | Number of plates at total reflux | — | — |
| $n_p$ | Number of actual plates | — | — |
| $P$ | Total pressure | N/m$^2$ | **ML$^{-1}$T$^{-2}$** |
| $P_A, P_B$ | Partial pressure of **A**, **B** | N/m$^2$ | **ML$^{-1}$T$^{-2}$** |
| $P_A^\circ, P_B^\circ$ | Vapour pressure of **A**, **B** | N/m$^2$ | **ML$^{-1}$T$^{-2}$** |
| $-\Delta P_{\mathrm{dry}}$ | Pressure drop over dry tray expressed as head | m | **L** |
| $-\Delta P_T$ | Total pressure drop expressed as head | m | **L** |
| $p$ | Constant in equation 11.174 | — | — |
| $Q$ | Quantity of heat | J | **ML$^2$T$^{-2}$** |
| $Q_B$ | Heat supplied to boiler for bottom product $W$ | W | **ML$^2$T$^{-3}$** |
| $Q_C$ | Heat removed in condenser for product $D$ | W | **ML$^2$T$^{-3}$** |
| $Q_R$ | Heat to still in batch distillation to provide reflux | J | **ML$^2$T$^{-2}$** |
| $Q'$ | Liquid rate | m$^3$/s | **L$^3$T$^{-1}$** |

| | | Units in SI System | Dimensions in **MLT**$\theta$ |
|---|---|---|---|
| $q$ | Heat to vaporise one mol of feed divided by molar latent heat | — | — |
| $q_c$ | $Q_C/D$ | W/kmol | $\mathbf{L^2T^{-2}}$ |
| $R$ | Reflux ratio | — | |
| $R_m$ | Minimum reflux ratio | — | — |
| $R'$ | Resistance force per unit area, as used in Chapter 4 | N/m$^2$ | $\mathbf{ML^{-1}T^{-2}}$ |
| **R** | Universal gas constant | J/kmol K | $\mathbf{L^2T^{-2}\theta^{-1}}$ |
| $r$ | Radius of an orifice | m | **L** |
| $r_m, r_n, r_f$ | Ratio of concentrations of key components on plates $m$, $n$ and in the feed | — | — |
| $S$ | Mols of material in the still | kmol | **M** |
| $S', S'', S'''$ | Mols or mass of material per unit time in sidestreams 1, 2, 3 | kmol/s | $\mathbf{MT^{-1}}$ |
| $T$ | Absolute temperature | K | $\theta$ |
| $T_B$ | Boiling point | K | $\theta$ |
| $t$ | Time | s | **T** |
| $u$ | Vapour velocity | m/s | $\mathbf{LT^{-1}}$ |
| $u_{dc}$ | Downcomer velocity | m/s | $\mathbf{LT^{-1}}$ |
| $u_e$ | Maximum vapour velocity | m/s | $\mathbf{LT^{-1}}$ |
| $u_n$ | Vapour velocity based on net area above tray (area of column less area of one downcomer) | m/s | $\mathbf{LT^{-1}}$ |
| $u_o$ | Velocity of vapour through perforations | m/s | $\mathbf{LT^{-1}}$ |
| $u_r$ | Velocity through riser | m/s | $\mathbf{LT^{-1}}$ |
| $u_z$ | Velocity of vapour at distance $z$ above bottom of slot | m/s | $\mathbf{LT^{-1}}$ |
| $V$ | Vapour flow in mols mass or volume per unit time | kmol/s, kg/s, m$^3$/s | $\mathbf{MT^{-1}, L^3T^{-1}}$ |
| $V''$ | Mols of vapour formed | kmol | **M** |
| $v$ | Volumetric vapour flowrate | m$^3$/s | $\mathbf{L^3T^{-1}}$ |
| $v_s$ | Volumetric rate of flow of vapour through slot | m$^3$/s | $\mathbf{L^3T^{-1}}$ |
| $W$ | Bottom product in mols or mass per unit time | kmol/s kg/s | $\mathbf{MT^{-1}}$ |
| $w$ | Slot width or width of weir | m | **L** |
| $x$ | Mol fraction of a component in the liquid phase | — | — |
| $y$ | Mol fraction of a component in the gas phase | — | — |
| $Z$ | Height of packed column | m | **L** |
| $Z_p$ | Plate spacing | m | **L** |
| $Z'_p$ | Depth of liquid in the downcomer | m | **L** |
| $Z_t$ | Height of equivalent theoretical plate | m | **L** |
| $z$ | Distance | m | **L** |
| $\alpha$ | Relative volatility or volatility relative to heavy key | — | — |
| $\alpha_{AB}$ | Volatility of **A** relative to **B** | — | — |
| $\beta$ | Aeration factor (equation 11.105) | — | — |
| $\gamma$ | Activity coefficient | — | — |
| $\varepsilon$ | $m(G'/L')$ | — | — |
| $\lambda$ | Latent heat per mol | J/kmol | $\mathbf{L^2T^{-2}}$ |
| $\mu_L$ | Viscosity of liquid | N s/m$^2$ | $\mathbf{ML^{-1}T^{-1}}$ |
| $\mu_v, \mu$ | Viscosity of vapour | N s/m$^2$ | $\mathbf{ML^{-1}T^{-1}}$ |
| $\rho_L$ | Density of liquid | kg/m$^3$ | $\mathbf{ML^{-3}}$ |
| $\rho_v, \rho$ | Density of vapour | kg/m$^3$ | $\mathbf{ML^{-3}}$ |
| $\rho_m$ | Density of metal | kg/m$^3$ | $\mathbf{ML^{-3}}$ |
| $\Sigma$ | Mols per unit time of steam for steam distillation | kmol/s | $\mathbf{MT^{-1}}$ |
| $\sigma$ | Surface tension | J/m$^2$ (or N/m) | $\mathbf{MT^{-2}}$ |
| $\theta$ | Root of equation 11.92 | — | — |
| $\phi$ | Intercept of operating line on $Y$-axis $x_d/(R+1)$ | — | — |
| $\psi$ | $r_m/r_n$ at minimum reflux or tractional entrainment | — | — |
| $\omega$ | Fraction of packing wetted | — | — |
| $Re$ | Reynolds number | — | — |
| $Sc$ | Schmidt number | — | — |

Suffixes

| | |
|---|---|
| 1, 2 | Inlet, outlet |
| $A, B, C, D$ | Materials **A**, **B**, **C**, **D** |
| $b$ | Bottom |

| $c$ | Intersection of equilibrium and operating lines at minimum reflux |
| --- | --- |
| $d$ | Top product |
| $e$ | Equilibrium |
| $f$ | Feed |
| $fs$ | Feed at its boiling point |
| $g$ | Azeotrope |
| $h$ | Component heavier than heavy key |
| $i$ | Interface |
| $L, l$ | Liquid |
| $m, n$ | Plates $m, n$ below and above feed plate respectively |
| $q$ | Intersection of operating lines |
| $S', S'', S'''$ | Sidestreams 1, 2, 3 |
| $s$ | Still |
| $t$ | Top |
| $v$ | Vapour |
| $w$ | Bottom product |
| av | Average |

CHAPTER 12

# Absorption of Gases

## 12.1. INTRODUCTION

The removal of one or more selected components from a mixture of gases by absorption into a suitable liquid is the second major operation of chemical engineering that is based on interphase mass transfer controlled largely by rates of diffusion. Thus, acetone can be recovered from an acetone–air mixture by passing the gas stream into water in which the acetone dissolves while the air passes out. Similarly, ammonia may be removed from an ammonia–air mixture by absorption in water. In each of these examples the process of absorption of the gas in the liquid can be treated as a physical process, the chemical reaction having no appreciable effect. However, when oxides of nitrogen are absorbed in water to give nitric acid, or when carbon dioxide is absorbed in a solution of sodium carbonate, a chemical reaction occurs, the nature of which influences the actual rate of absorption. Absorption processes are therefore conveniently divided into two groups, those in which the process is solely physical and those where a chemical reaction is occurring. In considering the design of equipment to achieve gas absorption, the main requirement is that the gas be brought into intimate contact with the liquid, and the effectiveness of the equipment will largely be determined by the success with which it promotes contact between the two phases. The general form of equipment is similar to that described earlier for distillation (Chapter 11), and packed and plate towers are generally used for large installations. The method of operation, as will be seen later, is not the same. In absorption, the feed is a gas and is introduced at the bottom of the column, and the solvent is fed to the top, as a liquid; the absorbed gas and solvent leave at the bottom, and the unabsorbed components leave as gas from the top. The essential difference between distillation and absorption is that in the former the vapour has to be produced in each stage by partial vaporisation of the liquid which is therefore at its boiling point, whereas in absorption the liquid is well below its boiling point. In distillation there is a diffusion of molecules in both directions so that for an ideal system equimolecular counter-diffusion exists but in absorption gas molecules are diffusing into the liquid, and the movement in the reverse direction is negligible (Volume 1, Chapter 8). In general, the ratio of the liquid to the gas flowrate is considerably greater in absorption than in distillation with the result that layout of the trays is different in the two cases. Furthermore, with the higher liquid rates in absorption, packed columns are much more commonly used.

## 12.2. CONDITIONS OF EQUILIBRIUM BETWEEN LIQUID AND GAS

The two phases when brought into contact tend to reach equilibrium. Thus, water in contact with air evaporates until the air is saturated with water vapour, and the air is absorbed by the water until it becomes saturated with the individual gases. In any mixture of gases, the degree to which each gas is absorbed is determined by its partial pressure. At a given temperature and concentration, each dissolved gas exerts a definite partial pressure. Three types of gases may be considered from this aspect: a very soluble one (ammonia), a moderately soluble one (sulphur dioxide), and a slightly soluble one (oxygen). The figures in Table 12.1 show the concentrations in kilograms per 1000 kg of water that are required

TABLE 12.1. *Partial Pressures and Concentrations of Aqueous Solutions of Gases at 303 K*

| Partial pressure of solute in gas phase (kN/m²) | Concentration of solute in water kg/1000 kg water | | |
|---|---|---|---|
| | Ammonia | Sulphur dioxide | Oxygen |
| 1·3 | 11 | 1·9 | — |
| 6·7 | 50 | 6·8 | — |
| 13·3 | 93 | 12 | 0·08 |
| 26·7 | 160 | 24·4 | 0·13 |
| 66·7 | 315 | 56 | 0·33 |

to develop a partial pressure of 1·3, 6·7, 13·3, 26·7, and 66·7 kN/m² at 303 K. It is seen that a slightly soluble gas requires a much higher partial pressure of the gas in contact with the liquid to give a solution of a given concentration. Conversely, with a very soluble gas a given concentration in the liquid phase is obtained with a lower partial pressure in the vapour phase. At 293 K a solution of 4 kg of sulphur dioxide per 1000 kg of water exerts a partial pressure of 2·7 kN/m². If a gas is in contact with this solution with a partial pressure $SO_2$ greater than 2·7 kN/m², sulphur dioxide will be absorbed. The most concentrated solution that can be obtained is one in which the partial pressure of the solute gas is equal to the partial pressure of the gas in the gas phase. These equilibrium conditions fix the limits of operation of an absorption unit. Thus, in an ammonia–air mixture containing 13·1 per cent of ammonia, the partial pressure of the ammonia is 13·3 kN/m² and the maximum concentration of the ammonia in the water at 303 K is 93 kg per 1000 kg of water.

Whilst the solubility of a gas is not substantially affected by the total pressure in the system for pressures up to about 500 kN/m², it is important to note that the solubility falls with a rise of temperature. Thus, for a concentration of 25 per cent by weight of ammonia in water, the equilibrium partial pressure of the ammonia is 30·3 kN/m² at 293 K and 46·9 kN/m² at 303 K.

In many instances the absorption is accompanied by the evolution of heat, and it is therefore necessary to fit coolers to the equipment to keep the temperature sufficiently low for an adequate degree of absorption to be obtained.

For dilute concentrations of most gases, and over a wide range for some gases, the equilibrium relationship is given by Henry's Law. This law, as used in Chapter 11, can be written as:

$$P_A = \mathscr{H} C_A \tag{12.1}$$

where $P_A$ is the partial pressure of the component $A$ in the gas phase,
     $C_A$ is the concentration of the component in the liquid, and
     $\mathscr{H}$ is Henry's constant.

## 12.3. THE MECHANISM OF ABSORPTION

### 12.3.1. The Two-film Theory

The most useful concept of the process of absorption is given by the two-film theory due to Whitman[1]. According to this theory, material is transferred in the bulk of the phases by convection currents, and concentration differences are regarded as negligible except in the vicinity of the interface between the phases. On either side of this interface it is supposed that the currents die out and that there exists a thin film of fluid through which the transfer is effected solely by molecular diffusion. This film will be slightly thicker than the laminar sub-layer, because it offers a resistance equivalent to that of the whole boundary layer. According to Fick's Law (Volume 1, equation 8.1), the rate of transfer by diffusion is proportional to the concentration gradient and to the area of interface over which the diffusion is occurring. This question has been discussed in Chapter 8 of Volume 1, but some of the important features will be given here.

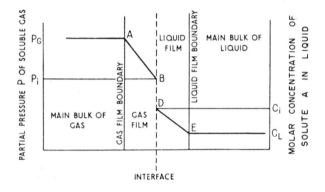

FIG. 12.1. Concentration profile for absorbed component $A$

The direction of transfer of material across the interface is, however, not dependent on the concentration difference, but on the equilibrium relationship. Thus, for a mixture of ammonia or hydrogen chloride and air which is in equilibrium with an aqueous solution, the concentration in the water is many times greater than that in the air. There is, therefore, a very big concentration gradient across the interface: but this is not the controlling factor in the mass transfer, as it is generally assumed that there is no resistance at the interface itself, where equilibrium conditions will exist. The controlling factor will be the rate of diffusion through the two films where all the resistance is considered to lie. The change in concentration of a component through the gas and liquid phases is illustrated in Fig. 12.1. $P_G$ represents the partial pressure in the bulk of the gas phase and $P_i$ the partial pressure at the interface. $C_L$ is the concentration in the bulk of the liquid phase and $C_i$ the concentration at the interface. Thus, according to the above theory, the concentrations at the interface are in equilibrium, and the resistance to transfer is centred in the thin films on either side. This kind of problem was encountered in heat transfer across a tube, where the main resistance to transfer was shown to lie in the thin films on either side of the wall; here the transfer was by conduction.

## 12.3.2. Diffusion of a Gas through a Stagnant Gas

The process of absorption may be regarded as the diffusion of a soluble gas **A** into a liquid. The molecules of **A** have to diffuse through a stagnant gas film and then through a stagnant liquid film before entering the main bulk of liquid. The absorption of a gas consisting of a soluble component **A** and an insoluble component **B** is a problem of mass transfer through a stationary gas to which Stefan's Law (Volume 1, Chapter 8) applies:

$$N'_A = -D_V \frac{C_T}{C_B} \frac{dC_A}{dz} \tag{12.2}$$

Here $N'_A$ is the overall rate of mass transfer (mols/unit area and unit time),

$\quad D_V$ is the gas-phase diffusivity,

$\quad z$ is distance in the direction of mass transfer, and

$\quad C_A$, $C_B$, and $C_T$ are the molar concentrations of **A**, **B**, and total gas, respectively.

Integrating over the whole thickness $z_G$ of the film, and representing concentrations at each side of the interface by suffixes 1 and 2:

$$N'_A = \frac{D_V C_T}{z_G} \ln \frac{C_{B2}}{C_{B1}} \tag{12.3}$$

$$= \frac{D_V P}{RT z_G} \ln \frac{P_{B2}}{P_{B1}} \text{ (for an ideal gas)} \tag{12.4}$$

since $C_T = P/RT$, where $R$ is the gas constant, $T$ the absolute temperature, and $P$ the total pressure.

Writing $P_{Bm}$ as the log mean of the partial pressures $P_{B1}$ and $P_{B2}$, then:

$$P_{Bm} = \frac{P_{B2} - P_{B1}}{\ln (P_{B2}/P_{B1})} \tag{12.5}$$

$$N'_A = \frac{D_V P}{RT z_G} \frac{P_{B2} - P_{B1}}{P_{Bm}}$$

$$= \frac{D_V P}{RT z_G} \left[ \frac{P_{A1} - P_{A2}}{P_{Bm}} \right] \tag{12.6}$$

Hence the rate of absorption of **A** per unit time over unit area:

$$N'_A = k'_G P \left[ \frac{P_{A1} - P_{A2}}{P_{Bm}} \right] \tag{12.7}$$

or

$$N'_A = k_G [P_{A1} - P_{A2}] \tag{12.8}$$

where

$$k'_G = \frac{D_V}{RT z_G}, \quad \text{and} \quad k_G = \frac{D_V P}{RT z_G P_{Bm}} = \frac{k'_G P}{P_{Bm}} \tag{12.9}$$

In the great majority of industrial processes the film thickness is not known, so that the rate equation of immediate use is 12.8 using $k_G$. $k_G$ is known as the gas film transfer coefficient for absorption and is a direct measure of the rate of absorption per unit area of interface with a driving force of unit partial pressure difference.

### 12.3.3. Diffusion in the Liquid Phase

The rate of diffusion in liquids is much slower than in gases, and mixtures of liquids may take a long time to reach equilibrium unless agitated. This is partly due to the much closer spacing of the molecules, as a result of which the molecular attractions are more important.

Whilst there is at present no theoretical basis for the rate of diffusion in liquids comparable with the kinetic theory for gases, the basic equation is taken as similar to that for gases: i.e. for dilute concentrates:

$$N'_A = -D_L \frac{dC_A}{dz} \qquad (12.10)$$

On integration:

$$N'_A = -D_L \left[ \frac{C_{A2} - C_{A1}}{z_L} \right] \qquad (12.11)$$

where $C_A$, $C_B$ are the molar concentrations of A and B,

$z_L$ is the thickness of liquid film through which diffusion occurs, and

$D_L$ is the diffusivity in the liquid phase.

Since the film thickness is rarely known, equation 12.11 is rewritten as:

$$N'_A = k_L [C_{A1} - C_{A2}] \qquad (12.12)$$

which is similar to equation 12.8 for gases.

Here $k_L$ is the liquid film transfer coefficient, which is usually expressed in kmol/s m$^2$ (kmol/m$^3$) = m/s. For dilute concentrations:

$$k_L = \frac{D_L}{z_L}$$

### 12.3.4. Rate of Absorption

In a steady state process of absorption, the rate of transfer of material through the gas film will be the same as that through the liquid film, and the general equation for mass transfer may be written as:

$$N'_A = k_G [P_G - P_i] = k_L [C_i - C_L] \qquad (12.13)$$

where $P_G$ is the partial pressure in the bulk of the gas, $C_L$ is the concentration in the bulk of the liquid, and $P_i$ and $C_i$ are the values of concentration at the interface where equilibrium conditions are assumed to exist. Therefore:

$$\frac{k_G}{k_L} = \frac{C_i - C_L}{P_G - P_i} \qquad (12.14)$$

These conditions may be illustrated graphically as in Fig. 12.2, where $ABF$ is the equilibrium curve.

Point $D$ $(C_L, P_G)$ represents conditions in the bulk of the gas and liquid,

$P_G$ is the partial pressure of A in the main bulk of the gas stream, and

$C_L$ is the average concentration of A in the main bulk of the liquid stream.

Point $A$ $(C_e, P_G)$   represents a concentration $C_e$ in the liquid in equilibrium with $P_G$ in the gas.

Point $B$ $(C_i, P_i)$   represents the concentration $C_i$ in the liquid in equilibrium with $P_i$ in the gas, and gives conditions at the interface.

Point $F$ $(C_L, P_e)$   represents a partial pressure $P_e$ in the gas phase in equilibrium with $C_L$ in the liquid.

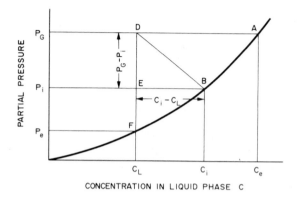

FIG. 12.2. Driving forces in the gas and liquid phases

Then the driving force causing transfer in the gas phase:

$$(P_G - P_i) \equiv DE$$

and the driving force causing transfer in the liquid phase:

$$(C_i - C_L) \equiv BE$$

Then

$$\frac{P_G - P_i}{C_i - C_L} = \frac{k_L}{k_G}$$

and the concentrations at the interface (point $B$) are found by drawing a line through $D$ of slope $-k_L/k_G$ to cut the equilibrium curve in $B$.

*Overall Coefficients*

To obtain a direct measurement of the values of $k_L$ and $k_G$ requires the measurement of the concentration at the interface. These values can only be obtained in very special circumstances, and it has been found of considerable value to use two overall coefficients $K_G$ and $K_L$ defined by the following equation:

$$N'_A = K_G[P_G - P_e] = K_L[C_e - C_L] \tag{12.15}$$

$K_G$ and $K_L$ are known as the overall gas and liquid phase coefficients, respectively.

*Relation between Film and Overall Coefficients*

The rate of transfer of **A** can now be written as:

$$N'_A = k_G[P_G - P_i] = k_L[C_i - C_L] = K_G[P_G - P_e] = K_L[C_e - C_L] \qquad (12.16)$$

$$\therefore \quad \frac{1}{K_G} = \frac{1}{k_G}\left[\frac{P_G - P_e}{P_G - P_i}\right]$$

$$= \frac{1}{k_G}\left[\frac{P_G - P_i}{P_G - P_i}\right] + \frac{1}{k_G}\left[\frac{P_i - P_e}{P_G - P_i}\right]$$

But from above:

$$\frac{1}{k_G} = \frac{1}{k_L}\left[\frac{P_G - P_i}{C_i - C_L}\right]$$

$$\therefore \quad \frac{1}{K_G} = \frac{1}{k_G} + \frac{1}{k_L}\left[\frac{P_G - P_i}{C_i - C_L}\right]\left[\frac{P_i - P_e}{P_G - P_i}\right]$$

$$= \frac{1}{k_G} + \frac{1}{k_L}\left[\frac{P_i - P_e}{C_i - C_L}\right]$$

But $(P_i - P_e)/(C_i - C_L)$ is the average slope of the equilibrium curve and, when the solution obeys Henry's Law, $\mathcal{H} = dP/dC \approx (P_i - P_e)/(C_i - C_L)$.

$$\therefore \quad \frac{1}{K_G} = \frac{1}{k_G} + \frac{\mathcal{H}}{k_L} \qquad (12.17)$$

Similarly:

$$\frac{1}{K_L} = \frac{1}{k_L} + \frac{1}{\mathcal{H}k_G} \qquad (12.18)$$

and:

$$\frac{1}{K_G} = \frac{\mathcal{H}}{K_L} \qquad (12.19)$$

The validity of using equations 12.17 and 12.18 in order to obtain an overall transfer coefficient has been examined in detail by King[2]. He pointed out that the equilibrium constant $\mathcal{H}$ must not vary over the equipment, there must be no significant interfacial resistance, and there must be no interdependence of the values of the two film coefficients.

*Rates of Absorption in Terms of Mol Fractions*

The mass transfer equations can be written:

$$N'_A = k''_G(y - y_i) = K''_G(y - y_e) \qquad (12.20)$$

and

$$N'_A = k''_L(x_i - x) = K''_L(x_e - x) \qquad (12.21)$$

where $x$, $y$ are the mol fractions of the soluble component **A** in the liquid and gas phases.

$k_G''$, $k_L''$, $K_G''$, $K_L''$ are transfer coefficients defined in terms of mol fractions by equations 12.20 and 12.21.

If $m$ is the slope of the equilibrium curve [approximately $(y_i - y_e)/(x_i - x)$], it can then be shown that:

$$\frac{1}{K_G''} = \frac{1}{k_G''} + \frac{m}{k_L''} \tag{12.22}$$

which is similar to the relation used for distillation 11.166. It should be noted that equations 12.20 and 12.21 are derived from equations 12.13 and 12.15, and that the total concentration in the liquid phase is assumed to remain constant.

### Factors Influencing the Transfer Coefficient

The general influence of the solubility of the gas on the shape of the equilibrium curve, and the resulting influence on film and overall coefficients, will be seen by considering very soluble, almost insoluble, and moderately soluble gases.

(a) *Very soluble gas.* Here the equilibrium curve lies close to the concentration-axis and the point $E$ (Fig. 12.2) approaches very close to point $F$. The driving force over the gas film ($DE$) is then approximately equal to the overall driving force ($DF$), so that $k_G$ is approximately equal to $K_G$.

(b) *Almost insoluble gas.* Here the equilibrium curve rises very steeply so that the driving force $(C_i - C_L)$ ($EB$) in the liquid film becomes approximately equal to the overall driving force $(C_e - C_L)$ ($AD$). In this case $k_L$ will be approximately equal to $K_L$.

(c) *Moderately soluble gas.* Here both films offer an appreciable resistance, and the point $B$ at the interface must be located by drawing a line through $D$ of slope $-(k_L/k_G) = -(P_G - P_i)/(C_i - C_L)$.

In most experimental work, the concentration at the interface cannot be measured directly, and only the overall coefficients will therefore be found. To obtain values for the film coefficients, the relations between $k_G$, $k_L$ and $K_G$ are utilised as discussed above.

## 12.4. APPLICATION OF MASS TRANSFER THEORIES

The analysis of the process of absorption so far given is based on the two-film theory of Whitman[1]. It is supposed that the two films have negligible capacity, but offer all the resistance to mass transfer. Any turbulence disappears at the interface or free surface, and the flow is thus considered to be laminar and parallel to the surface.

An alternative theory described in detail in Volume 1, Chapter 8, has been put forward by Higbie[3], and later extended by Danckwerts *et al.*[4,5], in which the liquid surface is considered to be composed of a large number of small elements each of which is exposed to the gas phase for an interval of time after which they are replaced by fresh elements arising from the bulk of the liquid.

All three of these proposals give the mass transfer rate $N_A'$ directly proportional to the concentration difference $(C_i - C_L)$ so that they do not directly enable a decision to be made between the theories. However, in the Higbie–Danckwerts theory $N'_A \propto \sqrt{D_L}$ whereas $N'_A \propto D_L$ in the two-film theory. Danckwerts[4] applied this theory to the problem of absorption coupled with chemical reaction but, although in this case the three proposals give somewhat different results, it has not been possible to distinguish between them.

The application of the penetration theory to the interpretation of experimental results obtained in wetted-wall columns has been studied by Lynn, Straatemeier, and Kramers[6]. They absorbed pure sulphur dioxide in water and various aqueous solutions of salts and found that, in the presence of a trace of Teepol which suppressed ripple formation, the rate of absorption was closely predicted by the theory. In very short columns, however, the rate was overestimated because of the formation of a region in which the surface was stagnant over the bottom one centimetre length of column. The studies were extended to columns containing spheres and again the penetration theory was found to hold, there being very little mixing of the surface layers with the bulk of the fluid as it flowed from one layer of spheres to the next.

Absorption experiments in columns packed with spheres (37·8 mm diam.) were also carried out by Davidson et al.[7] who absorbed pure carbon dioxide into water. When a small amount of surface active agent was present in the water no appreciable mixing was found between the layers of spheres. With pure water, however, the liquid was almost completely mixed in this region.

Davidson[8] has built up theoretical models of the surfaces existing in a packed bed. He assumed that the liquid ran down each surface in laminar flow and was then fully mixed before it commenced to run down the next surface. The angles of inclination of the surfaces were taken as random. In the first theory he assumed that all the surfaces were of equal length, and in the second that there was a random distribution of surface lengths up to a maximum. Thus the assumptions regarding age distribution of the liquid surfaces were similar to those of Higbie[3] and Danckwerts[4]. Experimental results were in good agreement with the second theory. All random packings of a given size appeared to be equivalent to a series of sloping surfaces and therefore the most effective packing would be that which gave the largest interfacial area.

In an attempt to test the surface renewal theory of gas absorption Danckwerts and Kennedy[9] measured the transient rate of absorption of carbon dioxide into various solutions by means of a rotating drum which carried a film of liquid through the gas. Results so obtained were compared with those for absorption in a packed column and it was shown that exposure times of at least 1 s were required to give a strict comparison; this was longer than could be obtained with the rotating drum. Roberts and Danckwerts[10] therefore used a wetted-wall column to extend the times of contact up to 1·3 s. The column was carefully designed to eliminate entry and exit effects and the formation of ripples. The experimental results and conclusions are reported by Danckwerts, Kennedy, and Roberts[11] who showed that they could be used, on the basis of the penetration theory model, to predict the performance of a packed column to within about 10 per cent.

There have been many recent studies of the mechanism of mass transfer in a gas absorption system. Many of these have been directed towards investigating whether there is a significant resistance to mass transfer at the interface itself. In order to obtain results which can readily be interpreted it is essential to operate with a system of simple geometry. For that reason a laminar jet has been used by a number of workers.

Cullen and Davidson[12] studied the absorption of carbon dioxide into a laminar jet of water. When the water issued with a uniform velocity over every cross-section the measured rate of absorption corresponded closely with the theoretical value. When the velocity profile in the water was parabolic, the measured rate was lower than the calculated value; this was attributed to a hydrodynamic entry effect.

The possible existence of an interface resistance in mass transfer has been examined by Raimondi and Toor[13] who absorbed carbon dioxide into a laminar jet of water with a flat velocity profile, using contact times down to 1 ms. They found that the rate of absorption was not more than 4 per cent less than that predicted on the assumption of instantaneous saturation of the surface layers of liquid. Thus, the effects of interfacial resistance could not have been significant. When the jet was formed at the outlet of a long capillary tube so that a parabolic velocity profile was established, absorption rates were lower than predicted because of the reduced surface velocity. The presence of surface-active agents appeared to cause an interfacial resistance, but this effect is probably attributable to a modification of the hydrodynamic pattern.

Sternling and Scriven[14] have examined interfacial phenomena in gas absorption and have explained the interfacial turbulence which has been noted by a number of workers in terms of the Marangoni effect which gives rise to movement at the interface due to local variations in interfacial tension. Some systems have been shown to give rise to stable interfaces when the solute is transferred in one direction, but instabilities develop during transfer in the reverse direction.

Goodridge and Robb[15] used a laminar jet to study the rate of absorption of carbon dioxide into sodium carbonate solutions containing a number of additives including glycerol, sucrose, glucose, and arsenites. For the short times of exposure used, absorption rates into sodium carbonate solution or aqueous glycerol corresponded to those predicted on the basis of pure physical absorption. In the presence of the additives, however, the process was accelerated as the result of chemical reaction.

Absorption of gases and vapour by drops has been studied by Garner et al.[16,17] who developed a vertical wind tunnel in which drops could be suspended for considerable periods of time in the rising gas stream. During the formation of each drop the rate of mass transfer was very high because of the high initial turbulence. After the initial turbulence had died out, the mass transfer rate fell towards the rate for molecular diffusion provided that the circulation had stopped completely. In a drop with stable natural circulation the rate was found to approach 2·5 times the rate for molecular diffusion.

## 12.5. VALUES OF TRANSFER COEFFICIENTS

In calculating the size of an absorption tower, the most important single factor is the value of the transfer coefficient or the height of the transfer unit. Whilst the total flowrates of the gas and liquid streams will be fixed by the process, it is necessary to determine the most suitable flow per unit area through the column. The gas flow is limited by the fact that the flooding rate must not be exceeded and there will be a serious drop in performance if the liquid rate is very low. It is convenient to examine the effects of flowrates of the gas and liquid on the transfer coefficients, and also to investigate the influence of variables such as temperature, pressure, and diffusivity.

In the laboratory, wetted-wall columns have been used by a number of workers and they have proved valuable in determining the importance of the various factors, and have served as a basis from which correlations have been developed for packed towers.

### 12.5.1. Wetted-wall Columns

Several workers have examined the vaporisation of liquids into an air stream in a wetted-wall column, using the apparatus as outlined in Fig. 12.3. Plots of $d/z_G$ and $du\rho/\mu$ on logarithmic axes give a series of approximately straight lines showing that $d/z_G = BRe^{0.83}$

where   $d$   is the diameter of tube,
      $z_G$ is the thickness of gas film,
      $u$   is the gas velocity,
      $\rho$   is the gas density,
      $\mu$   is the gas viscosity, and
      $B$   is a constant.

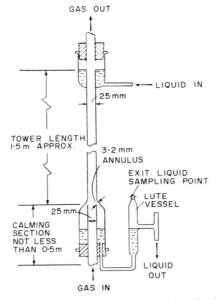

FIG. 12.3. Diagram of typical 25 mm wetted-wall column

It is convenient to eliminate the film thickness $z_G$ in the following way:

$$k_G = \frac{D_V P}{R T z_G P_{Bm}} \qquad \text{(equation 12.9)}$$

$$\therefore \quad \frac{k_G R T P_{Bm}}{D_V P} = \frac{1}{z_G} = \frac{B}{d} Re^{0.83}$$

or

$$\frac{h_D d P_{Bm}}{D_V P} = BRe^{0.83} \qquad (12.23)$$

where $h_D = k_G R T$ and is a mass transfer coefficient with the driving force expressed as a molar concentration difference.

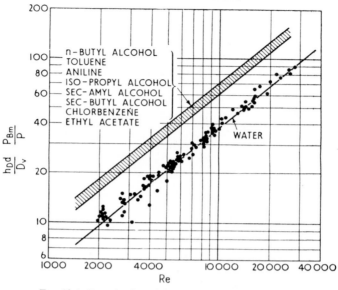

FIG. 12.4. Vaporisation of liquids in a wetted-wall column

Gilliland and Sherwood's results[18] expressed by equation 12.23 are shown in Fig. 12.4 for a number of systems. To allow for the variation in the physical properties, the Schmidt Group $Sc$, is introduced, and the general equation for mass transfer in a wetted-wall column is then given by:

$$\frac{h_D d}{D_V} \frac{P_{Bm}}{P} = B' Re^{0\cdot83} Sc^{0\cdot44} \tag{12.24}$$

Values of $B'$ from 0·021 to 0·027 have been reported and a mean value of 0·023 can be taken, which then makes equation 12.24 very similar to the general heat transfer equation (Volume 1, Chapter 7) for forced convection in tubes. The data shown in Fig. 12.4 are replotted as $(h_D d/D_V)(P_{Bm}/P) Sc^{-0\cdot44}$ in Fig. 12.5, and it will be noted that they can be correlated by means of a single line.

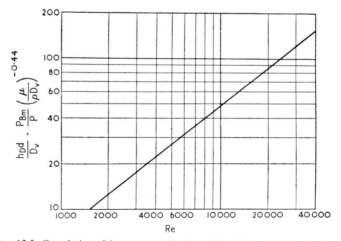

FIG. 12.5. Correlation of data on vaporisation of liquids in wetted-wall columns

In making comparison of the results from various workers, it is important to see that the inlet arrangements for the air are similar. Modifications of the inlet give rise (as found by Hollings and Silver[19]) to various values for the index on the Reynolds number. A good calming length is necessary before the inlet to the measuring section, if results are to be reproducible.

Equation 12.24 is frequently rearranged as:

$$\frac{h_D d}{D_V}\frac{P_{Bm}}{P}\frac{\mu}{du\rho}\left[\frac{\mu}{\rho D_V}\right]^{0.56} = B'Re^{-0.17}\frac{\mu}{\rho D_V}$$

i.e.

$$\frac{h_D}{u}\frac{P_{Bm}}{P}\left[\frac{\mu}{\rho D_V}\right]^{0.56} = B'Re^{-0.17} = j_d \tag{12.25}$$

(see Chapter 8 in Volume 1)

where $j_d$ is the $j$-factor for mass transfer as introduced by Chilton and Colburn[20] and discussed in Chapter 8 of Volume 1. The main interest of this type of work is that it suggests that $h_D \propto G'^{0.8}$, $D_V^{0.56}$ and $P/P_{Bm}$; this kind of relation will be the basis for correlating work on packed towers.

For wetted-wall columns, Morris and Jackson[21] have represented the experimental data for the mass transfer coefficient for the gas film $h_D$ in a form similar to equation 12.25, but with slightly different indices:

$$\frac{h_D}{u} = 0.04\left[\frac{ud\rho}{\mu}\right]^{-0.25}\left[\frac{\mu}{\rho D_V}\right]^{-0.5}\left[\frac{P}{P_{Bm}}\right] \tag{12.26}$$

The velocity $u$ of the gas is strictly the velocity relative to the surface of the falling liquid film, but little error is introduced if it is taken as the superficial velocity in the column.

For a 25 mm diameter tube at 293 K, equation 12.26 can be rearranged to give the coefficient $k_G[=(h_D/\mathbf{R}T)(P_{Bm}/P)]$ in g/s cm$^2$ atm (1 kg/m$^2$ s (kN/m$^2$) $\approx$ 0·1 g/s cm$^2$ atm):

$$k_G = 0.0317c_G\rho_{Ar}u^{0.75}P^{-0.25}\left(\frac{293}{T_f}\right)^{0.56} \tag{12.27}$$

where  $c_G$  is the gas-mixture constant $(\rho_r/\mu_r)^{0.25}(D_{Vr})^{0.5}$ in cgs units; it can be taken as equal to 0·6 as a first approximation, unless hydrogen is present,

$T_f$  is the temperature of the gas film in K,

$P$  is the total pressure in atmospheres (1 atm = 101·3 kN/m$^2$),

$\rho_{Ar}$ is the density of the soluble gas in cgs units at the reference temperature and pressure,

other quantities are in cgs units, and

suffix $r$ refers to the reference temperature and pressure (293 K, 101·3 kN/m$^2$).

*Compounding of Film Coefficients*

Assuming $k_G \propto G'^{0.8}$, equation 12.17 may be rearranged to give:

$$\frac{1}{K_G} = \frac{1}{k_G} + \frac{\mathscr{H}}{k_L} = \frac{1}{\psi u^{0.8}} + \frac{\mathscr{H}}{k_L} \tag{12.28}$$

If $k_L$ is assumed to be independent of the gas velocity, then a plot of $1/K_G$ versus $1/u^{0.8}$ will give a straight line with a positive intercept on the vertical axis representing the liquid film resistance $\mathscr{H}/k_L$. Data for ammonia and for sulphur dioxide are shown in this way in Fig. 12.6. It will be seen that in each case a straight line is obtained. The lines for ammonia

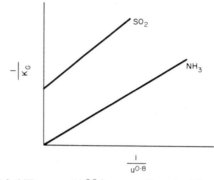

FIG. 12.6. $1/K_G$ versus $1/u^{0.8}$ for ammonia and sulphur dioxide

pass almost through the origin showing that the liquid film resistance is very small, but the line for sulphur dioxide gives a large intercept on the vertical axis, indicating a high value of the liquid film resistance.

For a constant value of $Re$, the film thickness $z_G$ should be independent of temperature, since $\mu/\rho D_V$ is almost independent of temperature. $k_G$ will then vary as $\sqrt{T}$, because $D_V \propto T^{3/2}$ and $k_G \propto D_V/T$. This is somewhat difficult to test accurately since the diffusivity in the liquid phase also depends on temperature. Thus, the data for sulphur dioxide, shown in Fig. 12.6, qualitatively support the theory for different temperatures, but the increase in value of $k_L$ masks the influence of temperature on $k_G$.

### 12.5.2. Coefficients in Packed Towers

The majority of published results for transfer coefficients in packed towers are for rather small laboratory units of 50–250 mm diameter, and there is still some uncertainty in extending the data for use in industrial units. One of the great difficulties in correlating the performance of packed towers is the problem of assessing the effective wetted area for interphase transfer. It is convenient to consider separately conditions where the gas film controls the process and then where the liquid film controls. The general method of expressing results is based on that used for wetted-wall columns.

*Gas-film Controlled Processes*

The absorption of ammonia in water has been extensively studied by a number of workers. Kowalke *et al.*[22] used a tower of 0.4 m internal diameter with a packing 1.2 m deep, and expressed their results as:

$$K_G a = \alpha G'^{0.8} \tag{12.29}$$

where $K_G$ is expressed in kmol/s m$^2$ (unit partial pressure difference in kN/m$^2$) and $a$ is the interfacial surface per unit volume of tower (m$^2$/m$^3$). Thus $K_G a$ is a transfer coefficient based on unit volume of tower. $G'$ is in kg/s m$^2$, and varies with the nature of the packing and the liquid rate. They noted that $K_G a$ increased with $L'$ for values up to $1 \cdot 1 \, \text{kg/s m}^2$, after which further increase gave no significant increase in $K_G a$. It was thought that the initial increase in the coefficient was occasioned by a more effective wetting of the packing. On increasing the liquid rate so that the column approached flooding conditions, it was found that $K_G a$ fell again. Reported values of $\alpha$ varied from $0 \cdot 00048$ for 25 mm balls to $0 \cdot 001$ for 6·4 mm crushed stone.

Borden and Squires[23] using a 250 mm tower gave their results in the form:

$$K_G a = 8 \cdot 0 \times 10^{-4} G'^{0 \cdot 5} L'^{0 \cdot 4} \, \text{kmol/m}^3 \, \text{s} \, (\text{kN/m}^2) \tag{12.30}$$

where $G'$ varied from $0 \cdot 07$ to $0 \cdot 69 \, \text{kg/m}^2 \, \text{s}$, and $L'$ from $0 \cdot 572$ to $2 \cdot 67 \, \text{kg/m}^2 \, \text{s}$.

Norman[24] carried out experiments on the absorption of ammonia in water and on the evaporation of water in an air stream, using a tower packed with carbon slates 150 mm long × 25 mm deep × 3 mm thick with shallow V-notches 6 mm deep cut in the lower edge to collect the liquid and distribute it on to the top edge of the lower grid. Such a unit gave good water distribution for values of $L'$ down to $1 \cdot 3 \, \text{kg/m}^2 \, \text{s}$ or $0 \cdot 012 \, \text{kg/s}$ per metre perimeter of grids.

He was able to show that:

$$K_G a = \alpha G'^{0 \cdot 8} \tag{12.31}$$

where $\alpha$ depended on the liquid rate and had a value $0 \cdot 000031$ for $L' = 3 \cdot 78 \, \text{kg/m}^2 \, \text{s}$, for absorption of ammonia. For the evaporation of water, he gave values of $\alpha$ in the range $0 \cdot 000027$ to $0 \cdot 000035$ as the water flowrate increased from $1 \cdot 2$ to $2 \cdot 7 \, \text{kg/m}^2 \, \text{s}$ with no further increase in $\alpha$ for higher liquid flows.

Fellinger[25] used a 450 mm diameter stoneware column in which a perforated packing support was fitted with 20 downcomers extending to within 25 mm of the bottom of the tower, and 120 risers were fitted extending 31 mm above the upper surface. In this way he avoided the problem of determining any entrance or exit effects. Some of his results for $\mathbf{H}_{OG}$ are shown in Table 12.2, taken from the graphs given by Perry[26] and further discussion on the use of transfer units is included in Section 12.8.8 and in Chapter 11.

TABLE 12.2. Height of the Transfer Unit $\mathbf{H}_{OG}$ in metres

| Raschig rings size (mm) | $G'$ kg/m² s | $\mathbf{H}_{OG}$ ($L' = 0 \cdot 65 \, \text{kg/m}^2 \, \text{s}$) | $\mathbf{H}_{OG}$ ($L' = 1 \cdot 95 \, \text{kg/m}^2 \, \text{s}$) |
|---|---|---|---|
| 9·5 | 0·26 | 0·37 | 0·23 |
|     | 0·78 | 0·60 | 0·32 |
| 25  | 0·26 | 0·40 | 0·22 |
|     | 0·78 | 0·64 | 0·34 |
| 50  | 0·26 | 0·60 | 0·34 |
|     | 0·78 | 1·04 | 0·58 |

Molstad et al.[27] have also measured the absorption of ammonia in water using a tower of 384 mm side packed with wood grids, or with rings or saddles. Their method of treatment was to measure $K_G a$ by direct experiment, and then to calculate the value of $k_G a$ from the following relation based on equation 12.17:

$$\frac{1}{K_G a} = \frac{1}{k_G a} + \frac{\mathscr{H}}{k_L a} \tag{12.32}$$

To find the value of $k_L a$ for ammonia, they used the data obtained for the absorption of oxygen in water, where the resistance is entirely in the liquid film, and obtained the value of $k_L a$ for ammonia from the relation:

$$(k_L a)_{NH_3} = (k_L a)_{O_2} \left( \frac{(D_L)_{NH_3}}{(D_L)_{O_2}} \right)^{0.5} \tag{12.33}$$

($k_L a$ is shown below to be proportional to $D_L^{0.5}$).

From these values of $(k_L a)_{NH_3}$ they could then find the value of $(k_G a)_{NH_3}$ using equation 12.32, and they were thus able to show that for absorption of ammonia in water some 20 per cent of the resistance was in the liquid phase.

The simplest method of representing data for gas-film coefficients is to relate the Sherwood number $[(h_D d/D_V)(P_{Bm}/P)]$ to the Reynolds number ($Re$) and the Schmidt number ($\mu/\rho D_V$). The indices used vary somewhat from one investigator to another, but van Krevelen and Hoftijzer[28] have given the following expression, which they claim to be valid over a wide range of Reynolds numbers:

$$\frac{h_D d}{D_V} \frac{P_{Bm}}{P} = 0.2 \, Re^{0.8} \left( \frac{\mu}{\rho D_V} \right)^{0.33} \tag{12.34}$$

Later work suggests that the coefficient of 0.2 is high and that 0.11 is a more realistic figure.

Morris and Jackson[21] have extended their own correlation for wetted-wall columns for use in connection with packed columns. The gas-film coefficient is then obtained by multiplying the coefficient obtained for a 25 mm diameter wetted-wall column by a packing factor $R_G$[2] to give:

$$k_G = 0.0317 R_G c_G \rho_{Ar} u^{0.75} P^{-0.25} \left( \frac{293}{T_f} \right)^{0.56} \tag{12.35}$$

The units of the quantities in equation 12.35 are as specified for equation 12.27. British or SI units can be used by rewriting equation 12.35 as equation 12.36 and using the numerical constants and physical quantities listed in Table 12.3.

$$k_G = \alpha' R_G c_G \rho_{Ar} u^{0.75} P^{-0.25} \left( \frac{T_r}{T_f} \right)^{0.56} \tag{12.36}$$

TABLE 12.3. *Significance of Quantities in Equation 12.36*

| Symbol | Significance | | | British units | SI units |
|---|---|---|---|---|---|
| $k_G$ | Gas-film coefficient | | | lb/hr ft² (atom) | kg/s m² (kN/m²) |
| $\alpha'$ | Numerical constant | | | 48·6 | 0·00031 |
| $R_G$ | Gas-film packing factor[21] | | | — | — |
| | Ceramic rings | (50 mm) | 2·7 | | |
| | (random) | (76 mm) | 2·5 | | |
| | Ceramic rings | (50 mm) | 1·4 | | |
| | (stacked) | | | | |
| | Metal rings | (50 mm) | 3·3 | | |
| | (random) | | | | |
| $c_G$ | Gas-mixture constant | | | — | — |
| $\rho_{Ar}$ | Density of soluble gas at 293 K and 101·3 kN/m² | | | lb/ft³ | kg/m³ |
| $u$ | Gas velocity | | | ft/s | m/s |
| $P$ | Total gas pressure | | | atm | atm |
| $T_r$ | Reference temperature (absolute) | | | 528°R | 293 K |
| $T_f$ | Absolute temperature of gas-film | | | °R | K |

Semmelbauer[29] has recommended the following correlation for $100 < (Re)_G$ $< 10,000$ and $0.01\,\text{m} < d_p < 0.05\,\text{m}$:

$$(Sh)_G = \beta (Re)_G^{0.59} (Sc)_G^{0.33} \tag{12.37}$$

where $\quad \beta = 0.69$ for Raschig rings and $0.86$ for Berl saddles,

$(Sh)_G = h_D d_p / G_G$,
$(Re)_G = G' d_p / \mu_G$,
$(Sc)_G = \mu_G / \rho_G D_G$, and
$d_p$ = packing size.

### Processes Controlled by Liquid-film

The absorption of carbon dioxide, oxygen, and hydrogen in water gives three instances in which most, if not all, of the resistance to transfer lies in the liquid phase. Sherwood and Holloway[30] have measured values of $k_L a$ for these systems using a tower of 500 mm diameter packed with 37 mm rings. They were able to express their results in the form:

$$\frac{k_L a}{D_L} = \beta \left[ \frac{L'}{\mu_L} \right]^{0.75} \left[ \frac{\mu_L}{\rho_L D_L} \right]^{0.50} \tag{12.38}$$

It should be noted that this equation has no term for characteristic length on the left-hand side and therefore is not a true dimensionless equation. If values of $k_L a$ are plotted

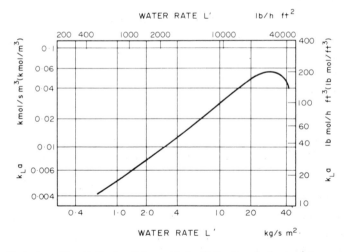

FIG. 12.7. Variation of liquid-film coefficient with liquid flow for absorption of oxygen in water

against value $L'$ on a logarithmic scale as shown in Fig. 12.7, a slope of about $0.75$ is obtained for values of $L'$ from $0.5$ to $13\,\text{kg/s}\,\text{m}^2$. Beyond this value of $L'$, they found that $k_L a$ tended to fall because the loading point for the column was reached. These values of $k_L a$ were found to be affected by the gas rate. More recently Cooper et al.[31] established that, at the high liquid rates and low gas rates used in practice, and with a tower packed to a depth of $2.2\,\text{m}$, the transfer rates were much lower than given by equation 12.38. This was

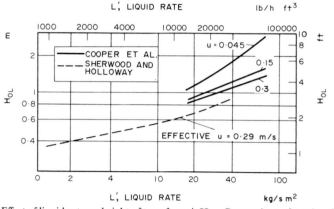

FIG. 12.8. Effect of liquid rate on height of transfer unit $H_{OL}$. Comparison of results of Sherwood and Holloway, and Cooper *et al.*

believed to arise as a result of maldistribution at gas velocities as low as 0·03 m/s; in fact it is doubtful if true countercurrent conditions operate under these conditions. The results of Cooper *et al.*[31] and Sherwood and Holloway[30] are compared in Fig. 12.8, where the height of the transfer unit $H_{OL}$ is plotted against the liquid rate for various gas velocities.

In an equation analogous to 12.37, Semmelbauer[29] produced the following correlation for the liquid film mass transfer coefficient $k_L$ for $3 < Re_L < 3000$ and $0·01\,\text{m} < d_p < 0·05\,\text{m}$:

$$(Sh)_L = \beta'(Re)_L^{0·59}(Sc)_L^{0·5}(Ga)^{0·17} \tag{12.39}$$

where      $\beta' = 0·32$ and $0·25$ for Raschig rings and Berl saddles, respectively,
$(Sh)_L = k_L d_p/D_L$,
$(Re)_L = L'd_p/\mu_L$,
$(Sc)_L = \mu_L/\rho_L D_L$,
$(Ga)_L = d_p^3 g\rho_L/\mu_L$.

It must be emphasised that values of the individual film mass transfer coefficients obtained from the above equations must be used with caution when designing large-scale towers and appropriately large safety factors should be incorporated[32].

## 12.5.3. Coefficients in Spray Towers

It is difficult to compare the performance of various spray towers since the type of spray distributor used will influence the results. Data from Hixson and Scott[33] and others show that $K_G a$ varies as $G'^{0·8}$, and is also affected by the liquid rate. Again their results for sulphur dioxide and ammonia are in fair agreement and show $K_G a$ varying as $\sqrt{D_V}$. One might expect to get more reliable data with spray columns if the liquid were introduced in the form of individual drops through a single jet into a tube full of gas. Unfortunately the drops tend to alter in size and shape and it is not possible to get the true interfacial area very accurately. This has been tried by Whitman *et al.*[34], who found that $k_G$ for the absorption of ammonia in water was about 0·035 kmol/s m² (N/m²), compared with 0·00025 for the absorption of carbon dioxide in water.

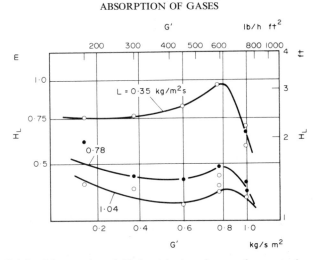

FIG. 12.9. Height of the transfer unit $H_L$ for stripping of oxygen from water in a spray tower

Some values for the height of a transfer unit $H_L$ for the stripping of oxygen from water are shown in Fig. 12.9[35]. There is reason to believe that, for short heights, the efficiency of the spray chamber approximates closely to that of a packed tower, but for heights greater than 1·2 m the efficiency of the spray tower drops off rather rapidly. Whilst it might be thought that it would be possible to obtain a very large active interface by producing small drops, in practice it is found impossible to prevent these coalescing, and hence the effective interfacial surface falls off with height, and spray towers are not extensively used.

## 12.6. ABSORPTION ASSOCIATED WITH CHEMICAL REACTION

In the instances so far considered, the process of absorption of the gas in the liquid is really entirely a physical one. There are, however, a number of cases in which the gas, on absorption, reacts chemically with a component of the liquid phase[36]. Thus, in the absorption of carbon dioxide by caustic soda, the carbon dioxide reacts directly with the caustic soda and the process of mass transfer is thus made much more complicated. Again, when carbon dioxide is absorbed in ethanolamine solution, there is direct chemical reaction between the amine and the gas. In such processes the conditions in the gas phase are similar to those already discussed, but in the liquid phase there is a liquid film followed by a reaction zone. The process of diffusion and chemical reaction can still be represented by an extension of the film theory by a method due to Hatta[37]. In the case considered, the chemical reaction is irreversible and of the type in which a solute gas **A** is absorbed from a mixture by a substance **B** in the liquid phase, which combines with **A** according to the equation $A + B \rightarrow AB$. As the gas approaches the liquid interface, it dissolves and reacts at once with **B**. The new product **AB**, thus formed, diffuses towards the main body of the liquid. The concentration of **B** at the interface falls; this results in diffusion of **B** from the bulk of the liquid phase to the interface. Since the chemical reaction is rapid, **B** is removed very quickly, so that it is necessary for the gas **A** to diffuse through part of the liquid film before meeting **B**. There is thus a zone of reaction between **A** and **B** which moves away

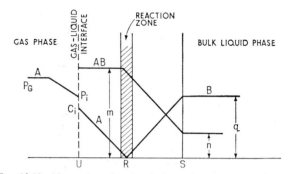

FIG. 12.10. Absorption with chemical reaction—concentration profiles

from the gas–liquid interface, taking up some position towards the bulk of the liquid. The final position of this reaction zone will be such that the rate of diffusion of **A** from the gas–liquid interface is equal to the rate of diffusion of **B** from the main body of the liquid. When this condition has been reached, the concentrations of **A**, **B**, and **AB** can be indicated as in Fig. 12.10, where the concentrations are shown as ordinates and the positions of a plane relative to the interface as abscissae. In the diagram, the plane of the interface between gas and liquid is shown by $U$, the reaction zone by $R$, and the outer boundary of liquid film by $S$. Then **A** diffuses through the gas film as a result of the driving force $P_G - P_i$, and diffuses to the reaction zone as a result of the driving force $C_i$ in the liquid phase. The component **B** diffuses from the main body of the liquid to the reaction zone under a driving force $q$, and the non-volatile product **AB** diffuses back to the main bulk of the liquid under a driving force $m - n$.

The difference between a physical absorption, and one in which a chemical reaction occurs, can also be shown by Figs. 12.11a and 12.11b, taken from a paper by van Krevelen and Hoftijzer[28]. Diagram $a$ shows the normal concentration profile for physical absorption whilst $b$ shows the profile modified by the chemical reaction. For transfer in the gas phase:

$$N'_A = k_G(P_G - P_i) \tag{12.40}$$

and in the liquid phase:

$$N'_A = k_L(C_i - C_L) \tag{12.41}$$

The effect of the chemical reaction is to accelerate the removal of **A** from the interface, and supposing that it is now $r$ times as great then:

$$N''_A = rk_L(C_i - C_L) \tag{12.42}$$

In Fig. 12.11a, the concentration profile through the liquid film of thickness $z_L$ is represented by a straight line such that $k_L = D_L/z_L$. In $b$, component **A** is removed by chemical reaction, so that the concentration profile is curved. The dotted line gives the concentration profile if—for the same rate of absorption—**A** were removed only by diffusion. The effective diffusion path is thus $1/r$ times the total film thickness $z_L$.

$$\therefore \quad N''_A = \frac{rD_L}{z_L}(C_i - C_L) = rk_L(C_i - C_L) \tag{12.43}$$

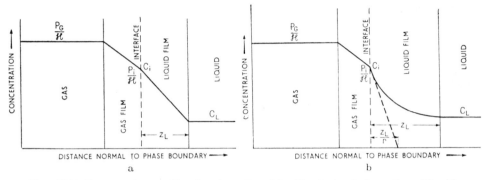

FIG. 12.11 Concentration profiles for absorption (a) without chemical reaction, (b) with chemical reaction

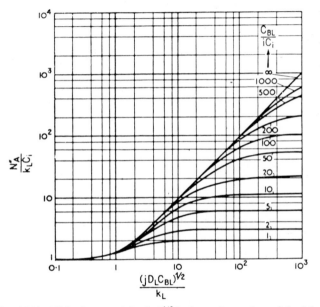

FIG. 12.12. $N''_A/k_L C_i$ versus $(jD_L C_{BL})^{1/2}/k_L$ for various values of $C_{BL}/iC_i$

van Krevelen has shown that this factor $r$ can be related to $C_i, D_L, k_L$, to the concentration of **B** in the bulk liquid $C_{BL}$, and to the reaction rate constant $j$. His suggested relationship is shown in Fig. 12.12, in which $r$, i.e. $N''_A/k_L C_i$, is plotted versus $(jD_L C_{BL})^{1/2}/k_L$ for various values of $C_{BL}/iC_i$ (where $i$ is the number of kmol of **B** combining with 1 kmol of **A**). This diagram brings out three conditions:

(a) If $j$ is very small, $r \simeq 1$, and conditions are those of physical absorption.

(b) If $j$ is very large, $r \simeq C_{BL}/iC_i$, and the rate of the process is determined by the transport of **B** towards the phase boundary.

(c) At moderate values of $j$, $r \simeq (jD_L C_{BL})^{1/2}/k_L$, and the rate of the process is determined by the rate of the chemical reaction.

Thus, from equation 12.43:

$$N''_A = k_L(C_i - C_L)\frac{(jD_LC_{BL})^{1/2}}{k_L} = (C_i - C_L)(jD_LC_{BL})^{1/2} \tag{12.44}$$

The controlling parameter is now $j$.

The above work is confirmed by Nijsing, Hendriksz, and Kramers[38].

As an illustration of combined absorption and chemical reaction, the results of Tepe and Dodge[39] on the absorption of carbon dioxide by sodium hydroxide solution may be considered. They used a 150 mm diameter tower filled to a depth of 915 mm with 12·5 mm carbon Raschig rings. Some of their results are indicated in Fig. 12.13. $K_Ga$

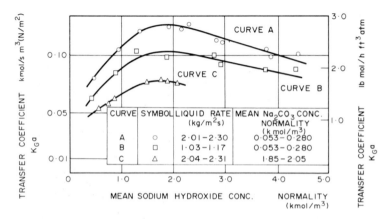

FIG. 12.13. Absorption of carbon dioxide in sodium hydroxide solution $G' = 0.24$–$0.25 \text{ kg/m}^2 \text{ s}$. Temperature $= 298 \text{ K}$

increases rapidly with increasing sodium hydroxide concentration up to a value of about 2 Normal. Changes in the gas rate were found to have negligible effect on $K_Ga$, indicating that the major resistance to absorption was in the liquid phase. The influence of the liquid rate was rather low, and corresponded to $L'^{0.28}$. It may be assumed that, in this case, the final rate of the process is controlled by the resistance to diffusion in the liquid, by the rate of the chemical reaction, or by both together.

Cryder and Maloney[40] have presented data on the absorption of carbon dioxide in diethanolamine solution, using a 203 mm tower filled with 19 mm rings. Some of their results are indicated in Fig. 12.14. The coefficient $K_Ga$ is found to be independent of the gas rate but to increase with the liquid rate, as expected in a process controlled by the resistance in the liquid phase.

It is very difficult at present to deduce the size of tower required for an absorption combined with a chemical reaction, and an experiment in some pilot plant should be carried out in all cases of this kind. Stephens and Morris[41] have suggested a small disc type tower illustrated in Fig. 12.15 which they have successfully used for preliminary experiments of this kind. They found that a simple wetted-wall column was unsatisfactory where chemical reactions took place. In their unit a series of discs, supported by means of a wire, is arranged one on top of the other as shown.

The absorption of carbon dioxide into aqueous amine solutions has been investigated

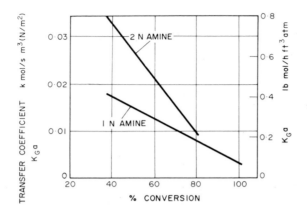

FIG. 12.14. Absorption of carbon dioxide in diethanolamine solutions
Liquid rate 1·85 kg/m² s

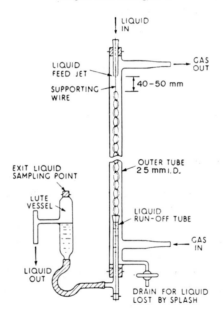

FIG. 12.15. Small disc tower for absorption tests

by Danckwerts and McNeil[42] using a stirred cell. They found that the reaction proceeded in two stages: first a fast reaction to give amine carbamate and secondly a slow reaction in the bulk of the liquid in which the carbamate was partially hydrolysed to bicarbonate. The use of sodium arsenite as catalyst considerably accelerated this second reaction. Hence the overall capacity of an absorber could be substantially increased by a suitable catalyst.

A comprehensive review of the work published to date on the absorption of carbon dioxide by alkaline solutions has been made by Danckwerts and Sharma[43] who have applied the results of recent research to the design of industrial scale equipment. More recently Sahay and Sharma[44] have shown that the mass transfer coefficient may be correlated with the gas and liquid rates and the gas and liquid compositions in the form:

$$K_G a = \text{const}.\ L'^{a_1} G'^{a_2} \exp(a_3 F' + a_4 y) \qquad (12.45)$$

where $a_1$, $a_2$, $a_3$, $a_4$ are experimentally determined constants,

$F' = $ fractional conversion of the liquid, and

$y = $ mol fraction of $CO_2$ in the gas.

Eckert[45], by using the same reaction, determined the mass transfer performance of packings in terms of $K_G a$ as:

$$K_G a = \frac{N}{V(\Delta P)_{lm}} \qquad (12.46)$$

where $N = $ number of mols of $CO_2$ absorbed,

$V = $ packed volume, and

$(\Delta P)_{lm} = $ log mean driving force.

Data obtained from the above work are limited by the conditions under which they were obtained and it is both difficult and dangerous to extrapolate over the entire range of conditions encountered on a full-scale plant.


## 12.7. ABSORPTION ACCOMPANIED BY THE LIBERATION OF HEAT

In some absorption processes, especially where a chemical reaction occurs, there is a liberation of heat. This generally gives rise to an increase in the temperature of the liquid, with the result that the position of the equilibrium curve is adversely affected.

In the case of plate columns, a heat balance may be performed over each plate and the resulting temperature determined. For adiabatic operation, where no heat is removed from the system, the temperature of the streams leaving the absorber will be higher than those entering due to the heat of solution. This rise in temperature lowers the solubility of the solute gas so that a larger value of $L_m/G_m$ and a larger number of trays will be required than for isothermal operation.

For packed columns, the temperature rise will affect the equilibrium curve and differential equations for heat and mass transfer, together with heat and mass balances, must be integrated numerically. An example of the procedure has been given in Chapter 11 of Volume 1 for the case of water cooling. For gas absorption under non-isothermal conditions, reference should be made to specialist texts[46,47] for a detailed description of the methods available. As an approximation, it is sometimes assumed that all the heat evolved is taken up by the liquid and the temperature rise of the gas is neglected. This method gives an overestimate of the rise in temperature of the liquid and results in the design of a tower which is taller than necessary. Figure 12.16 shows the effect of the temperature rise on the equilibrium curve for an adiabatic absorption process of ammonia in water. If the amount of heat liberated is very large, it may be necessary to cool the liquid. This is most conveniently done in a plate column, either with heat exchangers connected between consecutive plates, or with cooling coils on the plate, as shown in Fig. 12.17.

The overall heat transfer coefficient between the gas–liquid dispersion on the tray and the cooling medium in the tubes is dependent upon the gas velocity[48] but is usually in the range 500 to 2000 $W/m^2\,K$.

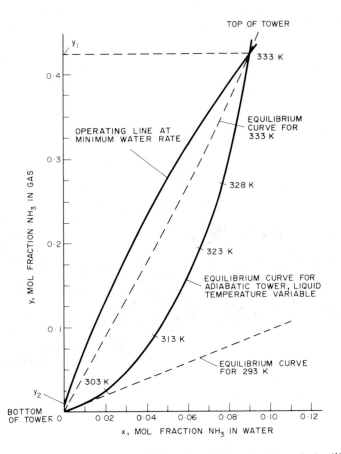

FIG. 12.16. Equilibrium curve modified to allow for heat of solution of solute[46]

With packed towers it is considerably more difficult to arrange for cooling and it is usually necessary to remove the liquid stream at intervals down the column and to cool externally. Coggan and Bourne[49] have presented a computer programme to enable the economic decision to be made between an adiabatic absorption tower or a smaller isothermal column with interstage cooling.

## 12.8. PACKED TOWERS

From the analysis given above of the diffusional nature of absorption, one of the outstanding requirements is to provide as large an interfacial area of contact as possible between the phases. For this purpose, columns similar to those used for distillation are suitable. In addition, equipment may be used in which gas is passed into a liquid which is agitated by a stirrer. A few special forms of units have also been used but it is the packed column which is most frequently used for gas absorption applications.

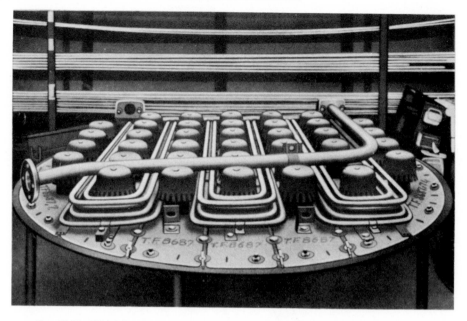

FIG. 12.17. Glitsch "truss type" bubble tray in stainless steel for 1·9 m absorption column

### 12.8.1. Construction

The essential features of a packed column as discussed in Chapter 4 are the shell, the arrangements for the gas and liquid inlets and outlets and the packing with its necessary supporting and redistributing systems. Reference should be made to Chapter 4 for details of these aspects whilst this section will largely be concerned with the determination of the height of packing for a particular duty. In installations where the gas is fed from a previous stage of a process where it is under pressure, there will be no need to use a blower for the transfer of the gas through the column. When this is not the case, a simple blower is commonly used; such blowers have been described in Chapter 6 of Volume 1. The pressure drop across the column may be calculated by the methods presented in Chapter 4 and the blower sized accordingly. A pressure drop exceeding 30 mm of water per metre of packing is said to improve gas distribution though process conditions may not permit a figure as high as this. The packed height should not normally exceed 6 m in any section of the tower and for some packings a much lower height must be used.

In the design of an absorption tower it is necessary to take into account the characteristics of the packing elements and the flow behaviour discussed in Chapter 4, together with the considerations given in the following sections concerning the performance of columns under operating conditions.

### 12.8.2. Mass Transport Coefficients and Specific Area in Packed Towers

Traditional methods of assessing the capacity of tower packings, which involved the use of the specific surface area $S$ and the voidage $e$, developed from the fact that these properties could be readily defined and measured for a packed bed of granular material such as granite, limestone, and coke which were some of the earliest forms of tower packings. The values of $S$ and $e$ enabled a reasonable prediction of hydraulic performance to be made. With the introduction of Raschig rings and other specially shaped packings, it was necessary to introduce a basis for comparing their relative efficiencies. Although the commonly published values of specific surface area $S$ provided a reasonable basis of comparison, papers such as that by Shulman et al.[50] in the early 1950s showed that the total area offered by Raschig rings was not used and varied considerably with hydraulic loading.

Further evidence of the importance of the wetted fraction of the total area came with the introduction of the Pall type ring. A Pall ring having the same surface area as a Raschig ring is up to 60 per cent more efficient, though many engineers still argue the relative merits of packings purely on the basis of surface area.

The selection of a tower packing is based on its hydraulic capacity, which determines the required cross-sectional area of the tower, and efficiency ($K_G a$ typically) which governs the packing height. Here $a$ is the area of surface per unit volume of column and is therefore equal to $S(1-e)$. Table 12.4[51] shows the capacity of the commonly available tower packings relative to 25 mm Raschig rings, for which a considerable amount of information is published in the literature. The table lists the packings in order of relative efficiency ($K_G a$) evaluated at the same approach to the hydraulic capacity limit (flooding) in each case.

### 12.8.3. Capacity of Packed Towers

The drop in pressure for the flow of gas and liquid over packings has been discussed in Chapter 4. It is important to see that, during operation, the tower does not reach flooding conditions. In addition, every effort should be made to have as high a liquid rate as possible, in order to attain satisfactory wetting of the packing.

With low liquid rates, the whole of the surface of the packing is not completely wetted. This can be seen very readily by allowing a coloured liquid to flow over packing contained in a glass tube. From the flow patterns, it is at once obvious how little of the surface is wetted until the rate is quite high. This difficulty of wetting can sometimes be overcome by having considerable recirculation of the liquid over the tower, but in other cases, such as vacuum distillation, poor wetting will have to be accepted because of the low volume of liquid available. In selecting a packing, it is desirable to choose the form which will give as near complete wetting as possible. The minimum liquid rate below which the packing will no longer perform satisfactorily is known as the minimum wetting rate, which has been discussed in Chapter 4.

Figure 12.18 illustrates the conditions that occur during the steady operation of a countercurrent gas–liquid absorption tower. It is convenient to express the concentrations of the streams in terms of mols of solute gas per mol of inert gas in the gas phase, and as mols of solute gas per mol of solute free liquid in the liquid phase. The actual area of interface between the two phases is not known, and the term $a$ is introduced as the interfacial area per unit volume of the column. On this basis the general equation (12.13) for mass transfer can be written as:

$$N'_A A \,dZ a = k_G a (P_G - P_i) A \,dZ$$
$$= k_L a (C_i - C_L) A \,dZ \tag{12.47}$$

TABLE 12.4. *Capacity of Commonly Available Packings Relative to 25 mm Raschig Rings* (Packing Factors given under size of packing) (1)

| Relative $K_G a$ | Raschig rings | Traditional saddles | Pall rings | Ceramic Pall rings | Ceramic cascade mini-ring (4) | Super Intalox saddles | Hypak (2) | Tellerettes (3) | Cascade mini-ring (4) |
|---|---|---|---|---|---|---|---|---|---|
| Materials available for this relative $K_G a$ | Ceramic | Ceramic | Metal (M) Plastic (P) | Ceramic | Ceramic | Ceramic Plastic | Metal | Plastic | Metal (M) Plastic (P) |
| 0·6–0·7 | 75 mm 32–36 | | | | | | | | |
| 0·7–0·8 | 50 mm 57–65 | | | | | | | | |
| 0·8–0·9 | 37 mm 83–95 | | | | | | | | |
| 0·9–1·0 | 25 mm 115–160 | | | | | | | | |
| 1·0–1·1 | 12 mm 300–400 | 75 mm 22 | 87 mm 16 | | No. 5 18 | No. 3 16 | | Size L 16 | |
| 1·1–1·2 | | 50 mm 40 | 50 mm 20 (M) | | | | No. 3 15 | | |
| 1·2–1·3 | | 37 mm 52 | 50 mm 25 (P) | | No. 3 24 | No. 2 21 | | | |
| 1·3–1·4 | | 25 mm 90 | 37 mm 28–30 | 50 mm 44 | | | No. 2 18 | | No. 4 10 (M) |
| 1·4–1·5 | | | | 37 mm 68 | | | | | No. 3 12 (P) |
| 1·5–1·6 | | | 25 mm 41 (M) | 25 mm 98 | No. 2 38 | No. 1 33 | | | |
| 1·6–1·7 | | | 25 mm 48 (P) | | | | No. 1 43 | Size S 36 | No. 3 14 (M) |
| 1·7–1·8 | | | 16 mm 71 | | | | | | No. 2 15 (P) |

| | |
|---|---|
| 1·8–1·9 | No. 2 22 (M) |
| 1·9–2·0 | No. 1 25 (P) |
| 2·0–2·1 | |
| 2·1–2·2 | No. 1 34 (M) |

Gas capacity before hydraulic limit (flooding) relative to 25 mm Raschig rings (also approx. the reciprocal of tower cross-sectional area relative to 25 mm Raschig rings for the same pressure drop throughout loading range). All relative capacity figures are valid for the same liquid to gas mass rate ratio:

(1) See Table 4.3.
(2) Trade Mark of Norton Company, U.S.A. (Hydronyl U.K.).
(3) Trade Mark of Ceilcote Company.
(4) Trade Mark of Mass Transfer Ltd. (& Inc.).

*Note:*

Relative $K_G a$ valid for all systems controlled by mass transfer coefficient ($K_G$) and wetted area (a) per unit volume of column. Some variation should be expected when liquid *reaction* rate is controlling (not liquid *diffusion* rate). In these cases liquid hold-up becomes more important. In general a packing having high liquid hold-up which is clearly greater than that in the falling film has poor capacity.

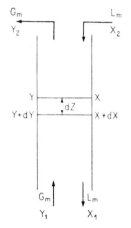

FIG. 12.18. Countercurrent absorption tower

where $N'_A$ = kmol of solute absorbed per unit time and unit interfacial area,
   $a$ = surface area of interface per unit volume of column,
   $A$ = cross-sectional area of column, and
   $Z$ = height of packed section.

$$\text{The interfacial area for transfer} = a\,dV = aA\,dZ \qquad (12.48)$$

### 12.8.4. Height of Column Based on Conditions in Gas Film

Let $G_m$ = mols of inert gas/(unit time) (unit cross-section of tower),
   $L_m$ = mols of solute-free liquor/(unit time) (unit cross-section of tower),
   $Y$ = mols of solute gas **A**/mol of inert gas **B** in gas phase, and
   $X$ = mols of solute **A**/mol of inert solvent in liquid phase.

Consider any plane at which the molar ratios of the diffusing material in the gas and liquid phases are $Y$ and $X$. Then over a small height $dZ$, the mols of gas leaving the gas phase will equal the mols taken up by the liquid. This gives:

$$AG_m\,dY = AL_m\,dX \qquad (12.49)$$

But

$$G_m A\,dY = N'_A(a\,dV) = k_G a[P_G - P_i]A\,dZ \qquad (12.50)$$

Since

$$P_G = \frac{Y}{1+Y}P$$

$$G_m\,dY = k_G aP\left[\frac{Y}{1+Y} - \frac{Y_i}{1+Y_i}\right]dZ$$

$$= k_G aP\left[\frac{Y-Y_i}{(1+Y)(1+Y_i)}\right]dZ$$

Hence the height of column $Z$ required to achieve a change in $Y$ from $Y_1$ at the bottom to $Y_2$ at the top of the column is given by:

$$\int_0^Z dZ = Z = \frac{G_m}{k_G aP} \int_{Y_2}^{Y_1} \frac{(1+Y)(1+Y_i)\,dY}{Y - Y_i} \qquad (12.51)$$

which for weak mixtures can be written as:

$$Z = \frac{G_m}{k_G aP} \int_{Y_2}^{Y_1} \frac{dY}{Y - Y_i} \qquad (12.52)$$

In this analysis it has been assumed that $k_G$ is a constant throughout the column, and provided the concentration changes are not too large this will be reasonably true.

### 12.8.5. Height of Column Based on Conditions in Liquid Film.

A similar analysis may be made in terms of the liquid film. Thus from equations 12.47 to 12.49:

$$AL_m\,dX = k_L a(C_i - C_L)A\,dZ \qquad (12.53)$$

where the concentrations $C$ are in terms of mols of solute per unit volume of liquor. If $C_T = $ (mols of solute + solvent)/(volume of liquid)

$$\frac{C}{C_T - C} = \frac{\text{mols of solute}}{\text{mols of solvent}} = X$$

whence

$$C = \frac{X}{1+X} C_T \qquad (12.54)$$

The transfer equation (12.53) may now be written:

$$L_m\,dX = k_L a C_T \left[ \frac{X_i}{1+X_i} - \frac{X}{1+X} \right] dZ$$

$$= k_L a C_T \left[ \frac{X_i - X}{(1+X_i)(1+X)} \right] dZ$$

$$\therefore \int_0^Z dZ = Z = \frac{L_m}{k_L a C_T} \int_{X_2}^{X_1} \frac{(1+X_i)(1+X)\,dX}{X_i - X} \qquad (12.55)$$

and for dilute concentrations this gives:

$$Z = \frac{L_m}{k_L a C_T} \int_{X_2}^{X_1} \frac{dX}{X_i - X} \qquad (12.56)$$

where $C_T$ and $k_L$ have been taken as constant over the column.

### 12.8.6. Height Based on Overall Coefficients

If the driving force based on the gas concentration is written as $Y - Y_e$ and the overall gas transfer coefficient as $K_G$, then the height of the tower for dilute concentrations becomes:

$$Z = \frac{G_m}{K_G a P} \int_{Y_2}^{Y_1} \frac{dY}{Y - Y_e} \tag{12.57}$$

or in terms of the liquor concentration as:

$$Z = \frac{L_m}{K_L a C_T} \int_{X_2}^{X_1} \frac{dX}{X_e - X} \tag{12.58}$$

*Equations for Dilute Concentrations*

As the mol fraction is approximately equal to the molar ratio at dilute concentrations then considering the gas film:

$$Z = \frac{G_m}{K_G a P} \int_{Y_2}^{Y_1} \frac{dY}{Y - Y_e} = \frac{G_m}{K_G a P} \int_{y_2}^{y_1} \frac{dy}{y - y_e} \tag{12.59}$$

and considering the liquid film:

$$Z = \frac{L_m}{K_L a C_T} \int_{X_2}^{X_1} \frac{dX}{X_e - X} = \frac{L_m}{K_L a C_T} \int_{x_2}^{x_1} \frac{dx}{x_e - x} \tag{12.60}$$

### 12.8.7. The Operating Line and Graphical Integration for Height of Column

Taking a material balance on the solute from the bottom of the column to any plane where the mol ratios are $Y$ and $X$ gives for unit area of cross-section:

$$G_m(Y_1 - Y) = L_m(X_1 - X) \tag{12.61}$$

or

$$Y_1 - Y = \frac{L_m}{G_m}(X_1 - X) \tag{12.62}$$

This is the equation of a straight line of slope $L_m/G_m$, which passes through the point $(X_1, Y_1)$. It can be seen by writing a material balance over the whole column that the same line passes through the point $(X_2, Y_2)$. This line is known as the operating line and represents the conditions at any point in the column; it is similar to the operating line used in Chapter 11. Figure 12.19 illustrates typical conditions for the case of moist air and sulphuric acid or caustic soda solution, where the main resistance lies in the gas phase.

The equilibrium curve is represented by the line $FR$, and the operating line is given by $AB$, $A$ corresponding to the concentrations at the bottom of the column and $B$ to those at the top of the column.

$D$ represents the condition of the bulk of the liquid and gas at any point in the column, and has coordinates $X$ and $Y$. Then, if the gas film is controlling the process, $Y_i$ equals $Y_e$, and is given by a point $F$ on the equilibrium curve, with coordinates $X$ and $Y_i$. The driving force causing transfer is then given by the distance $DF$. It is therefore possible to evaluate

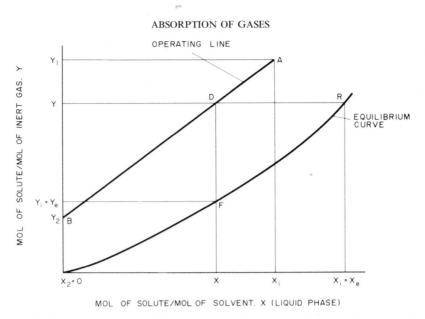

FIG. 12.19. Driving force in gas and liquid-film controlled processes. Diagram shows operating line BDA and equilibrium curve FR

the expression

$$\int_{Y_2}^{Y_1} \frac{dY}{Y - Y_i}$$

by selecting values of $Y$, reading off from the diagram the corresponding values of $Y_i$, and thus calculating $1/(Y - Y_i)$.

If the liquid film controls the process, $X_i$ equals $X_e$ and the driving force $X_i - X$ is given in the diagram by the line $DR$. Thus, the evaluation of

$$\int_{X_2}^{X_1} \frac{dX}{X_i - X}$$

may be effected in the same way as for the gas film.

### Special Case when Equilibrium Curve is a Straight Line

If over the range of concentrations considered the equilibrium curve is a straight line, it is permissible to use an average value of the driving force over the column. For dilute concentrations, over a small height $dZ$ of column, the absorption is given by:

$$N'_A A a \, dZ = G_m A \, dy = K_G a A P (y - y_e) dZ \tag{12.63}$$

If

$$y_e = mx + c \tag{12.64}$$

then:

$$y_{e2} = mx_2 + c$$

and

$$y_{e1} = mx_1 + c$$

so that

$$m = \frac{y_{e2} - y_{e1}}{x_2 - x_1} \qquad (12.65)$$

Further, taking a material balance over the lower portion of the columns gives:

$$L_m(x - x_1) = G_m(y - y_1)$$

and

$$x = x_1 + \frac{G_m}{L_m}(y - y_1) \qquad (12.66)$$

From equation 12.63:

$$\int_{y_2}^{y_1} \frac{dy}{y - y_e} = \int_0^Z \frac{K_G aP}{G_m} dZ \qquad (12.67)$$

$$= \int_{y_2}^{y_1} \frac{dy}{y - m[x_1 + (G_m/L_m)(y - y_1)] - c}$$

(from equations 12.64 and 12.66)

$$= \frac{1}{1 - (mG_m/L_m)} \ln \frac{y_1 - mx_1 - c}{y_2 - m[x_1 + (G_m/L_m)(y_2 - y_1)] - c}$$

$$= \frac{1}{1 - \dfrac{y_{e1} - y_{e2}}{x_1 - x_2} \cdot \dfrac{x_1 - x_2}{y_1 - y_2}} \ln \frac{(y - y_e)_1}{(y - y_e)_2}$$

(from equations 12.64, 12.65, and 12.66)

$$= \frac{y_1 - y_2}{(y - y_e)_1 - (y - y_e)_2} \ln \frac{(y - y_e)_1}{(y - y_e)_2}$$

$$= \frac{y_1 - y_2}{(y - y_e)_{lm}}$$

where $(y - y_e)_{lm}$ is the logarithmic mean value of $y - y_e$.

Substituting in equation 12.67:

$$\frac{K_G aP}{G_m} Z = \frac{y_1 - y_2}{(y - y_e)_{lm}}$$

$$\therefore \quad aAZN'_A = G_m(y_1 - y_2)A = K_G aAP(y - y_e)_{lm} Z \qquad (12.68)$$

and in terms of mol ratios:

$$aAZN'_A = G_m(Y_1 - Y_2)A = K_G aAP(Y - Y_e)_{lm} Z \qquad (12.69)$$

Thus, the logarithmic mean of the driving forces at the top and the bottom of the column may be used.

For concentrated solutions:

$$aAZN'_A = G_m(Y_1 - Y_2)A = K_G aA\phi P(Y - Y_e)_{lm} Z \qquad (12.70)$$

It is necessary to introduce the factor $\phi$ since $Y$ is not directly proportional to $P$. The value of $\phi$ may be found from the relation:

$$\phi Y = \frac{Y}{1 + Y} \qquad (12.71)$$

whence $\phi = 1/(1 + Y)$. Although the value of $\phi$ will change slightly over the column, an average value will generally be sufficiently accurate.

It is of interest to note from Fig. 12.20, that, as long as the ratio $k_L/k_G$ remains constant

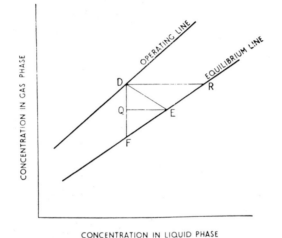

CONCENTRATION IN LIQUID PHASE

FIG. 12.20. Driving force when equilibrium curve is a straight line

(i.e. if the slope of $DE$ is constant), then the ratio of $DQ$, the driving force through the gas phase, divided by $DF$, the driving force assuming all the resistance to be in the gas phase, will be a constant. Thus, the use of the driving force $DF$ is satisfactory even if the resistance does not lie wholly in the gas phase. The coefficient $k_G$ on this basis is not an accurate value for the gas film coefficient, but is proportional to it. It follows that, if the equilibrium curve is straight, either the gas film or the liquid film coefficient may be used. This simplification is of considerable value.

### 12.8.8. Capacity of Tower in Terms of Partial Pressures for High Concentrations

A material balance taken between the bottom of the column and some plane where the partial pressure in the gas phase is $P_G$ and the concentration in the liquid is $X$ will give:

$$L_m(X_1 - X) = G_m \left[ \frac{P_{G1}}{P - P_{G1}} - \frac{P_G}{P - P_G} \right] \qquad (12.72)$$

Over a small height of the column $dZ$, therefore:

$$L_m\,dX = \frac{G_m P}{(P-P_G)^2}\,dP_G = k_G a(P_G - P_i)\,dZ \tag{12.73}$$

$$= \frac{k_G P_{Bm}}{P}\,aP\frac{(P_G - P_i)}{P_{Bm}}\,dZ$$

$$= k'_G aP\frac{(P_G - P_i)}{P_{Bm}}\,dZ$$

$$\therefore \quad \int_0^Z dZ = G_m \int_{P_{G_2}}^{P_{G_1}} \frac{P_{Bm}\,dP_G}{k'_G a(P-P_G)^2(P_G - P_i)} \tag{12.74}$$

The advantage of using $k'_G$ instead of $k_G$ is that $k'$ is independent of concentration, but this equation is almost unmanageable in practice. If a substantial amount of the gas is absorbed from a concentrated mixture, $k'_G$ will still change as a result of a reduced gas velocity, although it is independent of concentration.

### 12.8.9. The Transfer Unit

The group $\int(dy/y - y_e)$, which was used in Chapter 11, has been defined by Chilton and Colburn[52] as the number of overall gas transfer units $N_{OG}$. The application of this group to the countercurrent conditions in the absorption tower is given below.

Over a small height $dZ$, the partial pressure of the diffusing component $A$ will change by an amount $dP_G$. Then the mols of $A$ transferred are given by:

(change in mol fraction) $\times$ (total mols of gas)

Therefore

$$K_G a(P_G - P_e)\,dZ = \frac{dP_G}{P}G_m \tag{12.75}$$

(for dilute concentrations)

$$\therefore \quad \int_{P_{G_2}}^{P_{G_1}} \frac{dP_G}{P_G - P_e} = \int_0^Z \frac{K_G aP}{G_m}\,dZ \tag{12.76}$$

or in terms of mol fractions:

$$N_{OG} = \int_{y_2}^{y_1} \frac{dy}{y - y_e} = \int_0^Z K_G a\frac{P}{G_m}\,dZ = K_G a\frac{P}{G_m}Z \tag{12.77}$$

The number of overall gas transfer units $N_{OG}$ is an integrated value of the change in composition per unit driving force, and therefore represents the difficulty of the separation.

In many cases in gas absorption, $y - y_e$ is very small at the top of the column, and consequently $1/(y - y_e)$ is very much greater at the top than at the bottom of the column. Thus, equation 12.77 may lead to the use of an integral which is difficult to evaluate graphically because of the very steep slope of the curve. However:

$$\mathbf{N}_{OG} = \int_{y_2}^{y_1} \frac{dy}{y - y_e} = \int_{y_2}^{y_1} \frac{y\, d(\ln y)}{y - y_e} \qquad (12.78)$$

In these circumstances, the new form of the integral is much more readily evaluated, as pointed out by Rackett[53].

As in Chapter 11 (equation 11.156), equation 12.77 may be written as:

$$\mathbf{N}_{OG} = \frac{\text{Height of column}}{\text{Height of transfer unit}} = \frac{Z}{\mathbf{H}_{OG}} \qquad (12.79)$$

Then the height of the overall gas transfer unit $\mathbf{H}_{OG} = \dfrac{G_m}{PK_G a} \qquad (12.80)$

If the driving force is taken over the gas film only, the height of a gas-film transfer unit $\mathbf{H}_G = G_m/Pk_G a$ is obtained. Similarly for the liquid film, the height of the overall liquid transfer unit $\mathbf{H}_{OL}$ is given by:

$$\mathbf{H}_{OL} = \frac{L_m}{K_L a C_T} \qquad (12.81)$$

The height of the liquid film transfer unit is given by:

$$\mathbf{H}_L = \frac{L_m}{k_L a C_T} \qquad (12.82)$$

where $C_T$ is the mean molar density of the liquid.

In this analysis, it has been assumed that the total number of mols of gas and liquid remain the same. This is true in absorption only when a small change in concentration takes place. With distillation, the total number of mols of gas and liquid does remain approximately constant so that no difficulty arises in that case. In Chapter 11, the following relationships between individual and overall heights of transfer units were obtained and methods of obtaining the values of $\mathbf{H}_G$ and $\mathbf{H}_L$ were discussed:

$$\mathbf{H}_{OG} = \mathbf{H}_G + \frac{mG'}{L'} \mathbf{H}_L \qquad \text{(equation 11.163)}$$

$$\mathbf{H}_{OL} = \mathbf{H}_L + \frac{L'}{mG'} \mathbf{H}_G \qquad \text{(equation 11.164)}$$

For absorption duties, Semmelbauer[29] presented the following equations to evaluate $\mathbf{H}_G$ and $\mathbf{H}_L$ for Raschig rings and Berl saddles:

$$\mathbf{H}_G = \beta \left[ \frac{G'^{0.41} \mu_G^{0.26} \mu_L^{0.46} \sigma^{0.5}}{L^{0.46} \rho_G^{0.67} \rho_L^{0.5} D_G^{0.67} d_p^{0.05}} \right] \qquad (12.83)$$

$$\mathbf{H}_L = \beta \left[ \frac{\mu_L^{0.88} \sigma^{0.5}}{L'^{0.05} \rho_L^{1.33} D_L^{0.5} d_p^{0.55}} \right] \qquad (12.84)$$

where $\beta = 30$ and 21 for Raschig rings and Berl saddles respectively.

The limits of validity and the units for the terms in equations 12.83 and 12.84 are given in Table 12.5.

TABLE 12.5.  *Range of Application of Equations 12.83 and 12.84*

| $L'$ | 0·1–10 | kg/m$^2$ s | | $\mu_L$ | 0·2–2 | mN s/m$^2$ |
|---|---|---|---|---|---|---|
| $G'$ | 0·1–1·0 | kg/m$^2$ s | | $\mu_G$ | 0·005–0·03 | mN s/m$^2$ |
| $d_p$ | 0·006–0·06 | m | | $\sigma$ | $(20–200) \times 10^{-3}$ | J/m$^2$ |
| $\rho_L$ | 600–1400 | kg/m$^3$ | | $T$ | 273–373 | K |
| $\rho_G$ | 0·4–4 | kg/m$^3$ | | $d/d_p$ | 2·5–25 | — |
| $D_L$ | $(0·3–3·0) \times 10^{-4}$ | m$^2$/s | | $h_p/d_p$ | 10–100 | — |
| $D_G$ | $(0·3–3·0) \times 10^{-5}$ | m$^2$/s | | | | |

TABLE 12.6.  *Height of a Transfer Unit for Various Packings*

| Material | Size (mm) | | | Height of a transfer unit (m) | |
|---|---|---|---|---|---|
| Grids | Pitch | Height | Thickness | $H_G$ | $H_L$ |
| **Plain grids** | | | | | |
| Metal | 25 | 25 | 1·6 | 1 | 0·5 |
| | 25 | 50 | 1·6 | 1·2 | 0·6 |
| Wood | 25 | 25 | 6·4 | 0·9 | 0·5 |
| | 25 | 50 | 6·4 | 1·2 | 0·6 |
| **Serrated grids** | | | | | |
| Wood | 100 | 100 | 13 | 6·8 | 0·7 |
| | 50 | 50 | 9·5 | 1·8 | 0·6 |
| | 38 | 38 | 4·8 | 1·6 | 0·6 |
| **Solid material** | nominal size | | | | |
| Coke | 75 | | | 0·7 | 0·9 |
| | 38 | | | 0·25 | 0·8 |
| | 25 | | | 0·2 | 0·7 |
| Quartz | 50 | | | 0·5 | 0·8 |
| | 25 | | | 0·16 | 0·8 |

| | Diameter | Height | Thickness | | |
|---|---|---|---|---|---|
| **Stacked Raschig rings** | | | | | |
| Stoneware | 100 | 100 | 9·5 | 1·8 | 0·7 |
| | 75 | 75 | 9·5 | 1·1 | 0·6 |
| | 75 | 75 | 6·4 | 1·4 | 0·6 |
| | 50 | 50 | 6·4 | 0·7 | 0·6 |
| | 50 | 50 | 4·8 | 0·8 | 0·6 |
| **Random Raschig rings** | | | | | |
| Metal | 50 | 50 | 1·6 | 0·5 | 0·6 |
| | 25 | 25 | 1·6 | 0·2 | 0·5 |
| | 13 | 13 | 0·8 | 0·1 | 0·5 |
| | 75 | 75 | 9·5 | 0·8 | 0·7 |
| | 50 | 50 | 6·4 | 0·5 | 0·6 |
| Stoneware | 50 | 50 | 4·8 | 0·5 | 0·6 |
| | 38 | 38 | 4·8 | 0·3 | 0·6 |
| | 25 | 25 | 2·5 | 0·2 | 0·5 |
| | 19 | 19 | 2·5 | 0·15 | 0·5 |
| | 13 | 13 | 1·6 | 0·1 | 0·5 |
| Carbon | 50 | 50 | 6·4 | 0·5 | 0·6 |
| | 25 | 25 | 4·8 | 0·2 | 0·5 |
| | 13 | 13 | 3·2 | 0·1 | 0·5 |

$H_G$ ft

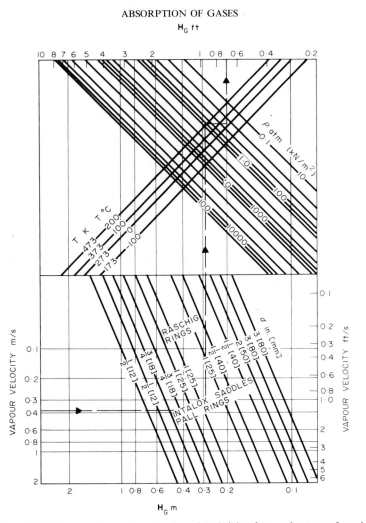

FIG. 12.21. Nomograph for the estimation of the height of a gas phase transfer unit

For a range of packings, Morris and Jackson[21] have presented values of the heights of the individual film transfer units as shown in Table 12.6. For Pall rings and Intalox saddles, the nomographs in Figs 12.21 and 12.22 may be used[54] though Fig. 12.22 must not be used to estimate $H_L$ for distillation applications. Table 11.10 gives the value as a function of size and type of packing. It is, however, satisfactory for absorption and stripping duties.

### Concentrated Solutions

With concentrated solutions, allowance must be made for the change in the total number of mols flowing, because the molar flow will decrease up the column if the amount of absorption is large. The following equation must be used in place of equation 12.75:

$$aA\, dZ N'_A = A\, d\left[\frac{P_G}{P} G_m\right] \qquad (12.85)$$

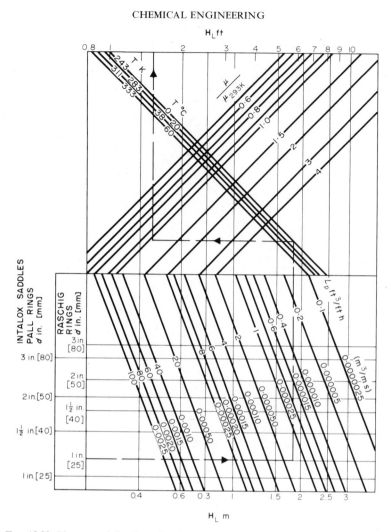

FIG. 12.22. Nomograph for the estimation of the height of a liquid phase transfer unit

Colburn[55] has shown that, under these conditions, the number of transfer units is given by:

$$N_{OG} = \int_{y_2}^{y_1} \frac{dy}{y - y_e} \frac{(1-y)_{lm}}{1-y}$$

(12.86)

where $(1-y)_{lm}$ is the logarithmic mean of $1-y$ and $1-y_i$.

### 12.8.10. The Importance of Liquid and Gas Flowrates and the Slope of the Equilibrium Curve

For a packed tower operating with dilute concentrations, since $x \simeq X_1$ and $y \simeq Y_1$:

$$G_m(y_1 - y_2) = L_m(x_1 - x_2)$$

(12.87)

where, as before, $x$ and $y$ are the mol fractions of solute in the liquid and gas phases, and $G_m$ and $L_m$ are the gas and liquid molar flowrates per unit area on a solute free basis.

A material balance between the top and some plane where the mol fractions are $x$, $y$ gives:

$$G_m(y - y_2) = L_m(x - x_2) \tag{12.88}$$

If the entering solvent is free from solute, then $x_2 = 0$ and:

$$x = \frac{G_m}{L_m}(y - y_2) \tag{12.89}$$

But the number of overall transfer units is given by:

$$\mathbf{N}_{OG} = \int_{y_2}^{y_1} \frac{dy}{y - y_e}$$

For dilute concentrations, Henry's Law holds and $y_e = mx$. Therefore:

$$\mathbf{N}_{OG} = \int_{y_2}^{y_1} \frac{dy}{y - \dfrac{mG_m}{L_m}(y - y_2)}$$

$$= \int_{y_2}^{y_1} \frac{dy}{y\left[1 - \dfrac{mG_m}{L_m}\right] + \dfrac{mG_m}{L_m}y_2}$$

$$\therefore \quad \mathbf{N}_{OG} = \frac{1}{1 - \dfrac{mG_m}{L_m}} \ln\left[\left(1 - \frac{mG_m}{L_m}\right)\frac{y_1}{y_2} + \frac{mG_m}{L_m}\right] \tag{12.90}$$

Colburn[55] has shown that this equation can usefully be plotted as shown in Fig. 12.23 which is taken from his paper. In this plot the number of transfer units $\mathbf{N}_{OG}$ is shown for values of $y_1/y_2$ using $mG_m/L_m$ as a parameter and it will be seen that the greater $mG_m/L_m$ is, the greater will be the value of $\mathbf{N}_{OG}$ for a given ratio of $y_1/y_2$. From equation 12.89:

$$\frac{L_m}{G_m} = \frac{y_1 - y_2}{x_1} = \frac{y_1 - y_2}{y_{e1}/m}$$

$$\therefore \quad \frac{mG_m}{L_m} = \frac{y_{e1}}{y_1 - y_2}$$

where $y_{e1}$ is the value of $y$ in equilibrium with $x_1$.

On this basis, the lower the value of $mG_m/L_m$, the lower will be $y_{e1}$, and hence the weaker the exit liquid. Colburn has suggested that the economic range for $mG_m/L_m$ lies from 0·7 to 0·8. If the value of $\mathbf{H}_{OG}$ is known, the quickest way of obtaining a good indication of the required height of the column is by using Fig. 12.23.

## Example 12.1

Gas, from a petroleum distillation column, has its concentration of $H_2S$ reduced from 0·03 kmol $H_2S$ per kmol of inert hydrocarbon gas to 1 per cent of this value by scrubbing

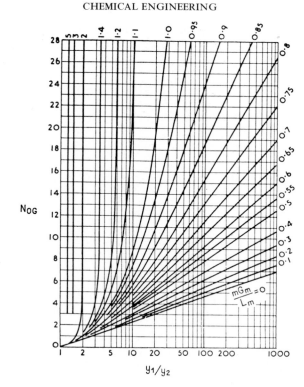

FIG. 12.23. Number of transfer units $N_{OG}$ as a function of $y_1/y_2$, with $mG_m/L_m$ as parameter

with a triethanolamine–water solvent in a countercurrent tower, operating at 300 K and at atmospheric pressure. The equilibrium relation for the solution may be taken as $Y_e = 2X$.

The solvent enters the tower free of $H_2S$ and leaves containing 0·013 kmol of $H_2S$ per kmol of solvent. If the flow of inert gas is 0·015 kmol/s m² of tower cross-section, calculate:

(a) the height of the absorber necessary, and

(b) the number of transfer units $N_{OG}$ required.

The overall coefficient for absorption $K''_G a$ may be taken as 0·04 kmol/s m³ (unit mol fraction driving force).

**Solution**

Driving force at top of column $= Y_2 - Y_{2e} = 0·0003$
Driving force at bottom of column $= Y_1 - Y_{1e} = 0·03 - 0·026 = 0·004$

Logarithmic mean driving force $= \dfrac{0·004 - 0·0003}{\ln \dfrac{0·004}{0·0003}}$

$$= \frac{0·0037}{2·59} = 0·00143$$

From equation 12.68

$$G_m(Y_1 - Y_2)S = K_G a P(Y - Y_e)_{\mathrm{lm}} SZ$$

i.e.

$$G_m(Y_1 - Y_2) = K_G'' a(Y - Y_e)_{lm} Z$$

$$\therefore \quad 0{\cdot}015(0{\cdot}03 - 0{\cdot}0003) = 0{\cdot}04 \times 0{\cdot}00143Z$$

$$\therefore \quad Z = \frac{0{\cdot}000446}{0{\cdot}0000572} = 7{\cdot}79 = 7{\cdot}8 \text{ m (say)}$$

Height of transfer unit $\mathbf{H}_{OG} = \dfrac{G_m}{K_G'' a}$

$$= \frac{0{\cdot}015}{0{\cdot}04} = 0{\cdot}375 \text{ m}$$

Number of transfer units $\mathbf{N}_{OG} = \dfrac{7{\cdot}79}{0{\cdot}375} = 20{\cdot}7 = \underline{\underline{21}} \text{ (say)}$

## 12.9. PLATE TOWERS

Bubble-cap columns or sieve trays, of similar construction to those described in the previous chapter for distillation, are sometimes used for gas absorption, particularly when the load is more than can be handled in a packed tower of about 1 m diameter and when there is any likelihood of deposition of solids which would quickly choke a packing. Plate towers are particularly useful when the liquid rate is sufficient to flood a packed tower. Since the ratio of liquid rate to gas rate is greater than with distillation, the slot area will be rather less and the downcomers rather larger. On the whole, plate efficiencies have been found to be less than with the distillation equipment, and to range from 20 to 80 per cent.

The plate column is a common type of equipment for large installations, but when the diameter of the column is less than 2 m, packed columns are more often used; for the handling of very corrosive fluids, packed columns are frequently preferred for larger units. The essential arrangement of such a unit is indicated in Fig. 12.24, where

$L_m$ is the molar rate of flow per unit area of solute free liquid,

$G_m$ is the molar rate of flow per unit area of inert gas,

$n$ refers to the plate numbered from the top downwards (and suffix $n$ refers to material leaving plate $n$),

$x$ is the mol fraction of the absorbed component in the liquid,

$y$ is the mol fraction of the absorbed component in the gas, and

$s$ is the total number of plates in the column.

It will be assumed that dilute solutions are used so that mol fractions and mol ratios are approximately equal. Each plate will be taken as an "ideal" unit, so that the gas leaving of composition $y_n$ is in equilibrium with the liquid of composition $x_n$ leaving the plate.

A material balance for the absorbed component from the bottom to a plane above plate $n$ will give:

$$G_m y_n + L_m x_s = G_m y_{s+1} + L_m x_{n-1} \tag{12.91}$$

$$\therefore \quad y_n = \frac{L_m}{G_m} x_{n-1} + y_{s+1} - \frac{L_m}{G_m} x_s \tag{12.92}$$

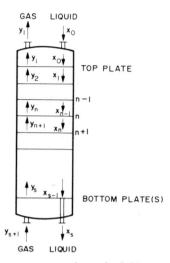

FIG. 12.24. Plate tower—nomenclature for fluid streams

This is the equation of a straight line of slope $L_m/G_m$, relating the composition of the gas entering a plate to the liquid leaving the plate, and is known as the *operating line*. Such a line passes through two points $A(x_s, y_{s+1})$ and $B(x_0, y_1)$ (Fig. 12.25) representing the terminal concentrations in the column. The equilibrium curve is shown in this figure as *pqr*.

Point $A$ represents conditions at the bottom of the tower. The gas rising from the bottom plate is in equilibrium with a liquid of concentration $x_s$ and is shown as point 3 on the operating line. Then point 4 indicates the concentration of the liquid on the second plate

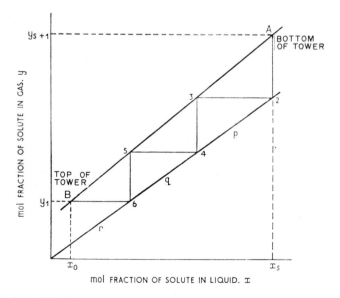

FIG. 12.25. Diagrammatic representation of changes in a plate column

from the bottom. In this way steps may be drawn to point $B$, giving the gas $y_1$ rising from the top plate and the liquid $x_0$ entering the top of the absorber.

### 12.9.1. Number of Plates by Use of Absorption Factor

If the equilibrium curve can be represented by the relation $y_e = mx$, then the number of plates required for a given degree of absorption can conveniently be found by a method due to Kremser[56] and Souders and Brown[57]. The same treatment is applicable for concentrated solutions provided concentrations are expressed as mol ratios, and if the equilibrium curve can be represented approximately by $Y_e = mX$.

A material balance over plate $n$ gives:

$$L_m(x_n - x_{n-1}) = G_m(y_{n+1} - y_n) \tag{12.93}$$

But for an ideal plate, $y_n = mx_n$;

$$\therefore \quad \frac{L_m}{mG_m}(y_n - y_{n-1}) = y_{n+1} - y_n \tag{12.94}$$

This group $L_m/mG_m$, which will be taken as constant, is called the absorption factor $\mathscr{A}$.

$$\therefore \quad y_n = \frac{y_{n+1} + \mathscr{A} y_{n-1}}{1 + \mathscr{A}} \tag{12.95}$$

Applying this relation to the top plate and letting $y_0$ be the mol fraction of absorbed component in the gas in equilibrium with the entering liquid of concentration $x_0$:

$$y_1 = \frac{y_2 + \mathscr{A} y_0}{1 + \mathscr{A}}$$

And for the second plate from the top:

$$y_2 = \frac{y_3 + \mathscr{A} y_1}{1 + \mathscr{A}}$$

$$= \frac{(1 + \mathscr{A})y_3 + \mathscr{A} y_2 + \mathscr{A}^2 y_0}{(1 + \mathscr{A})^2}$$

$$= \frac{y_3(1 + \mathscr{A}) + \mathscr{A}^2 y_0}{\mathscr{A}^2 + \mathscr{A} + 1}$$

And for the third plate from the top:

$$y_3 = \frac{y_4(1 + \mathscr{A} + \mathscr{A}^2) + \mathscr{A}^3 y_0}{\mathscr{A}^3 + \mathscr{A}^2 + \mathscr{A} + 1}$$

which can be written:

$$y_3 = \frac{[(\mathscr{A}^3 - 1)/(\mathscr{A} - 1)]y_4 + \mathscr{A}^3 y_0}{(\mathscr{A}^4 - 1)/(\mathscr{A} - 1)}$$

$$= \frac{(\mathscr{A}^3 - 1)y_4 + \mathscr{A}^3(\mathscr{A} - 1)y_0}{\mathscr{A}^4 - 1}$$

Proceeding thus until plate $n$ is reached:

$$y_n = \frac{(\mathscr{A}^n - 1)y_{n+1} + \mathscr{A}^n(\mathscr{A} - 1)y_0}{\mathscr{A}^{n+1} - 1}$$

Putting

$$x_n = \frac{y_n}{m}$$

$$x_n = \frac{(\mathscr{A}^n - 1)y_{n+1} + \mathscr{A}^n(\mathscr{A} - 1)y_0}{m(\mathscr{A}^{n+1} - 1)}$$

A material balance down to plate $n$ gives $\mathscr{A}(y_n - y_0) = y_{n+1} - y_1$ (see equation 12.95).

$$\therefore \quad \frac{y_{n+1} - y_1 + \mathscr{A}y_0}{\mathscr{A}} = y_n = \frac{(\mathscr{A}^n - 1)y_{n+1} + \mathscr{A}^n(\mathscr{A} - 1)y_0}{\mathscr{A}^{n+1} - 1}$$

$$\therefore \quad y_1 = y_{n+1}\left[\frac{\mathscr{A} - 1}{\mathscr{A}^{n+1} - 1}\right] + y_0\left[\frac{\mathscr{A}^{n+1} - \mathscr{A}}{\mathscr{A}^{n+1} - 1}\right]$$

Since

$$\frac{\mathscr{A} - 1}{\mathscr{A}^{n+1} - 1} = 1 - \frac{\mathscr{A}^{n+1} - \mathscr{A}}{\mathscr{A}^{n+1} - 1}$$

$$y_1 = y_{n+1} + (y_0 - y_{n+1})\frac{(\mathscr{A}^{n+1} - \mathscr{A})}{\mathscr{A}^{n+1} - 1}$$

or

$$\frac{y_{n+1} - y_1}{y_{n+1} - y_0} = \frac{\mathscr{A}^{n+1} - \mathscr{A}}{\mathscr{A}^{n+1} - 1}$$

But $y_{n+1} - y_1$ = actual change in composition of gas, and

$y_{n+1} - y_0$ = maximum possible change in composition of gas, i.e. if gas leaving absorber is in equilibrium with entering liquid.

If $y_1$, $y_{s+1}$, $x_s$, $x_0$ are the terminal concentrations, then $y_0 = mx_0$ and:

$$\frac{y_{s+1} - y_1}{y_{s+1} - mx_0} = \frac{(L_m/mG_m)^{s+1} - (L_m/mG_m)}{(L_m/mG_m)^{s+1} - 1} \qquad (12.96)$$

This equation is conveniently represented as suggested by Souders and Brown by Fig. 12.26, and it is easy to use such a diagram to determine the number of plates required.

A high degree of absorption can be obtained, either by using a large number of plates, or by using a high absorption factor $L_m/mG_m$. Since $m$ is fixed by the system, this means that $L_m/G_m$ must be large if a high degree of absorption is to be obtained, but this will result in a low value of $x$ for the liquid leaving at the bottom. This problem is to some extent met by recirculating the liquid over the tower, but the advantages of a countercurrent flow system are then lost. Colburn has suggested that a value of $mG_m/L_m$ of about 0·7–0·8 is probably the most economical, i.e. $L_m/mG_m = 1·3$.

It is important to note that, if $L_m/mG_m$ is less than 1, then a very large number of plates are required to achieve a high recovery, and even an infinite number will not give complete recovery. $L_m/mG_m$ is the ratio of the slope of the operating line $L_m/G_m$ to the slope of the

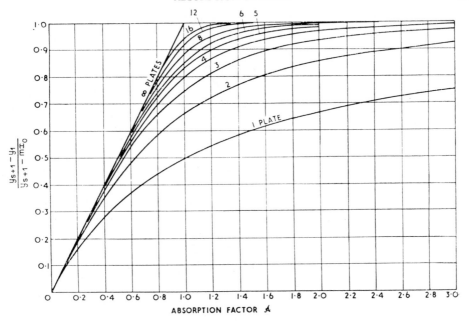

FIG. 12.26. Graphical representation of the effect of the absorption factor and the number of plates on the degree of absorption

equilibrium curve $m$, so that if $L_m/G_m < m$, i.e. $L_m/mG_m < 1$, then the operating line will never cut the equilibrium curve and the gas leaving the top of the column will not therefore reach equilibrium with the entering liquid.

### 12.9.2. Tray Types for Absorption

It has already been noted that trays which are suitable for distillation may be used for absorption duties though in general lower efficiencies will be obtained. In Chapter 11, the design of trays for common contacting devices was considered and the methods presented in that chapter are generally applicable. The most commonly used tray types are shown in Fig. 12.27[32] with the crossflow tray being the most popular.

At high liquid flowrates, the liquid gradient on the tray can become excessive and lead to poor vapour distribution across the plate. This problem may be overcome by the shortening of the liquid flow path as in the case of the double-pass and cascade trays.

Figure 12.28[32] enables a preliminary selection of the type of tray to be made, though only a complete tray design by the methods discussed in Chapter 11 will ascertain whether this initial choice is in fact the most suitable for the given duty.

### 12.10. MISCELLANEOUS EQUIPMENT

#### 12.10.1. The Use of Vessels with Agitators

A gas may be dissolved in a liquid by dispersing it through holes in a pipe immersed in the liquid which is stirred with some form of agitator (Fig. 12.29). This type of equipment

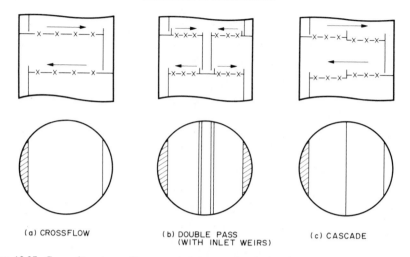

(a) CROSSFLOW          (b) DOUBLE PASS          (c) CASCADE
                       (WITH INLET WEIRS)

FIG. 12.27. General tray types. $X$ represents the contacting device, shaded areas denote downflow of liquid

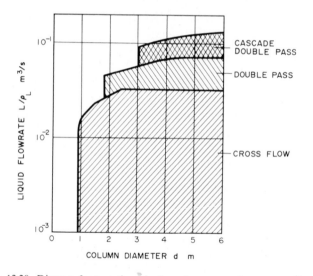

FIG. 12.28. Diagram for tentative selection of tray-type absorption tower

will give only one theoretical stage per unit, but it often provides a useful method of saturating a liquid with a gas. Cooper *et al.*[58] have studied the absorption of oxygen from the air in an aqueous solution of sodium sulphite. They used simple vessels of 0·15 to 2·44 m diameter fitted with four simple baffles, and introduced the air just below the agitator which was a vaned disc or flat-paddle. They found that the absorption coefficient $K_G a$ varied almost directly with $p_V$, the power input per unit volume. For constant values of $p_V$, they obtained the relation:

$$K_G a = \gamma u_s^{0.67} \tag{12.97}$$

where $u_s$ is the superficial gas velocity based on the volume of gas at inlet and the cross-

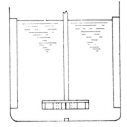

FIG. 12.29. Vessel with vaned-disc agitator

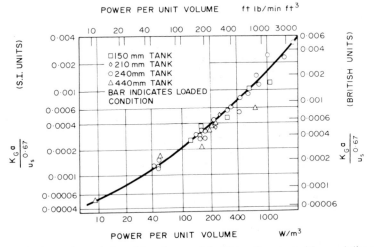

POWER PER UNIT VOLUME ft lb/min ft³

(S.I. UNITS)

□150 mm TANK
◇210 mm TANK
○240mm TANK
△440mm TANK
BAR INDICATES LOADED
CONDITION

$\dfrac{K_G a}{u_s^{0.67}}$

(BRITISH UNITS)

$\dfrac{K_G a}{u_s^{0.67}}$

POWER PER UNIT VOLUME W/m³

FIG. 12.30. General correlation of data for vessel (height = diameter) with vaned-disc agitator

section of tank. A general correlation was obtained by plotting $K_G a/u_s^{0.67}$ against the power input per unit volume, as shown in Fig. 12.30 taken from their work. Ayerst and Herbert[59] have given some data on the use of this type of unit for the absorption of carbon dioxide using ammoniacal solutions.

In general, there is little information on the prediction of $k_L a$ within agitated vessels though Sideman et al.[60] have produced an excellent review. For small bubbles (<2.5 mm diameter) produced in well-agitated vessels, Calderbank[61] suggests the following correlation for bubbles in agitated electrolytes:

$$k_L = 0.31\left(\frac{\Delta\rho\mu_L g}{\rho_L^2}\right)^{1/3}(Sc)^{-2/3} \tag{12.98}$$

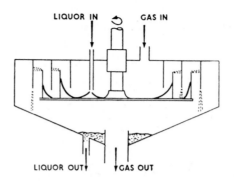

FIG. 12.31. The centrifugal absorber

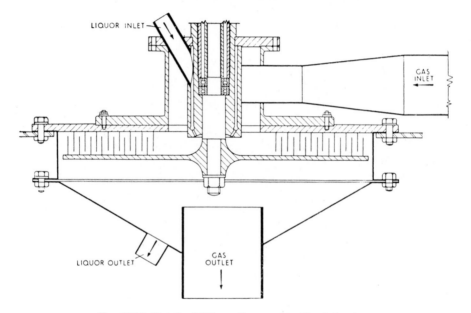

FIG. 12.32. Details of 510 mm diameter centrifugal absorber

where $\Delta\rho$ = density difference between gas and liquid,

$\rho_L, \mu_L$ = density and viscosity of the liquid, and

$Sc$ = Schmidt number for transport in the liquid.

For dilute aqueous systems at room temperature, values of $k_L$ in bubble swarms fall roughly within the range 0·08–0·5 mm/s. The interfacial area $a$ has been the subject of investigation by Westerterp et al.[62] though their correlations are complex. Maximum values of $a$ are about 1000 m²/m³.

### 12.10.2. The Centrifugal Absorber

In an attempt to obtain the benefits of repeated spray formations, a centrifugal type absorber has been developed from the ideas of Piazza for a still head. The principle of the unit is indicated in Fig. 12.31[63]. A set of stationary concentric rings intermeshes with a second set of rings attached to a rotating plate. Liquid fed to the centre of the plate is carried up the first ring, splashes over to the baffle and falls into the trough between the rings. It then runs up the second ring and in a similar way passes from ring to ring through the unit. The gas stream can be introduced at the top to give cocurrent flow, or at the bottom if countercurrent flow is desired. Some of the features of this unit are discussed by Ahmed[64] who found that the depth of the ring was not very important and that most of the transfer took place as the gas mixed with the liquid spray leaving the top of the rings. Chambers and Wall[63] have given some particulars of the performance of the 510 mm diameter unit shown in Fig. 12.32, for the absorption of carbon dioxide from air

TABLE 12.7. *Results for Absorption in Large Unit*

| Gas flow (m³/s) | Liquid flow (m³/s) | $CO_2$ in gas % | | Absorption (kg/s) |
|---|---|---|---|---|
| | | In | Out | |
| 0·016 | $1·07 \times 10^{-4}$ | 16·3 | 2·3 | 0·0044 |
| 0·024 | $1·07 \times 10^{-4}$ | 15·8 | 4·5 | 0·0055 |
| 0·031 | $1·07 \times 10^{-4}$ | 14·3 | 6·6 | 0·0051 |
| 0·039 | $1·07 \times 10^{-4}$ | 16·3 | 8·7 | 0·0065 |
| 0·032 | $1·0 \times 10^{-4}$ | 14·8 | 2·2 | 0·0082 (new monoethanolamine) |

containing 10–15 per cent of carbon dioxide, using mono-ethanolamine solution. Some figures are given in Table 12.7.

### 12.10.3. Spray Towers

In the spray tower, the gas enters at the bottom and the liquid is introduced through a series of sprays at the top. The performance of these units is generally rather poor, because the droplets tend to coalesce after they have fallen through a few feet, and the interfacial surface is thereby seriously reduced. Although there is considerable turbulence in the gas phase, there is little circulation of the liquid within the drops, and the resistance of the equivalent liquid film tends to be high. Spray towers are therefore useful only where the main resistance to mass transfer lies within the gas phase, and have consequently been used with moderate success for the absorption of ammonia in water. They are also used as air humidifiers; in this case the whole of the resistance lies within the gas phase.

*Centrifugal Spray Tower*

Figure 12.33, taken from the work of Kleinschmidt and Anthony[65], illustrates a spray tower in which the gas stream enters tangentially, so that the liquid drops receive the centrifugal force before they are taken out of the gas stream at the top.

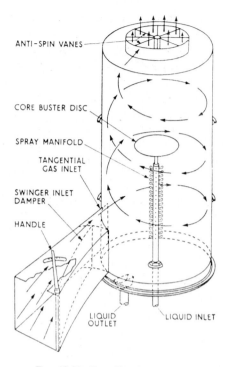

ANTI-SPIN VANES

CORE BUSTER DISC

SPRAY MANIFOLD

TANGENTIAL
GAS INLET

SWINGER INLET
DAMPER

HANDLE

LIQUID
OUTLET                    LIQUID INLET

Fig. 12.33.  Centrifugal spray tower

## 12.11.  FURTHER READING

HOBLER, T.: *Mass Transfer and Absorbers* (Pergamon Press, Oxford, 1966).
NORMAN, W. S.: *Absorption, Distillation and Cooling Towers* (Longmans, London, 1961).
SHERWOOD, T. K. and PIGFORD, R. L.: *Absorption and Extraction* (McGraw-Hill Book Co., New York, 1952).
SHERWOOD, T. K., PIGFORD, R. L., and WILKE, C. R.: *Mass Transfer* (McGraw-Hill Book Co., New York, 1975).
SMITH, B. D.: *Design of Equilibrium Stage Processes* (McGraw-Hill Book Co., New York, 1963).
TREYBAL, R. E.: *Mass Transfer Operations*, 2nd edn. (McGraw-Hill Book Co., New York, 1968).

## 12.12.  REFERENCES

1.  WHITMAN, W. G.: *Chem. Met. Eng.* **29** (1923) 147. The two-film theory of absorption.
2.  KING, C. J.: *A.I.Ch.E.Jl.* **10** (1964) 671. The additivity of individual phase resistances in mass transfer operations.
3.  HIGBIE, R.: *Trans. Am. Inst. Chem. Eng.* **31** (1935) 365. The rate of absorption of pure gas into a still liquid during short periods of exposure.
4.  DANCKWERTS, P. V.: *Ind. Eng. Chem.* **43** (1951) 1460. Significance of liquid-film coefficients in gas absorption.
5.  DANCKWERTS, P. V. and KENNEDY, A. M.: *Trans. Inst. Chem. Eng.* **32** (1954) S49. Kinetics of liquid-film processes in gas absorption.
6.  LYNN, S., STRAATEMEIER, J. R., and KRAMERS, H.: *Chem. Eng. Sci.* **4** (1955) 49, 58, 63. Absorption studies in the light of the penetration theory. I. Long wetted-wall columns. II. Absorption by short wetted-wall columns. III. Absorption by wetted-spheres, singly and in columns.
7.  DAVIDSON, J. F., CULLEN, E. J., HANSON, D., and ROBERTS, D.: *Trans. Inst. Chem. Eng.* **37** (1959) 122. The hold-up and liquid film coefficient of packed towers. Part I. Behaviour of a string of spheres.
8.  DAVIDSON, J. F.: *Trans. Inst. Chem. Eng.* **37** (1959) 131. The hold-up and liquid film coefficients of packed towers. Part II. Statistical models of the random packing.

9. DANCKWERTS, P. V. and KENNEDY, A. M.: *Chem. Eng. Sci.* **8** (1958) 201. The kinetics of absorption of carbon dioxide into neutral and alkaline solutions.
10. ROBERTS, D. and DANCKWERTS, P. V.: *Chem. Eng. Sci.* **17** (1962) 961. Kinetics of $CO_2$ absorption in alkaline solutions. I. Transient absorption rates and catalysis by arsenite.
11. DANCKWERTS, P. V., KENNEDY, A. M., and ROBERTS, D.: *Chem. Eng. Sci.* **18** (1963) 63. Kinetics of $CO_2$ absorption in alkaline solutions. II. Absorption in a packed column and tests of surface renewal models.
12. CULLEN, E. J. and DAVIDSON, J. F.: *Trans. Faraday Soc.* **53** (1957) 113. Absorption of gases in liquid jets.
13. RAIMONDI, P. and TOOR, H. L.: *A.I.Ch.E.Jl.* **5** (1959) 86. Interfacial resistance in gas absorption.
14. STERNLING, C. V. and SCRIVEN, L. E.: *A.I.Ch.E.Jl.* **5** (1959) 514. Interfacial turbulence: Hydrodynamic instability and the Marangoni effect.
15. GOODRIDGE, F. and ROBB, I. D.: *Ind. Eng. Chem. Fundamentals* **4** (1965) 49. Mechanism of interfacial resistance.
16. GARNER, F. H. and KENDRICK, P.: *Trans. Inst. Chem. Eng.* **37** (1959) 155. Mass transfer to drops of liquid suspended in a gas stream. Part I. A wind tunnel for the study of individual liquid drops.
17. GARNER, F. H. and LANE, J. J.: *Trans. Inst. Chem. Eng.* **37** (1959) 162. Mass transfer to drops of liquid suspended in a gas stream. Part II. Experimental work and results.
18. GILLILAND, E. R. and SHERWOOD, T. K.: *Ind. Eng. Chem.* **26** (1934) 516. Diffusion of vapours into air streams.
19. HOLLINGS, H. and SILVER, L.: *Trans. Inst. Chem. Eng.* **12** (1934) 49. The washing of gas.
20. CHILTON, T. H. and COLBURN, A. P.: *Ind. Eng. Chem.* **26** (1934) 1183. Mass transfer (absorption) coefficients—prediction from data on heat transfer and fluid friction.
21. MORRIS, G. A. and JACKSON, J.: *Absorption Towers* (Butterworths, 1953).
22. KOWALKE, O. L., HOUGEN, O. A., and WATSON, K. M.: *Bull. Univ. Wisconsin Eng. Sta.* Ser. No. 68 (1925). Transfer coefficients of ammonia in absorption towers.
23. BORDEN, H. M. and SQUIRES, W.: Massachusetts Institute of Technology, S.M. thesis (1937). Absorption of ammonia in a ring-packed tower.
24. NORMAN, W. S.: *Trans. Inst. Chem. Eng.* **29** (1951) 226. The performance of grid-packed towers.
25. FELLINGER, L. L.: Massachusetts Institute of Technology, D.Sc. thesis (1941). Absorption of ammonia by water and acids in various standard packings.
26. PERRY, J. H.: *Chemical Engineer's Handbook*, 5th edn. (McGraw-Hill, 1973) (revised by PERRY, R. and CHILTON, C.).
27. MOLSTAD, M. C., McKINNEY, J. F., and ABBEY, R. G.: *Trans. Am. Inst. Chem. Eng.* **39** (1943) 605, Performance of drip-point grid tower packings. III. Gas-film mass transfer coefficients; additional liquid-film mass transfer coefficients.
28. VAN KREVELEN, D. W. and HOFTIJZER, P. J.: *Chem. Eng. Sci.* **2** (1953) 45. Graphical design of gas–liquid reactors.
29. SEMMELBAUER, R.: *Chem. Eng. Sci.* **22** (1967) 1237. Die Berechnung der Schütthöhe bei Absorptionsvorgängen in Füllkörperkolonnen. (Calculation of the height of packing in packed towers.)
30. SHERWOOD, T. K. and HOLLOWAY, F. A. L.: *Trans. Am. Inst. Chem. Eng.* **36** (2940) 39, 181. Performance of packed towers—liquid film data for several packings.
31. COOPER, C. M., CHRISTL, R. J., and PERRY, L. C.: *Trans. Am. Inst. Chem. Eng.* **37** (1941) 979. Packed tower performance at high liquor rates—The effect of gas and liquor rates upon performance in a tower packed with two-inch rings.
32. NONHEBEL, G.: *Gas Purification Processes for Air Pollution Control*, 2nd edn. (Newnes–Butterworth, London, 1972).
33. HIXSON, A. W. and SCOTT, C. E.: *Ind. Eng. Chem.* **27** (1935) 307. Absorption of gases in spray towers.
34. WHITMAN, W. G., LONG, L., and WANG, H. Y.: *Ind. Eng. Chem.* **18** (1926) 363. Absorption of gases by a liquid drop.
35. PIGFORD, R. L. and PYLE, C.: *Ind. Eng. Chem.* **43** (1951) 1649. Performance characteristics of spray-type absorption equipment.
36. NORMAN, W. S.: *Absorption, Distillation and Cooling Towers* (Longmans, London, 1961).
37. HATTA, S.: *Tech. Repts. Tohoku Imp. Univ.* **10** (1932) 119. On the absorption velocity of gases by liquids. II. Theoretical considerations of gas absorption due to chemical reaction.
38. NIJSING, R. A. T. O., HENDRIKSZ, R. N., and KRAMERS, H.: *Chem. Eng. Sci.* **10** (1959) 88. Absorption of $CO_2$ in jets and falling films of electrolyte solutions, with and without chemical reaction.
39. TEPE, J. B. and DODGE, B. F.: *Trans. Am. Inst. Chem. Eng.* **39** (1943) 255. Absorption of carbon dioxide by sodium hydroxide solutions in a packed column.
40. CRYDER, D. S. and MALONEY, J. O.: *Trans. Am. Inst. Chem. Eng.* **37** (1941) 827. The rate of absorption of carbon dioxide in diethanolamine solutions.
41. STEPHENS, E. J. and MORRIS, G. A.: *Chem. Eng. Prog.* **47** (1951) 232. Determination of liquid-film absorption coefficients. A new type of column and its application to problems of absorption in presence of chemical reaction.
42. DANCKWERTS, P. V. and McNEIL, K. M.: *Trans. Inst. Chem. Eng.* **45** (1967) 32. The absorption of carbon dioxide into aqueous amine solutions and the effects of catalysis.

43. DANCKWERTS, P. V. and SHARMA, M. M.: *Chem. Engr. London* No. 202 (Oct. 1966) CE244. The absorption of carbon dioxide into solutions of alkalis and amines (with some notes on hydrogen sulphide and carbonyl sulphide).
44. SAHAY, B. N. and SHARMA, M. M.: *Chem. Eng. Sci.* 28 (1973) 41. Effective interfacial areas and liquid and gas side mass transfer coefficients in a packed column.
45. ECKERT, J. S.: *Chem. Engg.* 82 (14 April 1975) 70. How tower packings behave.
46. SHERWOOD, T. K. and PIGFORD, R. L.: *Absorption and Extraction*, 2nd edn. (McGraw-Hill, 1952).
47. TREYBAL, R. E.: *Mass Transfer Operations*, 2nd edn. (McGraw-Hill Book Co., New York, 1968).
48. POLL, A. and SMITH, W.: *Chem. Engg.* 71 (26 Oct. 1964) 111. Froth contact heat exchanger.
49. COGGAN, C. G. and BOURNE, J. R.: *Trans. I. Chem. E.* 47 (1969) T96, T160. The design of gas absorbers with heat effects.
50. SHULMAN, H. L., ULLRICH, C. F., PROULX, A. Z. and ZIMMERMAN, J. O.: *A.I.Ch.E.Jl.* 1 (1955) 2, 253. Interfacial areas—gas and liquid phase mass transfer rates.
51. EASTHAM, I. E.: Private communication (1977).
52. CHILTON, T. H. and COLBURN, A. P.: *Ind. Eng. Chem.* 27 (1935) 255. Distillation and absorption in packed columns.
53. RACKETT, H. G.: *Chem. Eng., Albany* 71 (21 Dec. 1964) 108. Modified graphical integration for determining transfer units.
54. TOWER PACKINGS (Hydronyl Ltd., Stoke-on-Trent, 1965).
55. COLBURN, A. P.: *Trans. Am. Inst. Chem. Eng.* 35 (1939) 211. The simplified calculation of diffusional processes. General consideration of two-film resistances.
56. KREMSER, A.: *Nat. Petroleum News* 22 (21 May 1930) 43. Theoretical analysis of absorption processes.
57. SOUDERS, M. and BROWN, G. G.: *Ind. Eng. Chem.* 24 (1932) 519. Fundamental design of high pressure equipment involving paraffin hydrocarbons. IV. Fundamental design of absorbing and stripping columns for complex vapours.
58. COOPER, C. M., FERNSTROM, G. A., and MILLER, S. A.: *Ind. Eng. Chem.* 36 (1944) 504. Performance of agitated gas–liquid contactors.
59. AYERST, R. R. and HERBERT, L. S.: *Trans. Inst. Chem. Eng.* 32 (1954) S68. A study of the absorption of carbon dioxide in ammonia solutions in agitated vessels.
60. SIDEMAN, S. O., HORTACSU, O., and FULTON, J. W.: *Ind. Eng. Chem.* 58 (July 1966) 32. Mass transfer in gas–liquid contacting systems.
61. CALDERBANK, P. H.: *Chem. Engnr.* No. 212 (Oct. 1967) CE 209. Gas absorption from bubbles.
62. WESTERTERP, K. R., VAN DIERENDONCK, L. L., and DE KRAA, J. R.: *Chem. Eng. Sci.* 18 (1963) 157. Interfacial areas in agitated gas–liquid contactors.
63. CHAMBERS, H. H. and WALL, R. C.: *Trans. Inst. Chem. Eng.* 32 (1954) S96. Some factors affecting the design of centrifugal gas absorbers.
64. AHMED, N.: University of London, Ph.D. thesis (1949). Design of gas scrubber based upon thin films and sprays.
65. KLEINSCHMIDT, R. V. and ANTHONY, A. W.: *Trans. Am. Soc. Mech. Eng.* 63 (1941) 349. Recent development of Pease–Anthony gas scrubber.

## 12.13. NOMENCLATURE

| | | Units in SI System | Dimensions in $\mathbf{MLT\theta}$ |
|---|---|---|---|
| $A$ | Cross-sectional area of column | $m^2$ | $\mathbf{L^2}$ |
| $\mathscr{A}$ | Absorption factor | — | — |
| $a$ | Surface of interface per unit volume of column | $m^2/m^3$ | $\mathbf{L^{-1}}$ |
| $a_1, a_2 \ldots$ | Constants in equation 12.45 | — | — |
| $B$ | A constant | — | — |
| $B'$ | A constant | — | — |
| $C$ | Molar concentration | $kmol/m^3$ | $\mathbf{ML^{-3}}$ |
| $C_A, C_B$ | Molar concentrations of $\mathbf{A}$, $\mathbf{B}$ | $kmol/m^3$ | $\mathbf{ML^{-3}}$ |
| $C_{BL}$ | Molar concentration of $\mathbf{B}$ in bulk of liquid phase | $kmol/m^3$ | $\mathbf{ML^{-3}}$ |
| $C_e$ | Molar concentration in liquid phase in equilibrium with partial pressure $P_G$ in gas phase | $kmol/m^3$ | $\mathbf{ML^{-3}}$ |
| $C_i$ | Molar concentration at interface | $kmol/m^3$ | $\mathbf{ML^{-3}}$ |
| $C_L$ | Molar concentration in bulk of liquid | $kmol/m^3$ | $\mathbf{ML^{-3}}$ |
| $C_T$ | Total molar concentration | $kmol/m^3$ | $\mathbf{ML^{-3}}$ |
| $c$ | Constant term in equation of equilibrium line | — | — |

| | | Units in SI System | Dimensions in $\mathbf{MLT\theta}$ |
|---|---|---|---|
| $c_G$ | Gas mixture constant $(\rho_r/\mu_r)^{0\cdot25}(D_{Vr})^{0\cdot5}$ in cgs units | — | $\mathbf{L}^{1/2}\mathbf{T}^{-1/4}$ |
| $D_L$ | Liquid phase diffusivity | m²/s | $\mathbf{L}^2\mathbf{T}^{-1}$ |
| $D_V$ | Vapour phase diffusivity | m²/s | $\mathbf{L}^2\mathbf{T}^{-1}$ |
| $d$ | Column diameter | m | $\mathbf{L}$ |
| $d_p$ | Packing size | m | $\mathbf{L}$ |
| $e$ | Voidage | — | — |
| $F$ | Packing factor (see equations 12.35 and 12.36) | — | — |
| $F'$ | Fractional conversion (equation 12.45) | — | — |
| $f$ | Fraction of surface renewed per unit time | 1/s | $\mathbf{T}^{-1}$ |
| $G_m$ | Molar rate of flow of inert gas per unit cross-section | kmol/m² s | $\mathbf{ML}^{-2}\mathbf{T}^{-1}$ |
| $G'$ | Gas flow rate (mass) per unit cross-section | kg/m² s | $\mathbf{ML}^{-2}\mathbf{T}^{-1}$ |
| $\mathbf{H}$ | Height of transfer unit | m | $\mathbf{L}$ |
| $h$ | Heat transfer coefficient | W/m² K | $\mathbf{MT}^{-3}\theta^{-1}$ |
| $h_D$ | Mass transfer coefficient $(D_V/z_G)$ | m/s | $\mathbf{LT}^{-1}$ |
| $h_p$ | Height of packing | m | $\mathbf{L}$ |
| $\mathscr{H}$ | Henry's constant | m²/s² | $\mathbf{L}^2\mathbf{T}^{-2}$ |
| $i$ | Number of kmol of $\mathbf{B}$ reacting with kmol of $\mathbf{A}$ | — | — |
| $j$ | Reaction rate constant | m³/kg s | $\mathbf{M}^{-1}\mathbf{L}^3\mathbf{T}^{-1}$ |
| $j_d$ | $j$-factor for mass transfer | — | — |
| $K_G$ | Overall gas-phase transfer coefficient | s/m | $\mathbf{L}^{-1}\mathbf{T}$ |
| $K_L$ | Overall liquid-phase transfer coefficient | m/s | $\mathbf{LT}^{-1}$ |
| $K_G''$ | Overall gas-phase transfer coefficient in terms of mol fractions | kmol/m² s | $\mathbf{ML}^{-2}\mathbf{T}^{-1}$ |
| $K_L''$ | Overall liquid-phase transfer coefficient in terms of mol fractions | kmol/m² s | $\mathbf{ML}^{-2}\mathbf{T}^{-1}$ |
| $k$ | Thermal conductivity | W/m K | $\mathbf{MLT}^{-3}\theta^{-1}$ |
| $k_G$ | Gas film transfer coefficient $(D_V P/\mathbf{R}Tz_G P_{Bm})$ | s/m | $\mathbf{L}^{-1}\mathbf{T}$ |
| $k_G'$ | Gas film transfer coefficient $(D_V/\mathbf{R}Tz_G)$ | s/m | $\mathbf{L}^{-1}\mathbf{T}$ |
| $k_G''$ | Gas film transfer coefficient in terms of mol fractions | kmol/m² s | $\mathbf{ML}^{-2}\mathbf{T}^{-1}$ |
| $k_L$ | Liquid film transfer coefficient | m/s | $\mathbf{LT}^{-1}$ |
| $k_L''$ | Liquid film transfer coefficient in terms of mol fractions | kmol/m² s | $\mathbf{ML}^{-2}\mathbf{T}^{-1}$ |
| $L_m$ | Molar rate of flow of solute-free liquor per unit cross-section | kmol/s m² | $\mathbf{ML}^{-2}\mathbf{T}^{-1}$ |
| $L'$ | Liquid flow rate (mass) per unit cross-section | kg/s m² | $\mathbf{ML}^{-2}\mathbf{T}^{-1}$ |
| $m$ | Slope of equilibrium line | — | — |
| $N_A$, $N_B$ | Molar rate of diffusion of $\mathbf{A}$, $\mathbf{B}$ per unit area | kmol/s m² | $\mathbf{ML}^{-2}\mathbf{T}^{-1}$ |
| $N_A'$, $N_B'$ | Molar rate of absorption of $\mathbf{A}$, $\mathbf{B}$ per unit area | kmol/s m² | $\mathbf{ML}^{-2}\mathbf{T}^{-1}$ |
| $N_A''$ | Molar rate of absorption of $\mathbf{A}$ per unit area with chemical reaction | kmol/s m² | $\mathbf{ML}^{-2}\mathbf{T}^{-1}$ |
| $\mathbf{N}$ | Number of transfer units | — | — |
| $n$ | Number of plates from top | — | — |
| $P$ | Total pressure | N/m² | $\mathbf{ML}^{-1}\mathbf{T}^{-2}$ |
| $P_A$, $P_B$ | Partial pressures of $\mathbf{A}$ and $\mathbf{B}$ | N/m² | $\mathbf{ML}^{-1}\mathbf{T}^{-2}$ |
| $P_{Bm}$ | Logarithmic mean value of $P_B$ | N/m² | $\mathbf{ML}^{-1}\mathbf{T}^{-2}$ |
| $P_e$ | Partial pressure of $\mathbf{A}$ in equilibrium with concentration $C_L$ in liquid phase | N/m² | $\mathbf{ML}^{-1}\mathbf{T}^{-2}$ |
| $P_G$ | Partial pressure of $\mathbf{A}$ in bulk of gas phase | N/m² | $\mathbf{ML}^{-1}\mathbf{T}^{-2}$ |
| $P_i$ | Partial pressure of $\mathbf{A}$ at interface | N/m² | $\mathbf{ML}^{-1}\mathbf{T}^{-2}$ |
| $\Delta P_{\mathrm{lm}}$ | Log mean driving force | N/m² | $\mathbf{ML}^{-1}\mathbf{T}^{-2}$ |
| $p_V$ | Power input per unit volume | W/m³ | $\mathbf{ML}^{-1}\mathbf{T}^{-3}$ |
| $\mathbf{R}$ | Universal gas constant | J/kmol K | $\mathbf{L}^2\mathbf{T}^{-2}\theta^{-1}$ |
| $R_G$ | Packing factor (see equations 12.35 and 12.36) | — | — |
| $r'$ | Ratio of effective film thickness for absorption without and with chemical reaction | — | — |
| $S$ | Specific surface of packing | 1/m | $\mathbf{L}^{-1}$ |
| $s$ | Total number of plates in column | — | — |
| $T$ | Absolute temperature | K | $\theta$ |
| $t$ | Time | s | $\mathbf{T}$ |
| $u$ | Gas velocity | m/s | $\mathbf{LT}^{-1}$ |
| $u_s$ | Superficial gas velocity based on inlet conditions | m/s | $\mathbf{LT}^{-1}$ |
| $V$ | Volume of packed section of column | m³ | $\mathbf{L}^3$ |

| | | Units in SI System | Dimensions in $\mathbf{MLT}\theta$ |
|---|---|---|---|
| $X$ | Mols of solute gas $\mathbf{A}$ per mol of solvent in liquid phase | — | — |
| $x$ | Mol fraction of $\mathbf{A}$ in liquid phase | — | — |
| $Y$ | Mols of solute gas $\mathbf{A}$ per mol of inert gas $\mathbf{B}$ in gas phase | — | — |
| $y$ | Mol fraction of $\mathbf{A}$ in gas phase | — | — |
| $Z$ | Height of packed column | m | $\mathbf{L}$ |
| $z$ | Distance in direction of mass transfer | m | $\mathbf{L}$ |
| $z_G$ | Thickness of gas film | m | $\mathbf{L}$ |
| $z_L$ | Thickness of liquid film | m | $\mathbf{L}$ |
| $\alpha$ | A coefficient | $\text{s}^{-1\cdot8}/\text{kg}^{0\cdot8}\,\text{m}^{0\cdot4}$ | $\mathbf{M}^{-0\cdot8}\mathbf{L}^{-0\cdot4}\mathbf{T}^{1\cdot8}$ |
| $\alpha'$ | Coefficient in equation 12.36 | — | — |
| $\beta$ | A coefficient | $1/\text{m}^{1\cdot25}$ | $\mathbf{L}^{-5/4}$ |
| $\beta'$ | A coefficient (equation 12.39) | — | — |
| $\gamma$ | A coefficient | $\text{t}^{1\cdot67}/\text{m}^{2\cdot67}$ | $\mathbf{L}^{-8/3}\mathbf{T}^{5/3}$ |
| $\mu$ | Viscosity of gas | $\text{Ns/m}^2$ | $\mathbf{ML}^{-1}\mathbf{T}^{-1}$ |
| $\mu_L$ | Viscosity of liquid | $\text{Ns/m}^2$ | $\mathbf{ML}^{-1}\mathbf{T}^{-1}$ |
| $\rho$ | Density of gas | $\text{kg/m}^3$ | $\mathbf{ML}^{-3}$ |
| $\rho_L$ | Density of liquid | $\text{kg/m}^3$ | $\mathbf{ML}^{-3}$ |
| $\sigma$ | Surface tension | $\text{J/m}^2$ | $\mathbf{MT}^{-2}$ |
| $\phi$ | Correction factor for concentrated solutions | — | — |
| $Ga$ | Galileo number | — | — |
| $Pr$ | Prandtl number | — | — |
| $Re$ | Reynolds number | — | — |
| $Sc$ | Schmidt number | — | — |
| $Sh$ | Sherwood number | — | — |

Suffixes

| | |
|---|---|
| 1 | denotes conditions at bottom of packed column, or at plane 1 |
| 2 | denotes conditions at top of packed column, or at plane 2 |
| $A$ | denotes soluble gas |
| $B$ | denotes insoluble gas |
| $e$ | denotes equilibrium value |
| $f$ | denotes film value |
| $i$ | denotes value at interface |
| $G$ | denotes gas phase |
| $L$ | denotes liquid phase |
| lm | denotes logarithmic mean value |
| $n$ | denotes values on plate $n$ |
| $r$ | denotes reference state (293 K, 101·3 kN/m$^2$) |

$G, OG, L, OL$ refer to gas film, overall gas, liquid film, and overall liquid transfer units

CHAPTER 13

# Liquid–Liquid Systems

## 13.1. INTRODUCTION

The separation of the components of a liquid mixture by treatment with a solvent in which one or more of the desired components is preferentially soluble is known as liquid–liquid extraction—an operation which is used, for example, in the processing of coal tar liquids and in the production of fuels in the nuclear industry, and which has been applied extensively to the separation of hydrocarbons in the petroleum industry. In the operation it is essential that the liquid-mixture feed and solvent are at least partially if not completely immiscible and, in essence, three stages are involved:

(a)  bringing the feed mixture and the solvent into intimate contact,
(b)  separation of the resulting two phases, and
(c)  removal and recovery of the solvent from each phase.

It is possible to combine stages (a) and (b) into a single piece of equipment such as a column which is then operated continuously—such an operation is known as differential contacting. Liquid–liquid extraction is also carried out in stagewise equipment, the prime example being a mixer–settler unit in which the main features are the mixing of the two liquid phases by agitation, followed by settling in a separate vessel by gravity. This mixing of two liquids by agitation is of considerable importance and it is dealt with before the general discussion of liquid–liquid extraction. It should be noted at the outset that a great deal of the discussion on the mixing of liquid–liquid systems also applies to the agitation of liquids and fine solids.

## 13.2. THE MIXING OF LIQUID–LIQUID SYSTEMS

### 13.2.1. Propellers and Turbines

For the great majority of reaction vessels, and for most operations involving liquid–liquid and to some extent liquid–solid mixing, the most commonly applied equipment involves a propeller or a turbine in a tank. Reavell[1] has suggested that mixing of this type can be divided into three classes:

(a)  Liquids, with or without solids, which remain free flowing when mixing is complete; e.g. water and salt, acid and sand, light or medium oils.
(b)  Liquids, with or without solids, which are viscous but still pourable when mixing is complete; e.g. heavy oils, paints, resins, syrups.
(c)  Liquids, with solids, which form stiff pastes; e.g. core sands and binders, oil-bound distempers, white lead and oil, putties.

The usual form of equipment is a vertical cylindrical tank, with a height one and a half times to twice the diameter, fitted with an agitator. When the thickness of the mix corresponds to class (c) above, it is necessary for the agitator to conform to the shape of the vessel, so that the action corresponds with that of a kneading machine described in Chapter 1. With thin liquids, high-speed propellers of diameter about one-third that of the vessel are suitable, and for thicker mixtures the propeller diameter is increased and its speed reduced. Thus, high-speed propellers are run at from 10–25 Hz, and scraping agitators at speeds down to 2 or even 1 Hz.

### Propeller in Cylindrical Tank

If a propeller is mounted centrally, there is a tendency for the lighter fluid (usually air) to be drawn in to form a vortex and for the degree of agitation to be reduced. The flow pattern should be as indicated in Fig. 13.1, where the stream leaving the propeller is moving with a

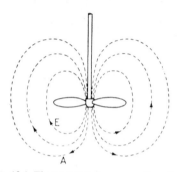

FIG. 13.1  Flow pattern from propeller mixer

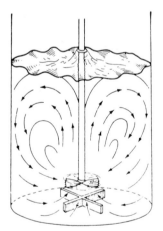

FIG. 13.2.  Flow pattern in vessel with cruciform baffle

high velocity and initially in a straight line. The outer part of the stream, shown as $E$, turns back on itself and re-enters the feed to the propeller, whilst the inner streams, as at $A$, are of much greater length. A particle in any one stream will enter the next at the inlet side of the propeller and effective mixing occurs, a considerable up and down motion being provided. The agitation is strongest near the propeller and dead spaces form at the bottom of the tank. With this arrangement the unsupported length of the propeller shaft should not exceed 2 m. If the contents of a very large vessel are to be stirred with a propeller of this kind, a foot-bearing is essential. Despite a considerable amount of practical experimentation, these foot-bearings usually give trouble, since corrosive liquids and solvents are frequently used so that it is very difficult to lubricate the bearings. It has been shown by Reavell[1] that the fitting of a cruciform baffle at the bottom of the vessel (Fig. 13.2) enables

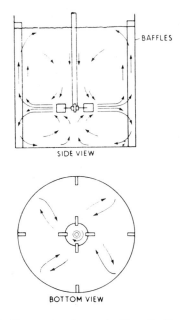

SIDE VIEW

BOTTOM VIEW

Fig. 13.3. Flow pattern in vessel with vertical baffles

much better dispersion to be obtained. The rotor is arranged to force the fluid upwards; this arrangement gives much better axial flow and avoids the development of rotational movements of the liquid. The great reduction in the side-thrust on the shaft enables longer shafts of up to 3 m to be used without foot-step bearings. To improve the rate of mixing and to minimise vortex formation, baffles are usually added. These take the form of thin vertical strips mounted against the walls of the vessel, as shown in Fig. 13.3. They considerably increase the power requirement, as discussed later. The off-setting of the agitator is another method of minimising vortex formation (Fig. 13.4).

*Portable Mixers*

For a wide range of applications, a portable mixer which can be clamped on the top or side of the vessel is now used. This is commonly fitted with two propeller blades so that the bottom rotor forces the liquid upwards and the top rotor forces the liquid downwards. This form of unit can be supplied with about 2 kW, though the size of the motor becomes too great at higher powers. To avoid excessive strain on the armature, some form of flexible coupling should be fitted between the motor and the main propeller shaft. Units

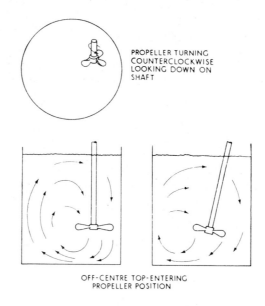

FIG. 13.4. Flow pattern with agitator offset from centre

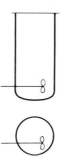

FIG. 13.5. Horizontally mounted propeller

of this kind are usually driven at a fairly high rate (15 Hz), and a reduction gear can be fitted to the unit fairly easily for low speed operation.

*Horizontally Mounted Propellers*

A propeller mounted on a horizontal shaft, positioned eccentrically as shown in Fig. 13.5, enables the contents of a very large tank to be stirred with a single propeller. This type of unit has been developed for very large power inputs (20 kW), though a good gland is needed where the shaft enters the vessel (Fig. 13.6).

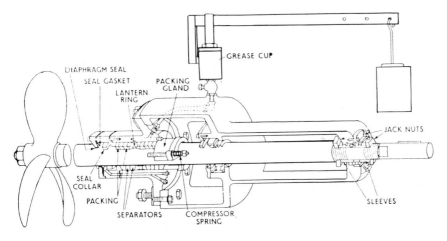

FIG. 13.6. Gland fittings for horizontally mounted propeller

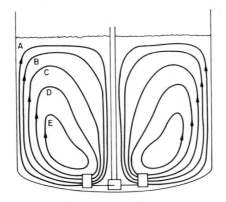

FIG. 13.7. Flow pattern with turbine rotor

## Turbine Rotors

The ordinary propeller may be replaced by a turbine which may be open or shrouded, the latter type being much more expensive. The flow pattern with a turbine, as shown in Fig. 13.7, is quite different from that obtained with a marine propeller. Turbines can be used for rather more viscous materials than propellers, though the power consumption is much greater. In comparing a propeller and a turbine, the following features should be noted:

Propellers:

(a) are self-cleaning in operation,
(b) can be used at a wide range of speeds,
(c) give excellent shearing effect at high speeds,
(d) do not damage dispersed particles at low speeds,

(e)  are reasonably economical in power, provided the pitch is adjusted according to the
     speed,
(f)  by offset mounting, avoid vortex formation,
(g)  if horizontally mounted, require a stuffing box in the liquid, and are not effective in
     viscous liquids.

Shrouded Turbines:
(a)  are excellent for providing circulation,
(b)  are normally mounted on a vertical shaft with the stuffing box above the liquid,
(c)  are effective in fluids of high viscosity,
(d)  are easily fouled or plugged by solid particles,
(e)  are expensive to fabricate,
(f)  are restricted to a narrow range of speeds, and
(g)  do not damage dispersed particles at economical speeds.

Open Impellers:
(a)  are less easily plugged than the shrouded type.
(b)  are less expensive, and
(c)  give a less well-controlled flow pattern.

*Propellers with Coil*

If a coil is used in the tank to provide cooling, then the geometrical arrangements
commonly adopted are as in Fig. 13.8. This set-up is very widespread for reaction vessels in
the organic chemical industry.

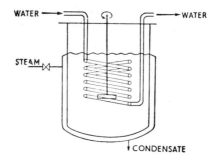

FIG. 13.8. Reaction vessel with jacket and coil

## 13.2.2. Power Consumption of Agitators

In considering the speed of rotation of an agitator, a compromise is usually made
between a high speed which gives rapid mixing and a lower speed where a smaller power is
required. One of the earliest publications in which these quantities were discussed is by
Wood, Whittemore, and Badger[2] who used a $2\,m^3$ vertical tank of 1·5 m diameter,
fitted with a simple paddle of 100 mm × 100 mm section. Strong brine was first run into the
tank, and water was then run in on top. The power for mixing was measured electrically,
though accurate results were difficult to obtain because of the difficulty of assessing the
power used in the belt drive and gears. Some of their results are shown in Fig. 13.9. From
curve 1, it is seen that the time of stirring fell off quite steadily with an increase in speed. The
addition of four simple baffles (25 mm × 100 mm positioned 50 mm from the wall), reduced

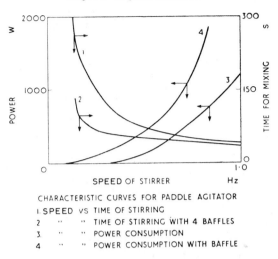

FIG. 13.9.  Power and time of mixing as a function of speed for paddle agitator

the time for stirring but increased the power requirement (curve 2). The degree of mixing was assessed by taking samples from various points and measuring their electrical conductivity. Curves 3 and 4 show the change in power consumption at various speeds with and without baffles.

Hixson and Wilkens[3] measured the power absorbed by a four-bladed propeller with a 45° pitch, operating in a cylindrical vessel. In general, they found that the power consumption per unit volume increased rapidly as the size of the system was increased.

Stoops and Lovell[4] examined the power consumption of a propeller agitator under various conditions. Three feasible methods for the determination of the power required were suggested:

(a) Fitting some form of Prony brake to the shaft between the motor and the propeller. In this way, the mechanical power output can be measured for various conditions of operation.

(b) Placing the mixing vessel on a turn-table and determining the torque necessary to prevent its rotation.

(c) Connecting the driving shaft to the propeller by a spring or dynamometer system, and measuring the torque by the relative displacement of the two shafts.

The first method is the simplest, but they found that reliable results were very difficult to obtain as the mechanical energy was always a small proportion of the electrical energy. The second method has been used by White and Brenner[5], but is rather difficult to apply to large installations. The third method, although more complicated, has been found by many workers to give the most satisfactory results. Various forms of dynamometers have been used, and these are described by Stoops and Lovell[4] and Black[6].

*Fluid Motion and Power Requirements*

One of the problems confronting the designer of agitating equipment is that of deducing from experimental work with small units what will be the most satisfactory arrangement for a larger unit. In order to achieve the same kind of flow pattern in two units, geometrical, kinematic, and dynamic similarity must be maintained, as well as similar boundary conditions. This problem has been discussed by a number of workers, including Rushton *et al.*[7] and Kramers *et al.*[8]. The latter authors have made virtually the only attempt to assess the relative merits of different arrangements of mixers. It has been found convenient to relate the power used by the agitator to the geometrical and mechanical arrangements of the mixer, and thus to obtain a direct indication of the change in power for any alteration to the mixer. The general method of attack is indicated in the following section, the nomenclature following that given in the diagram (Fig. 13.10).

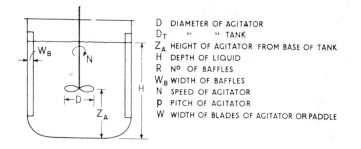

D   DIAMETER OF AGITATOR
$D_T$   "       " TANK
$Z_A$   HEIGHT OF AGITATOR FROM BASE OF TANK
H   DEPTH OF LIQUID
R   N° OF BAFFLES
$W_B$   WIDTH OF BAFFLES
N   SPEED OF AGITATOR
p   PITCH OF AGITATOR
W   WIDTH OF BLADES OF AGITATOR OR PADDLE

FIG. 13.10. Dimensions of agitator

For similarity in two mixing systems it is important to arrange for:

(a) *Geometrical similarity.* This will define the boundary conditions; corresponding dimensions will have the same ratio.

Thus the following ratios:

$$\frac{D_T}{D}; \frac{Z_A}{D}; \frac{W_B}{D}; \frac{h}{D}; \frac{W}{D}; \frac{H}{D}; \quad \text{must be the same in the two systems.}$$

(b) *Kinematic similarity.* This requires that velocities at corresponding points must have the same ratio as those at other corresponding points. The paths of motion must also be alike.

(c) *Dynamic similarity.* This requires that the ratio of forces at corresponding points is equal to that at other corresponding points.

If the boundary conditions are fixed, then one variable such as power **P** can be expressed in terms of a number of other independent variables:

$$\mathbf{P} = f(D, \mu, g, \rho, N) \tag{13.1}$$

The simplest form of function is the product of powers of the variables and then:

$$\mathbf{P} = K'(D^{n_1} \mu^{n_2} g^{n_3} \rho^{n_4} N^{n_5}) \tag{13.2}$$

Expressing these terms on the basis of the **LMT** system of dimensions:

$$\frac{\mathbf{ML}^2}{\mathbf{T}^3} = \left[ \mathbf{L}^{n_1} \left(\frac{\mathbf{M}}{\mathbf{LT}}\right)^{n_2} \left(\frac{\mathbf{L}}{\mathbf{T}^2}\right)^{n_3} \left(\frac{\mathbf{M}}{\mathbf{L}^3}\right)^{n_4} \left(\frac{1}{\mathbf{T}}\right)^{n_5} \right]$$

Equating the dimensions on each side:

for **L:** $\qquad\qquad\qquad 2 = n_1 - n_2 + n_3 - 3n_4$

**M:** $\qquad\qquad\qquad 1 = n_2 + n_4$

**T:** $\qquad\qquad\qquad -3 = -n_2 - 2n_3 - n_5$

Writing these indices in terms of $n_2$ and $n_3$:

$$n_4 = 1 - n_2$$

$$n_1 = 2 + n_2 - n_3 + 3 - 3n_2 = 5 - 2n_2 - n_3$$

$$n_5 = -n_2 - 2n_3 + 3$$

$$\therefore \quad \mathbf{P} = K'[D^{-2n_2 - n_3 + 5} \mu^{n_2} g^{n_3} \rho^{1-n_2} N^{-n_2 - 2n_3 + 3}]$$

$$= K'\left[(D^5 \rho N^3)\left(\frac{D^2 N \rho}{\mu}\right)^{-n_2}\left(\frac{DN^2}{g}\right)^{-n_3}\right]$$

$$\therefore \quad \frac{\mathbf{P}}{D^5 N^3 \rho} = K'\left[\frac{D^2 N \rho}{\mu}\right]^{-n_2}\left[\frac{DN^2}{g}\right]^{-n_3} \qquad (13.3)$$

or

$$\mathbf{N_P} = K' Re^b Fr^c \qquad (13.4)$$

In this analysis the Reynolds number $(D^2 N \rho/\mu)$ accounts for the viscous forces, and may be regarded as the ratio of the inertia to the viscous forces, and the Froude number $(DN^2/g)$ represents the influence of gravitation. Where, as is generally the case, the viscous forces are significant, then kinematic similarity will be obtained by arranging for $Re$ to be the same in the two systems. Rushton et al.[7] have given data for a propeller (Fig. 13.11), by plotting Power Number $\mathbf{N_P}$ vs $Re$.

For values of $Re < 300$, all the data fall on a single line indicating that the Froude number has no important effect. Thus

$$\mathbf{N_P} = K'(Re)^b$$

and

$$b = -1$$

so that

$$\mathbf{P} = K' \mu N^2 D^3 \qquad (13.5)$$

where $K$ depends on the impeller and surroundings. For marine-type three-bladed propellers with pitch equal to diameter, $K'$ is found to have a value of about 41.

For higher values of $Re$, the Froude number plays a part, and separate lines are drawn for various speeds. The Reynolds number was varied by using different viscosities as well as different speeds, and the slanting lines represent conditions of constant viscosity.

Figure 13.12 also taken from the work of Rushton, shows similar data for a 150 mm diameter turbine with six flat blades. The effect of introducing baffles is also brought out in this chart.

Bissell et al.[9] have given the data shown in Table 13.1 for the power consumption with different baffles for a turbine mixer situated one diameter from the bottom of a cylindrical container, and operating at a Reynolds number $Re$ of $5 \times 10^4$.

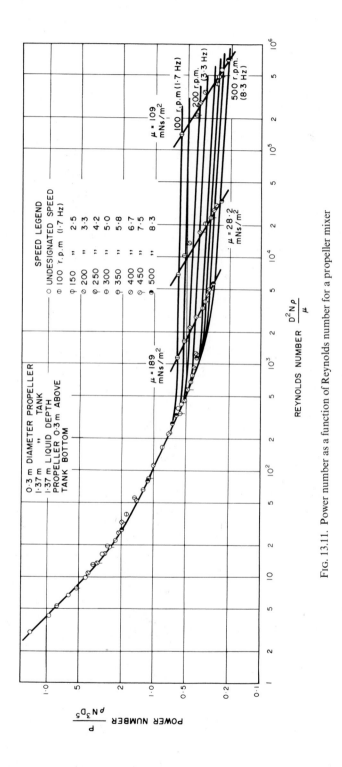

Fig. 13.11. Power number as a function of Reynolds number for a propeller mixer

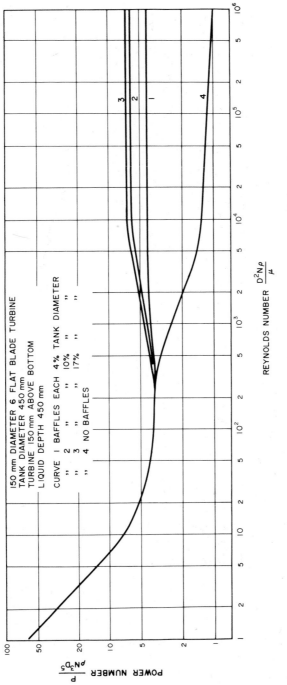

TABLE 13.1. *Effect of Width and Number of Baffles on Power*

| Baffle width as % of tank diameter | % power based on 4 baffles of $\frac{1}{12}$ tank diameter | | | | | | |
| --- | --- | --- | --- | --- | --- | --- | --- |
| | 1 | 2 | 3 | 4 | 5 | 6 | baffles |
| 2 | 30 | 52 | 63 | 72 | 76 | 78 | |
| 5·5 $(\frac{1}{18})$ | 40 | 64 | 78 | 87 | 92 | 94 | |
| 8·3 $(\frac{1}{12})$ | 50 | 78 | 92 | 100 | 102 | 104 | |
| 10·0 $(\frac{1}{10})$ | 58 | 82 | 95 | 103 | 105 | 106 | |

The power requirements of a turbine mixer operating at a peripheral speed of 3·5 m/s for different diameters is given by the figures below in Tables 13.2, 13.3 and 13.4.[10] The effects of peripheral speed and viscosity of the liquid are also given.

TABLE 13.2. *Effect of impeller Diameter on Power for Turbine Mixer*

| Diameter | | Power | |
| --- | --- | --- | --- |
| (in) | (m) | (hp) | (kW) |
| 6 | 0·15 | 0·1 | 0·075 |
| 10 | 0·25 | 0·3 | 0·224 |
| 20 | 0·51 | 1·1 | 0·814 |
| 40 | 1·02 | 4 | 2·98 |
| 80 | 2·04 | 15 | 11·17 |

TABLE 13.3. *Effect of Liquid Viscosity on Power for Turbine Mixer*

| $\mu$ (mN s/m$^2$) | Power (as percentage of that for 1 mN s/m$^2$) |
| --- | --- |
| 1 | 100 |
| 1,000 | 120 |
| 10,000 | 130 |
| 50,000 | 210 |

TABLE 13.4. *Effect of Peripheral Speed on Power for Turbine Mixer*

| Peripheral speed (m/s) | Power (as percentage of that for 3·5 m/s) |
| --- | --- |
| 0·66 | 1 |
| 1·63 | 10 |
| 2·79 | 50 |
| 3·56 | 100 |
| 4·57 | 200 |
| 6·10 | 520 |

Metzner et al.[11] give experimental results for viscous Newtonian and non-Newtonian fluids, agitated in vessels only slightly greater in diameter than the impeller of the mixer.

## 13.2.3. Rate of Mixing

In assessing the performance of a given unit, it is not only the power used that is important, but also the rate at which the desired degree of mixing is achieved. It has proved very difficult to define a scale for expressing the degree of mixing achieved at any time, and the final decision as to when a material is mixed is still left to the skill, experience and judgement of the operator. The essential test is the practical one of whether the mixture meets the requirement for which it is being prepared. This may be measured, for example, by the rate of burning of a blended nitrocellulose propellant, or by the colour reflection of a paint, or by the uniformity of conduction of an electrolyte in water. These are varying scales which cannot serve as a general criterion of performance. A few of the methods tried for assessing the effectiveness of agitation are:

(a) the rate of dispersion of an electrolyte, such as brine, in water,

(b) the rate of distribution of sand in water, and

(c) the rate of dissolution of solids in different solvents.

Method (a) was tried by Wood, Whittemore, and Badger[2] and more recently by Kramers et al.[8]. Kramers used two tanks of 0·32 and 0·64 m diameter and a three-bladed marine propeller of diameter one-quarter that of the tank. To a solution of potassium chloride containing about 3 mg/cm$^3$, a small quantity of concentrated solution containing 100 mg/cm$^3$ was added. By using a suitable electronic arrangement, a record was obtained of the change in concentration with time; the time to reach an arbitrary standard of 0·1 per cent deviation was taken as the mixing time. Typical records are illustrated in Fig. 13.13.

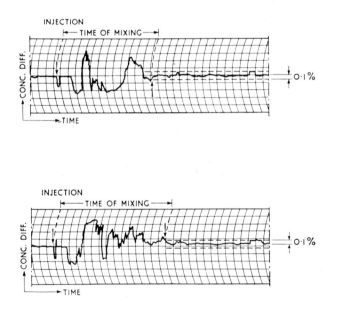

Fig. 13.13. Determination of time of mixing by method of Kramers

Method (b) has not been widely used but, as suggested by Reavell[1], if polyvinyl chloride particles are used, it is possible to observe the flow pattern of an agitator. In method (c), suggested by Hixson and Crowell[12], when the change in concentration is small then the rate of solution is:

$$\frac{dw}{dt} = -k'w^{2/3} \tag{13.6}$$

where $w$ is the mass of solid solute present at time $t$, i.e.

$$\frac{1}{3}k't = w_0^{1/3} - w^{1/3} \qquad (13.7)$$

## 13.3. LIQUID–LIQUID EXTRACTION

### 13.3.1. Applications

The principles implied in the operation of liquid–liquid extraction, which involves the three stages of contacting, separation and solvent recovery referred to in Section 13.1, are discussed further in this section together with several important applications.

The separation of aromatics from kerosene-based fuel oils to improve their burning qualities and the separation of aromatics from paraffin and naphthenic compounds to improve the temperature-viscosity characteristics of lubricating oils are important applications of this technique. It may also be used to obtain, for example, relatively pure compounds such as benzene, toluene, and xylene from catalytically produced reformates in the oil industry, in the production of anhydrous acetic acid, in the extraction of phenol from coal tar liquors, and in the purification of penicillin. The important feature is the selective nature of the solvent, in that the separation of compounds is based on differences in solubilities, rather than differences in volatilities as in distillation.

Extraction is in many ways complementary to distillation and is preferable in the following cases:

(a) where distillation would require excessive amounts of heat—for example when the relative volatility is near unity;
(b) when the formation of azeotropes limits the degree of separation obtainable in distillation;
(c) when heating must be avoided; and
(d) when the components to be separated are quite different in nature.

A recent and extremely important development lies in the application of the technique of liquid extraction to metallurgical processes. The successful development of methods for the purification of uranium fuel and for the recovery of spent fuel elements in the nuclear power industry by extraction methods, mainly based on packed, including pulsed, columns (Section 13.5.6) has led to their application to other metallurgical processes. Of these, the recovery of copper from acid leach liquors and subsequent electro-winning from these liquors is the most extensive yet in operation, though further applications to nickel and other metals are being actively examined. In many of these processes, some form of chemical complex is formed between the solute and the solvent so that the kinetics of the process become important. The extraction operation may be either a physical operation, as discussed previously, or a chemical operation. Chemical operations have been classified by Hanson[13] as follows:

(a) those involving cation exchange, for example, the extraction of metals by carboxylic acids;
(b) those involving anion exchange, such as the extraction of anions involving a metal with amines;
(c) those involving the formation of an additive compound, for example, extraction with

neutral organo-phosphorus compounds. An important operation of this type is the purification of uranium from the nitrate with tri-n-butyl phosphate.

This process of metal purification is of particular interest in that it involves the application of principles of both chemistry and of chemical engineering and necessitates the cost evaluation of alternatives. It is very important that the student should appreciate that selection of the process may frequently require the application of various disciplines and that plant fabrication will generally involve mechanical engineering.

### 13.3.2. Design Considerations

The three steps outlined in Section 13.1 which are necessary in a liquid–liquid extraction operation may be carried out either as a batch or as a continuous process.

In the single-stage batch process illustrated in Fig. 13.14, the solvent and solution are mixed together and then allowed to separate into the two phases: the *extract* E containing the required solute in the added solvent and the *raffinate* F the weaker solution with some associated solvent. With this simple arrangement mixing and separation occur in the same vessel.

A continuous two-stage operation is shown in Fig. 13.15, where the mixers and

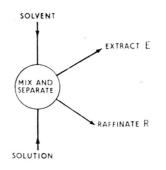

FIG. 13.14. Single stage batch extraction

separators are shown as separate vessels. There are three main forms of equipment: first the mixer-settler as shown in Fig. 13.14; secondly, the column type with trays or packing as in distillation; and thirdly, a variety of units incorporating rotating devices such as the Scheibel and the Podbielniak extractors. In all cases, the extraction units are followed by distillation or a similar operation in order to recover the solvent and the solute. Some indication of the form of these alternative arrangements can be seen by considering two of the processes referred to in the previous section.

One system for separating the benzene, toluene, and xylene groups from light feedstocks is shown in Fig. 13.16, where n-methylpryolidone (NMP) with the addition of some glycol is used as the solvent. Here the feed is passed to a multistage extractor arranged as a tower from which an aromatics-free raffinate is obtained at the top. The extract stream containing the solvent, aromatics, and low boiling non-aromatics is distilled to provide the extractor recycle stream as a top product and a mixture of aromatics and solvent at the bottom. This stream passes to a stripper from which the glycol and the aromatics are

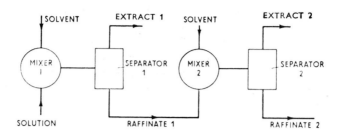

FIG. 13.15. Multiple contact system with fresh solvent

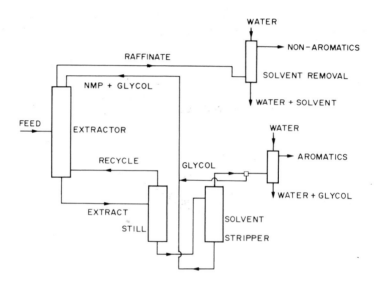

FIG. 13.16. Process for benzene, tolulene, xylene recovery

recovered. This is a complex system illustrating the need for careful recycling and recovery of solvent.

The concentration of acrylic acid by extraction with ethyl acetate[14] is a rather different illustration of this technique. As shown in Fig. 13.17, the dilute acrylic acid solution of concentration about 20 per cent is fed to the top of the extraction column 1, the ethyl acetate solvent being fed in at the base. The acetate containing the dissolved acrylic acid and water leaves from the top and is fed to the distillation column 2. Here the acetate is removed as an azeotrope with water and the dry acrylic acid is recovered as product from the bottom.

It may be seen from these illustrations that successful extraction processes are not to be judged simply by the performance of the extraction unit alone, but by assessment of the recovery achieved by the whole plant. This aspect of the process may be complex if chemical reactions are involved. The sections of the plant for mixing and for separation must be considered together when assessing capital cost; in addition, the cost of the organic solvents used in the metallurgical processes may also be high.

The mechanism of transfer of solute from one phase to the second is one of molecular and eddy diffusion and the concepts of phase equilibrium, interfacial area, and surface renewal are all similar in principle to those already met in distillation, though in liquid–liquid extraction, dispersion is effected by mechanical means (except in standard packed columns) including pumping and agitation.

In formulating design criteria for extraction equipment, it is necessary to take into account the equilibrium conditions for the distribution of solute between the phases as this will determine the maximum degree of separation possible in a single stage. The resistance to diffusion and, in the case of chemical effects, the kinetics are also important in that they

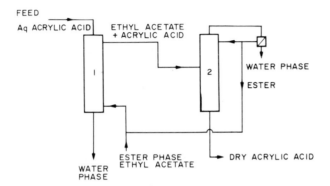

FIG. 13.17. Concentration of acrylic acid by extraction with ethyl acetate[14]

determine the residence time required to bring about near equilibrium in a stage-wise unit, or the height of a transfer unit in a differential contactor. The transfer rate is given by the usual equation:

$$\text{Rate per unit interfacial area} = k\Delta C \tag{13.8}$$

where $k$ is a mass transfer coefficient and $\Delta C$ the concentration driving force. A high value of $k$ can be obtained only if turbulent or eddy conditions prevail and, although these can be readily achieved in the continuous phase by some form of agitation, it is very difficult to generate eddies in the drops which constitute the dispersed phase.

### 13.3.3. Equilibrium Conditions

The equilibrium condition for the distribution of one solute between two liquid phases is conveniently considered in terms of the distribution law. Thus, at equilibrium, the ratio of the concentrations of the solute in the two phases is given by $C_E/C_R = K'$, where $K'$ is the distribution constant. This relation will only apply accurately if both solvents are immiscible, and if there is no association or dissociation of the solute. If the solute forms molecules of different molecular weight, then the distribution law holds for each molecular species. Where the concentrations are small, the distribution law usually holds provided no chemical reaction occurs.

*Ternary Systems*

The addition of a new solvent to a binary mixture of the solute in a solvent may lead to the formation of several types of mixture:

(a)  A homogeneous solution may be formed; then the selected solvent is unsuitable.

(b)  The solvent may be completely immiscible with the initial solvent.

(c)  The solvent may be partially miscible with the original solvent, resulting in the formation of one pair of partially miscible liquids.

(d)  The new solvent may lead to the formation of two or three partially miscible liquids.

Of these possibilities, types b, c, and d all give rise to systems that may be used, but those of types b and c are the most promising. With conditions of type b, the equilibrium relation is conveniently shown by a plot of the concentration of solute in one phase against the concentration in the second phase. Conditions given by c and d are usually represented by triangular diagrams. Equilateral triangles will be used, though it is also possible to employ right-angled isosceles triangles, which are discussed in Chapter 10.

*Use of Triangular Diagrams*

The system, acetone acetone(**A**)–water(**B**)–methyl isobutyl ketone(**C**), as indicated in Fig. 13.18, is of type c above. Here the solute **A** is completely miscible with the two solvents

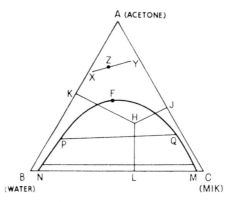

FIG. 13.18. Equilibrium relationship for acetone distributed between water and methyl isobutyl ketone

**B** and **C**, but the two solvents are only partially miscible with each other. Then a mixture indicated by point $H$ consists of the three components **A**, **B** and **C** in the ratio of the perpendiculars $HL$, $HJ$, $HK$. The distance $BN$ represents the solubility of solvent **C** in **B**, and $MC$ that of **B** in **C**. The area under the curved line $NPFQM$ (the binodal solubility curve) represents a two-phase region which will split up into two layers in equilibrium with each other. These layers have composition represented by points $P$ and $Q$, and $PQ$ is known as a "tie line". Such lines, two of which are shown in the diagram, connect together two phases in equilibrium with each other, and these points must be found by experiment. There will be one point on the binodal curve at $F$ which represents a single phase which does not split into two phases. $F$ is known as a "plait" point, and must also be found by experiment. The plait point is fixed if either the temperature or the pressure is fixed. Inside

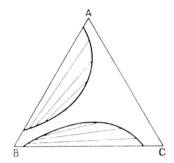

FIG. 13.19. Equilibrium relationship for aniline–water–phenol system

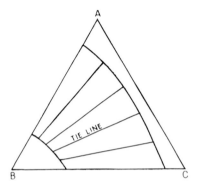

FIG. 13.20. Equilibrium relationship for aniline–water–phenol system

the area under the curve, the temperature and composition of one phase will fix the composition of the other. Applying the phase rule to the three-components system at constant temperature and pressure, the number of degrees of freedom is equal to 3 minus the number of phases. In the area where there is only one liquid phase, there are two degrees of freedom and two compositions must be stated. In a system where there are two liquid phases, there is only one degree of freedom.

One of the most useful features of this method of representation is that, if a solution of composition $X$ is mixed with one of composition $Y$ then the resulting mixture will have a composition shown by $Z$ on a line $XY$, such that:

$$XZ/ZY = (\text{amount of } Y)/(\text{amount of } X).$$

Similarly, if from a mixture $Z$ an extract $Y$ is removed the remaining liquor will have composition $X$.

In Fig. 13.19 two separate two-phase regions are formed, whilst in Fig. 13.20 the two-phase regions merge on altering the temperature. Aniline (**A**), water (**B**), and phenol (**C**) represent a system of the latter type. Under the conditions shown in Figs. 13.19 and 13.20, **A** and **C** are miscible in all proportions, but **B** and **A**, and **B** and **C** are only partially miscible.

Whilst these diagrams are of considerable use in presenting equilibrium data, Fig. 13.21 is in many ways more useful for determining the selectivity of the solvent, and the number of stages that are likely to be required. In Fig. 13.21, the percentage of solute in one phase is plotted against the percentage in the second phase in equilibrium with it. This is equivalent to plotting the compositions at either end of a tie line. The important factor in assessing the value of a solvent is the ratio of the concentrations of the desired component in the two phases, rather than the actual concentrations. A selectivity ratio can be defined as:

$$\beta = \left[\frac{x_A}{x_B}\right]_E / \left[\frac{x_A}{x_B}\right]_R \tag{13.9}$$

where $x_A$ and $x_B$ are the mass or mol fractions of **A** and **B** in the two phases $E$ and $R$.

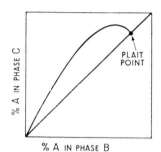

FIG. 13.21. Equilibrium distribution of solute $A$ in phases $B$ and $C$

For a few systems $\beta$ tends to be substantially constant, though more usually it varies with concentration. The selectivity ratio has the same significance in extraction as relative volatility has in distillation, so that the ease of separation is directly related to the numerical value of $\beta$. As $\beta$ approaches unity, a larger number of stages is necessary for a given degree of separation and the capital and operating costs increase correspondingly. When $\beta = 1$ any separation becomes impossible.

## 13.4. CALCULATION OF THE NUMBER OF THEORETICAL STAGES IN EXTRACTION OPERATIONS

### 13.4.1. Co-current Contact with Partially Miscible Solvents

For calculating the number of ideal stages required for a given degree of separation, the conditions of equilibrium expressed by one of the methods indicated in Section 13.3.3. will be used. The number of stages where single or multiple contact equipment is involved will be considered first and then the design of equipment where the concentration change is continuous will be discussed.

For the general case where the solvents are partially miscible, the feed solution $F$ is brought into contact with the selective solvent $S$, to give raffinate $R_1$ and an extract $E_1$. The addition of streams $F$ and $S$ is shown on the triangular diagram (Fig. 13.22), by the point $M$, where $FM/MS = S/F$. This mixture $M$ breaks down to give extract $E_1$ and raffinate $R_1$, at opposite ends of a tie line through $M$.

If a second stage is used, then the raffinate $R_1$ is treated with a further quantity of solvent $S$, and extract $E_2$ and raffinate $R_2$ are obtained as shown in the figure.

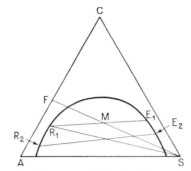

FIG. 13.22.  Multiple contact with fresh solvent used at each stage

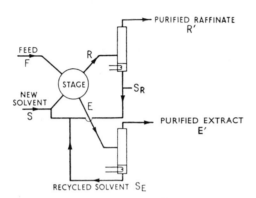

FIG. 13.23.  Single-stage process with solvent recovery

*Solvent recovery*

The complete process consists in carrying out the extraction, and recovering the solvent from the raffinate and extract obtained. Thus, for a single-stage system as indicated in Fig. 13.23, the raffinate $R$ is passed into the distillation column where it is separated to give purified raffinate $R'$ and solvent $S_R$. The extract $E$ is passed to another distillation unit to give extract $E'$ and a solvent stream $S_E$. These recovered solvents $S_R$ and $S_E$ are pumped back to the extraction process as shown. This cycle can be represented on the diagram (Fig. 13.24), by showing the removal of $S_R$ from $R$ to give composition $R'$, and the removal of $S_E$ from $E$ to give composition $E'$. It has been assumed in this case that perfect separation is obtained in the stills, so that pure solvent is obtained in the streams $S_R$ and $S_E$, though the same form of diagram can be used where imperfect separation is obtained. It should be noted that, when $ES$ is a tangent to the binodal curve, we have the maximum concentration of solute **C** in the extract $E'$. It also follows that $E'$ represents the maximum possible concentration of **C** in the feed. Sufficient solvent **S** must be used to bring the mixture $M$ within the two-phase area.

## 13.4.2.  Co-current Contact with Immiscible Solvents

In this case, as illustrated in Fig. 13.25, there is no need to use triangular diagrams. Suppose the initial solution contains $A$ kg of solvent **A** with a mass ratio $X_f$ of solute. Then

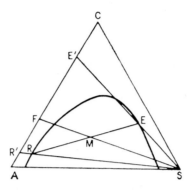

FIG. 13.24. Representation of process shown in Fig. 13.23

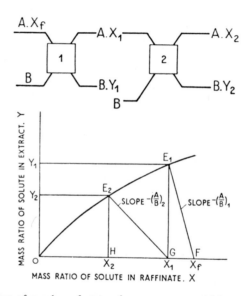

FIG. 13.25. Calculation of number of stages for co-current multiple-contact process, using immiscible solvents

the selective solvent to be added will be a mass $B$ of solvent **B**. On mixing and separating, a raffinate is obtained with the solvent **A** containing a mass ratio $X_1$ of solute, and an extract with the solvent **B** containing a mass ratio $Y_1$ of solute. A material balance on the solute gives:

$$AX_f = AX_1 + BY_1$$

$$\therefore \quad \frac{Y_1}{X_1 - X_f} = -A/B \tag{13.10}$$

This process can be illustrated by letting the point $F$ represent the feed solution and drawing a line $FE_1$, of slope $-(A/B)_1$, to cut the equilibrium curve in $E_1$. This will then give composition $Y_1$ of the extract and $X_1$ of the raffinate. If a further stage is then carried out by the addition of solvent **B** to the stream $AX_1$, then point $E_2$ is found on the equilibrium curve by drawing $GE_2$ of slope $-(A/B)_2$. Point $E_2$ then gives the compositions $X_2$ and $Y_2$ of the final extract and raffinate. This system can be used for any number of stages, with any assumed variation in the proportion of solvent **B** to raffinate from stage to stage.

*Conditions where the Distribution Law is followed.*

If the distribution law is followed, then the equilibrium curve becomes a straight line of equation $Y = mX$. The material balance on the solute may then be rewritten as:

$$AX_f = AX_1 + BY_1 = AX_1 + BmX_1 = (A + Bm)X_1$$

$$\therefore \quad X_1 = \left[\frac{A}{A + Bm}\right] X_f. \tag{13.11}$$

If a further mass $B$ of **B** is added to raffinate $AX_1$ to give extract of composition $Y_2$ and raffinate $X_2$ in a second stage, then:

$$AX_1 = AX_2 + BmX_2 = X_2(A + Bm)$$

$$\therefore \quad X_2 = \left[\frac{A}{A + Bm}\right] X_1 = \left[\frac{A}{A + Bm}\right]^2 X_f \tag{13.12}$$

And for $n$ stages:

$$X_n = \left[\frac{A}{A + Bm}\right]^n X_f \tag{13.13}$$

or the number of stages:

$$n = \frac{\log X_n/X_f}{\log\left[\dfrac{A}{A + Bm}\right]} \tag{13.14}$$

## 13.4.3. Countercurrent Contact with Immiscible Solvents

If a series of mixing and separating vessels is arranged so that the flow is countercurrent, then the conditions of flow can be represented as in Fig. 13.26, where each circle corresponds to a mixer and a separator. The initial solution $F$ of the solute **B** in solvent **A** is fed to the first unit and leaves as raffinate $R_1$. This stream passes through the units and leaves from the $n$th unit as stream $R_n$. The fresh solvent **S** enters at the $n$th unit and passes in the reverse direction through the units, leaving as extract $E_1$.

Let $X$ denote the ratio of solute to solvent in the raffinate streams, and
$Y$ denote the ratio of solute to solvent in the extract streams.

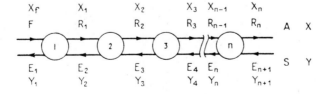

FIG. 13.26. Arrangement for multiple-contact extraction in countercurrent flow

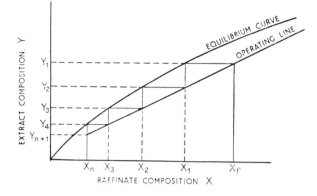

F<small>IG</small>. 13.27. Graphical method for determining the number of stages for process shown in Fig. 13.26, using immiscible solvents

Since the two solvents are immiscible, the solvent in the raffinate streams remains as $A$ units, and the added solvent in the extract streams as $S$. The following material balances for the solute may then be written:

(a) For the 1st stage

$$AX_f + SY_2 = AX_1 + SY_1$$

(b) For the $n$th stage

$$AX_{n-1} + SY_{n+1} = AX_n + SY_n$$

(c) For the whole unit

$$AX_f + SY_{n+1} = AX_n + SY_1$$

$$\therefore \quad Y_{n+1} = \frac{A}{S}(X_n - X_f) + Y_1 \tag{13.15}$$

This is the equation of a straight line of slope $A/S$, known as the *operating line*; from the above equalities, it is seen to pass through the points $(X_f, Y_1)$ and $(X_n, Y_{n+1})$. In Fig. 13.27 is drawn the equilibrium relation ($Y_n$ versus $X_n$) and the operating line, and the number of stages required to pass from $X_f$ to $X_n$ is found by drawing in steps between the operating line and the equilibrium curve; in the example shown four stages are required, and $(X_n, Y_{n+1})$ corresponds to $(X_4, Y_5)$. It should be noted that the operating line connects the compositions of the raffinate stream leaving and the fresh solvent stream entering a unit ($X_n$ and $Y_{n+1}$, respectively).

### 13.4.4. Countercurrent Contact with Partially Miscible Solvents

The arrangement of the equipment will be the same as for the previous case with immiscible solvents though, since the amounts of solvent in the extract and raffinate streams are changing, the material balance is taken for the total streams entering and leaving each stage.

With the notation as before (Fig. 13.26), suppose that the feed $F$, final extract $E_1$, fresh solvent $S(= \text{stream } E_{n+1})$ and final raffinate $R_n$ to be fixed. Then taking material balances:

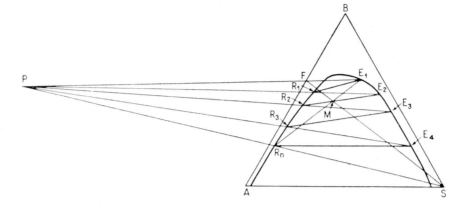

FIG. 13.28. Graphical method for determining the number of stages for process shown in Fig. 13.26, using partially miscible solvents

(a) *Over 1st Unit*

$$F + E_2 = R_1 + E_1$$

$$\therefore \quad F - E_1 = R_1 - E_2 = P, \; say \text{—the difference stream} \tag{13.16}$$

(b) *Over stages 1 to n*

$$F + E_{n+1} = R_n + E_1 = M, \; say \tag{13.17}$$

$$F - E_1 = R_n - E_{n+1} = P. \tag{13.18}$$

(c) *Over nth Unit*

$$R_{n-1} + E_{n+1} = E_n + R_n$$

$$\therefore \quad R_{n-1} - E_n = R_n - E_{n+1} = P \tag{13.19}$$

Thus the difference in quantity between the raffinate leaving a stage $R_n$, and the extract entering from next stage $E_{n+1}$, is constant. Similarly, it can be shown that the difference between the amounts of each component in the raffinate and the extract streams will be constant. This means that, with the notation of a triangular diagram, lines joining any two points representing $R_n$ and $E_{n+1}$ will pass through a common pole. The number of stages required to pass from an initial concentration $F$ to a final raffinate concentration $R_n$ can then be found using a triangular diagram, by the method shown in Fig. 13.28.

Join the points $F$ and $S$ representing the compositions of the feed and fresh solvent $S$. Then the composition of a mixture of $F$ and $S$ is shown by point $M$ where

$$\frac{MS}{MF} = \frac{\text{mass of } F}{\text{mass of } S}$$

Draw a line from $R_n$ through $M$ to give $E_1$ on the binodal curve. Draw $E_1F$ and $SR_n$ to meet at the pole $P$. It should be noted that $P$ represents an imaginary mixture, as already found for the leaching problems discussed in Chapter 10.

Then, in an ideal stage, the extract $E_1$ leaves in equilibrium with the raffinate $R_1$, so that the point $R_1$ is at the end of the tie line through $E_1$. To find the extract $E_2$, draw $PR_1$ to cut the binodal curve at $E_2$. Then the points $R_2$, $E_3$, $R_3$, $E_4$, etc., can be found in the same way. If the final tie line (say $ER_4$) does not pass through $R_n$, then the amount of solvent added is incorrect for the desired change in composition. This does not generally invalidate the method, since it will give the required number of ideal stages sufficiently accurately.

### 13.4.5. Continuous Extraction in Columns

The use of spray towers, packed towers or mechanical columns enables continuous countercurrent extraction to be obtained in a similar manner to gas absorption[15] or distillation. Applying the two-film theory of diffusion, the concentration gradients for transfer to a desired solute from a raffinate to an extract phase are as shown in Fig. 13.29, which is similar to Fig. 12.1 for gas absorption.

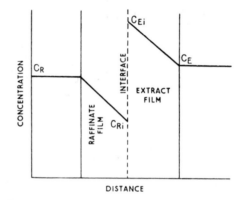

FIG. 13.29. Concentration profile near an interface

The transfer through the film on the raffinate side of the interface is brought about by a concentration difference $C_R - C_{Ri}$, and through the film on the extract side by a concentration difference $C_{Ei} - C_E$.

The rate of transfer across these films can be expressed, as in absorption, in the following way:

$$N' = k_R(C_R - C_{Ri}) = k_E(C_{Ei} - C_E) \tag{13.20}$$

where   $C_R$, $C_E$ are the concentrations of the solute in the raffinate and extract phases in mol per unit volume,

$k_R$, $k_E$ are the transfer coefficients for raffinate and extract films, respectively, and

$N'$ is the molar rate of transfer per unit area.

Then:

$$\frac{k_R}{k_E} = \frac{C_{Ei} - C_E}{C_R - C_{Ri}} = \frac{\Delta C_E}{\Delta C_R} \tag{13.21}$$

If the equilibrium curve can be taken as a straight line, then assuming equilibrium at the interface:

$$C_{Ei} = mC_{Ri} \tag{13.22}$$

Further:

$$C_E = mC_R^*$$
(13.23)

and:

$$C_E^* = mC_R$$
(13.24)

where $C_E^*$ is the concentration in phase $E$ in equilibrium with $C_R$ in phase $R$, and $C_R^*$ is the concentration in phase $R$ in equilibrium with $C_E$ in phase $E$.

The relations for mass transfer may also be written in terms of overall transfer coefficients $K_R$ and $K_E$ defined by the relations:

$$N' = K_R(C_R - C_R^*) = K_E(C_E^* - C_E)$$
(13.25)

and by similar reasoning to that for absorption (Chapter 12):

$$\frac{1}{K_R} = \frac{1}{k_R} + \frac{1}{mk_E}$$
(13.26)

and

$$\frac{1}{K_E} = \frac{1}{k_E} + \frac{m}{k_R}$$
(13.27)

### Capacity of Column Operating as Continuous Countercurrent Unit

The capacity of a column operating as a countercurrent extractor, as indicated in Fig. 13.30, can be derived as follows:

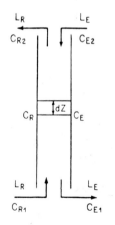

FIG. 13.30. Countercurrent flow in a packed column

If $L'_R$, $L'_E$ are the volumetric flow rates of raffinate and extract phases per unit area,
      $a$ is the interfacial surface per unit volume, and
      $Z$ is the height of packing.
then, over a small height $dZ$, a material balance gives:

$$L'_R \, dC_R = L'_E \, dC_E$$
(13.28)

From equation 13.20:

$$L'_R \, dC_R = k_R(C_R - C_{Ri})a \, dZ$$

$$\therefore \int_{C_{R2}}^{C_{R1}} \frac{dC_R}{C_R - C_{Ri}} = \frac{k_R a}{L'_R} Z \qquad (13.29)$$

The integral on the left-hand side of this equation is known as the number of raffinate film transfer units $\mathbf{N}_R$, and the height of the raffinate film transfer unit is:

$$\mathbf{H}_R = \frac{L'_R}{k_R a} \qquad (13.30)$$

In a similar manner, and by analogy with absorption:

$$\mathbf{H}_E = \frac{L'_E}{k_E a} = \text{height of extract film transfer unit} \qquad (13.31)$$

$$\mathbf{H}_{OR} = \frac{L'_R}{K_R a} = \begin{array}{l}\text{height of overall transfer unit based on}\\ \text{concentration in raffinate phase}\end{array} \qquad (13.32)$$

$$\mathbf{H}_{OE} = \frac{L'_E}{K_E a} = \begin{array}{l}\text{height of overall transfer unit based on}\\ \text{concentration in extract phase.}\end{array} \qquad (13.33)$$

Since

$$\frac{1}{K_R} = \frac{1}{k_R} + \frac{1}{mk_E} \qquad \text{(equation 13.26)}$$

$$\frac{L'_R}{K_R} = \frac{L'_R}{k_R} + \frac{L'_R}{mk_E} \times \frac{L'_E}{L'_E}$$

$$\therefore \quad \mathbf{H}_{OR} = \mathbf{H}_R + \frac{L'_R}{mL'_E} \mathbf{H}_E \qquad (13.34)$$

and

$$\mathbf{H}_{OE} = \mathbf{H}_E + \frac{mL'_E}{L'_R} \mathbf{H}_R \qquad (13.35)$$

This is the same form of relation as already obtained for absoprtion, but it is only with dilute solutions that the group $mL'_E/L'_R$ is constant. If equations 13.34 and 13.35 are combined, it will be seen that:

$$\mathbf{H}_{OR} = \frac{L'_R}{mL'_E} \mathbf{H}_{OE} \qquad (13.36)$$

The group $L'_R/mL'_E$ is the ratio of the slope of the operating line to that of the equilibrium curve so that when these two are parallel it follows that $\mathbf{H}_{OR}$ and $\mathbf{H}_{OE}$ are numerically equal.

In deriving these relationships, it has been assumed that $L'_R$ and $L'_E$ are constant throughout the tower; this will not be true if a large part of the solute is transferred from a concentrated solution to the other phase. Again, it is assumed that the transfer coefficients are independent of concentration. For dilute solutions and where the equilibrium relation is a straight line, a simple expression can be obtained for determining the required height of a column, by the same method as in Chapter 12.

Thus $L'_R dC_R = K_R (C_R - C_R^*) a dZ$ can be integrated over the height $Z$ and written as:

$$L'_R (C_{R1} - C_{R2}) = K_R (\Delta C_R)_{lm} aZ \qquad (13.37)$$

where $(\Delta C_R)_{lm}$ is the logarithmic mean of $(C_R - C_R^*)_1$ and $(C_R - C_R^*)_2$. This simple relation has been used by workers in the determination of $K_R$ or $K_E$ in small laboratory columns; care should be taken when applying these results to other conditions.

Equations 13.34 and 13.35 have been used as the basis of correlating mass transfer measurements in continuous countercurrent contactors. For example, Leibson and Beckmann[16] plotted $\mathbf{H}_{OE}$ against $mL'_E/L'_R$ (Fig. 13.31) for a variety of column packings

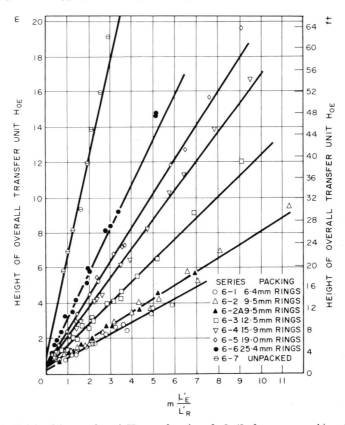

FIG. 13.31. Height of the transfer unit $\mathbf{H}_{OE}$ as a function of $mL_E/L_R$ for various packings. Transfer of diethylamine from water to dispersed toluene

and obtained good straight lines. Caution must, however, be exercised in drawing conclusions from such plots. Although equation 13.35 suggests that the intercepts and slopes of the lines are numerically equal to $\mathbf{H}_E$ and $\mathbf{H}_R$, this is only true provided that both these quantities are independent of the flow ratio $L'_E/L'_R$. This is not always so and, in the case of packed towers, the height of the continuous phase film transfer unit does in fact depend upon the flow ratio[17]. Under these conditions neither equation 13.34 nor 13.35 can be used to apportion the individual resistances to mass transfer between the two phases, and the film coefficients have to be determined by direct experiment.

## Example 13.1

In the extraction of acetic acid from an aqueous solution with benzene in a packed column of height 1·4 m and cross-sectional area 0·0045 m², the concentrations measured at the inlet and outlet of the column are as shown in Fig. 13.32.

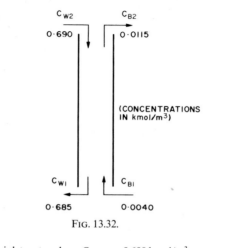

FIG. 13.32.

Acid concentration in inlet water phase $C_{W2}$ = 0·690 kmol/m³
Acid concentration in outlet water phase $C_{W1}$ = 0·685 kmol/m³
Flowrate of benzene phase = 5·7 cm³/s ≡ 1·27 × 10⁻³ m³/m² s
Inlet benzene phase concentration $C_{B1}$ = 0·0040 kmol/m³
Outlet benzene phase concentration $C_B$ = 0·0115 kmol/m³

Determine the overall transfer coefficient and the height of the transfer unit.

## Solution

The acid transferred to the benzene phase is

$$5·7 \times 10^{-6}(0·0115 - 0·0040) = 4·275 \times 10^{-8} \text{ kmol/s}$$

The equilibrium relationship for this system is

$$\frac{C_B^*}{C_W^*} = 0·0247$$

$$\therefore \quad C_{Bi}^* = 0·0247 \times 0·685 = 0·0169$$

$$C_{B2}^* = 0·0247 \times 0·690 = 0·0170$$

$$\therefore \quad \text{Driving force at bottom } \Delta C_1 = 0·0169 - 0·0040 = 0·0129$$

and

$$\text{Driving force at top } \Delta C_2 = 0.0170 - 0.0115 = 0.0055$$

$$\therefore \quad \text{Log mean driving force} = \Delta C_{\text{lm}} = 0.0087$$

$$\therefore \quad K_B a = \frac{4.275 \times 10^{-8}}{0.0063 \times 0.0087} = \frac{\text{mols transferred}}{\text{volume of packing} \times \Delta C_{\text{lm}}}$$

$$= 7.8 \times 10^{-4} \, \text{kmol/s m}^3 \, (\text{kmol/m}^3)$$

and

$$\mathbf{H}_{OB} = 1.27 \times 10^{-3} / 7.8 \times 10^{-4} = \underline{1.63 \, \text{m}}$$

## 13.5. CLASSIFICATION OF EXTRACTION EQUIPMENT

In most industrial applications, multistage countercurrent contacting is required. The hydrodynamic driving force necessary to induce countercurrent flow and subsequent phase separation may be derived from the differential effects of either gravity or centrifugal force on the two phases of different densities. Essentially there are two types of design by which effective multistage operation may be obtained:

(a) *stage-wise contactors*, in which the equipment includes a series of physical stages in which the phases are mixed and separated;

(b) *differential contactors*, in which the phases are continuously brought into contact with complete phase separation only at the exits from the unit.

The three factors, the inducement of countercurrent flow, stage-wise or differential contacting and the means of effecting phase separation are the basis of a classification of contactors proposed by Hanson[13] which is summarised in Table 13.5.

TABLE 13.5. *Classification of Contactors*

| Countercurrent flow produced by | Phase interdispersion by | Differential contactors | Stagewise contactors |
|---|---|---|---|
| Gravity | Gravity | GROUP A<br>Spray column<br>Packed column | GROUP B<br>Perforated plate column |
| | Pulsation | GROUP C<br>Pulsed packed column<br>Pulsating plate column | GROUP D<br>Pulsed sieve plate column<br>Controlled cycling column |
| | Mechanical agitation | GROUP E<br>Rotating disc contactor<br>Oldshue-Rushton column<br>Zeihl column<br>Graesser contactor | GROUP F<br>Scheibel column<br>Mixer-settlers |
| Centrifugal force | Centrifugal force | GROUP G<br>Podbielniak<br>Quadronic<br>De Laval | GROUP H<br>Westfalia<br>Robatel |

Typical regions for application of contactors of different types are given in Table 13.6. The choice of a contactor for a particular application requires an appreciation of several factors including chemical stability, the value of the products and the rate of phase separation. Occasionally, the extraction system may be chemically unstable and the contact time must then be kept to a minimum by using equipment such as a centrifugal contactor.

TABLE 13.6.  *Typical regions of application of contactor groups listed in Table 13.5*[13]

|  | System criterion | Modest throughput | High throughput |
|---|---|---|---|
| **Small number of stages required** | Chemically stable<br>Easy phase separation<br>Low value | A, B or mixer–settler | E or F |
|  | Chemically stable<br>Appreciable value | C or D | E or F (not mixer–settler) |
|  | Chemically unstable<br>Slow phase separation | G or H | G or H |
| **Appreciable number of stages required** | Chemically stable<br>Easy phase separation<br>Low value | B, C, D or mixer–settler | E or F |
|  | Chemically stable<br>Appreciable value | C or D | E or F (not mixer–settler) |
|  | Chemically unstable<br>Slow phase separation | G or H | G or H |

The more important types of stage-wise and differential contactors are now discussed in Sections 13.6 and 13.7 respectively.

## 13.6. STAGE-WISE EQUIPMENT FOR EXTRACTION

### 13.6.1. The Mixer–Settler

In this unit the solution and solvent are mixed by some form of agitator in the *mixer*, and then transferred to the *settler* where the two phases separate to give an extract and a raffinate. The mixer unit will usually be in the form of a circular or square vessel and it can then be designed on the principles given in Section 13.2. In the settler the separation is gravity controlled and the liquid densities and the form of the dispersion are important parameters. It is necessary to establish principles for determining the size of these units and an understanding of the criteria governing their internal construction. Whilst the mixer and settler are first considered as separate items, it is vitally important to realise that they are essential component parts of one processing unit.

*The mixer*. As a result of the agitation achieved in a mixer, the two phases will be brought to, or near to, equilibrium so that one theoretical stage is frequently obtained in a single mixer where a physical extraction process is taking place. Where a chemical reaction occurs, the kinetics must be established so that the residence time and the hold-up can be calculated. The hold-up is the key parameter in determining size, and scale-up is acceptably reliable though a reasonably accurate estimate of the power required is important with large units. For a circular vessel, baffles are required to obtain the optimum degree of agitation and the propeller, which should be about one-third of the diameter of the vessel, should be mounted just below the interface and have a tip speed in the range 3–15 m/s depending on the nature of the propeller or turbine. A shroud around the propeller helps to give good initial mixing of the streams, and also provides some pumping action and hence improves circulation.

As discussed in Section 13.2, the two key parameters determining power are the Reynolds number for the agitator and the Power number which are related as shown in Figs. 13.11 and 13.12. The Reynolds number should exceed $10^4$ for optimum agitation; this gives a power number of about 6 for a fully baffled tank. It is important to note that the Power number and the tip speed cannot both be kept constant in scale-up.

*The settler.* In this unit, gravitational settling occurs and, in addition, coalescence has to take place at the interface. Baffles are fitted at the inlet in order to aid distribution. The rates of sedimentation and coalescence increase with drop size, and therefore excessive agitation resulting in the formation of very small drops should be avoided. The height of the dispersion band $Z_B$ is influenced by the throughput since a minimum residence time is required for coalescence to occur. This height $Z_B$ is related to the dispersed and continuous phase superficial velocities, $u_d$ and $u_s$:

$$Z_B = \text{constant} \, (u_d + u_c) \qquad (13.38)$$

Pilot experiments may be necessary to achieve satisfactory design, though the sizing of the settler is a difficult problem when the throughputs are large.

*Combined Mixer–Settler Units*

Recent work has emphasised the need to look at the combined mixer–settler operation, particularly in metal extraction systems where the throughput may be very large. Thus Warwick[18] gives details of a recent design (Fig. 13.33) in which the two operations are effected in the one combined unit. Here the impeller has swept-back vanes with double

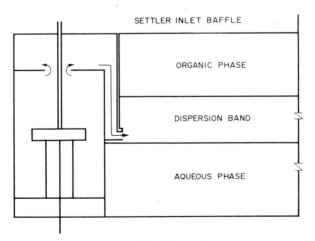

FIG. 13.33. Mixer settler

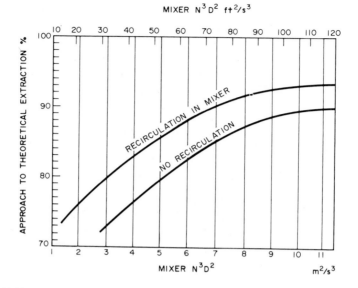

FIG. 13.34. The effect of variation of mixer internal recirculation on extraction efficiency

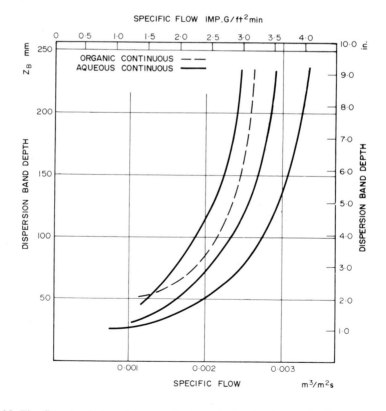

FIG. 13.35. The effect of variation of phase continuity and mixer $N^3D^2$ on settler dispersion band depth[18].

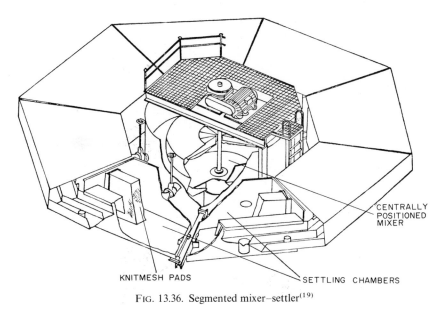

CENTRALLY
POSITIONED
MIXER

KNITMESH PADS

SETTLING CHAMBERS

FIG. 13.36. Segmented mixer–settler[19]

shrouds, and the two phases meet in the draught tube. A baffle on top of the agitator reduces air intake and a baffle on the inlet to the settler is important in controlling the flow pattern. This arrangement gives a good performance and is mechanically neat. Raising the impeller above the draught tube increases internal recirculation which in turn improves the stage efficiency, as seen in Fig. 13.34. The effect of agitation on the thickness of the dispersion band is indicated in Fig. 13.35 which is taken from the same reference. The depth of the dispersion band $Z_B$ will also vary with the flow per unit area. Whilst this work was primarily aimed at a design for copper-extraction processes, it is clear that there is scope for further important applications of these units.

*The segmented mixer–settler.* New ideas for a combined mixer–settler are incorporated in a unit from Davy International, described by Jackson *et al.*[19] and illustrated in Fig. 13.36. Specially designed knit-mesh pads are used to speed up the rate of coalescence, the effect of the packing being shown in Fig. 13.37.

The centrally situated mixer is designed to give the required hold-up, and its impeller and shroud pump the mixture at the required rate to the settler which is formed in segments around the mixer, each fed by individual pipework. The Knitmesh pads are positioned in each segment and are 0·75–1·5 m in depth. One key advantage of this design is that the holdup of the dispersed phase in the settler is reduced to about 20 per cent of that in the mixer, as compared with 50 per cent with simple gravity settlers.

### 13.6.2. Baffle-plate Columns

These are simple cylindrical columns provided with baffles to direct the flow of the dispersed phase. This is shown in Fig. 13.38. The efficiency of each plate is very low, though

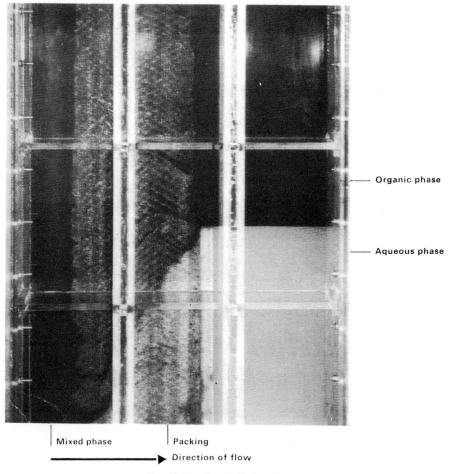

FIG. 13.37.  Flow in Knitmesh pads

since the baffles can be positioned very close together (75–150 mm), it is possible to obtain several theoretical stages in a reasonable height.


### 13.6.3. The Scheibel Column

One of the difficulties in using perforated plate and indeed packed columns is that redispersion of the liquids after each stage is very poor. To overcome this, Scheibel and Karr[20] introduced a unit, as indicated in Fig. 13.39, in which a series of agitators is mounted on a central rotating shaft. Between the agitators is fitted a wire mesh section which successfully breaks up any emulsions. Some results for a column 292 mm diameter, with 100 mm diameter agitators and with packing sections 230 and 343 mm, are shown in Fig. 13.40. It is found that one theoretical stage is obtained in a height of 0·45–0·75 m; this is a big improvement on that usually obtained in a packed column. Although there are few data on the use of large units, there should be no fall in efficiency as the diameter is increased.

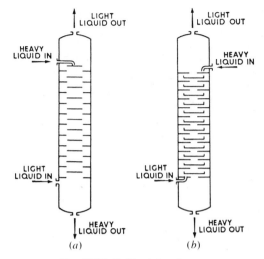

FIG. 13.38. Baffle-plate column

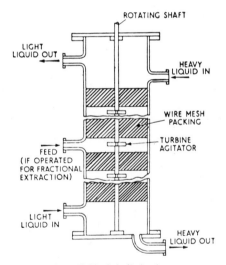

FIG. 13.39. Scheibel column

## 13.7. DIFFERENTIAL CONTACT EQUIPMENT FOR EXTRACTION

### 13.7.1. Spray Columns

Two methods of operating spray columns are shown in Fig. 13.41. Either the light or heavy phase may be dispersed. In the former case (a) the light phase enters from a distributor at the bottom of the column and the droplets rise through the heavier phase, finally coalescing to form a liquid-liquid interface at the top of the tower. Alternatively, the heavier phase may be dispersed, in which case the interface is held at the bottom of the tower as shown in (b). Although spray towers are simple in construction, they are

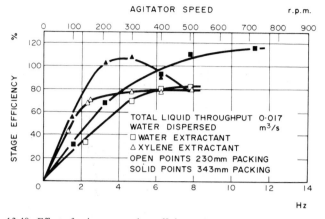

FIG. 13.40.  Effect of agitator speed on efficiency. System acetone–xylene–ester

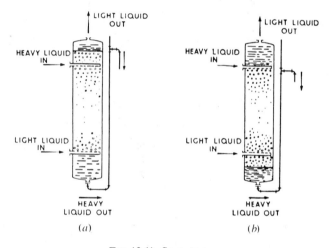

FIG. 13.41.  Spray towers

inefficient because considerable recirculation of the continuous phase takes place. As a result, true countercurrent flow is not maintained and up to 6 m may be required to obtain one theoretical stage. There is very little turbulence in the continuous phase and lack of interface renewal, and appreciable axial mixing results in poor performance.

Because the droplets of dispersed phase rise or fall through the continuous phase under the influence of gravity, it will be apparent that there is a limit to the amount of dispersed phase that can pass through the tower for any given flowrate of continuous phase. Thus referring to Fig. 13.41(a), any additional light phase fed to the bottom of the tower, in excess of that which can pass upwards under the influence of gravity, will be rejected from the bottom of the unit and the tower is then said to be flooded[21]. It is therefore important to be able to predict the conditions under which flooding will occur, so that the diameter of tower can be calculated for any required throughput. Although no complete analysis of this problem has, as yet, been achieved, it can be treated approximately in the manner as follows:

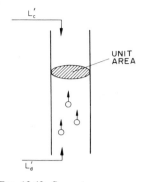

FIG. 13.42. Spray tower

*Dispersed phase hold-up.* Figure 13.42 represents a section of a spray tower of unit cross-sectional area. The light phase is assumed to be dispersed, and the volumetric flowrates of the two phases are $L'_d$ and $L'_c$ respectively. The superficial velocities $u_d$, $u_c$ of the phases are therefore also equal to $L'_d$ and $L'_c$. Under steady-state conditions the amount of dispersed phase held up in the tower in the form of droplets is conveniently expressed in terms of the fractional hold-up ($j$), i.e., the fractional volume of the two-phase dispersion occupied by the dispersed phase. This may also be thought of as the fraction of the cross-sectional area of the tower occupied by the dispersed phase. The velocity of the dispersed phase relative to the tower is therefore $L'_d/j$; similarly the relative velocity of the continuous phase is equal to $L'_c/(1-j)$. If the overall flow is regarded as strictly countercurrent, the sum of these two velocities will be equal to the velocity of the dispersed phase relative to the continuous phase, $u_r$:

$$\frac{L'_d}{j} + \frac{L'_c}{(1-j)} = u_r \tag{13.39}$$

In the case of spray towers it has been shown[22] that $u_r$ is well represented by $\bar{u}_0(1-j)$ where $\bar{u}_0$ is the velocity of a single droplet relative to the continuous phase, and is termed the droplet characteristic velocity. The term $(1-j)$ is a correction to $\bar{u}_0$ which takes into account the way in which the characteristic velocity is modified when there is a finite population of droplets present as opposed to a single droplet. It will be seen therefore that for very dilute dispersions (i.e. as $j \to 0$), $\bar{u}_0(1-j) \to \bar{u}_0$; on the other hand, as the fractional hold-up increases, the relative velocity of the dispersed phase decreases due to interactions between the droplets. Substituting for $u_r$, equation 13.39 may be written in the form:

$$\frac{u_d}{j} + \frac{u_c}{1-j} = \bar{u}_0(1-j) \tag{13.40}$$

This expression relates the hold-up to the flowrates of the phases and column diameter through the characteristic velocity, $\bar{u}_0$. It therefore affords a method of calculating the hold-up for a given set of flowrates if $\bar{u}_0$ is known. Conversely, equation 13.40 can be used to calculate $\bar{u}_0$ from experimental hold-up measurements performed at different flowrates. Thus, if hold-up data are plotted with $L'_d + (j/(1-j))L'_c$ as ordinate versus $j(1-j)$ as abscissa, a linear plot is obtained which passes through the origin and which has a slope equal to $\bar{u}_0$[22].

*Flooding-point condition.* A plot of equation 13.40 in the form of $u_d$ and $u_c$ against $j$ for a typical value of $\bar{u}_0 = 0.042$ m/s is shown in Fig. 13.43. Although equation 13.40 is cubic, only the root which lies between zero and the flooding-point values of $u_d$ and $u_c$ is

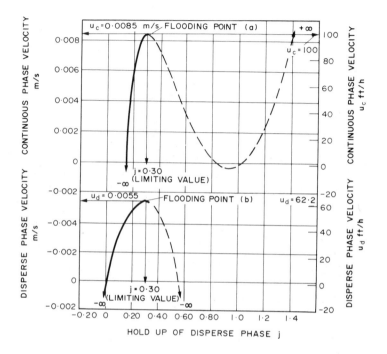

FIG. 13.43. Spray tower. (a) Continuous phase velocity ($u_c$) as function of hold-up of disperse phase ($j$) ($u_d = 0.0053$ m/s, $\bar{u}_0 = 0.042$ m/s. (b) Disperse phase velocity ($u_d$) as function of hold-up of disperse phase ($j$) ($u_c = 0.008$ m/s $\bar{u}_0 = 0.042$ m/s.

realisable in practice. These portions of the hold-up curves are shown by full lines in Fig. 13.43. If the flowrate of one of the phases is held constant, an increase in the flowrate of the other phase will result in an increased value of the hold-up until the flooding-point is reached. The latter corresponds to the maxima in the two curves shown in Fig. 13.43, so that the flooding-point condition is given by $du_d/dj = du_c/dj = 0$. Since those portions of the hold-up curves beyond the flooding-point are unrealistic in practice, the value of $j$ corresponding to the maximum flowrates also represents the limiting hold-up at flooding, although this condition is not obtainable mathematically from equation 13.40.

Performing the differentials described previously:

$$u_{df} = 2\bar{u}_0 j^2 f(1 - j_f) \qquad \text{when } du_c/dj = 0 \tag{13.41}$$

and

$$u_{cf} = \bar{u}_0 (1 - j_f)^2 (1 - 2j_f) \text{ when } du_d/dj = 0 \tag{13.42}$$

The value of $j_f$, the limiting hold-up at the flooding-point, may be obtained by eliminating

$\bar{u}_0$ between equations 13.41 and 13.42 and solving for $j_f$. Thus:

$$j_f = \frac{(r^2 + 8r)^{0.5} - 3r}{4(1-r)} \qquad (13.43)$$

where

$$r = u_{df}/u_{cf}. \qquad (13.44)$$

The derivation of equations 13.41 and 13.42 has been carried out assuming that $\bar{u}_0$ is constant and independent of the flowrates, up to and including the flooding-point. This in turn assumes that the droplet size is constant and that no coalescence occurs as the hold-

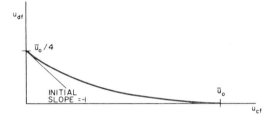

FIG. 13.44. Flooding-point curve for spraytower.

up increases. Whilst this assumption is essentially true in properly designed spray towers, this is certainly not the case with packed towers. Equations 13.41 and 13.42 cannot therefore be used to predict the flooding-point in packed towers and a more empirical procedure must be adopted.

A typical form of flooding-point curve for spray towers is shown in Fig. 13.44 where values of $u_{df}$ are plotted against $u_{cf}$. The limiting values of each flowrate as the other approaches zero can be determined readily from equations 13.41 and 13.42. Thus, when $u_{df} \rightarrow 0, j_f \rightarrow 0$ and $u_{cf} = \bar{u}_0$ in the limit. Similarly, as $u_{cf} \rightarrow 0, j_f \rightarrow 0.50$ and $u_{df} = \bar{u}_0/4$ in the limit. Furthermore, combining equations 13.41 and 13.42 and differentiating $u_{df}$ with respect to $u_{cf}$, it is seen that the slope of the flooding-point curve is given by:

$$\frac{du_{df}}{du_{cf}} = -\frac{j_f}{1-j_f} \qquad (13.45)$$

It is therefore apparent that the curve meets the abscissa tangentially as $u_{df} \rightarrow 0$, and meets the ordinate with a slope of $-1$ as $u_{cf} \rightarrow 0$[22].

Equations 13.41 and 13.42 may be used for correlating or extending incomplete flooding-point data. Thus, for example, if flooding-point hold-up data are available for a range of flowrates, a plot of $u_{df}$ as ordinate versus $j_f^2(1-j_f)$ as abscissa will result in a straight line through the origin, of slope $2\bar{u}_0$. Alternatively, if $\bar{u}_0$ is known, either equation 13.41 or 13.42 may be used to calculate the tower diameter for a prescribed throughput. The actual area of tower to be employed is then taken as twice this area which ensures that the unit operates at below 50 per cent of the flooding-point.

*Droplet characteristic velocity and droplet size.* The characteristic velocity may be calculated from hold-up measurements below the flooding-point using equation 13.40 or from hold-up measurements at the flooding-point using equations 13.41 or 13.42. In many instances, however, such data are not available and it is then necessary to be able to predict $\bar{u}_0$ from a knowledge of the physical properties of the extraction system. This involves predicting firstly the mean droplet size which will be present in the tower, and then the corresponding mean droplet velocity. A full discussion of the problem is beyond the scope of this chapter though basically, for low nozzle velocities, the mean droplet size can be established through the correlation proposed by Hayworth and Treybal[23] and the corresponding droplet velocity calculated by the methods of Hu and Kintner[24] or Klee and Treybal[25].

*Interfacial area.* The interfacial area per unit volume of tower (specific area) is given by:

$$a = \frac{\text{Total area}}{\text{Volume}} = \frac{\pi d_0^2}{d_0^3 \dfrac{1}{6} \dfrac{1}{j}} = \frac{6j}{d_0} \tag{13.46}$$

where $j$ is the fractional hold-up and $d_0$ is the mean droplet size, defined as:

$$d_0 = \frac{\sum n_1 d_1^3}{\sum n_1 d_1^2} \tag{13.47}$$

where $n_1$ is the number of droplets of diameter $d_1$ in a population.

## Example 13.2

The calculation of $d_0$ is illustrated by Table 13.7 below.

TABLE 13.7. *Calculation of Mean Droplet*

| $n_1$ | $d_1$ cm | $d_1^2$ | $d_1^3$ | $n_1 d_1^3$ | $n_1 d_1^2$ |
|---|---|---|---|---|---|
| 3 | 0·2 | 0·04 | 0·008 | 0·024 | 0·120 |
| 12 | 0·3 | 0·09 | 0·027 | 0·324 | 1·08 |
| 20 | 0·4 | 0·16 | 0·064 | 1·28 | 3·20 |
| 8 | 0·5 | 0·25 | 0·125 | 1·00 | 4·00 |
| 2 | 0·6 | 0·36 | 0·216 | 0·432 | 0·720 |

$$\sum n_1 d_1^3 = 3\cdot06 \quad \sum n_1 d_1^2 = 9\cdot120$$

$$\therefore \quad d_0 = \frac{3\cdot06}{9\cdot12} = 0\cdot336\,\text{cm}\,\underline{(3\cdot36\,\text{mm})}$$

*Mass transfer.* It is not yet possible to predict the mass transfer coefficient with a high further work is required before generalised correlations can be developed which take into imperfectly understood[26,27,28]. Furthermore, the flow in spray towers is not strictly countercurrent due to recirculation of the continuous phase, and consequently the effective overall driving force for mass transfer is not the same as that for true countercurrent flow.

To a first approximation, the droplets may be considered to be fully turbulent and the dispersed-phase film coefficients calculated on the Handlos–Baron model for circulating liquid spheres[29]. The continuous phase film coefficients may be estimated from the correlation of Ruby and Elgin[30], and the overall coefficients then computed using equations 13.27 or 13.28. Such procedures are, however, only approximate at best, and further work is required before generalised correlations can be developed which take into account the effect of recirculation and the different mass transfer rates at the dispersed phase entry nozzles, during droplet rise and during coalescence at the top of the tower. In the meantime the reader is referred to typical data[31–33] and standard texts[10,34,35] for H.T.U. values covering a range of extraction systems. In this respect, scale-up from pilot data raises problems. Reference should also be made to the work of Hanson[13] for further discussion on mass transfer in liquid–liquid systems.

Hitherto no mention has been made of interfacial effects accompanying the mass transfer process. Under certain conditions the presence of an undistributed solute gives rise to the Marangoni effect[36,37] which results in interfacial turbulence and droplet coalescence. These effects are generally more obvious when solute transfer takes place from an organic solvent droplet into an aqueous continuous phase. In the reverse direction of transfer, interfacial disturbances of this nature are frequently absent. The existence of interfacial phenomena of this kind presents yet one more obstacle to a quantitative interpretation of the mass transfer process. Furthermore, when the droplet coalescence is marked, the droplet size is no longer constant up the tower and the derivations of equations 13.41 and 13.42, which were based upon constant $\bar{u}_0$ values, are no longer valid. From a practical point of view, these equations may still be used for design purposes, since the effect of droplet coalescence is to enhance the flooding-point beyond the value predicted by equations 13.41 and 13.42, so that the tower operates in practice below 50 per cent of the flooding-point.

### 13.7.2. Packed Columns

These are similar to those used for distillation and absorption though it must be remembered that the flowrates of the phases are very different and two liquid phases are always present. The packing increases the interfacial area, and considerably increases mass transfer rates compared with those obtained with spray columns because of the continuous coalescence and break-up of the drops, though the H.T.U. values are still high. Packed columns are unsuitable for use with dirty liquids, suspensions, or high viscosity liquids. They have provided good results in the petroleum industry though at present they cannot be scaled-up to cope with the very high flows encountered in metallurgical processes. They are economical in the use of ground space.

*Minimum packing size.* For reproducible results the minimum packing size should be such that the mean void height is not less than the mean droplet diameter. This critical packing size is given by[38]:

$$d_{\text{crit}} = 2 \cdot 42 \left( \frac{\sigma}{\Delta \rho g} \right)^{0 \cdot 5} \tag{13.48}$$

*Dispersed phase hold-up.* Three regimes of flow can be distinguished with packings greater than the critical size:

(a) Region of linear hold-up. At low dispersed phase flowrates the droplets move freely within the interstices of the packing and the hold-up increases linearly with dispersed phase flowrate.

(b) Region of rapidly increasing hold-up. Above a hold-up of approximately 10 per cent, corresponding to the lower transition point, the hold-up increases more rapidly with increasing dispersed phase flow. This is due to the onset of hindered movement of the droplets within the packing voids.

(c) Region of constant hold-up. At higher dispersed phase flowrates, the upper transition point is encountered above which droplet coalescence occurs. In this region the hold-up remains constant as the flowrate is increased until the flooding-point is reached. It will be apparent therefore that a plot of $u_d$ versus $j$ will be of similar form to that shown in Fig. 13.43 except that the point corresponding to the flooding point for spray towers coincides with the upper transition point for packed towers.

Below the upper transition point, hold-up data can be correlated by an equation analogous to that used for spray towers, except that a correction must be introduced to take the packing voidage $(e)$ into account, thus[38,39]:

$$\frac{u_d}{j} + \frac{u_c}{1-j} = e\bar{u}_0(1-j) \tag{13.49}$$

*Droplet size and interfacial area.* In the absence of interfacial effects accompanying mass transfer, the droplets break down by impact with elements of packing and finally reach an equilibrium size which is independent of the packing size. Conversely, small droplets gradually coalesce until the equilibrium size is attained. Pratt and his co-workers[17,40] showed that the mean droplet size attained in the tower was well represented by the expression:

$$d_0 = 0.92\left(\frac{\sigma}{\Delta\rho g}\right)^{0.50}\left(\frac{\bar{u}_0 e j}{u_d}\right) \tag{13.50}$$

Correcting equation 13.46 for the packing voidage, the specific area in the tower is given by:

$$a = \frac{6ej}{d_0} \tag{13.51}$$

*Droplet characteristic velocity.* Gayler, Roberts and Pratt[38] have proposed a graphical correlation for $\bar{u}_0$ which takes into account the fact that the droplets are periodically halted by collisions with elements of packing and then accelerate to some fraction of their terminal velocity, $u_0$, before being deflected or stopped again by a further collision. Figure 13.45 enables $u_0$ to be determined from a knowledge of the physical properties of the system; this value is then used in conjunction with Fig. 13.46 to evaluate $\bar{u}_0$ for any given column diameter $d_c$ and nominal packing size $d_p$.

Once $\bar{u}_0$ has been determined, the hold-up for any particular set of flowrates can be calculated from equation 13.49, the mean droplet size from equation 13.50, and the specific area from equation 13.51.

*Flooding-point.* Because the flooding-point is no longer synonomous with that for spray towers, equations 13.41 and 13.42 only predict the upper transition point. Dell and

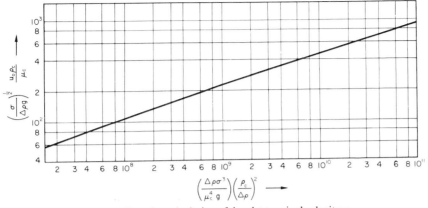

FIG. 13.45. Chart for calculation of droplet terminal velocity, $u_0$

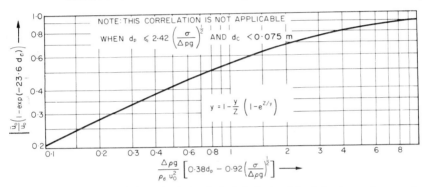

FIG. 13.46. Chart for calculation of velocity $(\bar{u}_0)$ of single droplet relative to continuous phase ($d_c$ in m)

Pratt[41] adopted a semi-empirical approach for the flooding-point by consideration of the forces acting on the separate dispersed and continuous phase channels which form when coalescence sets in just below the flooding-point. The following expression correlates data to within $\pm 20$ per cent.

$$1 + 0.835 \left(\frac{\rho_d}{\rho_c}\right)^{1/4} \left(\frac{u_d}{u_c}\right)^{1/2} = J\left[\frac{u_c^2 a_p}{g\, e^3}\left(\frac{\rho_c}{\Delta\rho}\right)\left(\frac{\sigma}{\sigma_T}\right)^{1/4}\right]^{-1/4} \tag{13.52}$$

A table of values of the constant $J$ is given in Table 13.8.

TABLE 13.8. Values of Constant J for use in Equation 13.52

| Type of packing | $d_p$ Nominal packing size, mm | | | | | |
|---|---|---|---|---|---|---|
| | 6 | 9 | 12·5 | 19 | 25 | 57·5 |
| Raschig rings* | 0·49 | 0·53 | 0·57 | 0·61 | 0·63 | 0·63 |
| Lessing rings | — | 0·61 | 0·61 | — | — | — |
| Berl saddles | 0·52 | — | 0·67 | 0·67 | — | — |
| Spheres | — | — | 0·61 | — | — | — |

* For values of the L.H.S. of equation 13.52 $< 1.6$ the exponent on the R.H.S. becomes $-0.20$ and the constant $1.10\, J^{0.8}$.

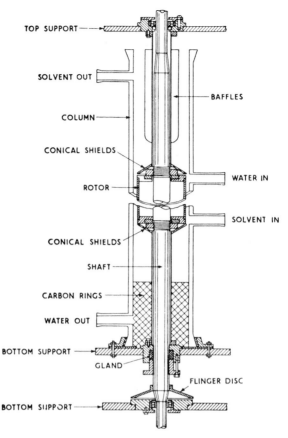

FIG. 13.47. Rotary annular column

*Mass transfer.* As in the case of spray columns, it is not yet possible to predict mass transfer rates from first principles. In the absence of any reliable correlations, use may be made of typical values of overall[32,42] and film[43,44] coefficients. A comprehensive summary is given by Perry[35].

### 13.7.3. Rotary Annular Columns[45] and Rotary Disc Columns[46]

With these columns, mechanical energy is provided to form the dispersed phase. The equipment is particularly suitable for installations where a moderate number of stages is required, and where the throughput is considerable. A well dispersed phase is obtained with this arrangement. Figure 13.47 shows a rotary annular column.

Flooding-point data may be correlated by equations 13.41 and 13.42 using the droplet characteristic velocity concept[45] since coalescence is absent.

### 13.7.4. Pulsed Columns

In order to prevent coalescence of the dispersed drops, Van Dijck[47] and others have devised methods of giving the whole of the continuous phase a pulsed motion. This can be

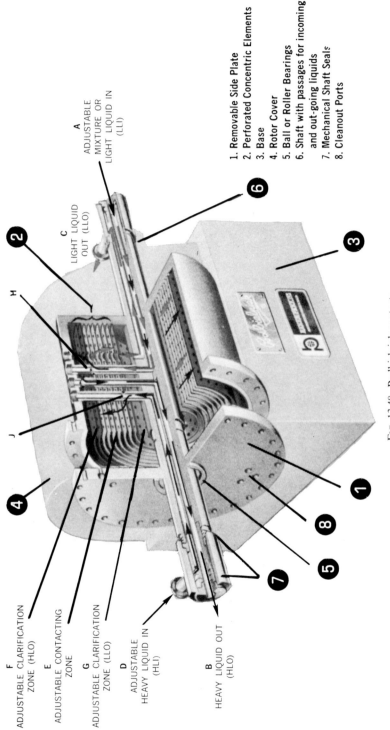

1. Removable Side Plate
2. Perforated Concentric Elements
3. Base
4. Rotor Cover
5. Ball or Roller Bearings
6. Shaft with passages for incoming and out-going liquids
7. Mechanical Shaft Seals
8. Cleanout Ports

A
ADJUSTABLE MIXTURE OR LIGHT LIQUID IN (LLI)

C
LIGHT LIQUID OUT (LLO)

F
ADJUSTABLE CLARIFICATION ZONE (HLO)

E
ADJUSTABLE CONTACTING ZONE

G
ADJUSTABLE CLARIFICATION ZONE (LLO)

D
ADJUSTABLE HEAVY LIQUID IN (HLI)

B
HEAVY LIQUID OUT (HLO)

FIG. 13.48. Podbielniak contactor

done either by some mechanical device, or by the introduction of compressed air.

The pulsation markedly improves performance of packed columns and the H.T.U. will be about half that of an unpulsed column. There are advantages in using gauze-type packings since the pulsation operation will often break ceramic rings. Perforated plates, as used in distillation, may also be used for pulsed extraction. Pulsed packed columns have been used in the nuclear industry though they are limited in size since the pulsation system is difficult to arrange and the pulsation itself demands strenghthening of the column.

Flooding-point data have been correlated by equations analogous to 13.41 and 13.42. This procedure is permissible since pulsed columns can be operated up to the flooding-point with no droplet coalescence[48,49].

### 13.7.5.  Centrifugal Extractors

If separation is difficult in the mixer–settler unit, the centrifugal extractor may be used; here the mixing and the separation stages are contained in the same unit which operates as a differential contactor.

*Podbielniak.* In this equipment, the first of the rotating machines to be developed, the heavy phase is driven outwards by centrifugal force and the light phase is displaced inwards. Referring to Fig. 13.48 which illustrates a unit produced by Baker Perkins, the heavy phases enters at $D$, passes to $J$ and is driven out at $B$. The light phase enters at $A$ and is

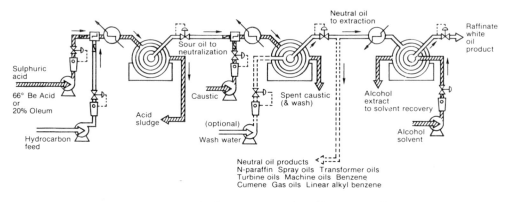

FIG. 13.49. Application of the Podbielniak contactor to the acid treatment of hydrocarbon feeds

displaced inwards towards the shaft and leaves at $C$. The two liquids intermix in zone $E$ where they are flowing countercurrently through the perforated concentric elements and are separated in the spaces between. In zones $F$ and $G$ the perforated elements are surfaces on which the small droplets of entrained liquid can coalesce, the large drops then being driven out by centrifugal force.

The contactor finds extensive use where high performance phase separation and countercurrent extraction or washing in the one unit are required; particularly important applications are the removal of acid sludges from hydrocarbons (Fig. 13.49), hydrogen peroxide extraction, sulphonate soap and antibiotics extraction, the extraction of rare earths such as uranium and vanadium from leach liquors, and the washing of refined edible oils.

*Alfa-Laval.* This machine has a vertical spindle and the rotor is fitted with concentric cylindrical inserts with helical wings forming a series of spiral passages. The two phases are fed into the bottom, the light phase being led to the periphery from which it flows inwards along the spiral, with the heavy phase flowing countercurrently. High shear forces are thus generated giving high extraction rates.

These units give many ideal stages, run continuously and do not take up much room.

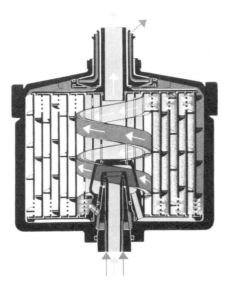

FIG. 13.50. Working principle of Alfa-Laval centrifugal extractor

For these reasons they have been adopted in many drug extractions, though they are unsuitable for medium or large throughputs.

*Scheibel column.* To some extent, the Scheibel column may be considered a centrifugal device, though as coalescence takes place between the layers of mesh packing, in this sense it is a stage-wise contactor, and has therefore been described in Section 13.6.3.

## 13.8. FURTHER READING

ALDERS, L.: *Liquid–Liquid Extraction* (Elsevier, New York, 1955).
BACKHURST, J. R., HARKER, J. H., and PORTER, J. E.: *Problems in Heat and Mass Transfer* (Edward Arnold, 1974).
BRODKEY, R. S.: *Turbulence in Mixing Operations* (Academic Press, London, 1975).
FRANCIS, A. W.: *Handbook for Components in Solvent Extraction* (Gordon & Breach, New York, 1972).
HANSON, C. (ed.): *Recent Advances in Liquid–Liquid Extraction* (Pergamon Press, Oxford, 1971).
HOLLAND, F. A. and CHAPMAN, F. S.: *Liquid Mixing and Processing in Stirred Tanks* (Reinhold, New York, 1966).
JAMRACK, W. D.: *Base Metal Extraction by Chemical Engineering Techniques* (Pergamon Press, Oxford, 1963).
KING, C. J.: *Separation Processes* (McGraw-Hill, New York, 1971).
MARCUS, Y. (ed): *Solvent Extraction Reviews* (Marcel Dekker, New York, 1971).
PRATT, H. R. C.: *Countercurrent Separation Processes* (Elsevier, Amsterdam, 1967).
SHERWOOD, T. K. and PIGFORD, R. L.: *Absorption and Extraction* (McGraw-Hill, New York, 1952).
SHERWOOD, T. K., PIGFORD, R. L., and WILKE, C. R.: *Mass Transfer* (McGraw-Hill, New York, 1975).

SMITH, B. D.: *Design of Equilibrium Stage Processes* (McGraw-Hill, New York, 1963).
TREYBAL, R. E.: *Liquid Extraction*, 2nd edn. (McGraw-Hill, New York, 1963).
Solvent Extraction, Proc. of the Int. Solvent Extraction Conference, The Hague, 1971. (Soc. Chem. Ind., London, 1971).

# 13.9. REFERENCES

1. REAVELL, B. N.: *Trans. Inst. Chem. Eng.* **29** (1951) 301. Practical aspects of liquid mixing and agitation.
2. WOOD, J. C., WHITTEMORE, E. R., and BADGER, W. L.: *Chem. Met. Eng.* **27** (1922) 1176. The measurement of stirrer performance.
3. HIXSON, A. W. and WILKENS, G. A.: *Ind. Eng. Chem.* **25** (1933) 1196. Performance of agitators in liquid–solid chemical systems.
4. STOOPS, C. E. and LOVELL, C. L.: *Ind. Eng. Chem.* **35** (1943) 845. Power consumption of propeller-type agitators.
5. WHITE, A. M. and BRENNER, E.: *Trans. Am. Inst. Chem. Eng.* **30** (1934) 585. Studies in agitation. V—The correlation of power data.
6. BLACK, C. R.: *J. Imp. Coll. Chem. Eng. Soc.* **4** (1948) 111. The unit operation of mixing. Part 1—Methods of assessing the performance of agitators.
7. RUSHTON, J. H., COSTICH, E. W., and EVERETT, H. J.: *Chem. Eng. Prog.* **46** (1950) 395, 467. Power characteristics of mixing impellers.
8. KRAMERS, H., BAARS, G. M. and KNOLL, W. H.: *Chem. Eng. Sci.* **2** (1953) 35. A comparative study on the rate of mixing in stirred tanks.
9. BISSELL, E. S., HESSE, H. C., EVERETT, H. J., and RUSHTON, J. H.: *Chem. Eng. Prog.* **43** (1947) 649. Design and utilisation of internal fittings for mixing vessels.
10. PERRY, J. H.: *Chemical Engineers' Handbook*, 5th end. (McGraw-Hill 1973) revised by R. Perry and C. Chilton.
11. METZNER, A. B., FEEHS, R. H., RAMOS, H. L., OTTO, R. E., and TUTHILL, J. D.: *A.I.Ch.E.Jl.* **7** (1961) 3–9. Agitation of viscous Newtonian and non-Newtonian fluids.
12. HIXSON, A. W. and CROWELL, J. H.: *Ind. Eng. Chem.* **23** (1931) 1160. Dependence of reaction velocity upon surface and agitation. III. Experimental procedure in study of agitation.
13. HANSON, C.: *Het Ingenieursblad.* **41**, 15–16 (Aug. 1972) 408–17. The technology of solvent extraction in metallurgical processes.
14. British Patent No. 995 472: Distillers Company Limited (29 April 1964). Acrylic acid recovery.
15. SHERWOOD, T. K. and PIGFORD, R. L.: *Absorption and Extraction*, 2nd edn. (McGraw-Hill, 1952).
16. LEIBSON, I. and BECKMANN, R. B.: *Chem. Eng. Prog.* **49** (1953) 405. The effect of packing size and column diameter on mass transfer in liquid–liquid extraction.
17. GAYLER, R. and PRATT, H. R. C.: *Trans. Inst. Chem. Eng.* **31** (1953) 69. Liquid–liquid extraction. Part V— Further studies of droplet behaviour in packed columns.
18. WARWICK, G. C. I. and SCUFFHAM, J. B.: *Het Ingenieursblad.* **41**, 15–16 (Aug. 1972) 442–449. The design for mixer-settlers for metallurgical duties.
19. JACKSON, I. D., NEWRICK, G. M., and WARWICK, G. C. I.: *I. Chem. E.* Symposium Series **42**, 15.1–15.8. A recent development in the design of hydrometallurgical mixer-settlers.
20. SCHEIBEL, E. G. and KARR, A. E.: *Ind. Eng. Chem.* **42** (1950) 1048. Semicommercial multistage extraction columns.
21. BLANDING, F. H. and ELGIN, J. C.: *Trans. Am. Inst. Chem. Eng.* **38** (1942) 305. Limiting flow in liquid–liquid extraction columns.
22. THORNTON, J. D.: *Chem. Eng. Sci.* **5** (1956) 201. Spray liquid–liquid extraction columns.
23. HAYWORTH, C. B. and TREYBAL, R. E.: *Ind. Eng. Chem.* **42** (1950) 1174. Drop formation in two-liquid-phase systems.
24. HU, S. and KINTNER, R. C.: *A.I.Ch.E.Jl.* **1** (1955) 42. The fall of single liquid drops through water.
25. KLEE, A. J. and TREYBAL, R. E.: *A.I.Ch.E.Jl.* **2** (1956) 44. Rate of rise or fall of liquid drops.
26. LICHT, W. and CONWAY, J. B.: *Ind. Eng. Chem.* **42** (1950) 1151. Mechanism of solute transfer in spray towers.
27. COULSON, J. M. and SKINNER, S. J.: *Chem. Eng. Sci.* **1** (1952) 197. The mechanism of liquid–liquid extraction across stationary and moving interfaces. Part 1. Mass transfer into single dispersed drops.
28. GARNER, F. H. and HALE, A. A.: *Chem. Eng. Sci.* **2** (1953) 157. The effect of surface agents in liquid extraction processes.
29. HANDLOS, A. E. and BARON, T.: *A.I.Ch.E.Jl.* **3** (1957) 127. Mass and heat transfer from drops in liquid–liquid extraction.
30. RUBY, C. L. and ELGIN, J. C.: *Chem. Eng. Prog.* Symp. Series No. 16, **51** (1955) 17. Mass transfer between liquid drops and a continuous liquid phase in a countercurrent fluidized system. Liquid–liquid extraction in a spray tower.

31. APPEL, F. J. and ELGIN, J. C.: *Ind. Eng. Chem.* **29** (1973) 451. Countercurrent extraction of benzoic acid between toluene and water.

32. SHERWOOD, T. K., EVANS, J. E., and LONGCOR, J. V. A.: *Trans. Am. Inst. Chem. Eng.* **35** (1939) 597. Extraction in spray and packed columns.

33. ROW, S. B., KOFFOLT, J. H., and WITHROW, J. R.: *Trans. Am. Inst. Chem. Eng.* **37** (1941) 559. Characteristics and performance of a nine-inch liquid–liquid extraction column.

34. TREYBAL, R. E.: *Liquid Extraction*, 2nd edn. (McGraw-Hill, 1963).

35. SHERWOOD, T. K., PIGFORD, R. L., WILKE, C. P.: *Mass Transfer* (McGraw-Hill, 1975).

36. DAVIES, J. T. and RIDEAL, E. K.: *Interfacial Phenomena* (Academic Press, 1961).

37. GROOTHUIS, H. and ZUIDERWEG, F. J.: *Chem. Eng. Sci.* **12** (1960) 288. Influence of mass transfer on coalescence of drops.

38. GAYLER, R., ROBERTS, N. W. and PRATT, H. R. C.: *Trans. Inst. Chem. Eng.* **31** (1953) 57. Liquid–liquid extraction. Part IV. A further study of hold-up in packed columns.

39. GAYLER, R. and PRATT, H. R. C.: *Trans. Inst. Chem. Eng.* **29** (1951) 110. Symposium on liquid–liquid extraction. Part II. Hold-up and pressure drop in packed columns.

40. LEWIS, J. B., JONES, I., and PRATT, H. R. C.: *Trans. Inst. Chem. Eng.* **29** (1951) 126. Symposium on liquid–liquid extraction. Part III. A study of droplet behaviour in packed columns.

41. DELL, F. R. and PRATT, H. R. C.: *Trans. Inst. Chem. Eng.* **29** (1951) 89, 270. Symposium on liquid–liquid extraction. Part I. Flooding rates for packed columns.

40. LEWIS, J. B., JONES, U., and PRATT, H. R. C.: *Trans. Inst. Chem. Eng.* **29** (1951) 126. Symposium on liquid–liquid extraction. Part III. A study of droplet behaviour in packed columns.

41. DELL, F. R. and PRATT, H. R. C.: *Trans. Inst. Chem. Eng.* **29** (1951) 89. Symposium on liquid–liquid extraction. Part I. Flooding rates for packed columns.

42. PRATT, H. R. C. and GLOVER, S. T.: *Trans. Inst. Chem. Eng.* **24** (1946) 54. Liquid–liquid extraction: Removal of acetone and acetaldehyde from vinyl acetate with water in a packed column.

43. COLBURN, A. P. and WELSH, D. G.: *Trans. Am. Inst. Chem. Eng.* **38** (1942) 179. Experimental study of individual transfer resistances in countercurrent liquid–liquid extraction.

44. LADDHA, G. S. and SMITH, J. M.: *Chem. Eng. Prog.* **46** (1950) 195. Mass transfer resistances in liquid–liquid extraction.

45. THORNTON, J. D. and PRATT, H. R. C.: *Trans. Inst. Chem. Eng.* **31** (1953) 289. Liquid–liquid extraction. Part VII. Flooding rates and mass transfer data for rotary annular columns.

46. VERMIJS, H. J. A. and KRAMERS, H.: *Chem. Eng. Sci.* **3** (1954) 55. Liquid–liquid extraction in a "rotating disc contactor".

47. VAN DIJCK, W. J. D.: *U.S. Patent* 2,011,186 (1935). Intimately contacting fluids (immiscible liquids).

48. THORNTON, J. D.: *Trans. Inst. Chem. Eng.* **35** (1957) 316. Liquid–liquid extraction. Part XIII. The effect of pulse wave-form and plate geometry on the performance and throughput of a pulsed column.

49. LOGSDAIL, D. H. and THORNTON, J. D.: *Trans. Inst. Chem. Eng.* **35** (1957) 331. Liquid–liquid extraction. Part XIV. The effect of column diameter upon the performance and throughput of pulsed plate columns.

## 13.10. NOMENCLATURE

| | | Units in SI System | Dimensions in **MLT** |
|---|---|---|---|
| $A$ | Mass of solvent **A** | kg | **M** |
| $a$ | Interfacial area per unit volume of tower | m²/m³ | $\mathbf{L}^{-1}$ |
| $a_p$ | Superficial area of packing per unit volume of tower | m²/m³ | $\mathbf{L}^{-1}$ |
| $B$ | Mass of solvent **B** | kg | **M** |
| $b$ | Index | — | — |
| $C$ | Concentration of solute | kg/m³ | $\mathbf{ML}^{-3}$ |
| $C^*$ | Value of concentration in equilibrium with second phase | kg/m³ | $\mathbf{ML}^{-3}$ |
| $c$ | Index | — | — |
| $D_T$ | Diameter of tank | m | **L** |
| $d$ | Particle size | m | **L** |
| $d_c$ | Diameter of column | m | **L** |
| $d_p$ | Nominal size of packing | m | **L** |
| $\mathbf{d}_{\text{crit}}$ | Critical packing size | m | **L** |
| $d_0$ | Surface mean (Sauter mean) diameter of drop | m | **L** |
| $E$ | Mass of extract | kg | **M** |
| $e$ | Voidage of packing | — | — |
| $F$ | Mass of feed | kg | **M** |
| $g$ | Acceleration due to gravity | m/s² | $\mathbf{LT}^{-2}$ |
| $H$ | Depth of liquid | m | **L** |
| $\mathbf{H}$ | Height of the transfer unit | m | **L** |
| $J$ | Coefficient in equation 13.52 for flooding data | — | — |

|  |  | Units in SI System | Dimension in **MLT** |
|---|---|---|---|
| $j$ | Fractional hold-up of dispersed phase | — | — |
| $K$ | Overall mass transfer coefficient | m/s | $\mathbf{LT^{-1}}$ |
| $K'$ | Constant | — | — |
| $K''$ | Distribution constant | — | — |
| $k$ | Mass transfer coefficient | m/s | $\mathbf{LT^{-1}}$ |
| $k'$ | Coefficient | $kg^{1/3}/s$ | $\mathbf{M^{1/3}T^{-1}}$ |
| $L'$ | Volumetric rate of flow per unit area | $m^3/m^2\,s$ | $\mathbf{LT^{-1}}$ |
| $M$ | Total flow of material through the system | kg | $\mathbf{M}$ |
| $m$ | Slope of the equilibrium line | — | — |
| $N$ | Revolutions per unit time | i/s | $\mathbf{T^{-1}}$ |
| $N'$ | Molar rate of transfer per unit area | $kmol/m^2\,s$ | $\mathbf{ML^{-2}T^{-1}}$ |
| $\mathbf{N}$ | Number of transfer units | — | — |
| $\mathbf{N_P}$ | Power number | — | — |
| $n$ | Number of stages | — | — |
| $n_1, n_2 \ldots$ | Indices or numbers of particles | — | — |
| $P$ | Difference stream | kg | $\mathbf{M}$ |
| $\mathbf{P}$ | Power | W | $\mathbf{ML^2T^{-3}}$ |
| $p$ | Pitch of agitator | m | $\mathbf{L}$ |
| $R$ | Mass of raffinate | kg | $\mathbf{M}$ |
| $r$ | Ratio $u_{df}/u_{cf}$ | — | — |
| $S$ | Added solvent in extract stream, | kg | $\mathbf{M}$ |
| $t$ | Time | s | $\mathbf{T}$ |
| $u$ | Volumetric flowrate per unit area | $m^3/m^2\,s$ | $\mathbf{LT^{-1}}$ |
| $u_r$ | Velocity of dispersed phase relative to continuous phase | m/s | $\mathbf{LT^{-1}}$ |
| $u_0$ | Terminal falling velocity of droplet | m/s | $\mathbf{LT^{-1}}$ |
| $\bar{u}_0$ | Velocity of single droplet relative to continuous phase | m/s | $\mathbf{LT^{-1}}$ |
| $W$ | Width of blades of agitator or paddle | m | $\mathbf{L}$ |
| $W_B$ | Width of baffles | m | $\mathbf{L}$ |
| $w$ | Mass of solute at time $t$ | kg | $\mathbf{M}$ |
| $w_0$ | Mass of solute at time $t = 0$ | kg | $\mathbf{M}$ |
| $X_f$ | Mass ratio of solute in feed | — | — |
| $X_1, X_2$ | Mass ratio of solute in raffinate | — | — |
| $x_A, x_B$ | Mass or mol fraction of **A**, **B** | — | — |
| $Y_1, Y_2$ | Mass ratio of solute in extract | — | — |
| $Z$ | Height of packing | m | $\mathbf{L}$ |
| $Z_A$ | Height of agitator from bottom of tank | m | $\mathbf{L}$ |
| $Z_B$ | Height of dispersion band | m | $\mathbf{L}$ |
| $Fr$ | Froude number | — | — |
| $Re$ | Reynolds number | — | — |
| $\beta$ | Selectivity ratio | i−1 | — |
| $\mu$ | Viscosity | $(N\,s/m^2)$ | $\mathbf{ML^{-1}T^{-1}}$ |
| $\rho$ | Density | $(kg/m^3)$ | $\mathbf{ML^{-3}}$ |
| $\Delta\rho$ | Density difference between phases | $kg/m^3$ | $\mathbf{ML^{-3}}$ |
| $\sigma$ | Interfacial tension | $J/m^2$ | $\mathbf{MT^{-2}}$ |
| $\sigma_T$ | Interfacial tension, water-air at 288 $K'$ (0·073 N/m) | $J/m^2$ | $\mathbf{MT^{-2}}$ |

Suffixes

| $c, d$ | continuous and disperse phases |
|---|---|
| $E, R$ | extract and raffinate phases |
| $O$ | overall (transfer units) |
| $f$ | limiting value at flooding point |
| $i$ | value at interface |
| lm | logarithmic mean value |
| 1, 2 | Values at bottom, top of column |

CHAPTER 14

# *Evaporation*

## 14.1. INTRODUCTION

Evaporation is one of the main methods used in the chemical industry for the concentration of aqueous solutions. The usual meaning of the term is the removal of water from a solution by boiling the liquor in a suitable vessel, the evaporator, and withdrawing the vapour. If the solution contains dissolved solids, the resulting strong liquor may become saturated so that crystals are deposited.

Various liquors which are to be evaporated may be classified in the following ways:

(a) Liquors that can be heated to high temperatures without decomposition, and those that can only be heated to a low temperature (*ca.* 330 K).

(b) Liquors which yield to solids on concentration, in which case crystal size and shape may be important and those which do not.

(c) Liquors which at any given pressure boil at about the same temperature as water, and those which have a much higher boiling point.

Evaporation is carried out by adding heat to the solution to vaporise the solvent. The heat is supplied principally to provide the latent heat of vaporisation, and, by adopting methods for recovery of heat from the vapour, the chemical engineer has been able to bring about great economy in heat utilisation. Whilst the normal heating medium is generally low pressure steam, for special purposes Dowtherm or flue gases may be adopted.

The design of an evaporation unit requires the practical application of data on heat transfer to boiling liquids, together with a realisation of what happens to the liquid during concentration. In addition to the three main features outlined above, liquors which have an inverse solubility curve and which are therefore likely to deposit scale on the heating surface, merit special attention.

## 14.2. HEAT TRANSFER IN EVAPORATORS

### 14.2.1. Heat Transfer Coefficients

The normal rate equation for heat transfer is given in the form:

$$Q = UA\Delta T \tag{14.1}$$

where $Q$ is the heat transferred per unit time,

$U$ is the overall heat transfer coefficient,

$A$ is the heat transfer surface, and

$\Delta T$ is the temperature difference between the two streams.

In applying this relation to evaporators, there is some difficulty in deciding the correct value for the temperature difference. This arises from what are known as the *boiling point*

rise (B.P.R.) and the *hydrostatic head*. If water is boiled in an evaporator under a given
pressure, then the temperature of the liquor can be determined from steam tables and the
temperature difference is readily calculated. At the same pressure, a solution has a boiling
point greater than that of water, and the difference between its boiling point and that of
water is the B.P.R. For example, at atmospheric pressure, a 25 per cent solution of
sodium chloride boils at 381 K and shows a B.P.R. of 8 deg K. Thus, if steam at 389 K
were used to concentrate the salt solution, the overall temperature difference would not
be 389–373 = 16 deg K, but only 389 − 381 = 8 deg K. Such solutions usually require

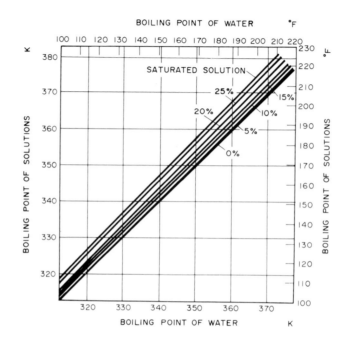

FIG. 14.1. Boiling point of solutions of sodium chloride as a function of the boiling point of
water (Dühring lines)

more heat to vaporise one kilogram of water, so that the reduction in capacity of a unit
may be considerable. The value of the B.P.R. cannot be calculated from physical data
of the liquor, though Dühring's rule is often used to find the change in B.P.R. with
pressure. If the boiling point of the solution is plotted against that of water at the same
pressure, then a straight line is obtained, as shown for sodium chloride in Fig. 14.1.
Thus, if the pressure is fixed, the boiling point of water is found from steam tables, and
the boiling point of the solution from Fig. 14.1. The boiling point rise is much greater
with strong electrolytes, such as salt and caustic soda.

  The effect of hydrostatic head may be considered by supposing the liquor to be boiling
at the top of a tube. Then the pressure of the liquid which is just at the top of the tubes is
that in the vapour space, and the boiling point can therefore be found. The liquor at the
bottom of the tubes is at a higher pressure due to the hydrostatic head ; and, if the liquor is
to boil, it must be heated to the higher temperature corresponding to the increased
pressure. Thus, the temperature difference in an evaporator between the steam outside the

tubes and the liquor will depend on where boiling starts, and there is no satisfactory way of determining this. In a natural circulation unit with vertical tubes, the liquid flows down the central downtake and enters the bottom of the tubes at a temperature below the boiling point at this depth. On rising, the liquor is heated and at the same time the pressure falls, so that at some point the liquid starts to boil. In calculating the transfer coefficient $U$ from equation 14.1, if the boiling point of the liquor is taken as that of water at that pressure, then the values of $U$ are known as the *apparent transfer coefficients*. If the boiling point is corrected for B.P.R., the values of $U$ are known as the *corrected coefficients*. Further correction for the hydrostatic head is not generally attempted. The real temperature difference, and $U$ in equation 14.1, will alter with the depth of liquor. This has been shown by Badger and Shepard[1], and from Fig. 14.2 it is seen that, after an initial sharp rise, $U$

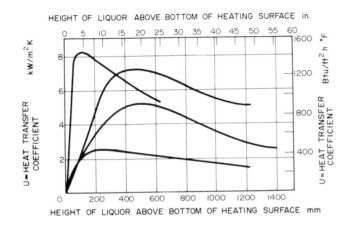

FIG. 14.2. Effect of liquor depth on heat transfer coefficient for various natural circulation evaporators

falls as the liquor level is increased. The problem is complicated by the fact that, as the level is reduced, the rate of circulation of the liquor rises.

The overall heat transfer coefficients for any form of evaporator will depend on the value of the film coefficients on the heating side and for the liquor, together with allowances for scale deposits and the tube wall. For condensing steam, a common heating medium, film coefficients are of the order of $6 \, \text{kW/m}^2 \, \text{K}$. There is no entirely satisfactory general method for calculating transfer coefficients for the boiling film; however, design equations of sufficient accuracy are available in the literature though this information should be applied with caution.

### 14.2.2. Boiling at a Submerged Surface

It is convenient to classify the heat transfer processes occurring in common evaporation equipment under two general headings. The first of these is concerned with boiling at a submerged surface. Typical examples in this category are the horizontal tube evaporator and the coil evaporator considered in Section 14.5. Here the basic heat transfer process is assumed to be nucleate boiling with convection induced predominantly by the growing and departing vapour bubbles. The second category embraces two-phase forced-convection boiling processes occurring in closed conduits. In this case convection is induced by the flow which results from natural or forced circulation effects.

### Nucleate Boiling from a Single Tube

The familiar heat flux-temperature difference characteristic observed when heat is transferred from a surface to a liquid at its boiling point, is shown in Fig. 14.3. In the range

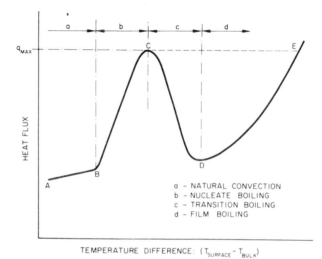

FIG. 14.3. Typical characteristic for boiling at a submerged surface

$AB$, although the liquid in the vicinity of the surface will be slightly superheated, there is no vapour formed and heat transfer is by natural convection with evaporation from the free surface. Boiling commences at $B$ with bubble columns initiated at preferred sites (nucleation centres) on the surface. Over the nucleate boiling region $BC$ the bubble sites become more numerous with increasing flux until at $C$ the surface is completely covered. In the majority of commercial evaporation processes the heating medium is a fluid and therefore the controlling parameter is the overall temperature difference. If an attempt is made to increase the heat flux beyond that at $C$, by increasing the temperature difference, the nucleate boiling mechanism will partially collapse and portions of the surface will be exposed to vapour blanketing. In the region of transition boiling $CD$ the average heat transfer coefficient, and therefore the heat flux, will decrease with increasing temperature difference, due to the increasing proportion of the surface exposed to vapour. This self-compensating behaviour is not exhibited if heat flux rather than temperature difference is the controlling parameter. In this case an attempt to increase the heat flux beyond point $C$ will cause the nucleate boiling regime to collapse completely, exposing the whole surface to a vapour film. The inferior heat transfer characteristics of the vapour mean that the surface temperature must rise (to $E$) in order to dissipate the heat. In many instances this temperature exceeds the melting point of the surface and results are disastrous. For obvious reasons the point $C$ is generally known as *burnout*, although the terms *departure from nucleate boiling (D.N.B. point)* and *maximum heat flux* are in common usage. For the evaporator designer, a method of predicting the heat transfer coefficient in nucleate boiling $h_b$ and the maximum heat flux which might be expected before $h_b$ begins to decrease, is of extreme importance. The complexity of the nucleate boiling process has made it the subject of many studies. In a review of the available correlations for nucleate boiling Westwater[2] has presented some fourteen equations. Palen and Taborek[3] reduced this bewildering array to seven and tested these against selected experimental data[4,5]. As a result of this study two equations, those due to McNelly[6] and Gilmour[7], were selected as the most accurate. Although the modified form of the Gilmour equation is somewhat more accurate, the relative simplicity of the McNelly equation commends its use. This equation is given in dimensionless form as:

$$\left[\frac{h_b d}{k}\right] = 0.225 \left[\frac{C_p \mu_L}{k}\right]^{0.69} \left[\frac{qd}{\lambda \mu_L}\right]^{0.69} \left[\frac{Pd}{\sigma}\right]^{0.31} \left[\frac{\rho_L}{\rho_v} - 1\right]^{0.31} \tag{14.2}$$

It will be noted that the inclusion of the characteristic dimension $d$ is necessary dimensionally though its value does not affect the result obtained for $h_b$.

This equation predicts the heat transfer coefficient for a single isolated tube and cannot be expected to be applicable to tube bundles. Indeed, Palen and Taborek[3] showed that the use of the above equation would have resulted in 50–250 per cent underdesign in a number of specific cases. The reason for this discrepancy can be explained as follows. In the case of a tube bundle, only the lowest tube in each vertical row is completely irrigated by the liquid; higher tubes will experience exposure to liquid–vapour mixtures. This partial vapour blanketing results in a lower average heat transfer coefficient for tube bundles than the value given by equation 14.2. In order to calculate these average values of $h$ for a tube bundle, equations of the form $h = C_s h_b$ have been suggested[3] where the surface factor $C_s$ is less than 1 and is, as might be expected, a function of the number of tubes in a vertical row, the pitch of the tubes, and the basic value of $h_b$. The factor $C_s$ can only be determined by statistical analysis of experimental data and further work will be necessary before it can be predicted from a physical model for the process.

*Maximum Heat Flux*

The single tube values for $h_b$ have been correlated by equation 14.2, which applies to the true nucleate boiling regime and takes no account of the factors which eventually lead to the maximum heat flux being approached. Several maximum flux equations have been tested by Palen and Taborek[3]. No improvements were suggested and the simplified Zuber equation[8], is recommended as follows:

$$q_{max} = \frac{\pi}{24} \lambda \rho_v \left[\frac{\sigma g(\rho_L - \rho_v)}{\rho_v^2}\right]^{1/4} \left[\frac{\rho_L + \rho_v}{\rho_L}\right]^{1/2} \tag{14.3}$$

where $q_{max}$ is the maximum heat flux,
    $\lambda$ is the latent heat of vaporisation,
    $\rho_L$ is the density of liquid,
    $\rho_v$ is the density of vapour,
    $\sigma$ is the interfacial tension, and
    $g$ is the acceleration due to gravity.

### 14.2.3. Forced Convection Boiling

In order to understand the functioning of evaporators operating with forced convection, it is convenient to consider what happens when a liquid is vaporised during flow through a vertical tube. If the liquid enters the tube below its boiling point, then the first section will operate as a normal heater and the heat transfer rates will be determined by the well-established equations for single phase flow. When the liquid temperature reaches the boiling point corresponding to the local pressure, boiling will commence. At this stage the vapour bubbles will be dispersed in the continuous liquid phase; however, progressive vaporisation of the liquid gives rise to a number of characteristic flow patterns which are shown in Fig. 14.4. Over the initial boiling section convective heat transfer

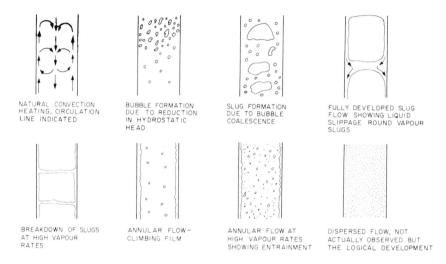

FIG. 14.4. The nature of two-phase flow in an evaporator tube

occurs with vapour bubbles dispersed in the liquid; higher up the tube bubbles become more numerous and elongated, and bubble coalescence occurs; eventually the bubbles form slugs which later collapse to give an annular flow regime in which vapour forms the central core with a thin film of liquid carried up the wall. In the final stage, dispersed flow with liquid entrainment in the vapour core occurs. In general, the conditions existing in the tube are those of annular flow. With further evaporation the rising liquid film becomes progressively thinner and this thinning, together with the increasing vapour core velocity, eventually causes breakdown of the liquid film, leading to dry wall conditions.

For boiling in a tube, there is therefore a contribution from nucleate boiling arising from bubble formation together with forced convection boiling due to the high velocity liquid–vapour mixture. Such a system is inherently complex since certain parameters influence these two basic processes in different ways.

Dengler and Addoms [9] measured heat transfer coefficients to water boiling in a 6 m tube. They found that the heat flux increased steadily up the tube as the percentage of vapour increased (Fig. 14.5).

For runs in which convection was predominant, their data were correlated using as the ordinate: the ratio of the observed two-phase heat transfer coefficient ($h_{tp}$) to that which would be obtained had the same total mass flow been all liquid ($h_L$). This ratio was plotted against the reciprocal of $X_{tt}$, the parameter for two-phase turbulent flow developed by Lockhart and Martinelli[10]. The liquid coefficient $h_L$ is given by:

$$h_L = 0.023\left[\frac{k}{d_t}\right]\left[\frac{4W}{\pi d_t \mu_L}\right]^{0.8}\left[\frac{C_p \mu_L}{k}\right]^{0.4} \tag{14.4}$$

where $W$ is the total mass rate of flow. The parameter $1/X_{tt}$ is given by:

$$\frac{1}{X_{tt}} = \left[\frac{y}{1-y}\right]^{0.9}\left[\frac{\rho_L}{\rho_v}\right]^{0.5}\left[\frac{\mu_v}{\mu_L}\right]^{0.1} \tag{14.5}$$

$1/X_{tt}$ is strongly dependent on the mass fraction of vapour $y$; the density and viscosity terms give a quantitative correction for the effect of pressure in the absence of nucleate boiling.

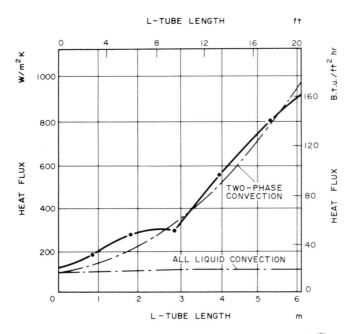

FIG. 14.5. Variation of heat flux to water up an evaporator tube[9]

Eighty-five per cent of the purely convective data for two-phase flow were correlated to within 20 per cent by the expression:

$$\frac{h_{tp}}{h_L} = 3.5\left[\frac{1}{X_{tt}}\right]^{0.5} \quad \left(0.25 < \frac{1}{X_{tt}} < 70\right) \tag{14.6}$$

Similar results for a range of organic liquids are reported by Guerrieri and Talty[11]. However, $h_L$ in this work is based on the point mass flowrate of the unvaporised part of the stream, i.e. $W$ is replaced by $W(1-y)$ in equation 14.4.

The processes discussed in this section are the subject of continuing research and as basic mechanisms become better understood the accuracy of the equations describing them will improve. Before concluding this section it is necessary to comment on the use of these equations for design purposes. An unusual characteristic of equation 14.2 is the dependence of $h_b$ on the heat flux $q$. The calculation of $h_b$ presents no difficulty in situations where the controlling parameter is the heat flux, as is the case with electrical heating. If a value of $q$ is selected, this together with a knowledge of operating conditions and the physical properties of the boiling liquid permits the direct calculation of $h_b$. The surface temperature of the heater can now be calculated from $q$ and $h_b$ and the process is described completely. Considering the evaluation of a process involving heat transfer from steam condensing at temperature $T_c$ to a liquid boiling at temperature $T_b$, assuming that the condensing coefficient is constant and specified as $h_c$, and also that the thermal resistance of the intervening wall is negligible, an initial estimate of the wall temperature $T_w$ can be made. The heat flux $q$ (for the condensing film) can now be calculated since $q = h_c(T_c - T_w)$, and the value of $h_b$ can then be determined from equation 14.2 using this value for the heat flux. A heat balance across the wall will now test the accuracy of the estimated value of $T_w$ since $h_c(T_c - T_w)$ must equal $h_b(T_w - T_b)$, assuming the intervening wall to be plane. If the error in this heat balance is unacceptable, further values of $T_w$ must be tried until the heat balance falls within specified limits of accuracy.

A more refined design procedure would include estimation of the steam-side coefficient $h_c$ by one of the methods discussed in Volume 1, Chapter 7. Whilst such iterative procedures are laborious when carried out by hand, they are ideally suited to machine computation so that rapid evaluation to any degree of accuracy is easily achieved.

## 14.2.4. Vacuum Operation

With a number of heat sensitive liquids it is necessary to work at low temperatures, and this is effected by boiling under a vacuum. In addition, as will be seen later, the last unit of a multi-effect system usually works under a vacuum. Operation under a vacuum increases the temperature difference between the steam and boiling liquid and therefore tends to increase the heat flux. At the same time, the reduced boiling point will usually result in a more viscous material and a lower film heat transfer coefficient.

Consider the operation of a standard evaporator using steam at $135\,\mathrm{kN/m^2}$ or $380\,\mathrm{K}$ with a total heat of $2685\,\mathrm{kJ/kg}$, evaporating a liquor such as water.

|  | Atmospheric pressure ($101.3\,\mathrm{kN/m^2}$) | $13.5\,\mathrm{kN/m^2}$ |
|---|---|---|
| Boiling point | 373 K | 325 K |
| Temp. drop to liquor | 7 deg K | 55 deg K |
| Heat lost in condensate | 419 kJ/kg | 216 kJ/kg |
| Heat used | 2266 kJ/kg | 2469 kJ/kg |

Thus the capacity under vacuum is $(101.3/13.5) = 7.5$ times as great as that at atmospheric pressure. The advantage in capacity for the same unit is therefore considerable though there is no real change in the consumption of steam in the unit. In practice, the advantages are not as great as indicated since operation at a lower boiling point will reduce the value of the heat transfer coefficient and additional energy is required to achieve and maintain the vacuum.

## 14.3. MULTIPLE-EFFECT EVAPORATORS

### 14.3.1. General Principles

If an evaporator is fed with steam at 399 K with a total heat of 2714 kJ/kg and it is evaporating water at 373 K, then each kilogram of water vapour produced will have a total heat of 2675 kJ. If this heat is allowed to go to waste, by condensing it in a tubular condenser or by direct contact in a jet condenser, such a system makes very poor use of steam. The vapour produced is, however, suitable for passing to the calandria of a similar unit, provided the boiling temperature in the second unit is reduced so that an adequate temperature difference is maintained. This, as discussed in Section 14.2.4, can be effected by applying a vacuum to the second effect in order to reduce the boiling point of the liquor. This is the principle reached in the multiple effect systems which were introduced by Rillieux in about 1830.

Considering three evaporators arranged as shown in Fig. 14.6, in which the

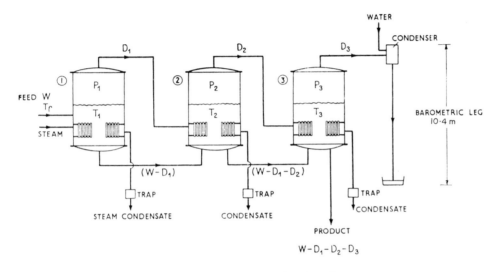

FIG. 14.6. Forward feed arrangement for triple effect evaporator

temperatures and pressures are shown as $T_1$, $T_2$, $T_3$, and $P_1$, $P_2$, $P_3$, respectively, in each unit. Suppose the liquor has no boiling point rise, and that the effects of hydrostatic head can be neglected. Then the heat transmitted per unit time across each effect will be:

Effect 1    $Q_1 = U_1 A_1 \Delta T_1$,    where    $\Delta T_1 = (T_0 - T_1)$,
Effect 2    $Q_2 = U_2 A_2 \Delta T_2$,    where    $\Delta T_2 = (T_1 - T_2)$,
Effect 3    $Q_3 = U_3 A_3 \Delta T_3$,    where    $\Delta T_3 = (T_2 - T_3)$.

Neglecting the heat required to heat the feed from $T_f$ to $T_1$, the heat $Q_1$ transferred across $A_1$ appears as latent heat in the vapour $D_1$ and is used as steam in the second effect. Hence

$$Q_1 = Q_2 = Q_3$$

So that

$$U_1 A_1 \Delta T_1 = U_2 A_2 \Delta T_2 = U_3 A_3 \Delta T_3 \qquad (14.7)$$

If, as is commonly the case, the individual effects are alike, $A_1 = A_2 = A_3$, so that:

$$U_1 \Delta T_1 = U_2 \Delta T_2 = U_3 \Delta T_3 \qquad (14.8)$$

On this analysis, the difference in temperature across each effect is inversely proportional to the heat transfer coefficient. This, however, represents a simplification, since

(a) the heat required to heat the feed from $T_f$ to $T_1$ has been neglected, and

(b) the liquor passing from 1 to 2 carries heat into the second effect, and this is responsible for some evaporation; similarly for the third effect.

The latent heat required to evaporate 1 kg of water in 1, is approximately equal to the heat obtained in condensing 1 kg of steam at $T_0$.

Thus 1 kg of steam fed to 1 evaporates 1 kg of water in 1. Again the 1 kg of steam from 1 evaporates about 1 kg of steam in 2.

Thus, in an $N$ effect system, 1 kg of steam fed to the first effect will evaporate in all about $N$ kilograms of liquid. This gives a simplified picture, as discussed later, though it does show that one of the great attractions of a multiple effect system in that considerably more evaporation per kilogram of steam is obtained than in a single effect unit. The economy of the system, measured by the kilograms of water vaporised per kilogram of steam condensed, increases with the number of effects.

The water evaporated in each effect is proportional to $Q$, since the latent heat is sensibly constant. Thus the total capacity,

$$\begin{aligned} Q &= Q_1 + Q_2 + Q_3 \\ &= U_1 A_1 \Delta T_1 + U_2 A_2 \Delta T_2 + U_3 A_3 \Delta T_3 \end{aligned} \qquad (14.9)$$

If an average value of the coefficients $U_{av}$ is taken, then:

$$Q = U_{av}(\Delta T_1 + \Delta T_2 + \Delta T_3)A \qquad (14.10)$$

assuming the area of each effect is the same. A single effect evaporator operating with a temperature difference $\Sigma \Delta T$, with this average coefficient $U_{av}$, would, however, have the same capacity $Q = U_{av} A \Sigma \Delta T$. Thus, it is seen that the capacity of a multiple effect system is the same as that of a single effect, operating with the same total temperature difference and having an area $A$ equal to that of one of the multiple effect units. The value of the multiple effect system is that better use is made of steam though, in order to bring this about, it is necessary to make a much bigger capital outlay for the increased number of units and accessories.

## 14.3.2. The Calculation of Multiple-effect Systems

In the equations considered in Section 14.3.1, various simplifying assumptions have been made which are now considered further in the calculation of a multiple-effect system; in particular, the temperature distribution in such a system and the heat transfer area required in each effect are to be determined. The method illustrated in Example 14.1 is essentially based on that of Hausbrand[12].

**Example 14.1**

4 kg/s of a liquor with 10 per cent solids is fed at 294 K to the first effect of a triple effect unit. Liquor with 50 per cent solids is to be withdrawn from the third effect, which is at a pressure of 13·3 kN/m². The liquor will be assumed to have a specific heat of 4·18 kJ/kg K and to have no B.P.R. Saturated dry steam at 205 kN/m² is fed to the heating element of the first effect, and the condensate is removed at the steam temperature in each effect (Fig. 14.6).

If the three units are to have equal areas, it is required to find the area, the temperature differences and the steam consumption. Assume heat transfer coefficients of 3·10, 2·00 and 1·10 kW/m² K for the first, second, and third effects respectively.

**Solution 1**

It should be realised that a precise theoretical solution is neither necessary nor possible. During the operation of the evaporator, variation of such points as the liquor levels will alter the heat transfer coefficients and hence the temperature distribution. In the first place, it is necessary to assume heat transfer coefficients. These, as noted previously, will only be approximate and will be based on practical experience with similar liquors in similar types of evaporators.

At a pressure of 13·3 kN/m², the boiling point of water is 325 K, so that the total temperature difference $\Sigma\Delta T = (394 - 325) = 69 \deg K$.

*First Approximation.* Assume the relation

$$U_1\Delta T_1 = U_2\Delta T_2 = U_3\Delta T_3 \qquad \text{(equation 14.8)}$$

Then substituting the values of $U_1$, $U_2$ and $U_3$ and $\Sigma\Delta T = 69 \deg K$:

$$\Delta T_1 = 13 \text{ K}, \quad \Delta T_2 = 20 \text{ K}, \quad \Delta T_3 = 36 \text{ K}$$

Since the feed is cold, it will be necessary to have a greater value of $\Delta T_1$ than given by this analysis. Assume therefore that the values of $\Delta T$ are $\Delta T_1 = 18 \deg K$, $\Delta T_2 = 17 \deg K$, $\Delta T_3 = 34 \deg K$.

If the latent heats are given by $\lambda_0$, $\lambda_1$, $\lambda_2$, $\lambda_3$ then,

| | | | |
|---|---|---|---|
| For steam to 1, | $T_0 = 394$ K | and | $\lambda_0 = 2200$ kJ/kg |
| For steam to 2, | $T_1 = 376$ K | and | $\lambda_1 = 2249$ kJ/kg |
| For steam to 3, | $T_2 = 359$ K | and | $\lambda_2 = 2293$ kJ/kg |
| | $T_3 = 325$ K | and | $\lambda_3 = 2377$ kJ/kg |

Assuming the condensate to leave at the steam temperature, heat balances across each effect may be made as follows:

Effect 1.   $D_0\lambda_0 = W(T_1 - T_f) + D_1\lambda_1$   or
$$2200D_0 = 4 \times 4\cdot18(376 - 294) + 2249D_1$$
Effect 2.   $D_1\lambda_1 + (W - D_1)(T_1 - T_2) = D_2\lambda_2$   or
$$2249D_1 + (4 - D_1)4\cdot18(376 - 359) = 2293D_2$$
Effect 3.   $D_2\lambda_2 + (W - D_1 - D_2)(T_2 - T_3) = D_3\lambda_3$   or
$$2293D_2 + (4 - D_1 - D_2)4\cdot18(359 - 325) = 2377D_3$$

where $W$ is the liquor fed to the system per unit time.

A material balance over the evaporator may be written as follows:

|         | Solids (kg/s) | Liquor (kg/s) | Total (kg/s) |
|---------|---------------|---------------|--------------|
| Feed    | 0·4           | 3·6           | 4·0          |
| Product | 0·4           | 0·4           | 0·8          |
| Evaporation |           | 3·2           | 3·2          |

Making use of the previous equations and the fact that $(D_1 + D_2 + D_3) = 3\cdot2\,\text{kg/s}$, the evaporation in each unit is, $D_1 \approx 0\cdot991$, $D_2 \approx 1\cdot065$, $D_3 \approx 1\cdot144$, $D_0 \approx 1\cdot635\,\text{kg/s}$. The area of the surface of each calandria necessary to transmit the necessary heat under the given temperature difference can then be obtained as:

$$A_1 = \frac{D_0 \lambda_0}{U_1 \Delta T_1} = \frac{1\cdot635 \times 2200}{3\cdot10 \times 18} = 64\cdot5\,\text{m}^2$$

$$A_2 = \frac{D_1 \lambda_1}{U_2 \Delta T_2} = \frac{0\cdot991 \times 2249}{2\cdot00 \times 17} = 65\cdot6\,\text{m}^2$$

$$A_3 = \frac{D_2 \lambda_2}{U_3 \Delta T_3} = \frac{1\cdot085 \times 2293}{1\cdot10 \times 34} = 65\cdot3\,\text{m}^2$$

The three areas, as calculated, are approximately equal, so that the temperature differences assumed may be taken as nearly correct. In practice, $\Delta T_1$ would have to be a little larger since $A_1$ is the smallest area. It may be noted that, on the basis of these calculations, the economy is given by $e = (3\cdot2/1\cdot635) = 2\cdot0$. Thus, a triple effect unit working under these conditions gives a reduction in steam utilisation compared with a single effect, though not as large an economy as would be expected.

### Solution 2

A simplified method of solving problems of multiple effect evaporation has been suggested by Storrow[13]. It is particularly useful for systems with a large number of effects because it obviates the necessity for solving many simultaneous equations. Essentially it depends on obtaining only a rough value for those heat quantities which are a small proportion of the whole. The same example will now be solved by this method.

From Fig. 14.6 it is seen that for a feed $W$ to the first effect, vapour $D_1$ and liquor $(W - D_1)$ are fed forward to the second effect. In the first effect steam is condensed partly in order to raise the feed to its boiling point and partly to effect evaporation. In the second effect further vapour is produced mainly as a result of condensation of the vapour from the first effect and to a smaller extent by flash vaporisation of the concentrated liquor which is fed forward. As the amount of vapour produced by the latter means is generally only comparatively small, it need be only roughly estimated. Similarly the vapour produced by flash evaporation in the third effect will be a small proportion of the total and only an approximate evaluation is required.

*Vapour Production by Flash Vaporisation—Approximate Evaluation*

If the heat transferred in each effect is the same:

$$U_1 \Delta T_1 = U_2 \Delta T_2 = U_3 \Delta T_3 \qquad \text{(equation 14.8)}$$

or

$$3 \cdot 10 \Delta T_1 = 2 \cdot 0 \Delta T_2 = 1 \cdot 10 \Delta T_3$$

Steam temperature $= 394\,\text{K}$. Temperature in condenser $= 325\,\text{K}$.

$$\therefore \quad \Sigma \Delta T = (394 - 325) = 69 \deg \text{K}$$

Solving:

$$\Delta T_1 = 13 \deg \text{K} \qquad \Delta T_2 = 20 \deg \text{K} \qquad \Delta T_3 = 36 \deg \text{K}$$

These values of $\Delta T$ will be valid provided the feed is approximately at its boiling point. Weighting the temperature differences since the feed enters at ambient temperature:

$$\Delta T_1 = 18 \deg \text{K} \qquad \Delta T_2 = 18 \deg \text{K} \qquad \Delta T_3 = 33 \deg \text{K}$$

and the temperatures in each effect are:

$$T_1 = 376\,\text{K} \qquad T_2 = 358\,\text{K} \quad \text{and} \quad T_3 = 325\,\text{K}$$

The total evaporation $(D_1 + D_2 + D_3)$ is obtained from a material balance as follows:

|            | Solids (kg/s) | Liquid (kg/s) | Total (kg/s) |
|------------|---------------|---------------|--------------|
| Feed       | 0·4           | 3·6           | 4·0          |
| Product    | 0·4           | 0·4           | 0·8          |
| Evaporation |              | 3·2           | 3·2          |

Assuming, as an approximation, equal evaporation in each effect, or $D_1 = D_2 = D_3$ $= 1 \cdot 07\,\text{kg/s}$. Thus the latent heat of flash vaporisation in the second effect is given by:

$$4 \cdot 18(4 \cdot 0 - 1 \cdot 07)(376 - 358) = 220 \cdot 5\,\text{kW}$$

and latent heat of flash vaporisation in third effect:

$$4 \cdot 18(4 \cdot 0 - 2 \times 1 \cdot 07)(358 - 325) = 256 \cdot 6\,\text{kW}$$

*Final Calculation of Temperature Differences*

Subsequent calculations are considerably simplified if it is assumed that the latent heat of vaporisation is the same at all temperatures in the multiple-effect system, for under these conditions the condensation of 1 kg of steam gives rise to the formation of 1 kg of vapour.

$$\text{At } 394\,\text{K the latent heat} = 2200\,\text{kJ/kg}$$
$$\text{At } 325\,\text{K the latent heat} = 2377\,\text{kJ/kg}$$
$$\text{Mean value, } \lambda = 2289\,\text{kJ/kg}$$

Now the amounts of heat transferred in each effect ($Q_1, Q_2, Q_3$) and in the condenser ($Q_c$) are related by the following equations:

$$Q_1 - W(T_1 - T_f) = Q_2 = (Q_3 - 220 \cdot 5) = (Q_c - 220 \cdot 5 - 256 \cdot 6)$$

i.e.

$$Q_1 - 4 \cdot 0 \times 4 \cdot 18(394 - \Delta T_1 - 294) = Q_2 = (Q_3 - 220 \cdot 5) = (Q_c - 477 \cdot 1)$$

$$\text{Total evaporation} = \frac{(Q_2 + Q_3 + Q_c)}{2289} = 3 \cdot 2$$

i.e.

$$Q_2 + (Q_2 + 220 \cdot 5) + (Q_2 + 477 \cdot 1) = 7325$$

i.e.

$$Q_2 = 2209 \,\text{kW}$$
$$Q_3 = 2430 \,\text{kW}$$
$$Q_1 = 2209 + 4 \cdot 0 \times 4 \cdot 18(394 - \Delta T_1 - 294)$$
$$= (3881 - 16 \cdot 72 \Delta T_1)$$

Thus, applying the heat transfer equations:

$$3881 - 16 \cdot 72 \Delta T_1 = 3 \cdot 1 A \Delta T_1, \quad \text{i.e. } A \Delta T_1 = (1252 - 5 \cdot 4 \Delta T_1)$$
$$2209 = 2 \cdot 0 A \Delta T_2, \quad \text{i.e. } A \Delta T_2 = 1105$$
$$2430 = 1 \cdot 1 A \Delta T_3, \quad \text{i.e. } A \Delta T_3 = 2209$$

Further,

$$\Delta T_1 + \Delta T_2 + \Delta T_3 = 69 \deg \text{K}$$

Values of $\Delta T_1, \Delta T_2, \Delta T_3$ must now be chosen to give equal values of $A$ in each effect. Trial and error solution is simplest.

| $\Delta T_1$ (deg K) | $A_1$ (m²) | $\Delta T_2$ (deg K) | $A_2$ (m²) | $\Delta T_3$ (deg K) | $A_3$ (m²) |
|---|---|---|---|---|---|
| 18 | 64·2 | 18 | 61·4 | 33 | 66·9 |
| 19 | 60·5 | 17 | 65·0 | 33 | 66·9 |
| 18 | 64·2 | 17·5 | 63·1 | 33·5 | 65·9 |
| 18 | 64·2 | 17 | 65·0 | 34 | 64·9 |

The areas, as calculated, are approximately equal so that the assumed temperature differences are approximately correct.

$$\text{Steam consumption} = \frac{Q_1}{2289} = \frac{3580}{2289} = 1 \cdot 56 \,\text{kg/s}$$

$$\text{Economy} = \frac{3 \cdot 2}{1 \cdot 56} \approx \underline{\underline{2 \cdot 0}}$$

The calculation of multiple-effect systems, whilst relatively straightforward for one or two configurations, become tedious in the extreme where a wide range of operating conditions are to be investigated. Fortunately the calculations involved lend themselves admirably to processing by computer and in this respect reference should be made to articles such as that by Stewart and Beveridge quoted in Section 14.7.

### 14.3.3. Comparison of Forward and Backward Feeds

In the unit discussed in Example 14.1, the weak liquor is fed to effect 1 and flows on to 2 and then to 3. The steam is also fed to 1, and the process is known as forward feed since the feed is to the same unit as the steam and travels down the unit in the same direction as the steam or vapour. It is possible, however, to introduce the weak liquor to effect 3 and cause it to travel from 3 to 2 to 1, whilst the steam and vapour still travel in the direction 1 to 2 to 3. This system, shown in Fig. 14.7, is known as backward feed. A further arrangement for

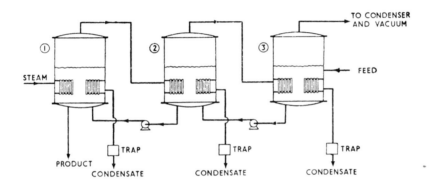

FIG. 14.7.  Backward feed arrangement for triple effect evaporator

the feed is known as parallel feed, which is shown in Fig. 14.8. In this case, the liquor is fed to each of the three effects in parallel, but the steam only to the first effect. This arrangement is commonly used in the concentration of salt solutions, where the deposition of crystals makes it rather difficult to use the standard forward feed arrangement.

It is of importance to examine the effect of backward feed on the temperature distribution, the areas of surface required, and the economy of the unit, and Example 14.1 will be considered with this flow arrangement. Since the weak liquor is now at the lowest temperature and the strong liquor at the highest, the heat transfer coefficients will not be the same as in the case of forward feed. In effect 1, the liquor is now much stronger than in the former case, and hence $U_1$ will not be as large as before. Again, on the same argument, $U_3$ will be larger than before. Although it is unlikely to be exactly the same, $U_2$ will be taken as being unaltered by the arrangement. Taking values of $U_1 = 2.5$, $U_2 = 2.0$ and $U_3 = 1.6\,\mathrm{kW/m^2\,K}$, the temperature distribution can be determined in the same manner as for forward feed by taking heat balances across each unit.

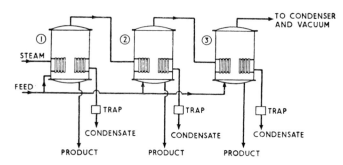

FIG. 14.8. Arrangement of triple effect evaporator for parallel feed

### Solution 1 for Backward Feed

Applying equation 14.8, and taking the temperature differences as:

$$\Delta T_1 = 18 \deg K, \quad \Delta T_2 = 23 \deg K, \quad \Delta T_3 = 28 \deg K$$

Modifying the figures to take account of the low feed temperature then:

$$\Delta T_1 = 20 \deg K, \quad \Delta T_2 = 24 \deg K, \quad \Delta T_3 = 25 \deg K$$

The temperatures in the effects and the corresponding latent heats will now be:

$$
\begin{aligned}
T_0 &= 394\,K \quad \text{and} \quad \lambda_0 = 2200\,kJ/kg \\
T_1 &= 374\,K \quad \text{and} \quad \lambda_1 = 2254\,kJ/kg \\
T_2 &= 350\,K \quad \text{and} \quad \lambda_2 = 2314\,kJ/kg \\
T_3 &= 325\,K \quad \text{and} \quad \lambda_3 = 2377\,kJ/kg
\end{aligned}
$$

The heat balance equations can be written as

Effect 3.   $D_2\lambda_2 = W(T_3 - T_f) + D_3\lambda_3$  or
$$2314D_2 = 4 \times 4 \cdot 18(325 - 294) + 2377D_3$$

Effect 2.   $D_1\lambda_1 = (W - D_3)(T_2 - T_3) + D_2\lambda_2$  or
$$2254D_1 = (4 - D_3)4 \cdot 18(350 - 325) + 2314D_2$$

Effect 1.   $D_0\lambda_0 = (W - D_3 - D_2)(T_1 - T_2) + D_1\lambda_1$  or
$$2200D_0 = (4 - D_3 - D_2)4 \cdot 18(374 - 350) + 2254D_1$$

Again remembering that $(D_1 + D_2 + D_3) = 3 \cdot 2\,kg/s$, these equations can be solved to give:

$$D_1 \approx 1 \cdot 261, \quad D_2 \approx 1 \cdot 086, \quad D_3 \approx 0 \cdot 853, \quad D_0 \approx 1 \cdot 387\,kg/s$$

The areas of transfer surface will then be obtained as:

$$A_1 = \frac{D_0\lambda_0}{U_1\Delta T_1} = \frac{(1 \cdot 387 \times 2200)}{(2 \cdot 5 \times 20)} = 61 \cdot 0\,m^2$$

$$A_2 = \frac{D_1\lambda_1}{U_2\Delta T_2} = \frac{(1 \cdot 261 \times 2254)}{(2 \cdot 00 \times 24)} = 59 \cdot 2\,m^2$$

$$A_3 = \frac{D_2\lambda_2}{U_3\Delta T_3} = \frac{(1 \cdot 086 \times 2314)}{(1 \cdot 6 \times 25)} = 62 \cdot 8\,m^2$$

These three areas are approximately equal, so that the temperature differences suggested are sufficiently near for design purposes. The economy for this system $e = (3\cdot2/1\cdot387) = \underline{\underline{2\cdot3}}$.

### Solution 2 for Backward Feed

The example will now be solved by Storrow's method.

As in the first method of solution, the temperatures in the effects will be taken as

$$T_1 = 374\,\text{K}, \quad T_2 = 350\,\text{K}, \quad T_3 = 325\,\text{K}$$

With backward feed (Fig. 14.7) the liquid has to be raised to its boiling point as it enters each effect.

Heat required to raise feed to second effect to boiling point

$$= 4\cdot18(4\cdot0 - 1\cdot07)(350 - 325)$$
$$= 306\cdot2\,\text{kW}$$

Heat required to raise feed to first effect to boiling point

$$= 4\cdot18(4\cdot0 - 2 \times 1\cdot07)(374 - 350)$$
$$= 186\cdot6\,\text{kW}$$

Assuming a constant value of 2289 kJ/kg for the latent heat at all stages, the relation between the heat transferred in each effect and in the condenser is given by:

$$Q_1 - 186\cdot6 = Q_2 = (Q_3 + 306\cdot2) = (Q_c + 306\cdot2 + 4 \times 4\cdot18(325 - 294))$$
$$= Q_c + 824\cdot5$$

$$\text{Total evaporation} = \frac{Q_2 + Q_3 + Q_c}{2289} = 3\cdot2\,\text{kg/s}$$

i.e.

$$Q_2 + (Q_2 - 306\cdot2) + (Q_2 - 824\cdot5) = 7325$$

i.e.

$$Q_2 = 2819 = A\Delta T_2 \times 2\cdot0$$
$$Q_3 = 2512 = A\Delta T_3 \times 1\cdot6$$
$$Q_1 = 3006 = A\Delta T_1 \times 2\cdot5$$

or

$$A\Delta T_1 = 1202$$
$$A\Delta T_2 = 1410$$
$$A\Delta T_3 = 1570$$
$$\Delta T_1 + \Delta T_2 + \Delta T_3 = 69\,\text{deg K}$$

| $\Delta T_1$ (deg K) | $A_1$ (m²) | $\Delta T_2$ (deg K) | $A_2$ (m²) | $\Delta T_3$ (deg K) | $A_3$ (m²) |
|---|---|---|---|---|---|
| 20 | 60·1 | 24 | 58·9 | 25 | 62·8 |

The areas are approximately equal and the assumed values of $\Delta T$ therefore correct.

$$\text{Economy} = \frac{3 \cdot 2}{3006 \div 2289} = \underline{\underline{2 \cdot 4}}$$

On the basis of heat transfer area and thermal considerations, a comparison of the two methods of feed is given as:

|                                            | Forward | Backward |
|--------------------------------------------|---------|----------|
| Total steam used $D_0$ (kg)                | 1·635   | 1·387    |
| Economy                                    | 2·0     | 2·3      |
| Condenser load $D_3$ (kg)                  | 1·44    | 0·853    |
| Heat transfer surface $A$ (m$^2$) per effect | 65·1    | 61·0     |

For the conditions of Example 14.1, the backward feed system shows a reduction in steam consumption, an improved economy, a reduction in condenser load, and a small reduction in heat transfer area.

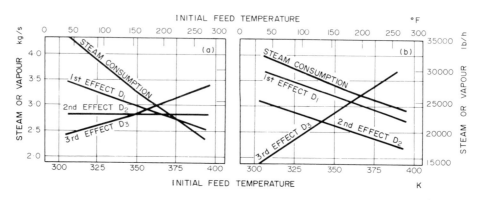

FIG. 14.9.  Effect of feed temperature on operation of triple effect evaporator. (a) Forward feed, (b) backward feed

In the case of forward feed systems, all the liquor has to be heated from $T_f$ to $T_1$ by steam but, in the case of backward feed, the heating of the feed in the last effect is done with steam that has already evaporated $(N-1)$ times its own weight of water, assuming ideal conditions. The feed temperature must therefore be regarded as a major feature in this class of problem. Webre[14] has examined the effect of feed temperature on the economy and the evaporation in each effect, for the case of a liquor fed at the rate of 12·5 kg/s to a triple effect evaporator where it was concentrated to 8·75 kg/s. Neglecting boiling-point rise and working with a fixed vacuum on the third effect, he prepared curves shown in Figs. 14.9 and 14.10 for the three methods of forward, backward and parallel feed.

Figure 14.9$a$ illustrates the drop in steam consumption as the feed temperature is increased with forward feed. It will also be seen that, for these conditions, $D_1$ falls, $D_2$ remains constant and $D_3$ rises with increase in the feed temperature $T_f$. With backward feed (Fig. 14.9$b$), the fall in steam consumption is not so marked and it will be seen that, whereas $D_1$ and $D_2$ fall, the load on the condenser $D_3$ increases. The results are conveniently interpreted in Fig. 14.10, which shows that the economy increases with $T_f$ for a forward feed system to a marked extent, whilst the corresponding increase with the backward feed system is relatively small. At low values of $T_f$, the backward feed gives the higher economy. At some intermediate value the two systems give the same value of economy, whilst for high values of $T_f$ the forward feed system is more economical in steam.

These results, whilst showing the influence of $T_f$ on the economy, must not be interpreted too rigidly, since the values for the coefficients for the two systems and the

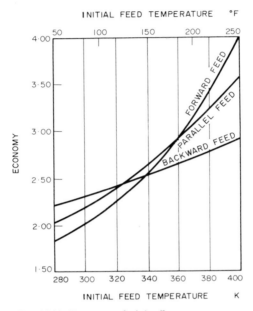

FIG. 14.10. Economy of triple effect evaporators

influencing of boiling-point rise may make a substantial difference to these curves. In general, however, it will be found that with cold feeds the backward feed system is more economical. Despite this fact, the forward feed system is the commonest, largely because it is the simplest to operate, the backward feed necessitating the use of pumps between each effect.

The main criticism of the forward feed system is that the most concentrated liquor is in the last effect, and the temperature is lowest there. The viscosity is therefore high and low values of $U$ are obtained; in order to compensate for this, a large temperature difference is required, though this limits the number of effects. It will sometimes be found, as in the sugar industry, that it is preferable to run a multiple effect system to a certain concentration, and to run a separate effect for the final stage where the crystals are formed.

## 14.4. IMPROVED EFFICIENCY IN EVAPORATORS

### 14.4.1. Vapour Compression Evaporators

The single effect evaporator uses rather more than 1 kg of steam to evaporate 1 kg of water. Three methods have been introduced which enable improved performance to be obtained, either by direct reduction in the steam consumption, or by improved energy efficiency of the whole unit:

(a)  Multiple effect operation. This has been considered in Section 14.3.

(b)  Recompression of the vapour rising from the evaporator.

(c)  Evaporation at low temperatures using a heat pump cycle.

Methods (b) and (c) will now be considered.

Consider an evaporator fed with saturated steam at 387 K, i.e. 165 kN/m² concentrating a liquor boiling at 373 K at atmospheric pressure. If the condensate leaves at 377 K, then:

1 kg of steam at 387 K has a total heat of 2698 kJ.

1 kg of condensate at 377 K has a total heat of 437 kJ.

∴   Heat given up per kg of steam is 2261 kJ.

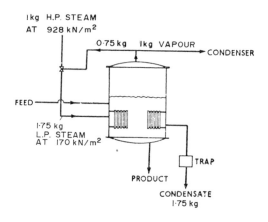

FIG. 14.11. Vapour compression evaporator with high pressure steam-jet compression

If this condensate is returned to the boiler house, then at least 2261 kJ/kg must be added to give 1 kg of steam to be fed back to the evaporator. In practice, of course, more heat per kilogram of condensate will be required. 2261 kJ will vaporise 1 kg of liquid at atmospheric pressure to give vapour with a total heat of 2675 kJ/kg. To regenerate 1 kg of steam in the original condition from this requires the addition of only 23 kJ. The idea of vapour compression is to make use of the vapour from the evaporator, and to upgrade it to the condition of the original steam. Such an idea offers enormous advantages in thermal economy, though it is by no means easy to add the 23 kJ to each kilogram of vapour in an economical manner. The two methods available are:

(a)  The use of steam-jet ejectors (Fig. 14.11).

(b)  The use of mechanical compressors (Fig. 14.12).

In selecting a compressor for this type of operation, the main difficulty is the very large volume of vapour to be handled. Rotary compressors of the Rootes type are suitable for small and medium size units, though they have not often been applied to large installations. Mechanical compressors have been used extensively for the purification of sea water.

The use of an ejector, fed with high-pressure steam, is illustrated in Fig. 14.11. High-pressure steam is injected through a nozzle and the low-pressure vapours are drawn in

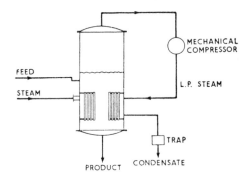

FIG. 14.12. Vapour compression evaporator with mechanical compressor

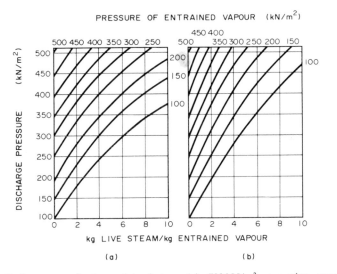

FIG. 14.13. Performance of steam jet ejector. (a) 790 kN/m² operating pressure, (b) 1135 kN/m² operating pressure

through a second inlet at right angles, the issuing jet of steam passing out to the calandria, as indicated. These units are relatively simple in construction and can be made of corrosion-resistant material; they have no moving parts and for this reason will have a long life. They have the great advantage over the mechanical compressor that they will handle large volumes of vapour and can therefore be arranged to operate at very low pressures. The disadvantage of the steam jet ejector is that it works at maximum efficiency at only one specific condition. Some indication of the performance of these units is shown in Fig. 14.13, where the pressure of the mixture, for different amounts of vapour compressed per kilogram of live steam, is shown for a series of different pressures. With an ejector of these characteristics using steam at 965 kN/m², 0·75 kg of vapour per kilogram of steam can be compressed to give 1·75 kg of vapour at 170 kN/m². An evaporator unit as shown in Fig. 14.11 will therefore give 1·75 kg of vapour per kilogram of high pressure steam. Of the 1·75 kg of vapour, 0·75 kg is taken to the compressor and the remaining 1 kg to the condenser. Ideally, this single effect unit gives an economy of 1·75, or approximately the economy of a double effect unit.

Two illustrations illustrate the application of this system. In the first, 1·2 kg/s of a potato waste liquor is concentrated from 5·5 to 45 per cent solids in a single effect fitted with vapour compression. 0·6 kg/s of steam at 710 kN/m² is used and 1·06 kg of water is evaporated per second, corresponding to an economy of 1·7.

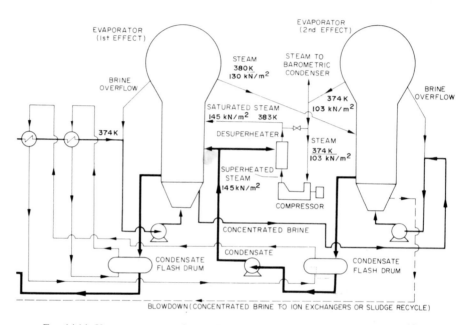

FIG. 14.14. Vapour recompression on the second effect of a two stage evaporator[15]

In a second installation, a single effect with vapour compression concentrates 0·104 kg/s of a glycerine–water liquor from 5 per cent to 88 per cent glycerine, by the evaporation of 0·10 kg of water per second, and uses 0·53 kg/s of steam at 1620 kN/m². This, too, corresponds to an economy of 1·7.

Vapour compression may be applied to the vapour from the first effect of a multiple effect system, thus giving increased utilisation of the steam (Fig. 14.14). Such a device is not suitable for use with liquors having a high boiling-point rise, for in these cases the vapour, although initially superheated, has to be compressed to such an extent, in order to give the desired temperature difference across the calandria, that the efficiency is reduced. The application of these compressors is bound up with the steam load of the factory. If there is plenty of low pressure steam available, then the use of vapour compression can scarcely be advocated. If, however, high-pressure steam is available, then it can be used to advantage in a vapour compression unit. It will, in fact, be far superior to the practice of passing high-pressure steam through a reducing valve to feed an evaporator.

### 14.4.2. The Heat Pump Cycle

The evaporation of citrus juices at temperatures up to 328 K, or of pharmaceutical products at even lower temperatures, has led to the development of the evaporator incorporating a heat pump cycle using a separate working fluid. Figure 14.15 shows the use of the heat pump cycle, with ammonia as the working fluid. In this arrangement, ammonia gas vaporises the feed liquor at a low temperature such as 288–313 K, the

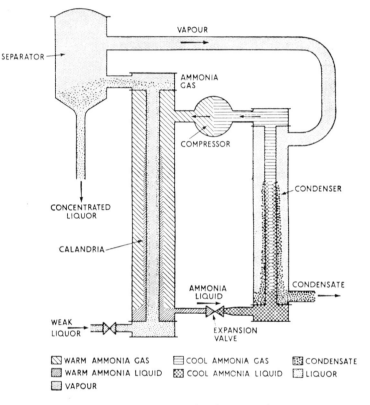

FIG. 14.15. Heat pump cycle using ammonia

ammonia being condensed. The liquid ammonia is then passed through an expansion valve, where it is cooled to a much lower temperature. The cooled ammonia liquor then enters the condenser where it condenses the vapour leaving the separator; it is itself vaporised and leaves as low pressure gas, to be compressed in a mechanical compressor and then passed to the evaporator for a second cycle. The excess heat introduced by the compressor must be removed from the ammonia by means of a cooler.

The main advantage of this form of unit lies in the very great reduction in the volume of gas handled by the compressor. Thus, 1 kg of water vapour at, say, 311 K, with a volume of $22 \, m^3$ and latent heat about 2560 kJ/kg, passes this heat to ammonia at a temperature of say 305 K. About 2·1 kg of ammonia will be vaporised to give a vapour with a volume of only about 0·22 $m^3$ at the high pressure used in the ammonia cycle.

Schwarz[16] gives a comparison of the various units used for low temperature evaporation. The three types in general use are the single-effect single-pass, the single-effect with recirculation, and the multiple effect with recirculation. Each of these may involve vapour compression or the addition of a second heat transfer medium. Schwarz suggests that the multiple effect units are the most economical, in terms of capital and operating costs. It is important to note that the single-effect, single-pass system offers the minimum hold-up, and hence a very short transit time. In view of work on film-type units, there seems little to be gained by recirculation, since over 70 per cent vaporisation can be achieved in one pass. The figures in Table 14.1 show the comparison between a double-effect unit with vapour compression on the first effect, and a unit with an ammonia refrigeration cycle, both units giving 1·25 kg/s of evaporation.

TABLE 14.1. *Comparison of Refrigeration and Vapour Compression Systems*

| System | Steam kg/s at 963 kN/m$^2$ | Water m$^3$/s at 300 K | Power (kW) |
|---|---|---|---|
| Refrigeration cycle | 0·062 | 0·019 | 320* |
| Vapour compression | 0·95 | 0·076 | 20 |
| Ratio of steam system to refrigeration | 15·1 | 4 | 0·06 |

*Includes 300 kW compressor.

The utilities, other than power, required for the refrigeration system are therefore very much less than for recompression with steam, though the capital cost and the cost of power will be much higher.

Figure 14.16 shows a double effect system with a capacity of 0·76 kg/s, using an ammonia refrigeration cycle; the first effect is heated by warm water obtained from the ammonia condenser. The first effect has a surface of 103 m$^2$, and the second of 195 m$^2$, the overall transfer coefficients being 425 and 664 W/m$^2$ K.

Reavell[17] has given a comparison of costs for concentration of a feed of a heat sensitive protein liquor at 1·70 kg/s from 10 per cent to 50 per cent solids, on the basis of a 160-hr (288 k s) week. It should be noted that, when using the double effect with vapour compression, a lower temperature can be used in the first effect than when a triple effect unit is used. In determining these figures no account has been taken of depreciation, though if this is taken as 15 per cent of the capital costs it will not make a significant difference to the comparison.

TABLE 14.2. *Comparison of Various Systems for Concentration of Protein Liquid*

| Type | Approx. installed cost | Cost of steam per year | Net saving per year compared with single effect |
|---|---|---|---|
| Single effect | £50,000 | £403,000 | — |
| Double effect | £70,000 | £214,000 | £189,000 |
| Double effect with vapour compression | £90,000 | £137,000 | £266,000 |
| Triple effect | £100,000 | £143,000 | £260,000 |

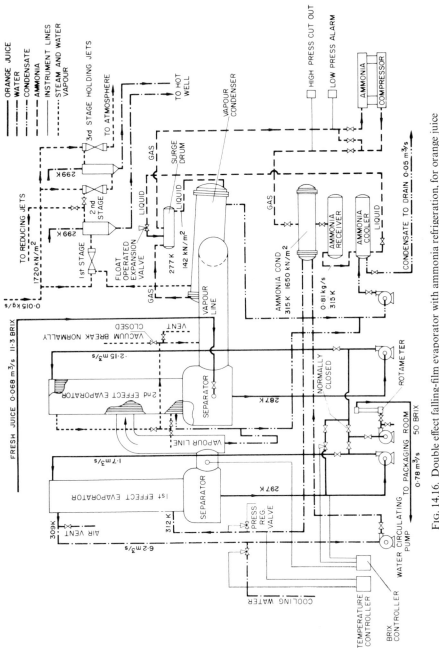

FIG. 14.16. Double effect falling-film evaporator with ammonia refrigeration, for orange juice

## 14.5. EQUIPMENT FOR EVAPORATION

### 14.5.1. Selection of Evaporators

The growth of the process industries and the rapid development of new products has provided liquids with a wide range of physical and chemical properties which require concentration by evaporation techniques. This has been the stimulus for the continued development of the evaporation equipment which is currently available and for the introduction of new techniques.

The type of equipment used depends largely on the method of applying heat to the liquor and the method of agitation. Heating may be either direct or indirect. Direct heating is represented by solar evaporation and by submerged combustion of a fuel. In indirect heating, the heat, generally provided by the condensation of steam, passes through the heating surface of the evaporator.

Some of the problems arising during evaporation are:

(a) High product viscosity.
(b) Heat sensitivity.
(c) Scale formation and deposition.

Consequently, new equipment has been designed in an attempt to overcome one or more of these problems. In view of the large number of types of evaporator which are available, the selection of equipment for a particular application can only be made after a detailed analysis of all relevant factors has been made. These will, of course, include the properties of the liquid to be evaporated, capital and running costs, capacity, holdup, and residence time characteristics. Evaporator selection has recently been discussed by Moore and Hesler[18] and Parker[19] although the former is orientated towards the food industry. Parker has attempted to test the suitability of each basic design for dealing with the problems encountered in practice. The basic information was presented in the form shown in Fig. 14.17. The factors which are considered include the ability to handle liquids in three viscosity ranges, to deal with foaming, scaling or fouling, crystal production, solids in suspension, and heat sensitive materials. A comparison of residence time and holding volume relative to the wiped film unit is also given. It is of interest to note that the agitated or wiped film evaporator is the only one which is shown to be applicable over the whole range of conditions covered.

### 14.5.2. Evaporators with Direct Heating

The use of solar heat for the production of Glauber's salt at Lake Chaplin has been described by Holland[20,21]. The brine is pumped in hot weather to reservoirs of a $100,000 \, m^2$ area to a depth of 3 to 5 m, and salt is deposited. Later in the year, the mother liquor is drained off and the salt stacked mechanically, and conveyed to special Holland evaporators. These are large vessels ($3 \times 3 \cdot 5 \times 4 \, m$); hot gases enter at 1150–1250 K through a suitable refractory duct and leave at about 330 K. The salt crystals melt in their own water of crystallisation and are then dried in the stream of hot gas. Bloch and associates[22] have examined the mechanism of evaporation of salt brines by direct solar energy. They found that the rate of evaporation increased with the depth of brine; thus, a 500 mm depth of brine gave a 16 per cent higher rate than a 200 mm depth. The addition of dyes, such as 2-naphthol green, enables the solar energy to be absorbed in a much

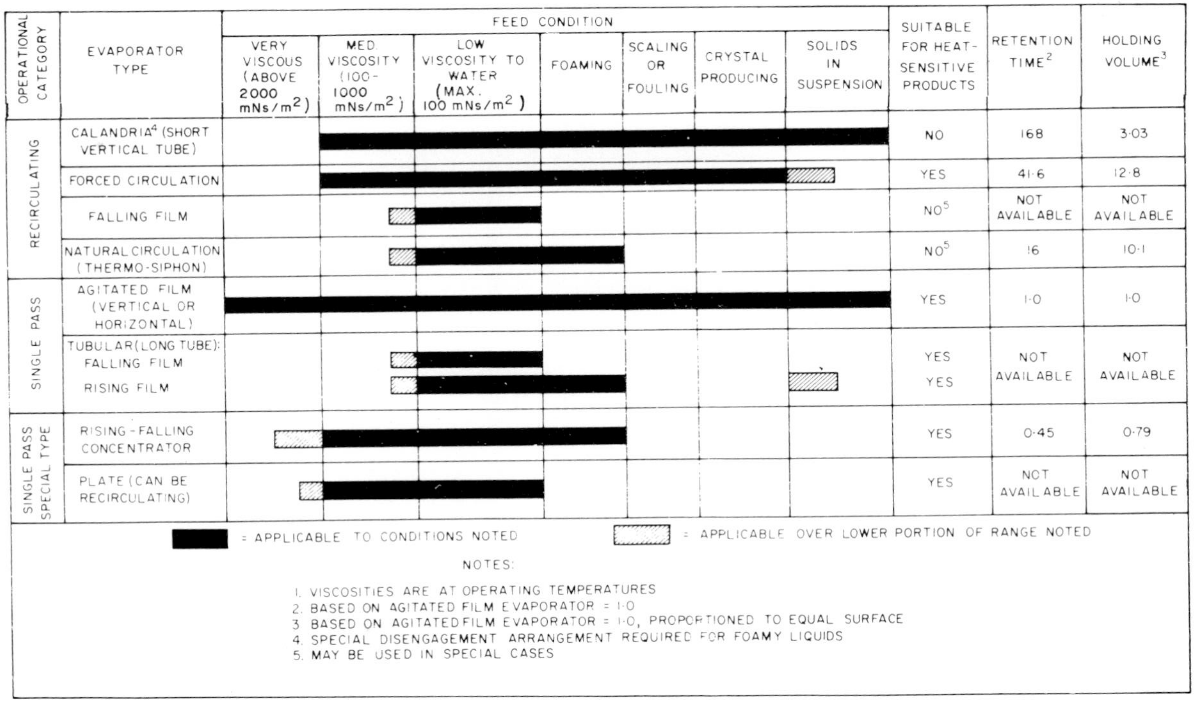

| OPERATIONAL CATEGORY | EVAPORATOR TYPE | FEED CONDITION | | | | | | | | SUITABLE FOR HEAT-SENSITIVE PRODUCTS | RETENTION TIME[2] | HOLDING VOLUME[3] |
| | | VERY VISCOUS (ABOVE 2000 mNs/m$^2$) | MED. VISCOSITY (100-1000 mNs/m$^2$) | LOW VISCOSITY TO WATER (MAX. 100 mNs/m$^2$) | FOAMING | SCALING OR FOULING | CRYSTAL PRODUCING | SOLIDS IN SUSPENSION | | | |
| RECIRCULATING | CALANDRIA[4] (SHORT VERTICAL TUBE) | | | | | | | | NO | 168 | 3·03 |
| | FORCED CIRCULATION | | | | | | | | YES | 41·6 | 12·8 |
| | FALLING FILM | | | | | | | | NO[5] | NOT AVAILABLE | NOT AVAILABLE |
| | NATURAL CIRCULATION (THERMO-SIPHON) | | | | | | | | NO[5] | 16 | 10·1 |
| SINGLE PASS | AGITATED FILM (VERTICAL OR HORIZONTAL) | | | | | | | | YES | 1·0 | 1·0 |
| | TUBULAR (LONG TUBE) FALLING FILM | | | | | | | | YES | NOT AVAILABLE | NOT AVAILABLE |
| | RISING FILM | | | | | | | | YES | | |
| SINGLE PASS | RISING-FALLING CONCENTRATOR | | | | | | | | YES | 0·45 | 0·79 |
| SPECIAL TYPE | PLATE (CAN BE RECIRCULATING) | | | | | | | | YES | NOT AVAILABLE | NOT AVAILABLE |

■ = APPLICABLE TO CONDITIONS NOTED   ▨ = APPLICABLE OVER LOWER PORTION OF RANGE NOTED

NOTES:

1. VISCOSITIES ARE AT OPERATING TEMPERATURES
2. BASED ON AGITATED FILM EVAPORATOR = 1·0
3. BASED ON AGITATED FILM EVAPORATOR = 1·0, PROPORTIONED TO EQUAL SURFACE
4. SPECIAL DISENGAGEMENT ARRANGEMENT REQUIRED FOR FOAMY LIQUIDS
5. MAY BE USED IN SPECIAL CASES

Fig. 14.17. Evaporator selection (after Parker[19])

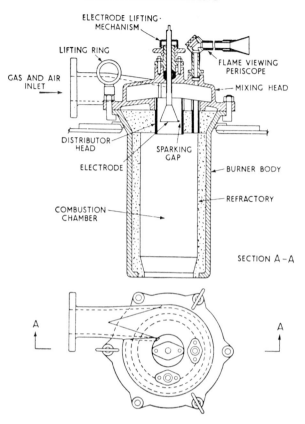

FIG. 14.18. Burner for submerged combustion

shallower depth of brine, and this technique has been used to obtain a marked increase in the rate of production in the Dead Sea area.

The submerged combustion of a gas, such as natural gas, has been used for the concentration of very corrosive liquors, including spent pickle liquors, weak phosphoric and sulphuric acids. A suitable burner for direct immersion in the liquor, as developed by Swindin[23], is shown in Fig. 14.18. The depth of immersion of the burner is determined by the time of heat absorption; thus, a 50 mm burner may be immersed 250 mm and a 175 mm burner about 450 mm. The efficiency of heat absorption is measured by the difference between the temperature of the liquid and that of the gases leaving the surface, values of 2–5 deg K being obtained in practice. The great attraction of this technique, apart from the ability to handle corrosive liquors, is the very great heat release obtained and the almost instantaneous transmission of the heat to the liquid. Thus Swindin quotes 70 MW/m$^3$ as a practical figure.

### 14.5.3. Natural Circulation Evaporators

Whilst each of the previous methods is of considerable importance in a given industry, it is the steam-heated evaporator that is the main type of unit in the chemical industry and this will be considered in detail. In this type of unit one is concerned with the transfer of the latent heat from steam to a boiling liquid. In Chapter 7 of Volume 1, on heat transfer, it was shown that the movement of the liquid over the heating surface had a marked influence on the rate of heat transfer, and it is thus convenient to classify evaporators according to the method of agitation or the nature of the circulation of the liquor over the heating surface. On this basis they may be divided into three main types:

(a) Natural circulation units.

(b) Forced circulation units.

(c) Film-type units.

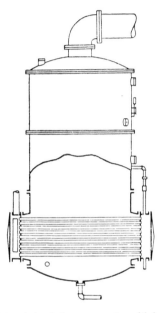

FIG. 14.19. Natural circulation evaporator with horizontal tubes

The developments that have taken place in design have, in the main, originated from the sugar and salt industries where the cost of evaporation represents a major item in the process. In recent years, particular attention has been given to obtaining the most efficient use of the heating medium, and the main techniques that have been developed are the use of the multiple effect unit, and of various forms of vapour compression units.

With natural circulation evaporators, circulation of the liquor is brought about by convection currents arising from the heating surface. This group may be subdivided according to whether the tubes are horizontal with the steam inside, or vertical with the steam outside.

### Horizontal Tubes

N. Rillieux is usually credited with the idea of using horizontal tubes, and a unit of this type is shown in Fig. 14.19. The horizontal tubes extend between two tube plates to which they are fastened either by packing plates or, more usually, by expansion. Above the heating section is a cylindrical portion in which separation of the vapour from the liquid takes place. The vapour leaves through some form of de-entraining device to prevent the carry-over of liquid droplets with the vapour stream. The steam enters one steam chest, passes through the tubes and out into the opposite chest, and the condensate leaves through some form of steam trap.

These horizontal evaporators are relatively cheap, require low head room, and are easy to install. They are suitable for handling liquors that do not crystallise, and give good heat

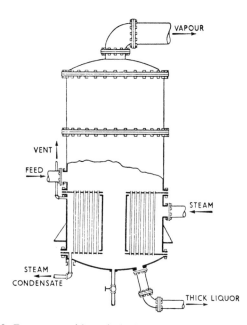

FIG. 14.20. Evaporator with vertical tubes and large central downcomer

transfer. They can be used either as batch or as continuous units. The body is generally from 0·9 to 3·6 m diameter and 2·4 to 3·9 m high, and the tubes generally range from 22 to 32 mm diameter. The liquor circulation is poor, and for this reason they are unsuitable for viscous liquors.

*Vertical Tubes*

The use of vertical tubes is associated with Robert, a director of a sugar factory in Austria, so that this type is sometimes known as the Robert or Standard Evaporator. A typical form of vertical evaporator is illustrated in Fig. 14.20, in which it will be seen that a vertical cylindrical body is used, with the tubes held between two horizontal tube plates which extend right across the body; the lower portion of the evaporator is frequently spoken of as the calandria section (Fig. 14.21). The early plants of this type consisted of a very large number of tubes covering the whole cross-section of the body. It was found, however, that the circulation was poor for, although it was easy to generate upward currents, it was difficult for the liquor to return to the bottom. This layout was improved by Kasalovsky who introduced the large central downtake tube which enables good circulation to be obtained. In this form of evaporator, the liquor level is usually maintained just above the top of the tubes. The area of this central tube varies from about 40 per cent to 100 per cent of the total cross-sectional area of the tubes. The Scott Company has used several tubes, of wider diameter than the main tubes, to serve as a

FIG. 14.21. Calandria of evaporator

number of downcomers. Again, other firms employ downcomers at the side of the body. The tubes range from 0·9 to 1·8 m in length and from 37 to 75 mm diameter, giving ratio of length to inside diameter of the tubes of between 20 and 40. In 1877, the basket type (Fig. 14.22) was introduced; in this, vertical tubes are still used with the steam outside, though the heating element is suspended in the body so as to give an annular downtake. The advantages claimed for this construction are that the heating unit is easily removed for repairs, and that crystals formed in the downcomer do not break up. As the circulation of the liquor in the tubes is better, the vertical tube evaporator is the standard type. It is widely used in the sugar, salt and heavy chemical industries (i.e. ammonium nitrate, caustic soda), where the throughputs are very large.

*Coil Evaporators*

For the final concentration of sugars, a coil heating unit has been found very suitable, and a modern form of this construction as described by Venton[24] is shown in Fig. 14.23. This pan has a capacity of $38.6 \, m^3$ and a heating surface of $4.72 \, m^2/m^3$, which is quite a high figure. The coils are of copper tubing of 125 mm outside diameter, and are flattened to

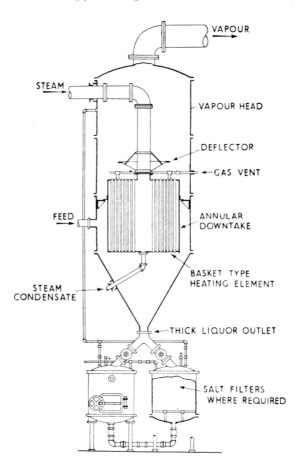

VAPOUR

STEAM

VAPOUR HEAD

DEFLECTOR

GAS VENT

FEED

ANNULAR
DOWNTAKE

STEAM
CONDENSATE

BASKET TYPE
HEATING ELEMENT

THICK LIQUOR OUTLET

SALT FILTERS
WHERE REQUIRED

FIG. 14.22. Basket type evaporator

give an oval shape as indicated. The spacing between the coils should be at least 75 mm, and the central well should be from 0.3 to 0.4 of the diameter of the pan. As an alternative, a vertical calandria may be used, as described by Shearon et al.[25]. This unit has tubes of 102 mm outside diameter, and 1.3 m high, set in a steam space of 3.0 m diameter at the bottom of the pan, and the disengaging space is of 3.6 m diameter. This type has a short overall height and a dished bottom; it is sometimes preferred to the coil type because of easier maintenance. Bosworth and Duloy[26] have measured the rate of circulation of the liquor and found figures of 100 mm/s for the weak sugars.

### 14.5.4. Forced Circulation Evaporators

Increasing the velocity of flow of the liquor through the tubes will bring about a marked increase in the liquid film transfer coefficient. This is achieved in the forced circulation units where a propeller or other impeller is mounted in the central downcomer, or a circulating pump is mounted outside the evaporator body. Thus, in the concentration of strong brines, an internal impeller is often fitted in the downtake, and this form of construction is particularly useful where crystallisation takes place. A turbine impeller, as

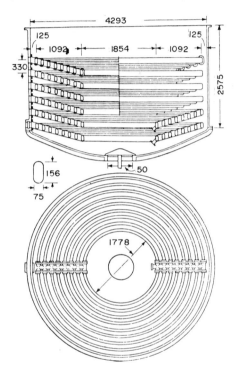

FIG. 14.23. Coil pan of 38·6 m³ (Dimensions in mm)

shown in Fig. 14.24, is sometimes used for this purpose. Forced circulation enables higher degrees of concentration to be achieved, since the heat transfer rate can be kept up in spite of the increased viscosity of the liquid. It also enables scale-forming liquors to be handled, since the scale may be cleared from the heating surface by the motion of the liquid. Because the pumping costs increase roughly as the cube of the velocity, the added cost of operation of this type of unit may make it uneconomic; though many forced circulation evaporators are running with a liquor flow through the tubes of from 2 to 5 m/s which is a marked increase on the value for natural circulation. Where stainless steel or expensive alloys such as Monel are to be used, forced circulation is favoured because the units can be made smaller and cheaper than those relying on natural circulation. In the type illustrated in Fig. 14.25, there is an external circulating pump; this is usually of the centrifugal type when crystals are present, though vane types may be used in the absence of crystals. The liquor is either introduced at the bottom and pumped straight through the calandria, or

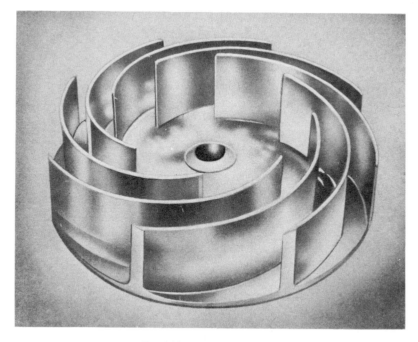

Fig. 14.24. Turbine agitator

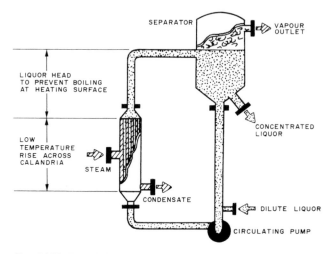

Fig. 14.25. Forced circulation evaporator with external pump

else it is introduced in the separating section. In most units, boiling does not take place in the tubes, because the hydrostatic head of liquid raises the boiling point above that in the separating space (Section 14.2.1). Thus the liquor enters the bottom of the tubes and is heated as it rises and at the same time the pressure falls. In the separator the pressure is sufficiently low for boiling to occur. Forced circulation evaporators will work well on such diverse materials as meat extracts, salt, caustic soda, alum or other crystallising materials, glues, alcohols, and foam-forming materials.

### 14.5.5. Film-type Units

In all the units so far discussed, the liquor remains some considerable time in the evaporator: with batch operation several hours, and with continuous operation perhaps an hour. This may be undesirable, as many liquors decompose if kept at temperatures at or near their boiling points for any length of time. The temperature can be reduced by operating under a vacuum, as discussed previously, though even so there are many liquors which are very heat-sensitive: for example, orange juice, blood plasma, liver extracts and

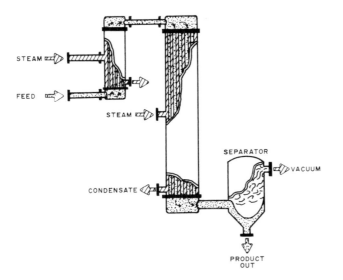

FIG. 14.26. Single effect falling film evaporator concentrating a heat sensitive chemical product

vitamins. If a unit is so constructed that the residence time is only a few seconds, then these dangers are very much reduced. This is the idea behind the Kestner long tube evaporator, introduced in 1909, which is fitted with tubes of 38 to 50 mm diameter, mounted in a simple vertical steam chest. The liquor enters at the bottom, and a mixture of vapour and entrained liquor leaves at the top and enters a separator, usually of the tangential type. The vapour passes out from the top and the liquid from the bottom of the separator. In the early models the thick liquid was recirculated through the unit, but the once-through system is now normally used.

An alternative name for the first Kestner long-tube evaporator is the climbing film evaporation, and this gives an accurate description of its method of operation. The progressive evaporation of a liquid, whilst it passes through a tube, gives rise to a number of flow regimes discussed in Section 14.2.3. In the long-tube evaporator the annular flow or climbing film regime is utilised throughout almost all the tube length, the climbing film being maintained by drag induced by the vapour core which moves at a high velocity relative to the liquid film. Kestner extended his thin film concept to more viscous materials, which could not be successfully concentrated in the climbing film unit, by developing an evaporator in which the material to be treated flows down the inner walls of a bank of vertical steam heated tubes as a falling film, Fig. 14.26. With many viscous materials, however, heat transfer rates in this unit are low. There is little turbulence in the film with such fluids and the thickness of the film is too great to permit much evaporation from the film as a result of conduction through it. (With these more viscous liquids it is usually necessary to resort to mechanically aided agitation of the liquid film, another topic which is discussed in Section 14.5.6.) In evaporators of this type it is essential that the feed should enter the tubes as nearly as possible at its boiling point. If the feed is subcooled the initial sections will act merely as a feed heater thus reducing the overall performance of the unit. Pressure drop over the tube length will be attributable to the hydrostatic heads of the single-phase and two-phase regions, friction losses in these regions, and losses due to the acceleration of the vapour phase. The first published analysis of the operation of this type of unit was given by Badger and his associates[27-29] who arranged for a small thermocouple to be fitted inside the experimental tube (32 mm outside diameter × 5·65 m) so that the couple could be moved up and down the centre of the tube. In this way, they found that the temperature rose slightly from the bottom of the tube to the point where boiling commenced, after which the change in temperature was relatively small. Applying this technique, it was possible to determine the heat transfer coefficients in the non-boiling and boiling sections of the tube.

### 14.5.6. Thin-layer Evaporators

This type of unit, known also as a thin-film evaporator, has been developed by Luwa S.A. of Zürich. As illustrated in Fig. 14.27, it consists of a vertical tube $A$, the lower portion of which is surrounded by a jacket $B$ which contains the heating medium. The upper part of the tube $A$ is not jacketed and acts as a separator. A rotor $D$, driven by an external motor, has blades which extend nearly to the bottom of the tube, so mounted that there is a clearance of only about 1·3 mm between their tips and the inner surface of $A$. The liquor to be concentrated enters at $a$ and is picked up by the rotating blades and thrown against the tube wall. This action provides a thin film of liquid and sufficient agitation to give good heat transfer, even with very viscous liquids. The film flows down by gravity, becoming concentrated as it falls. The concentrated liquor is taken off at the bottom by a pump and the vapour leaves the top of the unit, where it is passed to a condenser. Development of this basic design has been devoted mainly to the modification of the blade system. An early alternative was the use of a hinged blade. In this type of unit the blade is forced on to the wall under centrifugal action, the thickness of the film being governed by a balance between this force and the hydrodynamic forces produced in the liquid film on which the blade rides. The first experimental comparison of the fixed and hinged blade wiped-film evaporators was that of Bressler[30]. For each type of blade there appeared to be an optimum wiper speed beyond which an increase had no further effect on heat transfer. This optimum was reached at a lower speed for the hinged blade. Other agitator designs in which the blades (usually made from rubber, graphite or synthetic materials) actually scrape the wall have been studied. The use of nylon brushes as the active agitator elements has been thoroughly investigated by McManus[31] using a small steam heated evaporator (63 mm i.d. × 762 mm long). Water and various aqueous solutions of sucrose and glycerol were tested in the evaporator. A notable feature of the unit was the high heat fluxes obtained with the viscous solutions. Values as high as 70 kW/m² were obtained when concentrating a 60 per cent sucrose feedstock to 73 per

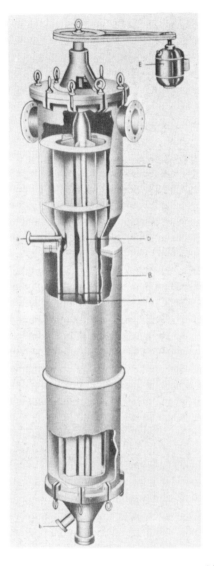

$A$ — vertical tube
$B$ — jacket
$C$ — separator
$D$ — rotor driven by
    external motor
$E$ — driving motor
$a$ — liquid inlet
$b$ — liquid outlet

FIG. 14.27. Luwa evaporator working on thin layers

cent, at a film temperature difference of 16·5 deg K with a wiper speed of 8·3 Hz. The fluxes obtained for the evaporation of water under similar conditions were nearly 4·5 times higher. A detailed analysis of the heat transfer mechanism, based on unsteady-state conduction to the frequently renewed film, was presented. Similar analyses are to be found in the work of Harriott[32] and Kool[33]. Close agreement between the theory and experimental data confirmed the accuracy of the assumed model for heat transfer; however, the theory has one main disadvantage in that a satisfactory method for the estimation of liquid film thickness is not available. Nevertheless, the analysis provided a rational basis for the presentation of an equation for the evaporating film coefficient $h_b$

given by:

$$h_b = 570 \left[ \frac{k^{0 \cdot 875}}{\mu_L^{0 \cdot 18}} \right] \left[ \frac{\rho_L C_p}{t} \right]^{0 \cdot 125} \tag{14.11}$$

where $k$, $\mu_L$, $\rho_L$, and $C_p$, the physical properties of the liquid, are in SI units, and $t$ seconds is the time available for film heating, i.e. the time between deposition of the film by one blade and its collection by the following blade. Equation 14.11 was claimed to correlate most of the data to within $\pm 6$ per cent, and clearly demonstrates that the most important factor influencing the evaporation coefficient is the thermal conductivity of the film material, and that the effects of viscosity and wiper speed (inversely proportional to the heating time $t$) are of less significance. A comprehensive discussion of the main aspects of the wiped-film evaporator technique covering thin film technology in general, the equipment, and its economics and process applications is given by Mutzenburg[34], Parker[35], Fischer[36], and Ryley[37]. An additional advantage of wiped-film evaporators, especially those producing a scraped surface, is the reduction or complete suppression of scale formation. However, in processes where the throughput is very high, this type of unit obviously becomes uneconomic and the traditional way of avoiding scale formation, by operating a flash evaporation process, is more suitable.

### 14.5.7. Plate-type Units

A plate evaporator consists of multiple gasketed plates mounted within a frame which utilises minimum headroom and floor space. The evaporator operates on a single pass,

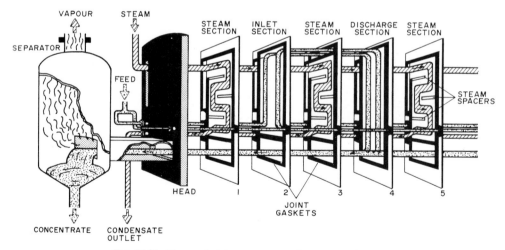

FIG. 14.28. Flow and plate arrangement for one complete pass

rising–falling film principle with the plates arranged in a series of processing units. Figure 14.28 shows the flow and plate arrangement for one complete pass. Each unit consists of a rising film, falling film and two adjacent steam sections.

As feed liquid passes simultaneously over the rising and falling film sections in each processing unit, it vaporises on contact with the adjacent heated plates and is discharged

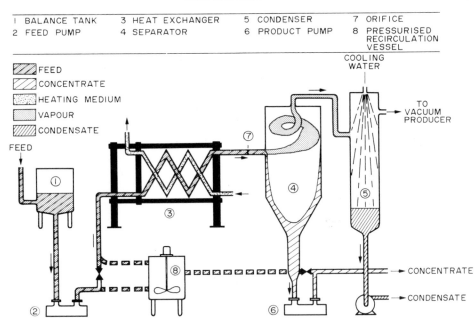

| 1 BALANCE TANK | 3 HEAT EXCHANGER | 5 CONDENSER | 7 ORIFICE |
| 2 FEED PUMP | 4 SEPARATOR | 6 PRODUCT PUMP | 8 PRESSURISED RECIRCULATION VESSEL |

FIG. 14.29. Layout of plant incorporating a plate-type evaporator

with its vapour to a horizontally or vertically mounted cyclone-type separator. Here the product is extracted and the vapour passed to a condenser or the next effect of the evaporator.

A typical schematic layout for a plant incorporating a plate evaporator is shown in Fig. 14.29. Feed liquor from balance tank 1 is pumped 2 to the heat exchanger 3 where it boils on contact with the heated plates. Under normal operation boiling is completed within the heat exchanger. Concentrated product and vapour are discharged into the separator 4, where vapour is removed to a condenser 5, product being extracted by positive pump 6. Alternatively, to satisfy the requirements of a particular product or processing condition, the plant can be arranged as a flash evaporator. An orifice plate 7, fitted between the heat exchanger and the separator, causes back pressure within the heat exchanger and prevents boiling until the liquid enters the separator.

The A.P.V. Company, whose *Paraflow* plate heat exchanger is illustrated in Volume 1, manufacture this type of evaporator with evaporative capacities of up to 8·5 kg/s. They offer maximum protection of product quality through high heat transfer rates, low liquid hold-up, and minimum residence time and thus find wide application in the evaporation of heat-sensitive materials. Capacity is easily changed by alteration of the number of plates in the stack. Where high viscosity or foaming liquids have to be processed or streams with a high solids content are produced, A.P.V. have developed the *Paravap*. Plate spacing in the Paravap is closer than in the normal evaporator and this results in a highly turbulent flow giving the high heat transfer coefficients required for the processing of viscous fluids. For low concentration ratios, single-pass operation is usually employed, whilst higher ratios can be obtained by incorporating recirculation of the partly evaporated product to the balance tank, or where mixing is difficult to achieve, to a pressurised vessel, 8 in Fig. 14.29, fitted with an agitator. Final viscosities of up to $1.5 \, Ns/m^2$ are obtainable, the limiting factor being the ability of the extraction pump to withdraw the concentrate from the separator. Control of evaporation rate is achieved by regulation of the steam supply or, alternatively, of the feed rate by means of a variable speed drive. When the plant is arranged as a flash evaporator, control may be

FIG. 14.30. Five type R145 APV Paraflow plate heat exchangers for Mobil North Sea Ltd. Now operating on Mobil Beryl "A" on closed-circuit cooling duties, they were designed to remove $58.6 \, MW$ by sea water. Their total heat-transfer area is $2400 \, m^2$ yet they occupy a total plot area of only $37 \, m^2$

by simple adjustment of the product recirculation rate. Maximum capacity, depending upon operating conditions, may reach $1.25 \, kg/s$ water evaporation.

### 14.5.8. Flash Evaporators

In the flash evaporator boiling in the actual tubes is suppressed and the superheated liquor is flashed into a separator operating at reduced pressure. Whilst the high heat transfer rates associated with boiling in tubes cannot be utilised, the thermodynamic and economical advantages of the system when operated in a multistage configuration outweigh this consideration. These advantages, stated independently by Frankel[38] and Silver[39], have been important in the past decade in the intensive effort to devise economic processes for the desalination of sea water. This topic is discussed further by Baker[40], who considers multistage flash evaporation with heat input supplied by a conventional steam boiler, by a gas turbine cycle, or by vapour recompression. The combined power–water plant is also considered. Attempts to reduce scale formation in flash evaporators to even lower levels have resulted in a number of novel developments. In one unit described by Woodward[41] sea water is heated by a countercurrent spray of hot immiscible oil; in this respect the process is similar to liquid–liquid extraction, the extracted quantity being heat in this case. The sea water is heated under pressure and subsequently flashed into a low pressure chamber. A similar direct contact system is discussed by Wilke *et al.*[42]. Yet another arrangement which avoids the intervening metallic wall of the conventional heat exchanger is described by Othmer *et al.*[43]. In this process direct mass transfer between brine and pure water is utilised in the desalination operation.

The formation of solids in evaporators is not, of course, always unwanted; indeed, this is precisely what is required in the evaporator-crystalliser. The evaporator-crystalliser is a unit in which crystallisation takes place largely as a result of the removal of solvent by evaporation. Cooling of the liquor may, in some cases, produce further crystallisation thus establishing conditions similar to vacuum crystallisation; however, the true evaporator-crystalliser is distinguished by its use of an external heat source. Crystallisation by evaporation is practised on those salt solutions having a small change of solubility with temperature, such as sodium chloride and ammonium sulphate, which cannot be dealt with economically by other means, as well as those with inverted solubility curves. It is also widely used in the production of many other crystalline materials, as outlined by Bamforth[44]. The problem of design for crystallising equipment is extremely complicated and consequently design data are extremely meagre and unreliable.

### 14.5.9. Foaming in Evaporators

The development of unwanted foams is one problem that evaporation has in common with a number of chemical processes, and a considerable amount of effort has been devoted to the study of defoaming techniques using chemical, thermal, or mechanical methods. Chemical techniques involve the addition of substances, called antifoams, to foam-producing solutions to eliminate completely, or at least reduce drastically, the resultant foam. Antifoams are in general poorly soluble in foaming solutions and can cause a decrease in surface tension; however, their ability to produce an expanded surface film is one explanation of their foam-inhibiting characteristic[45]. Foams may be caused to collapse by raising or lowering the temperature. Many foams collapse at high temperature due to a decrease in surface tension, solvent evaporation, or chemical degradation of the foam-producing agents; at low temperatures freezing or a reduction in surface elasticity may be responsible. All methods which are neither chemical nor thermal may be classified as mechanical. Tensile, shear, or compressive forces may be used to destroy foams; such methods are discussed in some detail by Goldberg and Rubin[46]. The ultimate choice of defoaming procedure depends on the process under consideration and the convenience with which a technique may be applied.

### 14.5.10. Typical Industrial Installations

An installation for the concentration of black liquor in the Kraft paper industry of America has been described by Sawyer[47]. The main evaporating unit, shown in Fig.

14.31, consists of a sextuple-effect system using long tube evaporators, each evaporator containing 552 tubes of 50 mm internal diameter and 7·2 m long in a steam shell of 1·2 m diameter with a dome of 2·4 m diameter. The material and heat balances for this unit are given in Table 14.3. The liquor leaving this unit is too viscous for further concentration in the film evaporator, and is further treated in two large disc units each 2·6 m in diameter with 33 discs spaced 67 mm apart. These discs pick up the liquor and evaporation is

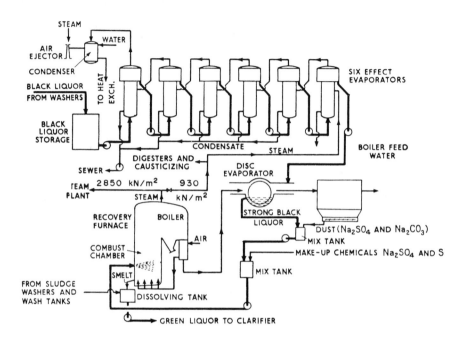

Fig. 14.31. Sextuple effect evaporator for the paper industry

effected by passing hot gases at 468 K over the discs. This arrangement gives a final concentration of 64 per cent solids. A combination of the long tube unit and the direct contact system is a method of making use of the best features of the two types. The backward feed system adopted again has technical advantages in this case.

A plant for the concentration of brine is installed at the works of the Murgatroyd Vacuum Salt Company[48]. The brine, after purification to remove the calcium and magnesium salts, passes through a triple effect vertical evaporator with parallel feed. The units are 3·35 m in diameter, and are of cast iron construction; they are fitted with a draw-off leg for the slurry at the bottom. Part of the brine feed is introduced at the bottom of this leg to wash the crystals free from mother liquor, and the remainder enters immediately below the calandria. This consists of a tube bundle of 1389 copper tubes (1·8 m long by 50 mm outside diameter), with a central downtake of 0·9 m diameter, and has a surface area of 306 m². The natural circulation is augmented by an internal agitator driven from the base, and heat transfer coefficients of 3·1, 2·0, and 1·4 kW/m² K are obtained for the various effects. The slurry drawn from the units is pumped to rotary drum filters, and the filtrate is returned to the evaporators.

TABLE 14.3. *Sextuple-effect Evaporator Test*

Material Balance (kg/s)

Feed of weak black liquor = $0.022 \, m^3/s \times 1.09 \,(sp.gr.) \times 1000 \, kg/m^3$ = 24·0

Discharge from evaporator = $24.0 \, kg/s \times \dfrac{17.5 \,(\% \text{ solids in inlet liquor})}{53.3 \,(\% \text{ solids in discharge liquor})}$ = 7·9

Total evaporation of water accomplished in evaporator = 16·1

kg $H_2O$ evaporation/kg steam = $\dfrac{16.1 \, kg/s \,(\text{total evaporation})}{3.28 \, kg/s \,(\text{steam discharge liquor})}$ = 4·90.

| | Flash tank | Evaporator test effect[a] | | | | | |
|---|---|---|---|---|---|---|---|
| | | I | II | III | IV | V | VI |
| Chest pressure (kN/m²) | — | 353 | 205 | 165 | 88 | 51 | 30 |
| Chest temperature (K) | — | 412 | 393 | 381 | 369 | 355 | 342 |
| Temp. diff. (deg K) | — | 12 | 7 | 7 | 8 | 6 | 17 |
| Liquor temp. calcd. (K) | 376 | 400 | 387 | 374 | 361 | 349 | 325 |
| Liquor temp. measured (K) | 370 | 401 | 385 | 372 | 359 | 346 | 327 |
| Boiling point rise K | 6 | 6 | 6 | 4 | 4 | 3 | 3 |
| Dome temperature (K) | 370 | 394 | 381 | 370 | 356 | 346 | 322 |
| Dome pressure (kN/m²) | — | 206 | 138·5 | 90 | 57 | 35 | 11 |
| Latent heat of vaporisation (kJ/kg) | 2266 | 2200 | 2235 | 2266 | 2303 | 2326 | 2387 |
| Total solids (%) | 53·3 | 45·7 | 37·9 | 31·3 | 25·9 | 22·4 | 24·3 |
| Feed (kg/s) | 8·3 | 11·3 | 13·7 | 16·5 | 18·3 | 12·00 | 12·4 |
| Discharge (kg/s) | 8·0 | 8·3 | 11·3 | 13·7 | 16·5 | 9·4 | 8·9 |
| Evaporation (kg/s) | 0·3 | 3·0 | 2·4 | 2·8 | 1·8 | 2·6 | 3·5 |
| Evaporation (heat bal.) (kg/s) | 0·28 | 2·96 | 2·73 | 2·78 | 2·49 | 2·81 | 3·59 |
| Overall temp. diff. (deg K) | — | 11·6 | 6·6 | 6·6 | 7·7 | 6·1 | 17·1 |

[a] Total temperature difference in 6 effects = 56 deg K; inlet weak black liquor temperature = 385 K; barometric pressure = 101·75 kN/m².

## 14.6. MECHANICAL DESIGN DETAILS

### 14.6.1. Materials of Construction

The various designs considered in Section 14.5 resulted from the necessity to concentrate materials having "difficult" properties. However, little has been said about mechanical design details, especially the important aspect of materials of construction. The specific problem of corrosion has not been considered as yet and, since the range of available materials is extensive, the present discussion will be limited to one of the more sophisticated constructional materials. The unique combination of chemical inertness and excellent thermal conductivity renders graphite important for constructional purposes in plant handling corrosive fluids. Graphite differs from most constructional materials in its high anisotropy resulting in directionally preferred thermal conductivity, and also in the difference between its relatively good compressive strength and its poor tensile or torsional strength. Although it is easily machinable, it is not ductile or malleable and cannot be cast or welded. The use of cements in assembly is undesirable because they are usually less chemically or thermally stable; there are also problems of differential expansion. In order to exploit the advantages of this material and avoid the foregoing problems, special

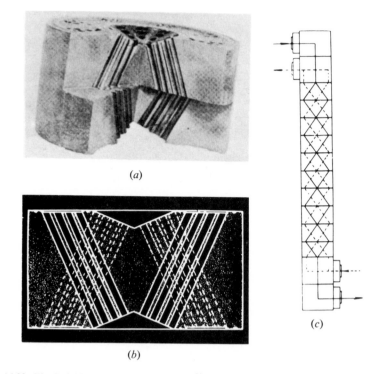

(a)

(b)

(c)

FIG. 14.32. The Polybloc system (after Hilliard[50]). (a) Cutaway section of x-flow block as used for two corrosive fluids. (b) Section through x-flow block. (c) Stacked Polybloc exchanger

constructional techniques are necessary. The Polybloc system, which is described by Hilliard[49−51], is based on the use of robust blocks assembled exclusively under compression. Heat transfer occurs between fluids passing through holes drilled in the blocks and so positioned as to exploit preferred anisotropic crystal orientation for the highest thermal conductivity in the direction of heat flow. Inert gaskets eliminate the need for cements and enable units of varying size to be assembled simply by stacking the required number of blocks as shown in Fig. 14.32. A similar form of construction has been adopted by the Powell Duffryn Company. In commercial installations they have given good values of overall transfer coefficients. Thus a value of $1\cdot1\,kW/m^2\,K$ has been obtained for concentrating thick fruit juice containing syrup, and also for concentrating 40 per cent sulphuric acid to 60 per cent. A somewhat lower figure of $0\cdot8$ is obtained for sulphuric acid concentration from 60 per cent to 74 per cent at a pressure of $1\cdot5\,kN/m^2$, and similar values are obtained with spinning-bath liquors and some pharmaceuticals.

Figure 14.33 shows a conventional climbing film evaporator with $1\cdot8\,m$ carbon tubes and a Keebush separator.

## 14.6.2 Condensing and Vacuum Equipment

The vapour leaving the last effect of a multiple effect unit must be condensed, either by direct contact with a jet of water, or in a normal tubular exchanger. If $M$ is the mass of

FIG. 14.33. Climbing film evaporator with carbon tubes and Keebush separator

cooling water used per unit mass of vapour in a jet condenser, and $H$ is the enthalpy per unit mass of vapour, a heat balance gives:

$$H + MC_pT_i = C_pT_e + MC_pT_e \qquad (14.12)$$
$$\underset{\text{(Heat in)}}{\phantom{H + MC_pT_i}} \quad \underset{\text{(Heat out)}}{\phantom{= C_pT_e + MC_pT_e}}$$

where $T_i$ and $T_e$ are the inlet and outlet temperatures of the water, reckoned above a standard datum temperature, and where the condensate is assumed to leave at the same temperature as the cooling water. From equation 14.12:

$$M = \frac{H - C_pT_e}{C_p(T_e - T_i)} \qquad (14.13)$$

For example, if $T_e$ is 316 K, $T_i$ is 302, and the pressure is $87 \cdot 8\,\text{kN/m}^2$:

$$M = \frac{2596 - 4 \cdot 18(316 - 273)}{4 \cdot 18(316 - 302)} = 41 \cdot 5\,\text{kg/kg}$$

The water is then conveniently discharged at atmospheric pressure, without the aid of a pump, by allowing it to flow down a vertical pipe, known as a barometric leg, of sufficient length for the pressure at the bottom to be slightly in excess of atmospheric pressure. For a jet condenser with a barometric leg, a chart for determining the water requirement has been prepared by Arrowsmith[52], and is shown in Fig. 14.34.

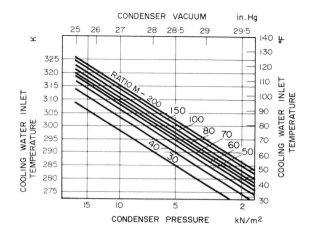

FIG. 14.34. Ratio ($M$) of cooling water to vapour required under various conditions

## Jet Condensers

Jet condensers may be either of the countercurrent or parallel flow type. In the countercurrent unit, the water leaves at the bottom through a barometric leg, and any entrained gases leave at the top. This provides what is known as a dry vacuum system, since the pump has to handle only the non-condensable gases. The cooling water will generally be heated to within 3 or 6 deg K of the vapour temperature. With the parallel flow system, the temperature difference will be rather greater and, therefore, more cooling water will be required. In this case, the water and gas will be withdrawn from the condenser and passed through a wet vacuum system. As there is no barometric leg, the unit can be mounted at floor level, although the pump displacement is about one and a half times that for the dry vacuum system.

## Vacuum Equipment

Air is introduced into a jet condenser from the cooling water, as a result of the evolution of non-condensable gases in the evaporator, and as a result of leakages. The volume of air to be removed is frequently about 15 per cent of that of the cooling water. The most convenient way of obtaining a vacuum is usually by means of a steam jet ejector (Fig. 14.35). Part of the momentum of a high velocity steam jet is transferred to the gas entering the ejector, and the mixture is then compressed in the diverging portion of the ejector by conversion of kinetic energy into pressure energy. $C$ represents the converging–diverging steam nozzle, giving rise to a high velocity jet of steam. In the converging section $B$, this steam entrains gas and, in the diffuser section $D$, part of the kinetic energy of the mixture is converted into pressure energy. Good performance by the ejector is obtained largely by correct proportioning of the steam nozzle and diffuser, and poor ejectors will use much more high pressure steam than a well designed unit. The amount of steam required increases with the compression ratio. Thus, a single-stage ejector will remove air from a system at a pressure of $17 \text{ kN/m}^2$; i.e. where a compression ratio of 6 is required. To remove air from a system at $3 \cdot 4 \text{ kN/m}^2$ would involve a compression ratio of 30, and a

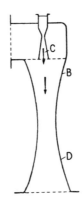

FIG. 14.35. Steam jet ejector

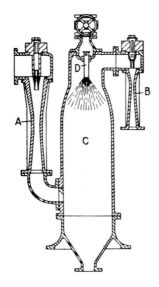

FIG. 14.36. Two-stage ejector with condenser: *A*—First stage. *B*—Second stage.
*C*—Condenser. *D*—Water spray

single-stage unit would be unecomonic in steam consumption. A two-stage ejector is shown in Fig. 14.36; the first stage withdraws air from the high vacuum vessel and compresses it to say $20 \text{ kN/m}^2$, and the second stage compresses the discharge from the first ejector to atmospheric pressure. A further improvement is obtained if a condenser is inserted after the first stage, as this will reduce the amount of vapour to be handled in the final stage. An indication of the number of stages required for various conditions is shown in Fig. 14.37[52]. The higher the steam pressure the smaller is the consumption, and pressures of 790 to $1135 \text{ kN/m}^2$ are commonly used in multistage units. Typical performance curves are shown in Fig. 14.38[52], where the air duty for a given steam consumption and given steam pressure is shown as a function of the number of stages and the operating pressure.

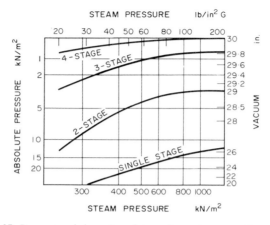

FIG. 14.37. Recommended number of stages for various operating conditions

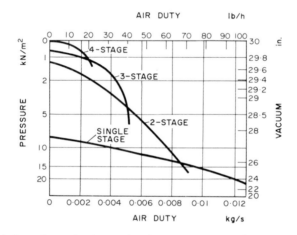

FIG. 14.38. Air duty of steam jet ejectors, for given steam consumption and steam pressure

### 14.6.3. Separators and Entrainment

In operating an evaporator, it is important to minimise entrainment of the liquid in the vapour passing over to the condenser. Entrainment is reduced by having a considerable headroom, of say 1·8 m, above the boiling liquid, though the addition of some form of de-entrainer is usually essential. Figure 14.39 shows three methods of reducing entrainment. The simplest is to take the vapour from an upturned pipe as in *a*, and this has been found to give quite good results in smaₗ units. The deflector type *b* is a common form of de-entrainer and the tangential separator *c* is the type usually fitted to climbing-film units. This problem is particularly important in the concentration of radioactive waste liquors and has been discussed by McCullough[53], who cites the case of a batch evaporator of the forced circulation type in which the vapours are passed to a 3·6 m diameter separator and then through four bubble cap trays to give complete elimination of entrained liquor. A

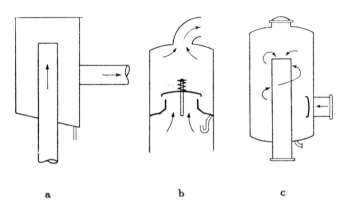

Fig. 14.39. Entrainment separators: (a) upturned pipe, (b) deflector type, (c) tangential type

good entrainment separator will reduce the amount of liquid carried over to between 10 and 20 kg per $10^6$ kg of vapour.

## 14.7. FURTHER READING

BACKHURST, J. R. and HARKER, J. H.: *Process Plant Design* (Heinemann, London, 1973).

BADGER, W. L.: *Heat Transfer and Evaporation* (Chemical Catalog Co., 1926).

HAUSBRAND, E.: *Evaporating, Condensing and Cooling Apparatus.* Translated from the second revised German edition by A. C. WRIGHT. Fifth English edition revised by B. HEASTIE (E. Benn, London, 1933).

HOLLAND, F. A., MOORES, R. M., WATSON, F. A., and WILKINSON, J. K.: *Heat Transfer* (Heinemann, London, 1970).

KERN, D. Q.: *Process Heat Transfer* (McGraw-Hill, New York, 1950).

KING, C. J.: *Separation Processes* (McGraw-Hill, New York, 1971).

KREITH, F.: *Principles of Heat Transfer*, 2nd edn. (International Textbook Co., London, 1965).

McADAMS, W. H.: *Heat Transmission*, 3rd edn. (McGraw-Hill, New York, 1954).

McCABE, W. L. and SMITH, J. C.: *Unit Operations of Chemical Engineering*, 3rd edn. (McGraw-Hill, New York, 1976).

PERRY, R. H. and CHILTON, C. H. (ED.): *Chemical Engineers Handbook*, 5th edn. (McGraw-Hill, New York, 1973).

PETERS, M. S. and TIMMERHAUS, K. D.: *Plant Design and Economics for Chemical Engineers*, 2nd edn. (McGraw-Hill, New York, 1968).

STEWART, G. and BEVERIDGE, G. S.: *Computers and Chemical Engineering* (Pergamon, London, 1977). Steady-state cascade simulation in multi-effect evaporation.

TYNER, M.: *Process Engineering Calculations* (Ronald Press, New York, 1960).

## 14.8. REFERENCES

1. BADGER, W. L. and SHEPARD, P. W.: *Trans. Am. Inst. Chem. Eng.* **13** (i) (1920) 139. Studies in evaporator design. II. The effect of hydrostatic head on heat transmission in vertical tube evaporators.

2. WESTWATER, J. W.: *Petro/Chem. Engr.* **33**, No. 9 (Aug. 1961) 186–9 and **33**, No. 10 (Sept. 1961) 219–26. Nucleate pool boiling (in two parts).

3. PALEN, J. W. and TABOREK, J. J.: *Chem. Eng. Prog.* **58**, No. 7 (July 1962) 37–46. Refinery kettle reboilers: proposed method for design and optimization.

4. CRYDER, D. S. and GILLILAND, E. R.: *Ind. Eng. Chem.* **24** (1932) 1382–7. Heat transmission from metal surfaces to boiling liquids. I. Effect of physical properties of boiling liquid on liquid film coefficient.

5. CICHELLI, M. T. and BONILLA, C. F.: *Trans. Am. Inst. Chem. Eng.* **41** (1945) 755–87. Heat transfer to liquids boiling under pressure.

6. McNELLY, M. J.: *J. Imp. Coll. Chem. Eng. Soc.* **7** (1953) 18. A correlation of the rates of heat transfer to nucleate boiling liquids.

7. GILMOUR, C. H.: *Chem. Eng. Prog.* Symp. Ser. No. 29, **55** (1959) 67–78. Performance of vaporizers: Heat transfer analysis of plant data.

8. ZUBER, N.: *Trans. Am. Soc. Mech. Eng.* **80** (1958) 711–20. On the stability of boiling heat transfer.

9. DENGLER, C. E. and ADDOMS, J. N.: *Chem. Eng. Prog.* Symp. Ser. No. 18, **52** (1956) 95–103. Heat transfer mechanism for vaporization of water in a vertical tube.

10. LOCKHART, R. W. and MARTINELLI, R. C.: *Chem. Eng. Prog.* **45** (1948) 39–48. Proposed correlation of data for two-phase, two-component flow in pipes.

11. GUERRIERI, S. A. and TALTY, R. D.: *Chem. Eng. Prog.* Symp. Ser. No. 18, **52** (1956) 69–77. A study of heat transfer to organic liquids in single-tube, natural-circulation, vertical-tube boilers.

12. HAUSBRAND, E.: *Evaporating, Condensing and Cooling Apparatus.* Translated from the second revised German edition by A. C. Wright. Fifth English edition revised by B. HEASTIE (E. Benn, London, 1933).

13. STORROW, J. A.: *Ind. Chemist* **27** (1951) 298. Design calculations for multiple-effect evaporators—Part 3.

14. WEBRE, A. L.: *Chem. Met. Eng.* **27** (1922) 1073. Evaporation—A study of the various operating cycles in triple effect units.

15. BRENNAN, P. J.: *Chem. Eng., Albany* **70**, No. 21 (14 Oct. 1963) 170–2. Fresh water from vapour-compression evaporation.

16. SCHWARZ, H. W.: *Food Technol.* **5** (1951) 476. Comparison of low temperature (e.g. 15–24° citrus juice) evaporators.

17. REAVELL, B. N.: *Ind. Chemist* **29** (1953) 475. Developments in evaporation with special reference to heat sensitive liquors.

18. MOORE, J. G. and HESLER, W. E.: *Chem. Eng. Prog.* **59**, No. 2 (Feb. 1963) 87–92. Equipment for the food industry—2: Evaporation of heat sensitive materials.

19. PARKER, N. H.: *Chem. Eng., Albany* **70**, No. 15 (22 July 1963) 135–40. How to specify evaporators.

20. HOLLAND, A. A.: *Chem. Eng., Albany* **55** (xii) (1948) 121. More Saskatchewan salt cake.

21. HOLLAND, A. A.: *Chem. Eng., Albany* **58** (i) (1951) 106. New type evaporator.

22. BLOCH, M. R., FARKAS, L., and SPIEGLER, K. S.: *Ind. Eng. Chem.* **43** (1951) 1544. Solar evaporation of salt brines.

23. SWINDIN, N.: *Trans. Inst. Chem. Eng.* **27** (1949) 209. Recent developments in submerged combustion.

24. VENTON, C. B.: *Intl. Sugar Journal* **53** (1951) 248, 281. The design of vacuum pans.

25. SHEARON, W. H., LOUVIERE, W. H., and LAPEROUSE, R. M.: *Ind. Eng. Chem.* **43** (1951) 552. Cane sugar refining.

26. BOSWORTH, R. C. L. and DULOY, J. S. K.: *Intl. Sugar Journal* **53** (1951) 165. The measurement of pan circulation.

27. BROOKS, C. H. and BADGER, W. L.: *Trans. Am. Inst. Chem. Eng.* **33** (1937) 392. Heat transfer coefficients in the boiling section of a long-tube, natural circulation evaporator.

28. STROEBE, G. W., BAKER, E. M., and BADGER, W. L.: *Trans. Am. Inst. Chem. Eng.* **35** (1939) 17. Boiling film heat transfer coefficients in a long-tube vertical evaporator.

29. CESSNA, O. C., LIENTZ, J. R., and BADGER, W. L.: *Trans. Am. Inst. Chem. Eng.* **36** (1940) 759. Heat transfer in a long-tube vertical evaporator.

30. BRESSLER, R.: *Z. Ver. deut. Ing.* **100**, No. 15 (1958) 630–8. Versuche über die Verdampfung von dünnen Flüssigkeitsfilmen.

31. MCMANUS, T.: University of Durham, Ph.D. Thesis (1963). The influence of agitation on the boiling of liquids in tubes.

32. HARRIOTT, P.: *Chem. Eng. Prog.* Symp. Ser. No. 29, **55** (1959) 137–9. Heat transfer in scraped surface exchangers.

33. KOOL, J.: *Trans. Inst. Chem. Eng.* **36** (1958) 253–8. Heat transfer in scraped vessels and pipes handling viscous materials.

34. MUTZENBURG, A. B.: *Chem. Eng., Albany* **72**, No. 19 (13 Sept. 1965) 175–8. Agitated thin film evaporators. Part 1. Thin film technology.

35. PARKER, N.: *Chem. Eng., Albany* **72**, No. 19 (13 Sept. 1965) 179–85. Agitated thin film evaporators. Part 2. Equipment and economics.

36. FISCHER, R.: *Chem. Eng., Albany* **72**, No. 19 (13 Sept. 1965) 186–90. Agitated thin film evaporators. Part 3. Process applications.

37. RYLEY, J. T.: *Ind. Chemist* **38** (1962) 311–19. Controlled film processing.

38. FRANKEL, A.: *Proc. Inst. Mech. Eng.* **174**, No. 7 (1960) 312–24. Flash evaporators for the distillation of sea-water.

39. SILVER, R. S.: *Proc. Inst. Mech. Eng.* **179**, Pt. 1, No. 5 (1964–5) 135–54. Nominated Lecture: Fresh water from the sea.

40. BAKER, R. A.: *Chem. Eng. Prog.* **59**, No. 6 (June 1963) 80–3. The flash evaporator.

41. WOODWARD, T.: *Chem. Eng. Prog.* **57**, No. 1 (Jan. 1961) 52–7. Heat transfer in a spray column.

42. WILKE, C. R., CHENG, C. T., LEDESMA, V. L., and PORTER, J. W.: *Chem. Eng. Prog.* **59**, No. 12 (Dec. 1963) 69–75. Direct contact heat transfer for sea water evaporation.

43. OTHMER, D. F., BENENATI, R. F., and GOULANDRIS, G. C.: *Chem. Eng. Prog.* **57**, No. 1 (Jan. 1961) 47–51. Vapour reheat flash evaporation without metallic surfaces.
44. BAMFORTH, A. W.: *Industrial Crystallization* (Leonard Hill, London, 1965).
45. BECKERMAN, J. J.: *Foams, Theory and Industrial Applications* (Reinhold, New York, 1953).
46. GOLDBERG, M. and RUBIN, E.: *Ind. Eng. Chem. Process Design and Development* **6** (1967) 195–200. Mechanical foam breaking.
47. SAWYER, F. G., HOLZER, W. F., and McGLOTHLIN, L. D.: *Ind. Eng. Chem.* **42** (1950) 756. Kraft pulp production.
48. —— *Ind. Chemist* **27** (1951) 115. An integrated alkali–chlorine plant.
49. HILLIARD, A.: *Brit. Chem. Eng.* **4** (1959) 138–43. Considerations on the design of graphite heat exchangers.
50. HILLIARD, A.: *Brit. Chem. Eng.* **8** (1963) 234–7. The X-flow Polybloc system of construction for graphite.
51. HILLIARD, A.: *Ind. Chemist* **39** (1963) 525–31. Effect of anisotropy on design considerations for graphite.
52. ARROWSMITH, G.: *Trans. Inst. Chem. Eng.* **27** (1949) 101. Production of vacuum for industrial chemical processes.
53. McCULLOUGH, G. E.: *Ind. Eng. Chem.* **43** (1951) 1505. Concentration of radioactive liquid waste by evaporation.

## 14.9. NOMENCLATURE

| | | Units in SI System | Dimensions in **MLT$\theta$** |
|---|---|---|---|
| $A$ | Heat transfer surface | m$^2$ | **L$^2$** |
| $C_p$ | Specific heat of liquid at constant pressure | J/kg K | **L$^2$T$^{-2}\theta^{-1}$** |
| $C_s$ | Surface factor | — | — |
| $D$ | Liquid evaporated or steam condensed per unit time | kg/s | **MT$^{-1}$** |
| $d$ | A characteristic dimension | m | **L** |
| $d_t$ | Tube diameter | m | **L** |
| $e$ | Economy | — | — |
| $g$ | Acceleration due to gravity | m/s$^2$ | **LT$^{-2}$** |
| $H$ | Enthalpy per unit mass of vapour | J/kg | **L$^2$T$^{-2}$** |
| $h$ | Average value of $h_b$ for a tube bundle | W/m$^2$ K | **MT$^{-3}\theta^{-1}$** |
| $h_b$ | Film heat transfer coefficient for boiling liquid | W/m$^2$ K | **MT$^{-3}\theta^{-1}$** |
| $h_c$ | Film heat transfer coefficient for condensing steam | W/m$^2$ K | **MT$^{-3}\theta^{-1}$** |
| $h_L$ | Liquid-film heat transfer coefficient | W/m$^2$ K | **MT$^{-3}\theta^{-1}$** |
| $h_{tp}$ | Heat transfer coefficient for two phase mixture | W/m$^2$ K | **MT$^{-3}\theta^{-1}$** |
| $k$ | Thermal conductivity of liquid | W/m K | **MLT$^{-3}\theta^{-1}$** |
| $M$ | Mass of cooling water per unit mass of vapour | kg/kg | — |
| $N$ | Number of effects | — | — |
| $P$ | Pressure | N/m$^2$ | **ML$^{-1}$T$^{-2}$** |
| $Q$ | Heat transferred per unit time | W | **ML$^2$T$^{-3}$** |
| $q$ | Heat flux | W/m$^2$ | **MT$^{-3}$** |
| $T$ | Temperature | K | $\theta$ |
| $T_b$ | Boiling temperature of liquid | K | $\theta$ |
| $T_c$ | Condensing temperature of steam | K | $\theta$ |
| $T_f$ | Feed temperature | K | $\theta$ |
| $T_w$ | Heater wall temperature | K | $\theta$ |
| $\Delta T$ | Temperature difference | K | $\theta$ |
| $t$ | Time | s | **T** |
| $U$ | Overall heat transfer coefficient | W/m$^2$ K | **MT$^{-3}\theta^{-1}$** |
| $W$ | Feed rate | kg/s | **MT$^{-1}$** |
| $X_{tt}$ | Lockhart and Martinelli's parameter (equations 14.5 and 14.6) | — | — |
| $y$ | Mass fraction of vapour | — | — |
| $Z$ | Hydrostatic head | m | **L** |
| $\lambda$ | Latent heat of vaporisation per unit mass | J/kg | **L$^2$T$^{-2}$** |
| $\mu_L$ | Viscosity of liquid | Ns/m$^2$ | **ML$^{-1}$T$^{-1}$** |
| $\mu_v$ | Viscosity of vapour | Ns/m$^2$ | **ML$^{-1}$T$^{-1}$** |
| $\rho_L$ | Density of liquid | kg/m$^3$ | **ML$^{-3}$** |
| $\rho_v$ | Density of vapour | kg/m$^3$ | **ML$^{-3}$** |
| $\sigma$ | Interfacial tension | J/m$^2$ | **MT$^{-2}$** |

Suffixes

$\quad\quad$ 0 refers to the steam side of the first effect

$\quad$ 1,2,3 refer to the first, second and third effects

$\quad\quad$ av refers to an average value

$\quad\quad\quad$ $c$ refers to the condenser

$i$ and $e$ refer to the inlet and exit cooling water

# CHAPTER 15

# *Crystallisation*

## 15.1. INTRODUCTION

Crystallisation is an important operation in the chemical industry, as a method of purification and as a method of providing crystalline materials in the desired size range. In a crystal the constituent molecules, ions or atoms are arranged in a regular manner with the result that the crystal shape is independent of size, and if a crystal grows, each of the faces develops in a regular manner. The presence of impurities will, however, usually result in the formation of an irregular crystal. Generally large regular crystals are a guarantee of the purity of the material, though a number of pairs of materials form "mixed crystals". In recent years considerable development has taken place in techniques for growing perfect single crystals which are used in the production of semiconductor devices, laser beams, and artificial gems.

The crystallisation process ranges from the production of the pure single crystal mentioned above to the production of crystalline sugar which is the purest bulk-produced chemical in the world. World production of salt at a rate of 2·6 Mg/s (80 Mtonne/year) gives an idea of the scale and importance of the operation.

In an energy-conscious environment, crystallisation can offer substantial savings as a method of separation when compared with distillation, though it must be recognised that it is more costly to effect cooling than to provide heating. Data provided in Table 15.1 show the energy savings which are possible[1].

The most significant figures in the table relate to water where common applications of

TABLE 15.1. *Energies of Crystallisation and Distillation*

| Substance | Crystallisation | | Distillation | |
|---|---|---|---|---|
| | Melting point (K) | Enthalpy of crystallisation (kJ/kg) | Boiling point (K) | Enthalpy of vaporisation (kJ/kg) |
| *o*-Cresol | 304 | 115 | 464 | 410 |
| *m*-Cresol | 285 | 117 | 476 | 423 |
| *p*-Cresol | 308 | 110 | 475 | 435 |
| *o*-Xylene | 248 | 128 | 414 | 347 |
| *m*-Xylene | 225 | 109 | 412 | 343 |
| *p*-Xylene | 286 | 161 | 411 | 340 |
| *o*-Nitrotoluene | 269 | 120 | 495 | 344 |
| *m*-Nitrotoluene | 289 | 109 | 506 | 364 |
| *p*-Nitrotoluene | 325 | 113 | 511 | 366 |
| Benzene | 278 | 126 | 353 | 394 |
| Water | 273 | 334 | 373 | 2260 |

the process include the production of drinking water from sea water by desalination and the freeze concentration of fruit juices.

Supersaturation alone is not sufficient for a system to begin to crystallise. A number of minute solid bodies known as nuclei must exist in the solution before the process can start and this topic is the subject of considerable study at the present time.

## 15.2. GROWTH AND PROPERTIES OF CRYSTALS

Crystallisation is normally carried out either from a solution or from a melt, though occasionally crystals are formed directly by condensation from the vapour phase. Many features of the process of condensation from a vapour to a liquid are also common to crystallisation, but only crystallisation is considered here. Crystals may also be grown from reagent solutions which produce and deposit material simultaneously, as in the hydrothermal crystallisation of quartz[2]. In some respects crystallisation can be regarded as the inverse of dissolution, which has already been discussed in Chapter 10, but there are important differences. Thus the number of particles present during dissolution will remain constant or decrease, whereas in crystallisation the number of nuclei on which material is deposited may continuously increase. Further, whereas in dissolution there is rarely an appreciable resistance to the transfer of material across the interface between the two phases, this is no longer true for crystallisation.

The crystallisation process consists essentially of two stages which generally proceed simultaneously but which can to some extent be independently controlled. The first stage is the formation of small particles or nuclei, and the second stage is the growth of the nuclei. If the number of nuclei can be controlled the size of the crystals ultimately formed can be regulated, and this forms one of the most important features of the crystallisation process.

### 15.2.1. Saturation

If a melt is slowly cooled, the temperature will gradually fall until the melting point is reached and further cooling will then produce, either super-cooling of the liquid, or solidification with the temperature remaining approximately constant. Generally a definite degree of supercooling must be achieved before change of phase will occur, and a metastable condition therefore exists at temperatures a little below the melting point because the rate of spontaneous nucleation is here negligible. On the other hand, if a small seed crystal or other nucleus is introduced, it will increase in size until the solution is no longer supercooled.

Similarly if a superheated vapour is cooled, its temperature will fall until it reaches the dry and saturated condition, and further cooling in the absence of suitable condensation nuclei will produce a supersaturated vapour, and condensation of liquid or solid will occur only after a finite degree of supersaturation has been achieved. Figure 15.1 shows the pressure–temperature diagram for a material, the line $AB$ representing the vapour pressure as a function of temperature. If a vapour is cooled from $a$ to $c$, it has a supersaturation $bc$ with respect to temperature. Condition $c$ can also be reached by increasing the pressure, at constant temperature, from $e$ to $c$, in which case the supersaturation, in terms of pressure, is $cf$. A vapour can thus become supersaturated by

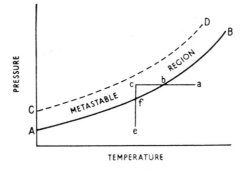

FIG. 15.1. Pressure–temperature relation for saturated vapour

increasing the pressure, by decreasing the temperature, or by both simultaneously. The boundary of the metastable region in which the spontaneous nucleation rate is negligible is represented by the curve $CD$ which is roughly parallel to $AB$.

Again for a solution, curve $AB$ in Fig. 15.2 is the solubility curve and the supersaturation of a solution represented by $c$ is $cb$ in terms of temperature and $cf$ in terms of concentration, $CD$ again represents the limit of the metastable region, in which the nucleation rate is very low. It was originally suggested by Miers[3,4] and others that there was no nucleation in the metastable region but this is now known to be untrue, and the line $CD$ therefore does not represent a sharp boundary.

### 15.2.2. Nucleation

Nuclei may form spontaneously if conditions are suitable, but in many cases small seed crystals may be added and small quantities of impurities may also act as nuclei. Spontaneous nucleation is a process which takes place with some reluctance and is thought to depend upon the existence of random variations in concentration or temperature on the molecular scale. Lord Kelvin has shown that the vapour pressure at a convex surface is greater than that at a plane surface under otherwise similar conditions, and is given as a first approximation by:

$$P_r = P_s + \frac{2\sigma}{r}\frac{\rho_V}{\rho_L} \tag{15.1}$$

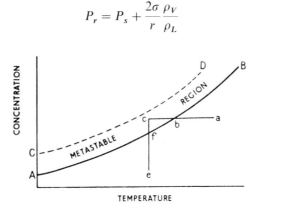

FIG. 15.2. Concentration–temperature relation for saturated solution

where $P_r$ is the vapour pressure at a convex surface of radius $r$,

  $P_s$ is the vapour pressure at a plane surface,

  $\sigma$ is the interfacial tension, and

  $\rho_V, \rho_L$ are the densities of the vapour and liquid.

This increase in vapour pressure occurs because the molecules in a large droplet are more firmly bound together than those in a small droplet, as can be seen in Fig. 15.3. Thus small droplets may re-evaporate in a slightly supersaturated vapour, and consequently large droplets will tend to grow at the expense of the smaller ones. Although this implies that very small droplets can never form, there are generally sufficiently large irregularities on the molecular scale for the production of a number of nuclei, sufficiently large to be stable. A similar situation arises in the case of a solution where the small particles have a greater solubility than the larger ones, and as a result a nucleus will grow only if it is sufficiently large for the solution to be supersaturated with respect to it.

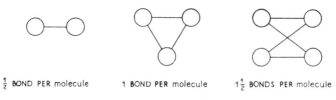

$\frac{1}{2}$ BOND PER molecule          1 BOND PER molecule          $1\frac{1}{2}$ BONDS PER molecule

FIG. 15.3. Groups of molecules

Study of the nucleation process is in most cases difficult because nucleation and growth usually occur simultaneously. Tammann[5], however, worked on the crystallisation of a number of somewhat complex organic compounds, such as piperine, from the melt and obtained conditions where the nucleation and growth could be studied virtually independently. Piperine is a compound with a relatively large molecule and melts at 403 K, but nucleation is appreciable only at temperatures below about 373 K. The material was melted in a small tube a few millimetres in diameter and then rapidly cooled to the required temperature for 600 s, during which time a finite number of nuclei was formed. The temperature was then rapidly raised to 373 K for 240 s to allow them to grow sufficiently to be counted. By this means the curve for the rate of nucleation as a function of temperature shown in Fig. 15.4a was obtained. The shape of this curve is readily explained qualitatively. As the temperature is reduced, the degree of supersaturation and the driving force causing nucleation are increased, and the curve rises steeply. Since, however, the kinetics of the process are adversely affected by lowering the temperature, the rate of nucleation attains a maximum value and then falls off as the temperature is further reduced. The effect of temperature on the kinetics of the process is more marked over the period of crystal growth than during nucleation, and the crystallisation rate reaches a maximum at a rather higher temperature than the nucleation. Figure 15.4b shows the effect of temperature on the linear rate of growth of crystals of triphenylguanidine. Whereas the rate of nucleation and of crystallisation can sometimes be studied separately for crystal growth from a melt, this is not possible in the case of a solution because the two rate curves overlap and the effects occur over a much narrower temperature range. It is thought, however, that the general picture is similar.

The general conclusion is that relatively large crystals will be obtained as a result of slow cooling, because the spontaneous nucleation is then reduced and the material is deposited

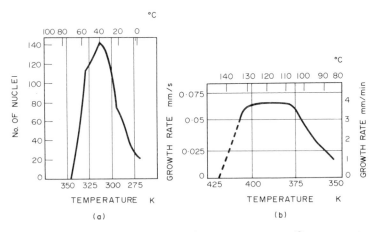

FIG. 15.4. (a) Formation of nuclei from $1 \cdot 2 \, cm^3$ of piperine in $600 \, s^{(5)}$. (b) Growth rate of triphenylguanidine crystals[5]

on a relatively few nuclei. This is borne out in practice. Conversely, rapid cooling produces a large crop of small crystals. Whereas with some materials nuclei form almost immediately supersaturation is achieved, others can be cooled without the deposition of an appreciable number of nuclei, so that crystallisation does not take place and the liquid gradually thickens to form a very viscous mass known as a glass as the temperature is reduced. Sugar syrup and glass are the most commonly encountered glasses, but many organic compounds with large molecules also form glasses very readily. That a glass is a liquid is shown by the fact that it has no crystal structure and that it can flow very gradually.

### 15.2.3. Crystallisation Rate

The rate of growth of a crystal in a solution is dependent on the temperature and concentration of the liquid at the crystal face. These conditions are not generally the same as those in the bulk of the solution because a concentration gradient is necessary for the transfer of solute towards the face, and a temperature gradient for the dissipation of the heat of crystallisation. The problem thus involves both heat transfer and mass transfer, though in most cases the heat transfer may be negligible. In the case of melts, the problem is one of heat transfer alone. Since the resistance to heat and mass transfer lies predominantly in the laminar sub-layer close to the surface of the crystal, the rate of growth of the crystal is improved by increasing the relative velocity between the solid and the liquid. Figure 15.5 shows the effect of increasing the speed of rotation of a crystal of sodium thiosulphate in an aqueous solution. It will be noted that the crystallisation rate initially rises very rapidly but a point is reached where further increase in the agitation produces no effect on the rate of growth. Had the only resistance been within the liquid, the transfer rate would have continued to increase; it therefore appears that there is an additional resistance which is independent of the agitation of the liquid.

The rate of crystallisation is a function of the degree of supersaturation. For crystallisation from a melt, the process is dependent on the rate of transfer of heat between

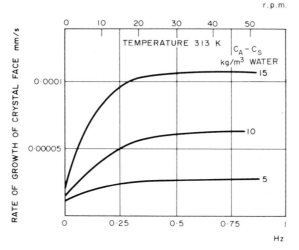

FIG. 15.5. Effect of speed of rotation on rate of growth of crystal of sodium thiosulphate

the crystal face and the surroundings. If $T_M$ is the melting point and $T_L$ is the temperature of the liquid, the apparent supercooling is $T_M - T_L$. Because of the conversion of latent heat into sensible heat as the material solidifies, the temperature at the interface rises to $T_i$ and the true supercooling is then only $T_M - T_i$ (Fig. 15.6). The temperature at the interface must clearly be greater than that of the bulk of the liquid since otherwise there would be no transfer of heat. The higher the heat transfer coefficient, the smaller is the required temperature difference $T_i - T_L$, and therefore a high rate of agitation will cause $T_i$ to approach $T_L$. Although there does not appear to be any regular relationship between the rate of growth of a single face of a crystal and the supercooling, there is a general tendency for a rise in the growth rate as the temperature difference is increased.

For crystallisation from a solution, very much less supercooling is possible—about 1 or 2 deg K as compared with 25 or 50 for melts—and the process involves mass transfer rather than heat transfer, as already indicated. In this case the two processes of nucleation and growth cannot be so conveniently separated. Figure 15.7 shows the concentration gradient in the neighbourhood of a crystal face. $C_S$ represents the

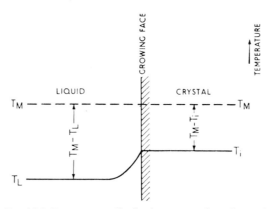

FIG. 15.6. Temperature distribution near surface of crystal

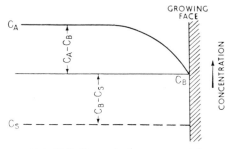

FIG. 15.7. Concentration near crystal

concentration of a saturated solution and $C_A$ the concentration in the bulk of the liquid. The concentration falls from $C_A$ to $C_B$ through the liquid, and the concentration difference $C_B - C_S$ is required in order to overcome the resistance at the interface. Thus, whereas $C_A - C_S$ is the apparent supersaturation, the concentration difference which is responsible for mass transfer is only $C_A - C_B$. In this case the heat effects at the face are generally insignificant.

The growth of a crystal face has been studied in detail by Bunn[6] who observed that the process was erratic in that there were irregular periods of rapid and of slow or zero rate. He also observed that the face grew by the building up of successive layers. Whereas the attachment of the first group of molecules to the surface is relatively difficult because of the weakness of the bonding, subsequent groups can add on quite readily as they are secured not only to the previous layer but also to the material already deposited in the new layer. Thus, the crystal grows layer by layer even though, as reported by Butler[7], the concentration of solution at the surface varies over the face of the crystal (Fig. 15.8). A recent theory of Frank[8] suggests that deposition occurs at a dislocation in the structure of the crystal and that growth takes place in a spiral manner.

Although there is no relation for the instantaneous rate of growth of a crystal face, one can obtain a statistical relation over a period of time. Noyes and Whitney[9] have assumed that the diffusion of material to the crystal face is controlled almost entirely by the resistance of the laminar layer near the face. Therefore:

$$\frac{dm}{dt} = \frac{DA}{\delta_b}(C_A - C_B) \tag{15.2}$$

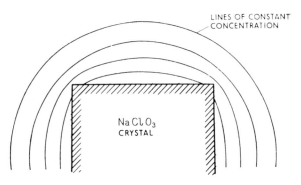

FIG. 15.8. Concentration of solution round crystal of $NaClO_3$

where $A$ is the area of the crystal surface,

$D$ is the diffusivity of the solute in the solution,

$\delta_b$ is the thickness of the laminar layer (it decreases as the degree of agitation is increased), and

$m$ is the mass of material deposited in time $t$.

Berthoud[10] assumed that the transfer at the interface was directly proportional to the supersaturation at the interface. Thus

$$\frac{dm}{dt} = KA(C_B - C_S) \tag{15.3}$$

where $K$ is the rate constant for the surface process.

Eliminating $C_B$ between equations 15.2 and 15.3:

$$\frac{dm}{dt} = \frac{A}{\dfrac{1}{K} + \dfrac{\delta_b}{D}}(C_A - C_S) \tag{15.4}$$

Butler[7] has shown that the surface process is more complicated than is suggested by equation 15.3, and that:

$$\frac{dm}{dt} \propto (C_A - C_S)^n \tag{15.5}$$

where $n$ is greater than unity.

Using the simplified equation (15.4), the linear rate of growth is given as:

$$\frac{dl}{dt} = \frac{1}{\rho_s}\frac{1}{\dfrac{1}{K} + \dfrac{\delta_b}{D}}(C_A - C_S) \tag{15.6}$$

Then if $1/K$ and $\delta_b/D$ are both constant, it follows that:

$$\frac{dl}{dt} = k\,\Delta C \tag{15.7}$$

Thus $dl/dt$ is independent of the actual size of the crystal. Subsequently, Bransom[11] has proposed an equation which takes account of crystal size, as a factor affecting the turbulence around the crystal, and thus affecting $\delta_b$. Using an equivalent Reynolds number $Re'$ he obtained:

$$\frac{dl}{dt} = a(Re')^b(C_A - C_S)^n \tag{15.8}$$

or equivalently:

$$\frac{dl}{dt} = a'l^b(C_A - C_S)^n \tag{15.9}$$

During growth, the shape of the crystal does not alter and therefore the large faces must grow less rapidly than the small ones. This is further evidence of a resistance to transfer at the crystal faces because, if all the resistance were in the liquid film, each face would grow at the same rate. It is the variation of $K$ from face to face which allows the crystal to maintain its shape.

The ratio of the volume, to the cube of the linear dimension, must remain constant as the shape does not alter, i.e.

$$V = k'l^3, \text{ say} \tag{15.10}$$

and

$$dV = 3k'l^2 \, dl$$

Thus, using the approximate relationship given by equation 15.7, the rate of increase of volume is given by:

$$\frac{dV}{dt} = 3k'kl^2 \, \Delta C$$

$$\propto l^2 \, \Delta C \tag{15.11}$$

From the above equations the rate of growth can be calculated approximately, but as in practice this may show considerable variation because, for instance, of the effect of small quantities of impurities, it is always necessary to check the results experimentally.

Because the linear rate of growth of a crystal is independent of its size, the change in the size distribution of a set of seed crystals can be calculated, on the assumption that the size range of these seeds is sufficiently small for the solubility of each crystal to be approximately the same. Thus if the seed crystals have dimensions in a particular direction of $l_1, l_2, \ldots$ etc., their volumes will be $k'l_1^3, k'l_2^3 \ldots$ etc. If during the period of growth of these crystals this dimension has increased by an amount $\Delta l$, the volumes of the crystals in the product will be $k' (l_1 + \Delta l)^3$, $k' (l_2 + \Delta l)^3 \ldots$ etc. Then if a total mass $M$ of solute is crystallised:

$$M = \rho_s k' \{ \Sigma (l_1 + \Delta l)^3 - \Sigma l_1^3 \} \tag{15.12}$$

This expression enables the change in the size distribution of the crystals to be calculated, after a known amount of solute has been removed from solution.

### 15.2.4. Effect of Impurities on Crystal Formation

Although in many cases a soluble impurity will remain in the liquid phase so that pure crystals are produced, there are several instances where both the rate of nucleation and the rate of crystal growth may be affected. The effect is more usually a retardation and is often stated to be due to adsorption of the impurity on the surface of the nucleus or crystal. Thus materials with relatively large molecules, such as glue, tannin, dextrin or sodium hexametaphosphate (Calgon), when added in small quantities to boiler feed water prevent nucleation and growth of calcium carbonate crystals and therefore reduce scaling. Further, addition of 0·1 per cent HCl and 0·1 per cent $PbCl_2$ will prevent the growth of sodium chloride crystals.

In certain cases the adsorption occurs preferentially on a particular face of the crystal with the result that the shape is modified. Thus sodium chloride, crystallised from solutions containing small quantities of urea, forms octohedral instead of the usual cubic crystals; these have a low packing density. A large number of dyes are preferentially adsorbed on inorganic crystals in this way[12]. Their action is quite specific and not generally predictable. Garrett[13] has described a number of industrial uses of additives as habit modifiers.

Solid impurities in the fluid act as condensation nuclei but generally cause some dislocation in the crystal structure.

### 15.2.5. Effect of Temperature on Solubility

Whereas increase in the temperature of the solution usually increases the solubility of the solute, there are instances where the temperature coefficient of solubility is negative, and sometimes it is zero. When the stable crystal-form changes as the temperature is altered, e.g. with hydrated salts, the curve is discontinuous; the coefficient may be positive over part of the range of temperature and negative over the remainder. A negative coefficient indicates a heat of reaction as the material passes into solution, which is greater than the heat absorbed due to the dissolution and dilution processes combined; anhydrous salts therefore tend to have negative coefficients. In Fig. 15.9 the solubility curves are given for a number of salts in water. Potassium chlorate has a large positive temperature coefficient and therefore is readily crystallised by cooling a saturated solution. Sodium chloride has a small coefficient and very little crystallisation occurs on cooling, and evaporation of solvent is therefore essential. The curves for sodium hydrogen phosphate and ferrous sulphate show a number of discontinuities which are attributable to changes in the stable form of the material. Thus crystallisation from a solution of ferrous

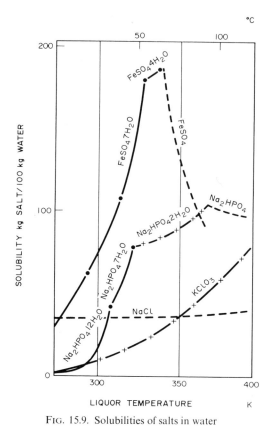

FIG. 15.9. Solubilities of salts in water

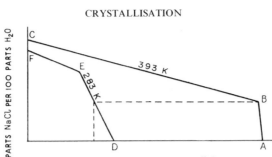

FIG. 15.10. Effect of sodium chloride on solubility of sodium nitrate

sulphate at temperatures below about 320 K results in the formation of the hydrate $FeSO_4. 7H_2O$. On the other hand, crystallisation at higher temperatures will yield either $FeSO_4 . 4H_2O$ or anhydrous $FeSO_4$. With careful regulation of temperature the material can be produced in the desired crystalline form. In many cases it is possible to obtain the hydrated salt only by use of comparatively low temperatures, and vacuum crystallisation is then frequently used.

### 15.2.6. Fractional Crystallisation

It is often possible to separate a mixture of two soluble salts by fractional crystallisation and, in order to understand the process, it is necessary to consider the effect of additions of one salt on the solubility of the second when one ion is common to both salts. Figure 15.10[14] illustrates the effect of additions of sodium chloride on the solubility of sodium nitrate at 393 K (line $AB$) and the effect of sodium nitrate on the solubility of sodium chloride (line $CB$). Line $AB$ then represents mixtures which are saturated with respect to sodium nitrate but unsaturated with respect to sodium chloride, and $CB$ represents the reverse condition. A mixture represented by $B$ is saturated with respect to both salts and therefore represents the composition of the solution obtained by mixing an excess of solid with solvent. Curves $DE$ and $FE$ represent the corresponding conditions at 283 K, and some line joining $B$ and $E$ therefore represents the composition of solutions saturated with both components at various temperatures.

Separation of a mixed salt into its constituents is then effected, for instance, by forming a saturated solution at 393 K, and allowing the solution to cool out of contact with the residual solid to 283 K. Since at this temperature a saturated solution contains more of sodium chloride and less of sodium nitrate, the latter will be deposited in pure form and the remaining solution will be unsaturated with respect to sodium chloride. The crystals can then be drained and washed, and a relatively pure product is obtained. The choice of conditions will affect the particle size and this in turn will affect the purity of the washed solids. If the crystals are relatively large they may tend to agglomerate so that mother liquor is occluded, whereas if they are too small drainage is difficult.

As a further illustration of fractional crystallisation, the separation of pure crystals of para-mononitrotoluene from a mixture of the three isomers, ortho, meta, and para may be considered[15]. The equilibrium diagram for this system is shown in Fig. 15.11. If the hot liquor has a composition, 3 per cent ortho, 8·5 per cent meta and 88·5 per cent para, then its composition is shown by point $P$ on the diagram. For a mixture of this composition no

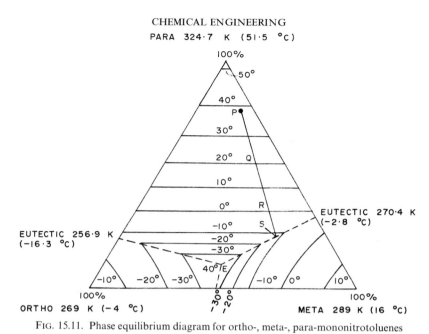

FIG. 15.11. Phase equilibrium diagram for ortho-, meta-, para-mononitrotoluenes

crystals are deposited until the liquor is cooled to 46°C (319 K). On further cooling in the crystalliser, the composition of the remaining liquor is shown by the line *PQRS* for various temperatures indicated by the isothermals on the diagram. On cooling, only the para isomer is deposited and the ratio of ortho to meta remains constant till the point *S* is reached. Further cooling results in the meta being deposited along the line *SE*, till at point *E* all sets solid as a ternary eutectic. It is therefore undesirable to cool the mixture below point *S* (260 K, −13°C). If cooling water is available, it should be possible to cool the liquor down to 293 K (20°C) and, as shown in Table 15.2, this will enable most of the para isomer to be removed.

TABLE 15.2. 100 kg of Liquor Cooled to 293 K (20°C)

| Component | Feed liquor | | Mother liquor | | Crystals |
|---|---|---|---|---|---|
| | % | Mass kg | % | Mass kg | Mass kg |
| Ortho | 3·0 | 3·0 | 12·0 | 3·0 | — |
| Meta | 8·5 | 8·5 | 34·0 | 8·5 | — |
| Para | 88·5 | 88·5 | 54·0 | 13·5 | 75 |
| | 100·0 | 100·0 | 100·0 | 25·0 | 75 |

Cooling to lower temperatures gives an increasing weight of para crystals, as shown in Table 15.3. It is clear that with this system comparatively little is gained by cooling below 293 K (20°C), whereas the cost of cooling to lower temperatures is considerable. If, however, it were decided to obtain the meta isomer as a pure material, cooling to a much lower temperature would be necessary in order to provide a liquor of sufficiently high concentration in meta to justify distillation. Thus, on cooling to 263 K (−10°C), the

TABLE 15.3.  100 kg of Liquor of Composition 3% ortho,
8·5% meta, 88·5% para

| Temperature on cooling | | Mass of para as crystals |
| (K) | (°C) | (kg) |
|---|---|---|
| 319 | 46 | 0 |
| 303 | 30 | 66·7 |
| 293 | 20 | 75·0 |
| 283 | 10 | 79·6 |
| 273 | 0 | 82·3 |
| 263 | −10 | 84·2 |
| 255 | −18 | 84·8 |

percentage of meta in the mother liquor would be nearly 54 and it would be possible to obtain pure meta by distillation.

### 15.2.7.  Caking of Crystals

The tendency for crystalline materials to cake is attributable to a small amount of dissolution taking place at the surface of the crystals and subsequent re-evaporation of the solvent. The crystals can then become very tightly bonded together.

Because the vapour pressure of a saturated solution of a crystalline solid is less than that of pure water at the same temperature, condensation can occur on the surfaces of the crystals even though the relative humidity of the atmosphere is less than 100 per cent. The solution so formed then penetrates into the pack of crystals by virtue of the capillary action of the small spaces between the crystals, and caking can occur if the absorbed moisture subsequently evaporates when the atmospheric humidity falls. Caking can also occur at a constant relative humidity, because the vapour pressure of a solution in a small capillary is lower than that in a large capillary, as a result of the Kelvin effect. As condensation takes place, the small particles are the first to dissolve and the average size of the capillaries is therefore increased, and the vapour pressure of the solution may increase sufficiently for evaporation to occur[16].

A crystalline material will cake more readily if the particle size is non-uniform, because the porosity is less in a bed of particles of mixed sizes and fine particles are more readily soluble. Thus, the propensity for caking can be reduced by forming crystals of relatively large and uniform sizes. Alternatively, small amounts of insoluble materials, such as calcium phosphate or talc, or of a water repellent agent, such as stearic acid or a stearate, may be added.

### 15.3.  CRYSTALLISERS

The solar evaporation of brine is a process which has been practised for over 5000 years and at the present time about one-ninth of the area of the Dead Sea, where the water is shallow, is used for the production of salts containing a high proportion of magnesium in the form of carnallite ($KCl.MgCl.6H_2O$). In this case, it is the sea shore itself which forms the crystalliser.

The more important types of plant used for producing crystals from solutions will now be described because this is the most common application in the chemical industry. Broadly, crystallisers may be classified according to whether they are batch or continuous in operation; continuous crystallisers can be divided into the linear and stirred types which will be referred to later. Crystallisers can also be classified according to the method by which supersaturation is achieved. In the evaporative crystalliser conditions are approximately isothermal, and supersaturation is achieved as a result of the removal of solvent. In the cooling crystalliser, supersaturation results from lowering of the temperature of the solution, and this can be effected either by means of exchange of sensible heat or by evaporative cooling; in the latter case, there is a small loss of solvent. Evaporative crystallisation must of course be used where the solubility shows little variation with temperature.

The main feature of the crystalliser is the method by which the size of the product is regulated, and this is almost entirely dependent on the control of the nucleation process. In general, slow cooling will result in the formation of relatively few and large crystals because spontaneous nucleation is then reduced, and rapid cooling will result in a high rate of nucleation and a small size of crystal. Crystallisers are generally simple in construction, the only moving parts being agitators and the scrapers which are used to keep the heat transfer surfaces free of solids.

### 15.3.1. Batch Crystallisers

*Tank Crystallisers*

The simplest and cheapest type of crystalliser consists of an open tank, which can be used either as an evaporative or as a cooling crystalliser. In the former case, it is generally heated by means of steam coils or from a jacket. In the latter case, the solvent is evaporated until the concentration has reached the required value, and cooling is then effected by transfer of sensible heat to the surroundings and evaporation at the free surface. The solution cools slowly and large crystals are therefore obtained but they are generally uneven because the nucleation is not controlled in any way. The tank is normally ridged to facilitate removal of the mother liquor, but even so drainage is usually incomplete. Labour costs are high for this type of plant. Salt for the fishing industry can be prepared using long shallow open pans heated by steam coils. As the solubility is almost independent of temperature, the crystals form in the liquid surface and are held there by surface tension forces until they exceed a certain weight, when they fall to the bottom and are removed by rakes. Hard saucer-shaped crystals are obtained by this method.

The capacity of the simple tank crystalliser can be increased, and the uniformity of the product improved, by using an agitator and a series of cooling coils. The main disadvantage is that crystals form on the surface of the coils and seriously impede the heat transfer, but it is sometimes possible to brush the surfaces free of crystals. Stainless steel construction is often used because particles do not adhere so firmly as to other materials.

*Evaporators*

Salting evaporators are frequently used for crystallisation and these have already been referred to in Chapter 14. They are generally calandria type evaporators with relatively

FIG. 15.12. Part of plant for recovery of sodium sulphate in viscose process. Crystallising evaporator on left. Rotary filter on right

short wide tubes and a large downcomer, so designed that the crystals do not cause obstruction. The solution is concentrated and then withdrawn, either using a salt box or through a barometric leg. The crystals are separated from the mother liquor which is then returned to the evaporator together with the feed, and evaporation is continued until the concentration of impurities in the solution becomes excessive. When used in this way evaporators may be operated either at atmospheric or at reduced pressure; reduced pressures are essential for the formation of many hydrated crystals. Figure 15.12 shows part of a plant using a crystallising evaporator.

### The Use of Vacuum

Crystallisers employing evaporative cooling often operate under vacuum, and the hot feed is introduced into the body which generally consists of a tall vertical cylinder; a low pressure is maintained by means of a steam jet ejector. Flash evaporation of the solution takes place and produces rapid cooling accompanied by a small increase in concentration. The suspension of crystals is then pumped out, either continuously or in batches. Because these plants operate at very low pressures, cooling water is not generally available at a low enough temperature to condense the vapour, and the vapour is therefore compressed before it enters the condenser; the size of the condenser is thereby considerably reduced in addition. In many cases the solution is agitated or continuously circulated by an external pump in order to prevent stratification.

Vacuum crystallisation is frequently used in the sugar industry where a concentrated sugar solution is fed to an evaporator operating under reduced pressure. Concentration is continued until nucleation commences, when the solution is suddenly cooled by flash evaporation by reducing the pressure. When nucleation is complete, fresh solution is added and evaporation is continued so as to produce crystals of the required size.

### 15.3.2. Continuous Crystallisers

There are two main classes of continuous crystallisers; the linear type in which the solution flows along a trough or pipe with very little longitudinal mixing, and the stirred type in which uniform conditions are maintained.

*Swenson-Walker Crystalliser*

The Swenson-Walker crystalliser (Fig. 15.13) is of the linear type and consists of a long open trough about 0·6 m across and divided into a number of sections, each of which can be independently cooled with water in an external jacket, so that it is possible to control

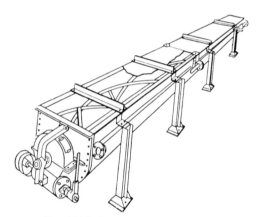

FIG. 15.13. Swenson-Walker crystalliser

the rate of cooling throughout. A spiral scraper keeps the cooling surfaces free of crystals. Sometimes small seed crystals are added to form the nuclei, but more often spontaneous nucleation is effected in the first section by suitable adjustment of the temperature. Slow cooling is obtained in subsequent sections, and the solution may be cooled below atmospheric temperature by using refrigerated brine in the jacket. This crystalliser is suitable only when supersaturation can be effected by cooling alone. It gives fairly uniform crystals but some breakage is caused by the scraper.

*Wulff-Bock Crystalliser*

The Wulff-Bock crystalliser (Fig. 15.14) has similar characteristics to the Swenson-Walker but relies on air cooling and gives more uniform crystals. It consists of a trough

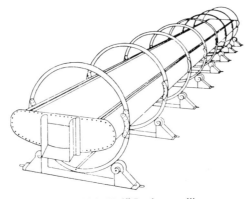

FIG. 15.14. Wulff-Bock crystalliser

which is slowly rocked from side to side; transverse baffles, fitted alternately to opposite sides, prevent surging. Because there are no moving parts, it is easily fabricated from corrosion-resistant materials, and sometimes it is rubber lined.

### The Double-pipe Cooler Crystalliser or Votator Apparatus

The Votator apparatus is a linear type cooler crystalliser in which the heat transfer medium flows through an annular jacket and the liquid undergoing treatment passes through the narrow space between the heat transfer surface and the shaft of the mutator which carries scraper blades (Fig. 15.15). By virtue of the high degree of turbulence and the

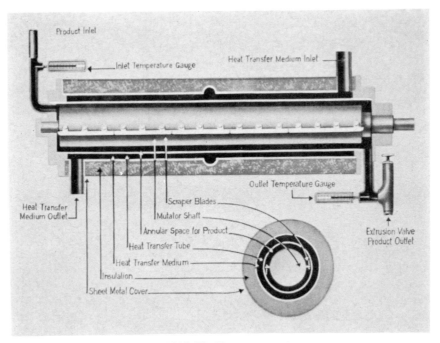

FIG. 15.15. The Votator apparatus

continuous scraping of the surface, very high heat transfer coefficients are obtained. The Votator apparatus is very suitable for highly viscous or heat-sensitive liquids. It is extensively used for crystallising paraffin wax in the petroleum industry and in the plasticising of margarine and cooking fats, and has many other applications.

### The Krystal or Oslo Crystalliser

The Oslo crystalliser was first developed in Norway[17], and is now extensively used where large quantities of crystals of controlled size are required. The plant is made in a number of different forms, but in each the basic principle is to pass a supersaturated solution upwards through a bed of crystals which are maintained in a fluidised state[18]. A uniform temperature is thereby attained, and the crystals segregate in the bed with the small ones at the top and the large ones at the bottom. A typical form of crystalliser is shown in Fig. 15.16 from which it is seen that mother liquor is withdrawn near the feed

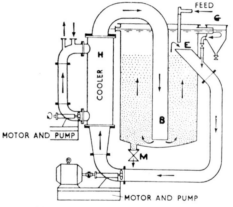

FIG. 15.16.  Oslo cooler crystalliser

point $E$ of the crystalliser by a circulating pump, and is passed through the cooler $H$ where it becomes supersaturated, and then is fed back to the bottom of the crystalliser through the central pipe $B$. The feed forms only a small proportion of the total liquid circulating, and nucleation does not generally take place in the cooler. Some nuclei form spontaneously in the crystal bed and some form as a result of breakage of the crystals. These nuclei circulate with the mother liquor until they have grown sufficiently large to be retained in the fluidised bed. The final product is removed entirely from the bottom of the crystalliser through a valve $M$, and a uniform product is thereby obtained because the crystals are not discharged until they have grown to the required size. A small amount of fines is often formed as a result of mechanical attrition, however. In some cases channelling of the bed occurs, but this can be prevented by introducing the solution through a perforated plate or gauze.

In the crystalliser just described there is no provision for evaporation, and therefore modifications are necessary for salts whose solubilities have small temperature coefficients. In the plant shown in Fig. 15.17, the solution is first passed through a heater

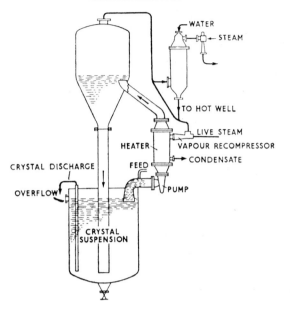

FIG. 15.17. Oslo evaporative crystalliser

and then to a flash evaporator before being returned to the crystalliser. This type can be operated under reduced pressures for the production of hydrated crystals, and multiple-effect evaporation is sometimes used.

Control of crystal formation is generally easier in the stirred than in the linear type of crystalliser. In the Oslo cooling crystalliser, the following are the main control variables:

    (a)  The feed rate.
    (b)  The feed temperature.
    (c)  The heat removed in the cooler.
    (d)  The circulation rate.
    (e)  The rate of removal of nuclei.

The effect of altering any one of these quantities may be quite complex, but the general picture is as follows:

    (a)  Increasing the feed rate causes the material to remain in the crystalliser for a shorter time. The capacity is increased but the crystals will be smaller, and the concentration of the discarded mother liquor will be greater.

    (b)  At high feed temperatures, it is possible to obtain a greater throughput, provided the additional heat is removed in the cooler, because a more concentrated feed can be used.

    (c)  If the heat removed in the cooler is increased, a greater yield is obtained. The crystals are smaller because there is a higher degree of supersaturation.

    (d)  The liquid must remain in contact with the crystals sufficiently long for it to lose its supersaturation. If the rate is too high, supersaturated liquid will be fed to the pump. If it is too low, all the supersaturation will be lost before the solution has reached the top of the crystal bed, and the upper portions of the bed will be effective.

    (e)  Nuclei may be withdrawn from the mother liquor before it enters the cooler, by means of a decanter ($G$ in Fig. 15.16). The solution obtained can then be reheated so that the nuclei dissolve before it is returned to the crystalliser. The fewer the nuclei present, the larger will be the final crystals.

Rumford and Bain[19] have made a detailed study of the crystallisation of sodium chloride in an Oslo crystalliser.

A typical Oslo crystalliser for the production of $MgSO_4 . 7H_2O$ is fed with saturated solution at 393 K. The temperature of the plant is 313 K. With a feed of $0.0005 \text{ m}^3/\text{s}$ (2000 kg/h) and a circulation of $0.0125 \text{ m}^3/\text{s}$ (50,000 kg/h), the circulating liquid becomes heated to 316 K.

Oslo crystallisers are made in very large sizes, with bodies up to 4·5 m in diameter and 6 m high. The internal surfaces must be carefully smoothed so that there is no possibility of nuclei becoming lodged.

## 15.4. FURTHER READING

BAMFORTH, A. W.: *Industrial Crystallisation* (Gresham Press, Surrey, England, 1965).
LAWSON, W. D. and NIELSEN, S.: *Preparation of Single Crystals* (Butterworths, London, 1958).
MULLEN, J. W.: *Crystallisation*, 2nd edn. (Butterworths, London, 1972).
NYVELT, J.: *Industrial Crystallisation from Solutions* (Butterworths, London, 1971).
STRICKLAND-CONSTABLE, R. F.: *Kinetics and Mechanism of Crystallisation* (Academic Press, London and New York, 1968).
Industrial Crystallisation Symposium. *I. Chem. E.* London, 1969.

## 15.5. REFERENCES

1. MULLEN, J. W.: Crystallisation. *Chem. and Ind.* (27 Feb. 1971) 237. A study in molecular engineering.
2. LAUDISE, R. A. and SULLIVAN, R. A.: *Chem. Eng. Prog.* **55,** No. 5 (May 1959) 55–59. Pilot plant production: synthetic quartz.
3. MIERS, H. A.: *Phil. Trans. Roy. Soc.* **202** (1903) 459. An enquiry into the variation of angles observed in crystals; especially of potassium alum and ammonium alum.
4. MIERS, H. A.: *J. Inst. Metals* **37** (1927) 331. The growth of crystals in supersaturated solutions.
5. TAMMANN, G.: *The States of Aggregation* (Constable, 1926). Translated from second German edition by R. T. MEHL.
6. BUNN, C. W.: *Discuss. Faraday Soc.* No. 5 (1949), Crystal Growth, p. 132. Crystal growth from solution—II. Concentration gradients and the rates of growth of crystals.
7. BUTLER, R. M.: University of London, Ph.D. Thesis (1950). Factors affecting the rate of growth of crystals from aqueous solutions.
8. FRANK, F. C.: *Discuss. Faraday Soc.* No. 5 (1949), Crystal Growth, p. 48. The influence of dislocations on crystal growth.
9. NOYES, A. A. and WHITNEY, W. R.: *J. Am. Chem. Soc.* **19** (1897) 930. The rate of solution of solid substances in their own solutions.
10. BERTHOUD, A.: *J. Chim. Phys.* **10** (1912) 624. Théorie de la formation des faces d'un cristal.
11. BRANSOM, S. H.: *Brit. Chem. Eng.* **5** (1960) 838–44. Factors in the design of continuous crystallisers.
12. WHETSTONE, J.: *Trans. Faraday Soc.* **51** (1955) 973–80 and 1142–53. The crystal habit modification of inorganic salts with dyes (in two parts).
13. GARRETT, D. E.: *Brit. Chem. Eng.* **4** (1959) 673–7. Industrial crystallization: influence of chemical environment.
14. BADGER, W. L. and BAKER, E. M.: *Inorganic Chemical Technology*, 1st edn. (McGraw-Hill, 1928).
15. COULSON, J. M. and WARNER, F. E.: *A Problem in Chemical Engineering Design: The Manufacture of Mononitrotoluene* (Institution of Chemical Engineers, 1949).
16. MORRIS, J. N.: University of London, Ph.D. Thesis (1953). Factors affecting the rate of growth of crystals from the melt.
17. SVANOE, H.: *Ind. Eng. Chem.* **32** (1940) 636. "Krystal" classifying crystallizer.
18. PULLEY, C. A.: *Ind. Chemist* **38** (1962) 63–66, 127–32, and 175–8. The Krystal crystallizer.
19. RUMFORD, F. and BAIN, J.: *Trans. Inst. Chem. Eng.* **38** (1960) 10–20. The controlled crystallisation of sodium chloride.

# 15.6. NOMENCLATURE

| | | Units in SI System | Dimensions in $\mathbf{MLT}\theta$ |
|---|---|---|---|
| $A$ | Area of crystal face | $m^2$ | $\mathbf{L}^2$ |
| $a$ | Empirical constant | — | — |
| $a'$ | Empirical constant | — | — |
| $b$ | Empirical constant | — | — |
| $C_A$ | Concentration in bulk of solution | $kg/m^3$ | $\mathbf{ML}^{-3}$ |
| $C_B$ | Concentration at interface | $kg/m^3$ | $\mathbf{ML}^{-3}$ |
| $C_S$ | Concentration of saturated solution | $kg/m^3$ | $\mathbf{ML}^{-3}$ |
| $D$ | Diffusivity of solute in solution | $m^2/s$ | $\mathbf{L}^2\mathbf{T}^{-1}$ |
| $K$ | Rate constant for surface process | $m/s$ | $\mathbf{LT}^{-1}$ |
| $k$ | Coefficient relating rate of growth and supersaturation | $m^4/kg\,s$ | $\mathbf{M}^{-1}\mathbf{L}^4\mathbf{T}^{-1}$ |
| $k'$ | Volume coefficient | — | — |
| $l$ | Linear dimension of crystal | $m$ | $\mathbf{L}$ |
| $M$ | Mass of solute crystallised | $kg$ | $\mathbf{M}$ |
| $m$ | Mass of solute deposited in time $t$ | $kg$ | $\mathbf{M}$ |
| $n$ | Index | — | — |
| $P_r$ | Vapour pressure at convex surface of radius $r$ | $N/m^2$ | $\mathbf{ML}^{-1}\mathbf{T}^{-2}$ |
| $P_s$ | Vapour pressure at plane surface | $N/m^2$ | $\mathbf{ML}^{-1}\mathbf{T}^{-2}$ |
| $r$ | Radius of droplet | $m$ | $\mathbf{L}$ |
| $T_i$ | Temperature of crystal face | $K$ | $\theta$ |
| $T_L$ | Temperature of bulk of liquid | $K$ | $\theta$ |
| $T_M$ | Melting point | $K$ | $\theta$ |
| $t$ | Time | $s$ | $\mathbf{T}$ |
| $V$ | Volume of crystal | $m$ | $\mathbf{L}^3$ |
| $\delta_b$ | Thickness of laminar layer | $m$ | $\mathbf{L}$ |
| $\rho_L$ | Density of liquid | $kg/m^3$ | $\mathbf{ML}^{-3}$ |
| $\rho_s$ | Density of crystal | $kg/m^3$ | $\mathbf{ML}^{-3}$ |
| $\rho_V$ | Density of vapour | $kg/m^3$ | $\mathbf{ML}^{-3}$ |
| $\sigma$ | Interfacial tension | $J/m^2$ | $\mathbf{MT}^{-2}$ |
| $Re'$ | Modified Reynolds number | — | — |

# CHAPTER 16

# *Drying*

## 16.1. INTRODUCTION

The drying of materials is often the final operation in a manufacturing process and is carried out immediately prior to packaging or dispatch. By drying, the final removal of water is referred to, and the operation often follows evaporation, filtration, or crystallisation. In some cases, drying is an essential part of the manufacturing process, as for instance in paper making or in the seasoning of timber, though, in the majority of processing industries, drying is carried out for one or more of the following reasons:

(a) To reduce the cost of transport.
(b) To make a material more suitable for handling, e.g. soap powders, dyestuffs, fertilisers.
(c) To provide definite properties, e.g. to maintain the free-flowing nature of salt.
(d) To avoid the presence of moisture which may lead to corrosion, e.g. the drying of coal gas or the drying of benzene prior to chlorination.

With a crystalline product it is necessary to ensure that the crystals are not damaged during drying, and, in the case of pharmaceutical products, care must be taken to avoid contamination. Shrinkage (as with paper), cracking (as with wood) or loss of flavour (as with fruit) must also be prevented. With the exception of the partial drying of a material by squeezing in a press or the removal of water by adsorption, almost all drying processes involve the removal of water by vaporisation, and thus require the addition of heat. In assessing the efficiency of a drying process, the effective utilisation of the heat is a major criterion.

## 16.2. GENERAL PRINCIPLES

The moisture content of a material is usually expressed as a percentage of the weight of the dry material. If a material is exposed to air at a given temperature and humidity, the material will either lose or gain water until an equilibrium condition is established. The equilibrium moisture content varies widely with the moisture content and the temperature of the air, as shown in Fig. 16.1. A non-porous insoluble solid, such as sand or China clay, will have an equilibrium moisture content approaching zero for all humidities and temperatures, though many organic materials, such as wood, textiles, and leather, show wide variations of equilibrium moisture content. The moisture may be present in the following two forms:

*Bound moisture.* This is water retained in such a way that it exerts a vapour pressure less than that of free water at the same temperature. Such water may be retained in small capillaries, adsorbed on surfaces, or as a solution in cell walls.

710

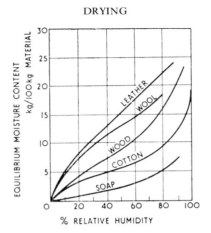

FIG. 16.1. Equilibrium moisture content of a solid as a function of relative humidity (293 K)

*Free moisture.* This is water which is in excess of the equilibrium moisture content.

The water removed by vaporisation is generally carried away by air or hot gases, and the ability of these gases to pick up the water will be determined by their temperature and humidity. In designing dryers using air, it is necessary to have available the properties of the air–water system in a convenient form. These properties have been discussed in detail in Volume 1, Chapter 11, where the development of the humidity chart is described. Some of the more important factors are briefly described below for the *air–water system.*

*Humidity.*$\mathscr{H}$. Mass of water carried per unit mass of dry air. Since

$$\frac{\text{kmol of water vapour}}{\text{kmol of dry air}} = \frac{P_w}{(P-P_w)}$$

$$\therefore \quad \mathscr{H} = \frac{18P_w}{29(P-P_w)}$$

where $P_w$ is the partial pressure of water vapour and $P$ is the total pressure.

*Humidity of saturated air* $\mathscr{H}_0$. Humidity of air when saturated with water vapour. The air is in equilibrium with water at the given temperature and pressure.

*Percentage humidity.*

$$\frac{\text{Humidity of air}}{\text{Humidity of saturated air}} \times 100$$

$$= \frac{\mathscr{H} \times 100}{\mathscr{H}_0}$$

*Percentage relative humidity.*

$$\frac{\text{Partial pressure of water vapour in air}}{\text{Vapour pressure of water at the same temperature}} \times 100$$

The difference between *percentage humidity* and *percentage relative humidity* should be noted.

*Humid volume.* Volume of unit mass of dry air and associated vapour. At atmospheric pressure, this is equal to:

$$\frac{22 \cdot 4}{29} \left( \frac{T}{273} \right) + \frac{22 \cdot 4 \mathscr{H}}{18} \left( \frac{T}{273} \right) \text{m}^3/\text{kg, where } T \text{ is in degrees Kelvin,}$$

or

$$\frac{359}{29} \left( \frac{T}{492} \right) + \frac{259 \mathscr{H}}{18} \left( \frac{T}{492} \right) \text{ft}^3/\text{lb, where } T \text{ is in degrees Rankine.}$$

*Saturated volume.* Volume of unit mass of dry air, together with the water vapour required to saturate it.

*Humid heat.* Heat required to raise unit mass of dry air and associated vapour through 1 degree K at constant pressure $(= 1 \cdot 00 + 1 \cdot 88 \mathscr{H}) \text{ kJ/kg K}$.

*Dew point.* Temperature at which condensation will first occur when air is cooled.

*Wet bulb temperature.* If a stream of air is passed rapidly over the surface of water, vaporisation occurs, provided the temperature of the water is above the dew point of the air. The temperature of the water falls and heat flows from the air to the water. If the surface is sufficiently small for the condition of the air to change inappreciably and if the velocity is in excess of about $5 \text{ m/s}$, the water reaches the wet bulb temperature $\theta_w$ at equilibrium.

Rate of heat transfer from gas to liquid

$$= hA(\theta - \theta_w) \qquad (16.1)$$

Mass rate of vaporisation

$$= \frac{h_D A M_w}{\mathbf{R} T} (P_{w0} - P_w)$$

$$= \frac{h_D A M_A}{\mathbf{R} T} (P - P_w)_{\text{mean}} (\mathscr{H}_w - \mathscr{H})$$

$$= h_D A \rho_A (\mathscr{H}_w - \mathscr{H}) \qquad (16.2)$$

Heat transfer required to effect vaporisation at this rate

$$= h_D A \rho_A (\mathscr{H}_w - \mathscr{H}) \lambda \qquad (16.3)$$

At equilibrium, the rates of heat transfer given by equations 16.1 and 16.3 must be equal. Hence

$$\mathscr{H} - \mathscr{H}_w = -\frac{h}{h_D \rho_A \lambda} (\theta - \theta_w) \qquad (16.4)$$

and the wet bulb temperature $\theta_w$ depends only on the temperature and humidity of the drying air.

In the above equations,

$h$ is the heat transfer coefficient,

$h_D$ is the mass transfer coefficient,

$A$ is the surface area,

$\theta$ is the temperature of the air stream,

$\theta_w$ is the wet bulb temperature,

$P_{w0}$ is the vapour pressure of water at the temperature $\theta_w$,

$M_A$ is the molecular weight of air,

$M_w$ is the molecular weight of water,

$\mathbf{R}$ is the universal gas constant,

$T$ is the absolute temperature,

$\mathscr{H}$ is the humidity of the gas stream,

$\mathscr{H}_w$ is the humidity of saturated air at a temperature $\theta_w$,

$\rho_A$ is the density of air at its mean partial pressure, and

$\lambda$ is the latent heat of vaporisation of water.

Equation 16.4 is identical with equation 11.8 in Volume 1, and reference should be made to that chapter for a more detailed discussion of the problem.

## 16.3. RATE OF DRYING

### 16.3.1. Drying Periods

In the drying of materials, it is necessary to remove free moisture from the surface and also moisture from the interior. If the change in moisture content with time for a material is determined, a smooth curve is obtained from which the rate of drying at any given moisture content can be evaluated. The form of the drying rate curve varies with the structure and type of material, and two typical curves are illustrated in Fig. 16.2. In curve 2 there are two well-defined zones: $AB$, where the rate of drying is constant; and $BC$, where there is a steady fall in the rate of drying as the percentage of moisture is reduced. The moisture content at the end of the constant rate period is represented by point $B$, and is known as the *critical moisture content*. Curve 1 which will be obtained for a rather different kind of material shows three stages: $DE$, $EF$ and $FC$. The stage $DE$ represents a constant rate period, and $EF$ and $FC$ are falling rate periods. In this case, however, the section $EF$ is a straight line and only the portion $FC$ is curved; section $EF$ is known as the first falling rate period and the final stage, shown as $FC$, as the second falling rate period. The drying of soap gives rise to a curve of type 2, and sand to a curve of type 1. A number of workers, notably Sherwood[1] and Newitt and co-workers[2-7], have made contributions to the theories on the rate of drying in these various stages.

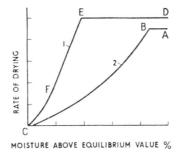

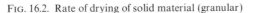

MOISTURE ABOVE EQUILIBRIUM VALUE %

FIG. 16.2. Rate of drying of solid material (granular)

*Constant Rate Period*

During the constant rate period, it is assumed that drying takes place from a saturated surface of the material by diffusion of the water vapour through a stationary air film into the air stream. Thus, Gilliland[8] has shown that the rates of drying of a variety of materials in this stage are substantially the same (Table 16.1).

TABLE 16.1. *Evaporation Rates for Various Materials under Constant Conditions* (Gilliland)

| Material | Rate of evaporation kg/m² s | Rate of evaporation g/m² s |
|---|---|---|
| Water | 0·00075 | 0·75 |
| Whiting pigment | 0·00058 | 0·58 |
| Brass filings | 0·00067 | 0·67 |
| Brass turnings | 0·00067 | 0·67 |
| Sand (fine) | 0·00055–0·00067 | 0·55–0·67 |
| Clays | 0·00064–0·00075 | 0·64–0·75 |

To calculate the rate of drying under these conditions, the relationships obtained in Volume 1 for diffusion of a vapour from a liquid surface into a gas may be used. The simplest equation of this type would be:

$$W = K_G A(P_s - P_w) \tag{16.5}$$

where $K_G$ represents the diffusional transfer coefficient.

Since the rate of transfer will depend on the velocity $u$ of the air stream, raised to a power of about 0·8, then:

$$W = K'A(P_s - P_w)u^{0·8} \tag{16.6}$$

In the above equations,

$A$  is the surface area,
$P_s$  is the vapour pressure of the water,
$P_w$  is the partial pressure of water vapour in the air stream, and
$W$  is the mass rate of evaporation.

This type of equation has been used in Volume 1 for studying the rate of vaporisation into an air stream, and simply states that the rate of transfer is equal to the transfer coefficient multiplied by the driving force. It should be noted, however, that $(P_s - P_w)$ is not only a driving force, but is also related to the capacity of the air stream to absorb moisture.

These equations suggest that the rate of drying is independent of the geometrical shape of the surface. Experiments by Powell and Griffiths[9] have shown, however, that the ratio of the length to the width of the surface is of some importance, and that the evaporation rate (kg/s) is given more accurately by equations 16.7 and 16.8, as follows:

(a)  For values of $u$ between 1 and 3 m/s:

$$W = 5·53 \times 10^{-9} L^{0·77} B(P_s - P_w)(1 + 61u^{0·85}) \tag{16.7}$$

(b)  For values of $u$ below 1 m/s:

$$W = 3·72 \times 10^{-9} L^{0·73} B^{0·8}(P_s - P_w)(1 + 61u^{0·85}) \tag{16.8}$$

where   $P_s$, the saturation pressure at the temperature of the surface, and
$\quad\quad$ $P_w$, the vapour pressure in the air stream (both in $N/m^2$),
$\quad\quad$ and $L$ and $B$ are the length and width of the surface in m.

For all normal engineering design work it may be assumed that the rate of drying is proportional to the transfer coefficient multiplied by $(P_s - P_w)$. Chakravorty[10] has shown that, if the temperature of the surface is greater than that of the air stream, then $P_w$ may easily reach a value corresponding to saturation of the air. Under these conditions, the capacity of the air to take up moisture is zero, while the force causing evaporation is $(P_s - P_w)$. As a result, a mist will form and water may be redeposited on the surface. In all drying equipment, care must therefore be taken to ensure that the air or gas used does not become saturated with moisture.

The rate of drying in the constant rate period is given by:

$$\frac{dw}{dt} = \frac{hA\Delta T}{\lambda} = K_G A(P_s - P_w) \tag{16.9}$$

where   $dw/dt$ is the rate of loss of water,
$\quad\quad$ $h$ is the heat transfer coefficient from air to the wet surface,
$\quad\quad$ $\Delta T$ is the temperature difference between the air and the surface,
$\quad\quad$ $\lambda$ is the latent heat of vaporisation,
$\quad\quad$ $K_G$ is the mass transfer coefficient for diffusion from the wet surface through the gas film,
$\quad\quad$ $A$ is the area of interface for heat and mass transfer, and
$\quad$ $(P_s - P_w)$ is the difference between the vapour pressure of water at the surface and the partial pressure in the air.

It is rather more convenient to express the mass transfer coefficient in terms of a humidity difference, so that $K_G A(P_s - P_w) \simeq kA(\mathscr{H}_s - \mathscr{H})$. The rate of drying is thus determined by the values of $h$, $\Delta T$ and $A$, and is not influenced by the conditions inside the solid. $h$ will depend on the air velocity and the direction of flow of the air, and it has been found that $h = CG'^{0.8}$ where $G'$ is the mass rate of flow of air in $kg/s\,m^2$. For air flow parallel to plane surfaces, Shepherd et al.[11] have given the value of $C$ as 14·5 for the heat transfer coefficient expressed in $W/m^2\,K$.

If the gas temperature is high, then a considerable proportion of the heat will pass to the solid by radiation, and the heat transfer coefficient will increase. This may result in the temperature of the solid rising above the wet bulb temperature.

*First Falling Rate Period*

The points $B$ and $E$ in Fig. 16.2 represent conditions where the surface is no longer capable of supplying sufficient free moisture to saturate the air in contact with it. Under these conditions, the real factor influencing the rate of drying is the mechanism by which the moisture from inside the material is transferred to the surface. In general, the curves in Fig. 16.2 will apply, but for type 2 a simplified expression for the rate of drying in this period can be obtained.

*Second Falling Rate Period*

At the conclusion of the first falling rate period it may be assumed that the surface is dry and that the plane of separation is moving into the solid. In this case, evaporation will be taking place from within the solid and the vapour reaches the surface by molecular diffusion through the material. The forces controlling the vapour diffusion will determine the final rate of drying, and these will be largely independent of the conditions outside the material.

## 16.3.2. Time for Drying

Suppose a material is dried by passing hot air over the surface which is initially wet. Then the rate of drying curve in its simplest form is represented by *BCE* as in Fig. 16.3.
In Fig. 16.3,

$w$ is the total moisture,
$w_e$ is the equilibrium moisture content $(E)$,
$w - w_e$ is the free moisture content,
and $w_c$ is the critical moisture content $(C)$.

### Constant Rate Period

During the period of drying from the initial moisture content $w_1$ to the critical moisture content $w_c$, the rate of drying is constant, so that the time of drying $t_c$ is given by:

$$t_c = \frac{w_1 - w_c}{R_c A} \tag{16.10}$$

where $R_c$ is the rate of drying per unit area in the constant rate period, and
$A$ is the area of exposed surface.

### Falling Rate Period

During this period the rate of drying is directly proportional to the free moisture content $(w - w_e)$, that is:

$$-\frac{1}{A}\frac{dw}{dt} = m(w - w_e) = mf \quad \text{(say)} \tag{16.11}$$

$$\therefore \quad -\frac{1}{mA}\int_{w_c}^{w}\frac{dw}{(w - w_e)} = \int_{0}^{t_f} dt$$

$$\therefore \quad \frac{1}{mA}\ln\left[\frac{w_c - w_e}{w - w_e}\right] = t_f$$

$$\therefore \quad t_f = \frac{1}{mA}\ln\frac{f_c}{f} \tag{16.12}$$

### Total Time of Drying

Total time of drying from $w_1$ to $w$ is $t = (t_c + t_f)$.

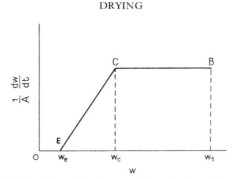

FIG. 16.3. Use of rate of drying curve for estimation of time drying

The rate of drying $R_c$ over the constant rate period is equal to the initial rate of drying in the falling rate period, so that $R_c = mf_c$.

$$\therefore \quad t_c = \frac{(w_1 - w_c)}{mf_c A} \tag{16.13}$$

$$\therefore \quad \text{Total drying time} \quad t = \frac{(w_1 - w_c)}{Amf_c} + \frac{1}{mA} \ln \frac{f_c}{f}$$

$$= \frac{1}{mA} \left[ \frac{(f_1 - f_c)}{f_c} + \ln \frac{f_c}{f} \right] \tag{16.14}$$

## 16.4. THE MECHANISM OF MOISTURE MOVEMENT DURING DRYING

### 16.4.1. Diffusion Theory of Drying

The general form of the curve for the rate of drying of a solid (Fig. 16.2), has shown that this takes the form of two or sometimes three distinct sections. During the constant rate period, moisture vaporises into the air stream and the controlling factor is the transfer coefficient for diffusion across the gas film. It is important to understand how the moisture moves to the drying surface during the falling rate period. Two ideas have been given of the physical nature of this process, one the diffusion theory and the second the capillary theory. In the diffusion theory, it is supposed that the rate of movement of water to the air interface is governed by rate equations similar to those for heat transfer, whilst in the capillary theory it is assumed that the forces controlling the movement of water are capillary in origin and arise from the minute pore spaces between the individual particles.

*Falling Rate Period. Diffusion Control*

In the falling rate period, the surface is no longer completely wetted and the rate of drying steadily falls. In the previous analysis, it has been assumed that the rate of drying per unit effective wetted area is a linear function of the water content, so that the rate of drying is given by:

$$\frac{1}{A}\frac{dw}{dt} = -m(w - w_e) \qquad \text{(equation 16.11)}$$

In many instances, however, the rate of drying is governed by the rate of internal movement of the moisture to the surface. It was initially assumed that this movement was a process of diffusion and would follow the same laws as heat transfer. This approach has been examined by a number of workers, and in particular by Sherwood[12] and Newman[13].

Let $w$ be the liquid content of the solid,

   $w_1$ be the initial content, and

   $w_e$ be the equilibrium content.

Then

$$\frac{(w - w_e)}{(w_1 - w_e)} = \frac{\text{Free liquid content at any point at any time}}{\text{Initial free liquid content}}$$

Consider a slab with the edges coated to prevent evaporation, and suppose the slab is dried by evaporation from two opposite faces, the $Y$-direction being taken perpendicular to the drying face, the central plane being taken as $y = 0$, and the slab thickness $2l$. Then, on drying, the moisture movement by diffusion will be in the $Y$-direction and hence from Chapter 8, Volume 1:

$$\frac{\partial w}{\partial t} = D_L \frac{\partial^2 w}{\partial y^2}$$

where $D_L$ is the coefficient of diffusion for the liquid. Sherwood and Newman have given the solution of this equation in the form:

$$\frac{w - w_e}{w_1 - w_e} = \frac{8}{\pi^2} \{ e^{-D_L t (\pi/2l)^2} + \tfrac{1}{9} e^{-9 D_L t (\pi/2l)^2} + \tfrac{1}{25} e^{-25 D_L t (\pi/2l)^2} + \ldots \} \qquad (16.15)$$

It should be noted that this equation assumes an initial uniform distribution of moisture, and that the drying is from both surfaces. When drying is from one surface, then $l$ is the total thickness. If the time of drying is long, then only the first term of the equation need be used and thus, on differentiating equation 16.15:

$$\frac{dw}{dt} = -\frac{\pi^2 D_L}{4l^2} (w - w_e) \qquad (16.16)$$

In the drying of materials such as wood or clay, the moisture concentration at the end of the constant rate period is not uniform, and is more nearly parabolic. Sherwood has developed an analysis for this case, and has given experimental values for the drying of brick clay.

In this solution it is assumed that the rate of movement of water is proportional to a concentration gradient, and capillary and gravitational forces are neglected. Water may, however, flow from regions of low concentration to those of high concentration if the pore sizes are suitable, and for this and other reasons Ceaglske and Hougen[14] suggested the capillary theory which is briefly outlined in the next section.

### 16.4.2. Capillary Theory of Drying

*Principles of the Theory*

The capillary theory of drying has been put forward in order to explain the movement of moisture in the bed during surface drying. The basic importance of the pore space between

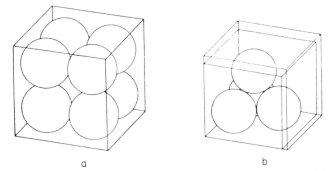

FIG. 16.4. Packing of spherical particles. (a) Cubic arrangement, one sphere touching six others. (b) Rhombohedral packing, one sphere touching twelve others, with layers arranged in rhombic formation

granular particles was first pointed out by Slichter[15] in connection with the movement of moisture in soils, and this work has now been modified and considerably expanded by Haines[16]. The principles will be given and then applied to the problem of drying. Considering a systematic packing of uniform spherical particles, these may be arranged in six different regular ways, ranging from the most open to the closest packing. In the former, the spheres are arranged as if at the corners of a cube, each sphere touching six others; whilst in the latter arrangement, each sphere rests in the hollow of three spheres in adjacent layers, and touches twelve other spheres. These configurations are shown in Fig. 16.4. The densities of packing of the other four arrangements will lie between those illustrated.

In each case, a regular group of spheres surrounds a space which is called a pore, and the bed is made up of a series of these elemental groupings. The pores are connected together by passages of various sizes, the smallest portions of which are known as waists. The size of a pore is defined as the diameter of the largest sphere which can be fitted into it, and the size of a waist as the diameter of the inscribed circle. The sizes of the pores and waists will differ for each form of packing, as indicated in Table 16.2.

TABLE 16.2. *Some Properties of Packing of Spheres of Radius r*

| Packing arrangement | Pore space % total volume | Radius of pore | Radius of waist | Value of $x$ in equation 16.20 for: | |
|---|---|---|---|---|---|
| | | | | limiting suction potential of pores | entry suction potential of waists |
| Cubical | 47·64 | 0·700r | 0·414r | 2·86 | 4·82 |
| Rhombohedral | 25·95 | 0·288r | 0·155r | 6·90 | 12·90 |

The continuous variation in the diameter of each passage is the essential difference between a granular packing and a series of capillary tubes. If a clean capillary of uniform diameter $2r'$ is placed in a liquid, the level will rise in the capillary to a height $h_s$ given by:

$$h_s = \frac{2\sigma}{r'\rho g} \cos \alpha \qquad (16.17)$$

where $\rho$ is the density of the liquid,

$\sigma$ is the surface tension,

$\alpha$ is the angle of contact, and

$g$ is the acceleration due to gravity.

A negative pressure, known as a suction potential, will exist in the liquid in the capillary. Immediately below the meniscus the suction potential will be equivalent to the height of the liquid column $h_s$ and, if water is used, will have the value:

$$h_s = \frac{2\sigma}{r'\rho g} \qquad (16.18)$$

If equilibrium conditions exist, the suction potential $h_1$ at any other level in the liquid, a distance $z_1$ below the meniscus, will be given by the equation:

$$h_s = h_1 + z_1 \qquad (16.19)$$

Similarly, if a uniform capillary is filled to a greater height than $h_s$, as given by equation 16.17, and its lower end is immersed, the liquid column will recede to this height.

The non-uniform passages in a porous material will also display the same characteristics as a uniform capillary, with the important difference that the rise of water in the passages will be limited by the pore size, whilst the depletion of saturated passages will be controlled by the size of the waists. The height of rise is controlled by the pore size, since this approximates to the largest section of a varying capillary, whilst the depletion of water is controlled by the narrow waists which are capable of a higher suction potential than the pores.

The theoretical suction potential of a pore or waist containing water can therefore be calculated from the equation:

$$h_t = \frac{x\sigma}{r\rho g} \qquad (16.20)$$

where $x$ is a factor depending on the type of packing, and

$r$ is the radius of the spheres.

For example, for an idealised bed of uniform rhombohedrally packed spheres of radius $r$, the waists are of radius $0.155r$ (from Table 16.2).

The maximum theoretical suction potential of which such a waist is capable

$$= \frac{2\sigma}{0.155r\rho g}$$

$$= \frac{12.9\sigma}{r\rho g}$$

from which it is seen $x = 12.9$ in this instance.

The maximum suction potential that can be developed by a waist is known as the "entry" suction potential. This is the controlling potential required to open a saturated pore protected by a meniscus in an adjoining waist. Some values for $x$ are given in Table 16.2.

When a bed is composed of granular material with particles of mixed sizes, the suction potential cannot be calculated but must be measured by methods such as those given by Haines[16] and Newitt et al.[3].

*Drying of a Granular Material according to Capillary Theory*

Suppose a bed of uniform spheres to be initially saturated and to be surface dried in a current of air of constant temperature, velocity and humidity. Then the rate of drying is given by:

$$\frac{dw}{dt} = K_G A(P_{w0} - P_w) \qquad (16.21)$$

where $P_{w0}$ is the saturation partial pressure of water vapour at the wet bulb temperature of the air, and $P_w$ is the partial pressure of the water vapour in the air stream. This rate, as indicated earlier, will remain constant so long as the inner surface of the "stationary" air film remains saturated.

As evaporation proceeds, the water surface recedes into the waists between the top layer of particles, and an increasing suction potential is developed in the liquid. When the menisci on the cubical waists (i.e. the largest) have receded to the narrowest section, the suction potential $h_s$ at the surface equals $4.82\sigma/r\rho g$ (Table 16.2). Further evaporation will result in $h_s$ increasing so that the menisci on the surface cubical waists will collapse, and the larger pores below will open. As $h_s$ steadily increases, the entry suction of progressively finer surface waists is reached, so that the menisci collapse into the adjacent pores which are thereby opened.

In considering the conditions below the surface, the suction potential $h_1$ a distance $z_1$ from the surface will be given by:

$$h_s = h_1 + z_1 \qquad (\text{equation } 16.19)$$

The flow of water through waists surrounding an open pore will be governed by the size of the waist in the following way:

(a) If the size of the waist is such that its entry suction potential exceeds the suction potential at that level within the bed, it will remain full by the establishment of a meniscus therein, in equilibrium with the effective suction potential to which it is subjected. This waist will then protect adjoining full pores which cannot be opened until one of the waists to which it is connected collapses.

(b) If the size of the waist is such that its entry suction potential is less than the suction potential existing at that level, it will in turn collapse and open the adjoining pore. Moreover, this successive collapse of pores and waists will progressively continue, so long as the pores so opened expose waists having entry suction potentials less than the suction potentials existing at that depth.

As drying proceeds, two processes take place simultaneously:

(a) The collapse of progressively finer surface waists, and the resulting opening of pores and waists connected thereto, which they previously protected.

(b) The collapse of further full waists within the bed adjoining opened pores, and the consequent opening of adjacent pores.

It should be noted that, even though the effective suction potential at a waist or pore within the bed may be in excess of its entry or limiting suction potential, this will not necessarily collapse or open. Such a waist can only collapse if it adjoins an opened pore, and the pore in question can only open upon the collapse of an adjoining waist.

*Effect of particle size.* Reducing the particle size of the bed will reduce the size of the pores and the waists, and will increase the entry suction potential of the waists. This increase will mean that the precentage variation in suction potentials with depth will be reduced, and the moisture distribution will be more uniform with small particles.

As the pore sizes are reduced, the frictional forces opposing the movement of water through these pores and waists may become significant, so that equation 16.19 is more accurately represented by:

$$h_s = h_1 + z_1 + h_f \tag{16.22}$$

where $h_f$ is the frictional head opposing the flow over a depth $z_1$ from the surface. The value of $h_f$ will depend on the particle size and it has been found[17] that, with coarse particles when only low suction potentials are found, the gravity effect is important but $h_f$ is small, whilst with fine particles $h_f$ becomes large. It has been found that:

(a) For particles of between 0·1 and 1 mm radius, the values of $h_1$ are independent of the rate of drying, but vary appreciably with depth. Frictional forces are, therefore, negligible: capillary and gravitational forces are in equilibrium throughout the bed and are the controlling forces. Under such circumstances the percentage moisture loss at the critical point at which the constant rate period ends is independent of the drying rate, but varies with the depth of bed.

(b) For particles of between 0·001 and 0·01 mm radius, the values of $h_1$ vary only slightly with rate of drying and depth, indicating that both gravitational and frictional forces are negligible; capillary forces are controlling. The critical point will here be independent of drying rate and depth of bed.

(c) For particles of less than 0·001 mm (1 μm) radius, gravitational forces are negligible, whilst frictional forces are of increasing importance; capillary and frictional forces may then be controlling. In such circumstances, the percentage moisture loss at the critical point will diminish with increased rate of drying and depth of bed. With beds of very fine particles an additional factor comes into play. The very high suction

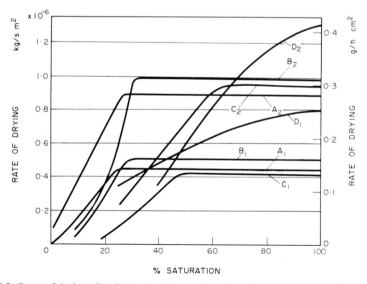

Fig. 16.5. Rates of drying of various materials as a function of percentage saturation. $A$ 60 μm glass spheres; bed 51 mm deep. $B$ 23·5 μm silica flour; bed 51 mm deep. $C$ 7·5 μm silica flour; bed 51 mm deep. $D$ 2·5 μm silica flour; bed 65 mm deep. 1. Low drying rate. 2. High drying rate

potentials which are developed cause a sufficient reduction of the pressure for vaporisation of water to take place inside the bed. This internal vaporisation results in a breaking up of the continuous liquid phase and a consequent interruption in the free flow of liquid by capillary action. Hence, the rate of drying is still further reduced.

Some of the experimental results of Newitt *et al.*[2] are illustrated in Fig. 16.5.

*Freeze Drying*

Special considerations apply to the movement of moisture in freeze drying. Since the water is frozen, liquid flow under capillary action is out of the question, and movement must be by vapour diffusion, analogous to the "second falling rate period" of the normal case.

Furthermore, at very low pressures the mean free path of the water molecules may be comparable with the pore size of the material. In these circumstances the flow is said to be of the Knudsen type.

## 16.5. CLASSIFICATION AND SELECTION OF DRYERS

There are many hundreds of dryer designs available on the market at the present time and this renders classification a virtually impossible task. Cronshaw[18] has made an attempt based on the form of the material being handled and Parker[19] takes into account the means by which material is transferred through the dryer as a basis of his classification with a view to presenting a guide to the selection of dryers. Probably the most thorough classification of dryer types has been made by Kröll[20] who has presented a decimalised system based on the following factors:

   (i) temperature and pressure in the dryer,
  (ii) the method of heating,
 (iii) the means by which moist material is transported through the dryer,
 (iv) any mechanical aids aimed at improving drying,
  (v) the method by which the air is circulated,
 (vi) the way in which the moist material is supported,
(vii) the heating medium, and
(viii) the nature of the wet feed and the way it is introduced into the dryer.

In selecting a dryer for a particular application two steps are of primary importance[21]:

 (i) a listing of the dryers which are capable of handling the material to be dried,
(ii) eliminating the more costly alternatives on the basis of annual costs (capital charges + operating costs). A summary of dryer types, together with cost data, have been presented by Backhurst and Harker[22].

Most dryers will accept free-flowing, particulate materials, though materials which are bulky or of irregular form will usually require equipment of a more specialised design. Once a group of possible dryers has been selected, the choice can be narrowed by considering the way in which the unit is to be operated, whether batch or continuous for example. In addition to the specific restraints imposed by the nature of the material, a further factor of some importance is the mode of heating, whether by contact with a solid surface or directly by convection and radiation.

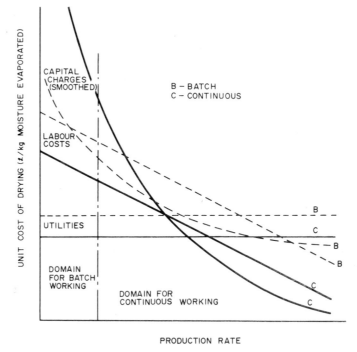

FIG. 16.6. Variation of unit costs of drying with production rate

### Batch or Continuous Operation

In general continuous operation has the important advantages of ease of integration into the rest of the process coupled with a lower unit cost of drying compared with batch processing. As the rate of throughput of material becomes smaller, however, the capital cost becomes the major component in the total running costs and the relative cheapness of batch plant becomes more attractive. This is illustrated in Fig. 16.6 which is taken from the work of Keey[23].

In simple terms, a throughput of around 5000 kg/day and under is best handled in batches whilst a throughput of 50,000 kg/day and over is more suited to continuous operation. There are further considerations, however, such as the ease of construction of a small batch dryer compared with the sophistication of the continuous dryer which generally demands the expertise of the plant manufacturer. In addition, a batch dryer is much more versatile and can often be used for different materials and also enables close control of humidity during an operation, especially in cases where the humidity has to be maintained at different levels for varying periods of time.

### Direct or Indirect Heating

Direct heating in which the material is heated primarily by convection from the surrounding air has several advantages. Firstly, directly heated dryers are, in general, less costly, mainly because of the absence of tubes or jackets within which the heating medium must be contained. Secondly, it is possible to control the temperature of the air within very fine limits, and indeed it is relatively simple to ensure that the material is not heated beyond a specified temperature. This is especially relevant with heat-sensitive materials. Against these advantages, the overall thermal efficiency of directly heated dryers is generally low due to the loss of energy in the exhaust air; furthermore, where an expensive solvent is evaporated from the solid the operation is often difficult and costly. Losses also occur in the case of fluffy and powdery materials and further problems are encountered where either the product or the solvent reacts with oxygen in the air.

Many of these disadvantages can be overcome by modifications to the design—however, though these again increase the cost, and in the event an indirectly heated dryer may prove to be more economical. This is especially the case when thermal efficiency, solvent recovery or maximum cleanliness is of paramount importance. One problem with indirectly heated dryers is the danger of overheating the product, bearing in mind the fact that heat is transferred by conduction through the material itself.

### Vacuum Operation

The maximum temperature at which the drying material may be held is fixed by the thermal sensitivity of the product and this temperature varies inversely with the time of retention. Where lengthy drying times are employed, as for example in a batch shelf dryer, it is necessary to operate under vacuum in order to maintain evaporative temperatures at acceptable levels. In most continuous dryers, the retention time is very low, however, and operation at atmospheric pressure is usually satisfactory.

## 16.6. TRAY AND TUNNEL DRYERS

### 16.6.1. Tray or Shelf Dryers

Tray or shelf dryers are commonly used for granular materials or for individual articles. The material is placed on a series of trays, as shown in Fig. 16.7, which may be heated from below by steam coils and drying is carried out by the circulation of air over the material. In some installations, the air is heated and then passed once through the oven, though, in the majority of dryers, some recirculation of air takes place, as indicated in Fig. 16.7, and the air reheated before it is passed over each shelf.

The condition of the air used on a tray or compartment dryer is conveniently shown by means of the humidity chart (see Volume 1, Chapter 11). If air of humidity $\mathscr{H}_1$ is passed over heating coils so that its temperature rises to $\theta_1$, the humidity staying constant, this operation can be represented by the line $AB$ on the diagram (Fig. 16.8). This air then passes over the wet material and leaves nearly saturated, say at 90 per cent relative humidity, its temperature falling to some value $\theta_2$. This change in the condition of the air is shown by the line $BC$, and the humidity has risen to $\mathscr{H}_2$. The wet bulb temperature of the air will not change appreciably and therefore $BC$ will coincide with an adiabatic cooling line. Each kilogram of air removes $(\mathscr{H}_2 - \mathscr{H}_1)$ kilogram of water, and the air required to remove a given amount of water from the material can easily be found. If the air at $\theta_2$ is now passed over a second series of heating coils and is heated to the original temperature $\theta_1$, the heating operation is shown by the line $CD$. This reheated air can then be passed

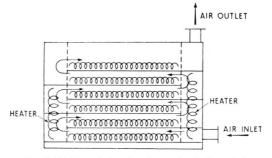

FIG. 16.7. Tray dryer, showing recirculation of air

over wet material on a second tray in the dryer, and pick up moisture until its relative humidity rises again to 90 per cent at a temperature $\theta_3$ (point $E$). In this way each kilogram of air has picked up water amounting to $(\mathscr{H}_3 - \mathscr{H}_1)$ kilogram. This type of reheating can be effected a number of times, as indicated in the diagram (Fig. 16.8), so that the moisture removed per kilogram of air can be considerably increased over that for a single pass. Thus, for three passes of air over the material, the total moisture removed is $(\mathscr{H}_4 - \mathscr{H}_1)$ kg/kg air.

If the air of humidity $\mathscr{H}_1$ had been heated initially to a temperature $\theta_5$, the same amount of moisture would have been removed by a single passage over the material, assuming that the air again leaves at a relative humidity of 90 per cent.

The reheating method has two main advantages. The first is that very much less air is required, because each kilogram of air picks up far more water than in a single stage system. The second is that, in order to pick up as much water in a single stage, it would be necessary to heat the air to a very much higher temperature. The reduction in the amount of air first simplifies the heating system, and secondly reduces the tendency of the air to carry away any small particles.

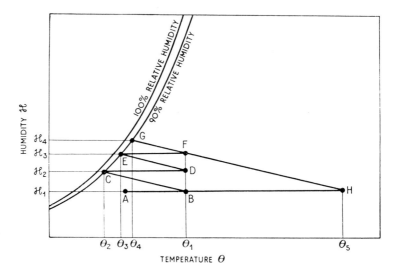

FIG. 16.8. Drying with reheating of air

A modern tray dryer consists of a well-insulated cabinet with integral fans and trays which are stacked on racks or are loaded on to trucks which are pushed into the dryer. Tray areas vary from 0·3 to 1 m² with a depth of material between 10 and 100 mm depending on the particle size of the product; air velocities of 1 to 10 m/s are used and in order to conserve heat 85–95 per cent of the air is recirculated. Even at these high figures, the steam consumption may be 2·5–3·0 kg/kg moisture removed. The capacity of tray dryers depends on many factors, such as the nature of the material, the loading and external conditions, though for dyestuffs a value to the evaporate capacity of $3 \times 10^{-5}$ to $30 \times 10^{-5}$ kg/m² s has been quoted with air at 300–360 K[23].

### 16.6.2. Tunnel Dryers

In the tunnel dryer, a series of trays or trolleys is moved slowly through a long tunnel, and drying takes place in a current of warm air; the tunnel itself may or may not be heated. Tunnel dryers are used for drying trays of paraffin wax, gelatine or soap, pottery ware, and in those cases where the throughput is so large that individual cabinet dryers would involve too much handling. In an alternative system, material is placed on a belt conveyor passing through the tunnel, an arrangement which is very suitable with a vacuum type unit.

Typical tunnel arrangements are shown in Fig. 16.9. The construction is usually of block or sheet-metal and the size varies over a wide range—a dryer for clayware being over 30 m long, for example.

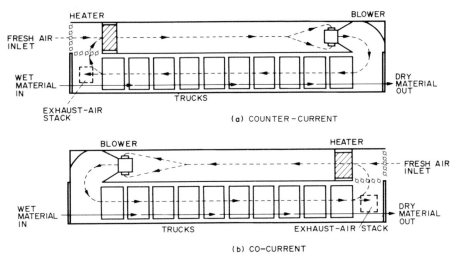

FIG. 16.9. Arrangements for drying tunnels

## 16·7. ROTARY DRYERS

### 16.7.1. Directly Heated Rotary Dryers

For the continuous drying of materials on a large scale (0·3 kg/s or greater), the rotary dryer is suitable. This consists of a relatively long cylindrical shell mounted on rollers, and

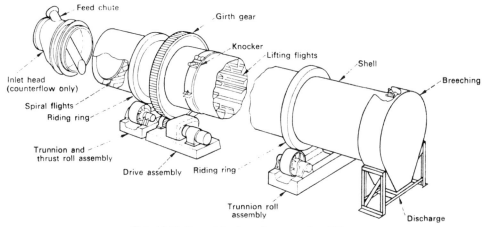

FIG. 16.10. Isometric view of a rotary dryer[21]

driven at a low speed—up to 0·4 Hz. The shell is arranged at a small angle to the horizontal so that material fed in at the higher end will travel through the dryer under gravity. The hot gases or air used as a drying medium are fed in either at the upper end of the dryer to give co-current flow, or at the discharge end of the machine to give countercurrent flow. One of two methods of heating is used:

(a) Direct heating, where the hot gases or air pass through the material in the dryer.
(b) Indirect heating, where the material is in an inner shell, heated externally by hot gases. Alternatively, steam is fed to a series of tubes inside the shell of the dryer.

The shell of the rotary dryer is usually made by welding rolled plate, and the thickness must be sufficient for transmission of the torque required to cause rotation, and to support its own weight and the weight of material in the dryer. The shell is usually supported on large tyres which run on wide rollers, as shown in Fig. 16.10. Although mild steel is the usual material of construction, alloy steels are used, and if necessary the shell may be coated with a plastics material to avoid contamination of the product.

With countercurrent operation, the gases are often exhausted by a fan, so that there is a slight vacuum in the dryer, and dust-laden gases are thereby prevented from escaping. This arrangement is suitable for sand, salt, ammonium nitrate and other inorganic salts, and is particularly convenient when the discharged product is at a high temperature. In this event, gas or oil firing is used where air is used as a drying medium, this may be filtered before heating, in order to minimise contamination of the product. As the gases leaving the dryer generally carry away very fine material, some form of cyclone or scrubber is usually fitted. Since the hot gases come into immediate contact with the dried material, the moisture content can be reduced to a minimum, though the charge may become excessively heated. Further, since the rate of heat transfer is a minimum at the feed end, a great deal of space is taken up with heating the material.

With co-current flow, the rate of passage of the material through the dryer tends to be greater since the gas is travelling in the same direction. Contact between the wet material and the inlet gases gives rise to rapid surface drying, and this is advantageous if the material tends to stick to the walls. This rapid surface drying is also helpful with materials containing water of crystallisation. The dried product will leave at a lower temperature than with countercurrent systems, and this may be an advantage. The rapid lowering of

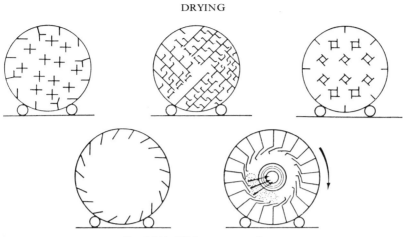

FIG. 16.11. Flights on rotary dryer

the gas temperature as a result of immediate contact with the wet material also enables heat sensitive materials to be handled rather more satisfactorily.

Since the drying action arises mainly from direct contact with hot gases, some form of lifter is essential to bring the material into the gas stream. This may take the form of flights, as shown in Fig. 16.11, or of louvres. In the former, the flights lift the material and then shower it across the gas stream, whilst in the latter the gas stream enters the shell along the louvres. From Figs. 16.12 and 16.13, it can be seen that in the rotary louvre dryer, the hot air enters through the louvres, and carries away the moisture at the end of the dryer. This is

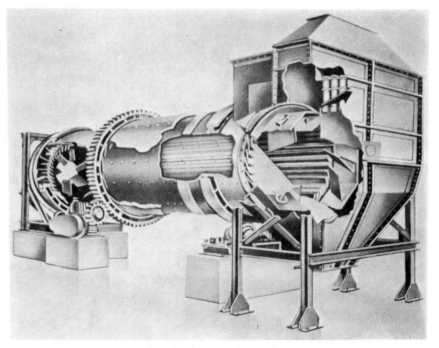

FIG. 16.12. Rotary louvre dryer

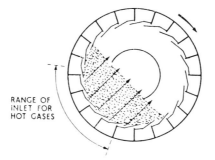

FIG. 16.13. Air flow through rotary louvre dryer

not strictly a co-current flow unit, but rather a through circulation unit, since the material continually meets fresh streams of air. The rotation of the shell, at about 0·05 Hz, maintains the material in agitation and conveys it through the dryer. These machines are built in sizes from 0·75 to 3·5 m in diameter and up to 9 m in length[24].

The thermal efficiency of rotary dryers is a function of the temperature levels and ranges from 30 per cent in the handling of crystalline foodstuffs to 60–80 per cent in the case of inert materials. Evaporative capacities in the range 0·0015–0·0080 kg/m$^3$ s may be expected and these may be increased by 50 per cent in the case of the louvre design.

### 16.7.2. Indirectly Heated Rotary Dryers

One form of indirectly heated dryer is shown in Fig. 16.14. The hot gases pass through the innermost cylinder, and then return through the annular space between the outer cylinders. This form of dryer can be arranged to give direct contact with the wet material during the return passage of the gases. Flights on the outer surface of the inner cylinder, and the inner surface of the outer cylinder, assist in moving the material along the dryer. This form of unit gives a better heat recovery than the single flow direct dryer, though it is more expensive. In a simpler arrangement, a single shell is mounted inside a brickwork chamber, through which the hot gases are introduced.

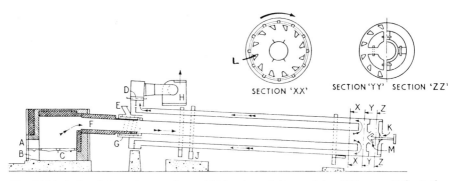

FIG. 16.14. Indirectly heated rotary dryer. A Firing door. B Air regulator. C Furnace. D Control valves. E Feed chute. F Furnace flue. G Feed screw. H Fan. J Driving gear. K Discharge bowl. L Duct lifters. M By-pass valve

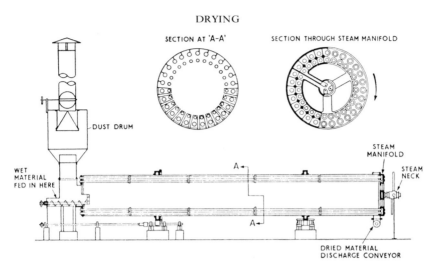

Fig. 16.15. Steam-tube rotary dryer

The steam-tube dryer, shown in Fig. 16.15, incorporates a series of steam tubes, fitted along the shell in concentric circles and rotating with the shell. These tubes may be fitted with fins to increase the heat transfer surface but material may then stick to the tubes. The solids pass along the inclined shell, and leave through suitable ports at the other end. A small current of air is passed through the dryer to carry away the moisture, and the air leaves almost saturated. In this arrangement, the wet material comes in contact with very humid air, and surface drying is therefore minimised. This type of unit has a high thermal efficiency, and can be made from corrosion resisting materials without difficulty.

### 16.7.3. Design Considerations

Many of the design problems associated with rotary dryers have been discussed by Friedman and Marshall[25] and by Miller et al.[26].

The heat from the air stream passes to the solid material during its fall through the air stream, and also by way of the hot walls of the shell, though, of these two mechanisms, the first is much the most important. The heat transfer equation may be written as:

$$Q = Ua V \Delta T \tag{16.23}$$

where $Q$ is the rate of heat transfer,
$\quad$ $U$ is the overall heat transfer coefficient,
$\quad$ $V$ is the volume of the dryer,
$\quad$ $a$ is the area of contact between the particles and the gas per
$\quad\quad$ unit volume of dryer, and
$\quad$ $\Delta T$ is the mean temperature difference between the gas and material.

The combined group $Ua$ has been shown[25] to be influenced by the feed rate of solids, the air rate and the properties of the material, and a useful approximation is given by an equation of the form:

$$Ua = \bar{\kappa} G'^{n}/D \tag{16.24}$$

where $\bar{\kappa}$ is a dimensional coefficient. Typical values obtained using a 300 mm diameter dryer revolving at 0·08–0·58 Hz (5–35 rpm) show that $n = 0·67$ for specific gas rates in the range 0·37–1·86 kg/m² s[27]. The coefficient $\bar{\kappa}$ is a function of the number of flights and, using SI units, is given approximately by:

$$\bar{\kappa} = 20(n_f - 1) \tag{16.25}$$

which was derived using a 200 mm diameter dryer with between 6 and 16 flights[28]. Combining equations 16.24 and 16.25,

$$Ua = 20(n_f - 1)G'^{0·67}/D \tag{16.26}$$

and hence for a 1 m diameter dryer with 8 flights, $Ua$ would be about 140 W/m³ K for a gas rate of 1 kg/m² s.

Saeman[27] has investigated the countercurrent drying of sand in a dryer of 0·3 m diameter and 2·0 m long with 8 flights rotating at 0·17 Hz (10 rpm) and has found that the volumetric heat transfer coefficient can be correlated in terms of the hold-up of solids, as shown in Fig. 16.16, and is independent of the gas rate in the range 0·25–20 kg/m² s.

The hold-up of a rotary dryer varies with the feed rate, the number of flights, the shell diameter and the air rate. For zero air flow, Friedman and Marshall[25] give the hold-up as $X$ per cent of the dryer volume, where

$$X = \frac{25·7F}{SN^{0·9}D} \tag{16.27}$$

Here $F$ is the feed rate in m³/s m²,
        $S$ is the slope of the dryer in m/m length, and
        $N$ is the rate of rotation in Hz.

As the air rate is increased, $X$ alters and an empirical relation for the hold-up $X_a$ with air flow is:

$$X_a = X \pm KG' \tag{16.28}$$

although the values of $K$ are poorly defined. It is usual for $X_a$ to have a value of about 3 per

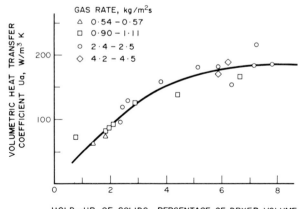

FIG. 16.16. Correlation of volumetric heat-transfer coefficients with hold-up[28]

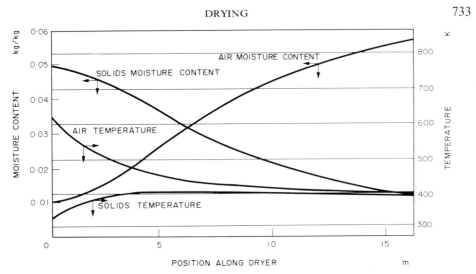

FIG. 16.17. Conditions in a 2·6 m diameter × 16 m long rotary dryer with 1° slope. Co-current drying of 11·3 kg/s of fertiliser granules with 9·1 kg/s of air[28]

cent, when working with a slope $S$ of about 0·1. In equation 16.28, the positive sign refers to countercurrent flow and the negative sign to co-current flow.

A more general approach has been made by Sharples[28] who has solved differential moisture and heat-balance equations coupled with expressions for the forward transport of solids, allowing for solids being cascaded out of lifting baffles. Typical results are shown in Fig. 16.17 which was obtained using a 2·6 m diameter dryer, 16 m long with a 1° slope (60 mm/m).

Bearing in mind that commercial equipment is available with diameters up to 3 m and lengths up to 30 m, the correlations outlined in this section must be used with caution outside the range used in the experimental investigations.

## 16.8. DRUM DRYERS

If a solution or slurry is run on to a steam-heated drum which is slowly rotating, evaporation will take place and the solids may be obtained in a dry form. This is the basic principle used in all drum dryers, some forms of which are illustrated in Figs. 16.18 *a, b, c*. The feed to a single drum dryer may be of the dip *a*, pan *b*, or splash *c* type. The dip-feed system was the earliest design and is still used where the liquid can be picked up from a shallow pan. The agitator prevents settling of any particles, and the spreader is sometimes used to produce a uniform coating on the drum. The knife which is employed for removing the dried material functions in a similar manner to the doctor blade on a rotary filter. If the material is dried to give a free flowing powder, this will come away from the drum quite easily. The splash feed *c* is used for materials, such as calcium arsenate, lead arsenate, and iron oxide, where a light fluffy type of material is desired from the dryer. The revolving cylinder throws the liquor against the drum, and a uniform coating is formed with materials which do not stick to the hot surface of the drum.

Double drum dryers may be used in much the same way, and Fig. 16.18 *d* and *e* show dip feed and top feed designs. Top feed gives a larger capacity, as a thicker coating is obtained.

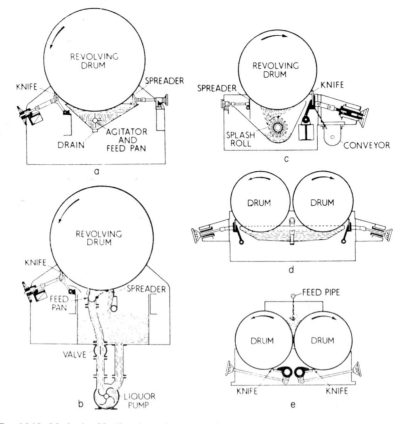

FIG. 16.18. Methods of feeding drum dryers. (a) Single drum, dip-feed. (b) Single drum, pan-feed.
(c) Single drum, splash-feed. (d) Double drum, dip-feed. (e) Double drum, top-feed

It is important to arrange for a uniform feed to a top feed machine, and this may be effected by using a perforated pipe for solutions and a travelling trough for suspensions.

The drums are usually made from cast iron, though chromium-plated steel or alloy steel is often used where contamination of the product must be avoided, such as with pharmaceuticals or food products. Arrangements must be made for accurate adjustment of the separation of the drums, and the driving gears should be totally enclosed. A range of speeds is usually obtained by selecting the gears, rather than by fitting a variable speed drive. Removal of the steam condensate is important, and an internal syphon is often fitted to keep the drum free of condensate. In some cases, it is better for the drums to be rotated upwards at the point of closest proximity, and the knives are then fitted at the bottom. By this means the dry material is kept away from the vapour evolved. Some indication of the sizes of this type of dryer is given in Table 16.3 and it should be noted that the surface of each drum is limited to about 35 m$^2$. This limitation, coupled with recent developments in the design of spray dryers, renders the latter more economically attractive especially where large throughputs have to be handled. For steam-heated drum dryers, normal evaporative capacities are in the range 0·003–0·02 kg/m$^2$ s though higher rates are claimed for grooved drum dryers.

TABLE 16.3. *Sizes of Double Drum Dryers*

| Drum dimensions Diameter (m) × Length (m) | Length (m) | Width (m) | Height (m) | Weight (kg) |
|---|---|---|---|---|
| 0·61 × 0·61 | 3·5 | 2·05 | 2·3 | 3,850 |
| 0·61 × 0·91 | 3·8 | 2·05 | 2·4 | 4,170 |
| 0·81 × 1·32 | 5·0 | 2·5 | 2·8 | 7,620 |
| 0·81 × 1·83 | 5·5 | 2·5 | 2·8 | 8,350 |
| 0·81 × 2·28 | 5·9 | 2·5 | 2·8 | 8,890 |
| 0·81 × 2·54 | 6·25 | 2·5 | 2·8 | 9,300 |
| 0·81 × 3·05 | 6·7 | 2·5 | 2·8 | 10,120 |
| 1·07 × 2·28 | 6·4 | 3·0 | 3·0 | 14,740 |
| 1·07 × 2·54 | 6·7 | 3·0 | 3·0 | 15,420 |
| 1·07 × 3·05 | 7·2 | 3·0 | 3·0 | 16,780 |
| 1·52 × 3·66 | 7·9 | 4·1 | 4·25 | 27,220 |

When it is desired to keep the temperature of the drying material as low as possible, vacuum drying is used, and Fig. 16.19 shows one form of vacuum dryer. The dried material is collected in two screw conveyors and carried usually to two receivers so that one can be filled while the other is emptied.

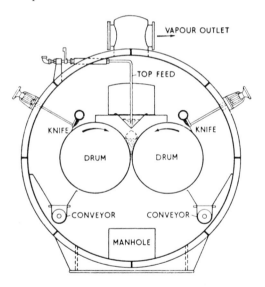

FIG. 16.19. Vacuum drum dryer

## 16.9. SPRAY DRYERS

### 16.9.1. Introduction

Water can be evaporated from a solution or a suspension of solid particles by spraying the mixture into a vessel through which a current of hot gases is passed. By this means a large interfacial area is produced and consequently a high rate of evaporation is obtained. Drop temperatures remain below the wet bulb temperature of the drying gas until drying is almost complete, and the process thus affords a convenient means of drying substances which may deteriorate if temperatures rise too high, such as milk, coffee, plasma, etc. Furthermore, because of the fine state of subdivision of the liquid, the dried material is obtained in a finely divided state.

In spray drying, it is necessary to atomise and distribute, under controlled conditions, a wide variety of liquids, the properties of which range from those of solutions, emulsions, and dispersions, to slurries and even gels. Most of the atomisers commonly employed in industry are designed for simple liquids, namely inviscid Newtonian liquids. When they are employed for slurries, pastes, and liquids having anomalous properties, there is a great deterioration in performance and, in many cases, they may be rapidly eroded and damaged so as to become quickly useless. Thus, much is to be gained by a study of the various types and designs of atomiser so that a suitable selection can be made for the duty under consideration.

The performance of a spray dryer or reaction system is critically dependent on the drop size produced by the atomiser and the manner in which the gaseous medium mixes with the drops. In this context an atomiser is defined as a device which causes liquid to be disintegrated into drops lying within a specified size range, and which controls their spatial distribution.

### 16.9.2. Atomisers

Atomisers are listed in Table 16.4 according to the three basic forms of energy commonly employed, namely pressure energy, centrifugal energy, and gaseous energy.

TABLE 16.4. *Classification of Atomisers*[29]

| Pressure energy used in pressure atomisers | Centrifugal energy used in rotary atomisers | Gaseous energy used in twin-fluid or blast atomisers |
| --- | --- | --- |
| Fan-spray nozzles $275-960\,kN/m^2$ | Spinning cups $6-30\,m/s$ | External mixing (fluid pressures independent) |
| Impact nozzles Impinging jet nozzles $240-1060\,kN/m^2$ Impact plate nozzles Up to $3000\,kN/m^2$ Deflector nozzles $7000\,kN/m^2$ | Spinning discs Flat discs Saucer-shaped discs Radial-vaned discs Multiple discs $30-180\,m/s$ | Internal mixing (fluid pressures interdependent) Low velocity Gas velocity $30-120\,m/s$ Gas–liquid ratio $2-25\,kg/kg$ |
| Swirl-spray nozzles Hollow cone, full cone $375-70,000\,kN/m^2$ | | Medium velocity Gas velocity $120-300\,m/s$ Gas–liquid ratio $0.2-1\,kg/kg$ |
| Divergent pintle nozzles $240-7000\,kN/m^2$ Fixed or vibrating pintle | | High velocity Gas velocity sonic or above Gas–liquid ratio $0.2-1\,kg/kg$ |

Where greater control is required over disintegration or spatial dispersion, combinations of atomiser types may be employed. For example, swirl-spray nozzles or spinning discs may be incorporated in a blast atomiser, their primary functions being to produce thin liquid sheets which are then eventually atomised by low, medium or high velocity gas streams.

The fundamental principle of disintegrating a liquid consists in increasing its surface area until it becomes unstable and disintegrates. The theoretical energy requirement is the increase in surface energy plus the energy required to overcome viscous forces, though in practice this is only a small fraction of the energy required. The process by which drops are produced from a liquid stream depends upon the nature of the flow in the atomiser, that is whether it is laminar or turbulent, the way in which energy is imparted to the liquid, the physical properties of the liquid, and the properties of the ambient atmosphere. The basic mechanism, however, is unaffected by these variables and consists essentially of the breaking down of unstable threads of liquid into rows of drops and conforms to the classical mechanism postulated by Lord Rayleigh[30]. The theory states that a free column of liquid is unstable if its length is greater than its circumference, and that for a non-viscous liquid the wavelength of that disturbance which will grow most rapidly in amplitude is 4·5 times the diameter; this corresponds to the formation of droplets of diameter approximately $1·89\,d_j$. Weber[31] has shown that the optimum wavelength for the disruption of jets of viscous liquid is:

$$\lambda_{\text{opt}} = \sqrt{2}\pi d_j \left[1 + \frac{3\mu}{\sqrt{\rho\sigma d_j}}\right]^{0·5} \tag{16.29}$$

A uniform thread will break down into a series of drops of uniform diameter, each separated by one or more satellite drops. Because of the heterogeneous character of the atomisation process, however, non-uniform threads are produced and this results in a wide range of drop sizes. An example of a part of a laminar sheet collapsing into a network of threads and drops is shown in Fig. 16.20. Only when the formation and disintegration of threads are controlled can a homogeneous spray cloud be produced. One method by which this can be achieved is by the use of a rotary cup atomiser operating within a critical range of liquid flow rates and rotor speeds, though, as shown later, this range falls outside that normally employed in practice.

Although certain features are unique to particular atomiser types, many of the detailed mechanisms of disintegration are common to most forms of atomiser[32]. The most effective way of utilising energy imparted to a liquid is to arrange that the liquid mass has a large specific surface before it commences to break down into drops. Thus the primary function of an atomiser is to transpose bulk liquid into thin liquid sheets.

Three modes of disintegration of such spray sheets have been established[33], namely "rim", "perforated sheet", and "wave".

Because of surface tension the free edge of any sheet contracts into a thick rim, and "rim" disintegration occurs as it breaks up by instabilities analogous to those of free jets. Figure 16.20a illustrates a fan spray sheet and shows that as the liquid in each edge moves along the curved boundary, the latter becomes disturbed and disintegrates. When this occurs, the resulting drops sustain the direction of flow of the edge at the point at which the drops are formed, and remain attached to the receding surface by thin threads which rapidly disintegrate into streams of drops.

In "perforated sheet" disintegration (Fig. 16.20b), small holes suddenly appear in the

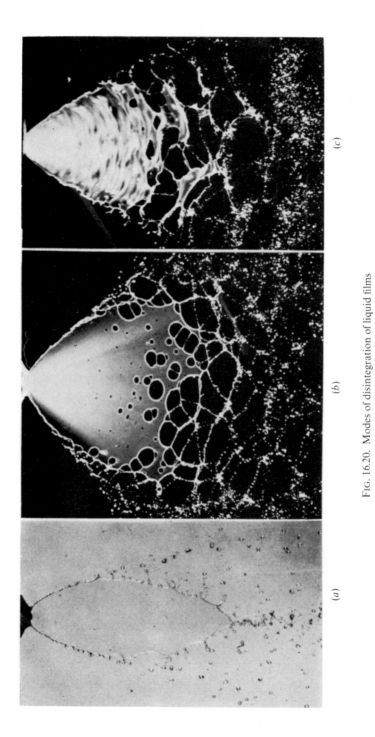

FIG. 16.20. Modes of disintegration of liquid films

sheet as it advances into the atmosphere. They rapidly grow in size[34] until the thickening rims of adjacent holes coalesce to form threads of varying diameter. The threads finally break down into drops.

Disintegration can also occur through the superimposition of a wave motion on the sheet (Fig. 16.20c). Sheets of liquid corresponding to half or full wavelengths of liquid are torn off and tend to draw up under the action of surface tension but may suffer disintegration by air action or liquid turbulence before a regular network of threads can be formed.

### Pressure Atomisers

In a pressure atomiser, liquid is forced under pressure through an orifice, and the form of the resulting liquid sheet can be controlled by varying the direction of flow towards the orifice. By this method, flat and conical spray sheets can be produced.

From the Bernoulli equation, the mass rate of flow through a nozzle may be derived as follows (cf. Volume 1, Chapter 5):

$$G = C_D \rho A_N \sqrt{2(-\Delta P)/\rho} \qquad (16.30)$$

Thus for a given nozzle and fluid, and an approximately constant coefficient of discharge, $C_D$ then

$$G = \text{Const.} \sqrt{(-\Delta P)} \qquad (16.31)$$

The capacity of a nozzle is conveniently described by the *flow number* (**FN**) a dimensional constant defined by the equation

$$\mathbf{FN} \equiv \frac{\text{volumetric flow}}{\sqrt{\text{pressure}}} = \frac{2 \cdot 23 \times 10^6 (\text{m}^3/\text{s})}{\sqrt{(\text{kN/m}^2)}} \qquad (16.32)$$

In the fan-spray drop nozzle (Fig. 16.21) two streams of liquid are made to impinge behind an orifice by specially designed approach passages, and a sheet is formed in a plane perpendicular to the plane of the streams.

In this type of nozzle, the orifice runs full, and since the functional portion is sharp-edged, high discharge coefficients are obtained which are substantially constant over wide ranges of Reynolds number.

The influence of conditions on the droplet size where the spray sheet disintegrates through aerodynamic wave motion can be represented by the following expression[32] for ambient densities around normal atmospheric conditions:

$$d_m = \frac{0 \cdot 000156}{C_D} \left[ \frac{\mathbf{FN} \sigma \rho}{\sin \phi (-\Delta P)} \right]^{1/3} \rho_A^{-1/6} \qquad (16.33)$$

where **FN** is given in equation 16.32, $-\Delta P$ is in kN/m² and all other quantities are expressed in SI units.

The principle of operation of the *impinging jet nozzle* resembles that of the fan spray nozzle with the exception that two or more independent jets are caused to impinge in the atmosphere. In *impact atomisers*, one jet is caused to strike against a solid surface, and for two jets impinging at 180°[32]:

$$d_m = 1 \cdot 73 \frac{d_j^{0 \cdot 75}}{u_l^{0 \cdot 5}} \left[ \frac{\sigma}{\rho} \right]^{0 \cdot 25} \qquad (16.34)$$

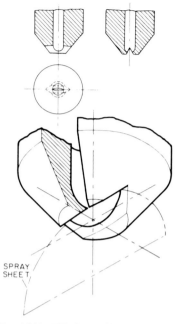

FIG. 16.21. Elliptical-orifice fan-spray nozzle

where all quantities are expressed in SI units. With this atomiser, drop size is effectively independent of viscosity, and the size spectrum is narrower than with other types of pressure nozzle.

When liquid is caused to flow through a narrow divergent annular orifice or around a pintle against a divergent surface on the end of the pintle, a conical sheet of liquid is produced where the liquid is flowing in radial lines. Such a sheet generally disintegrates by an aerodynamic wave motion. The angle of the cone and the root thickness of the sheet can

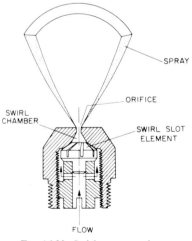

FIG. 16.22. Swirl-spray nozzle

be controlled by the divergence of the spreading surface and the width of the annulus. For small outputs this method is not so favourable because of difficulties in making an accurate annulus.

A conical sheet is also produced when the liquid is caused to emerge from an orifice with a tangential or swirling velocity resulting from its path through one or more tangential or helical passages before the orifice. Figure 16.22 shows a typical nozzle used for a spray dryer. In such swirl-spray nozzles the rotational velocity is sufficiently high to cause the formation of an air core throughout the nozzle, resulting in low discharge coefficients for this type of atomiser.

Several empirical relations have been formulated to express drop size in terms of the operating variables. One suitable for small atomisers with 85° spray cone angles, at atmospheric pressure is[32]:

$$d_m = 0.0134 \frac{\mathbf{FN}^{0.209}(\mu/\rho)^{0.215}}{(-\Delta P)^{0.348}} \tag{16.35}$$

where $-\Delta P$ is in $kN/m^2$, $\mathbf{FN}$ is given in equation 16.32, and all other quantities are in SI units.

Pressure nozzles are somewhat inflexible since large ranges of flowrate require excessive variations in differential pressure. For example, for an atomiser operating satisfactorily at $275\,kN/m^2$, a pressure differential of $17.25\,MN/m^2$ is required to increase the flow rate to ten times its initial value. These limitations inherent in all pressure-type nozzles have been overcome in swirl spray nozzles by the development of *spill, duplex, multi-orifice*, and *variable port atomisers*, in which ratios of maximum to minimum outputs in excess of 50 can be easily achieved[32].

### Rotary Atomisers

In a rotary atomiser, liquid is fed on to a rotating surface and spread out by centrifugal force. Under normal operating conditions the liquid extends from the periphery in the

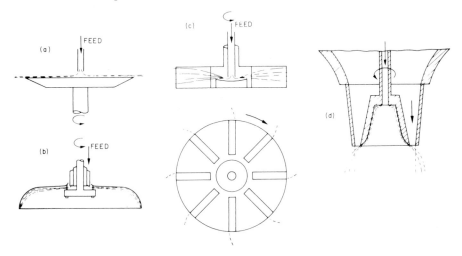

FIG. 16.23. Characteristic rotary atomisers: (a) Sharp-edge flat disc; (b) Bowl; (c) Vaned disc; (d) Air-blast bowl atomiser

form of a thin sheet which breaks down some distance away, either freely by aerodynamic action or by the action of an additional gas blast.

Since the accelerating force can be controlled independently, this type of atomiser is extremely versatile and it can handle successfully a wide range of feed rates with liquids having a wide range of properties. The rotating member may be a simple flat disc though slippage may then occur and consequently it is more usual to use bowls, vaned discs, and slotted wheels (Fig. 16.23). Diameters vary from 25 to 460 mm and small discs rotate at up to 1000 Hz while the larger discs rotate at up to 200 Hz with capacities of 1·5 kg/s. Where a coaxial gas blast is used to effect atomisation, lower speeds of the order of 50 Hz may be used.

At very low flow rates, e.g. 30 mg/s, the liquid spreads out towards the cup lip where it forms a ring. As liquid continues to flow into the ring, its inertia increases, overcomes the restraining surface tension and viscous forces and is centrifuged off as discrete drops of uniform size, which initially remain attached to the rim by a fine attenuating thread. When the drop is finally detached, the thread breaks down into a chain of small satellite drops. Since the satellite drops constitute only a small proportion of the total liquid flowrate, a cup operating under these conditions effectively produces a monodisperse spray (Fig. 16.24).

Under these conditions the drop size from sharp-edged discs is given by[35]:

$$d_m = \frac{0·52}{N}\sqrt{\frac{\sigma}{D\rho}} \qquad (16.36)$$

When the liquid flowrate is increased, the retaining threads grow in thickness and form long jets. As they extend into the atmosphere, these jets are stretched and finally break down into strings of drops (Fig. 16.25). Under more practical ranges of flowrate, the jets are unable to remove all the liquid, the ring is forced away and a thin sheet extends around the cup lip and eventually breaks up into a polydisperse spray (Fig. 16.26).

A far greater supply of energy for disintegration of the liquid jet can be provided by using a high-speed gas stream which impinges on a liquid jet or film. By this means a greater surface area is formed and drops of average size less than 20 μm can be produced; this is appreciably smaller than is possible by the methods previously described, but the

FIG. 16.24. Monodisperse spray from rotating cup

energy requirements are much greater. The range of flow rates which can be used is wide, because the supply of liquid and the energy for atomisation can be controlled independently. Gas velocities ranging from 50 m/s to sonic velocity are common and sometimes the gas is given a vortex motion.

Break-up of the jet occurs as follows: ligaments of liquid are torn off; these collapse to form drops, which may be subsequently blown out into films, which in turn further collapse to give a fine spray. Generally, this spray has a small cone angle and is capable of penetrating far greater distances than the pressure nozzle. Small atomisers of this type have been used in small spray-drying units.

FIG. 16.25. Strings of liquid breaking into drops from rotating cup

Where the gas is impacted on to a liquid jet the mean droplet size is given approximately by[32]:

$$d_s = \frac{0.585}{u_r} \sqrt{\frac{\sigma}{\rho}} + 0.0017 \left[ \frac{\mu}{\sqrt{\sigma\rho}} \right]^{0.45} \left[ \frac{1000}{j} \right]^{1.5} \qquad (16.37)$$

where $d_s$ is the surface mean diameter (m),
  $j$ is the volumetric ratio of gas to liquid rates at the pressure
    of the surrounding atmosphere,
  $u_r$ is the velocity of the gas relative to the liquid (m/s),
  $\mu$ is the liquid viscosity (Ns/m$^2$),
  $\rho$ is the liquid density (kg/m$^3$), and
  $\sigma$ is the surface tension (N/m).

More efficient atomisation is achieved if the liquid is spread out into a film before impact[34].

FIG. 16.26a. Spray sheet from cup rotating at low speed

FIG. 16.26b. Spray sheet from cup rotating at high speed

### 16.9.3. Drying of Drops

The amount of drying a drop undergoes depends upon the rate of evaporation and the contact time, the latter depending upon the velocity of fall and the length of path through the dryer. The terminal velocity and the transfer rate depend upon the flow conditions around the drop. Because of the nature of the flow pattern, the latter also varies with angular position around the drop but no practical design method has incorporated such detail and the drop is always treated as if it evaporates uniformly from all its surface.

There are two main periods of evaporation. When a drop is ejected from an atomiser its initial velocity relative to the surrounding gas is generally high and very high rates of transfer are achieved. However, it is rapidly decelerated to its terminal velocity, and the larger proportion of mass transfer takes place during the free-fall period. Little error is therefore incurred in basing the total evaporation time on this period.

An expression for the evaporation time for a pure liquid drop falling freely in air has been presented by Marshall[36,37]. For drop diameters less than 100 μm this may be simplified to give:

$$t = \frac{\rho\lambda(d_0^2 - d_t^2)}{8k_f\Delta T}$$ (16.38)

Sprays generally contain a wide range of drop sizes, and a stepwise procedure can be used[37] to determine the size spectrum as evaporation proceeds.

When single drops containing solids in suspension or solutions are suspended in hot gas streams it is found that evaporation initially proceeds in accordance with the foregoing equation for pure liquid, but that when solids deposition commences, a crust or solid film is rapidly formed which increases the resistance to transfer. Although this suggests the existence of a falling rate period similar to that found in tray drying, the available limited published data indicate that it has little effect on the total drying time. As a result of crust formation, the dried particles may be in the form of hollow spheres.

### 16.9.4. Industrial Spray Dryers

Spray dryers are used in a variety of industries where a fairly high grade product is to be made in granular form. In the drying chamber the gas and liquid streams are brought into contact, and the efficiency of mixing depends upon the flow patterns induced in the chamber. Rotating disc atomisers are most commonly used. Countercurrent dryers give the highest thermal efficiencies but result in high product temperatures. This limits their use to materials which are not affected by overheating. Co-current dryers suffer from relatively low efficiency but have the advantage of low product temperatures unless back-mixing occurs. In the case of materials which are extremely sensitive to heat great care has to be taken in the design of the chamber to eliminate this. Combustion gases are frequently used directly but, in some cases such as the preparation of food products, indirectly heated air is used; maximum temperatures are then normally limited to lower values than those with direct heating.

Typical flow arrangements in spray drying are illustrated in Fig. 16.27.

*Choice of Atomiser*

The drying time and size of the particles is directly related to the droplet size, and therefore the initial formation of the spray is of great importance. The factors which govern the choice of atomisers for any specific drying application are principally dependent upon the characteristics of the liquid feed and upon the required drying characteristics of the drying chamber.

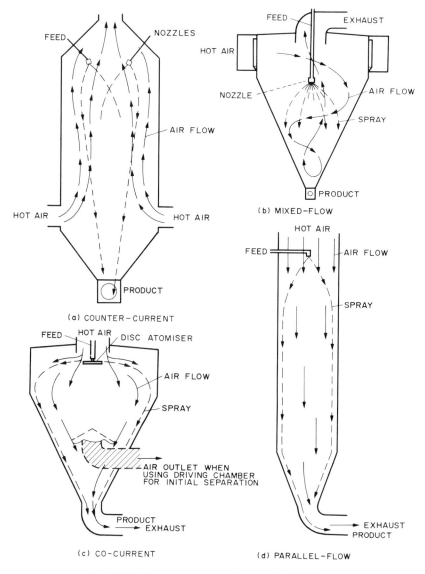

FIG. 16.27. Flow arrangements in spray-drying[21]

A guide to the choice of atomiser is given in Table 16.5.

Pressure nozzles are most suited to low viscosity liquids, and where possible, viscous liquids should be preheated to ensure the minimum viscosity at the nozzle. Because of their simplicity, pressure nozzles are also employed to atomise viscous liquids (up to $0.01 \text{ m}^2/\text{s}$, depending upon the nozzle capacity). Under these conditions, injection pressures of up to $0.14–70 \text{ MN/m}^2$ may be required to produce the required particle size. With slurries, the resulting high liquid velocities may cause severe erosion of the orifice and thus necessitate frequent replacement.

Spinning discs are very suitable for slurries and pastes while high viscosity liquids tend to produce a stringy product. Care must also be taken in the design to minimise incrustation around the lip and subsequent out-of-balance as drying takes place.

The simple gas atomiser is inherently fairly flexible although it has not yet found widespread application. This results from its tendency to produce a dusty product containing a large proportion of very small particles.

TABLE 16.5. *Choice of Atomiser*

| | Atomiser | | | |
|---|---|---|---|---|
| | Pressure | Rotary | Twin-fluid | Spinning-cup + gas blast |
| *Drying chamber* | | | | |
| Co-current | * | * | * | * |
| Countercurrent | * | | * | |
| *Feed* | | | | |
| Low viscosity solutions | * | | * | |
| High viscosity solutions | | | * | * |
| Slurries | | * | * | * |
| Pastes | | * | * | * |
| *Flexibility* | Flowrate $\alpha \sqrt{-\Delta P}$ | Flowrate independent of cup speed | Liquid flow independent of gas energy | Liquid flow independent of cup speed and gas energy |

## Product Recovery

Often no difficulty is experienced in removing the majority of the dried product but in most cases some effort must be taken to reclaim the smaller particles that may be carried over in the exit gases. Cyclones are the simplest form of separator but filter bags or even precipitators may be necessary. With heat-sensitive materials and in cases where sterility is of prime importance, more elaborate methods are required. For example, cooling streams of air may be used to aid the extraction of product while maintaining the required low temperature. Mechanical aids are often incorporated to prevent particles adhering to the chamber walls, and in one design the cooling air also operates a revolving device which sweeps the walls.

In some cases all the product is conveyed from the dryer by the exhaust gases and collected outside the drying chamber. This method is liable to cause breakage of the particles but is particularly suited for heat-sensitive materials which may deteriorate if left in contact with hot surfaces inside the dryer.

## Economics

Spray drying has generally been regarded as a relatively expensive process, especially when indirect heating is used. The data (Table 16.6), given by Grose and Duffield[38] in 1954, and updated to 1978 costs, illustrate the cost penalties associated with indirect heating or with low inlet temperatures in direct heating.

TABLE 16.6. *Operating Costs for a Spray Dryer Evaporating one tonne per hour (0·28 kg/s) of Water*

Basis: Air outlet temperature, 353 K
Steam, £5/Mg
Fuel oil, £75/Mg
Power, £0·027 per unit = £7·5 × 10⁻⁶/kJ

| Air heated | Air inlet temp. (K) | Steam (kg/s) | Oil (kg/s) | Power (kW) | Cost per 100 h (£) | | | Cost per Mg (£) |
|---|---|---|---|---|---|---|---|---|
| | | | | | Fuel | Power | Total | Total |
| Indirectly | 453 | 0·708 | — | 71 | 1275 | 192 | 1467 | 4075 |
| Direct combustion | 523 | — | 0·033 | 55 | 900 | 148 | 1048 | 2910 |
| Direct combustion | 603 | — | 0·031 | 47 | 825 | 127 | 952 | 2650 |

More recently Quinn[39] has drawn attention to the advantages being obtained with larger modern units using higher air-inlet temperatures (quoting 675 K for organic products and 925 K for inorganic products).

In spray dryers, using either a nozzle or rotating disc as the atomiser[39], volumetric evaporative capacities range from 0·0003–0·0014 kg/m³ s for cross and co-current flows with drying temperatures between 420 and 470 K. For handling large volumes of solutions, spray dryers are unsurpassed and it is only at feed rates below 0·1 kg/s that a drum dryer becomes more economic. Indeed the economy of spray drying improves with capacity until, at evaporative capacities of greater than 0·6 kg/s, the unit running cost is largely independent of scale.

In the jet spray dryer the cold feed is introduced[40] into preheated primary air which is blown through a nozzle at velocities up to 400 m/s. Very fine droplets are obtained with residence times of the order of 0·01 s, and with an air temperature of 620 K an evaporative capacity of 0·007 kg/m³ s has been claimed. The device has been used for evaporating milk without adverse effect on flavour and although operating costs are likely to be high the system is well suited to the handling of heat-sensitive materials.

## 16.10. PNEUMATIC DRYERS

In the pneumatic dryer the material is kept in a state of fine division, so that the surface per unit volume is high and a high rate of heat transfer is obtained. The solid is introduced into the dryer by some form of mechanical feeder, such as a rotating star wheel, or by an extrusion machine arranged with a high-speed guillotine to give short lengths of material, say 5 to 10 mm. Hot gases from a furnace, or more frequently from an oil burner, are passed into the bottom of the dryer, and these pick up the particles and carry them up the column. The stream of particles leaves the dryer through a cyclone separator and the hot gases pass out of the system. In some instances, final collection of the fine particles is by way of a series of bag filters. In this unit the time of contact of particles with the gases is small (typically 5 s even with a lengthy duct) and they do not approach the temperature of the hot gas stream. In some cases the material is recycled, especially where bound moisture is involved. Evaporative capacities are high and for normal materials fall in the range 0·003–0·06 kg/m³ s[41] though these are greatly affected by the solids–air ratio. Typical thermal and power requirements are 4·5 MJ/kg moisture evaporated and 0·2 MJ/kg respectively.

A typical installation showing the associated equipment is illustrated in Fig. 16.28. In this unit, wet feed is delivered on to a bed of previously dried material in a double paddle mixer to produce a friable mixture. It then passes to a cage mill where it comes into contact with hot gases from the furnace. Surface and some inherent moisture is immediately flash evaporated. The cage mill breaks up any agglomerates to ensure uniform drying of each individual particle. Gases and product are drawn up the drying column, where inherent moisture continues to be evaporated, before passing into the cyclone collector. Separated solids discharge through a rotary air lock into a dry divider which is set to recycle an adequate percentage of solids for conditioning the new wet feed in the double paddle mixer. Gases from the cyclone are vented to atmosphere through a suitable dust collector or wet scrubber. The system operates under suction and dust is therefore reduced to a minimum. A direct oil or gas fired furnace is generally employed, the heat input being controlled according to the vent stack gas temperature. Indirect heating can be used where contamination of the product is undesirable.

Materials handled include food products, chalk, coal, organic chemicals, clays, spent coffee grounds, sewage sludge and chicken manure. Where exhaust gases contain unpleasant odours "after burners" can be supplied to raise the temperature and burn off the organic and particulate content which cause the smell.

"Convex" dryers are continuously operating pneumatic dryers with an inherent classifying action in the drying chamber that produces residence times for the individual

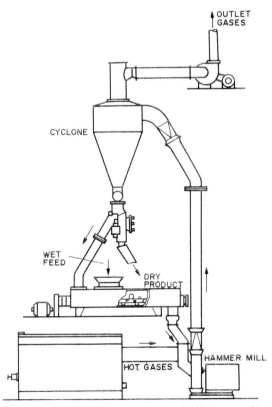

FIG. 16.28. Air-lift dryer with an integral mill (Raymond)

particles that differ according to particle size and moisture content. They offer all of the
processing advantages of short-time, co-current dryers and are used primarily for drying
reasonably to highly free-flowing moist products that can be conveyed pneumatically and
do not tend to stick together. By virtue of the pronounced classifying action, such dryers
are also well suited to the drying of thermally sensitive moist materials with widely
differing particle size—where the large particles have to be completely dried without any
overheating of the small ones. Basically, this form of pneumatic dryer consists of a
truncated cyclone with a bottom outlet that acts as a combined classifier and dryer.

## 16.11. FLUIDISED BED DRYERS

The principles of fluidisation, discussed in Chapter 6 are applied in this type of dryer.
Figure 16.29 shows a typical arrangement. Heated air, or hot gas from a burner, is passed
via a plenum chamber and a diffuser plate into the fluidised bed of material, from which it
passes to a dust separator. The diffuser plate is fitted with suitable nozzles to prevent backflow of solids. Wet material is fed continuously into the bed through a rotary valve, and
mixes immediately with the dry charge. Dry material overflows via a downcomer to an
integral after-cooler. An alternative design of this type of dryer is one in which a thin bed is
used.

Quinn[44], in an excellent account of the process, draws attention to a surprising aspect.
One would expect that it would be impossible to obtain very low product moisture levels
when the incoming feed is very wet. At least, one might think, care would be needed to site
the feed point well away from the discharge point and perhaps to use baffles to segregate
wetter and drier parts of the bed. These expectations are not borne out by operating
experience. Mixing in the bed is so rapid that it may be regarded as homogeneous, and
baffles or physical separation between feed and discharge points are largely without
effect. The very high mass-transfer rates achieved make it possible to maintain the
whole bed in a dry condition.

Many large-scale uses have been reported, for example on fertilisers, plastics materials,
foundry sand, inorganic salts, etc. Agarwal, Davis, and King[42] describe a plant consisting
of two units, each drying 10·5 kg/s of fine coal. Small fluidised-bed dryers are also finding
increasing use, for example in the drying of tablet granulations in the pharmaceutical
industry[43].

Dryers with grid areas of up to 14 m² have been built[44] and evaporative capacities vary
from 0·02 kg/s m² grid area for the low-temperature drying of food grains to 0·3 kg/s m² for
the drying of pulverised coal by direct contact with flue gases[42]. Specific air rates are
usually in the range 0·5–2·0 kg/s m² grid area and the total energy demand is
2·5–7·5 MJ/kg moisture evaporated.

## 16.12. TURBO-SHELF DRYERS

The handling of sticky materials presents many difficulties and one machine which is
useful for this type of material is the turbo dryer, made by the Buell Combustion
Company. As shown in Fig. 16.30, the wet solid is fed in a thin layer to the top member of a
series of annular shelves each made of a number of segmental plates with slots between
them. These shelves rotate and by means of suitably placed arms the material is pushed

through a slot on to a shelf below; after repeated movements, the solid leaves at the bottom of the dryer. The shelves are heated by a row of steam pipes, and in the centre there are three or more fans which suck the hot air over the material and remove it at the top.

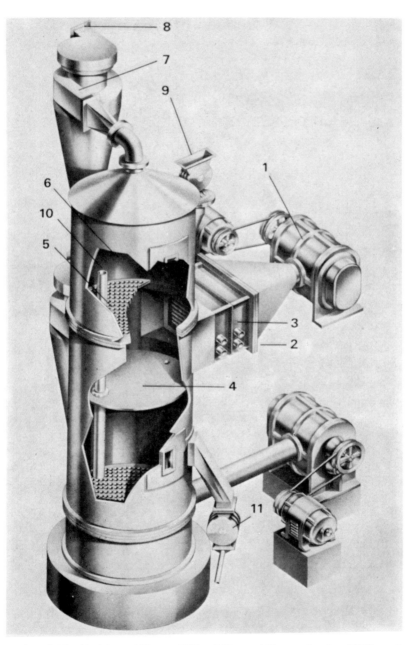

FIG. 16.29. Fluidised bed dryer. 1 Blower. 2 Filter. 3 Heater. 4 Plenum chamber. 5 Diffuser plate. 6 Fluidised bed. 7 Dust separator. 8 Air exhaust. 9 Rotary valve. 10 Downcomer. 11 Discharge rotary valve

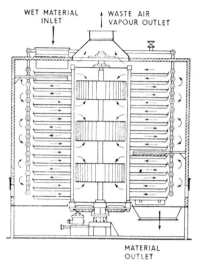

FIG. 16.30. Turbo-shelf dryer

The accelerated drying induced by the raking of the material results in evaporative capacities in the range 0·0002–0·0014 kg/s m² shelf area which are comparable with those obtained by through circulation on perforated belts. Shelf areas vary from 0·7 to 200 m² on a single unit and the dryer can easily be converted to closed-circuit operation either to prevent emission of fumes or where valuable solvents have to be recovered.

## 16.13. DISC DRYERS

A disc dryer provides a further way in which pasty and sticky materials can be handled; in addition, materials which tend to form a hard crust or pass through a rheologically difficult phase during the drying operation, can also be processed. The "Contivac dryer" is a good example (Fig. 16.31) of this class.

In principle, this is a single-agitator disc contact dryer, consisting of a heated cylindrical housing (1) assembled from unit sections, and a heated hollow agitating rotor (2) which has a simultaneous rotating and oscillating movement produced by means of a rotating drive (3) and a reciprocating drive (4).

The drive (3 + 4) and stuffing box (8) are located at the dry product end, and the stuffing box is protected by a reverse acting flight (12). The wet product is introduced at the far end and is drawn in by the screw flight (13) and continuously conveyed through the dryer.

The vapour passes through the vapour filter (15) to the condenser. This vapour filter system has a back scavenging action, and is specially designed for removing dust from vapour in vacuum drying plants.

The hollow agitator (2) is fitted, over its whole length, with heated flights, which are arranged equidistantly in pairs.

Between every two axially neighbouring agitator flights, there are, projecting inwards from the housing, fixed wiping pegs (10) or annular weir/kneading elements (11) which extend inward to the agitator core.

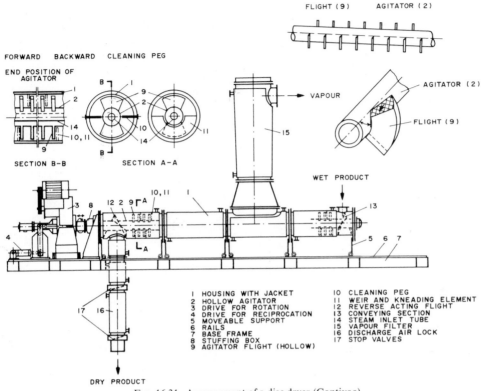

FIG. 16.31. Arrangement of a disc dryer (Contivac)

The self-cleaning of the heating surfaces is achieved by the combined rotating-reciprocating movement of the agitator. The stationary elements (10 and 11) clean the faces of the agitator flights, at the end of each oscillating movement, during one rotation of the agitator. By the forward and backward movement, the edges of the agitator flights (9) clean the inner surface of the housing and the fixed elements projecting inwards to the core of the agitator clean the agitator (2). In all, about 95 per cent of the heating surface is cleaned.

The rotating and reciprocating motions need not be synchronised, because the individual agitator flights (9) oscillate only between two adjacent fixed elements (10 and 11). This is so that the speed of rotation, the frequency of reciprocation and the forward and backward speed of reciprocation can be adjusted independently of one another over a wide range of settings.

The housing sections (1) are supported by frames (5) on rails and can be drawn forwards when cleaning is necessary.

The transport of pasty products through the Contivac dryer is achieved by the differential forward and backward oscillatory speed, combined with the action of the bevelled edges of the agitator flights.

A Contivac dryer with the housing removed is illustrated in Fig. 16.32, in which the agitator and flights can be clearly seen.

Contivac dryers have heating surfaces which range in size from 4 to 60 m$^2$ and volume between 0·1 and 3 m$^3$. They may be operated under vacuum or up to 400 kN/m$^2$

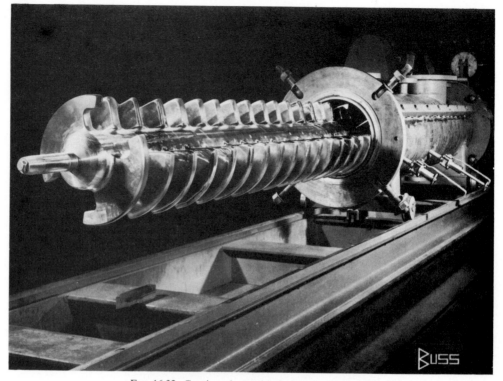

FIG. 16.32. Contivac dryer with the housing removed

with the temperature of heating fluids between 330 K to 670 K. The evaporative capacity ranges from 0·03 to 0·55 kg/s and the agitator speed from 0·1 Hz for rheologically difficult materials to 1 Hz for easier applications.

Operating on a similar principle is the *Buss paddle dryer* which effects batch drying of liquids, pasty and sandy materials. The paddle dryer constitutes the main element of the drying unit. It consists of a horizontal, cylindrical housing which contains a paddle agitator in the form of a hollow shaft carrying agitator arms. The jacket, the hollow shaft and the agitator arms are steam-heated. The paddle agitator is driven by an electric motor and gear unit. A built-on torque bracket with microswitch protects the paddle agitator against overloading. The product to be dried is filled in through the charging nozzle into the paddle dryer, where it is distributed uniformly by the rotating paddle agitator. The drying proceeds under vacuum while intensive intermixing by the agitator causes continual renewal of the product particles in contact with the heated surfaces. This guarantees efficient heat transfer and uniform product quality. The vapours are purged of dust in passing through a vapour filter and are then fed to a condenser. Non-condensable gases are drawn off by the vacuum system. The vapour filter is equipped with a removable filter insert, and to prevent excessive pressure drop across the fabric from thick (and possibly moist) dust build-up on the filter sacks, these are provided with a reverse jet arrangement, i.e. the individual filter sacks are cleaned in turn automatically during operation by short but powerful countercurrent blasts of steam blown through the filter sack. This serves to blow and shake off the dust layer and keep the filter sack dry. After the drying process is complete, the dried material can be cooled by applying cooling water to the dryer jacket and paddle agitator. The dryer is emptied by the arms of the paddle agitator, which are designed to shovel the material in the vessel towards the outlet when rotated backwards. The discharge outlet is specially constructed to prevent the formation of a plug of material. At 373 K, an evaporation rate of over 4 g/s m² can be achieved for a steam consumption of 1·5 kg/kg of evaporated water MJ/kg.

## 16.14. LESS COMMON DRYING METHODS

### 16.14.1. Solvent Drying

Two different processes may be considered under this heading:
*superheated-solvent drying* in which a material containing non-aqueous moisture is dried by contact with superheated vapours of its own moisture, and
*solvent dehydration* in which water-wet substances are exposed to an atmosphere of a saturated organic-solvent vapour.

The first of these has great attractions in applications involving material containing an inflammable liquid such as butanol, for example. Drying is effected with a gas with an air–moisture ratio of 90 kg/kg moisture in order to ensure that the composition is well below the lower explosive limit; the heat requirement is as great as that for superheated solvent drying at the same gas outlet temperature of 400 K[45]. Superheated solvent drying in a fluidised bed has been used for the drying of polypropylene pellets to eliminate the need for water washing and for fractionating; a flowsheet of the installation is shown in Fig. 16.33[46].

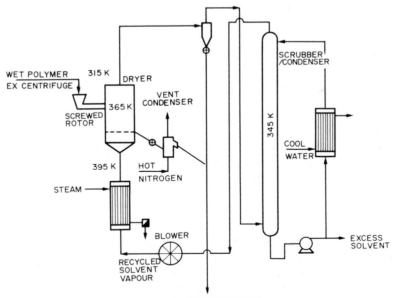

Fig. 16.33. Flowsheet for drying polypropylene by fluidising with a hot stream of solvent vapour[46]

The most important applications of solvent dehydration lie within the field of kiln drying methods for seasoning timber where substantial reductions in drying times have been achieved[47].

### 16.14.2. Superheated Steam Drying

The replacement of air by superheated steam to take up evaporating moisture is attractive in the drying of foodstuffs, in that the steam is completely clean and there is consequently less oxidation damage; furthermore, in the seasoning of timber drying times can be reduced significantly. Although the principle has been understood for many years, applications have been limited due to corrosion problems and the lack of suitable equipment[23]. The flowsheet of a batch dryer working on this principle is shown in Fig. 16.34. The dryer is initially filled with air, which is circulated with a blower together with evaporated moisture, and any excess pressure is vented to atmosphere so that the air is gradually replaced by water vapour. For an evaporation rate of $10 \, kg/m^3$ dryer, the chamber would contain 90 per cent of steam in about 600 s. Superheated steam drying is also achieved when a wet material is heated in a sealed autoclave with periodic release of the generated steam[48]. This pressure release causes flash evaporation of moisture throughout the material thus avoiding drying stresses and severe moisture gradients. The process has been applied to the drying of thermal insulating materials.

### 16.14.3. Freeze Drying

In this process, the material is first frozen and then dried by sublimation in a very high vacuum of the order of $10–40 \, N/m^2$, at a temperature in the range of 240–260 K. During the sublimation of the ice, a dry surface layer is left; this is not free to move because it has been frozen, and a honeycomb structure is formed. With normal vacuum drying of biological solutions containing dissolved salts, a high local salt concentration is formed at the skin, but with freeze drying this does not occur because of the freezing of the solid. Thus, freeze drying is useful not only as a method of working at low temperatures, but also as a method of avoiding surface hardening.

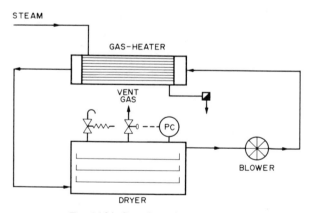

FIG. 16.34. Superheated-steam dryer

As indicated by Chambers[49], heat has to be supplied to the material to provide the heat of sublimation. The rate of supply of heat should be such that the highest possible water vapour pressure is obtained at the ice surface without danger of melting the material. During this stage, well over 95 per cent of the water present should be removed, and to complete the drying, the material should be allowed to rise in temperature, to say room temperature. The great attraction of this technique is that it does not harm heat-sensitive materials, and it has been widely used for the drying of penicillin, and other biological materials. Initially, high costs restricted the application of the process severely, but economic advances have reduced these considerably, and foodstuffs are now freeze-dried on a large scale. Maguire[50] suggests a total cost of £0·04/kg of water evaporated, in a plant handling 0·5–0·75 kg/s of meat.

A typical layout of a freeze-drying installation is shown in Fig. 16.35. Heat is

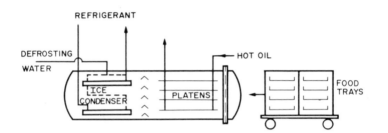

FIG. 16.35. Flowsheet of freeze-drying plant.

supplied either by conduction or by radiation from hot platens which interleave with trays containing the product and the sublimed moisture condenses on to a refrigerated coil at the far end of the drying chamber. The use of dielectric heating has been investigated[51] though uneven loading of the trays can lead to scorching, and ionisation of the residual gases in the dryer results in browning of the food.

Continuous freeze-drying equipment is being developed[52] and chopped meat and vegetables can be dried in a rotating steam-jacketed tube enclosed in the vacuum chamber. A model of the freeze-drying process has been presented by Dyer and Sunderland[53].

## 16.15. THE DRYING OF GASES

The drying of gases is carried out on a very large scale, the most important applications of the process being as follows:

(a) Drying gas in order to reduce its tendency to cause corrosion.
(b) Preparation of dry gas for use in a chemical reaction.
(c) Reduction of the humidity of air in air-conditioning plants.

The problems involved in the drying or dehumidification of gases have already been referred to in Chapter 11 of Volume 1. The most important methods available will now be summarised.

*Cooling.* A gas stream may be dehumidified by bringing it into contact with a cold liquid or a cold solid surface. If the temperature of the surface is lower than the dew point of the gas, condensation will take place, and the temperature of the surface will tend to rise by virtue of the liberation of latent heat; it is therefore necessary constantly to remove heat from the surface. Because a far larger interfacial surface can be produced with a liquid, it is usual to spray a liquid into the gas and then to cool it again before it is recycled. In many cases, countercurrent flow of the gas and liquid is obtained by introducing the liquid at the top of a column and allowing the gas to pass upwards.

*Compression.* The humidity of a gas may be reduced by compressing it, cooling it down to near its original temperature, and then draining off the water which has condensed. During compression, the partial pressure of the water vapour increases and condensation occurs as soon as the saturation value is exceeded.

*Liquid absorbents.* If the partial pressure of the water in the gas is greater than the equilibrium partial pressure at the surface of a liquid, condensation will take place as a result of contact between the gas and liquid. Thus, water vapour is frequently removed from a gas by bringing it into contact with concentrated sulphuric acid, phosphoric acid, or glycerol. Concentrated solutions of salts, such as calcium chloride, are also effective. The process may be carried out either in a packed column or in a spray chamber. Regeneration of the liquid is an essential part of the process, and this is usually effected by evaporation.

*Solid adsorbents and absorbents.* The use of silica gel or solid calcium chloride to remove water vapour from gases is a common operation in the laboratory. Moderately large units can be made, but the volume of packed space required is generally large because of the comparatively small transfer surface per unit volume. If the particle size is too small, the pressure drop through the material becomes excessive. The solid desiccants are also regenerated by heating.

Gas is frequently dried by using a calcium chloride liquor containing about 0·56 kg of calcium chloride per kg solution. The extent of recirculation of the liquor from the base of the tower is governed by heating effects, since the condensation of the water vapour gives rise to considerable heating. It is necessary to fit heat exchangers to cool the liquor leaving the base of the tower.

In the contact plant for the manufacture of sulphuric acid, sulphuric acid is itself used for drying the air for the oxidation of the sulphur. When drying hydrocarbons such as benzene, it is sometimes convenient to pass the material through a bed of solid caustic soda, but if the quantity is appreciable this method will be expensive.

The great advantage of the materials such as silica gel and activated alumina is that they enable the gas to be almost completely dried. Thus, with silica gel, air may be dried down to a dew point of 203 K. Small silica gel containers are frequently used to prevent moisture condensation in the low pressure lines of pneumatic control installations.

## 16.16. FURTHER READING

BACKHURST, J. R. and HARKER, J. H.: *Process Plant Design* (Heinmann, London, 1973).
BACKHURST, J. R., HARKER, J. H., and PORTER, J. E.: *Problems in Heat and Mass Transfer* (Arnold, London, 1974).
BARCLAY, S. F.: *A Study of Drying* (Inst. Fuel, London, 1957).
BRIDGMAN, M. J.: *Drying—Principles and Practice in the Process Industries* (Caxton, Christchurch, 1966).

Brown, W. H.: *An Introduction to the Seasoning of Timber* (Pergamon Press, Oxford, 1965).
Keey, R. B.: *Drying Principles and Practice* (Pergamon Press, Oxford, 1972).
Keey, R. B.: *Introduction to Industrial Drying Operations* (Pergamon Press, Oxford, 1978).
Luikov, A. V.: *Heat and Mass Transfer in Capillary-porous Bodies* (Pergamon Press, Oxford, 1966).
Masters, K.: *Spray Drying* (Godwin, London, 1976).
Nonhebel, G.: *Gas Purification Processes* (Newnes, London, 1964).
Nonhebel, G. and Moss, A. A. H.: *Drying of Solids in the Chemical Industry* (Butterworths, London, 1971).
Thijssen, H. A. C. and Rulkens, W. H.: *Recent Developments in Freeze Drying* (Int. Inst. Refrigeration, Lausanne, 1969).
Vaněček, J., Markvart, M. and Drbohlav, R.: *Fluidized Bed Drying* (Leonard Hill, London, 1966).
Williams-Gardner, A.: *Industrial Drying* (Godwin, London, 1971).

# 16.17. REFERENCES

1. Sherwood, T. K.: *Trans. Am. Inst. Chem. Eng.* **32** (1936) 150. The air drying of solids.
2. Pearse, J. F., Oliver, T. R., and Newitt, D. M.: *Trans. Inst. Chem. Eng.* **27** (1949) 1. The mechanism of the drying of solids. Part I. The forces giving rise to movement of water in granular beds, during drying.
3. Oliver, T. R. and Newitt, D. M.: *Trans. Inst. Chem. Eng.* **27** (1949) 9. The mechanism of drying of solids. Part II. The measurement of suction potentials and moisture distribution in drying granular solids.
4. Newitt, D. M. and Coleman, M.: *Trans. Inst. Chem. Eng.* **30** (1952) 28. The mechanism of drying of solids. Part III. The drying characteristics of China clay.
5. Corben, R. W. and Newitt, D. M.: *Trans. Inst. Chem. Eng.* **33** (1955) 52. The mechanism of drying of solids. Part VI. The drying characteristics of porous granular material.
6. King, A. R. and Newitt, D. M.: *Trans. Inst. Chem. Eng.* **33** (1955) 64. The mechanism of drying of solids. Part VII. Drying with heat transfer by conduction.
7. Corben, R. W.: *University of London*, Ph.D. Thesis (1955). The mechanism of drying of solids. Part IV. A study of the effect of granulation on the rate of drying of a porous solid.
8. Gilliland, E. R.: *Ind. Eng. Chem.* **30** (1938) 506. Fundamentals of drying and air conditioning.
9. Powell, R. W. and Griffiths, E.: *Trans. Inst. Chem. Eng.* **13** (1935) 175. The evaporation of water from plane and cylindrical surfaces.
10. Chakravorty, K. R.: *J. Imp. Coll. Chem. Eng. Soc.* **3** (1947) 46. Evaporation from free surfaces.
11. Shepherd, C. B., Hadlock, C., and Brewer, R. C.: *Ind. Eng. Chem.* **30** (1938) 388. Drying materials in trays. Evaporation of surface moisture.
12. Sherwood, T. K.: *Ind. Eng. Chem.* **21** (1929) 12, 976. The drying of solids I, II.
13. Newman, A. B.: *Trans. Am. Inst. Chem. Eng.* **27** (1931) 203. The drying of porous solids: diffusion and surface emission equations.
14. Ceaglske, N. H. and Hougen, O. A.: *Trans. Am. Inst. Chem. Eng.* **33** (1937) 283. The drying of granular solids. *Ind. Eng. Chem.* **29** (1937) 805. Drying granular solids.
15. Slichter, C. S.: *U.S. Geol. Survey, 19th Annual Report* (1897–8) Part 2, 301. Theoretical investigation of the motion of ground waters.
16. Haines, W. B.: *J. Agric. Science* **17** (1927) 264. Studies in the physical properties of soils. IV. A further contribution to the theory of capillary phenomena in soil.
17. Miller, C. O., Smith, B. A., and Schuette, W. H.: *Trans. Am. Inst. Chem. Eng.* **38** (1942) 841. Factors influencing the performance of rotary dryers.
18. Cronshaw, H. B.: *Modern Drying Machinery* (Benn, London, 1926).
19. Parker, N. H.: *Chem. Engg., Albany* **70**, No. 13 (1963) 115. Aids to dryer selection.
20. Kröll, K.: *Trockner, einteilen, ordnen, benennen, benumnern, Schilde Schriftenreihe* **6** (Schilde Bad-Hersfeld, 1965).
21. Sloan, C. E.: *Chem. Engg., Albany* **74**, No. 14 (1967) 169. Drying systems and equipment.
22. Backhurst, J. R. and Harker, J. H.: *Process Plant Design* (Heinemann, 1973).
23. Keey, R. B.: *Drying Principles and Practice* (Pergamon Press, Oxford, 1972).
24. Erisman, J. L.: *Ind. Eng. Chem.* **30** (1938) 996. Roto-louvre dryer.
25. Friedman, S. J. and Marshall, W. R.: *Chem. Eng. Prog.* **45** (1949) 482, 573. Studies in rotary drying.
26. Prutton, C. F., Miller, C. O., and Schuette, W. H.: *Trans. Am. Inst. Chem. Eng.* **38** (1942) 123, 251. Factors influencing the performance of rotary dryers.
27. Saeman, W. C.: *Chem. Eng. Prog.* **58**, No. 6 (June 1962) 49–56. Air–solids interaction in rotary dryers and coolers.
28. Sharples, K., Gliken, P. G. and Warne, R.: *Trans. Inst. Chem. Eng.* **42** (1964) T275. Complete simulation of rotary dryers.

29. Fraser, R. P., Eisenklam, P., and Dombrowski, N.: *Brit. Chem. Eng.* **2** (1957) 414–17, 496–501, 536–43, and 610–13. Liquid atomisation in chemical engineering.

30. Rayleigh, Lord: *Proc. Lond. Math. Soc.* **10** (1878–9) 4–13. On the instability of jets.

31. Weber, C.: *Z. angew. Math. Mech.* **11** (1931) 136–54. Zum Zerfall eines Flüssigkeitsstrahles.

32. Dombrowski, N. and Munday, G.: in *Biochemical and Biological Engineering Science*, Vol. 2, ch. 16 Spray drying (Academic Press, 1967).

33. Dombrowski, N. and Fraser, R. P.: *Phil. Trans.* **247A** (1954) 101. A photographic investigation into the disintegration of liquid sheets.

34. Fraser, R. P., Eisenklam, P., Dombrowski, N., and Hasson, D.: *A.I.Ch.E.Jl.* **8** (1962) 672–80. Drop formation from rapidly moving liquid sheets.

35. Walton, W. H. and Prewett, W. C.: *Proc. Phys. Soc.* **628** (1949) 341. The production of sprays and mists of uniform drop size by means of spinning disc type sprayers.

36. Marshall, W. R.: *Chem. Eng. Prog.* Monogr. Ser. No. 2, **50** (1954). Atomization and spray drying.

37. Marshall, W. R.: *Trans. Am. Soc. Eng.* **77** (1955) 1377. Heat and mass transfer in spray drying.

38. Grose, J. W. and Duffield, G. H.: *Chem. and Ind.* (1954) 1464. Chemical engineering methods in the food industry.

39. Quinn, J. J.: *Ind. Eng. Chem.* **57**, No. 1 (Jan. 1965) 35–37. The economics of spray-drying.

40. Bradford, P. and Briggs, S. W.: *Chem. Eng. Prog.* **59**, No. 3 (1963) 76. Jet spray drying equipment for the food industry.

41. Quinn, M. F.: *Ind. Eng. Chem.* **55**, No. 7 (July, 1963) 18–24. Fluidized bed dryers.

42. Agarwal, J. C., Davis, W. L., and King, D. T.: *Chem. Eng. Prog.* **58**, No. 11 (Nov. 1962) 85–90. Fluidized-bed coal dryer.

43. Scott, M. W., Lieberman, H. A., Rankell, A. S., Chow, F. S., and Johnston, G. W.: *J. Pharm. Sci.* **52**, No. 3 (Mar. 1963) 284–91. Drying as a unit operation in the pharmaceutical industry. I. Drying of tablet granulations in fluidized beds.

44. Vaněček, V., Markvart, M. and Drbohlav, R.: *Fluid Bed Drying* (Leonard Hill, 1966).

45. Chu Ju Chin, Lane, A. M. and Conker, D.: *Ind. Eng. Chem.* **45** (1953) 1586 Evaporation of liquids into their superheated vapours.

46. Basel, L. and Gray, E.: *Chem. Eng. Prog.* **58** (1962) 67. Superheated solvent-drying in a fluidised bed.

47. Ellwood, E. L., Gottstein, J. W., and Kauman, W. G.: A Laboratory Study of the Vapour Drying Process. CSIRO Forest Prod. Div., Paper 14 (1961) 111.

48. Yankelev, L. F. in Luikov, A. V.: *Heat and Mass Transfer in Capillary-porous Bodies* (Pergamon Press, Oxford 1966).

49. Chambers, H. H.: *Trans. Inst. Chem. Eng.* **27** (1949) 19. Vacuum freeze drying.

50. Maguire, J. F.: *Food Eng.* **34**, No. 8 (Aug. 1962) 54–5 and **34**, No. 9 (Sept. 1962) 48–52. Freeze drying moves ahead in U.S. (in two parts).

51. Harper, J. C., Chichester, C. O., and Roberts, T. E.: *Agric. Eng.* **43** (1962) 78, 90. Freeze-drying of foods.

52. Maister, H. G., Heger, E. N., and Bogard, W. M.: *Ind. Eng. Chem.* **50** (1958) 623. Continuous freeze-drying of *Serratia marcescens*.

53. Dyer, D. F. and Sunderland, J. E.: *Chem. Eng. Sci.* **23** (1968) 965. The role of convection in drying.

# 16.18. NOMENCLATURE

| | | Units in SI System | Dimensions in **MLT**$\theta$ |
|---|---|---|---|
| $A$ | Area for heat transfer or evaporation | $m^2$ | $L^2$ |
| $A_N$ | Area of nozzle or jet normal to direction of flow | $m^2$ | $L^2$ |
| $a$ | Surface area per unit volume | $m^2/m^3$ | $L^{-1}$ |
| $B$ | Width of surface | m | $L$ |
| $C$ | Coefficient | $kg^{0.2}/m^{0.4} s^{0.2}$ | $M^{0.2}L^{-0.4}T^{-0.2}$ |
| $C_D$ | Coefficient of discharge | — | — |
| $D$ | Diameter of drum or disc | m | $L$ |
| $D_L$ | Diffusion coefficient | $m^2/s$ | $L^2T^{-1}$ |
| $d_j$ | Nozzle or jet diameter | m | $L$ |
| $d_m$ | Main drop diameter | m | $L$ |
| $d_s$ | Surface-mean diameter of drop | m | $L$ |
| $d_t$ | Diameter of evaporating drop at time $t$ | m | $L$ |
| $d_0$ | Initial diameter of drop | m | $L$ |

|  |  | Units in SI System | Dimensions in $\mathbf{MLT}\theta$ |
|---|---|---|---|
| $F$ | Volumetric rate of feed per unit cross-section | $m^3/s\,m^2$ | $\mathbf{LT}^{-1}$ |
| $f$ | Free moisture content | kg | $\mathbf{M}$ |
| $f_c$ | Free moisture content at critical condition | kg | $\mathbf{M}$ |
| $f_1$ | Initial free moisture content | kg | $\mathbf{M}$ |
| $G$ | Mass flow in jet or nozzle | kg/s | $\mathbf{MT}^{-1}$ |
| $G'$ | Mass rate of flow of air per unit cross-section | $kg/m^2\,s$ | $\mathbf{ML}^{-2}\mathbf{T}^{-1}$ |
| $g$ | Acceleration due to gravity | $m/s^2$ | $\mathbf{LT}^{-2}$ |
| $\mathcal{H}$ | Humidity | kg/kg | — |
| $\mathcal{H}_s$ | Humidity of saturated air at surface temperature | kg/kg | — |
| $\mathcal{H}_w$ | Humidity of saturated air at temperature $\theta_w$ | kg/kg | — |
| $\mathcal{H}_0$ | Humidity of saturated air | kg/kg | — |
| $h$ | Heat transfer coefficient | $W/m^2K$ | $\mathbf{MT}^{-3}\theta^{-1}$ |
| $h_D$ | Mass transfer coefficient | m/s | $\mathbf{LT}^{-1}$ |
| $h_f$ | Friction head over a distance $z_1$ from surface | m | $\mathbf{L}$ |
| $h_s$ | Suction potential immediately below meniscus | m | $\mathbf{L}$ |
| $h_t$ | Theoretical suction potential of pore or waist | m | $\mathbf{L}$ |
| $h_1$ | Suction potential at distance $z_1$ below meniscus | m | $\mathbf{L}$ |
| $j$ | Volumetric gas/liquid ratio | — | — |
| $K$ | Coefficient | $m^2\,s/kg$ | $\mathbf{M}^{-1}\mathbf{L}^2\mathbf{T}$ |
| $K'$ | Coefficient | $s^{1.8}/m^{1.5}$ | $\mathbf{L}^{-1.8}\mathbf{T}^{1.8}$ |
| $K''$ | Transfer coefficient $(h_D\rho_A)$ | $kg/m^2\,s$ | $\mathbf{ML}^{-2}\mathbf{T}^{-1}$ |
| $K_G$ | Diffusion coefficient | s/m | $\mathbf{L}^{-1}\mathbf{T}$ |
| $k_f$ | Thermal conductivity of gas film at interface | W/mK | $\mathbf{MLT}^{-3}\theta^{-1}$ |
| $L$ | Length of surface | m | $\mathbf{L}$ |
| $l$ | Half thickness of slab | m | $\mathbf{L}$ |
| $M_A$ | Molecular weight of air | kg/kmol | — |
| $M_w$ | Molecular weight of water | kg/kmol | — |
| $m$ | Ratio of rate of drying per unit area to moisture content | $1/m^2\,s$ | $\mathbf{L}^{-2}\mathbf{T}^{-1}$ |
| $N$ | Revolutions per unit time | Hz | $\mathbf{T}^{-1}$ |
| $n$ | Index | — | — |
| $n_f$ | Number of flights | — | — |
| $P$ | Total pressure | $N/m^2$ | $\mathbf{ML}^{-1}\mathbf{T}^{-2}$ |
| $P_s$ | Vapour pressure of water at surface of material | $N/m^2$ | $\mathbf{ML}^{-1}\mathbf{T}^{-2}$ |
| $P_w$ | Partial pressure of water vapour | $N/m^2$ | $\mathbf{ML}^{-1}\mathbf{T}^{-2}$ |
| $P_{w0}$ | Partial pressure at surface of material at wet bulb temperature | $N/m^2$ | $\mathbf{ML}^{-1}\mathbf{T}^{-2}$ |
| $-\Delta P$ | Pressure drop across nozzle | $N/m^2$ | $\mathbf{ML}^{-1}\mathbf{T}^{-2}$ |
| $Q$ | Rate of heat transfer | W | $\mathbf{ML}^2\mathbf{T}^{-3}$ |
| $R_c$ | Rate of drying per unit area for constant rate period | $kg/m^2\,s$ | $\mathbf{ML}^{-2}\mathbf{T}^{-1}$ |
| $\mathbf{R}$ | Universal gas constant | J/kmol K | $\mathbf{L}^2\mathbf{T}^{-2}\theta^{-1}$ |
| $r$ | Radius of sphere | m | $\mathbf{L}$ |
| $r'$ | Radius of capillary | m | $\mathbf{L}$ |
| $S$ | Slope of drum | — | — |
| $T$ | Absolute temperature | K | $\theta$ |
| $\Delta T$ | Temperature difference | K | $\theta$ |
| $t$ | Time | s | $\mathbf{T}$ |
| $t_c$ | Time of constant rate period of drying | s | $\mathbf{T}$ |
| $t_f$ | Time of drying in falling rate period | s | $\mathbf{T}$ |
| $U$ | Overall heat transfer coefficient | $W/m^2K$ | $\mathbf{MT}^{-3}\theta^{-1}$ |
| $u$ | Gas velocity | m/s | $\mathbf{LT}^{-1}$ |
| $u_l$ | Liquid velocity in jet or spray | m/s | $\mathbf{LT}^{-1}$ |
| $u_r$ | Velocity of gas relative to liquid | m/s | $\mathbf{LT}^{-1}$ |
| $V$ | Volume | $m^3$ | $\mathbf{L}^3$ |
| $W$ | Mass rate of evaporation | kg/s | $\mathbf{MT}^{-1}$ |
| $w$ | Total moisture | kg | $\mathbf{M}$ |
| $w_c$ | Critical moisture content | kg | $\mathbf{M}$ |
| $w_e$ | Equilibrium moisture content | kg | $\mathbf{M}$ |
| $w_1$ | Initial moisture content | kg | $\mathbf{M}$ |
| $X$ | Hold-up of drum | — | — |
| $X_a$ | Hold-up of drum with air flow | — | — |
| $x$ | Factor depending on type of packing | — | — |

|  |  | Units in SI System | Dimensions in $\mathbf{MLT\theta}$ |
|---|---|---|---|
| $y$ | Distance in Direction of diffusion | m | $\mathbf{L}$ |
| $z_1$ | Distance below meniscus | m | $\mathbf{L}$ |
| $\alpha$ | Angle of contact | — | — |
| $\bar{\kappa}$ | Coefficient | $\text{kg}^{1-n}\,\text{m}^{2n}\,\text{s}^{n-3}/\text{K}$ | $\mathbf{M}^{1-n}\mathbf{L}^{2n}\mathbf{T}^{n-3}\mathbf{\theta}^{-1}$ |
| $\theta$ | Gas temperature | K | $\mathbf{\theta}$ |
| $\theta_s$ | Surface temperature | K | $\mathbf{\theta}$ |
| $\theta_w$ | Wet bulb temperature | K | $\mathbf{\theta}$ |
| $\lambda$ | Latent heat vaporisation per unit mass | J/kg | $\mathbf{L}^2\mathbf{T}^{-2}$ |
| $\lambda_{\text{opt}}$ | Optimum wavelength for jet disruption | m | $\mathbf{L}$ |
| $\mu$ | Absolute viscosity of liquid | N s/m$^2$ | $\mathbf{ML}^{-1}\mathbf{T}^{-1}$ |
| $\rho$ | Density of liquid | kg/m$^3$ | $\mathbf{ML}^{-3}$ |
| $\rho_A$ | Density of air at its mean partial pressure | kg/m$^3$ | $\mathbf{ML}^{-3}$ |
| $\sigma$ | Surface tension | J/m$^2$ | $\mathbf{MT}^{-2}$ |
| $\phi$ | Half-angle of spray cone or sheet | — | — |
| $\mathbf{FN}$ | Flow number of nozzle (see equation 16.32) | — | — |

# Problems

(A number of these questions are taken from examination papers)

**1.1.** The size analysis of a powdered material on a weight basis is represented by a straight line from 0 per cent weight at 1 μm particle size to 100 per cent weight at 101 μm particle size. Calculate the surface mean diameter of the particles constituting the system.

**1.2.** The equations giving the number distribution curve for a powdered material are $dn/dd = d$ for the size range 0–10 μm, and $dn/dd = 100,000/d^4$ for the size range 10–100 μm. Sketch the number, surface and weight distribution curves. Calculate the surface mean diameter for the powder.

Explain briefly how the data for the construction of these curves would be obtained experimentally.

**1.3.** The fineness characteristic of a powder on a cumulative basis is represented by a straight line from the origin to 100 per cent undersize at particle size 50 μm. If the powder is initially dispersed uniformly in a column of liquid, calculate the proportion by weight which remains in suspension at a time interval from commencement of settling such that a 40 μm particle would fall the total height of the column.

**1.4.** In a mixture of quartz, specific gravity 2·65, and galena, specific gravity 7·5, the sizes of the particles range from 0·0052 to 0·025 mm.

On separation in a hydraulic classifier under free settling conditions three fractions are obtained, one consisting of quartz only, one a mixture of quartz and galena, and one of galena only. What are the ranges of sizes of particles of the two substances in the mixed portion?

**1.5.** It is desired to separate into two pure fractions a mixture of quartz and galena of a size range from 0·015 mm to 0·065 mm by the use of a hindered settling process. What is the minimum apparent density of the fluid that will give this separation? How will the viscosity of the bed affect the minimum required density?

Specific gravity of galena = 7·5. Specific gravity of quartz = 2·65.

**1.6.** Write a short essay explaining the circumstances in which a particle size distribution would be determined by microscopical measurement or by sedimentation in a liquid. State the characteristics of these two methods of measurement.

The following table gives the size distribution of a dust as measured by the microscope. Convert these figures to obtain the distribution on a weight basis, and calculate the specific surface, assuming spherical particles of specific gravity 2·65.

**1.7.** The performance of a solids mixer has been assessed by calculating the variance occurring in the weight fraction of a component amongst a selection of samples withdrawn from the mixture. The quality was tested at intervals of 305 and the results are:

| sample variance | 0·025 | 0·006 | 0·015 | 0·018 | 0·019 |
|---|---|---|---|---|---|
| mixing time (s) | 30 | 60 | 90 | 120 | 150 |

If the component analyzed is estimated to represent 20% of the mixture by weight and each of the samples removed contained approximately 100 particles, comment on the quality of the mixture produced and present the data in graphical form showing the variation of mixing index with time.

**2.1.** A material is crushed in a Blake jaw crusher and the average size of particle reduced from 50 mm to 10 mm, with consumption of energy at the rate of 13·0 kW/(kg/s). What will be the consumption of energy needed to crush the same material of average size 75 mm to an average size of 25 mm,

(a) assuming Rittinger's Law applies, and
(b) assuming Kick's Law applies?

Which of these results would you regard as being more reliable and why?

763

**2.2.** A crusher was used to crush a material whose compressive strength was $22.5 \, MN/m^2$. The size of the feed was *minus* 50 mm, *plus* 40 mm and the power required was $13.0 \, kW/(kg/s)$. The screen analysis of the product was as follows:

| Size of aperture (mm) | Per cent of product |
|---|---|
| through 6·0 | all |
| on 4·0 | 26 |
| on 2·0 | 18 |
| on 0·75 | 23 |
| on 0·50 | 8 |
| on 0·25 | 17 |
| on 0·125 | 3 |
| through 0·125 | 5 |

What would be the power required to crush 1 kg/s of a material of compressive strength $45 \, MN/m^2$ from a feed *minus* 45 mm, *plus* 40 mm to a product of average size 0·50 mm?

**2.3.** A crusher, in reducing limestone of crushing strength $70 \, MN/m^2$ from 6 mm diameter average size to 0·1 mm diameter average size, requires 9 kW. The same machine is used to crush dolomite at the same rate of output from 6 mm diameter average size to a product which consists of 20 per cent with an average diameter of 0·25 mm, 60 per cent with an average diameter of 0·125 mm the balance having an average diameter of 0·085 mm. Estimate the power required to drive the crusher, assuming that the crushing strength of the dolomite is $100 \, MN/m^2$ and that crushing follows Rittinger's Law.

**2.4.** If crushing rolls 1 m diameter are set so that the crushing surfaces are 12·5 mm apart and the angle of nip is $31°$, what is the maximum size of particle which should be fed to the rolls?

If the actual capacity of the machine is 12 per cent of the theoretical, calculate the throughput in kg/s when running at 2 Hz if the working face of the rolls is 0·4 m long and the feed weighs $2500 \, kg/m^3$.

**2.5.** A crushing mill reduces limestone from a mean particle size of 45 mm to a product,

| size (mm) | *per cent* |
|---|---|
| 12·5 | 0·5 |
| 7·5 | 7·5 |
| 5·0 | 45·0 |
| 2·5 | 19·0 |
| 1·5 | 16·0 |
| 0·75 | 8·0 |
| 0·40 | 3·0 |
| 0·20 | 1·0 |

and in so doing requires 6 W/kg of material crushed.

Calculate the power required to crush the same material at the same rate, from a feed having a mean size of 25 mm to a product with a mean size of 1 mm.

**2.6.** A ball-mill 1·2 m in diameter is being run at 0·8 Hz; it is found that the mill is not working satisfactorily. Would you suggest any modification in the condition of operation?

**2.7.** 3 kW has to be supplied to a machine crushing material at the rate of 0·3 kg/s from 12·5 mm cubes to a product having the following sizes:

| | |
|---|---|
| 80% | 3·175 mm |
| 10% | 2·5 mm |
| 10% | 2·25 mm |

What would be the power which would have to be supplied to this machine to crush 0·3 kg/s of the same material from 7·5 mm cube to 2·0 mm cube?

| Size range in µm | Number of particles in range |
|---|---|
| 0–2 | 2000 |
| 2–4 | 600 |
| 4–8 | 140 |
| 8–12 | 40 |
| 12–16 | 15 |
| 16–20 | 5 |
| 20–24 | 2 |

**3.1.** A finely ground mixture of galena and limestone in the proportion of 1 to 4 by weight, is subjected to elutriation by an upwards current of water flowing at 5 mm/s. Assuming that the size distribution for each material is the same, and as shown by the following table, estimate the percentage of galena in the material carried away and in the material left behind. Take the absolute viscosity of water as 1 mN s/m$^2$ and use Stokes' equation.

| Diameter ($\mu$m) | 20 | 30 | 40 | 50 | 60 | 70 | 80 | 100 |
|---|---|---|---|---|---|---|---|---|
| Percentage weight of undersize | 15 | 28 | 48 | 54 | 64 | 72 | 78 | 88 |

Specific gravity of galena = 7·5; specific gravity of limestone = 2·7.

**3.2.** Calculate the terminal velocity of a steel ball, 2 mm diameter (density 7870 kg/m$^3$) in oil (density 900 kg/m$^3$, viscosity 50 mN s/m$^2$).

**3.3.** What will be the settling velocity of a spherical steel particle, 0·40 mm diameter, in an oil of specific gravity 0·82 and viscosity 10 mN s/m$^2$? The specific gravity of steel is 7·87.

**3.4.** What will be the settling velocities of mica plates, 1 mm thick and ranging in area from 6 to 600 mm$^2$, in an oil of specific gravity 0·82 and viscosity 10 mN s/m$^2$? The specific gravity of mica is 3·0.

**3.5.** A material of specific gravity 2·5 is fed to a size separation plant where the separating fluid is water which rises with a velocity of 1·2 m/s. The upward vertical component of the velocity of the particles is 6 m/s. How far will an approximately spherical particle, 6 mm diameter, rise relative to the walls of the plant before it comes to rest in the fluid?

**3.6.** A spherical glass particle is allowed to settle freely in water. If the particle starts initially from rest and if the value of the Reynolds number ($Re'$) with respect to the particle is 0·1 when it has attained its terminal falling velocity, calculate:

(a) the distance travelled before the particle reaches 90 per cent of its terminal falling velocity, and
(b) the time which has elapsed when the acceleration of the particle is one hundredth of its initial value.

**3.7.** In the hydraulic jig, a mixture of two solids is separated into its components by subjecting an aqueous slurry of the material to a pulsating motion, and allowing the particles to settle for a series of short time intervals such that they do not attain their terminal falling velocities. It is desired to separate materials of specific gravities 1·8 and 2·5 whose particle size ranges from 0·3 mm to 3 mm diameter. It may be assumed that the particles are approximately spherical and that Stokes' Law is applicable. Calculate approximately the maximum time interval for which the particles may be allowed to settle so that no particle of the less dense material falls a greater distance than any particle of the denser material.

$$\text{Viscosity of water} = 1 \text{ mN s/m}^2$$

**3.8.** Two spheres of equal terminal falling velocity settle in water starting from rest at the same horizontal level. How far apart vertically will the particles be when they have both reached their terminal falling velocities? Assume Stokes' law is valid and then check the assumption.

*Data:*

|  | Density (kg/m$^3$) | Viscosity (mN s/m$^2$) | Diameter ($\mu$m) |
|---|---|---|---|
| Particle 1 | 1500 | — | 40 |
| Particle 2 | 3000 | — | — |
| Water | 1000 | 1 | — |

**3.9.** The size of a powder is carried out by sedimentation in a vessel having the sampling point 180 mm below the liquid surface. If the viscosity of the liquid is 1·2 mN s/m$^2$, and the densities of the powder and liquid are 2650 and 1000 kg/m$^3$ respectively, determine the time which must elapse before any sample will exclude particles larger than 20 $\mu$m.

If turbulent conditions occur when the Reynolds number is greater than 0·2, what is the approximate maximum size of particle to which Stokes' Law can be applied under the above conditions?

**3.10.** Calculate the distance a spherical particle of lead shot of diameter ($d$) 0·1 mm will settle in a glycerol/water mixture before it reaches 99 per cent of its terminal falling velocity.

Density of lead = 11400 kg/m$^3$. Density of liquid = 1000 kg/m$^3$. Viscosity of liquid ($\mu$) = 10 mN s/m$^2$.

Assume that the resistance force can be calculated from Stokes' Law and is equal to $3\pi\mu du$, where $u$ is the velocity of the particle relative to the liquid.

**3.11.** Find the weight of a sphere of material of specific gravity 7·5 which falls with a steady velocity of 0·6 m/s in a large deep tank of water.

**3.12.** Two ores, of specific gravities 3·7 and 9·8, are to be separated in water by a hydraulic classification method. If the particles are all of approximately the same shape and each is sufficiently large for the drag force to be proportional to the square of its velocity in the fluid, calculate the maximum ratio of sizes which can be separated if the particles attain their terminal falling velocities. Explain why a wider range of sizes can be separated if the time of settling is so small that the particles do not reach their terminal velocities.

Obtain an explicit expression for the distance through which a particle will settle in a given time if it starts from rest and if the resistance force is proportional to the square of the velocity. The acceleration period is to be taken into account.

**3.13.** Salt, of specific gravity 2·35, is charged to the top of a reactor containing a 3 m depth of aqueous liquid (specific gravity 1·1 and viscosity ($\mu$) 2 mN s/m²) and the crystals must dissolve completely before reaching the bottom. If the rate of dissolution of the crystals is given by the relation:

$$-\frac{dd}{dt} = 3 \times 10^{-4} + 2 \times 10^{-4} u$$

where $d$ is the size of the crystal (cm) at time $t$ (s) and $u$ is its velocity in the fluid (cm/s), calculate the maximum size of crystal which can be charged. The inertia of the particles can be neglected and the resistance force can be taken as given by Stokes' Law ($3\pi\mu du$), $d$ being taken as the equivalent spherical diameter of the particle.

**3.14.** A balloon weighing 7 g is charged with hydrogen to a pressure of 104 kN/m². The balloon is released from ground level and, as it rises, hydrogen escapes in order to maintain a constant differential pressure of 2·7 kN/m², under which condition the diameter of the balloon is 0·3 m. If conditions are assumed to remain isothermal at 273 K as the balloon rises, what is the ultimate height reached and how long does it take to rise through the first 3000 m?

It may be assumed that the value of the Reynolds number with respect to the balloon exceeds 500 throughout, so that the resistance coefficient is constant at 0·22. Neglect the inertia of the balloon, i.e. assume that it is rising at its equilibrium velocity at any moment.

**3.15.** A mixture of quartz of specific gravity 3·7 and galena of specific gravity 9·8 whose size range is 0·3 to 1 mm is to be separated by a sedimentation process. If Stokes' Law is assumed to be applicable, what is the minimum density required for the liquid if the particles all settle at their terminal velocities?

Consideration was given to devising a separating system using water as the liquid. In this case the particles were to be allowed to settle for a series of short time intervals so that the smallest particle of galena settled a larger distance than the largest particle of quartz. What approximately is the maximum permissible settling period?

According to Stokes' Law, the resistance force $F$ acting on a particle of diameter $d$, settling at a velocity $u$ in a fluid of viscosity $\mu$ is given by

$$F = 3\pi\mu du$$
$$\text{Viscosity } (\mu) \text{ of water} = 1 \text{ mN s/m}^2$$

**4.1.** In a contact sulphuric acid plant the secondary converter is a tray type converter, 2·3 m in diameter with the catalyst arranged in three layers, each 0·45 m thick. The catalyst is in the form of cylindrical pellets 9·5 mm in diameter and 9·5 mm long. The void fraction is 0·35. The gas enters the converter at 675 K and leaves at 720 K. Its inlet composition is

$$SO_3 \, 6\cdot6, \quad SO_2 \, 1\cdot7, \quad O_2 \, 10\cdot0, \quad N_2 \, 81\cdot7 \, \text{mol\%}$$

and its exit composition

$$SO_3 \, 8\cdot2, \quad SO_2 \, 0\cdot2, \quad O_2 \, 9\cdot3, \quad N_2 \, 82\cdot3 \, \text{mol\%}$$

The gas flow rate is 0·68 kg/m² s. Calculate the pressure drop through the converter.

$$\mu = 0\cdot032 \, \text{mN s/m}^2.$$

**4.2.** Show how an equation for the pressure drop in a packed column can be modified to apply to cases where the total pressure and the pressure drop are of the same order of magnitude.

Two heat-sensitive organic liquids (average molecular weight = 155 kg/kmol) are to be separated by vacuum distillation in a 100 mm diameter column packed with 6 mm stoneware Raschig rings. The number of theoretical plates required is 16 and it has been found that the HETP is 150 mm. If the product rate is 5 g/s at a reflux ratio of 8, calculate the pressure in the condenser so that the temperature in the still does not exceed 395 K (equivalent to a pressure of 8 kN/m$^2$). Assume $a = 800 \text{ m}^2/\text{m}^3$, $\mu = 0.02$ mN s/m$^2$, $e = 0.72$ and neglect the temperature changes and the correction for liquid flow.

**4.3.** A column 0.6 m diameter and 4 m tall, packed with 25 mm ceramic Raschig rings, is used in a gas absorption process carried out at atmospheric pressure and 293 K. If the liquid and gas can be considered to have the properties of water and air, and their flowrates are 6.5 and 0.6 kg/m$^2$ s respectively, what will be the pressure drop across the column?

Use (a) Carman's method, (b) one other method, and compare the results obtained. How much can the liquid rate be increased before the column will flood?

**4.4.** A packed column, 1.2 m in diameter and 9 m tall, and packed with 25 mm Raschig rings, is used for the vacuum distillation of a mixture of isomers of molecular weight 155 kg/kmol. The mean temperature is 373 K, the pressure at the top of the column is maintained at 0.13 kN/m$^2$ and the still pressure ranges between 1.3 and 3.3 kN/m$^2$. Obtain an expression for the pressure drop on the assumption that it is not appreciably affected by the liquid flow and can be calculated using a modified form of Carman's equation (4.17). Show that, over the range of operating pressures used, the pressure drop is approximately directly proportional to the mass rate of flow of vapour, and calculate the pressure drop at a vapour rate of 0.125 kg/m$^2$.

Data:   Specific surface of packing, $S = 650 \text{ m}^2/\text{m}^3$
Mean voidage of bed, $e$     $= 0.71$
Viscosity of vapour, $\mu$     $= 0.018$ mN s/m$^2$
Molecular volume     $= 22.4 \text{ m}^3/\text{kmol}$

**4.5.** A packed column, 1.22 m in diameter and 9 m tall, and packed with 25 mm Raschig rings, is used for the vacuum distillation of a mixture of isomers of molecular weight 155 kg/kmol. The mean temperature is 373 K, the pressure at the top of the column is maintained at 0.13 kN/m$^2$, and the still pressure is 1.3 kN/m$^2$. Obtain an expression for the pressure drop on the assumption that it is not appreciably affected by the liquid flow and can be calculated using the modified form of Carman's equation.

Show that, over the range of operating pressures used, the pressure drop is approximately directly proportional to the mass rate of flow of vapour, and calculate approximately the flowrate of vapour.

Data:   Specific surface of packing, $S = 656 \text{ m}^2/\text{m}^3$
Mean voidage of bed, $e$     $= 0.71$
Viscosity of vapour, $\mu$     $= 0.018$ mN s/m$^2$
Kilogram molecular volume $= 22.4 \text{ m}^3/\text{kmol}$

**5.1.** A slurry containing 5 kg of water per kg of solids is to be thickened to a sludge containing 1.5 kg of water per kg of solids in a continuous operation. Laboratory tests using five different concentrations of the slurry yielded the following results:

| concentration (kg water/kg solid) | 5.0 | 4.2 | 3.7 | 3.1 | 2.5 |
|---|---|---|---|---|---|
| rate of sedimentation (mm/s) | 0.17 | 0.10 | 0.08 | 0.06 | 0.042 |

Calculate the minimum area of a thickener to effect the separation of 0.6 kg of solids per second.

**5.2.** If a centrifuge is 0.9 m diameter and rotates at 20 Hz, at what speed should a laboratory centrifuge of 150 mm diameter be run if it is to duplicate plant conditions?

**5.3.** What is the maximum safe speed of rotation of a phosphor-bronze centrifuge basket, 0.3 m diameter and 5 mm thick, when it contains a liquid of density 1000 kg/m$^3$ forming a layer 75 mm thick at the walls? Take the density of phosphor-bronze as 8900 kg/m$^3$ and the safe working stress as 55 MN/m$^2$.

**5.4.** A centrifuge with a phosphor-bronze basket 375 mm diameter is to be run at 30 Hz, with a 100 mm layer of solids of bulk density 2000 kg/m$^3$ at the walls. What should be the thickness of the walls of the basket if the perforations are so small that they have a negligible effect on strength?

Density of phosphor-bronze     $= 8900 \text{ kg/m}^3$
Maximum safe stress for phosphor-bronze $= 55 \text{ MN/m}^2$

**5.5.** An aqueous suspension consisting of particles of specific gravity 2·5 in the size range 1–10 $\mu$m is introduced into a centrifuge with a basket 450 mm diameter rotating at 80 Hz. If the suspension forms a layer 75 mm thick in the basket, approximately how long will it take to cause the smallest particle to settle out?

**5.6.** A centrifuge with a phosphor-bronze basket 375 mm diameter is to be run at 60 Hz with a 75 mm layer of liquid of specific gravity 1·2 in the basket. What thickness of walls is required in the basket?

$$\text{Density of phosphor-bronze} \qquad\qquad = 8900 \, \text{kg/m}^2$$
$$\text{Maximum safe stress for phosphor-bronze} = 55 \, \text{MN/m}^2$$

**5.7.** A centrifuge basket 600 mm long and 100 mm internal diameter has a discharge weir 25 mm diameter. What is the maximum volumetric flow of liquid through the centrifuge such that when the basket is rotated at 200 Hz all particles of diameter greater than 1 $\mu$m are retained on the centrifuge wall? The retarding force on a particle moving liquid can be taken as equal to $3\pi\mu du$, where

$$u = \text{particle velocity relative to the liquid}$$
$$\mu = \text{liquid viscosity, and}$$
$$d = \text{particle diameter.}$$
$$\text{Sp. gr. of liquid} = 1\cdot0$$
$$\text{Sp. gr. of solid} = 2\cdot0$$
$$\text{Viscosity of liquid } (\mu) = 1\cdot0 \, \text{mN s/m}^2$$

The inertia of the particle can be neglected.

**5.8.** Calculate the minimum area and diameter of a thickener with a circular basin to treat 0·1 m³/s of a slurry of solids with a concentration of 150 kg/m³. The results of batch settling tests are:

| solids concentration (kg/m³) | 100 | 200 | 300 | 400 | 500 | 600 | 700 | 800 | 900 | 1000 | 1100 |
|---|---|---|---|---|---|---|---|---|---|---|---|
| settling velocity ($\mu$m/s) | 148 | 91 | 55·3 | 33·3 | 21·4 | 14·5 | 10·3 | 7·4 | 5·6 | 4·2 | 3·3 |

A value of 1290 kg/m³ for the underflow concentration was selected from a retention time test. Estimate the underflow volumetric flowrate assuming total separation of all solids.

**6.1.** Oil, of specific gravity 0·9 and viscosity 3 mN s/m², passes vertically upwards through a bed of catalyst consisting of approximately spherical particles of diameter 0·1 mm and specific gravity 2·6. At approximately what mass rate of flow per unit area of bed will (a) fluidisation, and (b) transport of particles, occur?

**6.2.** Calculate the minimum velocity at which spherical particles (specific gravity 1·6) of diameter 1·5 mm will be fluidised by water in a tube of diameter 10 mm. Discuss the uncertainties in this calculation. (Viscosity of water = 1 mN s/m²; Kozeny's constant = 5.)

**6.3.** In a fluidised bed, iso-octane vapour is adsorbed from an air stream on to the surface of alumina microspheres. The mol fraction of iso-octane in the inlet gas is $1\cdot442 \times 10^{-2}$ and the mol fraction in the outlet gas is found to vary with time in the following manner:

| Time from start (s) | Mol fraction in outlet gas ($\times 10^2$) |
|---|---|
| 250 | 0·223 |
| 500 | 0·601 |
| 750 | 0·857 |
| 1000 | 1·062 |
| 1250 | 1·207 |
| 1500 | 1·287 |
| 1750 | 1·338 |
| 2000 | 1·373 |

Show that the results can be interpreted on the assumptions that the solids are completely mixed, that the gas leaves in equilibrium with the solids and that the adsorption isotherm is linear over the range considered. If the flowrate of gas is $0\cdot679 \times 10^{-6}$ kmol/s and the mass of solids in the bed is 4·66 g, calculate the slope of the adsorption isotherm. What evidence do the results provide concerning the flow pattern of the gas?

**6.4.** Discuss the reasons for the good heat transfer properties of fluidised beds. Cold particles of glass ballotini are fluidised with heated air in a bed in which a constant flow of particles is maintained in a horizontal direction. When steady conditions have been reached, the temperatures recorded by a bare thermocouple immersed in the bed are as follows:

| Distance above bed support (mm) | Temperature (K) |
|---|---|
| 0 | 339·5 |
| 0·64 | 337·7 |
| 1·27 | 335·0 |
| 1·91 | 333·6 |
| 2·54 | 333·3 |
| 3·81 | 333·2 |

Calculate the coefficient for heat transfer between the gas and the particles, and the corresponding values of the particle Reynolds and Nusselt numbers. Comment on the results and on any assumptions made.

| | |
|---|---|
| Gas flow rate | 0·2 kg/m² s; |
| Specific heat in air | 0·88 kJ/kg K |
| Viscosity of air | 0·015 mN s/m |
| Particle diameter | 0·25 mm |
| Thermal conductivity of air | 0·03 W/mK |

**6.5.** The relation between bed voidage $e$ and fluid velocity $u_c$ for particulate fluidisation of uniform particles small compared with the diameter of the containing vessel is given by

$$\frac{u_c}{u_0} = e^n$$

where $u_0$ is the free falling velocity.

Discuss the variation of the index $n$ with flow conditions, indicating why it is independent of the Reynolds number $Re$ with respect to the particle at very low and very high values of $Re$. When are appreciable deviations from this relation observed with liquid fluidised systems?

For particles of glass ballotini with free falling velocities of 10 and 20 mm/s the index $n$ has a value of 2·39. If a mixture of equal volumes of the two particles is fluidised, what will be the relation between the voidage and fluid velocity if it is assumed that complete segregation is obtained?

**6.6.** Obtain a relationship for the ratio of the terminal falling velocity of a particle to the minimum fluidising velocity for a bed of similar particles. Assume that Stoke's Law and the Carman-Kozeny equation are applicable. What is the value of the ratio if the bed voidage at the minimum fluidising velocity is 0·4?

Discuss the validity of using the Carman-Kozeny equation for calculation of pressure drop through a fluidised bed.

**6.7.** What is the significance of the term 'minimum fluidising velocity" and how would you determine it? Discuss its relevance to the behaviour of a practical fluidised bed.

A packed bed consisting of uniform spherical particles (diameter, $d = 3$ mm, and density, $\rho_s = 4200$ kg/m³) is fluidised by means of a liquid (viscosity, $\mu = 3$ mN s/m²·, and density, $\rho = 1100$ kg/m³). Using Ergun's equation for the pressure drop $(-\Delta P)$ through a bed of height $H$ and voidage $e$ as a function of superficial velocity, $u$, calculate the minimum fluidising velocity in terms of the settling velocity $(u_0)$ of the particles in the fluid.

State clearly any assumptions which you make and indicate how closely you would expect your result to be confirmed by an experiment.

Ergun's equation:

$$-\frac{\Delta P}{H} = 150\frac{(1-e)^2}{e^3}\frac{\mu u}{d^2} + 1·75\frac{(1-e)}{e^3}\frac{\rho u^2}{d}$$

**7.1.** It is required to transport sand of particle size 1·25 mm and density 2600 kg/m³ at the rate of 1 kg/s through a horizontal pipe, 200 m long. Estimate the air flowrate required, the pipe diameter and the pressure drop in the pipe-line.

**7.2.** Sand of mean diameter 0·2 mm is to be conveyed by water flowing at 0·5 kg/s in a 25 mm ID horizontal pipe, 100 m long. What is the maximum amount of sand which may be transported in this way if the head developed by the pump is limited to 300 kN/m²? Assume fully suspended heterogeneous flow.

**8.1.** The size distribution by weight of the dust carried in a gas is given in the following table, together with the efficiency of collection over each size range.

| Size range, μm | 0–5 | 5–10 | 10–20 | 20–40 | 40–80 | 80–160 |
|---|---|---|---|---|---|---|
| Per cent weight | 10 | 15 | 35 | 20 | 10 | 10 |
| Per cent efficiency | 20 | 40 | 80 | 90 | 95 | 100 |

Calculate the overall efficiency of the collector and the percentage by weight of the emitted dust which is smaller than 20 μm in diameter. If the dust burden is 18 g/m³ at entry and the gas flow 0·3 m³/s, calculate the weight of dust emitted in kg/s.

**8.2.** The collection efficiency of a cyclone is 45 per cent over the size range 0–5 μm, 80 per cent over the size range 5–10 μm, and 96 per cent for particles exceeding 10 μm. Calculate the efficiency of collection for the following dust:

$$\text{Weight distribution:} \quad \begin{array}{l} \text{50 per cent } 0\text{–}5\,\mu m \\ \text{30 per cent } 5\text{–}10\,\mu m \\ \text{20 per cent above } 10\,\mu m \end{array}$$

**8.3.** A sample of dust from the air in a factory has been collected on a glass slide. If the dust on the slide was deposited from one cubic centimetre of air, estimate the weight of dust in grams per cubic metre of air in the factory, given the number of particles in the various size ranges to be as shown in the following table:

| Size range in μm | 0–1 | 1–2 | 2–4 | 4–6 | 6–10 | 10–14 |
|---|---|---|---|---|---|---|
| Number of particles: | 2000 | 1000 | 500 | 200 | 100 | 40 |

Assume the specific gravity of the dust to be 2·6, and make an appropriate allowance for particle shape.

**8.4.** A cyclone separator 0·3 m in diameter and 1·2 m long, has a circular inlet 75 mm in diameter and an outlet of the same size. If the gas enters at 1·5 m/s, at what particle size will the theoretical cut occur?

$$\begin{array}{ll} \text{Viscosity of air} & 0\text{·}018\,\text{mN s/m}^2 \\ \text{Density of air} & 1\text{·}3\,\text{kg/m}^3 \\ \text{Density of particles} & 2700\,\text{kg/m}^3 \end{array}$$

**9.1.** A slurry, containing 0·2 kg of solid (specific gravity 3·0) per kilogram of water, is fed to a rotary drum filter, 0·6 m in diameter and 0·6 m long. The drum rotates at one revolution in 350 s and 20 per cent of the filtering surface is in contact with the slurry at any given instant. If filtrate is produced at the rate of 0·125 kg/s and the cake has a voidage of 0·5, what thickness of cake is formed when filtering at an absolute pressure of 35 kN/m²?

The rotary filter breaks down and the operation has to be carried out temporarily in a plate and frame press with frames 0·3 m square. The press takes 100 s to dismantle and 100 s to reassemble, and, in addition, 100 s is required to remove the cake from each frame. If filtration is to be carried out at the same overall rate as before, with an operating pressure of 275 kN/m², what is the minimum number of frames that must be used and what is the thickness of each? Assume the cakes to be incompressible and neglect the resistance of the filter media.

**9.2.** A slurry containing 100 kg of whiting, of specific gravity 3·0, per m³ of water, is filtered in a plate and frame press, which takes 900 s to dismantle, clean and re-assemble. If the filter cake is incompressible and has a voidage of 0·4, what is the optimum thickness of cake for a filtration pressure of 1000 kN/m²? If the cake is washed at 500 kN/m² and the total volume of wash water employed is one quarter of that of the filtrate, how is the optimum thickness of cake affected? Neglect the resistance of the filter medium and take the viscosity of water as 1 mN s/m². In an experiment, a pressure of 165 kN/m² produced a flow of water of 0·06 cm³/s through a centimetre cube of filter cake.

**9.3.** A plate and frame press, filtering a slurry, gave a total of 8 m³ of filtrate in 1800 s and 11·3 m³ in 3600 s when filtration was stopped. Estimate the washing time in seconds if 3 m³ of wash water are used. The resistance of the cloth can be neglected and a constant pressure is used throughout.

**9.4.** In the filtration of a certain sludge the initial period is effected at a constant rate with the feed pump at full capacity, till the pressure reaches 400 kN/m². The pressure is then maintained at this value for the remainder of the filtration. The constant rate operation requires 900 s and one-third of the total filtrate is obtained during this period.

Neglecting the resistance of the filter medium, determine (a) the total filtration time and (b) the filtration cycle with the existing pump for the maximum daily capacity, if the time for removing the cake and reassembling the press is 1200 s. The cake is not washed.

**9.5.** A rotary filter, operating at 0·03 Hz, filters 0·0075 m³/s. Operating under the same vacuum and neglecting the resistance of the filter cloth, at what speed must the filter be operated to give a filtration rate of 0·0150 m³/s?

**9.6.** A slurry is filtered in a plate and frame press containing 12 frames, each 0·3 m square and 25 mm thick. During the first 200 s, the filtration pressure is slowly raised to the final value of 500 kN/m² and, during this period, the rate of filtration is maintained constant. After the initial period, filtration is carried out at constant pressure and the cakes are completely formed in a further 900 s. The cakes are then washed at 375 kN/m² for 600 s, using "thorough washing". What is the volume of filtrate collected per cycle and how much wash water is used?

A sample of the slurry had previously been tested, using a vacuum leaf filter of 0·05 m² filtering surface and a vacuum equivalent to an absolute pressure of 30 kN/m². The volume of filtrate collected in the first 300 s was 250 cm³ and, after a further 300 s, an additional 150 cm³ was collected. Assume the cake to be incompressible and the cloth resistance to be the same in the leaf as in the filter press.

**9.7.** A sludge is filtered in a plate and frame press fitted with 25 mm frames. For the first 600 s the slurry pump runs at maximum capacity. During this period the pressure rises to 500 kN/m² and a quarter of the total filtrate is obtained. The filtration t ¹es a further 3600 s to complete at constant pressure and 900 s is required for emptying and resetting the press.

It is found that, if the cloths are precoated with filter aid to a depth of 1·6 mm, the cloth resistance is reduced to a quarter of its former value. What will be the increase in the overall throughput of the press if the precoat can be applied in 180 s?

**9.8.** Filtration is carried out in a plate and frame filter press, with 20 frames 0·3 m square and 50 mm thick, and the rate of filtration is maintained constant for the first 300 s. During this period, the pressure is raised to 350 kN/m², and one-quarter of the total filtrate per cycle is obtained. At the end of the constant rate period, filtration is continued at a constant pressure of 350 kN/m² for a further 1800 s, after which the frames are full. The total volume of filtrate per cycle is 0·7 m³ and dismantling and refitting of the press takes 900 s.

It is decided to use a rotary drum filter, 1·5 m long and 2·2 m in diameter, in place of the filter press. Assuming that the resistance of the cloth is the same in the two plants and that the filter cake is incompressible, calculate the speed of rotation of the drum which will result in the same overall rate of filtration as was obtained with the filter press. The filtration in the rotary filter is carried out at a constant pressure difference of 70 kN/m², and the filter operates with 25 per cent of the drum submerged in the slurry at any instant.

**9.9.** It is required to filter a certain slurry to produce 2·25 m³ of filtrate per working day of 8 h. The process is carried out in a plate and frame filter press with 0·45 m square frames and a working pressure of 450 kN/m². The pressure is built up slowly over a period of 300 s and, during this period, the rate of filtration is maintained constant.

When a sample of the slurry was filtered, using a pressure of 35 kN/m² on a single leaf filter of filtering area 0·05 m², 400 cm³ of filtrate were collected in the first 300 s of filtration and a further 400 cm³ was collected during the following 600 s. Assuming that the dismantling of the filter press, the removal of the cakes and the setting up again of the press takes an overall time of 300 s, plus an additional 180 s for each cake produced, what is the minimum number of frames that need be employed? Take the resistance of the filter cloth to be the same in the laboratory tests as on the plant.

**9.10.** The relation between flow and head for a certain slurry pump can be represented approximately by a straight line, the maximum flow at zero head being 0·0015 m³/s and the maximum head at zero flow 760 m of liquid. Using this pump to feed a particular slurry to a pressure leaf filter,

(a) how long will it take to produce 1 m³ of filtrate, and
(b) what will be the pressure across the filter after this time?

A sample of the slurry was filtered at a constant rate of 0·00015 m³/s through a leaf filter covered with a similar filter cloth but of one-tenth the area of the full scale unit and after 625 s the pressure across the filter was 360 m of liquid. After a further 480 s the pressure was 600 m.

**9.11.** A slurry containing 40 per cent by weight solid is to be filtered on a rotary drum filter 2 m diameter and 2 m long which normally operates with 40 per cent of its surface immersed in the slurry and under a pressure of 17 kN/m². A laboratory test on a sample of the slurry using a leaf filter of area 200 cm² and covered with a similar cloth to that on the drum, produced 300 cm³ of filtrate in the first 60 s and 140 cm³ in the next 60 s, when the leaf was under an absolute pressure of 17 kN/m². The bulk density of the dry cake was 1500 kg/m³ and the density of the filtrate 1000 kg/m³. The minimum thickness of cake which could be readily removed from the cloth was 5 mm.

At what speed should the drum rotate for maximum throughput and what is this throughput in terms of the weight of the slurry fed to the unit per unit time?

**9.12.** A continuous rotary filter is required for an industrial process for the filtration of a suspension to produce 0·002 m³/s of filtrate. A sample was tested on a small laboratory filter of area 0·023 m² to which it was fed by means of a slurry pump to give filtrate at a constant rate of 12·5 cm³/s. The pressure difference across the test filter increased from 14 kN/m² after 300 s filtration to 28 kN/m² after 900 s, at which time the cake thickness had reached 38 mm. Suggest suitable dimensions and operating conditions for the rotary filter, assuming that the resistance of the cloth used is one-half that on the test filter, and that the vacuum system is capable of maintaining a constant pressure difference of 70 kN/m² across the filter.

**9.13.** A rotary drum filter, 1·2 m diameter and 1·2 m long, can handle 6·0 kg/s of slurry containing 10 per cent of solids when rotated at 0·005 Hz. By increasing the speed to 0·008 Hz it is found that it can handle 7·2 kg/s. What will be the percentage change in the amount of wash water which can be applied to each kilogram of cake caused by the increased speed of rotation of the drum, and what is the theoretical maximum quantity of slurry which can be handled?

**9.14.** When an aqueous slurry is filtered in a plate and frame press, fitted with two 50 mm thick frames each 150 mm square at 450 kN/m², the frames are filled in 3500 s. The liquid in the slurry has the same density as water. The slurry is then filtered in a perforate basket centrifuge with a basket 300 mm diameter and 200 mm deep. If the radius of the inner surface of the slurry is maintained constant at 75 mm and the speed of rotation is 65 Hz, how long will it take to produce as much filtrate as was obtained from a single cycle of operations with the filter press?

Assume that the filter cake is incompressible and that the resistance of the cloth is equivalent to 3 mm of cake in both cases.

**9.15.** A rotary drum filter area of 3 m² operates with an internal pressure of 30 kN/m² and with 30 per cent of its surface submerged in the slurry. Calculate the rate of production of filtrate and the thickness of cake when it rotates at 0·0083 Hz (0·5 rpm), if the filter cake is incompressible and the filter cloth has a resistance equal to that of 1 mm of cake.

It is desired to increase the rate of filtration by raising the speed of rotation of the drum. If the thinnest cake that can be removed from the drum has a thickness of 5 mm, what is the maximum rate of filtration which can be achieved and what speed of rotation of the drum is required?

| | |
|---|---|
| Voidage of cake | $= 0·4$ |
| Specific resistance of cake | $= 2 \times 10^{12}/m^2$ |
| Density of solids | $= 2000 \, kg/m^3$ |
| Density of filtrate | $= 1000 \, kg/m^3$ |
| Viscosity of filtrate | $= 10^{-3} \, N \, s/m^2$ |
| Slurry concentration | $= 20\%$ by weight solids. |

**10.1.** 0·4 kg/s of dry sea-shore sand, containing 1 per cent by weight of salt, is to be washed with 0·4 kg/s of fresh water running countercurrent to the sand through two classifiers in series. Assume that perfect mixing of the sand and water occurs in each classifier and that the sand discharged from each classifier contains one part of water for every two of sand (by weight). If the washed sand is dried in a kiln dryer, what percentage of salt will it retain? What wash rate would be required in a single classifier in order to wash the sand equally well?

**10.2.** Caustic soda is manufactured by the lime-soda process according to the following equation:

$$Na_2CO_3 + Ca(OH)_2 = 2NaOH + CaCO_3$$

A solution of sodium carbonate in water (0·25 kg/s Na₂CO₃) is treated with the theoretical requirement of lime and, after the reaction is complete, the CaCO₃ sludge, containing by weight one part of CaCO₃ per nine parts of water is fed continuously to three thickeners in series and is washed countercurrently. Calculate the necessary rate of feed of neutral water to the thickeners, so that the calcium carbonate, on drying, contains only 1 per cent of sodium hydroxide. The solid discharged from each thickener contains one part by weight of calcium carbonate to three of water.

**10.3.** How many stages are required for 98 per cent extraction of a material containing 18 per cent of extractable matter (having a specific gravity of 2·7) and which requires 200 volumes of liquid per 100 volumes of solid for it to be capable of being pumped to the next stage? The strong solution is to have a concentration of 100 kg/m³.

**10.4.** Soda ash is mixed with lime and the liquor from the second of three thickeners and passes to the first thickener where separation is effected. The quantity of this caustic solution leaving the first thickener is such as to yield 10 Mg of caustic soda per day of 24 h. The solution contains 95 kg of caustic soda per 1000 kg of water, whilst the sludge leaving each of the thickeners consists of one part of solids to one of liquid.

Determine:

    (a) the weight of solids in the sludge;
    (b) the weight of water admitted to the third thickener;
    (c) the percentages of caustic soda in the sludges leaving the respective thickeners.

**10.5.** Seeds, containing 20 per cent by weight of oil, are extracted in a countercurrent plant and 90 per cent of the oil is recovered in a solution containing 50 per cent by weight of oil. If the seeds are extracted with fresh solvent and 1 kg of solution is removed in the underflow in association with every 2 kg of insoluble matter, how many ideal stages are required? Use a triangular diagram.

**10.6.** It is desired to recover precipitated chalk from the causticising of soda ash. After decanting the liquor from the precipitators the sludge has the composition 5 per cent $CaCO_3$, 0.1 per cent $NaOH$ and the balance water.

1000 Mg per day of this sludge is fed to two thickeners where it is washed with 200 Mg/day of neutral water. The pulp removed from the bottom of the thickeners contains 4 kg of water per kg of chalk. The pulp from the last thickener is taken to a rotary filter and concentrated to 50 per cent solids and the filtrate is returned to the system as wash water.

Calculate the net percentage of $CaCO_3$ in the product after drying.

**10.7.** Barium carbonate is to be made by the interaction of sodium carbonate and barium sulphide. The quantities that are fed to the reaction agitators per 24 h are as follows: 20 Mg of barium sulphide dissolved in 60 Mg of water together with the theoretically necessary amount of sodium carbonate.

There are three thickeners in series, run on a countercurrent decantation system. Overflow from the second thickener goes to the agitators and overflow from the first thickener is to be 10 per cent sodium sulphide. Sludge from all thickeners carries two parts water to one part barium carbonate (w/w).

How much sodium sulphide will remain in the dried barium carbonate precipitate?

**10.8.** In the production of caustic soda by the action of calcium hydroxide on sodium carbonate, 1 kg/s of sodium carbonate is treated with the theoretical quantity of lime. The sodium carbonate is made up as a 20 per cent solution. The material from the extractors is fed to a countercurrent washing system where it is treated with 2 kg/s of clean water. The washing thickeners are so arranged that the ratio of the volume of liquid discharged in the liquid offtake to that discharged with the solid is the same in all the thickeners and is equal to 4.0. How many thickeners must be arranged in series so that not more than 1 per cent of the sodium hydroxide discharged with the solid from the first thickener is wasted?

**10.9.** A plant produces 100 kg/s of titanium dioxide pigment which must be 99 per cent pure when dried. The pigment is produced by precipitation and the material, as prepared, is contaminated with 1 kg of salt solution containing 0.55 kg of salt per kg of pigment. The material is washed countercurrently with water in a number of thickeners arranged in series. How many thickeners will be required if water is added at the rate of 200 kg/s and the solid discharged from each thickener removes 0.5 kg of solvent per ton of pigment?

What will be the required number of thickeners if the amount of solution removed in association with the pigment varies in the following way with the concentration of the solution in the thickener?

| Concentration (solute/solution) | Solution/unit mass of pigment |
|---|---|
| 0 | 0.30 |
| 0.1 | 0.32 |
| 0.2 | 0.34 |
| 0.3 | 0.36 |
| 0.4 | 0.38 |
| 0.5 | 0.40 |

The concentrated wash liquor is mixed with the material fed to the first thickener.

**10.10.** Prepared cottonseed meats containing 35 per cent extractable oil are fed to a continuous countercurrent extractor of the intermittent drainage type using hexane as solvent. The extractor consists of ten sections, the section efficiency being 50 per cent. The entrainment, assumed constant, is 1 kg solution/kg solids. What will be the oil concentration in the outflowing solvent if the extractable oil content in the meats is to be reduced by 0.5 per cent by weight?

**10.11.** Seeds containing 25 per cent by weight of oil are extracted in a countercurrent plant and 90% of the oil is to be recovered in a solution containing 50% of oil. It has been found experimentally that the amount of solution removed in the underflow in association with every kilogram of insoluble matter is given by the equation:

$$k = 0.7 + 0.5y_s + 3y_s^2$$

where $y_s$ is the concentration of the overflow solution (wt. fraction of solute). If the seeds are extracted with fresh solvent how many ideal stages are required?

**10.12.** Halibut oil is extracted from granulated halibut livers in a countercurrent multibatch arrangement using ether as solvent. The solids charge contains 0·35 kg oil per kg of exhausted livers and it is desired to obtain a 90 per cent oil recovery. How many theoretical stages are required if 50 kg of ether are used per 100 kg of untreated solids. The entrainment data are as follows:

| Concentration of overflow (kg oil per kg solution) | 0 | 0·1 | 0·2 | 0·3 | 0·4 | 0·5 | 0·6 | 0·67 |
|---|---|---|---|---|---|---|---|---|
| Entrainment (kg solution per kg extracted livers) | 0·28 | 0·34 | 0·40 | 0·47 | 0·55 | 0·66 | 0·80 | 0·96 |

**11.1.** A liquor of four components, **A**, **B**, **C** and **D**, with 0·3 mol fraction each of **A**, **B** and **C**, is to be continuously fractionated to give a top product of 0·9 mol fraction **A** and 0·1 mol fraction **B**. The bottoms are to contain not more than 0·5 mol fraction **A**. Estimate the minimum reflux ratio required for this separation, if the relative volatility of **A** to **B** is 2·0.

**11.2.** During the batch distillation of a binary mixture in a packed column the product had 0·60 mol fraction of the more volatile component when the concentration in the still was 0·40 mol fraction. If the reflux ratio in use was 20:1, and the vapour composition $y$ is related to the liquor composition $x$ by the equation $y = 1·035x$ over the range of concentration concerned, determine the number of ideal plates represented by the column ($x$ and $y$ are in mol fractions).

**11.3.** A mixture of water and ethyl alcohol containing 0·16 mol fraction alcohol is continuously distilled in a plate fractionating column to give a product containing 0·77 mol fraction alcohol and a waste of 0·02 mol fraction alcohol. It is proposed to withdraw 25 per cent of the alcohol in the entering stream as a side stream with a mol fraction of 0·50 alcohol.

Determine the number of theoretical plates required and the plate from which the side stream should be withdrawn if the feed is liquor at the boiling point and a reflux ratio of 2 is used.

**11.4.** In a mixture to be fed to a continuous distillation column, the mol fraction of phenol is 0·35, of o-cresol 0·15, of m-cresol 0·30 and of xylenols 0·20. It is hoped to obtain a product with a mol fraction of phenol 0·952, of o-cresol 0·0474 and of m-cresol 0·0006. If the volatility relative to o-cresol of phenol is 1·26 and of m-cresol 0·70, estimate how many theoretical plates would be required at total reflux.

**11.5.** A continuous fractionating column, operating at atmospheric pressure, is to be designed to separate a mixture containing 15·67 per cent $CS_2$ and 84·33 per cent $CCl_4$ into an overhead product containing 91 per cent $CS_2$ and a waste of 97·3 per cent $CCl_4$ (all weight per cent). Assume a plate efficiency of 70 per cent and a reflux of 3·16 kmol per kmol of product.

Using the data below, determine the number of plates required.

Feed enters at 290 K with a specific heat of 1·7 kJ/kg K and boiling point of 336 K.

Latent heat of $CS_2$ and $CCl_4$ is 25900 kJ/kmol.

| Mol per cent $CS_2$ in vapour: 0 | 8·23 | 15·55 | 26·6 | 33·2 | 49·5 | 63·4 | 74·7 | 82·9 | 87·8 | 93·2 |
|---|---|---|---|---|---|---|---|---|---|---|
| Mol per cent $CS_2$ in liquor: 0 | 2·96 | 6·15 | 11·06 | 14·35 | 25·85 | 39·0 | 53·18 | 66·30 | 75·75 | 86·04 |

**11.6.** A batch fractionation is carried out in a small column which has the separating power of 6 theoretical plates. The mixture consists of benzene and toluene with 0·60 mol fraction of benzene. A distillate is required, of constant composition, of 0·98 mol fraction benzene, and the operation is discontinued when 83 per cent of the benzene charged has been removed as distillate. Estimate the reflux ratio needed at the start and finish of the distillation, if the relative volatility of benzene to toluene is taken as 2·46.

**11.7.** A continuous fractionating column is required to separate a mixture containing 0·695 mol fraction n-heptane ($C_7H_{16}$) and 0·305 mol fraction n-octane ($C_8H_{18}$) into products of 99 mol per cent purity. The column is to operate at a pressure of 101·3 kN/m² with a vapour velocity of 0·6 m/s. The feed is all liquid at its boiling-point, and is supplied to the column at 1·25 kg/s. The boiling-point at the top of the column may be taken as 372 K, and the equilibrium data are:

| $y$ = mol fraction heptane in vapour | 0·96 | 0·91 | 0·83 | 0·74 | 0·65 | 0·50 | 0·37 | 0·24 |
|---|---|---|---|---|---|---|---|---|
| $x$ = mol fraction heptane in liquid | 0·92 | 0·82 | 0·69 | 0·57 | 0·46 | 0·32 | 0·22 | 0·13 |

Determine the minimum reflux ratio that will be required. What diameter column would be required if the reflux used were twice the minimum possible?

**11.8.** The vapour pressures of chlorobenzene and water are as follows:

| Vapour pressure (kN/m²) | 13·3 | 6·7 | 4·0 | 2·7 |
|---|---|---|---|---|
| (mm Hg) | 100 | 50 | 30 | 20 |
| Temperatures, (K) | | | | |
| Chlorobenzene | 343·6 | 326·9 | 315·9 | 307·7 |
| Water | 324·9 | 311·7 | 303·1 | 295·7 |

A still is operated at a pressure of 18 kN/m², steam being blown continuously into it. Estimate the temperature of the boiling liquid and the composition of the distillate if liquid water is present in the still.

**11.9.** The following figures represent the equilibrium conditions in mol fraction of benzene in benzene-toluene mixtures at their boiling-point.

| Liquid | Vapour |
|---|---|
| 0·51 | 0·72 |
| 0·38 | 0·60 |
| 0·26 | 0·45 |
| 0·15 | 0·30 |

If the liquid compositions on four adjacent plates in a column were 0·18, 0·28, 0·41 and 0·57 under conditions of total reflux, determine the plate efficiencies.

**11.10.** A continuous rectifying column treats a mixture consisting of 40 per cent of benzene by weight and 60 per cent of toluene at the rate of 4 kg/s, and separates it into a product containing 97 per cent of benzene and a liquid containing 98 per cent toluene. The feed is liquid at its boiling-point.

(a) Calculate the weights of distillate and waste liquor per unit time.
(b) If a reflux ratio of 3·5 to 1 is employed, how many plates are required in the rectifying part of the column?
(c) What is the actual number of plates if the plate-efficiency is 60 per cent?

| Mol fraction of benzene in liquid | 0·1 | 0·2 | 0·3 | 0·4 | 0·5 | 0·6 | 0·7 | 0·8 | 0·9 |
|---|---|---|---|---|---|---|---|---|---|
| Mol fraction of benzene in vapour | 0·22 | 0·38 | 0·51 | 0·63 | 0·7 | 0·78 | 0·85 | 0·91 | 0·96 |

**11.11.** A compound rectifying column is fed with a mixture of benzene and toluene, in which the mol fraction of benzene is 0·35. The column is to yield a product in which the mol fraction of benzene is 0·95, when working with a reflux ratio of 3·2 to 1·0, and the waste from the column is not to exceed 0·05 mol fraction of benzene.

Assuming the plates used have an efficiency of 60 per cent, find the number of plates required and the position of the feed point.

The relation between the mol fraction of benzene in liquid and in vapour, is given by the following table:

| Mol fraction of benzene in liquid | 0·1 | 0·2 | 0·3 | 0·4 | 0·5 | 0·6 | 0·7 | 0·8 | 0·9 |
|---|---|---|---|---|---|---|---|---|---|
| Mol fraction of benzene in vapour | 0·20 | 0·38 | 0·51 | 0·63 | 0·71 | 0·78 | 0·85 | 0·91 | 0·96 |

**11.12.** The accompanying table gives the relationship between the mol fraction of carbon disulphide in the liquid and in the vapour evolved from the mixture during the distillation of a carbon disulphide–carbon tetrachloride mixture.

| x | 0 | 0·20 | 0·40 | 0·60 | 0·80 | 1·00 |
|---|---|---|---|---|---|---|
| y | 0 | 0·445 | 0·65 | 0·795 | 0·91 | 1·00 |

Determine graphically the theoretical number of plates required for the rectifying and stripping portions of the column, using the following data:

Reflux ratio, 3
Slope of fractionating line, 1·4.
Purity of product, 99 per cent.
Percentage carbon disulphide in waste liquors, 1 per cent.

What is the minimum slope of the rectifying line in this case?

**11.13.** A fractionating column is required to distil a liquid containing 25 per cent benzene and 75 per cent toluene by weight, so as to give a product of 90 per cent benzene. A reflux ratio of 3·5 is to be used, and the feed will enter at its boiling point.

If the plates used are 100 per cent efficient, calculate by the Lewis–Sorel method the composition of liquid on the third plate, and by McCabe and Thiele's method estimate the number of plates required.

**11.14.** A 50 mol per cent mixture of benzene and toluene is fractionated in a batch still which has the separating power of 8 theoretical plates. It is proposed to obtain a constant quality product with a mol per cent benzene of 95, and to continue the distillation till the still has a content of 10 mol per cent of benzene.

What will be the range of reflux ratios used in the process? Show graphically the relation between the required reflux ratio and the amount of distillate removed.

**11.15.** The vapour composition on a plate of a distillation column is:

|  |  |  |
| --- | --- | --- |
| $C_1$ | 0·025 mol fraction | 36·5 rel. volatility |
| $C_2$ | 0·205 | 7·4 |
| $i\text{-}C_3$ | 0·210 | 3·0 |
| $n\text{-}C_3$ | 0·465 | 2·7 |
| $i\text{-}C_4$ | 0·045 | 1·3 |
| $n\text{-}C_4$ | 0·050 | 1·0 |

What will be the composition of the liquid on the plate if it is in equilibrium with the vapour?

**11.16.** A liquor of 0·30 mol fraction of benzene and the rest toluene is fed to a continuous still to give a top product of 0·90 mol fraction benzene and a bottom product of 0·95 mol fraction toluene.

If the reflux ratio is 5·0, how many plates are required

(a) if the feed is saturated vapour?
(b) if the feed is liquid at 283 K?

**11.17.** A mixture of alcohol and water with 0·45 mol fraction of alcohol is to be continuously distilled in a column so as to give a top product of 0·825 mol fraction alcohol and a liquor at the bottom with 0·05 mol fraction alcohol.

How many theoretical plates are required if the reflux ratio used is 3? Indicate on a diagram what is meant by the Murphree plate efficiency.

**11.18.** It is desired to separate 1 kg/s of an ammonia solution containing 30 per cent $NH_3$ by weight into 99·5 per cent liquid $NH_3$ and a residual weak solution containing 10 per cent $NH_3$. Assuming the feed to be at its boiling point, a column pressure of 1013 $kN/m^2$, a plate efficiency of 60 per cent and that 8 per cent excess over minimum reflux requirements is used, how many plates must be used in the column and how much heat is removed in the condenser and added in the boiler?

**11.19.** A mixture of 60 mol per cent benzene, 30 per cent of toluene and 10 per cent xylene is run into a batch still. If the top product is to be 99 per cent benzene, determine

(a) the liquid composition on each plate at total reflux,
(b) the composition on the 2nd and 4th plates for $R = 1·5$,
(c) as for (b) but $R = 3$,
(d) as for (c) but $R = 5$,
(e) as for (d) but $R = 8$ and for the condition when the mol per cent benzene in the still is 10,
(f) as for (e) but with $R = 5$.

Take the relative volatility of benzene to toluene as 2·4, and xylene to toluene as 0·43.

**11.20.** A continuous still is fed with a mixture of mol fraction 0·5 of the more volatile component, and gives a top product of 0·9 mol fraction more volatile component and a bottom product of 0·10 mol fraction.

If the still operates with an $L_n/D$ ratio of 3·5:1, calculate by Sorel's method the composition of the liquid on the third theoretical plate from the top.

(a) for benzene-toluene, and
(b) for $n$-heptane-toluene.

**11.21.** A mixture of 40 mol per cent benzene with toluene is distilled in a column to give a product of 95 mol per cent benzene and a waste of 5 mol per cent benzene, using a reflux ratio of 4.

(a) Calculate by Sorel's method the composition on the second plate from the top.
(b) Using McCabe and Thiele's method, determine the number of plates required and the position of the feed if supplied to the column as liquid at the boiling point.
(c) Find the minimum reflux ratio possible.
(d) Find the minimum number of plates.
(e) If the feed is passed in at 288 K, find the number of plates required using the same reflux ratio.

**11.22.** 7·5 kg/s of *n*-propanol is obtained off the top of a distillation column working at 205 kN/m². The propanol is totally condensed and part taken for product and part returned for reflux. Under these conditions the boiling point is 390 K and the liquid density 775 kg/m³.

Determine the diameter and plate spacing based on entrainment and estimate the height of liquor in the downcomer. The liquor flow down may be taken as 5·0 kg/s.

What size do you suggest for the downcomer, the weir, the liquid seal and the area of the risers to the bubble caps? The $F_w$-factor for the height over the weir may be taken as 1.

**11.23.** Determine the minimum reflux ratio by the following two methods for the three systems given below:

(a) using Fenske's equation, and
(b) using Colburn's rigorous method.

1. $C_6$ (0·60 mol fraction), $C_7$ (0·30), $C_8$ (0·10) to give a product of 0·99 mol fraction $C_6$.

2.

| | | Mol. Frac. | Rel. Vol. $\alpha$ | $x_d$ |
|---|---|---|---|---|
| Components | **A** | 0·3 | 2 | 1·0 |
| | **B** | 0·3 | 1 | — |
| | **C** | 0·4 | $\frac{1}{2}$ | — |

3.

| | | | | |
|---|---|---|---|---|
| Components | **A** | 0·25 | 2 | 1·0 |
| | **B** | 0·25 | 1 | — |
| | **C** | 0·25 | $\frac{1}{2}$ | — |
| | **D** | 0·25 | $\frac{1}{4}$ | — |

**11.24.** A liquor consisting of phenol and cresols with some xylends is fractionated to give a top product of 95·3 mol per cent phenol. The compositions of the top product and of the phenol in the bottoms are given. A reflux ratio of 10 will be used.

(a) Complete the material balance over the still for a feed of 100 kmol.
(b) Calculate the composition on the second plate from the top.
(c) Calculate the composition on the the second plate from the bottom.
(d) Calculate the minimum reflux ratio by Underwood's equation and by Colburn's approximation.

| | Compositions in mol % | | |
|---|---|---|---|
| | Feed | Top | Bottom |
| phenol | 35 | 95·3 | 5·24 |
| *o*-cresol | 15 | 4·55 | |
| *m*-cresol | 30 | 0·15 | |
| xylenols | 20 | — | |
| | 100 | 100 | |

Heavy key is *m*-cresol, light key is phenol.

**11.25.** A continuous fractionating column is to be designed to separate 2·5 kg/s of a mixture of 60 per cent toluene and 40 per cent benzene, so as to give an overhead of 97 per cent benzene and a waste of 98 per cent toluene by weight.

A reflux ratio of 3·5 kmol of reflux per kmol of product is to be used and the molar latent heat of benzene and toluene may be taken as 30 MJ/kmol.

Calculate:

(a) The weight of product and waste per unit time.
(b) The number of theoretical plates and position of feed if the feed is liquid at 295 K, of specific heat 1·84 kJ/kg K.
(c) How much steam at 240 kN/m$^2$ is required in the still.
(d) What will be the required diameter of the column if it operates at atmospheric pressure and a vapour velocity of 1 m/s.
(e) If the vapour velocity is to be 0·75 m/s, based on free area of column, the necessary diameter of the column.
(f) The minimum possible reflux ratio, and the minimum number of plates for a feed entering at its boiling-point.

**11.26.** For a system that obeys Raoult's law show that the relative volatility $\alpha_{AB}$ is $P^0_A/P^0_B$, where $P^0_A$ and $P^0_B$ are the vapour pressures of the components **A** and **B** at the given temperature.

From the vapour pressure curves of benzene, toluene, ethyl benzene and of o-, m- and p-xylenes, obtain a plot of the volatilities of each of the materials relative to meta-xylene over the range of 340–430 K.

**11.27.** A still has a liquor composition of o-xylene 10 per cent, m-xylene 65 per cent, p-xylene 17 per cent, benzene 4 per cent and ethyl benzene 4 per cent. How many plates at total reflux are required to give a product of 80 per cent m-xylene, and 14 per cent p-xylene? The data are given as weight per cent.

**11.28.** The vapour pressure of n-pentane and of n-hexane is given in the table below.

| Pressure (kN/m$^2$) | 1·3 | 2·6 | 5·3 | 8·0 | 13·3 | 26·6 | 53·2 | 101·3 |
|---|---|---|---|---|---|---|---|---|
| (mm Hg) | 10 | 20 | 40 | 60 | 100 | 200 | 400 | 760 |
| C$_5$H$_{12}$ Temp. | 223·1 | 233·0 | 244·0 | 257·0 | 260·6 | 275·1 | 291·7 | 309·3 |
| C$_6$H$_{14}$ (K) | 248·2 | 259·1 | 270·9 | 278·6 | 289·0 | 304·8 | 322·8 | 341·9 |

Equilibrium curve at atmospheric pressure:

| $x = 0·1$ | 0·2 | 0·3 | 0·4 | 0·5 | 0·6 | 0·7 | 0·8 | 0·9 |
|---|---|---|---|---|---|---|---|---|
| $y = 0·21$ | 0·41 | 0·54 | 0·66 | 0·745 | 0·82 | 0·875 | 0·925 | 0·975 |

(a) Determine the relative volatility of pentane to hexane at temperatures of 273, 293 and 313 K.
(b) A mixture containing 0·52 mol fraction pentane is to be distilled continuously to give a top product of 0·95 mol fraction pentane and a bottom of 0·1 mol fraction pentane. Determine the minimum number of plates (i.e. the number of plates at total reflux) by the graphical McCabe–Thiele method, and analytically by using the relative volatility method.
(c) Using the conditions as in (b), determine the liquid composition on the second plate from the top by Lewis's method, if a reflux ratio of 2 is used.
(d) Using the conditions as in (b), determine by the McCabe–Thiele method the total number of plates required, and the position of the feed.

It may be assumed that the feed is all liquid at the boiling-point.

**11.29.** The vapour pressures of n-pentane and n-hexane are given in the data for Problem 11.28. Assuming that both Raoult's and Dalton's Laws are obeyed:

(a) Plot the equilibrium curve for a total pressure of 13·3 kN/m$^2$.
(b) Determine the relative volatility of pentane to hexane as a function of liquid composition for a total pressure of 13·3 kN/m$^2$.
(c) Would the error caused by assuming the relative volatility constant at its mean value be considerable?
(d) Would it be more advantageous to distil this mixture at higher pressures?

**11.30.** It is desired to separate a binary mixture by simple distillation. If the feed mixture has a composition of 0·5 mol fraction, calculate the fraction it is necessary to vaporise in order to obtain:

(a) a product of composition 0·75 mol fraction, when using a continuous process, and
(b) a product whose composition is not less than 0·75 mol fraction at any instant, when using a batch process.

If the product batch distillation is all collected in a single receiver, what is its mean composition?

Assume that the equilibrium curve is given by

$$y = 1 \cdot 2x + 0 \cdot 3$$

within the liquid composition range 0·3–0·8.

**11.31.** A liquor, consisting of phenol and cresols with some xylenol, is separated in a plate column. Given the following compositions, complete the material balance:

| Mol % | Feed | Top | Bottom |
|-------|------|-----|--------|
| $C_6H_5OH$ | 35 | 95·3 | 5·24 |
| $o\text{-}C_7H_7OH$ | 15 | 4·55 | |
| $m\text{-}C_7H_7OH$ | 30 | 0·15 | |
| $C_8H_9OH$ | 20 | — | |
| | 100 | 100 | |

Also, calculate:

  (a) the composition on the second plate from the top,
  (b) the composition on the second plate from the bottom.

A reflux ratio of 4:1 is used.

**11.32.** A mixture of 60, 30, and 10 mol per cent benzene, toluene, and xylene respectively is separated by a plate–column to give a top product containing at least 90 mol per cent benzene and negligible xylene, and a waste containing not more than 60 mol per cent toluene.

Using a reflux ratio of 4, and assuming that the feed is boiling liquid, determine the number of plates required in the column, and the approximate position of the feed.

Take the relative volatility of benzene to toluene as 2·4 and of xylene to toluene as 0·45, and assume these values are constant throughout the column.

**11.33.** It is desired to concentrate a mixture of ethyl alcohol and water from 40 mol per cent to 70 mol per cent ethanol. A continuous fractionating column, 1·2 m in diameter and having 10 plates is available. It is known that the optimum superficial vapour velocity in the column at atmosphere pressure is 1 m/s, giving an overall plate efficiency of 50 per cent.

Assuming that the mixture is fed to the column as a boiling liquid and using a reflux ratio of twice the minimum value possible, determine the feed plate and the rate at which the mixture can be separated.

Equilibria

| Mol fraction alcohol in liquid | 0·1 | 0·2 | 0·3 | 0·4 | 0·5 | 0·6 | 0·7 | 0·8 | 0·89 |
|--------------------------------|-----|-----|-----|-----|-----|-----|-----|-----|------|
| Mol fraction alcohol in vapour | 0·43 | 0·526 | 0·577 | 0·615 | 0·655 | 0·70 | 0·754 | 0·82 | 0·89 |

**12.1.** Some experiments are made on the absorption of carbon dioxide from a carbon dioxide–air mixture in 2·5 normal caustic soda, using a 250 mm diameter tower packed to a height of 3 m with 19 mm Raschig rings.

In one experiment at atmospheric pressure, the results obtained were:

Gas rate $G'$: 0·34 kg/m$^2$ s. Liquid $L'$: 3·94 kg/m$^2$ s.

The carbon dioxide in the inlet gas is 315 parts per million and in the exit gas 31 parts per million.

What is the value of the overall gas transfer coefficient $K_G a$?

**12.2.** An acetone–air mixture containing 0·015 mol fraction of acetone has the mol fraction reduced to 1 per cent of this value by countercurrent absorption with water in a packed tower. The gas flowrate $G$ is 1 kg/m$^2$ s of air and the water entering is 1·6 kg/m$^2$ s. For this system, Henry's law holds and $y_e = 1·75x$, where $y_e$ is the mol fraction of acetone in the vapour in equilibrium with a mol fraction $x$ in the liquid. How many overall transfer units are required?

**12.3.** An oil containing 2·55 mol per cent of a hydrocarbon is stripped by running the oil down a column up which live steam is passed, so that 4 kmol of steam are used per 100 kmol of oil stripped. Determine the number of theoretical plates required to reduce the hydrocarbon content to 0·05 mol per cent, assuming that the oil is non-volatile. The vapour–liquid relation of the hydrocarbon in the oil is given by $y_e = 33x$, where $y_e$ is the mol fraction in the vapour and $x$ the mol fraction in the liquid. The temperature is maintained constant by internal heating, so that the steam does not condense in the tower.

**12.4.** Gas, from a petroleum distillation column, has its concentration of $H_2S$ reduced from $0.03$ kmol $H_2S$/kmol of inert hydrocarbon gas to 1 per cent of this value, by scrubbing with a triethanolamine–water solvent in a countercurrent tower, operating at 300 K and at atmospheric pressure.

$H_2S$ is soluble in such a solution and the equilibrium relation may be taken as $Y = 2X$, where $Y$ is kmol of $H_2S$/kmol inert gas and $X$ is kmol of $H_2S$/kmol of solvent.

The solvent enters the tower free of $H_2S$ and leaves containing $0.013$ kmol of $H_2S$/kmol of solvent. If the flow of inert hydrocarbon gas is $0.015$ kmol/m$^2$s of tower cross-section and the gas-phase resistance controls the process, calculate:

(a) the height of the absorber necessary, and
(b) the number of transfer units required.

The overall coefficient for absorption $K_G''a$ may be taken as $0.04$ kmol/s m$^3$ of tower volume (unit driving force in $Y$).

**12.5.** It is known that the overall liquid transfer coefficient $K_L a$ for absorption of $SO_2$ in water in a column is $0.003$ kmol/s m$^3$ (kmol/m$^3$). Obtain an expression for the overall liquid film coefficient $K_G a$ for absorption of $NH_3$ in water in the same apparatus at the same water and gas rates. The diffusivities of $SO_2$ and $NH_3$ in air at 273 K are $0.103$ and $0.170$ cm$^2$/s. $SO_2$ dissolves in water, so that Henry's constant $\mathscr{H}$ is equal to 50 (kN/m$^2$)/(kmol/m$^3$). All data are expressed for the same temperature.

**12.6.** A packed tower is used for absorbing sulphur dioxide from air by means of a N/2 caustic soda solution. At an air flow of 2 kg/m$^2$ s, corresponding to a Reynolds number of 5160, the friction factor $R/\rho u^2$ is found to be $0.0200$.

Calculate the mass transfer coefficient in kg $SO_2$/s m$^2$ (kN/m$^2$) under these conditions if the tower is at atmospheric pressure. At the temperature of absorption the following values may be taken:

$$\text{Diffusion coefficient } SO_2 = 0.116 \text{ cm}^2/\text{s}$$
$$\text{Viscosity of gas} = 0.018 \text{ mN s/m}^2$$
$$\text{Density of gas stream} = 1.154 \text{ kg/m}^3$$

**12.7.** In an absorption tower, ammonia is being absorbed from air at atmospheric pressure by acetic acid. The flowrate of 2 kg/m$^2$ s in an experiment corresponds to a Reynolds number of 5100 and hence a friction factor $R/\rho u^2$ of $0.0199$. At the temperature of absorption the viscosity of the gas stream $(\mu)$ is $0.018$ mN s/m$^2$, the density $(\rho)$ is $1.154$ kg/m$^2$ and the diffusion coefficient of ammonia in air $(D)$ is $0.196$ cm$^2$/s.

Determine the mass transfer coefficient through the gas film in kg/m$^2$ s (kN/m$^2$).

**12.8.** Acetone is to be recovered from a 5 per cent acetone–air mixture by scrubbing with water in a packed tower using countercurrent flow. The liquid rate is $0.85$ kg/m$^2$ s and the gas rate is $0.5$ kg/m$^2$ s.

The overall absorption coefficient $K_G$ may be taken as $1.5 \times 10^{-4}$ kmol/s m$^3$ (kN/m$^2$) partial pressure difference, and the gas film resistance controls the process.

What should be the height of the tower to remove 98 per cent of the acetone? The equilibrium data for the mixture are:

| Mol fraction acetone in gas | 0.0099 | 0.0196 | 0.0361 | 0.0400 |
|---|---|---|---|---|
| Mol fraction acetone in liquid | 0.0076 | 0.0156 | 0.0306 | 0.0333 |

**12.9.** Ammonia is to be removed from 10 per cent ammonia–air mixture by countercurrent scrubbing with water in a packed tower at 293 K so that 99 per cent of the ammonia is removed when working at a total pressure of $101.3$ kN/m$^2$.

If the gas rate is $0.95$ kg/m$^2$ s of tower cross-section and the liquid rate is $0.65$ kg/m$^2$ s, find the necessary height of the tower if the absorption coefficient $K_G a = 0.001$ kmol/m$^3$ s (kN/m$^2$) partial pressure difference. The equilibrium data are:

| kmol $NH_3$/kmol water | 0.021 | 0.031 | 0.042 | 0.053 | 0.079 | 0.106 | 0.159 |
|---|---|---|---|---|---|---|---|
| Partial pressure $NH_3$ | | | | | | | |
| (mm Hg) | 12.0 | 18.2 | 24.9 | 31.7 | 50.0 | 69.6 | 114.0 |
| (kN/m$^2$) | 1.6 | 2.4 | 3.3 | 4.2 | 6.7 | 9.3 | 15.2 |

**12.10.** Sulphur dioxide is recovered from a smelter gas containing 3·5 per cent by volume of $SO_2$, by scrubbing with water in a countercurrent absorption tower. The gas is fed into the bottom of the tower, and in the exit gas from the top the $SO_2$ exerts a partial pressure of 1·14 kN/m². The water fed to the top of the tower is free from $SO_2$, and the exit liquor from the base contains 0·001145 kmol $SO_2$ per kmol water. The process takes place at 293 K, at which the vapour pressure of water is 2·3 kN/m². The water rate is 0·43 kmol/s.

If the area of the tower is 1·85 m² and the overall coefficient of absorption for these conditions $K''_L a$ is 0·19 kmol $SO_2$/s m³ (kmol of $SO_2$ per kmol $H_2O$), calculate the necessary height of the column.

The equilibrium data for $SO_2$ and water at 293 K are:

| kmol $SO_2$ per 1000 kmol $H_2O$ | | | | | | |
|---|---|---|---|---|---|---|
| 0·056 | 0·14 | 0·28 | 0·42 | 0·56 | 0·84 | 1·405 |

| kmol $SO_2$ per 1000 kmol inert gas | | | | | | |
|---|---|---|---|---|---|---|
| 0·7 | 1·6 | 4·3 | 7·9 | 11·6 | 19·4 | 36·3 |

**12.11.** Ammonia is removed from a 10 per cent ammonia–air mixture by scrubbing with water in a packed tower, so that 99·9 per cent of the ammonia is removed. What is the required height of tower?

| Entering gas | 1·2 kg/m² s |
|---|---|
| Water rate | 0·94 kg/m² s |
| $K_G a$ | 0·0008 kmol/s m³ (kN/m²) |

**12.12.** A soluble gas is absorbed from a dilute gas–air mixture by countercurrent scrubbing with a solvent in a packed tower. If the liquid led to the top of the tower contains no solute, show that the number of transfer units required is given by:

$$N = \frac{1}{\left[1 - \dfrac{m G_m}{L_m}\right]} \ln\left[\left(1 - \frac{m G_m}{L_m}\right)\frac{y_1}{y_2} + \frac{m G_m}{L_m}\right]$$

where $G_m$ and $L_m$ are the flowrates of the gas and liquid in kmol/s m² of tower area, and $y_1$ and $y_2$ the mol fraction of the gas at the inlet and outlet of the column. The equilibrium relation between the gas and liquid is represented by a straight line with the equation $y_e = mx$, where $y_e$ is the mol fraction in the gas in equilibrium with mol fraction $x$ in the liquid.

In a given process, it is desired to recover 90 per cent of the solute by using 50 per cent more liquid than the minimum necessary. If the H.T.U. of the proposed tower is 0·6 m, what height of packing will be required?

**12.13.** A paraffin hydrocarbon of molecular weight 114 kg/kmol at a temperature of 373 K, is to be separated from a mixture with a non-volatile organic compound of molecular weight 135 kg/kmol by stripping with steam. The liquor contains 8 per cent of the paraffin by weight and this is to be reduced to 0·08 per cent using an upward flow of steam saturated at 373 K. If three times the minimum amount of steam is used, how many theoretical stages will be required?

The vapour pressure of the paraffin at 373 K is 53 kN/m² and the process takes place at atmospheric pressure. It may be assumed that the system obeys Raoult's law.

**12.14.** Benzene is to be absorbed from coal gas by means of a wash oil. The inlet gas contains 3 per cent by volume of benzene, and the exit gas should not contain more than 0·02 per cent benzene by volume. The suggested oil circulation rate is 480 kg oil per 100 m³ of inlet gas measured at STP. The wash oil enters the tower solute-free. If the overall height of a transfer unit based on the gas phase is 1·4 m, determine the minimum height of the tower which is required to carry out the absorption. Use the following equilibrium data:

| Benzene in oil (% by wt.) | 0·05 | 0·01 | 0·50 | 1·0 | 2·0 | 3·0 |
|---|---|---|---|---|---|---|
| Equilibrium partial pressure of benzene in gas (kN/m²) | 0·013 | 0·033 | 0·20 | 0·53 | 1·33 | 3·33 |
| (mm Hg) | 0·1 | 0·25 | 1·5 | 4·0 | 10·0 | 25·0 |

**12.15.** Ammonia is to be recovered from a 5 per cent by volume ammonia–air mixture by scrubbing with water in a packed tower. The gas rate is 1·25 m³/s m² measured at NTP and the liquid rate is 1·95 kg/m² s. The temperature of the inlet gas is 298 K and of the inlet water 293 K. The mass transfer coefficient is $K_G a$ = 0·113 kmol/m³ s (mol ratio difference) and the total pressure is 101·3 kN/m². Find the height of the tower to remove 95 per cent of the ammonia. The equilibrium data and the heats of solutions are as follows:

| Mol fraction in liquid | 0·005 | 0·01 | 0·015 | 0·02 | 0·03 |
|---|---|---|---|---|---|
| Integral heat of solution | | | | | |
| (kJ/kmol of solution) | 181 | 363 | 544 | 723 | 1084 |

Equilibrium partial pressure:
mm Hg and (kN/m²)

|  |  |  |  |  |  |
|---|---|---|---|---|---|
| at 293 K | 3·0 (0·4) | 5·8 (0·77) | 8·7 (1·16) | 11·6 (1·55) | 17·5 (2.33) |
| at 298 K | 3·6 (0·48) | 7·3 (0·97) | 10·7 (1·43) | 14·4 (1·92) | 22·0 (2·93) |
| at 303 K | 4·6 (0·61) | 9·6 (1·28) | 13·7 (1·83) | 18·5 (2·47) | 29·0 (3·86) |

Assume adiabatic conditions and neglect the heat transfer between phases.

**12.16.** A thirty-plate bubble-cap column is to be used to remove $n$-pentane from solvent oil by means of steam stripping. The inlet oil contains 6 kmol of $n$-pentane per 100 kmol of pure oil and it is desired to reduce the solute content to 0·1 kmol per 100 kmol of solvent. Assuming isothermal operation and an overall plate efficiency of 30 per cent, find the specific steam consumption, i.e. kmol of steam required per kmol of solvent oil treated, and the ratio of the specific and minimum steam consumptions. How many plates would be required if this ratio were 2·0?

The equilibrium relation for the system may be taken as $Y_e = 3·0 X$, where $Y_e$ and $X$ are expressed in mol ratios of pentane in the gas and liquid phases respectively.

**12.17.** A mixture of ammonia and air is scrubbed in a plate column with fresh water.

If the ammonia concentration is reduced from 5 per cent to 0·01 per cent, and the water and air rates are respectively 0·65 and 0·40 kg/m² s, how many theoretical plates are required? The equilibrium relationship can be written $Y = X$, where $X$ is the mol ratio in the liquid phase.

**13.1.** Tests are made on the extraction of acetic acid from a dilute aqueous solution by means of a ketone in a small spray tower of diameter 46 mm and effective height 1090 mm, the aqueous phase being run into the top of the tower. The ketone enters free from acid at a rate of 0·0014 m³/s m², and leaves with an acid concentration of 0·38 kmol/m³. The concentration in the aqueous phase falls from 1·19 to 0·82 kmol/m³.

Calculate the overall extraction coefficient based on the concentrations in the ketone phase, and the height of the corresponding overall transfer unit.

The equilibrium conditions are expressed by the concentration of acid in the ketone phase being 0·548 times that in the aqueous phase.

**13.2.** A laboratory examination is made of the extraction of acetic acid from dilute aqueous solution, by means of methyl iso-butyl ketone, using a spray tower of 47 mm diameter and 1080 mm high. The aqueous liquor is run into the top of the tower and the ketone enters at the bottom.

The ketone enters at a rate of 0·0022 m³/s m² of tower cross-section, containing no acetic acid, and leaves with a concentration of 0·21 kmol/m³. The aqueous phase flows at a rate of 0·0013 m³/s m² of tower cross-section, and enters containing 0·68 kmol acid/m³.

Calculate the overall extraction coefficient based on the driving force in the ketone phase. What is the corresponding value of the overall H.T.U., based on the ketone phase?

Using units of kmol/m³, the equilibrium relationship under these conditions may be taken as:

Concentration of acid in the ketone phase equals 0·548 times the concentration in the aqueous phase.

**13.3.** Propionic acid is extracted with water from a dilute solution in benzene, by bubbling the benzene phase into the bottom of a tower to which water is fed at the top.

The tower is 1·2 m high and 0·14 m² in area, the drop volume is 0·12 cm³, and the velocity of rise 12 cm/s. From experiments in the laboratory the value of $K_w$ for forming drops is $7·6 \times 10^{-5}$ kmol/s m² (kmol/m³) and for rising drops $K_w = 4·2 \times 10^{-5}$ kmol/s m² (kmol/m³).

What do you expect to be the value of $K_w a$ for the tower in kmol/s m³ (kmol/m³)?

**13.4.** A 50 per cent solution of solute **C** in solvent **A** is extracted with a second solvent **B** in a countercurrent multiple contact extraction unit. The weight of **B** is 25 per cent that of the feed solution, and the equilibrium data are as given in the diagram (page 783).

Determine the number of ideal stages required, and the weight and concentration of the first extract if the final raffinate has 15 per cent of solute **C**.

**13.5.** Acetaldehyde (5 per cent) is in solution in toluene and is to be extracted with water in a five stage cocurrent

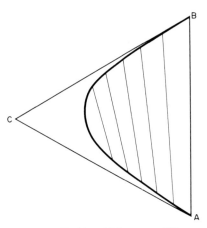

Problem 13.4 on page 782

unit. If 25 kg of water are used per 100 kg of feed, find the amount of acetaldehyde extracted and the final concentration.

If $Y$ = kg acetaldehyde/kg water and $X$ = kg acetaldehyde/kg toluene, then $Y_e = 2\cdot20X$ represents the equilibrium relation.

**13.6.** If a drop is formed in an immiscible liquid, show that the average surface available during formation of the drop is $\frac{12}{5}\pi r^2$, where $r$ is the radius of the drop, and that the average time of exposure of the interface is $\frac{3}{5}t_f$, where $t_f$ is the time of formation of the drop.

**13.7.** In the extraction of acetic acid from an aqueous solution in benzene in a packed column of height 1·4 m and cross-sectional area 0·0045 m$^2$, the concentrations measured at the inlet and outlet of the column are as follows:

$$
\begin{aligned}
&\text{acid concentration in the inlet water phase, } C_{W2} &&= 0\cdot69\,\text{kmol/m}^3\\
&\text{acid concentration in the outlet water phase, } C_{W1} &&= 0\cdot684\,\text{kmol/m}^3\\
&\text{flowrate of benzene phase} = 5\cdot6\times10^{-6}\,\text{m}^3/\text{s} &&= 1\cdot24\times10^{-3}\,\text{m}^3/\text{m}^2\,\text{s}\\
&\text{inlet benzene phase concentration, } C_{B1} &&= 0\cdot0040\,\text{kmol/m}^3\\
&\text{outlet benzene phase concentration, } C_{B2} &&= 0\cdot0115\,\text{kmol/m}^3
\end{aligned}
$$

Determine the overall transfer coefficient and the height of the transfer unit.

**13.8.** It is required to design a spray tower for the extraction of benzoic acid from solution in benzene.

Some experiments have been made on the rate of extraction of benzoic acid from a dilute solution in benzene to water, in which the benzene phase was bubbled into the base of a 25 mm diameter column and the water fed to the top of the column. Arrangements were made to measure the rate of mass transfer during the formation of the bubbles in the water phase and during the rise of the bubbles up the column. For conditions where the drop volume was 0·12 cm$^3$, the velocity of rise 12·5 cm/s, the value of $K_W$ for the period of drop formation was 0·000075 kmol/s m$^2$ (kmol/m$^3$), and for the period of rise 0·000046 in the same units.

If these conditions of drop formation and rise are reproduced in a spray tower of 1·8 m height and 0·04 m$^2$ cross-sectional area, what value would you expect for the transfer coefficient, $K_W a$, kmol/s m$^3$ (kmol/m$^3$), where $a$ represents the interfacial area in m$^2$ per unit volume of the column? The benzene phase enters at 38 cm$^3$/s.

**13.9.** It is proposed to reduce the concentration of acetaldehyde in aqueous solution from 50 per cent to 5 per cent by weight, by extraction with solvent **S** at 293 K. If a countercurrent multiple contact process is adopted and 0·025 kg/s of the solution is treated with an equal quantity of solvent, determine the number of theoretical stages required and the weight and concentration of the extract from the first stage.

The equilibrium relationship for this system at 293 K is given on the diagram (page 784).

**13.10.** 160 cm$^3$/s of a solvent **S** is used to treat 400 cm$^3$/s of a 10 per cent by wt. solution of **A** in **B**, in a three-stage countercurrent multiple contact liquid–liquid extraction plant. What is the composition of the final raffinate?

Using the same total amount of solvent, evenly distributed between the three stages, what would be the

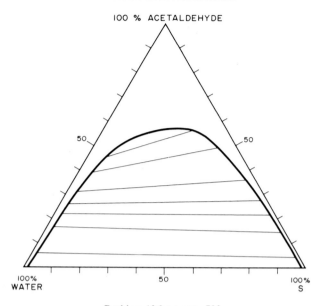

100 % ACETALDEHYDE

50

50

100%
WATER

50

100%
S

Problem 13.9 on page 783

composition of the final raffinate, if the equipment were used in a simple multiple contact arrangement? Equilibrium data:

| $\dfrac{\text{kg A}}{\text{kg B}}$ | 0·05 | 0·10 | 0·15 |
|---|---|---|---|
| $\dfrac{\text{kg A}}{\text{kg S}}$ | 0·069 | 0·159 | 0·258 |

Densities:     $\rho_A = 1200$,   $\rho_B = 1000$,   $\rho_c = 800\,\text{kg/m}^3$

**13.11.** In order to extract acetic acid from dilute aqueous solution, with isopropyl ether, the two immiscible phases are passed countercurrently through a packed column 3 m in length and 75 mm in diameter.

It is found that if $0.5\,\text{kg/m}^2$ of the pure ether is used to extract $0.25\,\text{kg/m}^2\,\text{s}$ of 4·0 per cent acid by wt., then the ether phase leaves the column with a concentration of 1·0 per cent acid by weight. Calculate:

(a) the number of overall transfer units, based on the raffinate phase, and
(b) the overall extraction coefficient, based on the raffinate phase.

The equilibrium relationship is given by weight per cent acid in isopropyl ether phase = 0·3 times the weight per cent acid in water phase.

**13.12.** A reaction is to be carried out in an agitated vessel. Pilot-plant experiments were performed under fully turbulent conditions in a tank 0·6 m in diameter, fitted with baffles and provided with a flat-bladed turbine. It was found that the satisfactory mixing was obtained at a rotor speed of 4 Hz, when the power consumption was 0·15 kW and the Reynolds number 160,000. What should be the rotor speed in order to retain the same mixing performance if the linear scale of the equipment is increased 6 times? What will be the power consumption and the Reynolds number?

**13.13.** It is proposed to recover material $A$ from an aqueous effluent by washing with a solvent $S$ and separating the resulting two phases. The light product phase will contain $A$ and the solvent $S$ and the heavy phase will contain $A$ and water. Show that the most economical solvent rate, $W$ kg/s, is given by:

$$W = ((F^2 a x_0)/(mb))^{0.5} - F/m$$

where:

The feedrate is $F$ kg/s water containing $x_0$ kg $A$/kg water.
The value of $A$ in the solvent product phase $= £a$/kg $A$.
The cost of solvent $S = £b$/kg $S$ and the equilibrium data are given by:

$(\text{kg } A/\text{kg } S)_{\text{product phase}} = m(\text{kg } A/\text{kg water})_{\text{water phase}}.$
$a, b$ and $m$ are constants.

**14.1.** A single-effect evaporater is used to concentrate 7 kg/s of a solution from 10 to 50 per cent of solids. Steam is available at 205 kN/m² and evaporation takes place at 13·5 kN/m².

If the overall heat transfer coefficient is 3 kW/m² K, calculate the heating surface required and the amount of steam used if the feed to the evaporator is at 294 K and the condensate leaves the heating space at 352·7 K.

Specific heat of 10 per cent solution 3·76 kJ/kg K
Specific heat of 50 per cent solution 3·14 kJ/kg K

**14.2.** A solution containing 10 per cent of caustic soda has to be concentrated to a 35 per cent solution at the rate of 180,000 kg/day during a year of 300 working days. A suitable single-effect evaporator for this purpose (neglecting condensing plant) costs £1600 and for a multiple-effect evaporator the cost may be taken as £1600 $N$, where $N$ is the number of effects.

Boiler steam can be purchased at £0·2 per 1000 kg and the vapour produced may be assumed to be 0·85 $N$ kg/kg of boiler steam. Assuming that interest on capital, depreciation, and other fixed charges amount to 45 per cent of the capital involved per annum, and that the cost of labour is constant and independent of the number of effects employed, determine the number of effects which, on the data supplied, will give the maximum economy.

**14.3.** Saturated steam is given off from an evaporator at atmospheric pressure and is being compressed by means of saturated steam at 1135 kN/m² in a steam jet to a pressure of 135 kN/m². If 1 kg of the high pressure steam compresses 1·6 kg of the evaporated atmospheric steam, comment on the efficiency of the compressor.

**14.4.** A single effect evaporator operates at 13 kN/m². What will be the heating surface necessary to concentrate 1·25 kg/s of 10 per cent caustic soda to 41 per cent, assuming a value of $U$ of 1·25 kW/m² K, using steam at 390 K? The heating surface is 1·2 m below the liquid level.

Boiling-point rise of solution = 30 deg K
Feed temperature = 291 K
Specific heat of the feed = 4·0 kJ/kg K
Specific heat of the product = 3·26 kJ/kg K
Specific gravity of boiling liquid = 1·39

**14.5.** Distilled water is produced from sea-water by evaporation in a single-effect evaporator, working on the vapour compression system. The vapour produced is compressed by a mechanical compressor of 50 per cent efficiency, and then returned to the calandria of the evaporator. Extra steam, dry and saturated at 650 kN/m², is bled into the steam space through a throttling valve. The distilled water is withdrawn as condensate from the steam space. Half the sea-water is evaporated in the plant. The energy supplied in addition to that necessary to compress the vapour may be assumed to appear as superheat in the vapour.

Using the data given below, calculate the quantity of extra steam required in kg/s.

Production of distillate                              0·125 kg/s
Pressure in vapour space                             atmospheric
Temperature difference from steam to liquor          8 deg K
Boiling point rise of sea-water                      1·1 deg K
Specific heat of sea-water                           4·18 kJ/kg K

Sea-water enters the evaporator at 344 K from an external heater.

**14.6.** It is claimed that a jet booster requires 0·06 kg/s of dry and saturated steam at 700 kN/m² to compress 0·125 kg/s of dry and saturated vapour from 3·5 kN/m² to 14·0 kN/m². Show whether this claim is reasonable.

**14.7.** A forward-feed double-effect evaporator, having 10 m² of heating surface in each effect, is used to concentrate 0·4 kg/s of caustic soda solution from 10 per cent to 50 per cent by weight. During a particular run, when the feed is at 328 K, the pressures in the two calandrias are 375 and 180 kN/m² respectively, while the condenser operates at 15 kN/m². Under these conditions, find:

(a) the load on the condenser,
(b) the steam economy, and
(c) the overall heat transfer coefficient in each effect.

Would there have been any advantages in using backward feed in this case?

*Data.* Assume negligible heat losses to the surroundings.

Physical properties of caustic soda solutions:

| Per cent solids by weight | B.pt. rise (deg K) | Sp. ht. (kJ/kg K) | Heat pf dilution (kJ/kg) |
|---|---|---|---|
| 10 | 1·6 | 3·85 | 0 |
| 20 | 6·1 | 3·72 | 2·3 |
| 30 | 15·0 | 3·64 | 9·3 |
| 50 | 41·6 | 3·22 | 220 |

**14.8.** A 12 per cent glycerol-water mixture is produced as a secondary product in a continuous process plant and flows from the reactor at 4·5 MN/m² and at a temperature of 525 K. Suggest, with preliminary calculations, a method of concentration to 75 per cent glycerol, in a factory which has no low-pressure steam available.

**14.9.** A forward feed double-effect standard vertical evaporator, with equal heating areas in each effect, is fed with 5 kg/s of a liquor of specific heat capacity 4·18 kJ/kg K, and with no boiling point rise, so that half of the feed liquor is evaporated. The overall heat transfer coefficient in the second effect is three-quarters of that in the first. Steam is fed in at 395 K and the boiling point in the second effect is 373 K. The feed is heated by external heater to the boiling point in the first effect.

It is decided to bleed off 0·25 kg/s of vapour from the vapour line to the second effect for use in another process. If the feed is still heated to the boiling point of the first effect by external means, what will be the change in steam consumption of the evaporator unit?

For the purpose of calculation, the latent heat of the vapours and of the live steam may be taken as 2230 kJ/kg.

**14.10.** A liquor containing 15 per cent solids is concentrated to 55 per cent solids in a double-effect evaporator, operating under a pressure in the second effect of 18 kN/m². No crystals are formed. The feed is 2·5 kg/s at a temperature of 375 K with a specific heat of 3·75 kJ/kg K. The boiling point rise of the concentrated liquor is 6 deg K and the steam fed to the first effect is 240 kN/m².

The overall heat transfer coefficients in the first and second effects are 1·8 and 0·63 kW/m² K, respectively. If the heat transfer area is to be the same in each effect, determine its value.

**14.11.** Liquor containing 5 per cent solids is fed at 340 K to a four effect evaporator. Forward feed is used to give a product containing 28·5 per cent solids. Do the following figures indicate normal operation? If not, suggest a reason for the abnormality.

| Effect | I | II | III | IV |
|---|---|---|---|---|
| % solids entering | 5·0 | 6·6 | 9·1 | 13·1 |
| Temperature in steam chest (K) | 382 | 374 | 367 | 357·5 |
| Temperature of boiling solution (K) | 369·5 | 364·5 | 359·6 | 336·6 |

**14.12.** 1·25 kg/s of a solution is concentrated from 10 to 50 per cent solids in a triple-effect evaporator using steam at 393 K and a vacuum such that the boiling point in the last effect is 325 K. If the feed is initially at 297 K and a backward feed is used, what will be the steam consumption, the temperature distribution in the system and the heat transfer area in each effect; each effect being identical?

For the purpose of calculation, assume that the specific heat is 4·18 kJ/kg K, that there is no boiling point rise, and that the latent heat of vaporisation is constant at 2330 kJ/kg K over the temperature range in the system. The overall heat transfer coefficient may be taken as 2·5, 2·0 and 1·6 kW/m² K in the first, second and third effects.

**14.13.** A triple-effect evaporator concentrates a liquid with no appreciable elevation of boiling point. If the temperature of the steam to the first effect is 395 K, and vacuum is applied to the third effect so that the boiling point is 325 K, what are the approximate boiling points in the three effects? The overall heat transfer coefficients may be taken as 3·1, 2·3, 1·1 kW/m² K in the three effects.

**14.14.** A 3-stage evaporator is fed with $1·25$ kg/s of a liquor which is concentrated from 10–40 per cent solids. The heat transfer coefficients may be taken as $3·1$, $2·5$ and $1·7$ kW/m$^2$ K.

Calculate the steam at 170 kN/m$^2$ required per second and the temperature distribution in the three effects.

 (a) if the feed is at 294 K, and
 (b) if the feed is at 355 K.

Forward feed is used in each instance and the values of $U$ may be considered the same for the two systems. The boiling point in the third effect is 325 K; the liquor has no boiling point rise.

**14.15.** An evaporator operating on the thermo-recompression principle employs a steam ejector to maintain atmospheric pressure over the boiling liquid. This ejector uses $0·04$ kg/s of steam at 650 kN/m$^2$ superheated 100 deg K, and produces a pressure in the steam chest of 205 kN/m$^2$. A condenser removes surplus vapour from the atmospheric pressure line.

What is the capacity and economy of the system?

How could the economy be improved?

*Data.* Properties of the ejector:

| | |
|---|---|
| Nozzle efficiency | 0·95 |
| Efficiency of momentum transfer | 0·80 |
| Efficiency of compression | 0·90 |

The feed enters the evaporator at 293 K and thick liquor is withdrawn at a rate of $0·025$ kg/s. The concentrated liquor exhibits a boiling point rise of 10 deg K. The plant is lagged sufficiently to render negligible the heat losses to the surroundings.

**14.16.** A single-effect evaporator is used to concentrate $0·075$ kg/s of a 10 per cent caustic soda liquor to 30 per cent. The unit employs forced circulation in which the liquor is pumped through the vertical tubes of the calandria which are 32 mm o.d. by 28 mm i.d., and $1·2$ m long.

Steam is supplied at 394 K, dry and saturated, and the boiling point rise of the 30 per cent solution is 15 deg K. If the overall heat transfer coefficient is $1·75$ kW/m$^2$ K, how many tubes do you suggest should be used, and what material of construction would you advise for the evaporator? The latent heat of vaporisation under these conditions is 2270 kJ/kg.

**14.17.** A steam jet booster compresses $0·1$ kg/s of dry and saturated vapour from $3·4$ kN/m$^2$ to $13·4$ kN/m$^2$. H.P. steam consumption is $0·05$ kg/s at 700 kN/m$^2$. (a) What must be the condition of the H.P. steam for the booster discharge to be superheated 20 deg K? (b) What is the overall efficiency of the booster if the compression efficiency is assumed equal to 1?

**14.18.** A triple-effect backward-feed evaporator concentrates 5 kg/s of liquor from 10 per cent to 50 per cent solids. Steam at 375 kN/m$^2$ is available and the condenser operates at $13·5$ kN/m$^2$. Find the area required in each effect (areas to be equal) and the economy of the unit.

Assume that the specific heat capacity is $4·18$ kJ/kg K at all concentrations and that there is no boiling-point rise. The overall heat transfer coefficients are $2·3$, $2·0$ and $1·7$ kW/m$^2$ K respectively in the three effects. The feed enters the third effect at 300 K.

**14.19.** A double-effect evaporator of the climbing film type is connected so that the feed passes through two preheaters, one heated by vapour from the first effect and the other by vapour from the second effect. The condensate from the first effect is passed into the steam space of the second. The temperature of the feed is initially 288 K, after the first heater 348 K and after the second heater 383 K. The vapour temperature in the last effect is 398 K and in the second 373 K. The feed is $0·25$ kg/s and the steam is dry and saturated at 413 K. Find the economy of the unit if the evaporation rate is $0·125$ kg/s.

**14.20.** A triple-effect evaporator is fed with 5 kg/s of a liquor containing 15 per cent solids. The concentration in the last effect, which operates at $13·5$ kN/m$^2$, is 60 per cent solids.

If the overall heat transfer coefficients are $2·5$, $2·0$ and $1·1$ kW/m$^2$ K respectively and the steam is fed at 388 K to the first effect, determine the temperature distribution and the area of heating surface required in each effect. The calandrias are to be identical. What is the economy and what is the heat load on the condenser?

       Feed temperature 294 K
       Specific heat of all liquors is $4·18$ kJ/kg K

If the unit is run as a backward feed system, the coefficients are $2·3$, $2·0$ and $1·6$ kW/m$^2$ K. Under these conditions, determine the new temperatures, heat economy and heating surface required.

**14.21.** A double-effect forward-feed evaporator is required to give a product which contains 50·0 per cent by weight of solids. Each effect has 10 m² of heating surface and the heat transfer coefficients are known to be 2·8 and 1·7 kW/m² K in the first and second effects respectively. Dry and saturated steam is available at 375 kN/m² and the condenser operates at 13·5 kN/m². The concentrated solution exhibits a boiling-point rise of 3 deg K.

What is the maximum feed rate if the feed contains 10 per cent solids and is at a temperature of 310 K?

Assume a latent heat of 2330 kJ/kg and a specific heat capacity of 4·18 kJ/kg K under all conditions.

**14.22.** You are required to consider proposals for concentrating fruit juice by evaporation. It is proposed to use a falling film evaporator and to incorporate a heat pump cycle with ammonia as the medium. The ammonia in vapour form enters the evaporator at 312 K and the water is evaporated from the juices at 287 K. The ammonia in the vapour-liquid mixture enters the condenser at 278 K and the vapour then passes to the compressor. It is estimated that the work in compressing the ammonia will be 150 kJ/kg of ammonia and that 1·05 kg of ammonia is cycled per kilogram of water evaporated. The following proposals are made for driving the compressor:

    (a) To use a diesel engine drive taking 0·4 kg of fuel per MJ, the calorific value of the fuel being 42 MJ/kg, and the cost £0·02/kg.

    (b) To pass steam, costing £0·01 for 10 kg, through a turbine which operates at 70 per cent isentropic efficiency, between 700 and 101·3 kN/m².

Explain by means of a diagram how this plant will work, including all necessary major items of equipment. Which method would you suggest for driving the compressor?

**14.23.** A double-effect forward-feed evaporator is required to give a product consisting of 30 per cent crystals and a mother liquor containing 40 per cent by weight of dissolved solids. Heat transfer coefficients are 2·8 and 1·7 kW/m² K in the first and second effects respectively. Dry saturated steam is supplied at 375 kN/m² and the condenser operates at 13·5 kN/m².

    (a) What area of heating surface is required in each effect (both effects to be identical) if the feed rate is 0·6 kg/s of liquor, containing 20 per cent by weight of dissolved solids, and the feed temperature 313 K?

    (b) What is the pressure above the boiling liquid in the first effect?

**14.24.** 1·9 kg/s of a liquid containing 10 per cent by weight of dissolved solids is fed at 338 K to a forward-feed double-effect evaporator. The product consists of 25 per cent by weight of solids and a mother liquor containing 25 per cent by weight of dissolved solids. The steam fed to the first effect is dry and saturated at 240 kN/m² and the pressure in the second effect of this evaporator is 20 kN/m². The specific heat capacity of the solid can be taken as 2·5 kJ/kg K whether it is in solid form or in solution and the heat of solution may be neglected. The mother liquor exhibits a boiling point rise of 6 deg K. If the two effects are identical, what area is required if the heat transfer coefficients in the first and second effects are 1·7 and 1·1 kW/m² K respectively?

**14.25.** 2·5 kg/s of a solution at 288 K containing 10 per cent of dissolved solids is fed to a forward-feed double-effect evaporator, operating a pressure of 14 kN/m² in the last effect. If the product is to consist of a liquid containing 50 per cent by weight of dissolved solids and dry saturated steam is fed to the steam coils, what should be the pressure of the steam? The surface in each effect is 50 m² and the coefficients for heat transfer in the first and second effects are 2·8 and 1·7 kW/m² K respectively. Assume that the concentrated solution exhibits a boiling-point rise of 5 deg K, that the latent heat has a constant value of 2260 kJ/kg, and that the specific heat capacity of the liquid stream is constant at 3·75 kJ/kg K.

**14.26.** A salt solution at 293 K is fed at the rate of 6·3 kg/s to a forward-feed triple-effect evaporator and is concentrated from 2 per cent to 10 per cent of solids. Saturated steam at 170 kN/m² is introduced into the calandria of the first effect and a pressure of 34 kN/m² is maintained on the last effect. If the heat transfer coefficients in the three effects are respectively 1·7, 1·4 and 1·1 and the specific heat capacity of the liquid is approximately 4 kJ/kg K, what area is required if each effect is identical? Condensate may be assumed to leave at the vapour temperature at each stage, and the effects of boiling point rise may be neglected. The latent heat of vaporisation may be taken as constant throughout.

**14.27.** A single effect evaporator with 10 m² of heating surface is used to concentrate NaOH solution from 10 per cent to 33·33 per cent, the weight of entering feed being 0·38 kg/s. The feed enters at 338 K; its specific heat capacity is 3·2 kJ/kg K. The pressure in the vapour spaces is 13·5 kN/m². 0·3 kg/s of stream is used from a supply at 375 K. Calculate:

(a) The apparent overall heat transfer coefficient.

(b) The coefficient corrected for boiling point rise of dissolved solids.

(c) The corrected coefficient if the depth of liquid is 1·5 m.

**14.28.** An evaporator, working at atmospheric pressure, is to concentrate a solution from 5 per cent to 20 per cent solids at the rate of 1·25 kg/s. The solution, which has a specific heat capacity of 4·18 kJ/kg K, is fed to the evaporator at 295 K and boils at 380 K. Dry saturated steam at 240 kN/m² is fed to the calandria, and the condensate leaves at the temperature of the condensing stream. If the heat transfer coefficient is 2·3 kW/m² K, what is the required area of heat transfer surface and how much steam is required? The latent heat of vaporisation of the solution is equal to that of water.

**15.1.** A saturated solution containing 1500 kg of potassium chloride at 360 K is cooled in an open tank to 290 K. If the specific gravity of the solution is 1·2, the solubility of potassium chloride per 100 parts of water is 53·55 at 360 K and 34·5 at 290 K s calculate:

(a) the capacity of the tank required, and

(b) the weight of crystals obtained, neglecting loss of water by evaporation.

**15.2.** What do you understand by the term "fractional crystallisation"? Explain with the aid of a diagram how the operation can be applied to a mixture of sodium chloride and sodium nitrate, given the following data. At 290 K, the solubility of sodium chloride is 36 parts per 100 of water and of sodium nitrate 88 parts per 100 of water. Whilst at this temperature, a saturated solution comprising both salts will contain 25 parts of sodium chloride and 59 parts of sodium nitrate per 100 parts of water. At 375 K these figures, per 100 parts of water, are 40, 176 and 17 and 160 respectively.

**16.1.** A wet solid is dried from 35 per cent of 10 per cent moisture, under constant drying conditions in 18 ks. If the equilibrium moisture content is 4 per cent and the critical moisture content is 14 per cent, how long will it take to dry to 6 per cent moisture under the same conditions?

**16.2.** Strips of material 10 mm thick are dried under constant drying conditions from 28 per cent to 13 per cent moisture in 25 ks. If the equilibrium moisture content is 7 per cent, find the time taken to dry 60 mm planks from 22 per cent to 10 per cent moisture under the same conditions, assuming no loss from the edges. All moistures are given on the wet basis.

The relation between $E$, the ratio of the average free moisture content at time $t$ to the initial free moisture content, and the parameter $j$ is given by:

| $E$ | 1 | 0·64 | 0·49 | 0·38 | 0·295 | 0·22 | 0·14 |
|-----|---|------|------|------|-------|------|------|
| $j$ | 0 | 0·1 | 0·2 | 0·3 | 0·5 | 0·6 | 0·7 |

Note that $j = kt/l^2$, where $k$ is a constant, $t$ the time in ks and $2l$ the thickness of the sheet of material in mm.

**16.3.** A granular material containing 40 per cent moisture is fed to a countercurrent rotary dryer at a temperature of 295 K and is withdrawn at 305 K containing 5 per cent moisture. The air supplied, which contains 0·006 kg water vapour per kg of dry air, enters at 385 K and leaves at 310 K. The dryer handles 0·125 kg/s wet stock.

Assuming that radiation losses amount to 20 kJ/kg of dry air used, determine the weight of dry air supplied to the dryer per second and the humidity of the air leaving it.

| | |
|---|---|
| Latent heat of water vapour at 295 K | = 2449 kJ/kg |
| Specific heat capacity of dried material | = 0·88 kJ/kg K |
| Specific heat capacity of dry air | = 1·00 kJ/kg K |
| Specific heat capacity of water vapour | = 2·01 kJ/kg K |

**16.4.** 1 Mg (dry weight) of a non-porous solid is dried under constant drying conditions with an air velocity of 0·75 m/s; the area of surface drying is 55 m². If the initial rate of drying is 0·3 g/m² s, how long will it take to dry the material from 0·15 to 0·025 kg of water/kg of dry solid? The critical moisture content of the material may be taken as 0·125 kg of water/kg of dry solid.

If the air velocity were increased to 4·0 m/s, what would be the anticipated saving in time if surface evaporation controlled?

**16.5.** A 100 kg batch of granular solids containing 30 per cent of moisture is to be dried in a tray dryer to 15·5 per cent of moisture by passing a current of air at 350 K tangentially across its surface at a velocity of 1·8 m/s. If the constant rate of drying under these conditions is 0·7 g/s m² and the critical moisture content is 15 per cent, calculate approximately the drying time.

Assume the drying surface to be 0·03 m²/kg dry weight.

# Conversion Factors for Some Common SI Units

An asterisk (*) denotes an exact relationship.

| | | | |
|---|---|---|---|
| Length | *1 in. | : | 25·4 mm |
| | *1 ft | : | 0·304 8 m |
| | *1 yd | : | 0·914 4 m |
| | 1 mile | : | 1·609 3 km |
| | *1 Å (angstrom) | : | $10^{-10}$ m |
| Time | *1 min | : | 60 s |
| | *1 h | : | 3·6 ks |
| | *1 day | : | 86·4 ks |
| | 1 year | : | 31·5 Ms |
| Area | *1 in.$^2$ | : | 645·16 mm$^2$ |
| | 1 ft$^2$ | : | 0·092 903 m$^2$ |
| | 1 yd$^2$ | : | 0·836 13 m$^2$ |
| | 1 acre | : | 4046·9 m$^2$ |
| | 1 mile$^2$ | : | 2·590 km$^2$ |
| Volume | 1 in.$^3$ | : | 16·387 cm$^3$ |
| | 1 ft$^3$ | : | 0·028 32 m$^3$ |
| | 1 yd$^3$ | : | 0·764 53 m$^3$ |
| | 1 UK gal | : | 4546·1 cm$^3$ |
| | 1 US gal | : | 3785·4 cm$^3$ |
| Mass | 1 oz | : | 28·352 g |
| | *1 lb | : | 0·453 592 37 kg |
| | 1 cwt | : | 50·802 3 kg |
| | 1 ton | : | 1016·06 kg |
| Force | 1 pdl | : | 0·138 26 N |
| | 1 lbf | : | 4·448 2 N |
| | 1 kgf | : | 9·806 7 N |
| | 1 tonf | : | 9·964 0 kN |
| | *1 dyn | : | $10^{-5}$ N |
| Temperature difference | *1 deg F (deg R) | : | $\frac{5}{9}$ deg C (deg K) |
| Energy (work, heat) | 1 ft lbf | : | 1·355 8 J |
| | 1 ft pdl | : | 0·042 14 J |
| | *1 cal (internat. table) | : | 4·186 8 J |
| | 1 erg | : | $10^{-7}$ J |
| | 1 Btu | : | 1·055 06 kJ |
| | 1 hp h | : | 2·684 5 MJ |
| | *1 kW h | : | 3·6 MJ |
| | 1 therm | : | 105·51 MJ |
| | 1 thermie | : | 4·185 5 MJ |
| Calorific value (volumetric) | 1 Btu/ft$^3$ | : | 37·259 kJ/m$^3$ |
| Velocity | 1 ft/s | : | 0·304 8 m/s |
| | 1 mile/h | : | 0·447 04 m/s |
| Volumetric flow | 1 ft$^3$/s | : | 0·028 316 m$^3$/s |
| | 1 ft$^3$/h | : | 7·865 8 cm$^3$/s |
| | 1 UK gal/h | : | 1·262 8 cm$^3$/s |
| | 1 US gal/h | : | 1·051 5 cm$^3$/s |

| | | | |
|---|---|---|---|
| Mass flow | 1 lb/h | : | $0 \cdot 126 \, 00 \, \text{g/s}$ |
| | 1 ton/h | : | $0 \cdot 282 \, 24 \, \text{kg/s}$ |
| Mass per unit area | 1 lb/in.$^2$ | : | $703 \cdot 07 \, \text{kg/m}^2$ |
| | 1 lb/ft$^2$ | : | $4 \cdot 882 \, 4 \, \text{kg/m}^2$ |
| | 1 ton/sq mile | : | $392 \cdot 30 \, \text{kg/km}^2$ |
| Density | 1 lb/in$^3$ | : | $27 \cdot 680 \, \text{g/cm}^3$ |
| | 1 lb/ft$^3$ | : | $16 \cdot 019 \, \text{kg/m}^3$ |
| | 1 lb/UK gal | : | $99 \cdot 776 \, \text{kg/m}^3$ |
| | 1 lb/US gal | : | $119 \cdot 83 \, \text{kg/m}^3$ |
| Pressure | 1 lbf/in.$^2$ | : | $6 \cdot 894 \, 8 \, \text{kN/m}^2$ |
| | 1 tonf/in.$^2$ | : | $15 \cdot 444 \, \text{MN/m}^2$ |
| | 1 lbf/ft$^2$ | : | $47 \cdot 880 \, \text{N/m}^2$ |
| | *1 standard atm | : | $101 \cdot 325 \, \text{kN/m}^2$ |
| | *1 atm (1 kgf/cm$^2$) | : | $98 \cdot 066 \, 5 \, \text{kN/m}^2$ |
| | *1 bar | : | $10^5 \, \text{N/m}^2$ |
| | 1 ft water | : | $2 \cdot 989 \, 1 \, \text{kN/m}^2$ |
| | 1 in. water | : | $249 \cdot 09 \, \text{N/m}^2$ |
| | 1 in. Hg | : | $3 \cdot 386 \, 4 \, \text{kN/m}^2$ |
| | 1 mm Hg (1 torr) | : | $133 \cdot 32 \, \text{N/m}^2$ |
| Power (heat flow) | 1 hp (British) | : | $745 \cdot 70 \, \text{W}$ |
| | 1 hp (metric) | : | $735 \cdot 50 \, \text{W}$ |
| | 1 erg/s | : | $10^{-7} \, \text{W}$ |
| | 1 ft lbf/s | : | $1 \cdot 355 \, 8 \, \text{W}$ |
| | 1 Btu/h | : | $0 \cdot 293 \, 07 \, \text{W}$ |
| | 1 ton of refrigeration | : | $3516 \cdot 9 \, \text{W}$ |
| Moment of inertia | 1 lb ft$^2$ | : | $0 \cdot 042 \, 140 \, \text{kg m}^2$ |
| Momentum | 1 lb ft/s | : | $0 \cdot 138 \, 26 \, \text{kg m/s}$ |
| Angular momentum | 1 lb ft$^2$/s | : | $0 \cdot 042 \, 140 \, \text{kg m}^2/\text{s}$ |
| Viscosity, dynamic | *1 P (Poise) | : | $0 \cdot 1 \, \text{N* s/m}^2$ |
| | 1 lb/ft h | : | $0 \cdot 413 \, 38 \, \text{mN s/m}^2$ |
| | 1 lb/ft s | : | $1 \cdot 488 \, 2 \, \text{N s/m}^2$ |
| Viscosity, kinematic | *1 S (Stokes) | : | $10^{-4} \, \text{m}^2/\text{s}$ |
| | 1 ft$^2$/h | : | $0 \cdot 258 \, 06 \, \text{cm}^2/\text{s}$ |
| Surface energy (surface tension) | 1 erg/cm$^2$ (1 dyn/cm) | : | $10^{-3} \, \text{J/m}^2$ $(10^{-3} \, \text{N/m})$ |
| Mass flux density | 1 lb/h ft$^2$ | : | $1 \cdot 356 \, 2 \, \text{g/s m}^2$ |
| Heat flux density | 1 Btu/h ft$^2$ | : | $3 \cdot 154 \, 6 \, \text{W/m}^2$ |
| | *1 kcal/h m$^2$ | : | $1 \cdot 163 \, \text{W/m}^2$ |
| Heat transfer coefficient | 1 Btu/h ft$^2$ °F | : | $5 \cdot 678 \, 3 \, \text{W/m}^2 \, \text{K}$ |
| Specific enthalpy (latent heat, etc.) | *1 Btu/lb | : | $2 \cdot 326 \, \text{kJ/kg}$ |
| Specific heat capacity | *1 Btu/lb °F | : | $4 \cdot 186 \, 8 \, \text{kJ/kg K}$ |
| Thermal conductivity | 1 Btu/h ft °F | : | $1 \cdot 730 \, 7 \, \text{W/m K}$ |
| | 1 kcal/h m °C | : | $1 \cdot 163 \, \text{W/m K}$ |

(Taken from MULLIN, J. W.: *The Chemical Engineer* No. 211 (Sept. 1967), 176. SI units in chemical engineering.)

# Index

793